Wake Conference 2017

Journal of Physics: Conference Series Volume 854

Visby, Sweden
30 May – 1 June 2017

ISBN: 978-1-5108-4268-7
ISSN: 1742-6588

Printed from e-media with permission by:

Curran Associates, Inc.
57 Morehouse Lane
Red Hook, NY 12571

CURRAN ASSOCIATES INC.
proceedings.com

Some format issues inherent in the e-media version may also appear in this print version.

Printed by Curran Associates, Inc. (2017)

Institute of Physics
Dirac House, Temple Back
Bristol BS1 6BE UK

Phone: 44 1 17 929 7481
Fax: 44 1 17 920 0979

techtracking@iop.org

Additional copies of this publication are available from:

Curran Associates, Inc.
57 Morehouse Lane
Red Hook, NY 12571 USA
Phone: 845-758-0400
Fax: 845-758-2633
Email: curran@proceedings.com
Web: www.proceedings.com

Table of contents

Volume 854

Wake Conference 2017

30 May to 1 June 2017, Visby, Sweden

Accepted papers received: 22 May 2017
Published online: 13 June 2017

Preface

Papers

IOP Conf. Series: Journal of Physics: Conf. Series **854** (2017) 011001 doi:10.1088/1742-6596/854/1/011001

The 52 papers in this volume constitute the proceedings of the 2017 Wake Conference, held in Visby on the island of Gotland, Sweden. The Wake Conference series began in Visby, where it was held in 2009 and 2011. In 2013 the conference took place in Copenhagen where it was combined with the International Conference on Offshore Wind Energy and Ocean Energy. In 2015 it went back to where it started, Visby, and this time it once again takes place at Uppsala University's Gotland campus, May 30th - June 1st.

Modern wind turbines are today clustered in large farms with a total production capacity reaching those of a nuclear power plant. When placed in a wind farm, the turbines will be fully or partially influenced by the wake of upstream turbines. This wake interaction results in a decreased power production, caused by the lower kinetic energy in the wind, and an increase in the turbulence intensity. Therefore, understanding the physical nature of vortices and their dynamics in the wake of a turbine is important for the optimal design of wind farms.

The increased importance and interest in the field of wake and wind farm aerodynamics can be seen in the increased number of scientific articles on the subject. For example, on the Web of Science citation index, the number citations on the topic 'wind turbine wakes' increased from about 50 in 2006 to more than 3800 in 2016. This citation growth essentially shows that the growth in the global production of electrical energy has become a scientific problem to be solved by scientists and engineers. In order to make a substantial impact on one of the most significant challenges of our time, global climate change, the wind industry's growth must continue. A part of making this growth possible will require research into the physics of wind turbine wakes and wind farms.

This conference is aimed at scientists and PhD students working in the field of wake dynamics. The conference covers the following subject areas: Wake and vortex dynamics, instabilities in trailing vortices and wakes, simulation and measurements of wakes, analytical approaches for modeling wakes, wake interaction and other wind farm investigations.

We would like to thank all those that assisted in the review process for these proceedings and we are also thankful for the help we have recieved from the editorial team at IOP Publishing during the process.

Visby, Sweden May/June 2017

Andrew Barney, Stefan Ivanell, Maria Klemm and Jens Nørkær Sørensen

Peer review statement

All papers published in this volume of *Journal of Physics: Conference Series* have been peer reviewed through processes administered by the proceedings Editors. Reviews were conducted by expert referees to the professional and scientific standards expected of a proceedings journal published by IOP Publishing.

IOP Conf. Series: Journal of Physics: Conf. Series **854** (2017) 012001 doi:10.1088/1742-6596/854/1/012001

Performance and Equivalent Loads of Wind Turbines in Large Wind Farms

Søren J. Andersen, Jens N. Sørensen, Robert F. Mikkelsen
DTU Wind Energy, Technical University of Denmark, 2800 Lyngby, Denmark
E-mail: sjan@dtu.dk

Abstract.
Ten simulations of large wind farms have been performed using a fully coupled LES and aero-elastic framework to form a database of full turbine operational conditions in terms of both production and loads. The performance is examined in terms of averaged power production and thrust, as well as 10min equivalent flapwise bending, yaw, and tilt moment loads. Certain scenarios operating below rated wind speed shows unexpected peaks in the loads. The influence on the operating conditions are examined for various parameters and compared relative to an effective power production per area.

1. Introduction

State-of-the-art in numerical simulations of wind farms is based on a combination of Large Eddy Simulations (LES) for solving the flow and the actuator line method by Sørensen and Shen [1] to model the turbines. LES have proven capable of providing detailed flow results and analysis of large wind farms, *e.g.* Calaf et al. [2], Meyers and Meneveau [3]-[4], Porté-Agel et al. [5], and Nilsson et al. [6]. Similarly, the actuator line method has proven good for modelling the turbine(s), *e.g.* Troldborg [7] and Jha et al. [8]. However, work including full coupling of the flow modelling and aero-elastic modelling for large wind farms is still scarce, but see Sørensen et al. [9], Andersen [10], Storey et al. [11], Lee et al. [12] as well as Vitsas and Meyers [10], who used a multibody approach rather than actuator lines.

However, such state-of-the-art simulations are still very computational expensive, but could potentially provide calibration data and insight into lower fidelity models, for instance the Dynamic Wake Meandering model by Larsen et al. [13]-[14].

Here, a collection of large wind farms simulations is examined in terms of both power production and aero-elastic load estimations to elucidate on the operating conditions of wind turbines deep inside large wind farms and how these conditions relate to various parameters, such as atmospheric turbulence, freestream velocity, and turbine spacing.

2. Methodology

2.1. Numerical Implementation

The simulations are performed using EllipSys3D, which solves the discretized incompressible Navier-Stokes equations with a finite volume approach in general curvilinear coordinates. EllipSys3D is formulated in pressure-velocity variables, and the pressure correction is solved through the PISO algorithm. Rhie-Chow interpolation technique is employed to avoid pressure decoupling. The convective terms are discretized in a combination of the third order QUICK scheme and the fourth order CDS scheme in order to prevent numerical wiggles. EllipSys3D is developed at DTU, see Michelsen [15]) and Risø (Sørensen[16]).

The wind turbines are modelled by the actuator line (AL) method, see Sørensen and Shen [1]. AL applies body forces ($\boldsymbol{f}_{WT}$) along rotating lines representing the turbine blades. The imposed body forces are calculated by the aero-elastic tool, Flex5, see e.g. Øye [17].

The actuator lines are fully coupled to Flex5, hence the turbines respond directly to the turbulent inflow, see Sørensen et al. [9] for details.

The atmospheric boundary layer (ABL) is prescribed explicitly by imposing additional body forces ($\boldsymbol{f}_{pbl}$) in the flow as described by Troldborg et al. [18]. The prescribed boundary layers are here taken as a combination of a parabolic and a power law profile described by:

$$U_{pbl}(z) = \begin{cases} U_0 \cdot (c_2 z^2 + c_1 z) & z \leq \Delta_{PBL} \\ U_0 \cdot \left(\frac{z}{H_{hub}}\right)^{\alpha_{PBL}} & z > \Delta_{PBL} \end{cases} \tag{1}$$

where Δ_{PBL} determines the height, where the velocity profile changes from parabolic to power law. H_{hub} is the hub height, c_1, c_2, and α_{PBL} are shape parameters. c_1 and c_2 are calculated to ensure a smooth transition between the parabolic and the power law expression.

Atmospheric turbulence is also added through body forces ($\boldsymbol{f}_{turb}$) in a Simulations O, B, C, F, and I. The forces are added in a plane $4R$ upstream the turbine. The body forces are derived based on stochastically generated turbulence using the Mann model, see Mann [19]-[20]. The advantage of the Mann model is that second-order statistics (variance, cross-spectra, etc.) are matched to those occurring in a neutral atmosphere. The turbulence box has an extent of $179,000\text{m} \times 6400\text{m} \times 6400\text{m}$, which is larger than a standard Mann box in order to include very long length scales. The advantage of using body forces is that no precursor simulation is needed and it is possible to apply different turbulence intensities for the same shear.

Hence, the filtered flow field is found by solving the 3D incompressible Navier-Stokes equations with added body forces:

$$\frac{\partial \overline{\boldsymbol{V}}}{\partial t} + \overline{\boldsymbol{V}} \cdot \nabla \overline{\boldsymbol{V}} = -\frac{1}{\rho} \nabla \overline{p} + \nabla[(\nu + \nu_{SGS}) \nabla \overline{\boldsymbol{V}}] + \frac{1}{\rho} \boldsymbol{f}_{WT} + \frac{1}{\rho} \boldsymbol{f}_{pbl} + \frac{1}{\rho} \boldsymbol{f}_{turb}. \tag{2}$$

The sub-grid scale (SGS) viscosity (ν_{SGS}) used in the LES is modeled using the vorticity based mixing scale by Ta Phuoc et al. [21].

2.2. Turbine

The modeled turbine is the NM80 turbine, see Madsen et al. [22]. NM80 has a radius $R = 40m$ and is rated to $2.75MW$ for $U_{hub} = 14$m/s.

The turbine includes an active controller consisting of a variable speed P-controller for small wind speeds ($U_{hub} < 14m/s$) and a PI-pitch angle controller for higher wind speeds. The controller responds continuously to the dynamic inflow as it would for a real turbine, which means that the modelled rotors are not constantly loaded and the blades pitch when the velocity is above rated.

2.3. Simulations Overview

A total of ten large wind farms have been modeled and summarized in Table 1. Each wind farm includes 16 turbines in the streamwise direction with hub height of $Z_{hub} = 2R$. Simulation O is a reference case, and identical to Simulation C except the turbines do not exert any forces on the flow, *i.e.* the turbines do not influence the flow, but statistics are extracted at the same locations for direct comparison. Cyclic boundary conditions are imposed in the lateral direction to mimic infinitely wide farms. The main parameters are streamwise and lateral spacings (S_X, S_Y), atmospheric turbulence intensity (TI) and freestream velocity at hub height (U_0), whereas shear is kept constant in the present work. Furthermore, the blade resolution and the number of cells in the computational domain are provided. The used grids have very comparable resolution, and no stretching is applied to prevent smearing of the wakes. Five of the ten simulations have been run with an atmospheric turbulence intensity of 0%, which is an idealized situation, but enables the study of the self-generated turbulence stemming only from the turbines themselves. The simulations have been run for at least two flow throughs to flush the domain of any transients and statistics are based on $60mins$, except Simulation H, which is based on $30min$ statistics. Simulations A-C, F, and H-I constitute the same wind farm under different inflow conditions. Several of the simulations have previously been reported in Andersen et al. [23], which showed that higher order flow statistics are comparable when scaled by a representative velocity, which can be assessed directly from the power curve, both instanteneously and averaged. Simulations O, A, B, and C have previously been investigated in detail in Andersen et al. [24], who concluded that the dominant length scales responsible for the energy entrainment in large wind farms is limited by the turbine spacing.

Table 1. Overview of simulations. *No forces are applied from the turbines to the flow, but aero-elastic results are extracted similar to simulation C.

Name	U_0	TI	Shear	Spacing ($S_X \times S_Y$)	Time step	Blade resolution	Cells in grid
O	$8m/s$	15%	0.14	$12R \times 20R^*$	$0.025s$	17	$327 \cdot 10^6$
A	$8m/s$	0%	0.14	$12R \times 20R$	$0.025s$	17	$327 \cdot 10^6$
B	$8m/s$	3%	0.14	$12R \times 20R$	$0.025s$	17	$327 \cdot 10^6$
C	$8m/s$	15%	0.14	$12R \times 20R$	$0.025s$	17	$327 \cdot 10^6$
D	$8m/s$	0%	0.14	$12R \times 12R$	$0.025s$	17	$120 \cdot 10^6$
E	$15m/s$	0%	0.14	$12R \times 12R$	$0.013s$	17	$120 \cdot 10^6$
F	$15m/s$	15%	0.14	$12R \times 20R$	$0.013s$	18	$235 \cdot 10^6$
G	$8m/s$	0%	0.14	$20R \times 20R$	$0.025s$	17	$524 \cdot 10^6$
H	$20m/s$	0%	0.14	$12R \times 20R$	$0.010s$	17	$327 \cdot 10^6$
I	$20m/s$	15%	0.14	$12R \times 20R$	$0.010s$	17	$327 \cdot 10^6$

3. Results

3.1. Statistical treatment

The present analyses focus on the following quantities:

- Velocity extracted $1R$ upstream at hub height, V_{hub}
- Power production, P
- Flapwise bending moment, M_F
- Yaw moment, M_Y
- Tilt moment, M_T
- Thrust force, T

where the statistics are calculated for different 10min periods shifted by 1min within the total simulation time, *i.e.* a total of 51 periods within 60min. The equivalent loads are computed using rainflow counting with Wöhler exponents of $m = 10$ for flap and $m = 6$ for yaw and tilt.

IOP Conf. Series: Journal of Physics: Conf. Series **854** (2017) 012001 doi:10.1088/1742-6596/854/1/012001

The data have been scaled by the following quantities for all the simulations to enable a direct comparison of the relative change. Hence, all quantities are denoted by a supercript * to indicate the scaling. Table 2 contain the scalings parameters.

Table 2. Overview of scaling parameters.

Quantity	Scaling parameter
V_{hub}^*	Maximum mean velocity of all simulations.
P^*	Rated power
M_F^*	Maximum of the mean equivalent flapwise bending moment of all simulations
M_Y^*	Maximum of the mean equivalent yaw moment of all simulations
M_T^*	Maximum of the mean equivalent tilt moment of all simulations
T^*	Maximum mean thrust force of all simulations

3.2. Spatial Variability

As previously shown by Andersen et al. [25] and Andersen et al. [23] the flow statitics have generally converged after the $4th$ or $5th$ turbine, although there can be large temporal varibility. Certain simulations also yield significant spatial variability, albeit not continuously increasing or decreasing through the farm, but rather alternating around the converged level. Figure 1 shows boxplot distributions for each of the 16 turbines in Simulations A-C. The lines connect the median of the distributions to make the trend even clearer. The velocity distributions are generally converged with only minor deviations in terms of medians and spread of the distributions. The standard deviation of V_{hub} shows a larger variation with distinct peaks, *e.g.* around the $9th$ turbine. Whether the standard deviation of the velocity at a single point is representative is questionable, but it shows how numerical power estimates should include the fluctuating part, *i.e.* $P \propto (\overline{U} + u')^3$, as the turbine is capable of extracting some of the larger turbulent scales. Peaks are even more pronounced in power, flapwise bending, yaw, tilt(not shown) moments, and thrust, and they are consistently located at the $5th$, $7th$, $10th$, $12th$, and $15th$ turbine across all three simulations. In addition to the consistent location, it does not stem from the imposed atmospheric turbulence as it also occurs for Simulation A and the peaks are not consistent in the standard deviation of the hub height velocity.

Figure 2 shows contour of the mean streamwise velocity and inplane streamlines for simulation A. The particular turbines show a significant smaller wake and higher velocity in a thin, horizontal band at hub height, but there appears to be no increased lateral movement. The repeating pattern could resemble a large scale harmonic effect, and it has recently been shown by Newman et al. [26] and Andersen et al. [24] that the largest turbulent structures are limited by the turbine spacing, yet grow further into the farms.

Boxplots of power, yaw, and thrust for the remaining simulations are shown in Figure 3. Here, the effect is less pronounced, although it occurs in the yaw (and tilt) moments for a number of the simulations, *e.g.* for turbines 8, 10, and 13 in simulation D. However, it is hardly seen in power and thrust. This "anomaly" requires additional investigation to determine if it is a physical or numerical artifact.

3.3. Aggregated Statistics

The data is aggregated for the $7th - 16th$ turbines to yield statistics of turbines operating deep inside large wind farms, and not only statistics of the $8th$ or $11th$ turbine. Hence, a total of 510

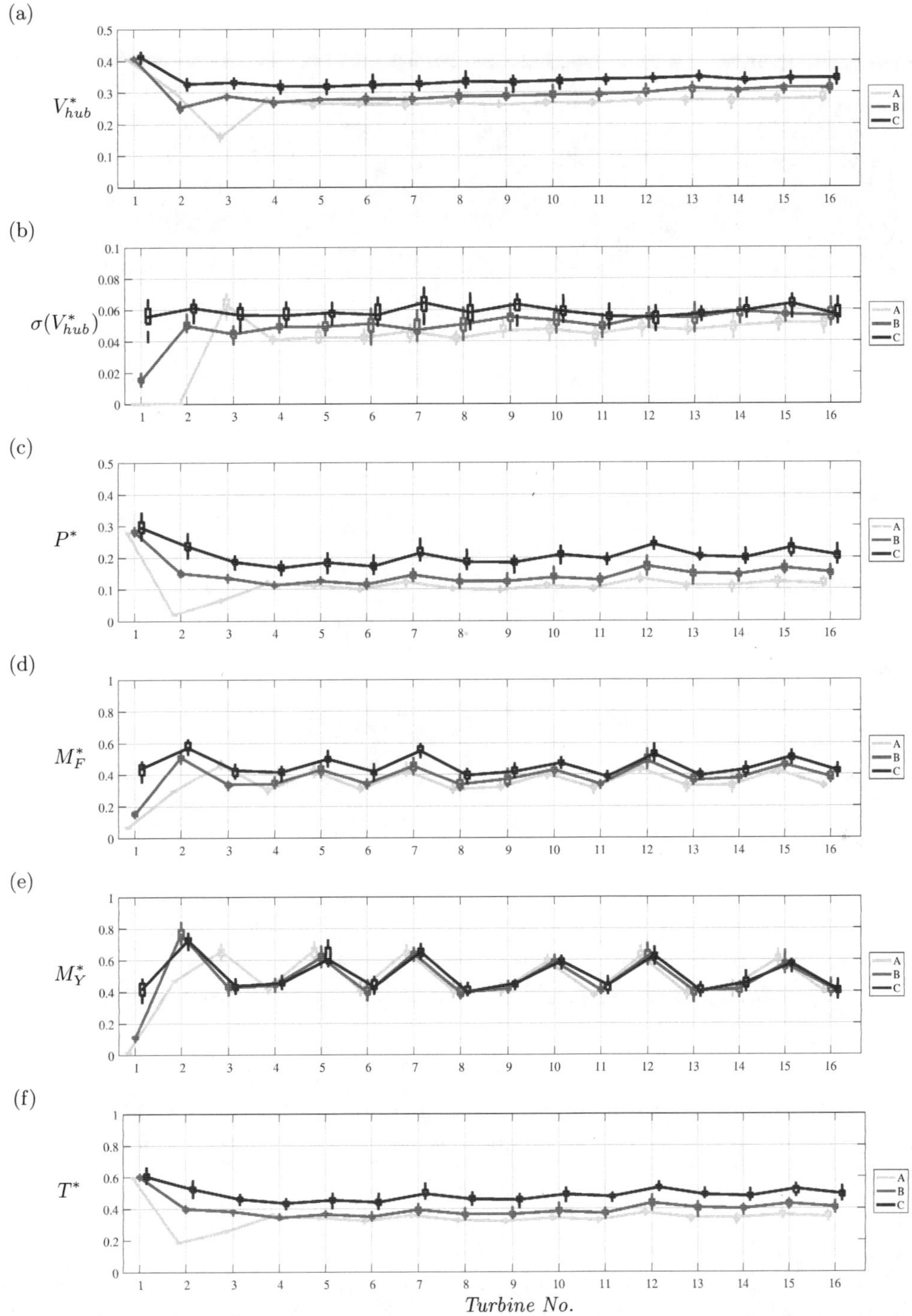

Figure 1. Boxplots of normalised 10min averages of (a) V^*_{hub}, standard deviation of (b) $\sigma(V^*_{hub})$, average of (c) P^* and 10min equivalent loads for (d) M^*_F, (e) M^*_Y, and (f) T^* for the 16 turbines in Simulations *A*-*C*. Lines connect the median of the distributions.

different 10min periods are included and reported for each simulation in the following, except

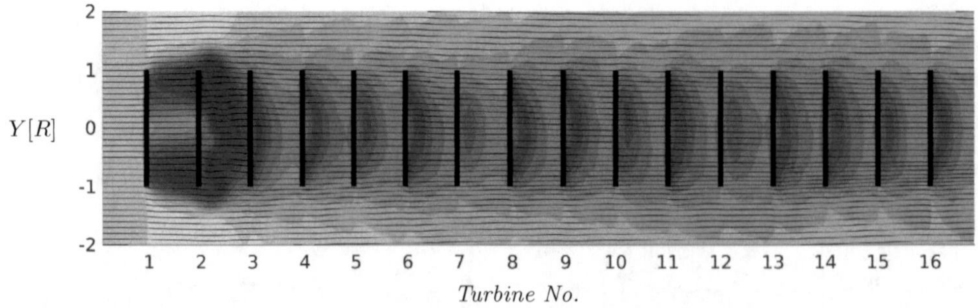

Figure 2. 10mins mean streamwise velocity and streamlines for simulation A.

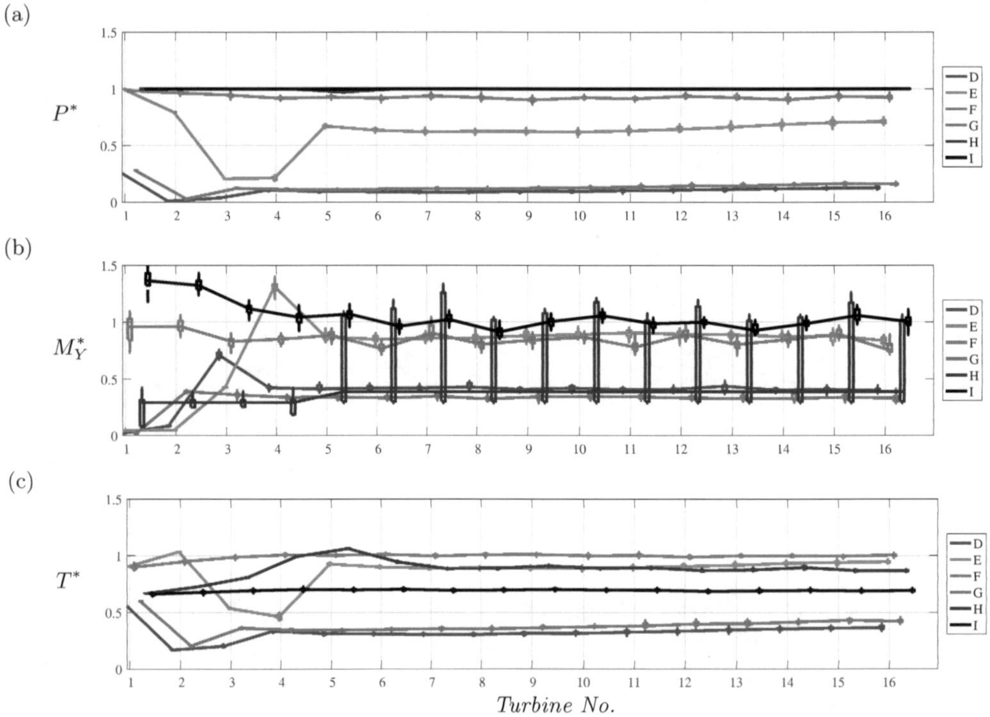

Figure 3. Boxplots of normalised 10min averages of (a) P^* and 10min equivalent loads for (b) M_Y^*, and (c) T^* for the 16 turbines in Simulations D-I. Lines connect the median of the distributions.

only 210 periods for Simulation H.

Figure 4 shows the aggregated statistics for each of the simulations (blue) in a polar plot, where each individual quantity is plotted at a given azimuthal position and the radial position shows the relative magnitude of the quantity compared to scaling parameters as described in Table 2. The standard deviations of the 510 periods are plotted as dashed lines. Simulation 0 without the effect of turbines is included in all plots for reference (black). The plots comprise a significant amount of information, but also give an immediate visual indication of the operating conditions for a given scenario, *i.e.*is the operation well balanced between power and loads or is it skewed either positively towards power or negatively towards loads. The reference is naturally skewed with higher velocities as well as higher ficticious power and thrust and very low loads.

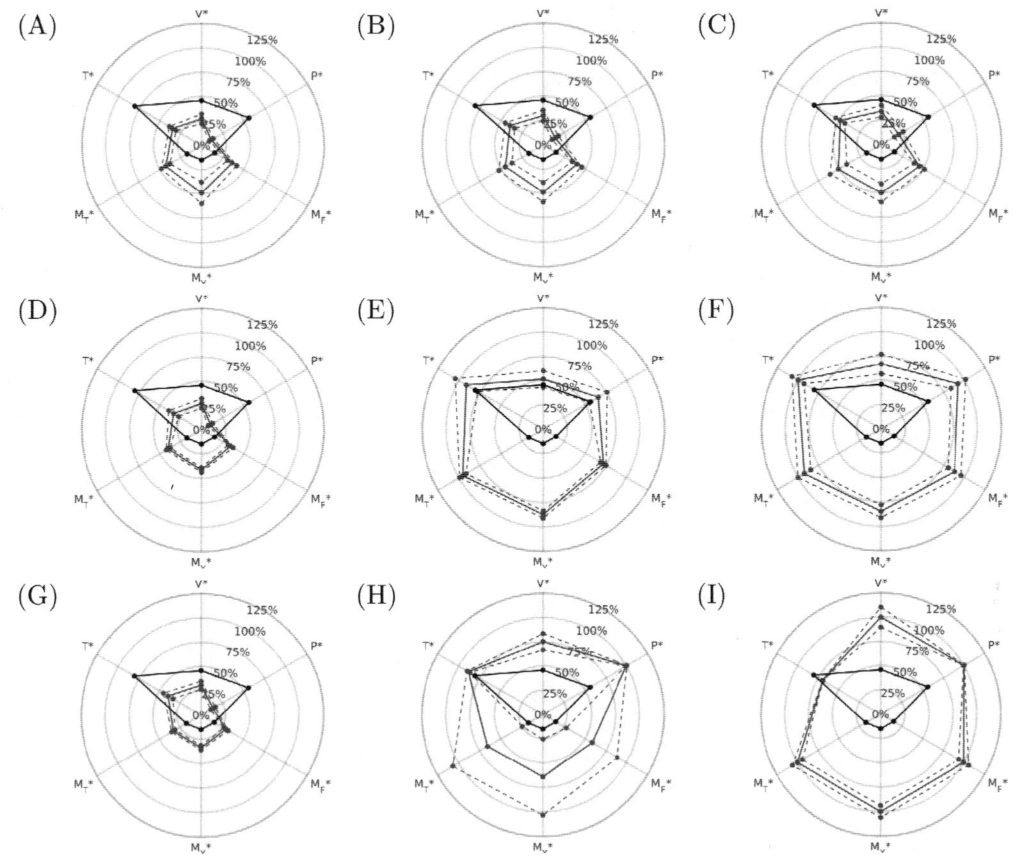

Figure 4. Polar plots showing mean velocity at hub height extracted $1R$ upstream, mean power production, mean 10min equivalent loads of flap, yaw, and tilt moments, and mean thrust. Statistics are based on the $7th$-$16th$ turbine. Quantities have been scaled by maximum values of all simulation for direct comparison. Broken lines show $\pm 1\sigma$. Simulation O with no turbines is shown for reference black.

Clearly, the majority of the scaling parameters comes from Simulation I, where the atmospheric turbulence intensity is large and the velocity is above rated, so the turbines are producing rated power. However, the highest thrust is experienced in Simulation F as the turbines operate around rated wind speed of 14m/s. The aforementioned increased spatial variability in Simulations A-C is evident in the larger standard deviations compared to Simulation D, which is very similar to Simulation A except a smaller lateral spacing. The largest variations in power occur for Simulation E and F with a freestream velocity just above rated, meaning that the waked turbines are constantly operating below rated. However, the largest variation in loads occur for Simulation H with high velocity and only self-generated turbulence. The variation could be related to large variations in the wake breakdown as described by Sørensen et al. [9].

Equivalent simulations, where only one parameter varies, are assembled in Figure 5. The effect of atmospheric turbulence in Simulations A-C shows (as expected) how increased atmospheric turbulence results in larger wake recovery, *i.e.* higher velocity and power, but also larger loads. However, the overall shape of the operation condition is maintained. Increased freestream velocity alters the operating condition significantly. The turbines below rated produce about

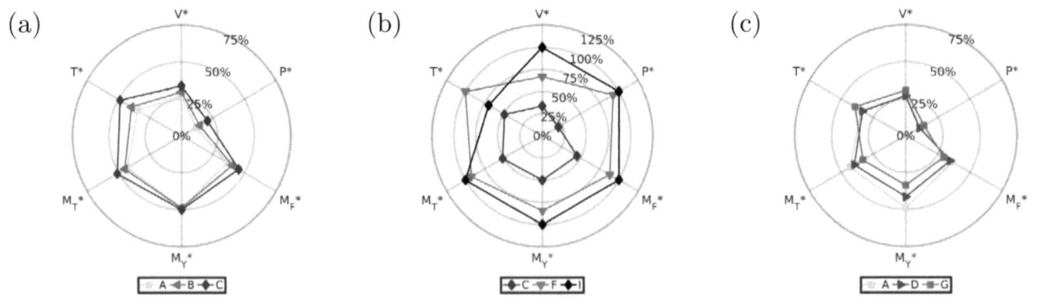

Figure 5. Comparing polar plots from Figure 4 showing effect of (a) atmospheric turbulence, (b) freestream velocity, and (c) spacing. Statistics are based on the $7th$-$16th$ turbine. Quantities have been scaled by maximum values of all simulation for direct comparison. Only mean normalised values are shown for clarity. Note, that the radial extent is changed in (a) and (c).

25% of rated power, but are exposed to approximately 50% of the maximum loading in the dataset. Turbines operating around rated experience the highest thrust and produced 92% of rated power with similar flapwise, yaw, and tilt loading around $85-92\%$ of maximum. For very high wind speed, the power and loads are at maximum, while thrust is reduced to 70%. Finally, the effect of spacing from Simulations A, D, and G show a slight increase in loads, particular yaw, when the lateral spacing is increased (A compared to D). However, increasing both streamwise and lateral spacing (D vs G) shows how the power production is increased, while loading is reduced.

3.4. Farm Layout

The full operational conditions can be used to elucidate on different farm layout scenarios similar to the work of *e.g.* Stevens et al. [27], who compared power production for different effective turbine spacings. Here, the equivalent loading and thrust on the turbines are added as a new dimension as the actual power is normalised by turbine spacing to yield a measure of the effective land use in $[W/m^2]$, see Figure 6. The five simulations with the same turbine arrangement are connected by a dashed line for reference as it covers a range of different operational conditions.

The data is concentrated in two regions, one below rated where power production is approximately $0.6-1.5W/m^2$ with corresponding loads of $30-50\%$ of the maximum equivalent load encountered in the total dataset for this specific turbine. The above rated region has a power production on the order of $6.6-8W/m^2$ and larger spread in the equivalent loads as the turbines operate around or above rated.

4. Conclusions

Ten simulations of large wind farms have been performed using a fully coupled LES and aero-elastic framework to form a database of full turbine operational conditions in terms of both production and loads. The performance is examined in terms of averaged power production and thrust, as well as 10min equivalent flapwise bending, yaw, and tilt moment loads. Certain scenarios operating below rated wind speed shows unexpected peaks in the loads, which occurs consistently across several simulations, but not immediately apparent or consistent in the flow statistics. The operating conditions of the farms are compared and the influence of atmospheric turbulence, freestream velocity, and turbine spacing is examined. The changes in equivalent loads and mean thrust are compared to the effective power production per area, which shows two distinct regions of waked turbine operation, corresponding to one below and one around/above

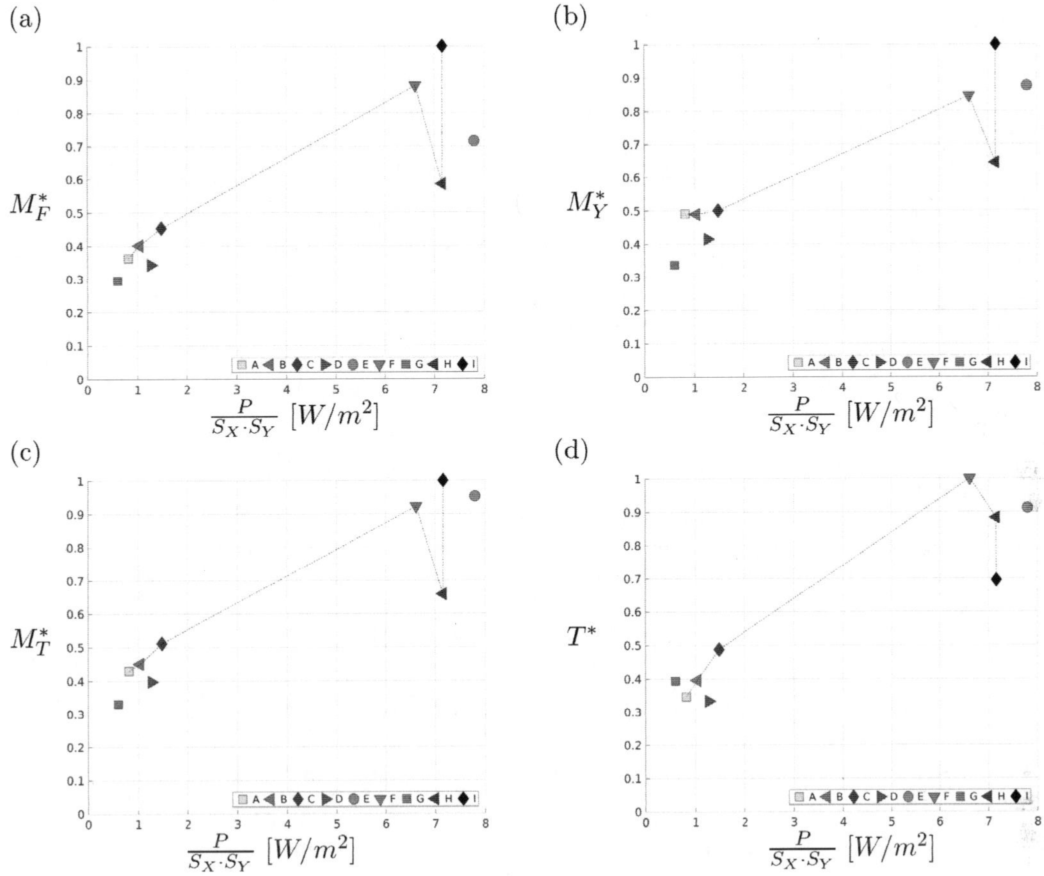

Figure 6. Equivalent (a) flapwise bending moment, (b) yaw moments, (c) tilt moment, and (d) mean thrust against actual power normalised by area. Simulations *A-C*, *F*, and *H-I* constituting the same wind farm is connected by a dashed line.

rated wind speeds. The database, albeit far from complete at present, provides detailed insights for designing large wind farms, *e.g.* the potential benefits of changing the farm layout to improve the levelized cost of energy by including loads. Furthermore, the database could be employed to alleviate particular unfavorable conditions by designing turbines and/or their controllers specifically to operate inside large wind farms under particular atmospheric conditions.

4.1. Acknowledgments

Acknowledgments
The analysed data has been gather during and as part of the following projects: Center for Computational Wind Turbine Aerodynamics and Atmospheric Turbulence (grant 2104-09-067216/DSF) (COMWIND: http://www.comwind.org), Nordic Consortium on Optimization and Control of Wind Farms funded by the Swedish Energy Agency, Eurotech Greentech Wind project, Winds2Loads, and CCA LES. Computer time was granted by the Swedish National Infrastructure for Computing (SNIC) and as well as on JESS (DTU). Furthermore, the proprietary data for Vestas' NM80 turbine has been used.

References

[1] Sørensen J N and Shen W Z 2002 *Journal of Fluids Engineering* **124** 393–399
[2] Calaf M, Meneveau C and Meyers J 2010 *Phys. Fluids* **22**
[3] Meyers J and Meneveau C 2012 *Wind Energy* **15** 305–317 ISSN 10991824, 10954244
[4] Meyers J and Meneveau C 2013 *J. Fluid Mech.* **715** 335–358
[5] Porté-Agel F, Wu Y T, Lu H and Conzemius R J 2011 *Journal of Wind Engineering and Industrial Aerodynamics* **99** 154–168
[6] Nilsson K, Ivanell S, Hansen M H, Mikkelsen R, Breton S P and Henningson D 2014 *Wind Energy*
[7] Troldborg N 2008 *Actuator Line Modeling of Wind Turbine Wakes* Ph.D. thesis DTU Mechanical Engineering, Technical University of Denmark, DTU, DK-2800 Kgs. Lyngby, Denmark Denmark
[8] Jha P K, Churchfield M J, Moriarty P J and Schmitz S 2014 *Journal of Solar Energy Engineering* **136** 031003–031003
[9] Sørensen J N, Mikkelsen R, Henningson D S, Ivanell S, Sarmast S and Andersen S J 2015 *Royal Society of London. Philosophical Transactions A. Mathematical, Physical and Engineering Sciences* **373** 20140071–20140071 ISSN 14712962, 1364503x
[10] Andersen S J 2013 *Simulation and Prediction of Wakes and Wake Interaction in Wind Farms* Ph.D. thesis Technical University of Denmark, Wind Energy.
[11] Storey R, Cater J and Norris S 2016 *Renewable Energy* **95** 31 – 42 ISSN 0960-1481 URL `http://www.sciencedirect.com/science/article/pii/S0960148116302518`
[12] Lee S, Churchfield M J, Moriarty P J, Jonkman J and Michalakes J 2013 *Journal of Solar Energy Engineering-transactions of the Asme* **135** 31001 ISSN 15288986, 01996231
[13] Larsen G C, Madsen H A, Bingöl F, Mann J, Ott S, Sørensen J N, Okulov V, Troldborg N, Nielsen M, Thomsen K, Larsen T J and Mikkelsen R 2007 Dynamic wake meandering modeling Tech. Rep. R-1607(EN) Risø-DTU Roskilde, Denmark
[14] Larsen G C, Madsen H A, Thomsen K and Larsen T J 2008 *Wind Energy* **11** 377–395
[15] Michelsen J A 1992 Basis3D – a Platform for Development of Multiblock PDE Solvers Tech. rep. Danmarks Tekniske Universitet
[16] Sørensen N N 1995 *General Purpose Flow Solver Applied to Flow over Hills* Ph.D. thesis Technical University of Denmark
[17] Øye S 11-12 April 1996 *Proceedings of 28th IEA Meeting of Experts Concerning State of the Art of Aeroelastic Codes for Wind Turbine Calculations. Available through International Energy Agency.* (Lyngby, Denmark: Danmarks Tekniske Universitet) pp 71–76
[18] Troldborg N, Sørensen J, Mikkelsen R and Sørensen N 2013 *Wind Energy* ISSN 1095-4244
[19] Mann J 1994 *Journal of Fluid Mechanics* **273** 141–168 ISSN 1469-7645
[20] Mann J 1998 *Probabilistic Engineering Mechanics* **13** 269–282 ISSN 0266-8920
[21] Ta Phuoc L, Lardat R, Coutanceau M and Pineau G 1994 Rechereche et analyse de modeles de turbulence de sous maille adaptes aux ecoulements instationnaires decolles.
[22] Aagaard Madsen H, Bak C, Schmidt Paulsen U, Gaunaa M, Fuglsang P, Romblad J, Olesen N, Enevoldsen P, Laursen J and Jensen L 2010 *The DAN-AERO MW Experiments* Denmark. Forskningscenter Risø. Risø-R (Danmarks Tekniske Universitet, Risø Nationallaboratoriet for Bæredygtig Energi) ISBN 978-87-550-3809-7
[23] Andersen S J, Sørensen J N, Mikkelsen R and Ivanell S 2016 *Journal of Physics: Conference Series. The Science of Making Torque from Wind 2016.* **753** ISSN 1742-6596
[24] Andersen S J, Sørensen J N and Mikkelsen R F 2017 *Philosophical Transactions of the Royal Society of London A: Mathematical, Physical and Engineering Sciences* **375** ISSN 1364-503X (*Preprint* `http://rsta.royalsocietypublishing.org/content/375/2091/20160107.full.pdf`) URL `http://rsta.royalsocietypublishing.org/content/375/2091/20160107`
[25] Andersen S J, Witha B, Breton S P, Sørensen J N, Mikkelsen R F and Ivanell S 2015 *Journal of Physics: Conference Series* **625** 012027 URL `http://stacks.iop.org/1742-6596/625/i=1/a=012027`
[26] Newman A J, Drew D A and Castillo L 2014 *Renewable Energy* **70** 129–141 ISSN 18790682, 09601481
[27] Stevens R J A M, Gayme D F and Meneveau C 2015 *Journal of Renewable and Sustainable Energy* **7**

IOP Conf. Series: Journal of Physics: Conf. Series **854** (2017) 012002 doi:10.1088/1742-6596/854/1/012002

A CFD framework for offshore and onshore wind farm simulation

Matias Avila, Abel Gargallo-Peiró and Arnau Folch

Barcelona Supercomputing Center, Jordi Girona 29, 08034 Barcelona, Spain

E-mail: matias.avila@bsc.es

Abstract. We present a wind simulation framework for offshore and onshore wind farms. The simulation framework involves an automatic hybrid high-quality mesh generation process, a pre-processing to impose initial and boundary conditions, and a solver for the Reynolds Averaged Navier-Stokes (RANS) equations with two different turbulence models, a modified standard k-ε model and a realizable k-ε model in which we included the Coriolis effects. Wind turbines are modeled as actuator discs. The wind farm simulation framework has been implemented in Alya, an in-house High Performance Computing (HPC) multi-physics finite element parallel solver. An application example is shown for an onshore wind farm composed of 165 turbines.

1. Introduction

Simulation of wind farms with Computational Fluid Dynamics (CFD) models involves the resolution of the turbulent Atmospheric Boundary Layer (ABL) and the effects induced by wind turbines, including wind speed deficit, increase of turbulent kinetic energy, and interaction among wakes. Commonly, wind turbines are modeled as uniformly loaded actuator discs [1], an option which allows a compromise between a reasonable model accuracy and a low computational cost.

However, it is well known that the Reynolds-Averaged Navier-Stokes (RANS) standard k-ε model significantly overestimates Reynolds stresses [2] behind actuator discs, resulting on a significant underestimation of velocities and on an excessive wake damping. Several modifications of the standard k-ε model have been proposed for delaying the simulated wake flow recovery [3], but most of them require the adjustment of model parameters that depend on the characteristics of the wind turbine at stake. Shi et al.[4] proposed a realizable k-ε model that enhances the wake predictions of actuator discs [5] and needs no adjustment of parameters. On the other hand, it is also well-established that the k-ε model needs an additional mixing length limitation model [6] to accurately predict wind intensity profiles and wind veering with height in the ABL when considering Coriolis forces. However, no attempts have been made to modify the realizable model for the simulation of Coriolis forces in the ABL.

Several aspects of wind farm modeling pose requirements on the computational mesh. Firstly, the existence of a boundary layer imposes stretching requirements in order to capture near-surface sharp gradients. Secondly and for onshore farms, the underlying terrain must be discretized and properly approximated by the mesh. Finally, the actuator discs have to be embedded in the mesh with a subsequent mesh refinement around and downstream the turbines in order to capture near wake flow effects properly. This is a constrain for conformal structured meshes, which do not allow to prescribe finer resolutions around turbines without extending the finer discretization across all directions of the computational domain, thereby increasing dramatically and unnecessarily the number of computational cells (nodes). Several codes circumvent this problem by loosing mesh conformity (*e.g.* use of hexahedral cells/elements with hanging nodes). In contrast, we propose an alternative based on hybrid element meshes, which allow local mesh refinements while keeping mesh conformity. The objective is to present a High Performance Computing (HPC) simulation framework for offshore and onshore wind farms with an automatic high-quality mesh generation process.

The rest of the manuscript is arranged as follows. Section 2 presents the two RANS physical models; a standard k-ε adapted to ABL flows [6] (*i.e.* with Coriolis effects, mixing length limitation and appropriate model constants), and the k-ε realizable model of [4] modified to account for Coriolis forces and with appropriate coefficients for the simulation of ABL flows. In both cases, wind turbines are simulated under the actuator disc theory [7]. Section 3 focus on numerical aspects including the hybrid mesh generation, the implementation of the actuator disc model, the numerical method and its implementation in the multi-physics parallel solver (Alya) and, finally, a succinct model validation. Section 4 shows an application example for an onshore wind farm consisting on 165 turbines deployed on complex terrain. The effect of Coriolis terms is analyzed for both (k-ε modified and k-ε realizable) RANS models. Finally, Section 5 wraps-up and briefly discusses model limitations and concluding remarks.

2. RANS models governing equations

2.1. k-ε modified model

Considering the flow as incompressible and isothermal (neutral stability), the modified k-ε RANS model accounting for Coriolis effects and using the Apsley and Castro correction for the mixing length limitation [6] are written as:

$$\begin{aligned} \nabla \cdot \boldsymbol{u} &= 0 && (1)\\ \frac{\partial \boldsymbol{u}}{\partial t} + \boldsymbol{u}\cdot\nabla\boldsymbol{u} - \nabla\cdot(\nu_t \nabla^s \boldsymbol{u}) + \nabla p + 2\boldsymbol{\omega}\times\boldsymbol{u} + \frac{1}{2}\frac{C_t}{\Delta}U_\infty^2 \boldsymbol{n}_d &= \boldsymbol{0} && (2)\\ \frac{\partial k}{\partial t} + \boldsymbol{u}\cdot\nabla k - \nabla\cdot\left(\frac{\nu_t}{\sigma_k}\nabla k\right) + \frac{C_\mu}{\nu_t}k^2 &= P_k && (3)\\ \frac{\partial \varepsilon}{\partial t} + \boldsymbol{u}\cdot\nabla \varepsilon - \nabla\cdot\left(\frac{\nu_t}{\sigma_\varepsilon}\nabla \varepsilon\right) + \frac{C_2}{k}\varepsilon^2 &= C_1' C_\mu k S && (4)\\ \nu_t &= C_\mu \frac{k^2}{\varepsilon}. && (5) \end{aligned}$$

where the unknowns are the velocity field $\boldsymbol{u}$, pressure p, turbulent kinetic energy k, dissipation rate of turbulent kinetic energy ε, and turbulent viscosity ν_t (computed with the diagnostic equation (5)). The fourth term on the left hand side (LHS) of momentum equation (2) models the Coriolis force being $\boldsymbol{\omega}$ the Earth's angular velocity. The sixth term on the LHS of equation (2) is the actuator disc force, which is active only inside the disc volume, where C_t is the thrust coefficient, U_∞ is the free-stream velocity at hub height, $\boldsymbol{n}_d$ the disc normal unit vector (pointing opposite to inflow), and Δ is the thickness of the disc. The forces inside each disc volume are uniformly distributed. In the turbulence equations (3)-(4), the term $P_k = \nu_t S$ is the kinetic energy production due to shear stress, with $S = \boldsymbol{\nabla}^s\boldsymbol{u} : \boldsymbol{\nabla}^s\boldsymbol{u}$ ($\boldsymbol{\nabla}^s$ denotes the symmetrical gradient operator). For the coefficients of the k-ε modified model we follow Panofsky and Dutton [8] and adopt:

$$C_\mu = 0.0333; \quad C_1 = 1.176; \quad C_2 = 1.92; \quad \sigma_k = 1.0; \quad \sigma_\varepsilon = \frac{\kappa^2}{C_\mu^{1/2}(C_2 - C_1)} \qquad (6)$$

where κ is the Von Karman constant. The coefficient C_1' in the RHS of the dissipation equation (4) is a modified coefficient, originally proposed by Apsley and Castro [6], to prevent the increase of mixing length $l_m = C_\mu^{3/4} k^{3/2}/\varepsilon$ above a maximum value $l_{\max}$ when accounting for Coriolis effects:

$$C_1' = C_1 + (C_2 - C_1)\frac{l_m}{l_{\max}} \tag{7}$$

where $l_{\max}$ is calculated as [9] :

$$l_{\max} = 0.00027\frac{|\boldsymbol{u}_g|}{2|\boldsymbol{\omega}|\sin\lambda} \tag{8}$$

being $\boldsymbol{u}_g$ the geostrophic wind velocity and λ the latitude. Note that, if no Coriolis forces are considered (*i.e.* $|\boldsymbol{\omega}| = 0$), then $l_{\max} \to \infty$ and $C_1' = C_1$.

2.2. k-ε realizable model with Coriolis effects

The k-ε realizable model proposed by Shi et al. [4] presents the advantage of satisfying realizability conditions on the Reynolds stresses. This model is known to improve the accuracy of flows involving detachment and re-circulations, and specifically, to enhance the prediction of wakes when using actuator disc models [5]. The realizable model shares the same turbulent kinetic energy equation (3) with the standard k-ε model. However, differences exist for the dissipation rate (4) and turbulent viscosity (5) equations. The turbulent viscosity ν_t is calculated from (5) with a variable C_μ that depends on the local values of k, ε and velocity gradients [4]. In turn, the dissipation equation is derived from the vorticity fluctuation transport equation, which is modified in the present work to use the model coefficients for ABL flows (6) and corrected with the Apsley and Castro limitation model to account for Coriolis effects in the ABL:

$$\frac{\partial \varepsilon}{\partial t} + \boldsymbol{u}\cdot\nabla\varepsilon - \nabla\cdot\left(\frac{\nu_t}{\sigma_\varepsilon}\nabla\varepsilon\right) + \frac{C_2}{k+\sqrt{\nu\varepsilon}}\varepsilon^2 = C_1'' S\varepsilon \tag{9}$$

$$C_1'' = C_{1r} + \left(C_2 C_{\mu 0}^{1/2} - C_{1r}\right)\frac{l_m}{l_{\max}} \tag{10}$$

$$C_{1r} = f\max\left(0.43, \frac{\eta}{\eta + 5/f}\right) \quad \text{with} \quad f = \left(\frac{C_{\mu 0}}{0.09}\right)^{1/2}, \quad \eta = (2S)^{1/2}\frac{k}{\varepsilon} \tag{11}$$

where ν is the laminar viscosity and $C_{\mu 0}$ is the value taken by coefficient C_μ under homogeneous shear, boundary layer logarithmic profile.

2.3. Boundary conditions

Proper boundary conditions need to be added to the Navier Stokes (1)-(2) and turbulence k-ε (3)-(4) ((3) and (9) for the realizable model) equations. The boundaries of the computational domain are split into inflow, outflow, bottom and top.

- **On the inflow boundary** a vertical profile is imposed for inflow velocity $\boldsymbol{u}$ and turbulence unknowns k and ε. The profiles are generated from a single-column (1D) precursor simulation (*i.e.* flat terrain and uniform roughness) .
- **On the outflow boundary** geostrophic pressure and no shear stress are imposed for the momentum equation and symmetric boundary conditions (no gradient) are imposed for the turbulence unknowns.
- **On the top boundary** symmetry boundary conditions are imposed for tangential velocity and turbulence unknowns. The normal velocity component is fixed to zero (*i.e.* $\boldsymbol{u}\cdot\boldsymbol{n} = 0$) and pressure is set to geostrophic.

- **On the bottom boundary** a wall law satisfying the Monin-Obukhov equilibrium is imposed to the momentum and turbulence equations removing a boundary layer of thickness δ_w. The imposed shear stress $\boldsymbol{\tau}_w$ tangent to the wall is expressed in terms of two velocity scales, namely u_{*v} and u_{*k}, based on the tangent velocity and the turbulent kinetic energy respectively:

$$\boldsymbol{\tau}_w = -u_{*v} u_{*k} \frac{\boldsymbol{u}}{|\boldsymbol{u}|}; \quad u_{*v} = \frac{|\boldsymbol{u}|\kappa}{\ln\left(1 + \frac{\delta_w}{z_0}\right)}; \quad u_{*k} = k^{1/2} C_\mu^{1/4} \tag{12}$$

where $\boldsymbol{u}$ is the component of the velocity tangent to the wall and $|\boldsymbol{u}|$ denotes its norm. The friction velocity u_{*v} is obtained from the neutral atmospheric velocity profile at a distance δ_w from the wall, being κ the Von Karman constant and z_0 the roughness length of the terrain. Finally, zero diffusion through the wall is imposed for the turbulent kinetic energy ($\boldsymbol{\nabla} k \cdot \boldsymbol{n} = 0$) and ε is imposed as:

$$\varepsilon = \frac{u_{*k}^3}{\kappa\left(\delta_w + z_0\right)} = \frac{k^{3/2} C_\mu^{3/4}}{\kappa\left(\delta_w + z_0\right)} \tag{13}$$

3. Numerical aspects and implementation

3.1. Hybrid mesh generation

The objective is to generate high-quality conformal hybrid meshes specifically designed for the simulation of ABL flows over complex terrains using actuator discs. The required mesh inputs include topography and terrain roughness, turbine characteristics (location, diameter and hub height), and wind inflow direction to determine the orientation of the actuator discs. To compute the wind inflow direction for each turbine we perform a precursor simulation using the background ABL mesh without turbines. Given these data and parameters, the hybrid mesh generation procedure is fully automatic and consists on 3 main steps (see [10] for further details):

i) Surface mesh generation. First, a 2D flat surface quadrilateral mesh is generated with three differentiated zones, namely *farm*, *transition* and *buffer* zones. The *farm* zone contains the area of interest and has a smaller element size. The *transition* zone surrounds the *farm* and has elements of increasing size outwards. Finally, the external *buffer* zone has coarser elements and is designed to accommodate the inflow and outflow conditions. This 2D flat mesh is then projected to the topography to obtain a 3D surface mesh. The underlying topography and terrain roughness are assimilated from a data file having any of the standard formats (point cloud, cartesian grid, contour levels, etc.) and filtered to remove data noise using the signal processing smoothing method presented in [11]. In certain zones with high topographic gradients, this simple projection can result on low quality elements (see Fig. 1(a)). For this reason, the 3D surface mesh is optimized maximizing the elemental quality presented in [12, 13] and imposing the nodes position over the exact topography to avoid lost of real geometry representation, using the process presented in [14, 15]. Fig. 1(b) shows the surface mesh after optimization, where a significant quality improvement is clearly observed. Only the *transition* and *farm* zones contain assimilated topography and roughness data whereas, in contrast, the *buffer* zone is left flat to guarantee consistence with the inflow profiles.

ii) Semi-structured volume mesh generation. Second, a semi-structured volume mesh of hexahedral elements is generated extruding the surface mesh into layers using a desired geometrical growing ratio to have higher vertical resolution near the ground. The extruding direction for a given node is computed as the *pseudo-normal* direction [16], using an average

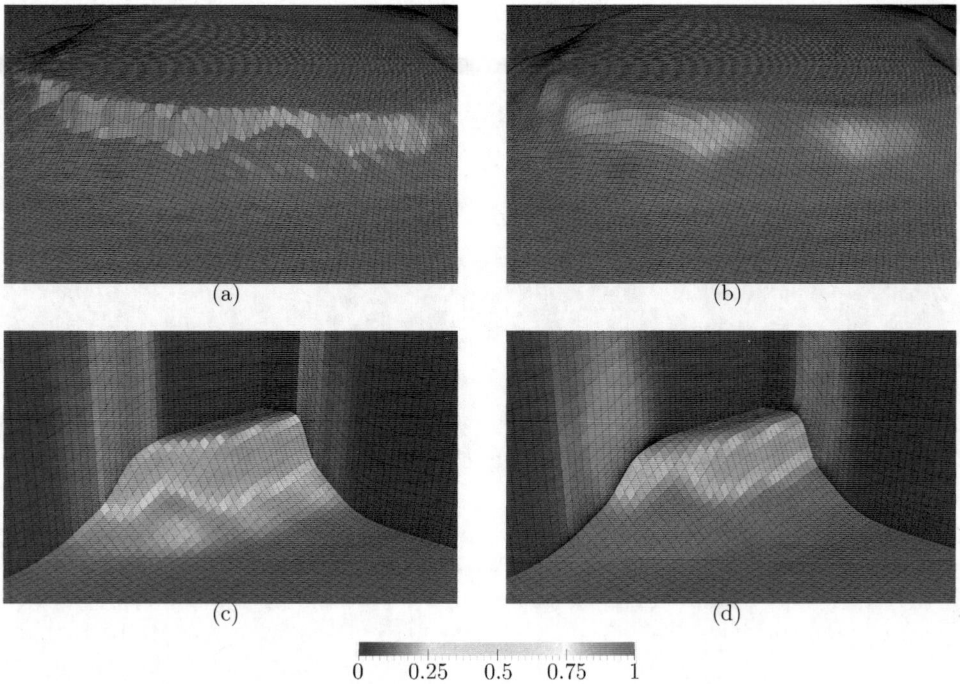

Figure 1. Mesh generation steps for the Bolund hill: (a) not optimized surface mesh, (b) optimized surface mesh, (c) not optimized volume mesh, and (d) optimized volume mesh. Elements are colored depending on their quality (0 degenerated, 1 orthogonal elements with the desired size). The quality of the elements improves from a minimum of 0.18 up to 0.49.

normal of the adjacent elements, that maximizes the orthogonality of the new generated layer. Each extruding step is combined with a non-linear optimization [17, 18] of the element quality to improve the mesh configuration before generating a new layer of hexahedra (see Fig. 1(d)).

iii) Actuator disc insertion and hybrid mesh. Finally, the third step consists on inserting the discs and refining the mesh around and downstream each disc conserving conformity. Firstly, the area surrounding each turbine is emptied. The disc is discretized using hexahedra, and the resulting disc mesh is extruded upstream, downstream and radially. Different element growing factors are used across each direction in order to smoothly tend to the element size of the background mesh in the different regions. At this stage, the background mesh and the mesh surrounding each disc are disconnected and pyramids and tetrahedra are used to perform a conformal union, thus leading to a final hybrid mesh. Pyramid elements are generated at the faces of the hexahedra facing other hexahedra, while tetrahedra are generated at the faces facing the gap between meshes. In this way, the gap between the different meshes is now bounded by triangles, which it is exploded to conform using the tetrahedral mesher TetGen [19]. The process is depicted in Figure 2 for a single turbine case.

3.2. Actuator disc model

The force exerted by the wind turbine over the flow is modeled as a uniformly loaded disc by means of the force term $\frac{C_t}{2\Delta}U_\infty^2\boldsymbol{n}_d$, where the thrust coefficient C_t is supplied by the manufacturers as a thrust coefficient curve depending on the undisturbed wind velocity $C_{tm}(U_\infty)$. However, in the case of wind farms, there is no obvious approach to estimate the free stream

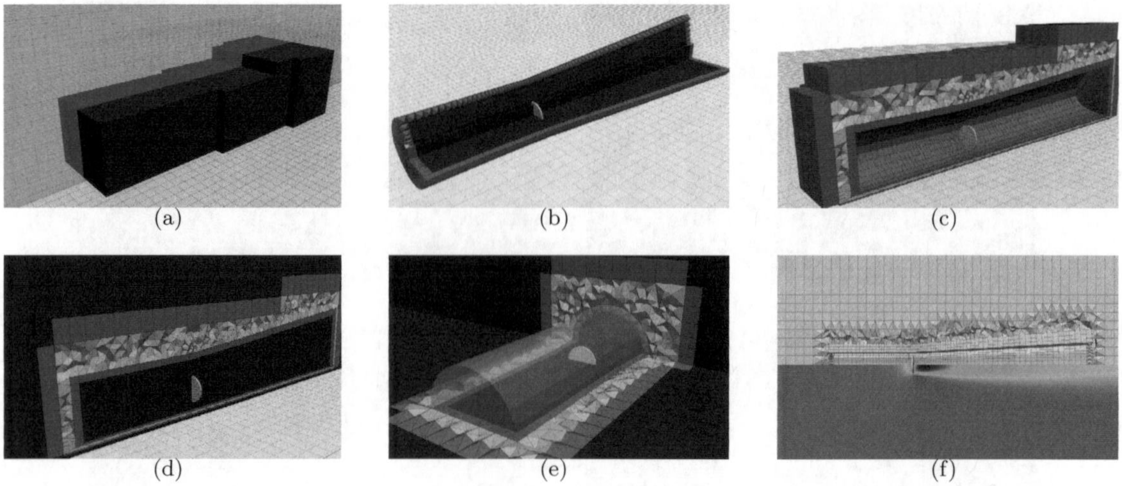

Figure 2. Hybrid mesh generation process for a single actuator disc. (a) Background structured mesh (grey) and region to be removed. (b) Wake and upwind mesh around the disc. (c) Transition pyramids and tetrahedra. (d,e) Final hybrid mesh. (f) Final mesh and computed velocity speedup (zenith view). In (a)-(e), elements are colored with respect to their element type: blue for hexahedra, green for tetrahedra, and red for pyramids. In (f), the velocity speedup color scale ranges from 0.66 (dark blue) to 1.05 (dark red).

velocity U_∞ (and therefore C_t) because the wind turbine power and thrust curves are usually provided for single-machine operation rather than operation in the wake of another turbine. For this reason, we relate the free stream velocity U_∞ to the velocity at hub height U_{hub} in terms of the thrust coefficient using one dimensional momentum theory. For high thrust coefficients momentum theory is no longer valid and we use the empirical relationship developed by Glauert [20] to obtain the theoretical thrust coefficient C_{ta}:

$$C_{ta}(a) = \begin{cases} 4a(1-a) & a < 0.4 \\ 0.889 - \left(0.0203 - (a-0.143)^2\right)/0.6427 & a \geq 0.4 \end{cases} \tag{14}$$

$$a = 1 - \frac{U_{hub}}{U_\infty} \tag{15}$$

where a is the axial induction factor. The velocity at hub height U_{hub} is calculated as the wind velocity component perpendicular to the disc surface averaged over the entire disc volume. To compute the proper value of U_∞ (and C_t) it is standardly posed an iterative procedure [3]. Given an initial guess for U_∞ it is first computed $C_{tm}(U_\infty)$, following, the induction factor a is updated in terms of $C_{ta} = C_{tm}(U_\infty)$ (inverse of Eq. (14)) and, finally, a new U_∞ is computed in terms of a (from Eq. (15)) until U_∞ converges to a fixed value (verifying $C_{tm} = C_{ta}$). However, this iterative scheme diverged in some of the tested complex cases, even using relaxation methods. Herein, introducing (15) in (14) we rewrite the iterative problem in terms of U_∞ and we translate it into solving the non-linear equation

$$f(U_\infty) = C_{tm}(U_\infty) - C_{ta}\left(1 - \frac{U_{hub}}{U_\infty}\right) = 0. \tag{16}$$

To solve (16) we use the bisection method, exploiting that the equation is one-dimensional and avoiding to compute the derivatives of the target function f. We found this nonlinear method to always converge to a proper value of U_∞ in all the tested cases.

3.3. Numerical method: linearization, discretization scheme and linear solvers

The RANS models have been implemented in Alya, an in-house high performance computing (HPC) code able to run large-scale applications. The code was recently tested on 100,000 processors with a parallel efficiency above 90% [21].

The Navier-Stokes and turbulence equations are discretized using a stabilized finite element method using equal interpolation for all the unknowns. As stabilization scheme we used the Algebraical Subgrid Scale method (ASGS) [22] extended for nonlinear equations [23], which gives stability to convection and Coriolis dominating terms in the momentum equation and to convection and reactive terms in the turbulence equations, removing spurious oscillations. The ASGS stabilization method gives also stability to pressure, allowing equal interpolation spaces for pressure and velocity. The velocity-pressure problem is decoupled using an Orthomin solver [24] that converges to the monolithic scheme.

A robust finite element scheme written in block-triangular form [25] is obtained for the k-ε equations (3)-(4)/(9). A priori, the k-ε equations are "well behaved" since the diffusion and reaction coefficients are positive. However, the numerical scheme cannot always guarantee positiveness (numerical oscillations occur) and sign variations in the reactive terms leading to loss of stability. In order to avoid instabilities and numerical convergence issues, ε and k are not allowed to drop below a predefined limit by applying a clipping. In addition, the innermost iterative loops of the k and ε equations (3)-(4)/(9) are linearized using a Newton-Raphson scheme for the quadratic terms, considering ν_t and P_k constants within the innermost loops. The stabilization of the reactive terms of the dissipation equation (4)/(9) has been found essential to achieve convergence of the unknowns. Once the algebraical system of equations are obtained, a Deflated Conjugate Gradient [26] solver with a linelet pre-conditioner [27] is used to solve the pressure, and a Generalized Minimizing Residual (GMRES) solver is used for the velocity and turbulence unknowns, leading to un-symmetric problems.

3.4. Validation

Figure 3 shows a basic model validation for the well-known Sexbierum benchmark case [5], in which wind velocity deficits and added turbulence intensities are analyzed at different transects downstream of a single isolated turbine. Four different simulations were performed using the standard modified and realizable RANS models accounting and not accounting for Coriolis forces. The obtained results are in very good agreement with those presented in [5]. As expected, the realizable model gives more accurate results than the standard modified model downstream of the actuator disc. Note also how the effect of the Coriolis force is to increase the wind velocity deficit along the wake.

4. Application to an onshore wind farm

As an illustrative application example, this section shows results for an onshore wind farm located in Spain and consisting on 165 turbines having a diameter of 77 m and a hub height of 80 m (Fig. 4(a)). The computational domain is 17×14 km^2 with the top at 2 km above the highest terrain elevation. The final horizontal mesh resolution was determined after a preliminary mesh convergence analysis without considering the presence of turbines. We found that an horizontal mesh resolution of 25 m guarantees a numeric accuracy of L_2-norm of the error ($\|e\|_2 := (\int e^2)^{1/2}$) below the 0.05% (discretization errors were computed against the solution from a finer mesh). Note from Fig. 4(b) how the mesh convergence of the solver is quadratic even for complex terrain. After introducing the actuator discs, the final hybrid mesh (25 m resolution at surface) is composed of 21.7 M elements (16M hexahedra, 4M tetrahedra, 1M pyramids). The actuator discs were meshed with elements of 11.6 m (15% of the turbine diameter) on the disc plane and assuming a disc thickness of 4 m (6% of its diameter).

IOP Conf. Series: Journal of Physics: Conf. Series **854** (2017) 012002 doi:10.1088/1742-6596/854/1/012002

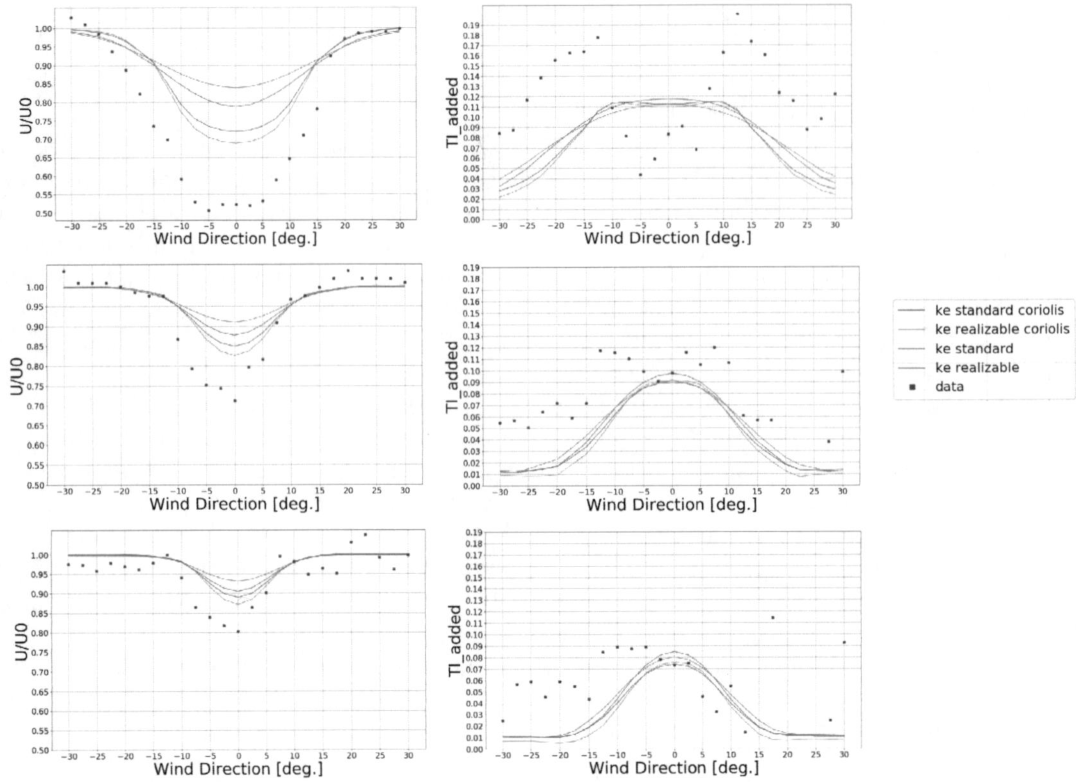

Figure 3. Wind speed deficit (left) and added turbulence intensity (right) in the flow direction at a downstream distance of 2.5 (first row), 5.5 (second row) and 8 (third row) diameters.

The CPU time required to complete the entire meshing process was of 284 seconds using a single thread on a MacBookPro with core i7-2.7GHz. Figures 4(c) and 4(d) show the velocity speed up and the Turbulent Kinetic Energy (TKE) at hub height, respectively. Note that, for this particular wind direction, many wind turbines are affected by the upstream turbines.

Figure 5 compares hub wind velocities and TKE from 4 different simulations (2 RANS models with and without Coriolis terms). Note how, when the Coriolis effect is not accounted for, the standard modified model predicts higher hub velocities than the realizable model. This is due to the over-prediction of the wake velocity in the standard model [5]. However, when accounting for Coriolis force, both models predict similar hub velocities and TKE at hub height. The mixing length limitation model counteracts the over-prediction of Reynolds stresses in the standard model behind the discs, leading to larger wakes. When accounting for Coriolis force the use of the mixing length limitation model decreases the obtained TKE at hub height for each RANS model in almost all wind turbines. This effect is stronger for the standard modified model, which predicts the lowers TKE values at hub height. Note also how, when using the realizable model, the effect of accounting for Coriolis force over the obtained wind velocity and TKE at hub height is much smaller than in the standard k-ε model.

5. Conclusions

A wind simulation framework for offshore and onshore wind farms has been implemented in the Alya HPC solver considering two different RANS models and simulating wind turbines as actuator discs. This includes an automatic hybrid mesh generator tailored to wind farms,

IOP Conf. Series: Journal of Physics: Conf. Series **854** (2017) 012002 doi:10.1088/1742-6596/854/1/012002

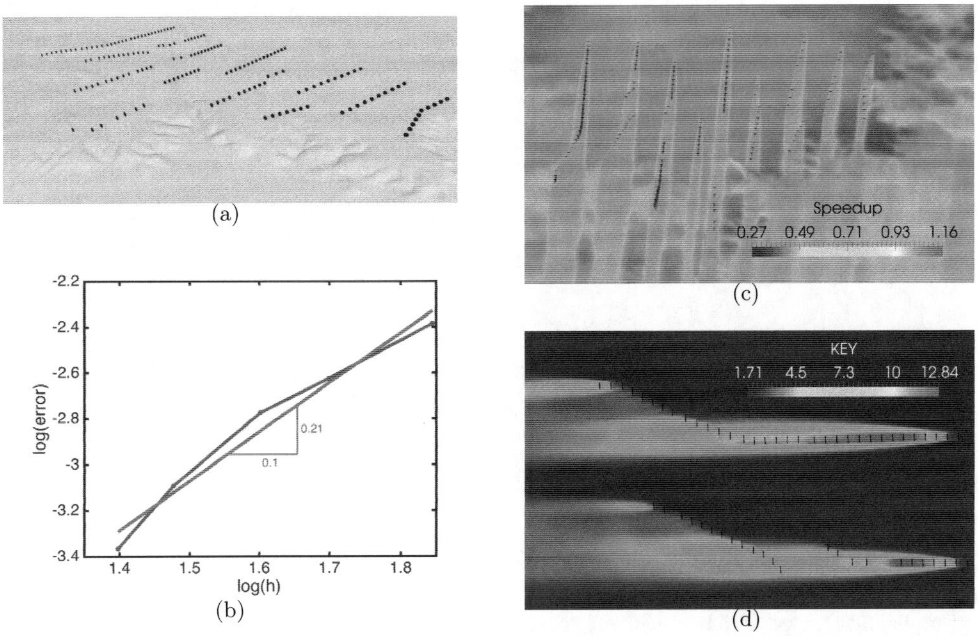

Figure 4. Onshore wind farm example. (a) Topography and turbine location. (b) Horizontal resolution mesh convergence analysis. The blue line indicates the error values (with respect to the finest mesh) depending on the surface mesh size h_s. For reference, the red line shows the quadratic slope. (c) Wind speedup at hub height (80 m) using the realizable model with Coriolis effect and an inflow velocity profile with velocity 11 m/s at 82 m above terrain. (d) Same for turbulent kinetic energy.

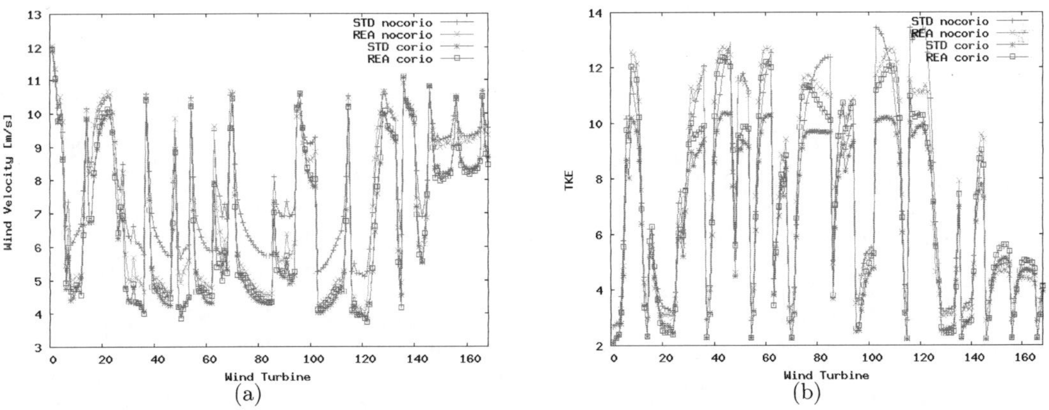

Figure 5. Simulated hub height wind velocities (a) and turbulent kinetic energy (b) at 165 turbines for the standard modified (STD) and realizable (REA) $k-\varepsilon$ RANS models with and without Coriolis terms.

i.e. able to reproduce the ABL, capture the underlying terrain, and refine areas upstream and downstream turbines in a conformal way. The presented mesher is fully automatic and features quadratic convergence to the topography.

Regarding the actuator disc model, a new robust nonlinear method has been presented to calculate the thrust coefficient and free stream velocity for each wind turbine. This new approach has proved to converge in all the tested cases, in contrast with the standard iterative process, which was not convergent for some tested complex onshore wind farm configurations.

The k-ε realizable model has been modified to account for Coriolis effects in the atmospheric boundary layer. The realizable and the standard modified k-ε models obtain very similar results over flat and homogeneous terrains without wind turbines. However, on complex terrains, the realizable model predicts lower velocities in the wakes of wind turbines, specially when not accounting for Coriolis effects. The obtained results of the standard and realizable models are in closer agreement when accounting for Coriolis force.

Acknowledgments

This work has been supported by the EU H2020 projects New European Wind Atlas ERA-NET PLUS (NEWA), High Performance Computing for Energy (HPC4E, grant agreement 689772), and the Energy oriented Centre of Excellence (EoCoE, grant agreement 676629). We also thank two anonymous reviewers for their constructive comments on the first version of the manuscript.

References

[1] Sanderse B, van der Pijl S and Koren B 2011 *Wind Energy* **14** 799–819
[2] Rethore P E 2009 *Wind Turbine Wake in Atmospheric Turbulence* Ph.D. thesis Riso National Laboratory for Sutanaible Energy Technical University of Denmark
[3] Prospathopoulos J M, Politis E S, Rados K G and Chaviaropoulos P K 2011 *Wind Energy* **14** 285–300 ISSN 1099-1824
[4] Shih T H, Liou W W, Shabbir A, Yang Z and Zhu J 1995 *Computers & Fluids* **24** 227–238
[5] Cabezón D, Migoya E and Crespo A 2011 *Wind Energy* **14** 909–921
[6] Apsley D and Castro I 1997 *Boundary-Layer Meteorology* **83** 75–98
[7] Avila M, Folch A, Houzeaux G, Eguzkitza B, Prieto L and Cabezón D 2013 *Procedia Computer Science* **18** 2157 – 2166 2013 International Conference on Computational Science
[8] Panofsky H A and Dutton J 1984 *Atmospheric turbulence, models and methods for engineering applications* (Wiley)
[9] AK B 1962 *J. Geophysic. Res* **97** 3095–3102
[10] Gargallo-Peiró A, Avila M, Owen H, Prieto L and Folch A 2015 *Procedia Engineering* **124** 239–251
[11] Taubin G 1995 *Proc. 22nd Annual Conference on Computer Graphics and Interactive Techniques* SIGGRAPH '95 (New York, NY, USA: ACM) pp 351–358 ISBN 0-89791-701-4
[12] Roca X, Gargallo-Peiró A and Sarrate J 2012 *Proc. 20th Int. Meshing Roundtable* (Springer International Publishing) pp 365–383
[13] Gargallo-Peiró A, Roca X, Peraire J and Sarrate J 2013 *Proc. 21st Int. Meshing Roundtable* (Springer International Publishing) pp 85–102
[14] Gargallo-Peiró A, Roca X and Sarrate J 2014 *Comput. Mech.* **53** 587–609 ISSN 0178-7675
[15] Gargallo-Peiró A, Roca X, Peraire J and Sarrate J 2015 *Int. J. Numer. Meth. Eng.* **106** 1100–1130 ISSN 1097-0207
[16] Roca X and Sarrate J 2010 *Eng. Comput.* **26** 391–406
[17] Gargallo-Peiró A, Roca X, Peraire J and Sarrate J 2015 *Int. J. Numer. Meth. Eng.* **103**(5) 342–363
[18] Gargallo-Peiró A, Roca X, Peraire J and Sarrate J 2015 *Eng. Comput.* **31**(3) 423–437
[19] Si H 2015 *ACM Trans. Math. Softw.* **41** 11:1–11:36 ISSN 0098-3500
[20] M E and S S 1987 *Wind Turbine Engineering Design* (Van Nostrand Reinhold Company)
[21] et al M V 2016 *Journal of Computational Science* **14** 15 – 27 ISSN 1877-7503 the Route to Exascale: Novel Mathematical Methods, Scalable Algorithms and Computational Science Skills
[22] Codina R 1998 *Computer Methods in Applied Mechanics and Engineering* **156** 185 – 210 ISSN 0045-7825
[23] Avila M, Codina R and Principe J 2015 *International Journal of Numerical Methods for Heat & Fluid Flow* **25** 1361–1384 ISSN 0961-5539
[24] Houzeaux G, Aubry R and Vazquez M 2011 *Computers & Fluids* **44** 297 – 313 ISSN 0045-7930
[25] Codina R and Soto O 1999 *Int. J. Num. Meth. Fluids* **30** 309–334
[26] Lohner R, Mut F, Cebral J R, Aubry R and Houzeaux G 2011 *International Journal for Numerical Methods in Engineering* **87** 2–14 ISSN 1097-0207
[27] Soto O, Lohner R and Camelli F 2003 *International Journal of Numerical Methods for Heat & Fluid Flow* **13** 133–147

Large-Eddy Simulation of turbine wake in complex terrain

J. Berg[1], N. Troldborg[1], N.N. Sørensen[1], E. G. Patton[2] and P. P. Sullivan[2]

[1]DTU Wind Energy, Frederiksborgvej 399, 4000 Roskilde, Denmark jbej@dtu.dk
[2] National Center for Atmospheric Research, Boulder, CO, US.

E-mail: `jbej@dtu.dk`

Abstract. We present Large-Eddy Simulation results of a turbine wake in realistic complex terrain with slopes above 0.5. By comparing simulations including and without the wind turbine we can estimate the induction factor, a, and we show how the presence of a strong recirculation zone in the terrain dictates the positioning of the wake. This last finding is in contrast to what would happen in gentle terrain with no substantial increase of turbulent kinetic energy in the terrain induced wakes.

1. Introduction

Wind turbine wakes are characterized by reduced wind speed and increased levels of turbulent kinetic energy. While wind turbine wakes have been studied extensively in homogeneous terrain (and in the absence of terrain) using both Reynolds Averaged Navier-Stokes methods (RANS) [21, 22, 23] and Large-Eddy Simulation methods (LES) [2, 13, 9, 20, 3, 4], only very few modelling studies exist of wakes in complex terrain.

Most of the numerical investigations of rotor wakes in complex terrain are carried out using RANS [12, 5], whereas the only systematic study of accuracy of LES in topography, to our knowledge, has been carried out recently by Shamsodddin and Porté-Agel [15]. In the latter study LES was carried out on five turbines located on a Gaussian hill. Their predictions were compared to measurements obtained in a wind tunnel and rather good agreement was achieved. In the study the hill height to hill half length ratio was 0.5 and the hill height was only about 12% higher than the diameter of the considered turbines. Thus, the complexity of the considered topography in this study is rather limited.

In the present contribution we present LES simulations of the wake of a wind turbine rotor positioned at Perdigão in Portugal - a complex site characterized by a double ridge configuration, and currently the target of a large international measurement campaign under the EU New European Wind Atlas Project[1]. The focus of our work is not the flow around Perdigão as such, but rather the interaction between the terrain and the turbine wake.

2. LES Setup

The LES code is presented in [17, 19]. It is a pseudo-spectral model; wave number representation in the two horizontal directions and finite differencing in the vertical direction. The Boussinesq approximation is adopted and the governing equations are integrated forward in time using adaptive time stepping (in the current simulations, Δt vary between 0.4 s and 0.7 s) in a terrain following grid: Whereas the horizontal coordinate lines are Cartesian the vertical z-coordinate is stretched exponentially from the ground and up as a function of the local terrain height. The pressure is solved during an iterative procedure of a general Poisson equation. Compared to running the model on flat terrain ([18]) this last step slows down the code substantially, depending on the complexity of the terrain. We use the Deardorff SGS model in the form as presented by [10] and [16].

2.1. *Turbine representaion*

In complex terrain we do not have a well defined upstream undisturbed velocity, U_∞, which is often used when estimating the total force the turbine exerts on the flow. Instead we exploit the classical expression [8],

$$U_\infty = \frac{U_d}{1-a}, \tag{1}$$

where a is the induction factor and U_d is the velocity on the rotor disk, which can easily be estimated in LES simulations. As in the JHU-LES code [6] turbines are represented by disks with thrust force

$$F_t = -\frac{1}{2}\rho C_T' \langle \bar{u} \rangle_d^2 \frac{\pi}{4} D^2, \tag{2}$$

where C_T' is related through 1-d momentum theory to the thrust coefficient, C_T, and induction factor, a, through

$$C_T' = \frac{C_T}{(1-a)^2} = \frac{4a}{1-a}. \tag{3}$$

D is the rotor diameter and $\langle \bar{u} \rangle_d$ is the rotor area averaged streamwise wind speed. Since a wind turbine act as a low pass filter we also averaged in time: a 10-minute exponential time filter is applied, and its effect on $\langle \bar{u} \rangle_d$ from a LES simulation can be observed in Figure 1. Its effect on the wake loss, however, is almost negligible (not shown in this paper).

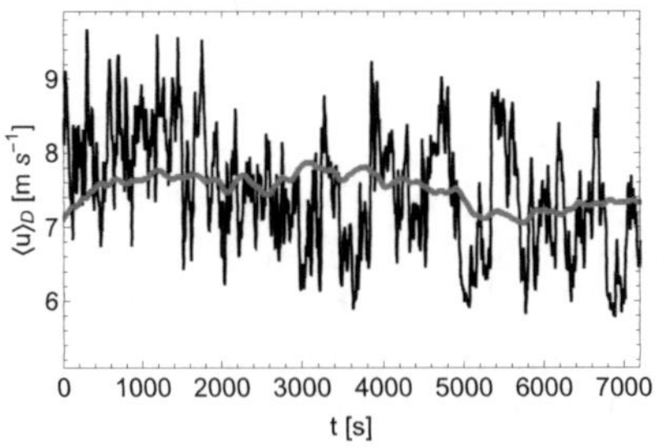

Figure 1: Rotor disk velocity, $\langle u \rangle_d$ with (*red*) and without time averaging (*black*)

In this paper we choose $a = 1/4$ corresponding to $C_T = 3/4$ and finally $C_T' = 4/3$.

The force is distributed proportional to the fractional cross area, ξ, of a given cell covered by the rotor. The force per unit volume in the x direction (perpendicular to the rotor plane) applied to a given cell with indices, (i, j, k), then becomes

$$f_x(i,j,k) = -\frac{1}{2}\rho C_T' \langle \bar{u} \rangle_d^2 \frac{\xi(j,k)}{\Delta x}, \tag{4}$$

where Δx is the length of the cell.

2.2. Cases and model spin-up

The simulations are carried out on a 256 × 128 × 128 grid spanning 5120 × 2560 × 3000 km^3 on a terrain representing the double ridge Perdigão in Portugal ([24]). The terrain is smoothed (maximum slope is 0.57) with an exponential filter in wave number space in order to reduce the number of pressure iterations in the code, and hence reduce computer time. Since the focus of this contribution is on turbine wake interaction and not on the precise details of the flow around Perdigão, we do not consider this an issue.

The driving pressure gradient is held constant throughout the study at $dP/dx = -u_\star^2/H$, where $u_\star = 0.6\,\mathrm{m\,s^{-1}}$ and $H = 3000\,\mathrm{m}$ is the height of the computational domain. With $z_0 = 0.5\,\mathrm{m}$ we have $z_0/H = 1.67 \times 10^{-4}$. The simulations were started from an initial random incompressible velocity fields and ran for approximately $T_S = 100\,T_E$, with $T_E = H/u_\star$ before time stationarity was obtained. The periodic boundary conditions are maintained by adding a small buffer zone to the terrain. In this way we can ensure that there are no large gradients in the terrain height between the last row and the first row and between the last column and the first column of the terrain file. We have also tried to increase the buffer zone considerably, but did not observe any significant changes to the results presented in this paper (not shown). Since Perdigão is a highly complex site (there is always another bump on the road upstream) running with periodic boundary conditions might even be a more realistic option compared to a well defined inlet condition as is often the case in complex terrain studies [7]. Alternatively, precursor simulations similar to what was introduced by Munters et al. in [11] could be used, although the non-homogeneity of the terrain would add another layer of complexity.

In Figure 2 we show the terrain around Perdigão and in Figure 3 a snapshot of the instantaneous streamwise velocity component along the main transect. The wind direction is from the south west at 235 degrees along the main transect.

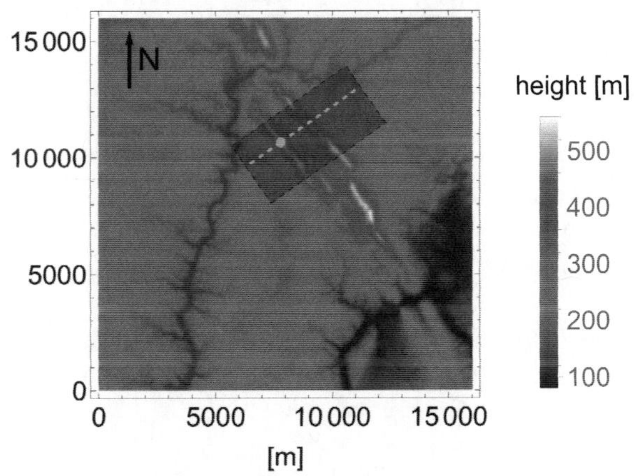

Figure 2: Perdigão terrain heights. The simulated area is represented by the shaded rectangle and the yellow line is the main transect.

3. Data

We analyse averaged LES data. The wind turbine is located on the first ridge as in reality. The average streamwise velocity and resolved turbulent kinetic energy from simulation without the presence of the wind turbine are plotted in Figure 4. The average is taken 10 hours of simulated real-time. With slopes above 0.5 the terrain is challenging and flow separation is

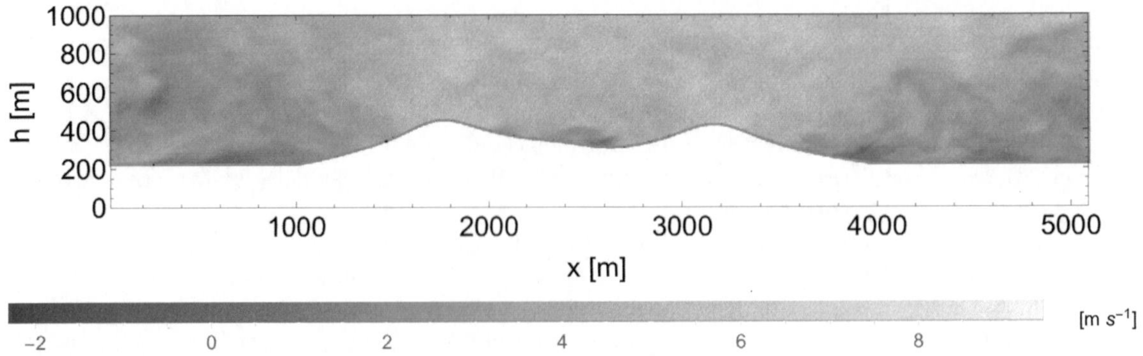

Figure 3: Instantaneous streamwise velocity along the main transect presented in Figure 2 from LES simulation without wind turbine.

present behind the ridge with associated recirculation zone and enhanced turbulence. The exact size and position of the recirculation zone is strongly dependent of the specific smoothing technique performed to the terrain as well as on the spatial resolution (not shown).

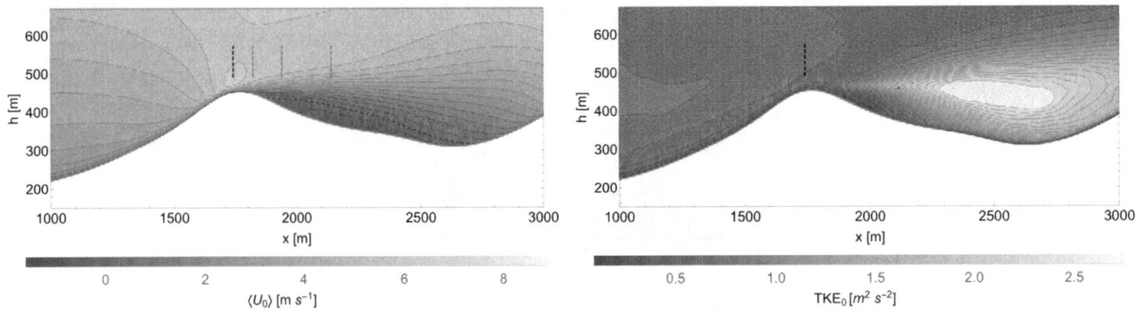

Figure 4: Average streamwise velocity without turbine along main transect (*left*) and resolved turbulent kinetic energy (*right*). The turbine position (not present in the simulation here) and downstream distances $1D$, $2.5D$ and $5D$ are shown for illustration.

In Figure 5 we compare the flow field around the first ridge with and without a wind turbine. The trouble in estimating the undisturbed velocity, U_∞ in eq. 1 motivates the definition of an alternative induction factor, $\tilde{a}$, defined as

$$\tilde{a} = \frac{\langle U_0 \rangle - \langle U_1 \rangle}{\langle U_{0,hub} \rangle}, \tag{5}$$

where $\langle U_0 \rangle$ and $\langle U_1 \rangle$ are the velocities from the LES simulations without and including the wind turbine, respectively. The denominator velocity is evaluated at hub height equal to 80 m. Since LES simulations are random in nature (here understood in the sense that individual realizations - here 30-min simulations - are different [25]), the quantities in eq. 5 are time dependent, and we have therefore used the velocity in the simulation without the wind turbine in the same time 30-min interval as the one including the wind turbine at the position of the wind turbine, i.e. on top of the first ridge.

In the left panel we observe induction factors on the disk slightly above the target value of 0.25, which is dictated by classical one dimensional momentum theory. The resolved turbulent kinetic energy presented in the right panel indicate expected enhanced production in the shear layers of the turbine wake, especially above the turbine, where a very strong narrow band is observed. The far wake is slightly pointing upwards. This fits the observations in Figure 4 where the narrowing of streamlines behind the hill above the flow separation zone enhances the turbulent kinetic energy extending to higher elevations compared to the hill height.

In Figure 6 we present plots of $\tilde{a}$ at fixed positions downstream. The error bars indicate the standard deviation of the many 30-min runs analysed. We observe quite large deviations in $\tilde{a}$ within the different 30-min runs.

In Figure 7 we observe how the wake stays more or less at fixed positions in the $y-z$ plane with only a slight move to the right for distances above $2.5D$. It thus follows the geometry of the terrain, which is almost symmetric around the main transect.

In Figure 13 and 15 of [15] it is observed how the wakes more or less follow the terrain, which in all cases are more gentle than the terrain in our study. Here, we show in contrary that the complexity of the terrain and the strong recirculation zone behind the ridge creates a strong shear layer which prevents the wake from following the terrain and blend with the wake of the ridge (besides a small region of excess turbulent kinetic energy present around $x \sim 2000\,\mathrm{m}$).

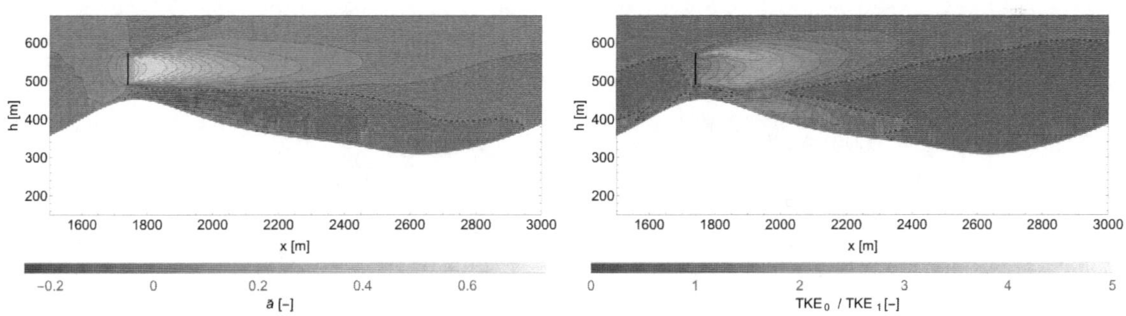

Figure 5: 30-minutes average plots of $\tilde{a}$ (*left*) and ratio of turbulent kinetic energy, TKE between simulations with and without the wind turbine (*right*). The dashed lines indicate where $\tilde{a}=0$ (*left*) and where the TKE ratio equals one (*right*).

4. Outlook

In the near future there will be measurements of the turbine wake itself at Perdigão. This will provide us with data that can be used to validate the results presented in this contribution. Another aspect that needs attention is whether the turbine representation used in this paper is too simplified: in complex terrain the mean wind is rarely horizontal and do not always follow the largest gradients in the terrain. In addition the inclusion of Coriolis force, stratification and inversion strength will force us to use a PID controller somewhat similar to what is presented in [14]. Perdigão also poses a challenge when it comes to resolution and domain size. The complexity and size of the terrain features of interest makes it computational very expensive if smaller scales are to be resolved while keeping the size of the domain constant. It is expected that the flow around Perdigão will change when we obtain a higher resolution and less smoothing (work in progress). The interaction with the wake will also change as a consequence. Thus in order to really numerically study the interaction between turbine wakes and realistic complex terrain as present at Perdigão much computer power is needed.

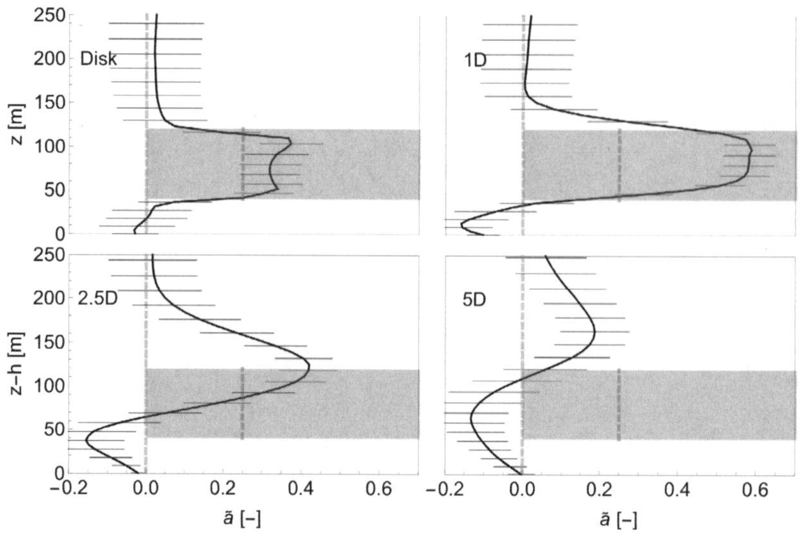

Figure 6: Profiles of the induction factor, $\tilde{a}$, at fixed positions downstream from the turbine. The rotor area is indicated by the gray shading and the red line indicates the $a = 0.25$. Error bars indicate the standard deviation.

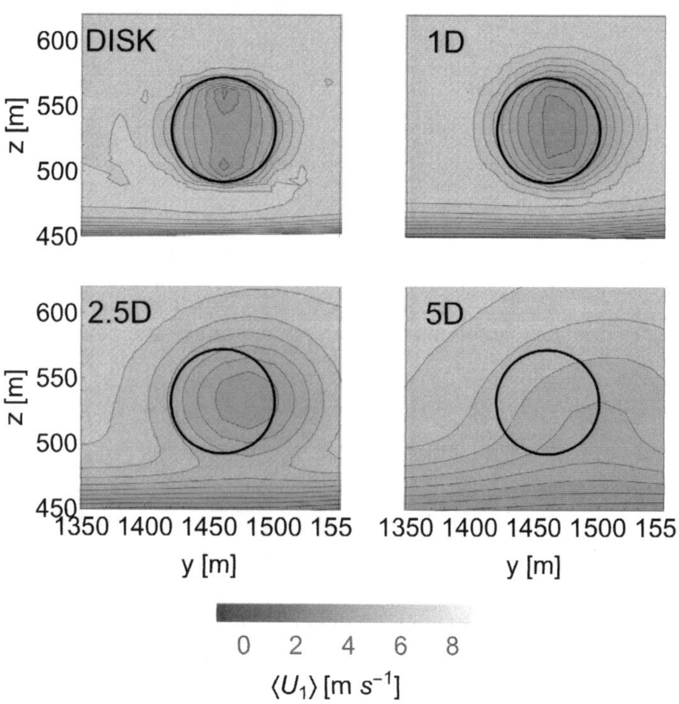

Figure 7: $\langle U_1 \rangle$ at fixed distances downstream of the rotor. The black circle represent the position and size of the rotor

References

[1] `http://www.neweuropeanwindatlas.eu`.
[2] M. Abkar, A. Sharifi, and F. Porté-Agel. Wake flow in a wind farm during a diurnal cycle. *J. Turbulence*, 17:420–441, 2016.

[3] D. Allaerts and J. Meyers. Large eddy simulation of a large wind-turbine array in a conventionally neutral atmospheric boundary layer. *Physics of Fluids*, 27:065108, 2015.

[4] D. Allaerts and J. Meyers. Boundary-layer development and gravity waves in conventionally neutral wind farms. *J. Fluid Mech.*, 814:95130, 2017.

[5] D. Cábezon, K.S. Hansen, and R.J. Barthelmie. Analysis and validation of CFD wind farm models in complex terrain . Effects induced by topography and wind turbines. *EWEC 2010*, pages 1–4, 2008.

[6] M. Calaf, C. Meneveau, and J. Meyers. Large eddy simulation study of fully developed wind-turbine array boundary layers. *Physics of Fluids*, 22(1):1–16, 2010.

[7] D. Cavar, P.-E. Réthoré, A. Bechmann, N. N. Sørensen, B. Martinez, F. Zahle, J. Berg, and M. C. Kelly. Comparison of openfoam and ellipsys3d for neutral atmospheric flow over complex terrain. *Wind Energy Science*, 1(1):55–70, 2016.

[8] M. O. L. Hansen. *Aerodynamics of Wind Turbines, 3.ed.* Routledge, Taylor Francis Group, 2015.

[9] S. Lee, M. Churchfield, P. Moriarty, J. Jonkman, and J. Michalakes. A Numerical Study of the Effects of Atmospheric and Wake Turbulence on Wind Turbine Dynamics. *J. Turbulence*, 13:1–32, 1989.

[10] C.-H. Moeng. A large-eddy-simulation model for the study of planetary boundary-layer turbulence. *Journal of the Atmospheric Sciences*, 41(13):2052–2062, 1984.

[11] W. Munters, C. Meneveau, and J. Meyers. Turbulent inflow precursor method with time-varying direction for large-eddy simulations and applications to wind farms. *Boundary-Layer Meteorology*, 159(2):305–328, 2016.

[12] E.S. Politis, J. Prospathopoulos, D. Cabezon, Kurt Schaldemose Hansen, P.K. Chaviaropoulos, and R.J. Barthelmie. Modeling wake effects in large wind farms in complex terrain: the problem, the methods and the issues. *Wind Energy*, 15(1):161–182, 2012.

[13] F. Porté-Agel, Y-T. Wu, H. Lu, and R.J. Conzemius. Large-eddy simulation of atmospheric boundary layer flow through wind turbines and wind farms. *J. Wind Eng. Ind. Aerodyn*, 99:154–168, 2011.

[14] A. Sescu and C. Meneveau. A control algorithm for statistically stationary large-eddy simulations of thermally stratified boundary layers. *Q. J. R. Meteorol. Soc.*, 140(683):2017–2022, 2014.

[15] S. Shamsoddin and F. Porté-Agel. Large-eddy simulation of atmospheric boundary-layer flow through a wind farm sited on topography. *Boundary-Layer Meteorol.*, 163(1):1–17, 2017.

[16] P. P. Sullivan, J. C. McWilliams, and Chin-Hoh Moeng. A subgrid-scale model for large-eddy simulation of planetary boundary-layer flows. *Boundary-Layer Meteorol.*, 71:247–276, 1994.

[17] P. P. Sullivan, J. C. McWilliams, and E. G. Patton. Large eddy simulation of marine atmospheric boundary layers above a spectrum of moving waves. *J. Atmos. Sci.*, 71:4001–4027, 2014.

[18] P. P. Sullivan and E. G. Patton. The effect of mesh resolution on convective boundary layer statistics and structures generated by large-eddy simulation. *J. Atmos. Sci.*, 68:2395–2415, 2011.

[19] P.P. Sullivan, E.G. Patton, and K.W. Ayotte. Turbulent flow over and around sinusoidal bumps, hills, gaps and craters derived from large eddy simulations. In *19th Conference on Boundary Layer and Turbulence, Keystone, CO*, 2010.

[20] N. Troldborg, F. Zahle, P-E Réthoré, and N.N Sørensen. Comparison of wind turbine wake properties in non-sheared inflow predicted by different computational fluid dynamics rotor models. *Wind Energy*, 2014.

[21] M. P. van der Laan, N. N. Sørensen, P.-E. Réthoré, J. Mann, M. C. Kelly, and N. Troldborg. The k-ε-f_p model applied to double wind turbine wakes using different actuator disk force methods. *Wind Energy*, 18(12):2223, December 2015.

[22] M. P. van der Laan, N. N. Sørensen, P.-E. Réthoré, J. Mann, M. C. Kelly, N. Troldborg, K. S. Hansen, and J. P. Murcia. The k-ε-f_p model applied to wind farms. *Wind Energy*, 18(12):2065, December 2015.

[23] M. P. van der Laan, N. N. Sørensen, P.-E. Réthoré, J. Mann, M. C. Kelly, N. Troldborg, J. G. Schepers, and E. Machefaux. An improved k-ε model applied to a wind turbine wake in atmospheric turbulence. *Wind Energy*, 18(5):889, May 2015.

[24] N. Vasiljević, J. M. L. M. Palma, N. Angelou, J. C. Matos, R. Menke, G. Lea, J. Mann, M. Courtney, L. F. Ribeiro, and V. M. M. G. C. Gomes. Perdigão 2015: methodology for atmospheric multi-doppler lidar experiments. *Atmos. Meas. Tech. Discussions*, 2017:1–28, 2017.

[25] John C. Wyngaard. *Turbulence in the Atmosphere.* Cambridge University Press, 2010.

Analysis of the reduced wake effect for available wind power calculation during curtailment

M P C Bontekoning[1], S Sanchez Perez-Moreno[1], B C Ummels[2] and M B Zaaijer[1]

[1] Delft University of Technology, [2] Ventolines BV

E-mail: contact@mpcbontekoning.nl

Abstract. With the increase of installed wind power capacity, the contribution of wind power curtailment to power balancing becomes more relevant. Determining the available power during curtailment at the wind farm level is not trivial, as curtailment changes the wake effects in a wind farm. Current best practice to estimate the available power is to sum the available power calculated by every wind turbine. However, during curtailment the changed local wind conditions at the wind turbines lead to inaccurate results at the wind farm level. This paper presents an algorithm to determine the available power of a wind farm during curtailment. Moreover, results of curtailment experiments are discussed that were performed on nearshore wind farm Westermeerwind to validate the algorithm. For the case where a single turbine is being curtailed, it is shown that the algorithm reduces the estimation error for the first downstream turbine significantly. Further development of the algorithm is required for accurate estimation of the second turbine. All further downstream turbines did not experience a change in wake conditions.

1. Introduction

The electricity market is currently experiencing a paradigm shift from a market dominated by generators running on fossil fuels, towards a more varied market with multiple renewable power sources. Wind power plays an important role in this transition. Notwithstanding the benefits of wind power, the transition also poses new challenges concerning the power balance in the electrical system. These challenges are mostly related to the variability and partial unpredictability of wind power. [1] This makes it difficult for the power supply to follow the demand, both on the short-term and long-term. From the two cases of power imbalances that can be identified — either the supply being higher than the demand, or vice versa — this research considers a surplus of power production. A method of mitigating this surplus is to (temporarily) reduce the power production.

It is important to continuously determine how much power is being curtailed, to allow settlement of the curtailed energy afterwards and to optimise the amount of curtailed power that provides a buffer for peaks in the demand. In order to calculate the curtailed energy ($E_{curtailed}$) during a certain period (T), equation 1 can be used.

$$E_{curtailed} = \int_T \left(P_{available}(t) - P_{produced}(t)\right) dt \tag{1}$$

The actual power produced ($P_{produced}$) is measured directly. For conventional power systems the maximum available power ($P_{available}$) is equal to the nominal capacity. However, for wind farms determining the available power is not trivial.

The available power of a wind farm depends on the installed capacity, the availability of each wind turbine and the wind conditions. The latter determines the production of upstream turbines directly, but also influences the wake development behind those turbines. This in turn, influences the power production of downstream turbines in the wind farm. During curtailment, less power is extracted from the wind and thus the wake effects are reduced. This leads to a wind speed increase at the downstream turbine and therefore to an apparent increase of its available power. [2] [3] This phenomenon is referred to as the *reduced wake effect*. Summing the individual available power estimations of all turbines, will therefore lead to an overestimation of the total wind farm available power during curtailment. However, this is currently the industry's best practice.

This paper presents and validates an algorithm to determine the available wind power of a wind farm during curtailment. For the validation, wind turbines in the nearshore wind farm Westermeerwind have been curtailed to measure the reduced wake effect and determine the reduction of the error of the algorithm with respect to the error of the current best practice. Westermeerwind is located in the middle of The Netherlands and operated by Ventolines BV.

2. The algorithm

The algorithm has two requirements: 1) deliver more accurate results than the current best practice and 2) be fast, allowing direct estimation of available power and settlement shortly after curtailment. Due to the validation study of Westermeerwind, the implementation of the algorithm could only use data signals from the turbines, so no data from external meteorological instruments were included, besides an air density correction.

2.1. Selection of wake models

In order to quantify the difference in wake losses during the normal and curtailed operation modes, proven wake models are used. Models that require computational fluid dynamics are excluded, as they are too computationally expensive. For this research, only the kinematic Jensen [4] and Larsen [5] wake models are considered. These wake models are quick to evaluate, but also yield a less accurate physical representation of the wake. It is part of this research to find if these wake models are useful for the determination of the available power. For the Jensen wake model the wake decay parameter is defined as in equation 2 [2], where TI refers to the local turbulence intensity measured at each turbine. Using the local turbulence intensity resulted in better estimations than using a single value for the entire wind farm. These two wake models were selected for their robustness, widely available validation studies in literature, and simple implementation.

$$k = 0.4 \cdot TI \tag{2}$$

2.2. Modelling of wakes during curtailment

The mentioned wake models are designed for the evaluation of the wind speed deficit at a certain position behind a single turbine. The boundary conditions for the application of the wake models are determined by the turbine state. Under curtailment this state is different from normal operation. Therefore, the changed state needs to be modelled to obtain the correct inputs for the wake models. This will be called the *curtailment model* and together with the thrust curve, the wake model and a method of mixing multiple wakes, it forms the *wake – curtailment model*, see figure 1. This model allows the calculation of the wind speed deficit at each turbine taking into account the curtailment of each turbine.

IOP Conf. Series: Journal of Physics: Conf. Series **854** (2017) 012004 doi:10.1088/1742-6596/854/1/012004

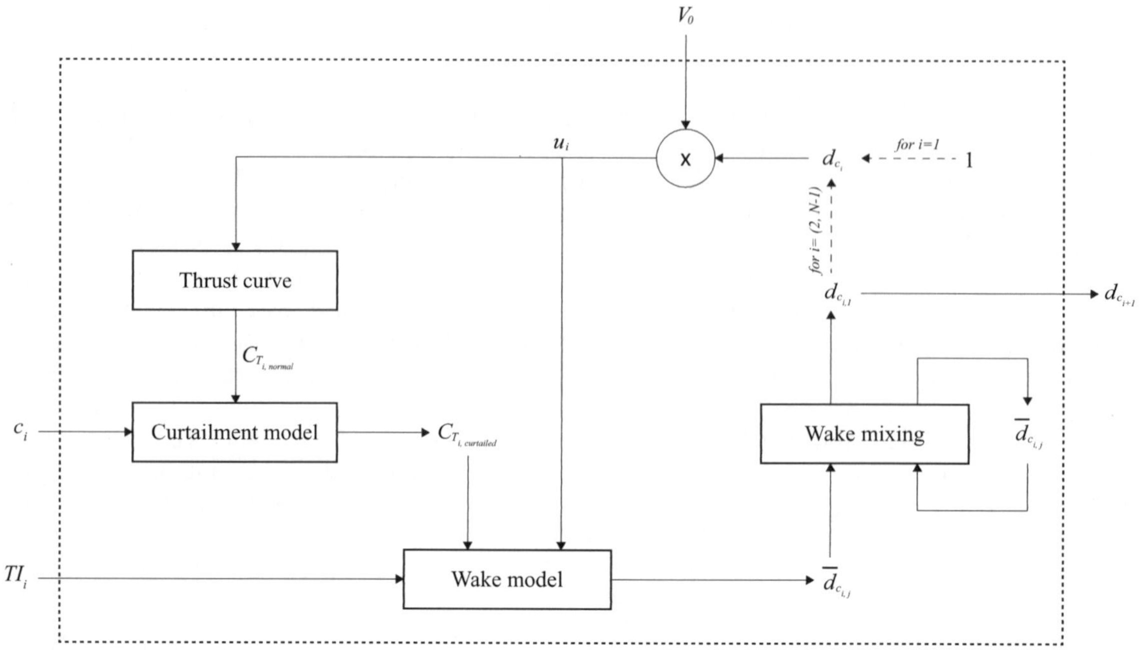

Figure 1. Flowchart of the wake – curtailment model, which calculates the wind speed deficit at each wind turbine taking into account the curtailment of each turbine, see table 1 for the description of each symbol.

Symbol	Description
$C_{T_{i_{normal}}}$	The thrust coefficient of turbine i if the turbine would not curtail its power
$C_{T_{i_{curtailed}}}$	The thrust coefficient of turbine i including curtailment
c_i	Curtailment factor of turbine i
d_{c_i}	Wind speed deficit at turbine i during curtailment
$\bar{d}_{c_{i,j}}$	Vector of wind speed deficits at downstream turbines j due to turbine i only
$d_{c_{i,1}}$	The wind speed deficit at the turbine directly downstream of turbine i
j	A turbine downstream of turbine i, i.e. $j \in (i+1, N)$
N	Number of turbines
TI_i	Turbulence intensity at turbine i
u_i	Wind speed at turbine i
V_0	Free stream wind speed

Table 1. Symbol descriptions of figure 1

2.2.1. Curtailment model. Both wake models require the thrust coefficient of the turbine as an input parameter. The implementation of the curtailment model depends on the selected turbine and available data. For this research curtailed thrust coefficient curves were available for different curtailment factors $c_i = 0.1, 0.2, \ldots, 0.9$. For the applicable wind speed, the thrust coefficient as a function of curtailment factor is linearly interpolated with the actual curtailment factor (equation 3), which is the ratio between the measured power production and the estimated available power.

$$c_i = 1 - \frac{\hat{P}_{c_i}}{P_{APE_i}} \tag{3}$$

2.2.2. Wake mixing. Some turbines are positioned inside the wakes of multiple turbines. In order to determine the total wind speed deficit at those turbines, the total effective wake deficit is calculated using the root mean square method, see equation 4 where d_j refers to the wake deficit due to the operation of upstream turbine j.

$$d_{total} = 1 - \sqrt{\left(\sum_{j}(1-d_j)^2\right)} \tag{4}$$

2.2.3. Partial wake incidence. Considering the layout of Westermeerwind, see figure 3, it was possible to consider straight rows of turbines. Validation was performed for the wind direction aligned with these rows. Implementation of rotor averaging for partial wakes was therefore not required. It is noted that in order to evaluate the Larsen wake model, the equations were discretized over the rotor radius with 100 sections. This is necessary as in the Larsen wake model the wind speed deficit varies over the rotor area of the turbine.

2.3. Full algorithm

Besides the wake – curtailment model, there are more modules in the algorithm, as can be seen in figure 2. Part 1 of the algorithm is evaluated twice: once where the inputs of the wake models are set to the values in normal operation, and once with the curtailed equivalents.

2.3.1. Turbulence model. Both the Jensen and the Larsen model for wind speed deficit depend on turbulence intensity (see e.g. equation 2). During curtailment, the hypothetical turbulence intensity that would exist without curtailment is not known and needs to be modelled. In this research the turbulence intensity is measured just before and just after curtailment and these values are used as the hypothetical turbulence intensity that would need to be modelled during curtailment. Such measurements without curtailment are only representative for the hypothetical turbulence intensity during curtailment periods, if they are taken not to long before or after the curtailed operation. During continuous curtailment, such measurements would not be available.

2.3.2. Reduced wake effect. The reduced wake effect is the difference between the power of a turbine in the wake of a curtailed turbine and the power it would have if the upstream turbine was not curtailed. This effect is the result of the increased wind speed at the downstream turbines, due to the curtailment of upstream turbines, effectively reducing the wake loss. It is calculated as the difference between the modelled power for curtailed condition (P_{c_i}) and the power for the modelled normal condition (P_i).

2.3.3. Available power estimator. The available power estimator determines the available power at a single turbine during curtailment for the local wind conditions. The methodology generally uses a combination of the pitch-power curve and rotational speed - power curve. [6] For this research the turbines were provided with an available power estimator by the manufacturer.

2.3.4. Available wind farm power The total available power of the wind farm equals the sum of the individual available power signals minus the reduced wake effects of each turbine, see equation 5.

$$P_{AWF} = \sum_{i=1}^{N}(P_{APE_i} - \Delta P_{rw_i}) \tag{5}$$

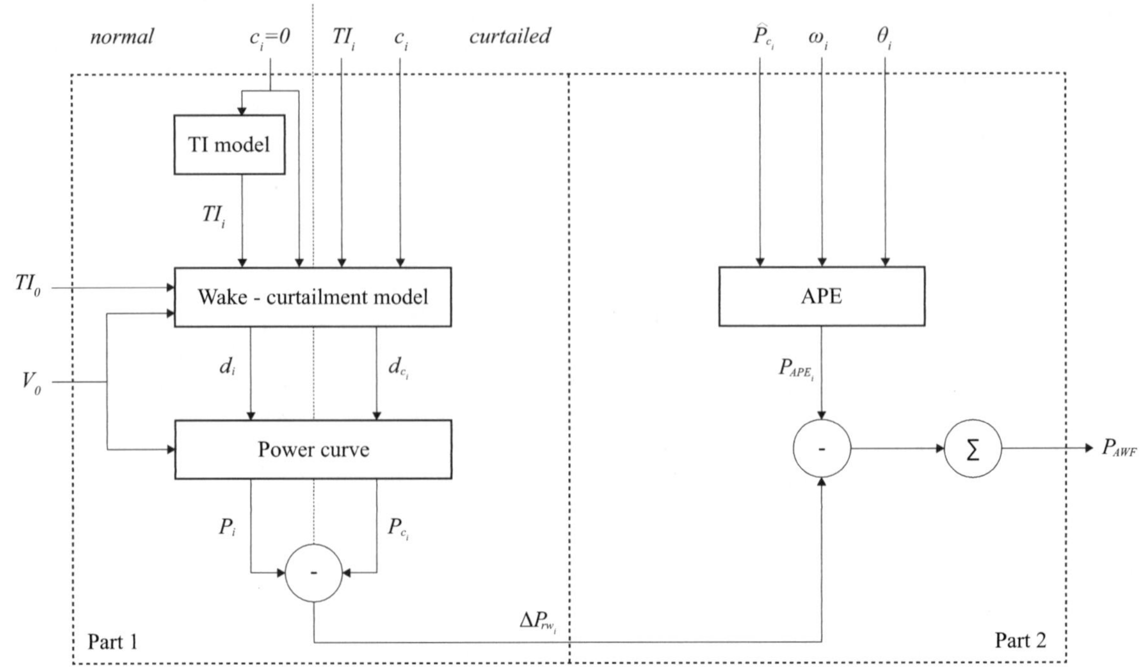

Figure 2. Flowchart of the algorithm, see table 2 for description of each symbol in addition to table 1.

Symbol	Description
θ_i	Blade pitch angle of turbine i
ω_i	Rotational speed of turbine i
APE	Available power estimator
d_i	Wind speed deficit at turbine i in normal operation mode
ΔP_{rwi}	Reduced power loss of turbine i due to wake reduction
P_{AWF}	Available wind farm power, including the reduced wake effect
P_{APE_i}	Estimated available power of turbine i, excluding the reduced wake effect
P_i	Modelled power production of turbine i in normal operation
P_{c_i}	Modelled power production of turbine i during curtailment
$\hat{P}_{c_i}$	Measured power production of turbine i during curtailment
TI_0	Free stream turbulence intensity

Table 2. Symbol descriptions of figure 2 in addition to table 1

2.4. Uncertainties of the algorithm

The main uncertainties of the algorithm are 1) the wake models; 2) using the wake models instantaneously; 3) applying the wake models during curtailment and 4) using the power curve instantaneously. Most validation studies available in literature test wake models for large time spans and on a 10-minute average basis. It is unknown if the wake models are able to produce valid results instantaneously during curtailment. The power curves provided by the turbine manufacturers are generally only guaranteed for 10-minute periods too. The preliminary validation of this study will indicate the combined effect of these uncertainties on the overall performance of the algorithm.

IOP Conf. Series: Journal of Physics: Conf. Series **854** (2017) 012004 doi:10.1088/1742-6596/854/1/012004

3. Experimental methodology

In order to validate the presented algorithm, curtailment experiments have been performed on wind farm Westermeerwind.

3.1. Wind farm Westermeerwind

The nearshore Westermeerwind project in the Netherlands consists of 48 3.0-108 MW wind turbines from Siemens Wind Power. The layout consists of three straight rows of turbines of which only the northern row is used in the experiment, see figure 3. This row contains thirteen turbines, and is 30 deg inclined with respect to North. Just south from the turbines of interest, the other rows of turbines are located. The relevant turbine specifications can be found in table 3. The turbines are spaced $520m$ apart, which equals 4.8 times the rotor diameter.

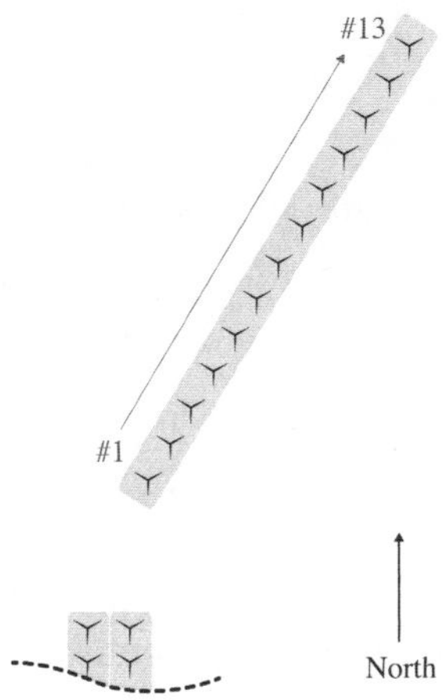

Figure 3. Layout wind farm Westermeerwind

SWT 3.0 MW D3-10	
Rated power	$3MW$
Cut-in wind speed	$3\frac{m}{s}$
Cut-out wind speed	$25\frac{m}{s}$
Rated wind speed	$12\frac{m}{s}$
Rotor diameter	$108m$
Hub height	$95m$

Table 3. Wind turbine specifications [7]

3.2. Curtailment experiments

In the experiments turbine #1 from figure 3 is being curtailed by pitch control. In each experiment a different power production limit is set, resulting in different curtailment factors. Between each consecutive experiment, a reference period without curtailment is reserved, to reacquire the normal wind conditions at each turbine. Both the curtailment experiment and the reference period had a duration of four minutes.

3.3. Using wind speed data

The accuracy of the sonic wind speed sensor, placed on top of the nacelle behind the rotor, proved to be too low for power calculations as the documentation states a maximum error of 5%. Therefore it was decided to use the power production measurements — with a maximum error of 0.2% — together with the power curve to determine wind speed. This resulted in much more consistent wind speed calculations between turbines. However, this does limit the maximum allowable wind speed for the validation to the rated wind speed.

However the sonic sensor is used to determine the turbulence intensity, as it does provide a good indication of the short-term fluctuation of wind speed. In order to determine the free stream wind conditions for the wake models the data signals of turbine #1 are used. A density correction is applied to adjust for temperature, pressure and humidity changes between the power curve specifications and the conditions during the experiments.

3.4. Time delay
In order to compare the wind speeds of different turbines, the time delay for wind conditions to travel over the wind farm needs to be accounted for. Gebraad et al. [8] achieved good results using the mean of the wind speed behind the first turbine based on the actuator disk theory and the wind speed at the direct downstream turbine (u_{i+1}), see equation 6. Note that the distance between turbines (Δx) is $520m$ for this research. It was found that this methodology yields good results in this research as well, using a fixed induction factor of $\frac{1}{3}$.

$$t_{delay} = \frac{\Delta x}{\frac{1}{2}\left(u_i(1-2a_i)+u_{i+1}\right)} \tag{6}$$

3.5. Stability criteria experiments
The results from the curtailment experiments are only relevant when the environmental conditions do not change significantly during the experiment. The following criteria were set to ensure meaningful results:

(i) The wind speed cannot reduce to a point where less than $200kW$ is being curtailed.
(ii) The wind direction cannot change to the extend of modifying the wake profile of the turbines. Due to the absence of high quality wind direction data, this criterion was monitored manually by observing the nacelle directions and wind speed deficits.
(iii) None of the turbines can operate above rated power.

4. Experimental results
Applying the stability criteria lead to five accepted experiments out of more than twenty performed experiments. This shows that the reduced wake effect is only measurable under specific stable wind conditions. See table 4 for an overview of the experiments, where $P1_{normal}$ and $P1_{curtailed}$ refer to the power of turbine #1 in the reference period and during the experiment.

Experiment	$V_0[\frac{m}{s}]$	$P1_{normal}[kW]$	$P1_{curtailed}[kW]$	*curtailment*
1	8.9	1,700	700	60%
2	8.8	1,700	500	70%
3	8.4	1,400	800	40%
4	7.8	1,100	600	45%
5	10.3	2,600	1,200	55%

Table 4. Overview accepted experiments

Figure 4 provides an overview of the experiments by showing the difference between mean power produced during the experiment and the reference period. When curtailing the leading turbine, the second turbine experiences an increase in power production due to the reduced wake effect; the third turbine has a decrease in power production smaller than that increase and all further downstream turbines have no significant change in power production. This confirms the premises of this research and the findings of Göçmen et al. [2], that it is important to estimate the changes in wake losses during curtailment. In general it was found that the accumulated effect of the curtailment of turbine #1 on the row of turbines, is a net decrease of power in the range of 0% to 60% of the power curtailed by the turbine #1.

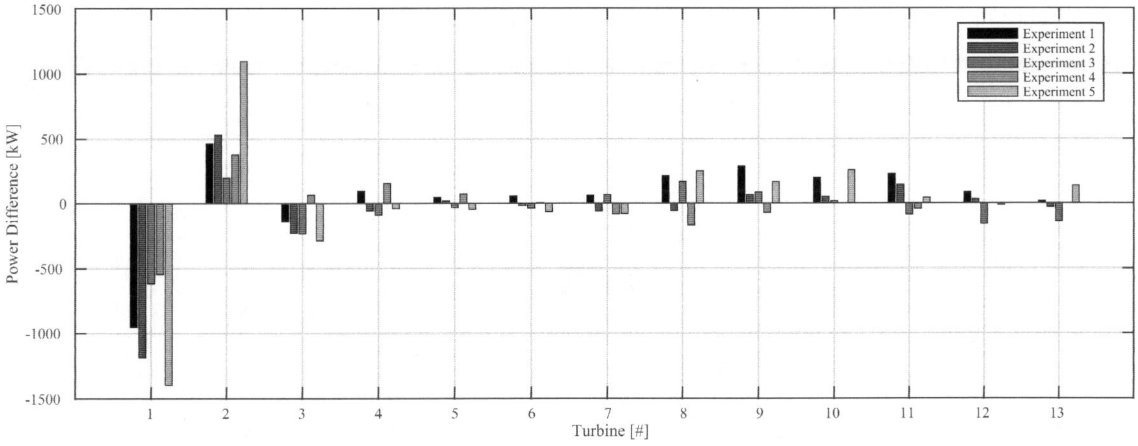

Figure 4. Mean power difference between normal and curtailed operation. The bars of turbine #1 indicate the curtailed power, whereas the bars for turbine #2 show the reduced wake effect.

4.1. Results algorithm validation turbine #2

Figure 5 shows the algorithm applied using the measurements of experiment 1, which is a typical result. First, a reference period can be seen without curtailment. Then, turbine #1 is curtailed to $700kW$. Finally, turbine #1 returns to the normal operation state for a second reference period.

The second solid curve indicates the power estimation for turbine #2 during curtailment and allows validation of the wake model. The estimations of the two wake models have a maximum error of $300kW$ on average. The reduced wake effect during curtailment is modelled with a similar accuracy.

The dashed curves in figure 5 for the models indicate the power production estimations as it would be without curtailment, i.e. the hypothetical normal operation. These estimations look reasonable as well, as they stay between the situations in the reference periods. Together with the solid curves, they allow validation of the performance of the algorithm to estimate the reduced wake effect.

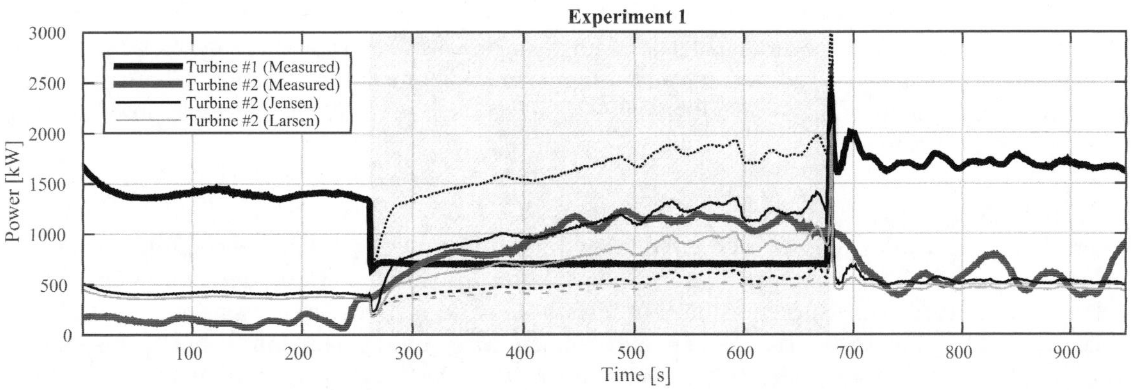

Figure 5. Power production signal of turbine #1 and #2. The dotted line of turbine #1 indicates the available power estimation. The dashed lines of the wake models indicate the estimated power production of turbine #2 if turbine #1 would not curtail its power.

In figure 6 the mean results of the algorithm with respect to the current best practice — summing the available power signals — and the measured available power during curtailment can be seen for turbine #2. The error of the approximation reduces from 60%–100% using the current best practice to 10%–40% for the algorithm with the Larsen wake model and to 1%–20% using the Jensen variant.

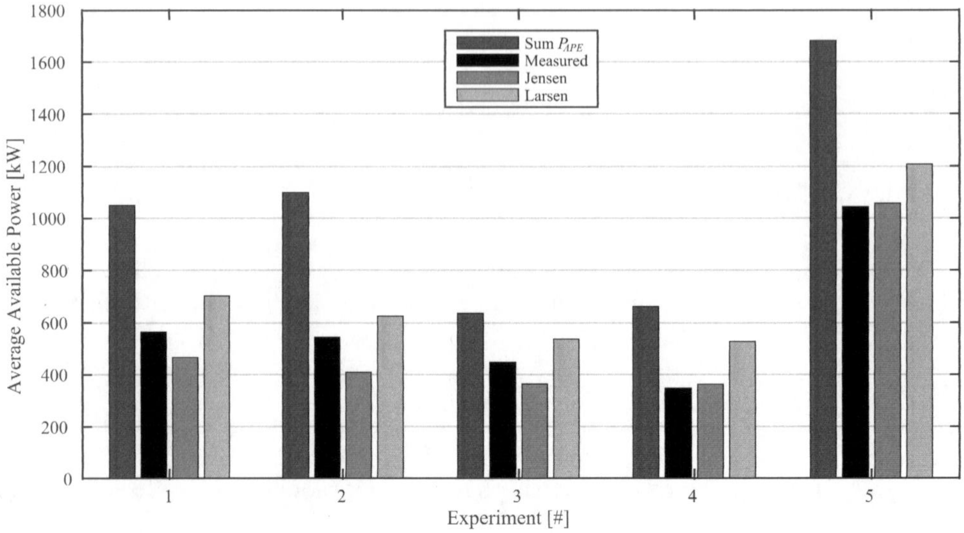

Figure 6. Average available power of turbine #2 for the current best practice; measured as the difference between normal and the reference period; using the algorithm with the two wake models.

4.2. Results algorithm validation turbine #3

For turbine #3 an increase of the wake loss is observed in figure 4. This can be explained by the fact that the wake development behind turbine #2 is increased, due to its increase in power production. The wake models were not able to accurately estimate the power at turbine #3 and further downstream for both normal and curtailed operation — the error was larger than 30% in most cases. The accuracy of the algorithm depends on the accuracy of wake model estimations for turbine #3 for both normal operation and during curtailment. Therefore, with the current implementation, it was not possible to accurately quantify the available power during curtailment for turbine #3.

5. Conclusion

Determining the available power of a wind farm during curtailed operation is complicated as the wake losses are different than during normal operation. In this paper an algorithm is presented and validated against measurement data of nearshore wind farm Westermeerwind.

It was found that using the Jensen and Larsen wake models instantaneously, instead of averaged over a longer period, still delivers accurate power predictions for turbine #2. The instantaneous use of the power curve in the algorithm was also found to be acceptable, despite the fact that it is defined for longer averaging periods. Finally, It was found that changing the thrust coefficient to model curtailment with the existing wake models proved accurate. This resulted in an improved estimation of the available power during curtailment for turbine #2, with an error reduction from 60%–100% using the current best practice to 10%–40% for the algorithm with the Larsen wake model and to 1%–20% using the Jensen variant. For available power estimation of turbine #3 additional research is required.

There are still several steps to take to arrive at a full algorithm, but based on the findings it can be expected that the algorithm could be developed to correctly determine the available wind power for wind farms as a whole during curtailed operation of specific or all wind turbines.

6. Recommendations

In order to expand the validation of the algorithm as currently presented, the following recommendations are made:

- Perform curtailment experiments on wind farms with a layout different from a straight row.
- Perform curtailment experiments with multiple turbines under curtailment.
- Perform curtailment experiments above rated power. This will require a alternative method of determining the wind speed, as the power curve does not have a one-on-one relation above rated power.

In order to further develop the presented algorithm, the following recommendations are made:

- Validate the algorithm against simulations with computation fluid dynamics to reduce uncertainties regarding meteorological conditions, delay time, turbine characteristics, etc.
- Implement a turbulence model to model the turbulence intensity during the experiment as it would be without any curtailment.
- Add advanced wake phenomena like wake meandering and the wind shear.
- Test more wake models, like the wake models by Ainslie [9], Frandsen [10], Compagnolo et al. [11] and Bastankhah et al. [12].

References

[1] Ummels BC, Gibescu M, Kling WL and Paap GC 2006 Integration of Wind Power in the Liberalized Dutch Electricity Market *Wind Energy* **9(6)** 579–90

[2] Göçmen T, Giebel G, Sørensen PE, Poulsen NK 2016 Possible Power Estimation of Down-Regulated Offshore Wind Power Plant *Technical University of Denmark*

[3] Bartl J, Sætran LR 2016 Experimental testing of axial induction based control strategies for wake control and wind farm optimization J *J. Phys.: Conf. Ser.* **753** 032035

[4] Jensen NO 1983 A note on wind generator interaction *Risø-M* **2411**

[5] Larsen GC 2009 A simple stationary semi-analytical wake model *Risø National Laboratory for Sustainable Energy, Technical University of Denmark* **1713**

[6] Esbensen T, Krishna R, Scheurich F and Stoettrup M 2015 *System for automatic power estimation adjustment* EP20130187918

[7] 2016 Wind Turbine SWT-3.0-108 Technical Specifications *Siemens AG* `https://www.siemens.com/content/dam/internet/siemens-com/global/market-specific-solutions/wind/data_sheets/data-sheet-wind-turbine-swt-3.0-108.pdf` *accessed on 2017-03-07*

[8] Gebraad PMO and van Wingerden JW 2015 Maximum power-point tracking control for wind farms *Wind Energy* **18** 429–47

[9] Ainslie JF 1988 Calculating the flowfield in the wake of wind turbines *J. Wind Eng. and Industrial Aer.* **27** 213–24 (Elsevier Science Publishers B.V., Amsterdam)

[10] Frandsen S, Barthelmie R, Pryor S, Rathmann O, Larsen S, Højstrup J and Thøgersen M 2006 Analytical modelling of wind speed deficit in large offshore wind farms *Wind Energ.* **9** 39–53

[11] Campagnolo F, Petrović V, Bottasso CL and Croce A 2016 Wind tunnel testing of wake control strategies *American Control Conference (ACC), Boston, MA* 513–8.

[12] Bastankhah M and Porté-Agel F 2014 A new analytical model for wind-turbine wakes *Renewable Energy* **70** 116–23

IOP Conf. Series: Journal of Physics: Conf. Series **854** (2017) 012005 doi:10.1088/1742-6596/854/1/012005

Validation of the actuator disc and actuator line techniques for yawed rotor flows using the New MEXICO experimental data

S.-P. Breton[1], W.Z. Shen[2], S. Ivanell[1]

[1]Uppsala University Campus Gotland, Cramérgatan 3, 62 157 Visby, Sweden

[2]Technical University of Denmark, Energivej, Building 414, 2800 Kgs. Lyngby, Denmark

E-mail: `wzsh@dtu.dk`

Abstract. Experimental data acquired in the New MEXICO experiment on a yawed 4.5m diameter rotor model turbine are used here to validate the actuator line (AL) and actuator disc (AD) models implemented in the Large Eddy Simulation code EllipSys3D in terms of loading and velocity field. Even without modelling the geometry of the hub and nacelle, the AL and AD models produce similar results that are generally in good agreement with the experimental data under the various configurations considered. As expected, the AL model does better at capturing the induction effects from the individual blade tip vortices, while the AD model can reproduce the averaged features of the flow. The importance of using high quality airfoil data (including 3D corrections) as well as a fine grid resolution is highlighted by the results obtained. Overall, it is found that both models can satisfactorily predict the 3D velocity field and blade loading of the New MEXICO rotor under yawed inflow.

1. Introduction

Wind turbines are often operated in conditions where the wind does not flow perpendicularly into the rotor plane, but with a so-called yaw angle. This configuration leads to higher fatigue loads for the rotor blades as well a more complex development of the flow downstream from the rotor. This configuration being often met in reality, it is important to verify if simulation tools are able to correctly predict the blade loading and velocity field associated to it. Wind tunnel measurements, being performed under controlled inflow conditions, allow to validate simulation methods in this regard. The New MEXICO experiments [1] performed in 2014 in Europe's largest wind tunnel on a three-bladed 4.5m diameter wind turbine will be used here to validate the blade loading and velocity fields under different configurations as predicted using the Ellipsys3D Large Eddy Simulation where the rotor plane is modelled using either the actuator disc (AD) or actuator line (AL) model. While both the AD and AL methods in EllipSys3D have recently been used with success to model the New MEXICO rotor under axial inflow [2], the current study will test if they also perform satisfactorily under yawed inflow. Let us mention that yawed inflow has been a subject of investigation for the AL model in EllipSys3D following the first round of the MEXICO wind tunnel experiments [3]. The experimental conditions of the New MEXICO setup differing relatively to the first MEXICO experiment, they allow a further validation of this method.

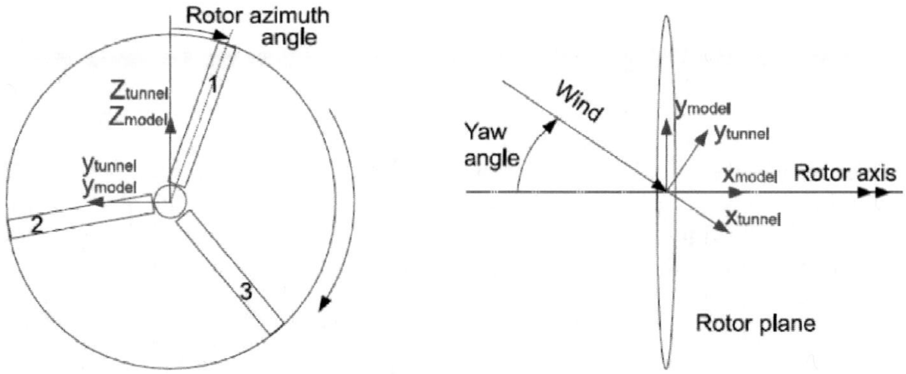

Figure 1. Convention used in the New MEXICO experiments for the coordinate system and yaw angle [4]. Front (left) and top (right) view

In this paper, the numerical and experimental model and setup will first be presented, followed by a presentation and discussion of results regarding the velocity field and blade loading of the New MEXICO rotor under various configurations.

2. Numerical and experimental model and setup

2.1. Experimental setup

The New MEXICO [1] experiments were performed on a three-bladed, 4.5 m diameter upwind wind turbine in the German Dutch Wind Tunnel (DNW). The rotational velocity of the rotor was 425.1rpm, and three incoming velocities (V_0) were considered, i.e. 10.03, 15.01 (design wind velocity) and 24.08m/s. For the case under consideration here, the rotor was yawed at an angle of 30 degrees relatively to the incoming flow. Figure 1 [4] illustrates the conventions used for the coordinate system, yaw and azimuth angles. One can see from this figure that a 0 degree azimuth angle corresponds to the reference blade (blade 1) in the vertical position.

The blades were equipped with Kulite pressure sensors, distributed around the airfoil sections at five radial positions, namely at $r/R = 0.25$, 0.35, 0.60, 0.82, and 0.92, with r the spanwise position, and R the 2.25m radius of the blade. Loads on the blades were calculated as a function of the rotor azimuth angle by integrating the pressure forces around the airfoil section. The blades consisted of three airfoils, namely the DU 91-W2-250 at the root (20 to 45.6 % span), the RISOE A2-21 at the centre of the blade (54.4 to 65.6 % span), and the NACA 64-418 at the tip of the blade (outboard of 74.4% span), with transition zones in between [5]. A cylinder was used from the root (9.3% span) of the blade to 13.3% span. As opposed to the MEXICO first experiments [6], the blades in the current experiments were not tripped with zig-zag strips in the tip region [1].

As in the first MEXICO experiment, the velocity field was also measured in this case, using Particle Image Velocimetry (PIV). The measurements were performed at the 9 o'clock plane relative to the rotor using windows disposed under two configurations and are rotor-phase locked averaged. Under the first configuration, windows were disposed radially in the y_{model} direction shown in Figure 1 to perform so-called radial traverse measurements 0.3m upstream and 0.3m downstream from the rotor plane (i.e. $x_{model} = \pm 0.3$m), leading to results being available for $-3.0m < y_{model} < 3.0m$ for a yaw angle of 30 deg. Under the second configuration, windows were positioned axially along the x_{tunnel} direction shown in Figure 1 to create axial traverses, whose range covers the interval -4.5m< x_{tunnel} <5.9m, leading to results being available at $y_{tunnel} = \pm1.5$m for the same yaw angle. Let us mention that the velocities considered in this work are always defined in the tunnel coordinate system.

While the same spatial resolution as in the first MEXICO measurements was used, the size of the PIV windows increased from 337×394 mm to 380×610 mm in the axial and lateral directions respectively, allowing to cover a larger measurement area. More information about the first and New MEXICO measurement campaigns can be found in [6] and [1] respectively.

2.2. Numerical model and setup

Large-eddy simulations are performed using the Computational Fluid Dynamics (CFD) multi-block finite volume solver EllipSys3D. The rotor is modeled using either the AL or AD technique [7]. Using the AL and AD techniques avoids the need for resolving the development of the boundary layer over the blades, allowing the use of a higher resolution to capture the details of the wake. In the AL model, body forces representing the aerodynamic loads on the rotor are distributed along points on rotating lines representing the individual blades, which allows information about the force on the blades to be directly provided as a function of the blade azimuth angle. In the AD model, the body forces are distributed on points over a rotating disc represented on a local polar mesh, which can provide information about the local forces on the disc as a function of the azimuth angle. The body forces in both techniques are smeared into their imposition points with a three-dimensional Gaussian function [8]. They are calculated through the use of so-called airfoil data, i.e. values of lift and drag coefficients as a function of the local angle of attack, obtained in the LES. Furthermore, the tip loss model by Shen et al. was used in the computations [9, 10].

The airfoil data were obtained from 2D wind tunnel measurements of the RISOE-A1-21 airfoil and provided to the Mexnext participants. It was recently realized that those were not obtained at the right Reynolds number and tripping condition for the RISOE-A2-21 airfoil used in the middle of the blade. However, let us note that the difference in lift-drag polar of the two airfoils is not so different. Work is currently underway to perform measurements under the appropriate conditions. In both the AD and AL simulations, the airfoil data are linearly interpolated in the transition regions between the different airfoil sections. In the AD simulations, the airfoil data associated with a cylinder (i.e., a zero lift coeffcient and a value of 1.2 for the drag coefficient) are used for the hub region, while a hole is considered in the AL calculations. Neither the presence of the hub nor of the nacelle is modelled in the EllipSys3D simulations.

The grid used in the AD and AL simulation is the same in order to make a fair comparison possible. It is made up of about 11.8 million points and covers the range [-20R,20R] × [-16R,16R] × [-16R,16R] in the x , y and z directions respectively, with the finest resolution given by $R/40$. More information regarding the numerical setup for the calculations performed with the AD and AL models can be found in [2].

3. Results and discussion

An overview of the available simulated and experimental results in terms of velocities in the wake and blade loading, representative of the overall results obtained, is provided and discussed in this section.

3.1. Velocities in the wake

Velocities shown are averaged over the last fourth of the simulations performed with the AD, while the velocities at the last simulation iteration are shown for the AL. Please note that the uncertainty on the experimental velocities, although not included in the graphs, is reported to be around $\pm$ 0.2m/s [11].

Figure 2 shows the axial (u, along x), crosswise (v, along y) and vertical velocities (w, along z) on the radial traverses located respectively 0.3m upstream and downstream from the rotor in the model coordinate system. Results are shown for the LES using the AD and AL representations of the rotor and are compared to the experimental results. The AD results are shown, upstream from the rotor, to slightly underestimate the downwash effect from the turbine wake as seen in the overestimated axial velocity. It is seen to also underestimate to a certain extent the velocity deficit in terms of axial velocity downstream from the rotor. The AL performs better in this sense and produces results closer to the experimental data. Both the AL and AD simulations compare rather well with the experimental results in terms of crosswise and vertical velocities. The AL and AD results also agree quite well, except for small differences in the hub region due to their different treatment of this region.

Figure 3 shows the same quantities as Figure 2, but for an incoming velocity of 15.01m/s. Observations similar to those related to an incoming velocity of 10.03m/s can be made here. The crosswise velocity which was slightly overestimated upstream from the rotor with the AD model at V_0=10.03m/s for negative y values is so in a more important way here, and so is the underestimation of the vertical velocity as predicted by the AL model downstream from the rotor for positive y values. Let us notice that the discrepancies observed are similar for these two wind speeds when non dimensionalized with respect to the incoming wind speed. They could result from the use of a slightly too coarse mesh and uncorrected 2D airfoil data.

Figure 4 shows the same quantities as the two previous figures, but for an incoming velocity of 24.08m/s. It is to be noted here that experimental values were available for comparison only for negative y values. In this region, the agreement of both the AD and AL results with the experimental data is quite satisfactory, except for a slight overestimation of the the crosswise velocity component for the AD model both upstream and downstream from the rotor.

The three velocity component as simulated with the AL and AD models and measured experimentally along the axial traverse located at y_{tunnel}=-1.5m as a function of the axial position x_{tunnel} are shown in Figure 5 for the three incoming wind velocities considered in the experiments. In all cases shown, it can be observed that both the results associated to the AL model agree quite well with the experimental data. This model is in general able to predict the velocity variations obtained in the experimental data due to the presence of the tip vortices, although with a smaller amplitude. Indeed, for the 10m/s and 15m/s cases, the fine fluctuation details associated to the vertical velocity observed in the experimental data are not reproduced by this AL simulation, while more large variations are produced at a wind velocity of 24m/s [12]. It is expected that the use of a finer resolution mesh would have allowed to reproduce such detailed features of the flow at lower wind velocities, as has been observed before [13]. The AD model, due to its average nature, cannot reproduce these features, but is able to predict the average behavior in the flow quite satisfactorily. Slight overestimations are however obtained as regards the axial velocity downstream from the rotor for incoming velocities of 10.03 and 15.01m/s.

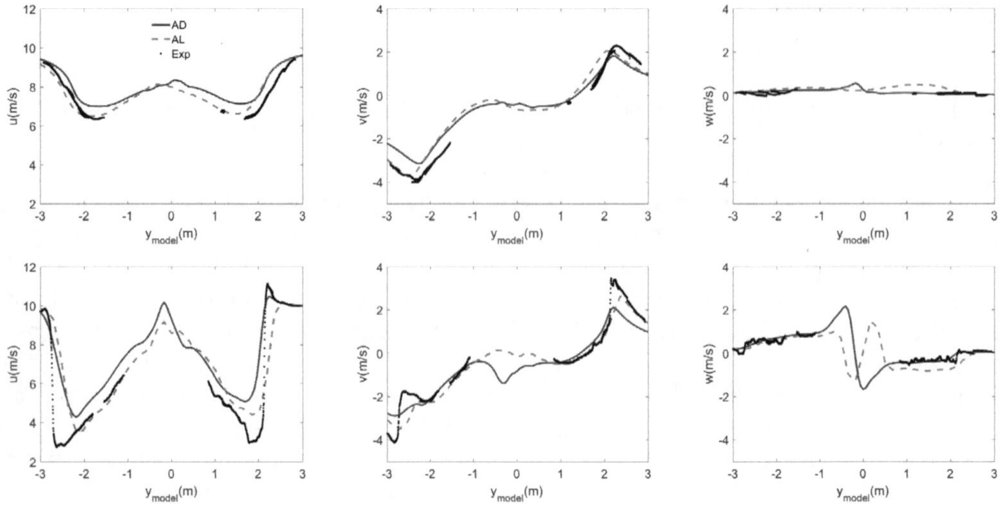

Figure 2. Streamwise, crosswise and vertical wind velocities obtained using the AD and AL in EllipSys3D compared with experimental data. V_0=10.03m/s, yaw angle=30deg, 0.3m upstream (top figures) and downstream (bottom figures) from the rotor plane.

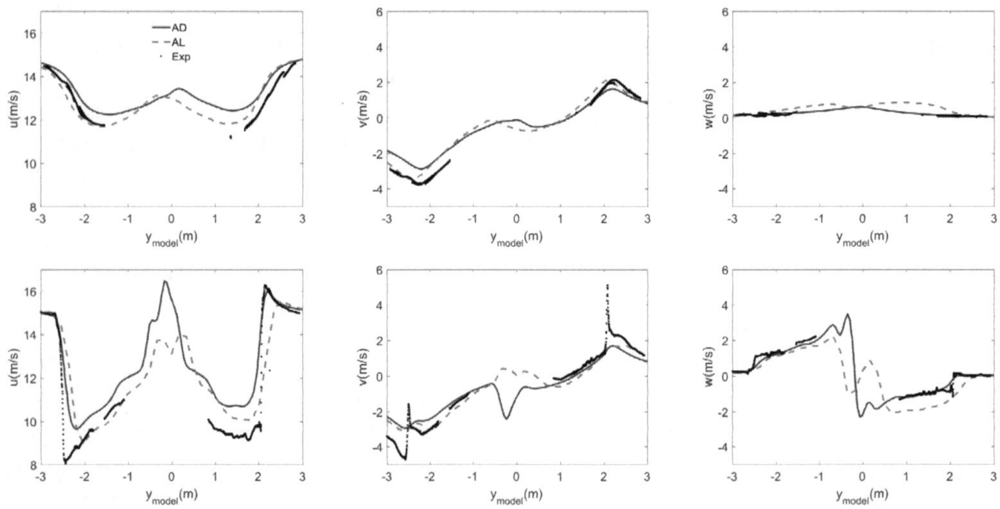

Figure 3. Streamwise, crosswise and vertical wind velocities obtained using the AD and ALin EllipSys3D compared with experimental data. V_0=15.01m/s, yaw angle=30deg, 0.3m upstream (top figures) and downstream (bottom figures) from the rotor plane.

The AL and AD method used in this work, although they do not explicitly model complex phenomena such as flow detachment, which is more important for increasing incoming wind velocities, are shown to be able to satisfactory reproduce the main flow features. Since the structures of the flow, as tip vortices, move more slowly at low incoming velocities, the use of a finer resolution at low wind velocities is expected to lead to a better reproduction of the details of the flow [13].

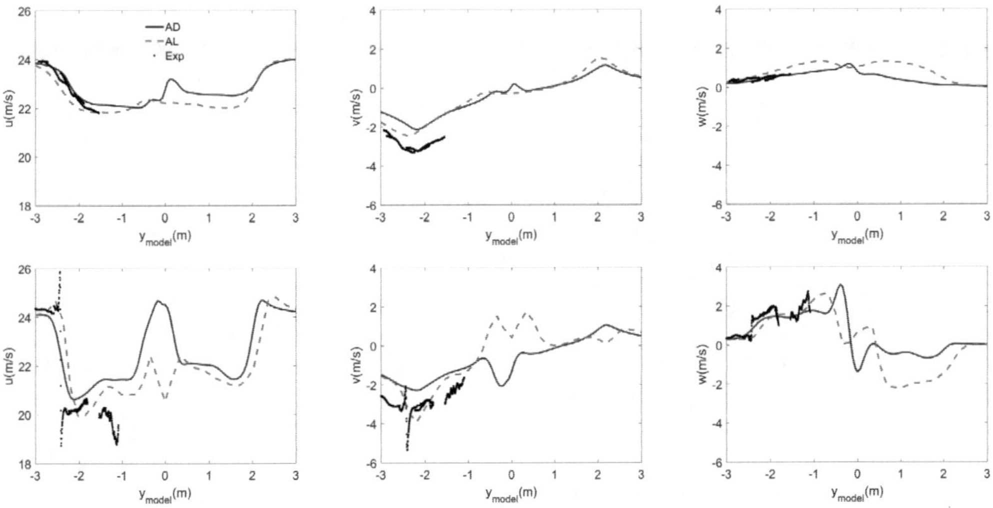

Figure 4. Streamwise, crosswise and vertical wind velocities obtained using the AD and AL in EllipSys3D compared with experimental data. V_0=24.08m/s, yaw angle=30deg, 0.3m upstream (top figures) and downstream (bottom figures) from the rotor plane.

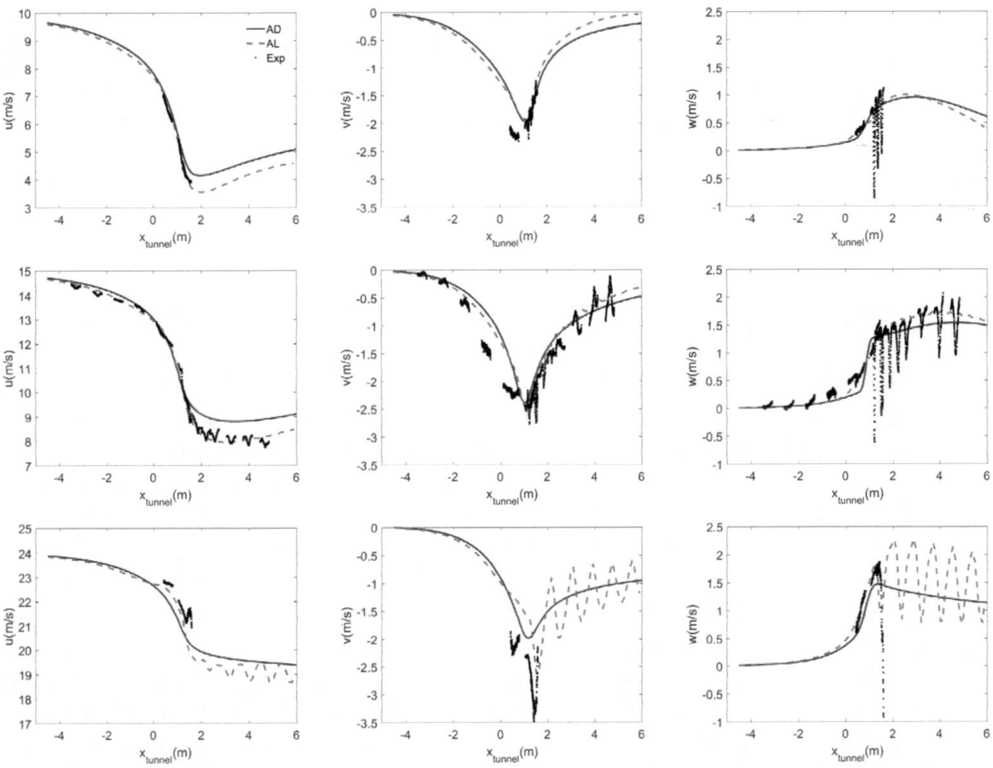

Figure 5. Streamwise (u), crosswise (v) and vertical (w) wind velocities obtained using the AD and AL in EllipSys3D compared with experimental data for the axial traverse located at y=-1.5m. V_0=10.03m/s (top), 15.01m/s (middle), and 24.08m/s (bottom).

3.2. Blade loading

Figure 6 shows the force normal to the blade chord as simulated using the AD and AL models and measured experimentally the three incoming velocities of 10.03, 15.01 and 24.08m/s. Four positions along the blade representative of the root, center and tip are considered. In general, the AD and AL simulations produce very similar results, and they capture well the azimuthal trend of the normal force. The normal forces are seen to be slightly underestimated in the root region, which is expected to be due to the use of 2D airfoil data which neglect stall delay effects. 2D airfoil data were used in the computations performed within the Mexnext consortium to prevent the use of various 3D correction models that would make a direct comparison between the simulation codes more difficult. Note however that the maximal uncertainty on the force at that position is rather large. The largest differences in terms of amplitude of the forces are seen at a radial position of $0.6R$, near the middle of the blade. These differences could be explained by the fact that the experimental airfoil data used in this section to represent the RISOE airfoil section were not available at the correct Reynolds number and tripping conditions, which could lead to incorrect force values to be computed. The best overall agreement is found for the $0.82R$ position for which the complex influence the tip, known to be difficult to accurately model, is not as important as for the $0.92R$ position.

Figure 7 shows the force tangential to the local blade chord measured and obtained using the AD and AL models for the same radial positions and incoming velocities as in Figure 6. Once again, the AD and AL models lead to very similar results, and are able to reproduce in a satisfactory way the experiments results in terms of trends and absolute values (notice the much smaller y-axis range than for the normal force shown in Figure 6). The tangential force is found to be slightly underestimated in both the AD and AL cases in the root region, which can be explained, as for the normal force, by the neglect of stall delay effects. The differences as regards the amplitude of the tangential force observed at a radial position of $0.6R$ could, as for the normal force, be explained by the issue outlined above regarding the RISOE airfoil in this region. One can notice an interesting feature of the measured tangential force at a radial position of $0.6R$, which does not show the same azimuthal trend as for the other velocities and radial positions. This is expected to result from dynamic stall effects affecting the yawed turbine at this high incoming velocity, a phenomenon that is not specifically accounted for in the current simulations. It would be interesting to investigate whether full CFD simulations modeling in detail the development of the boundary layer over the blades would reproduce this flow behavior.

4. Conclusion

Simulations of the New MEXICO rotor tested under controlled conditions in a wind tunnel were performed under yawed inflow with the Large Eddy Simulation code EllipSys3D, where the disc was modelled using either an actuator disc (AD) or actuator lines (ALs). Comparisons in terms of loading and velocity field with experimental data were made under various configurations. Satisfactory agreement was in general obtained between the experiments and simulations. As regards the velocities, the AL, as expected, was shown in general to be able to capture the induction effect from the individual blade tip vortices, however with a smaller amplitude for low incoming velocities. Although not able to capture this effect, the AD was shown to be capable of reproducing the averaged flow in the near wake quite satisfactorily. It would be interesting in a next step to model the geometry of the nacelle and hub under yawed inflow. This was recently performed under axial flow using both the AD and AL [2] and shown to improve the agreement of the near wake in the blade root region.

The general trends in blade loading were well captured with both the AD and AL models. Small differences were obtained in terms of force amplitude, which were explained by the non-modelling

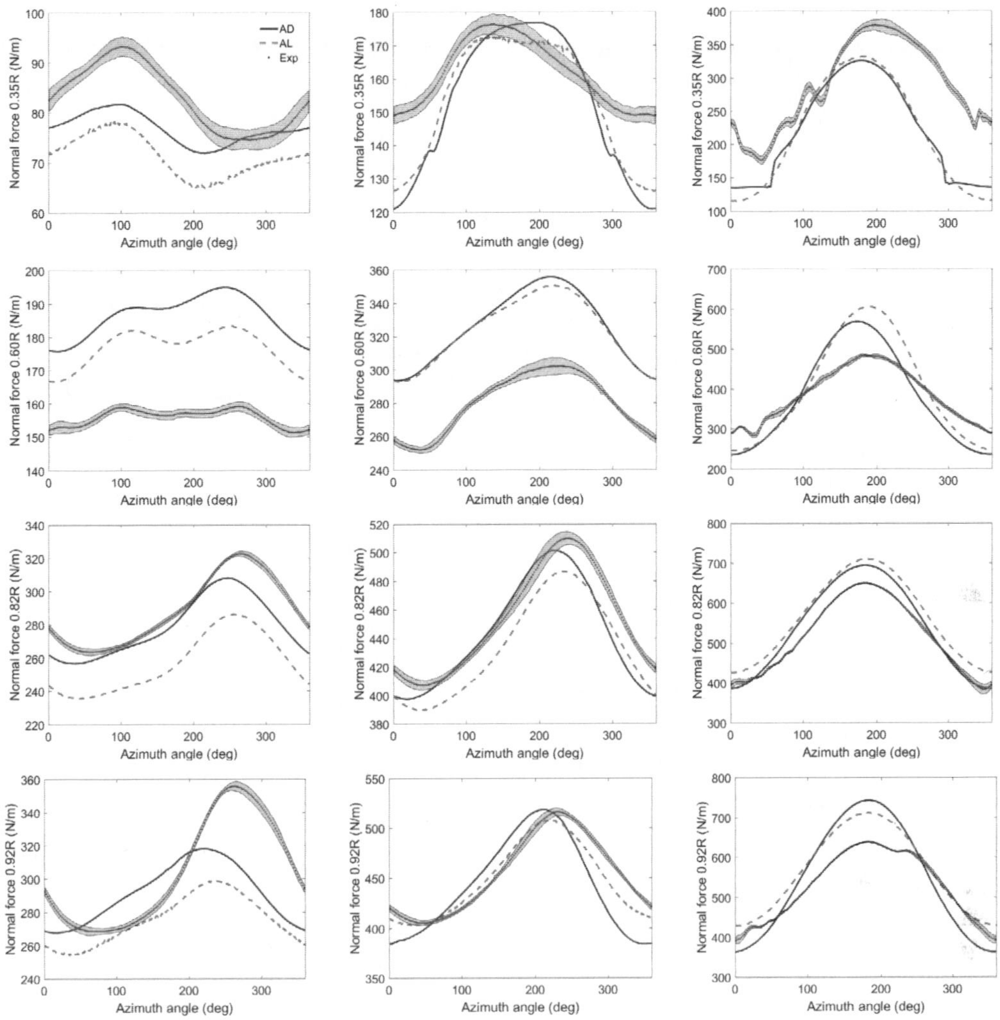

Figure 6. Force per unit length normal to the blade local chord at four radial positions, i.e. $0.35R$ (top), $0.6R$ (2nd row), $0.82R$ (3rd row) and $0.92R$ (bottom) as a function of the blade azimuth angle obtained using the AD and AL in EllipSys3D compared with measurement. Three incoming wind velocities are shown, i.e. V_0=10.03m/s (left), 15.01m/s (middle), and 24.08m/s (right). The shaded area around the experimental curve shows the uncertainty as $\pm$ one standard deviation, indicating the repeatability between the data points.

of stall delay effects in the root region and use of incorrect airfoil data in the middle part of the blade. This emphasizes the importance of employing as accurate airfoil data as possible for such simulations. This is especially true of the tangential force whose small amplitude makes it very dependent on the validity of the airfoil data. The nature of the airfoil data is found to be more important as regards the prediction of the loading on the blade than for that of the velocity field. It would also be of great interest to compare the present simulations with calculations where the full rotor geometry would be considered to see if experimental trends not totally captured by the current simulations could be reproduced. Nevertheless, both the AD and AL model tested in this work, even though they do not model the development of the boundary layer on the blades or the geometry of the hub and nacelle, are shown to be capable of reproducing the main flow

IOP Conf. Series: Journal of Physics: Conf. Series **854** (2017) 012005 doi:10.1088/1742-6596/854/1/012005

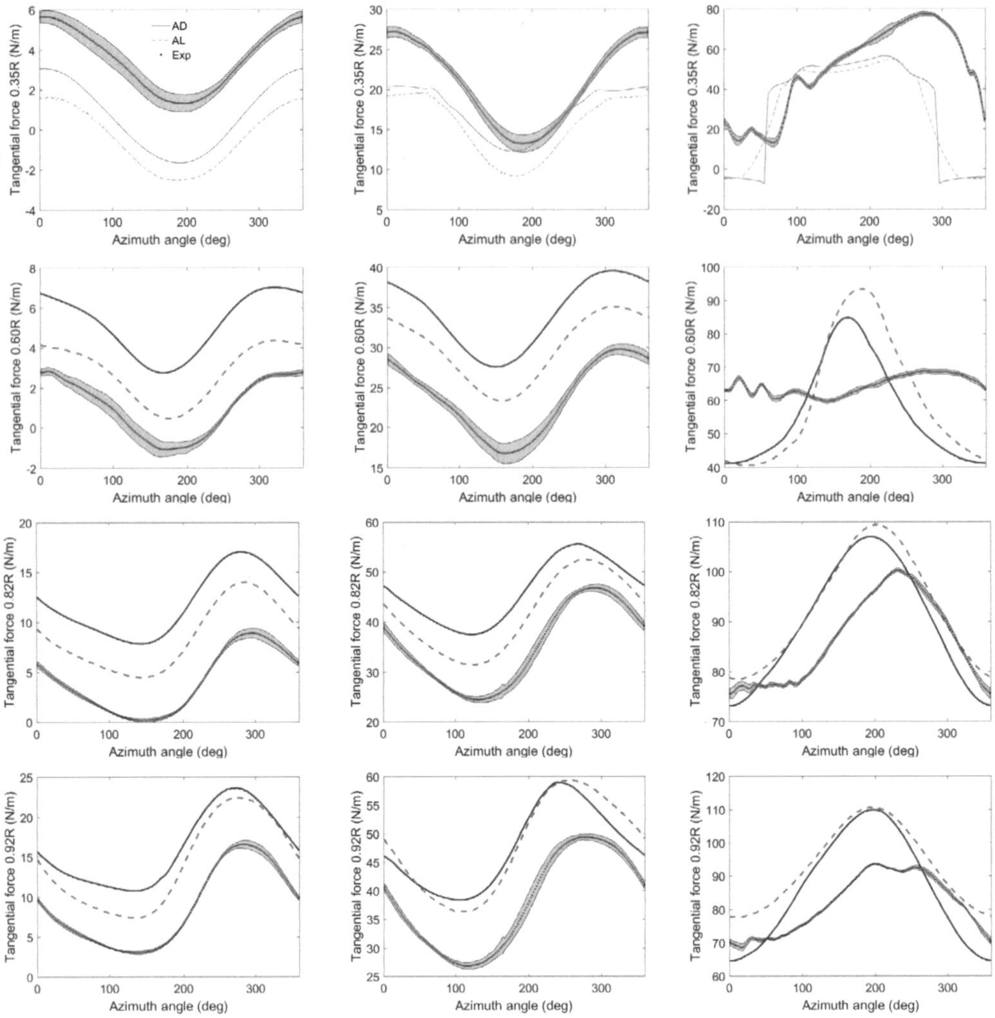

Figure 7. Force per unit length tangential to the blade local chord at four radial positions, i.e. $0.35R$ (top), $0.6R$ (2nd row), $0.82R$ (3rd row) and $0.92R$ (bottom) as a function of the blade azimuth angle obtained using the AD and AL in EllipSys3D compared with measurement. Three incoming wind velocities are shown, i.e. V_0=10.03m/s (left), 15.01m/s (middle), and 24.08m/s (right). The shaded area around the experimental curve shows the uncertainty as $\pm$ one standard deviation, indicating the repeatability between the data points.

features and blade loading characteristics under yawed inflow.

Acknowledgments
The Swedish Energy Agency is acknowledged for providing research funds for this work. This work was also supported by the Energy Technology Development and Demonstration Program (EUDP2014, J. nr. 64014-0543) under the Danish Energy Agency. The simulations were performed on resources provided by the Swedish National Infrastructure for Computing (SNIC) within the project SNIC 2016/10-33. The authors wish to thank the international partners for their collaborations in the MexNext project coordinated by Energy research Centre of the Netherlands (ECN), within the framework of IEA Task 29. S. Sarmast is acknowledged for help

in setting up the AD simulations.

References

[1] Boorsma K and Schepers J 2014 New MEXICO experiment - preliminary overview with initial validation. ECN-E–14-048, The Netherlands
[2] Sarmast S, Shen W, Zhu W, Mikkelsen R, Breton S P and Ivanell S 2016 *Journal of Physics: Conference Series* **753** 032026
[3] Shen W, Zhu W and Yang H 2015 *Journal of Power and Energy Engineering* **3** 7–13
[4] Boorsma K and Schepers J 2016 mexnext III: Definition of second round of calculations. Document internal to the Mexnext consortium, ECN, The Netherlands
[5] Schepers J, Boorsma K, Cho T, Gomez-Iradi S, Schaffarczyk P, Jeromin A, Shen W, Lutz T, Meister K, Stoevesandt B, Schreck S, Micallef D, Pereira R, Sant T, Madsen H and Sørensen N 2012 Final report of IEA task 29, mexnext (phase 1): Analysis of mexico wind tunnel measurements. ECN-E–12-004, The Netherlands
[6] Schepers J and Snel H 2007 Model experiments in controlled conditions - final report. ECN-E–07-042, The Netherlands
[7] Sørensen J and Shen W 2002 *Journal of fluids engineering* **124** 393–399
[8] Mikkelsen R 2003 Ph.D. thesis Technical University of Denmark, Denmark
[9] Shen W, Mikkelsen R, Sørensen J and Bak C 2005 *Wind Energy* **8** 457–475
[10] Shen W, Zhang J and Sørensen J 2009 *Journal of Solar Energy Engineering* **131** 011002
[11] Boorsma K and Schepers J 2015 Description of experimental setup new MEXICO Experiment. ECN-X–15-093-v1, The Netherlands
[12] Shen W, Zhu W and Sørensen J 2012 *Wind Energy* **15** 811–825
[13] Nilsson K, Shen W, Sørensen J, Breton S P and Ivanell S 2015 *Wind Energy* **18** 499–514

IOP Conf. Series: Journal of Physics: Conf. Series **854** (2017) 012006 doi:10.1088/1742-6596/854/1/012006

Wake losses from averaged and time-resolved power measurements at full scale wind turbines

Francesco Castellani[1], Davide Astolfi[1], Matteo Mana[1], Matteo Becchetti[1], Antonio Segalini[2]

[1] Department of Engineering, University of Perugia, Perugia, Italy
[2] KTH Royal Institute of Technology, Linné Flow Centre, STandUP for Wind, Stockholm, Sweden

E-mail: francesco.castellani@unipg.it

Abstract. This work deals with the experimental analysis of wake losses fluctuations at full-scale wind turbines. The test case is a wind farm sited on a moderately complex terrain: 4 turbines are installed, having 2 MW of rated power each. The sources of information are the time-resolved data, as collected from the OPC server, and the 10-minutes averaged SCADA data. The objective is to compare the statistical distributions of wake losses for far and middle wakes, as can be observed through the "fast" lens of time-resolved data, for certain selected test-case time series, and through the "slow" lens of SCADA data, on a much longer time basis that allow to set the standards of the mean wake losses along the wind farm. Further, time-resolved data are used for an insight into the spectral properties of wake fluctuations, highlighting the role of the wind turbine as low-pass filter. Summarizing, the wind rose, the layout of the site and the structure of the data sets at disposal allow to study middle and far wake behavior, with a "slow" and "fast" perspective.

1. Introduction

The comprehension of the behavior of wind turbines under wakes is a stimulating subject [1, 2, 3, 4]. Studying wake fluctuations is also a matter of time scales, because it involves the control system and the wind-turbine ability in reacting to the variability of the wind, especially when wake-added turbulence and veering flow are relevant. For the sake of saving space and computational cost, the wind-energy industry commonly employs Supervisory Control And Data Acquisition (SCADA) data sets having 10 minutes of averaging time for performance assessment and condition monitoring. Actually, significant achievements in the comprehension of wakes have been reached by using numerical modeling and 10-minute based SCADA data [5, 6, 7, 8, 9]. On the other hand, some phenomena must still be described only partially due to a too coarse time resolution: for example, the alignment of wind turbines to the incoming wind [10, 11, 12, 13] and mechanical fatigue loads [14, 15, 16]. For this reason, the attention of the scientific community towards high-frequency wind turbine data has recently increased. In [17], for example, SCADA data having 1 minute time resolution are employed for validating LES simulations over the Lillgrund wind farm [18]. In [19], high-frequency time series measurements obtained with remote scanning systems are analyzed, with focus on the identification of wind-turbine wake properties in complex terrain. In [20], high-frequency data are studied and it is shown that the complex structure of turbulence dominates the power output fluctuation for one single wind turbine as well as for an entire wind farm.

On these grounds, the present work aims at investigating the time-resolved wake dynamics on a wind farm sited in southern Italy over a moderately complex terrain. The test case is particularly valuable because the usual 10-minutes data and time-resolved data, as collected from the OPC (Object Linking and Embedding for Process Control) server, are available. This allows to exploit the large amount of time-wise information to get a statistical description of the wake behavior, with particular focus on the dynamics of the wind turbines. The layout of the test case is characterized by two characteristics inter-turbine distances: approximately 4 and 8 rotor diameters. Therefore, testing grounds are available for observing middle and far wake, with their associated "slow" and "fast" dynamics. The idea is therefore to compare the statistical distributions of wake losses for far and middle wakes, as can be observed through the "fast" lens of time-resolved data and through the "slow" lens of SCADA data. Subsequently, time-resolved data are used for an insight into the spectral properties of wake losses. The structure of the paper is as follows: in Section 2 the test-case wind farm is briefly described, while in Section 3 the data sets and the techniques are introduced. In Section 4 the results are collected, and finally in Section 5 some concluding remarks are proposed and further directions of the present work are discussed.

2. The Wind Farm

The wind farm, selected as test case for the present work, is sited in southern Italy and features 4 wind turbines having 2 MW of rated power each. The layout and the inter-turbine distances (in units of the rotor diameter D) are sketched in Figure 1.

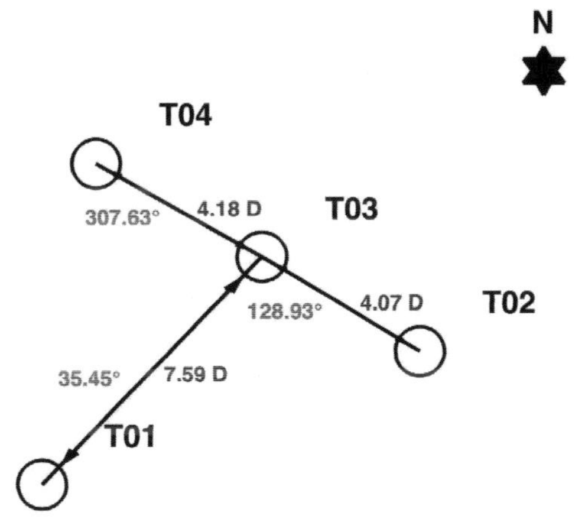

Figure 1. The layout of the test-case wind farm with turbines distance in rotor diameters (blue) and the relative waked direction of minimum distance (red).

From Figure 1, it is highlighted that the subcluster T2-T3-T4 is an interesting test case for investigating the mid-wake regime when the wind blows from approximately 310° and 130°. When the wind blows from approximately 40° (respectively 220°), the far-wake effect of turbine T3 on T1 (respectively, T1 on T3) can be studied.

3. The method

The "slow" (10-minutes based SCADA data) and "fast" (time-resolved) data sets are analyzed in parallel, as follows: a vast data set of 10-minutes SCADA (57168 data, during which all the four turbine have been all producing) and acquisitions of time-resolved data are filtered on similar wind conditions and the statistics are compared. The two sources are not synchronized: the reason for this is to show the time-resolved "fast" wake fluctuations during certain spot events and the average behavior over a catalog of similar situations (spanned on a much longer period in time), observed through the "slow" lens of SCADA data. This approach has pros and cons: the main pro is that one analyzes time-resolved data during consecutive events and therefore one can appreciate how the wind conditions and the response of the wind turbines evolve in time. The drawback is that one can't control strictly the wake regime during the time-resolved events: due to the fluctuations, full wakes and partial wakes can occur during the same event. For this reason, peculiar attention has been devoted to understand the wind conditions, especially as regards directions, as shall be explained further on. The wind speed and direction intervals are selected based on the criteria of interest for investigating the near and far wakes. The time series are characterized by low-moderate wind speed, and therefore high thrust.

Couples of turbines are compared by analyzing mainly the statistical distribution of the power ratio between upstream and downstream turbine. The power ratios between upstream and downstream turbine are computed at the same time interval. This is obvious in the case of 10-minutes data, it is less obvious in the case of time-resolved data because at those sampling times (cited in Table 1), the delay between upstream and downstream is a visible phenomenon. The way one considers the transport of the wake from upstream to downstream inevitably includes a certain degree of arbitrariness if approached theoretically. If approached experimentally, a cross-correlation between signals upstream and downstream would in principle detect the delay. There are two reasons why this can't be applicable in the selected case: the first reason is that the turbulence induced by the turbine upstream alters dramatically the signal downstream and the second reason is that the selected wake events are too short to circumvent the first reason through a large statistics. Therefore, in this work it has been decided to watch unambiguously the picture at the same time. In doing this, one is reassured by the fact that the selected time series all belong to a same "wake event".

The time-resolved acquisitions, selected for the present work, are summarized in Table 1. The sampling frequency of the time-resolved series depends on buffering issues and must be reconstructed from the data themselves[1]. For this reason, in Table 1, for each time series, the sampling frequency is indicated.

Series	Sampling frequency (Hz)	Wind Speed (m/s)	Wind direction (o)
FW1	0.20	4.7 ± 1.3	17.4 ± 20
FW2	0.13	5.4 ± 1.9	27.4 ± 30
MW1	0.57	5.1 ± 1.8	318.7 ± 24
MW2	0.20	3.8 ± 1.8	300.6 ± 48

Table 1. Wind conditions (with 2σ interval) at the upstream wind turbine, from the time-resolved data sets employed for the analysis.

A critical issue as regards the objective stated above is filtering the master SCADA data set on reasonably similar conditions with respect to the time series, despite it has different sampling

[1] The "run total" is employed for reconstructing the sampling frequency: it stores the elapsed time, with respect to a reference, according to discrete jumps of 600 seconds (i.e. the averaging time of the SCADA data). Identifying the number of time-resolved measurements inside each block of data, displaying the same "run total" value, allows to compute the actual acquisition rate.

time with respect to the time-resolved series. As regards wind intensity, the procedure is as follows: the standard deviation of each time-resolved wind-intensity series has been computed and the SCADA data are filtered around the average values of Table 1, on an interval having two standard deviations as total amplitude. As regards wind directions, the procedure is as follows: it is based on the standards from the International Electrotechnical Commission [21] defining the amplitude of the wake sector of a wind turbine on a neighboring one, on the basis of their relative distance L in units of rotor diameter D. The formula is given in Equation 1.

$$\alpha = 1.3\frac{180\arctan\left(2.5D/L + 0.15\right)}{\pi} + 10\,. \qquad (1)$$

In the case of FW1 and FW2 time series, the focus is on the wake of turbine T3 on turbine T1. By applying Equation 1 between turbines T3 and T1, the amplitude of the waked sector is 48.13°. Therefore, the average wind direction values of the time series FW1 and FW2, as reported in Table 1, are well inside the waked sector. Similarly, as regards the wake between turbines T3 and T4, the amplitude of the waked sector is 64.52° and the average wind direction values for time series MW1 and MW2 are well inside the waked sector too. Basing on this criterion, the 10-minutes SCADA data sets are filtered on wind directions corresponding to the waked sector. The time-resolved fluctuations of wind direction during the four time series of Table 1 are analyzed too: the occurrence outside the waked sector is quantified. On these grounds, despite the different nature of the data sets, it is reasonable to juxtapose the "slow" and "fast" fluctuations and statistics because the regimes are fairly corresponding. In the following Section 4, it is shown that this is indeed the case.

Further interesting information about wakes dynamics can be discovered through the analysis in the frequency domain. The OPC server can give real-time measurement with a good resolution in time. Unfortunately, technical problems regarding the buffering and the transfer rate did not allow a robust and reliable frequency when asking to the system very fast acquisition (up to 2 Hz) for the whole wind farm. The use of "run hour" counter, as explained above, allows anyway to reconstruct the acquisition frequency and this allows to set the scale for frequency-domain analysis. Therefore, the spectral density of wind speed and power production of some sample time series are shown and discussed in Section 4.

4. Results

From the histograms in Figures 2, 4, 6 and 8, it arises that the very vast majority of data from the time-resolved series belongs to the waked sector of, respectively, the T01-T03 and T03-T04 couple of wind turbines. Therefore, the procedure for filtering the 10-minutes SCADA data set (described in Section 3) is ex post supported.

Comparing Figures 3 and 5 to Figures 7 and 9, it arises that the distribution of power ratios of the SCADA data differs more from the one of the time-resolved series in the cases of mid wake. In particular, in the mid-wake cases, for the time resolved data sets, a considerable population at the lower and higher tails of the distribution is evident. In order to have an insight into this phenomenon, for each time series, data have been grouped in tertiles with respect to the downstream - upstream power ratio. For each subset, the distribution of wind directions (relative to the center of the wake) is analyzed. They are shown in Figures 10 and 11 for the FW2 and MW2 time series, respectively. From these Figures, a clear dependency of the power ratio on the wake condition (full or partial) doesn't arise (except some expected peak of low power ratios around 0°).

In Table 2, the average power ratios between upstream and downstream turbine are shown for the time-resolved series of Table 1 and for each corresponding SCADA milestone: considerable deviations occur in the mid-wake case.

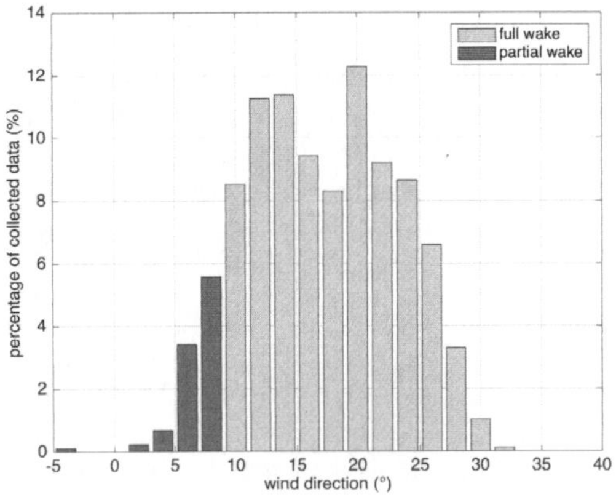

Figure 2. Distribution of wind direction: time-resolved data, FW1 time series. Data inside the waked sector between T01 and T03 in gray, in red the data outside the waked sector.

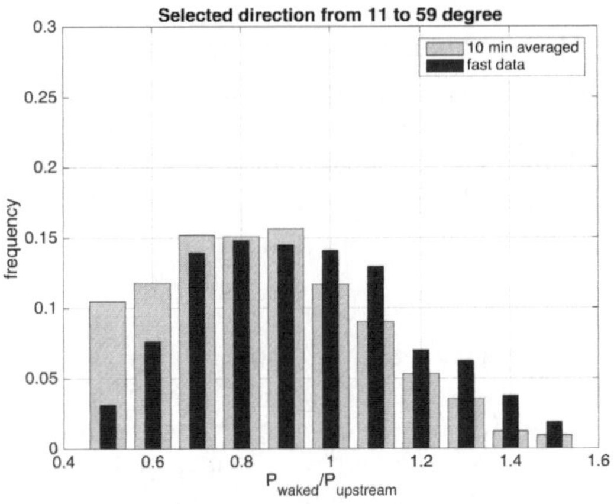

Figure 3. Distribution of power ratios (P_{T01}/P_{T03}): time-resolved vs. SCADA, FW1 time series.

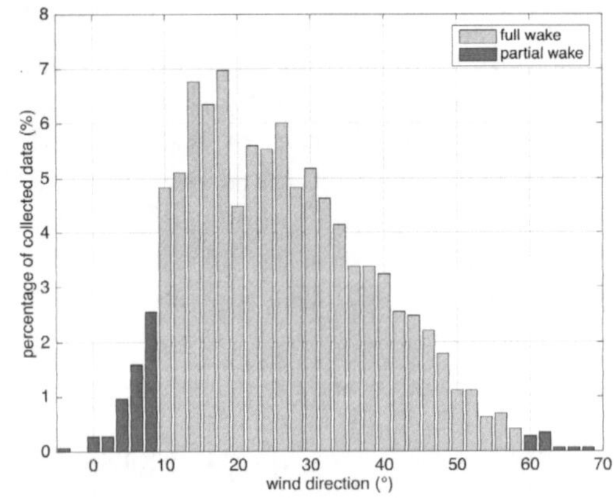

Figure 4. Distribution of wind direction: time-resolved data, FW2 time series. Data inside the waked sector between T01 and T03 in gray, in red the data outside the waked sector.

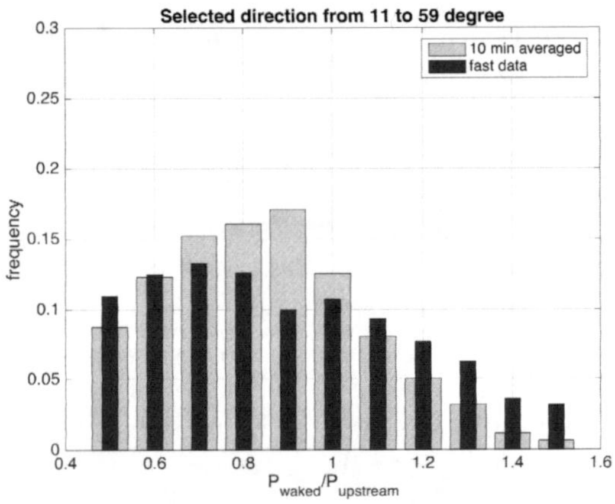

Figure 5. Distribution of power ratios (P_{T01}/P_{T03}): time-resolved vs. SCADA, FW2 time series.

Table 2 suggests that in the far-wake region the results from steady numerical simulation, as RANS (Reynold Averaged Navier Stokes) or linearized models, have more chances of reproducing with a good approximation the wake fluctuations. Actually, the far-wake behavior is less turbulent and more steady and this supports the fact that the distributions of Figures 3 and 5 show a good agreement between "slow" and "fast" data. In any case, the fluctuations observed in Figures 10 and 11 invoke a deeper comprehension on the dynamics of the wake fluctuations and this is why a spectral analysis should be carried on.

Figures 12 and 13 show the spectral density of power production and wind speed for the

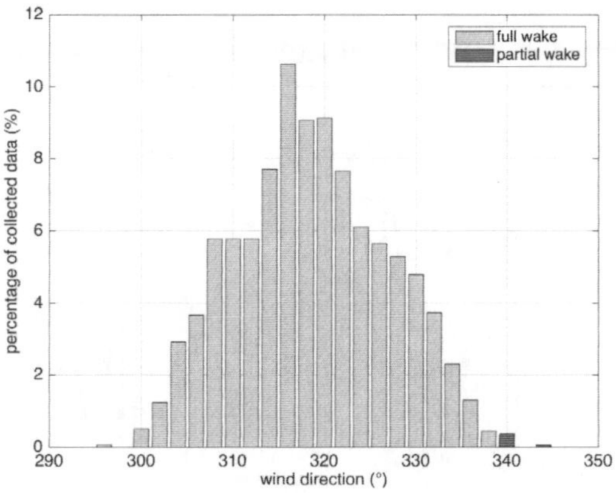

Figure 6. Distribution of wind direction: time-resolved data, MW1 time series. Data inside the waked sector between T03 and T04 in gray, in red the data outside the waked sector.

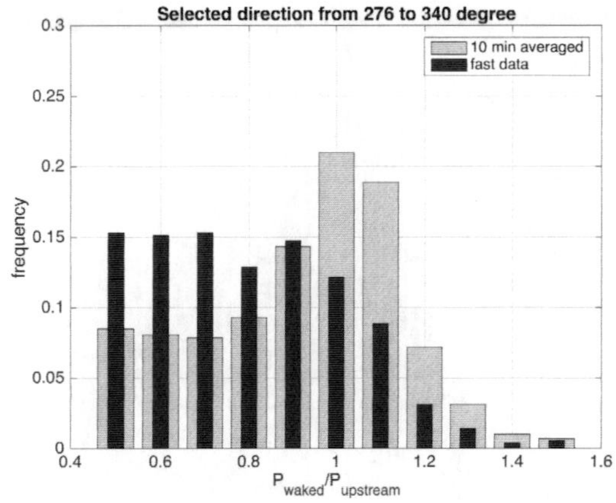

Figure 7. Distribution of power ratios (P_{T03}/P_{T04}): time-resolved vs. SCADA, MW1 time series.

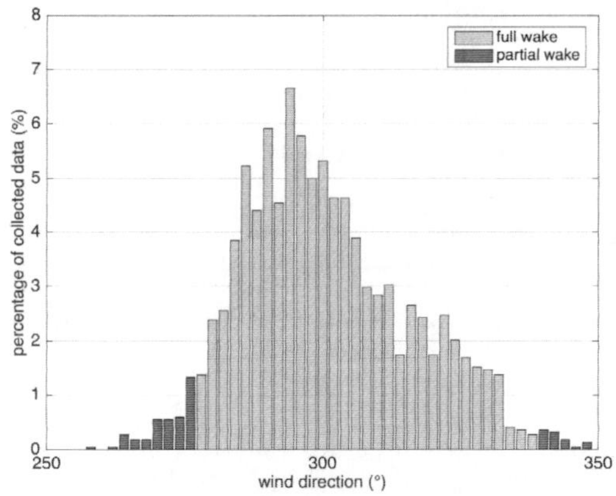

Figure 8. Distribution of wind direction: time-resolved data, MW2 time series. Data inside the waked sector between T03 and T04 in gray, in red the data outside the waked sector.

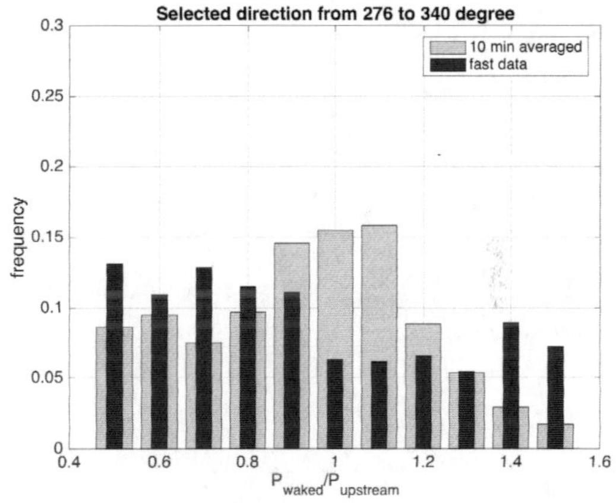

Figure 9. Distribution of power ratios (P_{T03}/P_{T04}): time-resolved vs. SCADA, MW2 time series.

upstream T4 and the downstream T3 wind turbines, during the MW1 acquisition time series. The MW1 wake event was chosen for this analysis as it is characterized by the highest sampling rate. It arises that the downstream turbine T3 shows higher oscillation amplitudes and is more noisy in the high-frequency domain, as regards wind speed (Figure 13). This is due to the additional turbulent effects induced by the wake. Comparing Figure 12 to 13, the role of the wind turbine as low-pass filter arises: the turbine cuts out frequencies and absorbs unsteady mechanical loads. Up to the highest frequencies at disposal, the PSD of the wind speed fairly agrees with the $-\frac{5}{3}$ law of turbulence from Kolmogorov [22]. The presence of the wind turbine cuts this behavior and acts as low-pass filter, as the difference between Figures 12 and 13 shows.

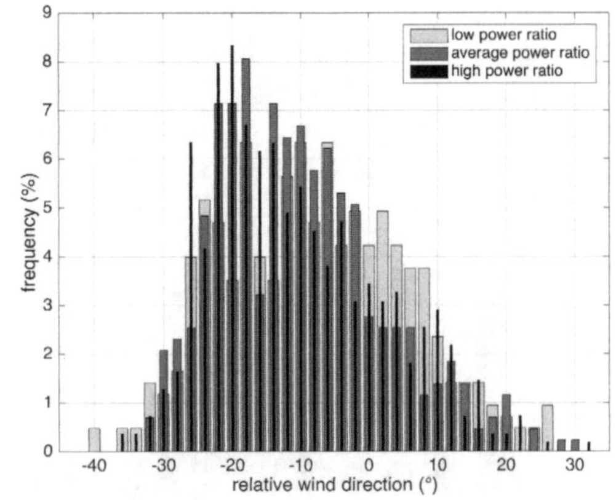

Figure 10. Distribution of wind directions of turbine T03, FW2 time series. Data are grouped in tertiles with respect to the P_{T01}/P_{T03} ratio.

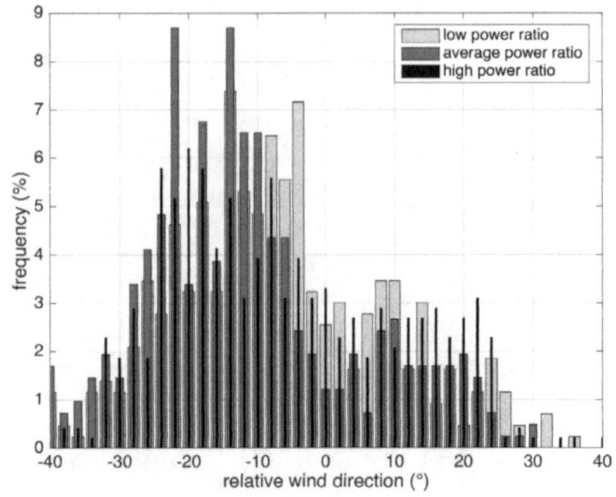

Figure 11. Distribution of wind directions of turbine T04, MW2 time series. Data are grouped in tertiles with respect to the P_{T03}/P_{T04} ratio.

Series	Time-resolved power ratio	SCADA power ratio
FW1	0.98	0.80
FW2	0.95	0.81
MW1	0.68	0.75
MW2	0.42	0.74

Table 2. Average power ratios for the high-frequency time series of Table 1 and their corresponding SCADA milestones.

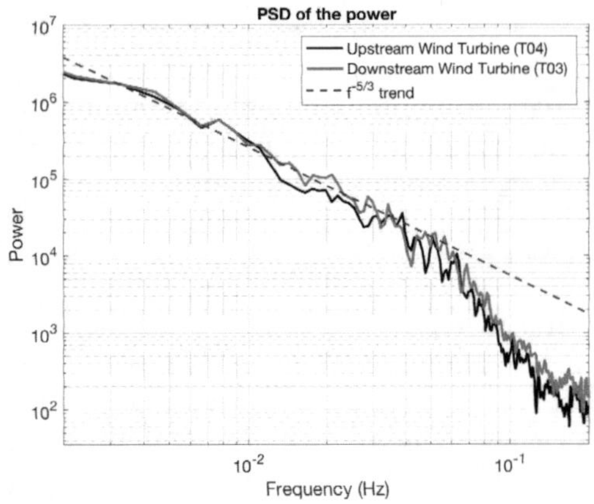

Figure 12. PSD of power production: upstream T4 and downstream T3 wind turbines, MW1 time series.

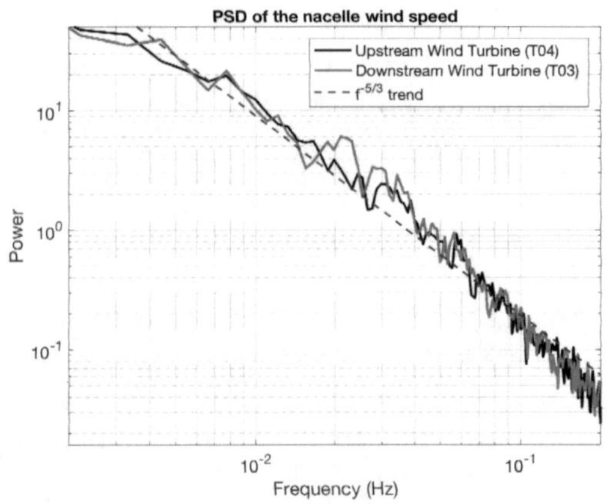

Figure 13. PSD of wind speed: upstream T4 and downstream T3 wind turbines, MW1 time series.

5. Discussion and Further Directions

In this work, wake fluctuations have been analyzed through the experimental data of a test wind farm sited in Italy in a moderately complex terrain. The key point of the work is that the analysis has been based on SCADA data, with their usual 10-minute sampling time, and high-frequency time resolved data as well. Thus it has been possible to observe wakes using two lenses: a "slow" one (SCADA) on a catalog of wake events spanned over a long period, and a "fast" one on some spot time series of wake events. Upon filtering the SCADA data in order to resemble reasonably the wind conditions of each spot wake event, the "slow" and "fast" fluctuations have been juxtaposed and analyzed, in order to appreciate their different features. Further, the layout of the wind farm (Figure 1) makes it possible to observe two wake regimes: mid-wake (approximately 4 rotor diameters) and far-wake (approximately 8 rotor diameters). This allowed to observe, for example, the characteristics of power-ratios fluctuation in the mid- and far-wake region. The results of this work are promising about the possibility of employing high-frequency data of wind turbines for validating LES frameworks, that are currently at their early stages especially in complex terrain [23], and for cross-validation against independent averaged informations from the turbines. Figures 12 and 13 suggest another direction of this work, because they highlight the role of the wind turbine in smoothing the high-frequency oscillations coming from the wind: it would be interesting to analyze how the wind turbine, acting de-facto as a low-pass filter, is affected by the unsteady mechanical loads [24].

Acknowledgments

The authors acknowledge the Italian company Lucky Wind spa for providing the production data and all the technical information used in present work.

References

[1] Matthew J Churchfield, Sang Lee, John Michalakes, and Patrick J Moriarty. A numerical study of the effects of atmospheric and wake turbulence on wind turbine dynamics. *Journal of turbulence*, (13):N14, 2012.

[2] B Subramanian, N Chokani, and RS Abhari. Aerodynamics of wind turbine wakes in flat and complex terrains. *Renewable Energy*, 85:454–463, 2016.

[3] Ann Hyvärinen and Antonio Segalini. Effects from complex terrain on wind-turbine performance. *Journal of Energy Resources Technology*, 2017.

[4] Fredrik Seim, Arne R Gravdahl, and Muyiwa S Adaramola. Validation of kinematic wind turbine wake models in complex terrain using actual windfarm production data. *Energy*, 123:742–753, 2017.

[5] Niels Otto Jensen. A note on wind generator interaction. Technical Report Ris-M-2411, Ris National Laboratory, 1983.

[6] Sten T Frandsen, Hans E Jørgensen, Rebecca Barthelmie, Ole Rathmann, Jake Badger, Kurt Hansen, Søren Ott, Pierre-Elouan Rethore, Søren E Larsen, and Leo E Jensen. The making of a second-generation wind farm efficiency model complex. *Wind Energy*, 12(5):445–458, 2009.

[7] Rebecca Jane Barthelmie, K Hansen, Sten Tronæs Frandsen, Ole Rathmann, JG Schepers, W Schlez, J Phillips, K Rados, A Zervos, ES Politis, et al. Modelling and measuring flow and wind turbine wakes in large wind farms offshore. *Wind Energy*, 12(5):431–444, 2009.

[8] Rebecca Jane Barthelmie and LE Jensen. Evaluation of wind farm efficiency and wind turbine wakes at the nysted offshore wind farm. *Wind Energy*, 13(6):573–586, 2010.

[9] Kurt S Hansen, Rebecca J Barthelmie, Leo E Jensen, and Anders Sommer. The impact of turbulence intensity and atmospheric stability on power deficits due to wind turbine wakes at horns rev wind farm. *Wind Energy*, 15(1):183–196, 2012.

[10] Francesco Castellani, Davide Astolfi, Alberto Garinei, Stefania Proietti, Paolo Sdringola, Ludovico Terzi, and Umberto Desideri. How wind turbines alignment to wind direction affects efficiency? a case study through scada data mining. *Energy Procedia*, 75:697–703, 2015.

[11] Francesco Castellani, Davide Astolfi, Emanuele Piccioni, and Ludovico Terzi. Numerical and experimental methods for wake flow analysis in complex terrain. In *Journal of Physics: Conference Series*, volume 625, page 012042. IOP Publishing, 2015.

[12] Francesco Castellani, Davide Astolfi, Paolo Sdringola, Stefania Proietti, and Ludovico Terzi. Analyzing wind

turbine directional behavior: Scada data mining techniques for efficiency and power assessment. *Applied Energy*, 185:1076–1086, 2017.

[13] Francesco Castellani, Davide Astolfi, Matteo Mana, Emanuele Piccioni, Matteo Becchetti, and Ludovico Terzi. Investigation of terrain and wake effects on the performance of wind farms in complex terrain using numerical and experimental data. *Wind Energy*, 2017.

[14] Torben J Larsen, Helge Aa Madsen, Gunner C Larsen, and Kurt S Hansen. Validation of the dynamic wake meander model for loads and power production in the egmond aan zee wind farm. *Wind Energy*, 16(4):605–624, 2013.

[15] Soo-Hyun Kim, Hyung-Ki Shin, Young-Chul Joo, and Keon-Hoon Kim. A study of the wake effects on the wind characteristics and fatigue loads for the turbines in a wind farm. *Renewable Energy*, 74:536–543, 2015.

[16] Luis Vera-Tudela and Martin Kühn. Analysing wind turbine fatigue load prediction: The impact of wind farm flow conditions. *Renewable Energy*, 107:352–360, 2017.

[17] Wolf-Gerrit Früh, Angus CW Creech, and A Eoghan Maguire. Turbulence characteristics in offshore wind farms from les simulations of lillgrund wind farm. *Energy Procedia*, 59:182–189, 2014.

[18] Jan-Ake Dahlberg. Assessment of the lillgrund wind farm: power performance wake effects. *Vattenfall Vindkraft AB, 6_1 LG Pilot Report, http://www. vattenfall. se/sv/file/15_Assessment_of_the_Lillgrund_W. pdf_16596737. pdf (cited March 30, 2012)*, 2009.

[19] Kurt Schaldemose Hansen, Gunner Chr Larsen, Robert Menke, Nikola Vasiljevic, Nikolas Angelou, Ju Feng, Wei Jun Zhu, Andrea Vignaroli, C Xu, Wen Zhong Shen, et al. Wind turbine wake measurement in complex terrain. In *Journal of Physics: Conference Series*, volume 753, page 032013. IOP Publishing, 2016.

[20] Patrick Milan, Matthias Wächter, and Joachim Peinke. Turbulent character of wind energy. *Physical review letters*, 110(13):138701, 2013.

[21] Robert Gasch and Jochen Twele. *Wind power plants: fundamentals, design, construction and operation.* Springer Science & Business Media, 2011.

[22] H Eugene Stanley. Turbulence: The legacy of an kolmogorov. *Journal of Statistical Physics*, 88(1):521–523, 1997.

[23] Nina Zhou, Jun Chen, Douglas E Adams, and Sanford Fleeter. Influence of inflow conditions on turbine loading and wake structures predicted by large eddy simulations using exact geometry. *Wind Energy*, 2015.

[24] C Braccesi, F Cianetti, and L Tomassini. Random fatigue. a new frequency domain criterion for the damage evaluation of mechanical components. *International Journal of Fatigue*, 70:417–427, 2015.

Numerical Study of the Impact of Atmospheric Stratification on a Wind-Turbine Performance

A. Chaudhari[1], O. Agafonova[1], A. Hellsten[2,1] and J. Sorvari[1]

[1] CEID, School of Engineering Science, Lappeenranta University of Technology, P.O. Box 20, 53851 Lappeenranta, Finland

[2] Finnish Meteorological Institute, P.O. Box 503, 00101 Helsinki, Finland

E-mail: Ashvinkumar.Chaudhari@lut.fi

Abstract. This paper is aimed to obtain a better understanding of how a stably stratified atmospheric condition involving low-level jet, in comparison to neutral condition, affects to a wind-turbine performance. In this work, large-eddy simulation approach in combination with actuator-line model is used to simulate the wake flows in an infinitely long wind farm. Here, numerical simulations of the wind-turbine wakes are carried out under the influence of neutral and stable conditions. The numerical results are compared between the two atmospheric conditions. It has been seen that the wind-turbine performance is highly influenced under the stable conditions. For an infinitely long wind-farm scenario, the power produced by a turbine under the stable condition is smaller by 81% than that under the neutral condition.

1. Introduction

Diurnal and seasonal variations induce drastic changes to the Atmospheric Boundary-Layer (ABL), which in turn, affect wind energy production. Especially, stably stratified ABL-conditions, typical in wintertime, are a challenge to wind energy production as such conditions are characterized by relatively shallow ABL depth, lower and sometimes intermittent turbulence levels, strong wind shear, and a relatively high wind near the top of the boundary layer. At this height, the local wind becomes super-geostrophic and can form a Low-Level Jet (LLJ). In this way, a LLJ is a localized maximum in a vertical profile of wind speed and that typically occurs at a height between 80–1000 m [1, 2]. This region is very important for wind energy production as modern wind turbines are operating in this region and thus LLJs have a direct impact over the wind turbines performance. As such LLJs can be found in every continent, their occurrences are frequently been reported for example in the mid-latitudes, especially in Southern Great Plains, and in the polar regions [1, 2]. According to the authors [2], LLJs occur more frequently (70%) in the cold season than in the warm season. According to [3], about 75% of LLJs in the Great Plains occur at nights in summers.

Due to the localized maximum wind velocity in LLJ, stable conditions involving LLJs may provide larger energy potential compared to those of neutral and convective ABLs having the same geostrophic wind speed. However, Stable Boundary Layer (SBL) may induce high deformation strain and great fatigue loads on wind-turbine structure due to strong shear and veer inherited with stable conditions [4, 5, 6, 3, 7]. Further, Coriolis effects associated with the planetary rotation are responsible for changes in wind direction with height (i.e. veering) and the formation of an Ekman spiral [8]. Such effects can lead to a skewed wind turbine wake and they are more pronounced in stable conditions [9].

This paper is aimed to obtain a better understanding of how a stably stratified atmospheric condition involving LLJ, in comparison to neutral condition, affects to a wind-turbine performance. Previously, several numerical studies [10, 11, 12, 13, 14, 15, 16, 17, 8, 18, 19] attempted to study the influence of thermal and especially stable stratifications on the wind-turbine wakes. However, most of the previous studies [10, 11, 13, 15, 16, 18, 19] have simulated stable conditions in the absence of LLJs. Apart from that, a number of previous numerical attempts [12, 13, 14, 18, 19] made for stable conditions were focused on a single turbine wake, therefore, their findings are only limited to a stand-alone turbine and not to a large wind farm. Nevertheless, some recent studies [8, 14, 17] have focused on wake development under the influence of real-time stable atmospheric conditions, in which, proper representations of LLJs do occur.

In this paper, we carry out Large-Eddy Simulations (LES) for infinitely long wind-farm wakes under the influence of neutral and stable stratifications. To consider real-time stable atmospheric conditions in this study, we have adopted the stably-stratified wind conditions from the GABLS (Global Energy and Water Cycle Experiment Atmospheric Boundary Layer Study) [20] case study. This case represents a typical quasi-equilibrium moderately SBL with a LLJ (at a height between 115–255 m), similar to those commonly observed over the polar regions and equilibrium night-time conditions over land in the mid-latitudes [17, 14]. In this study, firstly simulations are performed without any wind turbine for both, stable and neutral, atmospheric conditions in order to generate their reference cases. Then, the simulations are repeated by including a 5-MW wind turbine. The results from the both conditions are compared against each other to see the impact of the stably-stratified condition with respect to that from the neutral condition. The result comparison is shown into perspective for wind energy applications.

2. Computational approach

In this work, the incompressible filtered Navier-Stokes equations including the Coriolis and buoyancy forces are solved along with the continuity equation and potential-temperature transport equation. The Boussinesq approximation is used for the buoyancy forces. Further, Lagrangian-Averaged Scale-Independent (LASI) dynamic Smagorinsky model [21] is employed to model the effect of the sub-grid scales. Here, the LES calculations are carried out using the OpenFOAM-based SOWFA (Simulator for On/Offshore Wind Farm Applications) package developed by the NREL (National Renewable Energy Laboratory) researchers [5]. Recently, the SOWFA package has been successfully used in the ABL flow simulations for wind energy applications [5, 14, 22, 23, 24, 25, 26].

Following [8, 14, 17, 24, 27, 28, 23], we have chosen to model the turbine effects into the simulations rather than resolving the turbine geometry. The turbine-induced forces are parametrized in the simulations using the actuator line model developed by Sørensen and Shen [29]. The turbine used in this study is the NREL 5-MW reference turbine [30]. It is a three-bladed horizontal axis wind turbine with a rotor diameter D of 126 m and a hub height H of 90 m. More information on the NREL 5-MW turbine can be found in [30].

The simulations are performed over a computational domain of the size of $7D \times 7D \times 5D$ in the stream-wise (x), span-wise (y) and vertical (z) directions, respectively. Here, $D = 126$ m is the diameter of the rotor. The computational domain is discretized uniformly into $221 \times 221 \times 158$ number of grid-cells in x, y and z directions, respectively. This leads to a grid resolution Δx_i ($i = 1, 2, 3$) of nearly 4 m in all directions. This resolution ensures that there are about 31 finite-volume grid cells within the rotor area, which indicates finer grid resolution than those utilized in many previous studies [8, 17, 28, 15, 27, 13, 31]. It is obvious that one would desired to use even finer grid resolutions than what we used here, such as up to 1 m in the wake region.

However, due to higher CPU resources requirement we limited our grid resolution to 4 m. Also, Basu and Porté-Agel [32], who simulated the same stable condition (the GABLS case), showed that it is sufficient to have grid resolution of 4 m to simulate this particular stable condition.

To simulate an infinitely long wind farm, periodic boundary conditions are utilized in both horizontal directions in all simulations. The turbine was placed in the middle of the computational domain. Such arrangement implies a seven diameter (7D) distance between individual turbines in horizontal (stream-wise and span-wise) directions. The stably stratified condition simulated here is adopted from the GABLS case [20] in order to examine the influence of especially LLJ to the wind-turbine performance. In all simulations, the flow is driven by fixing a pressure gradient force which adjusts a desired geostrophic wind U_g near the domain's top boundary. For this reason, in all simulations the height of the computational domain is kept high (at $z = 5D = 630$ m), so that the flow at the height where the U_g is fixed is not affected by the turbine wakes.

In the stable case, the geostrophic wind is specified as $U_g = (8.0, 0.0, 0.0)$ m/s at $z = 628$ m, and a cooling rate of 0.25 K/h is specified at the surface boundary. The velocity is initialized to the geostrophic wind throughout the domain. From the surface to up 100 m, the potential temperature θ is initialized to 265 K, and above that a capping inversion with a strength of 10 K/km is applied. Also, below 100 m, random fluctuations with amplitudes of 0.15 K and 0.5 m/s are added to the initial potential-temperature and horizontal-velocity fields, respectively. The reference temperature θ_0 is set to 263.5 K. Using these initial and boundary conditions, the stable case attains a quasi-steady state in roughly 8–9 h with a boundary-layer depth ranging between 190–200 m.

In the neutral case, the geostrophic wind (at $z = 628$ m) is adjusted such that in the *absence* of the turbine, the wind speed at hub height is the same as that in the stable case. This condition insures that in the absence of the turbine there is a similar amount of potential power (due to the similar wind-speeds at hub height) available in both atmospheric conditions. Like in the stable case, the velocity is initialized to the geostrophic wind in the entire domain with 0.5 m/s of random fluctuations.

In all simulations, the aerodynamic roughness length z_0 is set to 0.1 m, and all the simulations take place at 73° north latitude. The LASI sub-grid scale model requires to solve two partial-differential equations for the model quantities: $\pounds_{LM}$ and $\pounds_{MM}$. These quantities are initially set to $\pounds_{LM} = 2.56 \times 10^6$ m^4/s^4 and $\pounds_{MM} = 1.0 \times 10^4$ m^4/s^4 throughout the entire domain such that the Smagorinsky constant C_s is 0.16. The boundary conditions for $\pounds_{LM}$ and $\pounds_{MM}$ quantities are zero normal-gradient at the surface and top boundaries.

Following [8, 17], who carried out wind-turbine wake simulations using the GABLS case, the simulation with stable condition is run for 9 h of flow physical time, and the turbine was introduced in the simulations only during the last two hours, i.e. from the 7th hour on-wards till the end of 9th hour. The same is also done in the neutral case. The numerical results are time averaged over the last hour and they are also space averaged in the horizontal directions.

3. Numerical results and discussions

We start this section by briefly discussing the validation of especially the stable simulation from the reference cases, that it, in the absence of the wind turbine. After that we discuss the impact of the two (stable and neutral) atmospheric conditions in the reference simulations as well as in the turbine simulations.

3.1. Validation of the simulations

Figure 1 shows the LES predicted vertical profiles of the time-and-space averaged x-directional velocity U_x (a, b), x-directional velocity fluctuations σ_x (c) and potential temperature θ (d). The figure also compares the predicted results with their corresponding reference data. To validate our stable case we have chosen to compare our results with another LES data by Basu and Porté-Agel [32]. For the neutral case, we compare the x-directional velocity U_x with the standard logarithmic law: $U = \frac{u_*}{0.4} ln(\frac{z}{z_0})$. Here u_* is the frictional velocity. In both cases, the results generally agree well with their respective reference data. In the stable case, the boundary layer shows the existence of a LLJ between z =115–240 m. The upper height of the LLJ from our LES is slightly shorter (by 10 m) than that in [32]. Both LES profiles shows the same magnitude of the maximum velocity in the LLJ region. The boundary layer height (δ) and the Obukhov length (L) from our stable simulation are 190 m and 118 m, respectively. These parameter are also in close agreement with the reference LES [32].

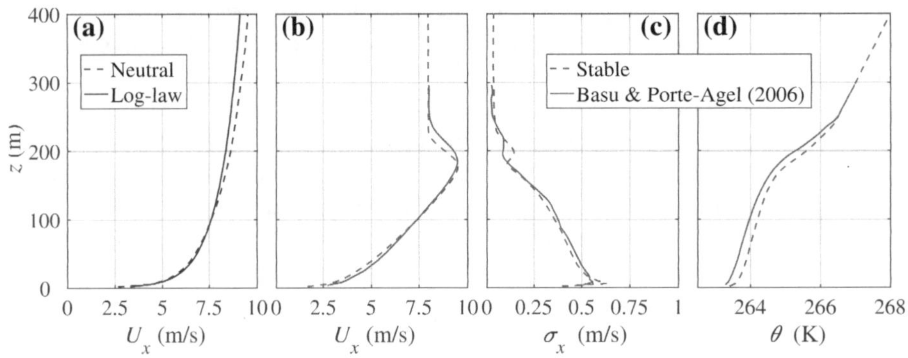

Figure 1. Vertical profiles of the LES results (dashed lines) of the x-directional wind-speed U_x (a, b), x-directional velocity fluctuations σ_x (c), and potential temperature θ (d) obtained from the reference cases (without turbine). The solid lines represents their corresponding reference data for validating the simulations.

3.2. Impact of stratifications on the reference (without turbine) wind conditions

Here we first compare and study the impact of stratifications on the reference (without the wind turbine) wind conditions. In the following, Figure 2 compares various wind statistics from the reference cases of the two conditions. The vertical profiles of the normalized horizontal wind-speed U calculated as $U = \sqrt{U_x^2 + U_y^2}$ are shown in Figure 2(a). Here, U_x and U_y are the mean-velocity components in x and y directions, respectively. The results are normalized with the hub-height velocity U_H. It can be observed that the upper-half area of the rotor seems to be in high influence of the LLJ. However the center of the LLJ appears at $z \approx 2H$, and it is just outside of the rotor area. The boundary-layer height in the stable case is about 190 m ($z/H \approx 2.1$), which is only the 30% of the neutral boundary-layer height. Because the neutral boundary-layer grows up to the top boundary, the wind-speed increases with height. As a result, the geostrophic wind in the neutral case is higher by 22% compared to that of the stable case.

Figure 2(b) shows the horizontal turbulence intensity I_M for the two cases. Here, I_M is the magnitude of the turbulence intensities from the two horizontal directions, and it is calculated as

$$I_M \text{ (in \%)} = 100\sqrt{\left(\frac{\sigma_x}{U_H}\right)^2 + \left(\frac{\sigma_y}{U_H}\right)^2} \quad (1)$$

where, σ_x and σ_y are the root-mean-square of the velocity fluctuations in x and y directions, respectively. The negative heat-flux in stable stratification has a tendency to reduce the turbulence level. As a consequence the turbulence intensity is smaller thoroughly in the stable boundary layer as compared to that in the neutral boundary layer. At the hub height, the

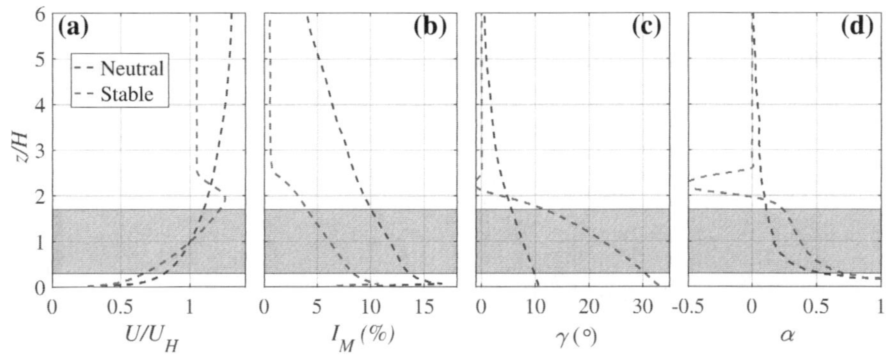

Figure 2. Vertical profiles of the normalized horizontal wind-speed U/U_H (a), horizontal turbulence intensity I_M (b), wind direction γ (c), and wind shear α obtained from the reference (without turbine) cases for both atmospheric conditions. The light-gray background depicts the rotor area.

intensity in the stable case is smaller by 48% compared that from the neutral condition. Figure 2(c) explains that the change in the wind direction (wind veer) is stronger in the stable case compared to that in the other case. The wind direction changes about 19° across the rotor diameter in the stable case, while only 4° of change is observed in the neutral case. This indicates asymmetric load to wind turbine under the influence of stable condition.

Figure 2(d) presents the wind shear α, and it is calculated as $\alpha = \frac{H}{U_H}\left(\frac{dU}{dz}\right)$. It is observed that in the neutral case the shear is 0.52 at the rotor-lowest position and then decreases to 0.11 at the rotor-highest position leading to the reduction of 79% within the rotor. The shear reduction in the stable case is 65% within the rotor area. Importantly, the shear observed especially in the stable case, is too high for safe turbine operations according to the IEC (International Electrotechnical Commission) standards [33]. For example, IEC suggests $\alpha = 0.2$ as one of the design criteria for the strongest class of wind turbines [33]. The shear values at the hub height are found to be 0.17 and 0.37 in the neutral and stable cases respectively. Thus the IEC design criterion for even the strongest class of wind turbines is exceeded under the stable condition. Generally high shear is very likely the case in almost all stable conditions. The increased shear in the stable case is likely to affect the pitching angle of the turbine. Also, the high wind veer under the stable condition are expected to affect power production and to increase the yawing moments with possibly harmful consequences for turbine lifetime.

3.3. Impact of stratifications in the presence of the wind turbine

Next we look the results from the turbine simulations. Figure 3 shows the contours of the time-averaged velocity magnitude at a horizontal $(x - y)$ plane at the hub height from the two cases. Because the wind directions at the hub height are different in both cases, the turbine yaw-angles were aligned to their respective wind directions at the hub height, which were known from their reference cases. The velocity distributions in Figure 3 explain that the stable stratification has a significant impact on the wind-turbine wake. The wake is stronger under the influence of the stable condition, as it prolongs longer as well as the hub-height wind speed is smaller in the entire domain compared to that in the neutral case. These findings are in agreement with other studies [10, 13, 18, 19, 16]. Due to less vertical turbulent mixing in stable conditions in general, the wake recovery is slower in stable condition than for example in convective or neutral conditions. In addition, the wake is also triggered by the use of periodic boundary conditions in the horizontal directions (infinitely long wind farm) and the effects are more profound in the

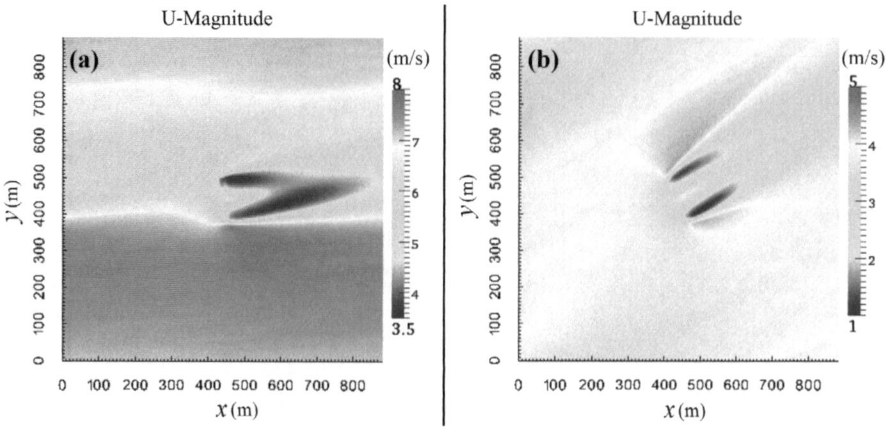

Figure 3. Contours of 1-h time-averaged velocity magnitude at the hub height from the neutral (a) and stable (b) cases.

stable case due to less turbulent mixing.

The time-and-space averaged results from the turbine simulations are presented in Figure 4. The results from the turbine simulations (solid lines) are also compared with their reference cases (dashed lines). The horizontal wind speed in Figure 4(a) suggests that due to the stronger wake effects in the stable case the mean wind is smaller by 46% at the hub height than that in the neutral case. This was however expected from the velocity distributions in Figure 3. In the mean wind-speed (Figure 4(a)), the wake effects extend vertically till $z = 3.4H$ and $z = 5.1H$ under the stable and neutral conditions, respectively. In both cases, the turbulence intensity in the wake is found to be increased thoroughly in the boundary layers. In the neutral case, the turbulence intensity does not vary significantly across the rotor, whereas in the stable case it increases from 7.7% at the lower-tip to 10.2% at the upper-tip, and it maintains till $z = 2.9H$. In both cases, the wind veer changes from their respective reference conditions, however the changes are not significant in the neutral case according to Figure 4(c). The veer across the rotor has been reduced from 19° in the reference case to 15° in the wake for the stable case. For the neutral case, the reduction is from 4° to 2°. Despite having the large difference (112% at the hub height) between the two shear values of the two reference cases (neutral and stable), the shear values from the turbine cases seem to close to each other at the hub height (Figure 4(d)). Apart from that, the shear behaves differently between the two cases. Table 1 summarize the hub-height values of the various wind statistics from the reference and turbine simulations. The table also reports differences between the two simulations for each atmospheric condition.

Next we discuss the velocity deficit. Figure 5 shows the velocity deficit ΔU profiles for the two cases and compares with another LES results of Porté-Agel et al. [8]. The deficit ΔU is calculated using the spaced-averaged velocity as $\Delta U = ((U - U_{ref})/U_{ref})$, where U_{ref} is the velocity from the reference case. In [8], they simulated wind turbine wakes using the same stable condition (i.e. the GABLS case) for infinitely long wind farm case with two arrangements: $5D$ and $8D$ distance between individual turbines in the horizontal directions. In our case, the corresponding distance is $7D$. It has to be mentioned that Porté-Agel et al. [8] used different turbine than what we have used in this work. They used so-called V112-3.0MW wind turbine with a hub height of 119 m and a rotor diameter of 112 m. Apart from that their turbine's aerodynamic properties are much more different than the turbine (NREL-5MW) used here. Therefore, a direct comparison is not possible between the present results and the results from [8], rather we use their results as a qualitative validation.

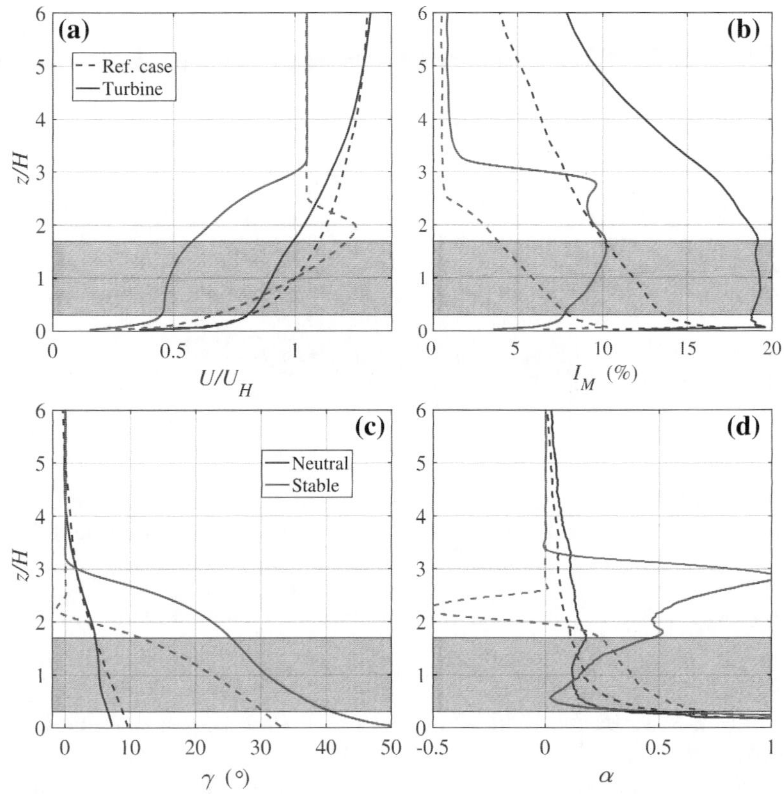

Figure 4. Vertical profiles of the normalized horizontal wind-speed U/U_H (a), horizontal turbulence intensity I_M (b), wind direction γ (c), and wind shear α. The solid lines represent results from the turbine simulations and the dashed lines from their respective reference cases. The light-gray background depicts the rotor area.

Table 1. Wind statistics from the reference and turbine simulations at the hub-height. The difference is calculated with respect to their reference values. The sign, + or −, in the difference shows either increment or decrement in the respective value.

		Neutral			Stable	
	reference	wake	difference	reference	wake	difference
U (m/s)	7.5	6.8	−9 %	7.5	3.7	−51 %
I_M (%)	11.8	19.3	+64 %	5.9	9.5	+61 %
γ (°)	6.7	5.2	+22 %	22.6	30.9	+37 %
α	0.17	0.12	−29 %	0.36	0.14	−61 %

According to Figure 5, the velocity deficit at the hub height in our stable and neutral cases are 51% and 9%, respectively. In the results of [8], the deficit at the corresponding hub height for $5D$ and $8D$ cases are 41% and 31%, respectively. This suggests that the wake deficit for infinitely long wind farm scenario is generally higher (more than 30%) under stable stable conditions. Further, the deficit seems to be so sensitive to the wind farm arrangement (the distance between

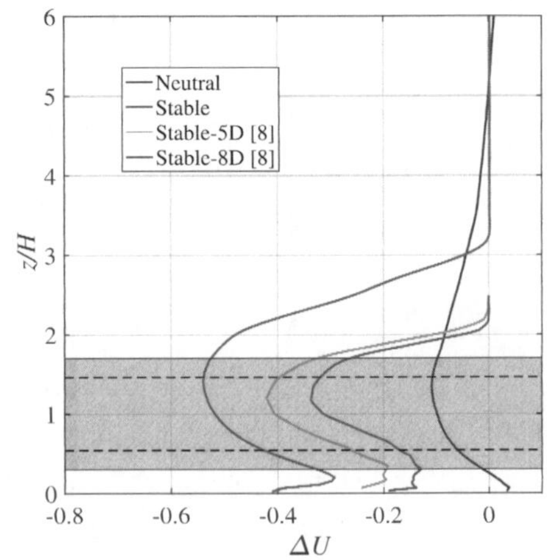

Figure 5. Vertical profiles of the velocity deficit ΔU compared with another LES results of [8]. The black dashed lines in the gray background shows the turbine-rotor area of the turbine used by [8].

turbines), as there is already 10% difference between the two cases of [8]. In our stable case, the deficit is even higher by for example 10% compared to their $5D$ case and 20% compared to the $8D$ case. This difference can be justified by the bigger turbine that we have used here, which has 33% of bigger rotor area than the turbine used in [8]. Bigger rotor area is also responsible for higher vertical expansion of the wake in our case. In addition to that different turbine and its aerodynamic properties, such as airfoil profile, twist angle, chord distribution, tip-speed ratio, yaw angle, etc., should also be responsible at a certain level for the higher velocity deficit (i.e. stronger wake) in our case.

3.4. Power output and losses

In this sub-section, we discuss the power output and the associated statistics from the present work. Figure 6 shows the time series of the power P produced by the turbine during the last 1-h of the flow time in both cases. At first, it is observed that the power production under the stable condition is far smaller than that under the neutral condition. The last 1-h time-averaged powers from the two cases are 1.37 MW and 0.26 MW in the neutral and stable cases respectively. In other words, the mean power output in the stable case is 81% smaller than that in the neutral case.

Table 2 gives the statistics on the power output and power loss (deficit) for the neutral and stable cases. The table also reports the difference between the two conditions for each statistics, which is calculated with respect to the neutral condition. Following [34, 35], we define the power loss as

$$P_{loss}\ (\text{in }\%) = 100\left(\frac{P_{avail} - P}{P_{avail}}\right), \tag{2}$$

which is a loss in the mechanical power with respect to the total-available power P_{avail} from the reference wind (potential power). Here the total-available power P_{avail} is calculated as [34, 35]

$$P_{avail} = \frac{1}{2}\rho\pi R^2 U_{avail}^3, \tag{3}$$

IOP Conf. Series: Journal of Physics: Conf. Series **854** (2017) 012007 doi:10.1088/1742-6596/854/1/012007

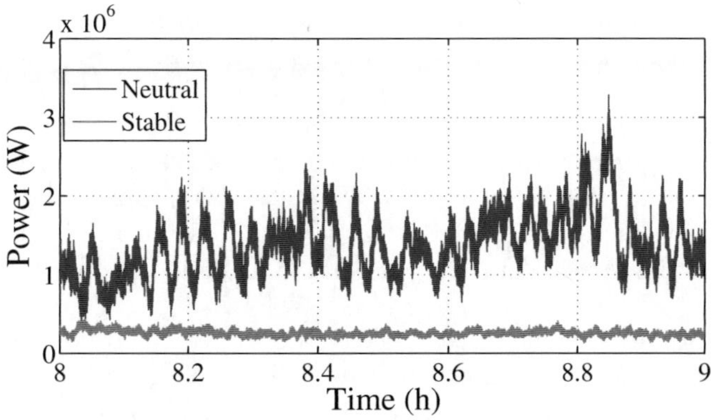

Figure 6. Time series of the power produced by the turbine during the last 1-h of the flow time.

in which U^3_{avail} is defined as

$$U^3_{avail} = \frac{1}{2R}\int_{H-R}^{H+R} U^3(z)dz, \tag{4}$$

where $R = D/2$ is the radius of the rotor.

Table 2. Wind-turbine power and related statistics: total-available power (P_{avail}), power output (P), power loss (P_{loss}) and standard deviation (P_{std}) of the power from the neutral and stable cases. The difference between the two conditions is calculated with respect to the neutral case.

-	Neutral	Stable	difference
P_{avail} (MW)	3.19	3.15	-1%
P (MW)	1.37	0.26	-81%
P_{loss} (%)	57	92	+61 %
P_{std} (MW)	0.39	0.04	-90%

From the table one can see that despite having the negligible difference (of 1%) between the total-available powers of the two conditions, the actual (mechanical) power (P) produced by the turbine under the stable condition differs by 81% from the power obtained in the neutral case. Further it is noticed that the stable condition loses 92% (= P_{loss}) of its total-available power, while in the neutral case the power loss is 57%. Moreover, the standard deviation from the last 1-h of power output P_{std} is 0.39 MW and 0.04 MW from the neutral and stable cases, respectively. The standard deviation is smaller (by 90%) in the stable case than in the neutral case which suggests that the power fluctuates less (due to the less turbulent fluctuations in wind) and that the load is more uniformly distributed in time in the stable case than in the neutral case.

The poor power production is a direct consequence of the stronger wake effects associated with the stable case. This reduces the wind speed more in the stable case than in the neutral case. According to Table 1, the wind speeds at the hub height in the wake simulations are 6.8 (m/s) and 3.7 (m/s) in the neutral and stable cases, respectively. Nevertheless, if one would take the power curve of the NREL-5MW turbine [30] into the account, the difference in the powers between the two conditions is acceptable. For example, according to the NREL-5MW turbine data (e.g. power curve) the rotor powers at the wind speeds of 3.7 (m/s) and 6.8 (m/s) are about 0.15 MW and 1.2 MW, respectively [30], leading to a difference of 87% itself in the turbine's power curve. This difference is in the close agreement with the difference (81%) obtained by our simulations. In terms of power production, this turbine looks very sensitive with the wind speed variation, as the power is varying from 0.043 MW to 4.8 MW (approximately 110 times) within the wind-speed variation from 3 (m/s) to 11 (m/s) [30]. This also indicate that this turbine is generally not efficient at lower wind speeds (less than 6 m/s) as the power is too small compared to its rated power.

Of course the question is why the wind speed is too low in the wake simulation of the stable case? There can be a number of reasons for this. Firstly, less turbulence mixing due to the negative buoyancy flux in the stable case which does not allow the wake to recover faster, and hence the wake is stronger (i.e. lower wind speed) compared to that in the neutral case. Recently, Abkar and Porté-Agel [11] studied the effect of free atmosphere on a large wind-farm power output. They pointed out that 10 times stronger stratification in the free atmosphere can decrease the power up to 30%. The second reason is the use of periodic boundary conditions (infinitely long wind farm). With respect to a single turbine, the wake effects in an infinitely long wind farm are always stronger as the wake gets stronger towards infinite number of turbines and the effects are more profound in the stable conditions for the above mentioned reason.

4. Conclusions

In this paper, we carried out LES calculations for an infinitely long wind-farm wakes under the influence of neutral and stable stratifications. The stable condition simulated here is characterized by a low-level jet, which appears at a lower ABL depth (115-240 m) in which modern wind turbines are operating. The turbine is modeled using actuator line model, which parametrizes the turbine-induced forces (e.g., lift and drag) into the governing equations. For the both atmospheric conditions, the simulations are also carried in the absence of the wind turbine in order to generate their reference cases. The LES results from the both atmospheric conditions are compared against each other in order to study the impact of the stably-stratified condition with respect to that from the neutral condition.

The results from the reference cases (in the absence of wind turbine) demonstrated that for the similar wind speed at the hub height the stably-stratified wind condition provides lower (e.g. by 35%) turbulence intensity but higher wind-shear (by 112% at the hub height) than that observed under the neutrally-stratified wind condition. The results from the wake simulations show that the stable condition has significant impact on the turbine wake and power statistics. The wake under the stable condition is far stronger than that seen under the neutral condition due to the less turbulent mixing associated with the stable condition. The velocity deficit in the stable case is higher by 47% compared to that in the neutral case. In the wake, the vertical variation of the wind statistics within the rotor is also high in the stable case, while the profiles in the neutral case are vertically less profound (weaker vertical variation). With respect to the neutral case, the vertical expansion of the wake is smaller by 33% in the stable case, as the negative buoyancy flux acts against it. Due to the wake, the boundary-layer depth in the stable case is increased by 35% compared to its reference case.

Concerning the power statistics, we noticed that the power production under the stable condition is 81% smaller than the power obtained in the neutral case. With respect to the total-available power, the power loss (deficit) is higher in the stable case. The power loss is found to be 57% and 92% in the stable and neutral cases respectively. However, for the obtained wind speeds in the wake simulations, the power statistic (production and losses) obtained in this study is in agreement with the turbine's power curve. In the future, we will extend this work by simulating a single-turbine wake as well as finite-size wind farm using the same atmospheric conditions and the results will be compared with the existing results. This comparison will bring further insight into the impact of stable conditions involving low-level jets on the wind turbine performance.

Acknowledgements
The computational resources for the numerical simulations were provided by CSC–Finnish IT Center for Science, Espoo (Finland). The authors would like to thank Lappeenranta University of Technology (LUT) for funding this research. The first author, A. Chaudhari, acknowledges with thanks the support of the TU1304 action "WINERCOST".

References

[1] Tuononen M, Sinclair V A and Vihma T 2015 *Atmos. Sci. Lett.* **16** 492–499 ISSN 1530-261X
[2] Karipot A, Leclerc M Y and Zhang G 2009 *Mon. Weather Rev* **137** 2605–2621
[3] Gutierrez W, Araya G, Basu S, Ruiz-Columbie A and Castillo L 2014 *J. Phys.: Conf. Ser.* **524** 012008
[4] Porté-Agel F, Lu H and Wu Y T 2014 *Procedia IUTAM* **10** 307–318
[5] Churchfield M J, Lee S and Moriarty P J 2014 *ITM Web of Conferences* vol 2 (EDP Sciences) p 02001
[6] Sathe A, Mann J, Barlas T, Bierbooms W and van Bussel G 2013 *Wind Energ.* **16** 1013–1032 ISSN 1099-1824
[7] Gutierrez W, Araya G, Kiliyanpilakkil P *et al.* 2016 *J Renew & Sustain Energ* **8** 023308
[8] Porté-Agel F, Lu H and Wu Y T 2010 *Proceedings of the 5th international symposium on computational wind engineering (CWE2010)* May 23-27, Chapel Hill, North Carolina
[9] van der Laan M P and Srensen N N 2017 *Wind Energ. Sci. Discuss.* **in review**
[10] Abkar M, Sharifi A and Porté-Agel F 2016 *J. Turbul.* **17** 420–441
[11] Abkar M and Porté-Agel F 2013 *Energies* **6** 2338–2361
[12] Park J, Basu S and Manuel L 2014 *Wind Energy* **17** 359–384
[13] Abkar M and Porté-Agel F 2014 *J. Phys.: Conf. Ser.* **524** 012138
[14] Bhaganagar K and Debnath M 2014 *Energies* **7** 5740–5763
[15] Chivaee H S and Sørensen J N 2014 *J. Phys.: Conf. Ser.* **555** 012087
[16] Dörenkämper M, Witha B *et al.* 2015 *J. Wind Eng. Ind. Aerodyn.* **144** 146–153
[17] Lu H and Porté-Agel F 2011 *Physics of Fluids* **23** 065101
[18] Abkar M and Port-Agel F 2015 *Physics of Fluids* **27** 035104
[19] Machefaux E, Larsen G C, Koblitz T, Troldborg N *et al.* 2015 *Wind Energy* DOI: 10.1002/we.1950
[20] Beare R J *et al.* 2006 *Boundary-Layer Meteorol.* **118** 247–272
[21] Meneveau C, Lund T S and Cabot W H 1996 *J. Fluid Mech.* **319** 353–385
[22] Chaudhari A, Conan B, Aubrun S and Hellsten A 2016 *J. Phys.: Conf. Ser.* **753** 032037
[23] Churchfield M J, Lee S, Michalakes J and Moriarty P 2012 *J. Turbul.* **13** 1–32
[24] Churchfield M J, Li Y and Moriarty P 2013 *Phil Trans R Soc A* **371** 20120421
[25] Chaudhari A, Conan B and Hellsten A 2017 *Proceedings of the 1st international conference on Wind energy Harvesting (WINERCOST-2017)* April 21-22, Coimbra, Portugal
[26] Agafonova O, Avramenko A, Chaudhari A and Hellsten A 2016 *AIP Conf. Procee.* vol 1738 p 480082
[27] Porté-Agel F, Wu Y T, Lu H and Conzemius R J 2011 *J. Wind Eng. Ind. Aerodyn.* **99** 154–168
[28] Tabib M, Rasheed A and Kvamsdal T 2015 *J. Phys.: Conf. Ser.* vol 625 p 012032
[29] Sørensen J S W 2002 *J. Fluids Eng.* **124** 393–399
[30] Jonkman J, Butterfield S, Musial W and Scott G 2009 Definition of a 5-MW reference wind turbine for offshore system development, Tech. Rep. NREL/TP-500-38060 NREL Golden, CO 80401
[31] Wu Y T and Porté-Agel F 2011 *Boundary-Layer Meteorol.* **138** 345–366
[32] Basu S and Porté-Agel F 2006 *J. Atmos. Sci.* **63** 2074–2091
[33] Commission I E *et al.* 2007 Wind turbines-part 1: Design requirements Tech. rep. IEC-61400-1
[34] Odemark Y and Segalini A 2014 *J. Phys.: Conf. Ser.* **555** 012079
[35] Agafonova O, Avramenko A, Chaudhari A and Hellsten A 2016 *J. Phys.: Conf. Ser.* **753** 032001

Experimental study on the wind-turbine wake meandering inside a scale model wind farm placed in an atmospheric-boundary-layer wind tunnel

N. Coudou[1,2,3], S. Buckingham[1], J. van Beeck[1]

[1]Environmental and Applied Fluid Dynamics Department, von Karman Institute for Fluid Dynamics, Waterloosesteenweg 72, 1640 Sint-Genesius-Rode, Belgium
[2]Fluids-Machines Department, Université de Mons, Rue du Joncquois 53, 7000 Mons, Belgium
[3]Institute of Mechanics, Materials and Civil Engineering, Université catholique de Louvain, Place du Levant 2, 1348 Louvain-la-Neuve, Belgium

E-mail: `nicolas.coudou@vki.ac.be`

Abstract. Increasing use of wind energy over the years results in more and larger clustered wind farms. It is therefore fundamental to have an in-depth knowledge of wind-turbine wakes, and especially a better understanding of the well-known but less understood wake-meandering phenomenon which causes the wake to move as a whole in both horizontal and vertical directions as it is convected downstream. This oscillatory motion of the wake is crucial for loading on downstream turbines because it increases fatigue loads and in particular yaw loads. In order to address this phenomenon, experimental investigations were carried out in an atmospheric-boundary-layer wind tunnel using a 3×3 scaled wind farm composed of three-bladed rotating wind-turbine models subject to a neutral atmospheric boundary layer (ABL) corresponding to a *slightly rough* terrain, i.e. to offshore conditions. Particle Image Velocimetry (PIV) measurements were performed in a horizontal plane, at hub height, in the wake of the three wind turbines in the wind-farm centreline. From the PIV velocity fields obtained, the wake-centrelines were determined and a spectral analysis was performed to obtain the characteristics of the wake-meandering phenomenon. In addition, Hot-Wire Anemometry (HWA) measurements were performed in the wakes of the same wind turbines to validate the PIV results. The spectral analysis performed with the spatial and temporal signals obtained from PIV and HWA measurements respectively, led to Strouhal numbers $St = fD/U_{hub} \simeq 0.20 - 0.22$.

1. Introduction

Between the years 2000 and 2015, the cumulative wind-power installations in the European Union increased from 12.9 GW to 141.6 GW [1]. Based on the European Wind Energy Association's central scenario [2], wind-energy installations could amount to over 320 GW in 2030. This important rise in the use of wind energy over the years results in more and larger clustered wind farms. It is therefore fundamental to have an in-depth knowledge of wind-turbine wakes, and especially a better understanding of the well-known but less understood wake-meandering phenomenon. Because this phenomenon causes the wake to be swept in and out of the rotor disk of downstream turbines, it is crucial for loading on these turbines since it might considerably increase fatigue loads and in particular yaw loads.

In their study performed with a two-bladed rotating wind-turbine scale model submitted to a uniform free-stream, Medici and Alfredsson [3] argued that the low-frequency vortex shedding might be responsible for the wake-meandering motion. Their measurements were supported by a theoretical study performed by Okulov and Sørensen [4]. However, as mentioned by Larsen *et al.* [5], notable differences exist between the experimental conditions of Medici and Alfredsson's study [3], and the conditions present for a full-scale wind turbine located in the atmospheric boundary layer (ABL).

Larsen *et al.* [5] described the wake meandering process as a series of consecutive wake "releases", each advected downstream with the mean wind field and displaced, in the plane perpendicular to the mean wind direction, by the large-scale turbulent structures. In relation to the large-scale turbulent structures, the individual wake "releases" are thus presumed to behave like passive tracers [5]. España *et al.* provided experimental evidence of the role of the large eddies on the wake-meandering phenomenon behind a wind turbine modelled with a static porous disk [6–8]. The results showed that the meandering process is only generated when the turbulent length scales contained in the ABL are larger than the disk diameter, and that the meandering phenomenon cannot be attributed to the periodic vortex shedding. Further studies performed with a static porous disk located in a neutral ABL [9, 10] showed that the instantaneous horizontal wake position correlates well with the upstream transverse velocity for wavelengths larger than three times the disk diameter, whereas it is not at all the case for wavelengths smaller than twice the disk diameter. Furthermore, using two porous disks located $5D$ apart from each other highlighted an amplification of the wake meandering for the downstream turbine because of the upstream wake.

The wake of a three-bladed rotating wind-turbine scale model in onshore (*moderately rough* terrain) and offshore (*slightly rough* terrain) neutral ABL conditions was investigated by Barlas *et al.* [11]. For offshore conditions, i.e. low turbulence level ($I = 7.5\%$ at hub height), a wake meandering mechanism was detected. It was found that the Strouhal number related to this meandering was independent of the inflow velocity and was always $\mathcal{O}(0.25)$. Additional investigations were performed in offshore and onshore conditions downstream of a 3×3 wind farm but as for the single wind turbine in onshore conditions ($I = 17.5\%$ at hub height), the incoming turbulence level was too high for the meandering to survive [12].

Additional studies performed on rotating wind-turbine scale models [13–15] as well as on a hydrokinetic turbine [16] provided evidence of wake oscillations corresponding to a Strouhal number St ranging from 0.1 to 0.3. In some of these studies [14, 16], it is reported that wake meandering is caused by instabilities of the hub or tip vortices.

While most of the experimental studies about wind-turbine wake meandering concern an isolated wind-turbine scale model or two models aligned with the mean flow in a few cases, this study focusses on the three wind turbines constituting the mid row (in stream-wise direction) of a 3 x 3 scaled wind farm. In particular, the aim of the current study is to investigate if the wake meandering persists in a wind farm and to determine the characteristics of this phenomenon. In order to address that, state-of-the-art experimental techniques combined with spectral analysis are used.

2. Experimental setup

2.1. Wind turbine scale model

The three-bladed wind-turbine scale models used for the experiments have a rotor diameter of 0.15 m and a hub height of 0.13 m (Fig. 1); they can therefore be considered as representative of a scaled 2 MW offshore wind turbine (e.g. scale 1/440 of Vestas V66-2MW). The rotors, made of a HAM-STD HS1-606 airfoil, are designed by Blade Element Momentum (BEM) theory [17] with an incoming velocity of 8 m s^{-1} at hub height. Each rotor is attached to a direct current (DC) motor used as a generator which can be "counter-loaded" to extract power from the incoming wind and to control the turbine tip-speed ratio

$$TSR = \frac{\Omega \cdot R}{U_{hub}}, \tag{1}$$

where Ω is the angular velocity and U_{hub} is the mean velocity at hub height. Based on the incoming velocity at hub height ($U_{hub} = 8.3$ m s^{-1}), the tip-speed ratio of the first turbine was equal to 7.2.

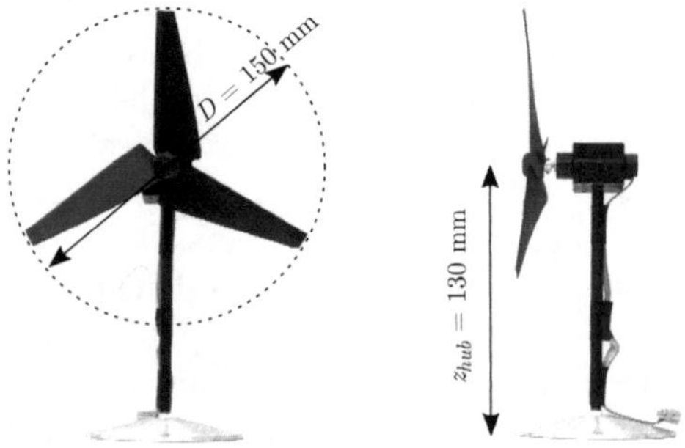

Figure 1. Wind-turbine scale model dimensions.

Figure 2. Wind farm position in configuration 1 inside the VKI Wind Engineering facility L-1B.

2.2. Wind farm setup

The 3×3 wind farm was located at the end of the test section of the VKI Wind Engineering facility L-1B (Fig. 2). The wind-turbine scale models forming the wind farm were spaced by $3D$ and $5D$ in the span-wise and stream-wise direction, respectively. Due to the limitations in the positioning of the CCD camera underneath the wind tunnel, the Particle Image Velocimetry (PIV) system was kept at its best position and the wind farm was shifted longitudinally in order to perform measurements successively in the wake of the first, the second, and the third wind turbines as depicted in Fig. 3. These changes in wind-farm positioning have no influence on the reliability of the results since the entire testing area is submitted to a fully developed ABL. Furthermore, since the PIV system remains in the same position, the comparison of the wind-turbine wakes is not affected by changes in the PIV setup.

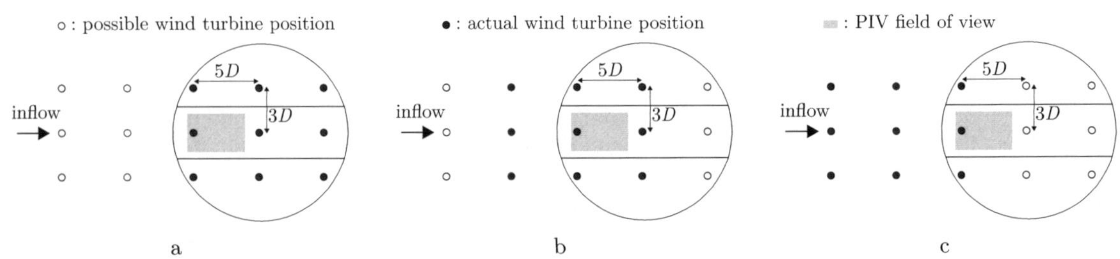

Figure 3. Wind farm position in the VKI Wind Engineering facility L-1B : sketch of configurations 1 (a), 2 (b), and 3 (c).

2.3. Incoming ABL flow

A neutral ABL was modelled in the VKI Wind Engineering facility L-1B which is a low-speed, closed-loop wind tunnel with a 2-m high, 3-m wide, and 20-m long test section. The length of the flow development section was equal to 11 m for the shortest scenario corresponding to the wind farm in configuration 3, this fetch being long enough to obtain a fully developed ABL. More information concerning the modelling of atmospheric flows in the wind tunnel L-1B can be found in [18, 19].

The ABL modelled was characterised using a one-component hot-wire anemometer and is representative of a *slightly rough* terrain, with a friction velocity $u_* = 0.33$ m s^{-1}, a full-scale aerodynamic roughness length $z_0 = 3.7 \times 10^{-3}$ m (full scale $= 440 \times$ wind-tunnel scale) and a power law coefficient $\alpha = 0.11$ using the hub height as reference height. The main properties of the ABL modelled are compared to the VDI [20], Eurocode [21] and ESDU [22] standards for a *slightly rough* terrain (category of *type 0*) in Fig. 4a,c. The integral length scale at hub height, computed by applying the autocorrelation method on the time series and the Taylor's frozen turbulence hypothesis, is about three rotor diameters. This characteristic size of the energy-containing eddies therefore fulfils the condition presented by España *et al.* [6, 8] to observe the wake-meandering phenomenon, i.e. to have modelled turbulent scales larger than the wind-turbine rotor diameter in the inflow ABL. With regard to the spectral characteristics of the ABL, the normalised turbulent spectrum obtained at hub height (Fig. 4d) matches well with the Kansas spectrum for stream-wise component [23]. Like in [19], there is a slight overshoot in the amplitude of the peak. This overshoot is most probably caused by a low-frequency instability ($f \simeq 1.5Hz$) produced by the two contra-rotating fans of the wind tunnel. Note that this frequency is about ten times smaller than the meandering frequency.

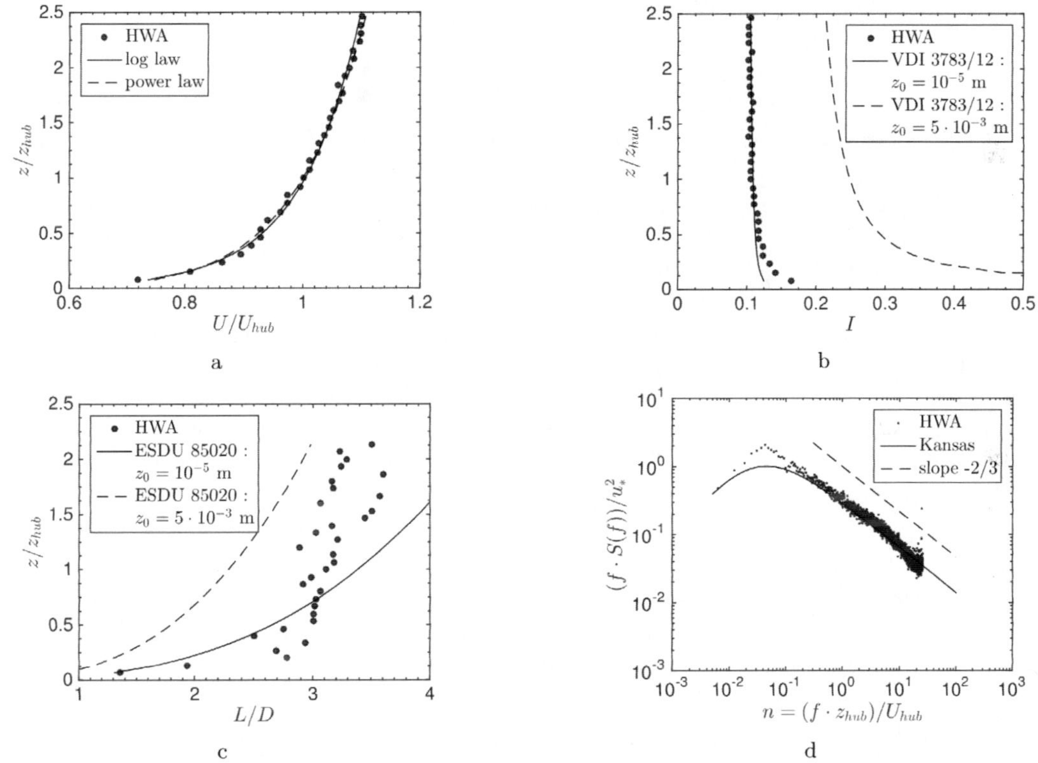

Figure 4. Vertical profiles and normalised spectrum of the atmospheric boundary layer modelled in the wind tunnel L-1B. Normalised mean velocity (a), turbulence intensity (b), integral length scales (c), normalised spectrum at hub height and Kansas spectrum [23] (d). Note that these quantities are based on the stream-wise velocity obtained by 1C-HWA measurements.

2.4. *Measurement techniques*

2.4.1. Particle Image velocimetry (PIV) The 2D2C-PIV technique was used to perform measurements in a horizontal plane at hub height in the wake of the wind-turbine scale models. The PIV system was set up to obtain a field of view larger than $4D$ in the stream-wise direction (Fig. 5). The seeding particles were oil droplets with a diameter of 1-5 μm. Measurements were performed using LaVision DAVIS 8 software. For each measurement, 800 pair images were acquired at a frequency of 2.5 Hz. The processing of the images was carried out using the in-house software WIDIM (WIndow Displacement and Distortion Iterative Multigrid) [24]. The parameters used are 64 × 64 pixels interrogation areas and an overlap of 75%. For all fields of views, the signal-to-noise ratio was of the order of 6, what enables to have full confidence in the quality of the measurements and processing. On the basis of the uncertainty analysis procedure proposed by the International Towing Tank Conference [25] and considering a confidence interval of 95%, the uncertainty on the mean velocity was determined to equal 1.5% for hub-height velocities of 4.9 m s^{-1} and 8.3 m s^{-1}.

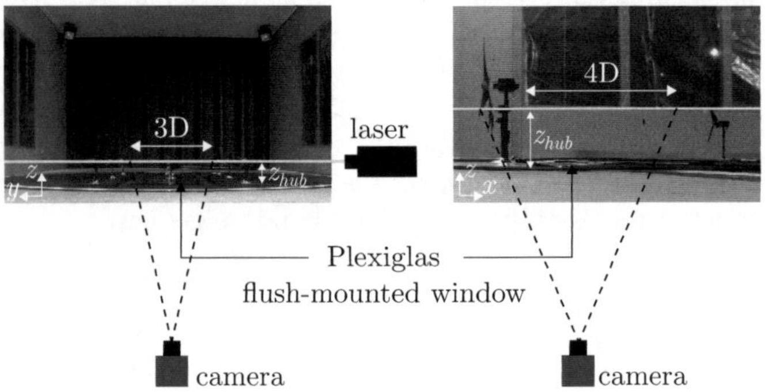

Figure 5. PIV setup in the VKI Wind Engineering facility L-1B.

2.4.2. Hot-Wire Anemometry (HWA) The hot-wire anemometer was calibrated using a low-turbulent and uniform jet created by a nozzle and controlled by a pressure difference. The calibration was carried out before and after each experimental campaign to verify the validity of the calibration throughout the experiments. The calibration curve was then fitted with a third-order polynomial in order to convert the instantaneous voltage signature into instantaneous wind velocity. The measurements were performed at $x = 2D$ downstream of each rotor disk at hub and upper tip heights. For each location, the HWA signal was acquired during 120 s with a sampling frequency of 3 kHz and low-pass filtered at 1 kHz. Based on classical uncertainty analysis [26] and considering a confidence interval of 95%, the uncertainty on the velocity was determined to be equal to 4.1% for a hub-height velocity of 7.8 m s^{-1}.

3. Estimation of the wake centreline

From the PIV measurements of the scaled wind-turbine wakes, the instantaneous deviations of the wake-centreline from its time-average location was determined using a specific processing technique based on the determination of a Gaussian function

$$f(y) = \frac{A}{2\pi\sigma} e^{-\frac{(y_i - \mu_y)^2}{2\sigma^2}} \qquad (2)$$

which optimally fits the instantaneous wake velocity deficits downstream of the wind turbine [27–30]. The independent Cartesian spatial variable of the function is referred to as y_i and the shape of the function is parameterised in terms of three parameters, namely: a position parameter of the profile μ_y, a width parameter of the profile σ, and a scaling parameter A [30]. For each snapshot, the Gaussian function was fitted, through a least-squares approach, to the instantaneous wake velocity deficit profiles (Fig. 6) for stream-wise locations x_i going from $x = 0.5D$ to $x = 4D$. The wake-centre position at each location x_i was then determined from the location of the maximum of the Gaussian function fitting each wake velocity deficit profile.

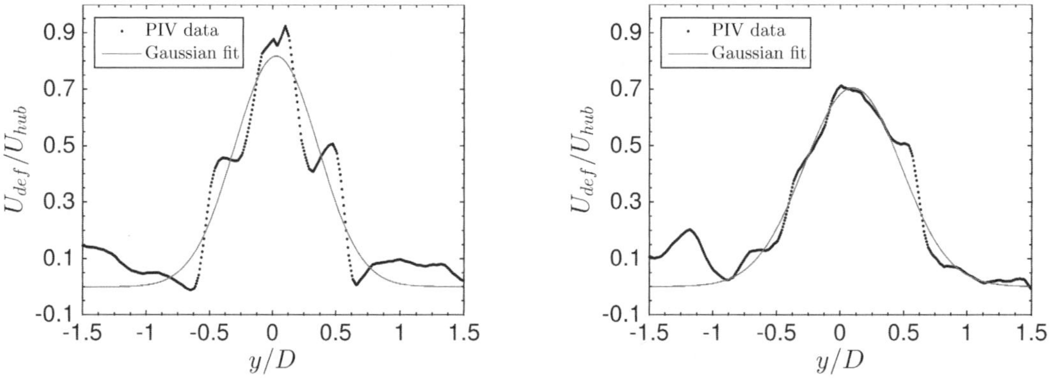

Figure 6. Gaussian fit of two normalised instantaneous wake velocity deficit profiles measured by PIV at hub height : $x = 0.5D$ (left) and $x = 3D$ (right).

4. Results and discussion

The normalised instantaneous horizontal wind speed fields obtained from PIV measurements at hub height (Fig. 7) reveal the presence of wake meandering for the three wind turbines constituting the mid row (in stream-wise direction) of the wind farm. Figure 8 presents, for each instantaneous velocity field processed, the instantaneous normalised wake-centre positions at hub height as a function of the stream-wise location. The wake-centre displacements in the y-direction appear to increase when moving downstream and reach, for some cases, the field of view limit of the CCD camera. This maximum value of $1.5D$ is not obtained because of post-processing inaccuracy but is due to "extreme deflections" of the wake (Fig. 7, bottom).

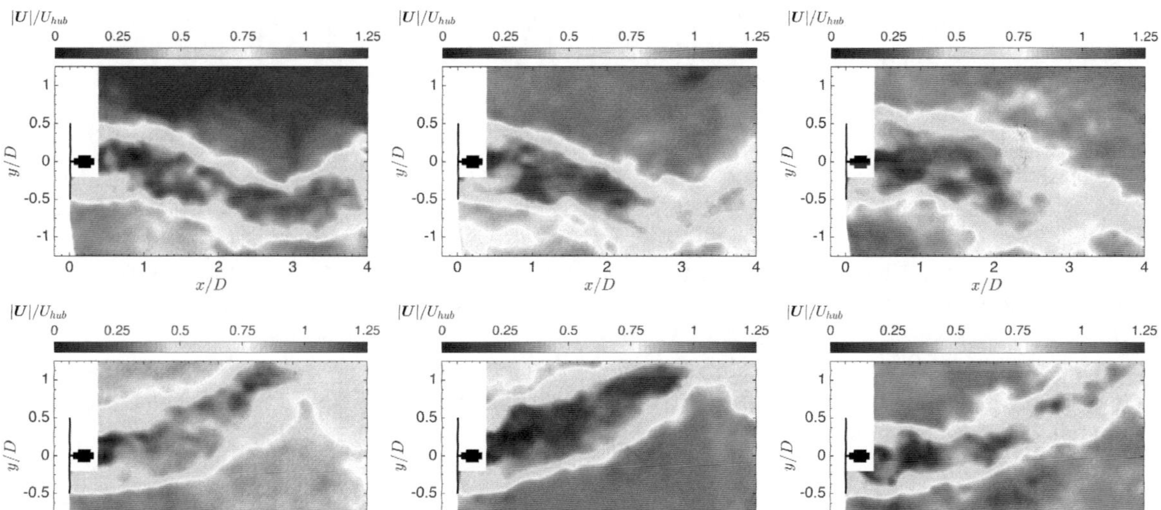

Figure 7. Normalised instantaneous horizontal wind speed ($|\boldsymbol{U}| = \sqrt{u^2 + v^2}$) fields obtained by PIV measurements at hub height at two different times revealing wake deflection : first wind turbine (left), second wind turbine (centre), and third wind turbine (right).

For each instantaneous velocity field, the wake-centreline position with reference to its time-averaged position was determined and a Fast Fourier Transform (FFT) was applied, after zero-padding, in order to obtain the characteristics of the wake-meandering phenomenon. Since the FFT was applied on a spatial signal, the resulting spectrum was obtained as a function of the wavenumber k. The spectrum was then expressed as a function of the wavelength $\lambda = \frac{1}{k}$ and as a function of the Strouhal number by using the Taylor's frozen turbulence hypothesis

$$\left.\begin{array}{l} St = \frac{f \cdot D}{U_{hub}} \\ f = \frac{U_{hub}}{\lambda} \end{array}\right\} St = \frac{D}{\lambda}. \quad (3)$$

In order to investigate the influence of incoming wind speed on the meandering phenomenon, the same methodology was applied for measurements performed for a velocity of 4.9 m s^{-1} at hub height. The average spectra of the wake-centrelines as a function of the Strouhal number and of the frequency normalised by the wind-turbine rotational frequency f_T are showed in Fig. 9. The wake meandering characteristics, obtained from the spectra, are summarised in Table 1. Note that the meandering wavelength ($\lambda = D/St$) is of the order of $4.5D - 4.75D$, thus justifying the large field of view required for the PIV measurements.

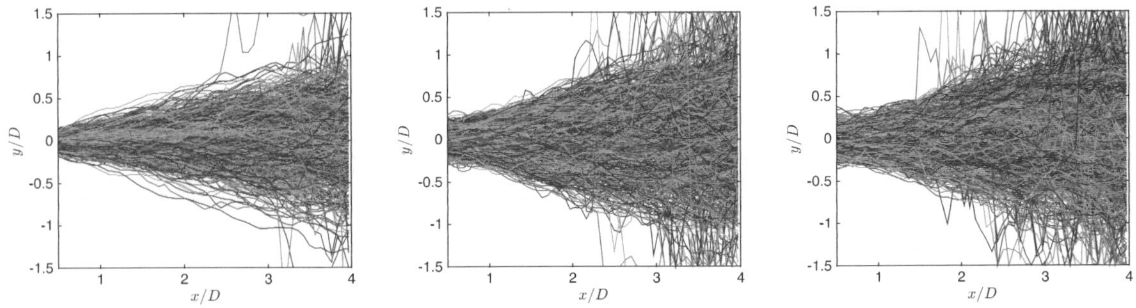

Figure 8. Normalised wake-centre positions obtained from PIV measurements at hub height as a function of the stream-wise position for each instantaneous velocity field processed : first wind turbine (left), second wind turbine (centre), and third wind turbine (right).

In order to validate the concept of the wake-centreline algorithm used in this study, the meandering characteristics obtained via PIV (spatial approach) at hub height were compared to the ones obtained from HWA (temporal approach) measurements at $x = 2D$ and at the same height (Fig.10, bottom). PIV and HWA results agree well together, particularly for the second and third turbines (Table 1). However, in the first wind-turbine wake the comparison is slightly worse because the Taylor's frozen turbulence hypothesis is probably less valid at the origin of the wake meandering. However, the excellent agreement observed between the PIV and HWA results provides an excellent proof of concept of the wake-centreline algorithm. In addition, it is interesting to note that, in contrast to what was observed by Barlas *et al.* [11, 12], the meandering phenomenon is not characterised by a well-pronounced narrow peak in the spectra but rather by a bumb spread over a larger low-frequencies range, especially at tip height (Fig.10).

		WT1		WT2		WT3	
		St	f/f_T	St	f/f_T	St	f/f_T
PIV	$U_{hub} = 4.9$ m s^{-1}	0.22	0.11	0.22	0.14	0.22	0.13
	$U_{hub} = 8.3$ m s^{-1}	0.21	0.09	0.22	0.12	0.21	0.11
HWA	$U_{hub} = 7.8$ m s^{-1}	0.14	0.06	0.22	0.11	0.20	0.10

Table 1. Meandering characteristics, Strouhal and frequency normalised by the rotor rotational frequency f_T, obtained via PIV and HWA measurements at hub height.

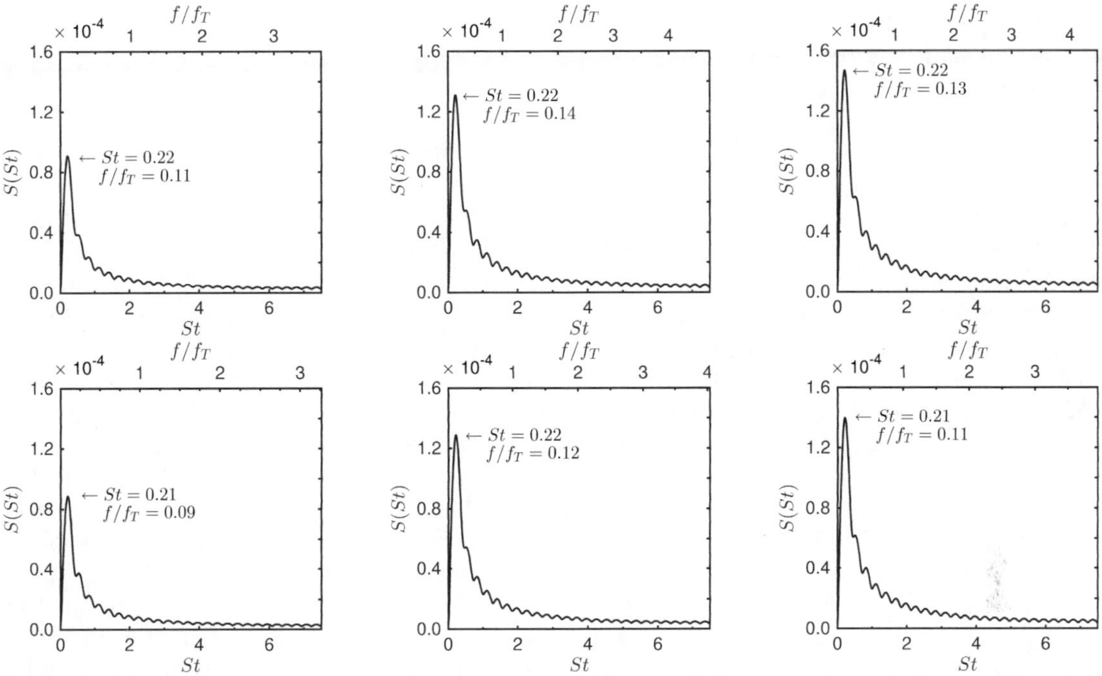

Figure 9. Average zero-padded spectra of the wake-centreline position with reference to its time-averaged position obtained from PIV measurements at hub height for $U_{hub} = 4.9$ m s^{-1} (top) and $U_{hub} = 8.3$ m s^{-1} (bottom) : first wind turbine (left), second wind turbine (centre), and third wind turbine (right). For $U_{hub} = 4.9$ m s^{-1} : $f_{T_1} = 66.7$ Hz, $f_{T_2} = 51.5$ Hz, $f_{T_3} = 54.6$ Hz and for $U_{hub} = 8.3$ m s^{-1} : $f_{T_1} = 127.2$ Hz, $f_{T_2} = 102.9$ Hz, $f_{T_3} = 105.1$ Hz.

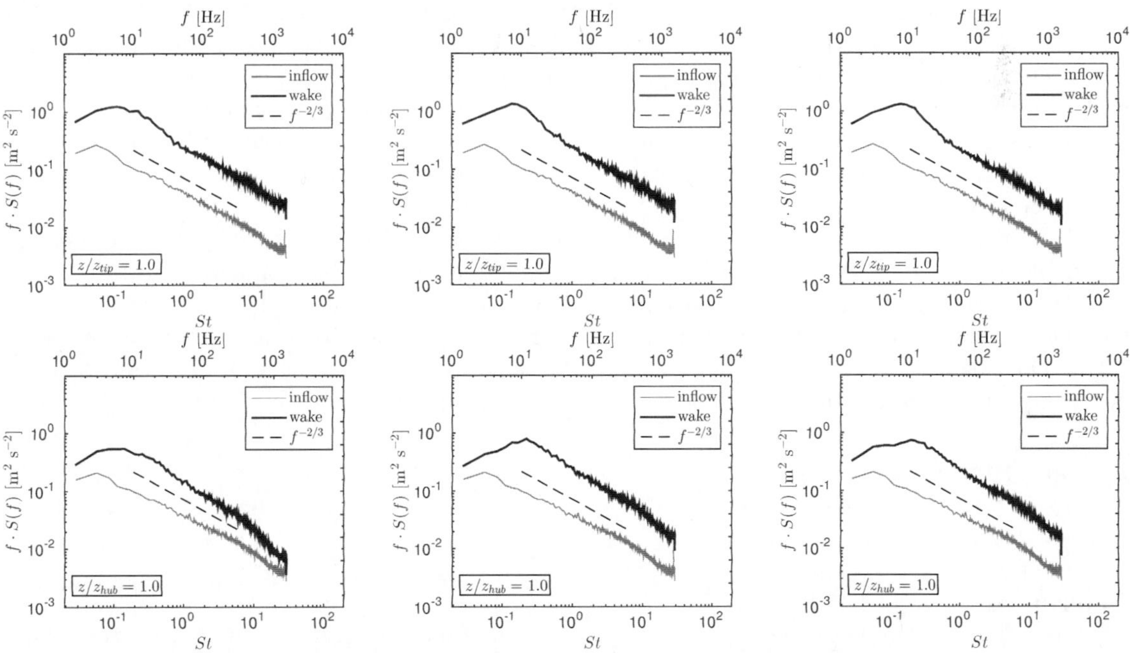

Figure 10. Comparison between the energy spectra of the inflow and at $x = 2D$ obtained from HWA measurements in the wake of the three wind turbines at $z/z_{tip} = 1.0$ (top) and $z/z_{hub} = 1.0$ (bottom) : first wind turbine (left), second wind turbine (centre), and third wind turbine (right).

5. Discussion and Conclusion

The wake-meandering phenomenon was investigated experimentally inside a 3×3 scaled wind farm placed in a neutral ABL wind tunnel inside which an ABL corresponding to a *slightly rough* terrain was modelled. PIV measurements were performed in a horizontal plane, at hub height, in the wake of the three turbines constituting the mid row (in stream-wise direction) of the scaled wind farm. Even if the PIV system employed was not able to resolve wake meandering in time, the large field of view set for the PIV camera made possible to perform a spatial analysis of the meandering structure from the velocity fields acquired. The fact that the Strouhal numbers obtained from HWA ($St = fD/U_{hub}$) and PIV ($St = D/\lambda$) are both $St \simeq 0.20 - 0.22$ proves the validity of the wake-centreline algorithm employed and the validity of the Taylor's frozen turbulence hypothesis for the wake-meandering phenomenon.

The results fall into the Strouhal numbers range $0.1 - 0.3$ obtained in various studies of wind-turbines [11–15] or hydrokinetic-turbines wakes [16]. Like in [11–14], the Strouhal number corresponding to wake meandering seems to be independent of the incoming wind speed. Furthermore, in contrast to what has been observed by Barlas *et al.* [12], the wake-meandering phenomenon persists in the wind farm and is characterised by a bumb spread over a large low-frequencies range in the HWA spectra. The differences observed are most probably due to the higher turbulence intensity of the ABL modelled in the present study : $I = 10.5\%$ vs $I = 7.5\%$ in [11, 12] at hub height. The major influence of the turbulence intensity is reinforced by the fact that, in [11, 12], no meandering was observed in the wake of a single wind turbine located in an ABL corresponding to a *rough* terrain and presenting therefore very high turbulence intensity ($I = 17.5\%$ at hub height).

Compared to literature, the wind-turbine wake meandering formation seems neither to be entirely due to intrinsic instabilities of the wake that could be associated with a periodic vortex shedding within the wake nor to be entirely due to the effects of the large scale turbulent eddies contained in the atmospheric boundary layer. The wake meandering formation seems in fact to be due to a combination of both of these propositions. Indeed, the Strouhal number obtained is in favour of the role of the intrinsic instabilities of the wake, and the fact that the phenomenon does not appear when integral length scales are smaller than the rotor diameter [6] underlines the influence of the large-scale eddies contained in the atmospheric boundary layer. The wake-meandering phenomenon seems therefore to be caused by the amplification of the intrinsic instabilities of the wake by large-scale turbulent eddies contained in the atmospheric boundary layer.

The perspectives of this work are many. The influence of the terrain roughness as well as the influence of the stability of the ABL on the wake-meandering phenomenon will be further studied. In particular, offshore and onshore ABL with various turbulence intensity levels will be modelled in the VKI Wind Engineering facility L-1B to investigate in detail the influence of the inflow on the wake-meandering phenomenon. Furthermore, the influence of the wind-turbine operating conditions on the wake meandering characteristics will be investigated.

Acknowledgments
Nicolas Coudou is funded by the *Fonds pour la Formation à la Recherche dans l'Industrie et dans l'Agriculture (FRIA)*, Belgium.

References

[1] Corbetta G, Mbistrova A, Ho A, Pineda I and Ruby K 2016 Wind in power - 2015 European statistics http://www.ewea.org/fileadmin/files/library/ publications/statistics/EWEA-Annual-Statistics-2015.pdf The European Wind Energy Association
[2] Corbetta G, Ho A, Pineda I, Ruby K and Van de Velde L 2015 Wind energy scenarios for 2030

http://www.ewea.org/fileadmin/files/library/publications/ reports/EWEA-Wind-energy-scenarios-2030.pdf The European Wind Energy Association

[3] Medici D and Alfredsson P H 2006 *Wind Energy* **9** 219–236

[4] Okulov V L and Sørensen J N 2007 *J Fluid Mech* **576** 1–25

[5] Larsen G C, Madsen H A, Thomsen K and Larsen T J 2008 *Wind Energy* **11** 377–395

[6] España G, Aubrun S, Loyer S and Devinant P 2012 *J Wind Eng Ind Aerodyn* 24–33

[7] Aubrun S, Loyer S, España G, Hayden P and Hancock P 2011 Experimental study on the wind turbine wake meandering with the help of a non-rotating simplified model and of a rotating model AIAA 2011-0460, 49th AIAA Aerospace Sciences Meeting (Orlando, FL)

[8] España G, Aubrun S, Loyer S and Devinant P 2011 *Wind Energy* **14** 923–937

[9] Muller Y A, Aubrun S and Masson C 2015 *Exp Fluids* **56** 1–11

[10] Aubrun S, Muller Y A and Masson C 2015 *Journal of Physics: Conference Series* **625**

[11] Barlas E, Buckingham S and van Beeck J 2016 *Boundary-Layer Meteorol* **158** 27–42

[12] Barlas E, Buckingham S, Glabeke G and van Beeck J 2015 Offshore and onshore wind turbine wake meandering studied in an ABL wind tunnel, International Workshop on Physical Modeling of Flow and Dispersion Phenomena (PHYSMOD) (Zürich)

[13] Medici D and Alfredsson P H 2008 *Wind Energy* **11** 211–217

[14] Okulov V, Naumov I, Mikkelsen R, Kabardin I and Sørensen J 2014 *J Fluid Mech* **747** 369–380

[15] Howard K B, Singh A, Sotiropoulos F and Guala M 2015 *Phys Fluids* **27**

[16] Chamorro L P, Hill C, Morton S, Ellis C, Arndt R E A and Sotiropoulos F 2013 *J Fluid Mech* **716** 658–670

[17] Burton T, Sharpe D, Jenkins N and Bossanyi E 2001 *Wind Energy Handbook* (John Wiley & Sons)

[18] Buckingham S 2010 Wind park siting in complex terrains assessed by wind tunnel simulations VKI PR 2010-04 von Karman Institute for Fluid Dynamics

[19] Conan B 2012 *Wind resource accessment in complex terrain by wind tunnel modelling* VKI PHDT 2013-04 Université d'Orléans - von Karman Institute for Fluid Dynamics

[20] VDI Guidelines 3783/12 2000 Physical modelling of flow and dispersion processes in the atmospheric boundary layer - Application for wind tunnels (Beuth Verlag)

[21] Eurocode 2005 EN 1991-1-4:2005 : Actions on structures - Part 1-4: General actions - Wind actions (CEN)

[22] ESDU 1985 Characteristics of atmospheric turbulence near the ground. Part II: single point data for strong winds (neutral atmosphere) Engineering Sciences Data Unit (London)

[23] Kaimal J and Finnigan J 1994 *Atmospheric Boundary Layer Flows: Their Structure and Measurement* (Oxford University Press)

[24] Scarano F 2000 *Particle Image Velocimetry - Development and Application* VKI PHDT 2000-06 von Karman Institute for Fluid Dynamics

[25] International Towing Tank Conference 2008 Recommended Procedures and Guidelines : Uncertainty Analysis - Particle Image Velocimetry 7.5-01-03-03

[26] Anthoine J, Arts T *et al* 1994 Measurements techniques in Fluid Dynamics: An Introduction Reprint of VKI LS 1994-01 Third revised edition (Von Karman Institute for Fluid Dynamics)

[27] Bingöl F, Mann J and Larsen G C 2010 *Wind Energy* **13** 51–61

[28] Trujillo J J, Bingöl F, Larsen G C, Mann J and Kühn M 2011 *Wind Energy* **14** 6175

[29] Aubrun S, Tchouaké T F, España G, Larsen G, Mann J and Bingöl F 2012 *Progress in Turbulence and Wind Energy IV: Proceedings of the iTi Conference in Turbulence 2010* (Springer Berlin Heidelberg)

[30] Vollmer L, Steinfeld G, Heinemann D and Kühn M 2016 *Wind Energy Science* **1** 129–141

IOP Conf. Series: Journal of Physics: Conf. Series **854** (2017) 012009 doi:10.1088/1742-6596/854/1/012009

Momentum balance in a fully developed boundary layer over a staggered array of NREL 5MW rotors

PL Delafin, T Nishino

Centre for Offshore Renewable Energy Engineering, Cranfield University, Cranfield, MK43 0AL, UK

E-mail: `pierreluc.delafin@gmail.com`

Abstract. 3D Reynolds-averaged Navier-Stokes (RANS) simulations of a fully developed wind farm boundary layer over a staggered array of NREL 5MW turbines are presented. The turbine is modeled as an actuator disk and as a fully resolved rotor to compare the effect of the turbine model on the wind farm aerodynamics, in particular the streamwise momentum balance across the farm. Results show that the difference in the turbine model affects the average wind speed through the farm as well as the local flow pattern around each turbine; both contributing to the difference in the prediction of farm performance. Results are also compared with a simple theoretical model of very large wind farms proposed recently. The actuator disk simulations agree very well with the theoretical model, whereas the fully resolved rotor simulations show some consistent and expected differences from the model.

1. Introduction

As wind energy is gaining more and more interest for its ability to produce clean, affordable and reliable electricity [1, 2], modeling and simulations of large wind farms are becoming an important subject [3, 4, 5], especially for offshore applications. Actuator disk models are often used in such large wind farm modeling and simulations [3, 5, 6]. There are many earlier studies comparing actuator disk simulations and blade-resolved rotor simulations, but these studies are either for an isolated turbine [7, 8] or a very small number (usually two) of turbines [9, 10], not for large wind farms. Therefore in this study we perform actuator disk (AD) and fully resolved rotor (FRR) simulations of an NREL 5MW horizontal-axis wind turbine in an infinitely large wind farm. The main objective is to analyze how the difference between the AD and FRR simulations will affect the momentum balance across the farm and thus the prediction of farm performance. The results of these simulations will be analyzed in association with the theoretical model of very large wind farms proposed recently by Nishino [5].

2. Theoretical model

The theoretical model proposed by Nishino [5] is a simple quasi-one-dimensional model to predict a practical upper limit to the efficiency of a very large wind farm. The basic concept underlying the Nishino model is similar to the so-called 'top-down' models [11, 12]. However, the Nishino model is based on a simplified 'two-scale coupled' momentum conservation argument and does not consider the details of wind profile across the farm boundary layer explicitly. On one hand, the 'farm-scale' wind speed reduction factor, $\beta = U_F/U_{F0}$ (where U_F is the wind speed averaged over a nominal farm layer height, H_F, and U_{F0} is its undisturbed equivalent), is calculated from the momentum balance between the driving force of the atmospheric boundary layer and the resistance due to the turbines' thrust and bottom friction. On the other hand, the 'turbine-scale' wind speed reduction factor, $\alpha = U_T/U_F$ (with U_T the face-averaged wind speed over the turbine rotor swept area), is calculated from the classical actuator disk theory. These two wind speed reduction factors α and β satisfy the (simplified) momentum conservation equation [5]:

$$1 - \beta^{\gamma} = \lambda \cdot \frac{1}{C_{f0}} \cdot \beta^2 \cdot 4\alpha(1-\alpha) \tag{1}$$

where $\lambda = A/S$ is the ratio of the turbine swept area A to the land area S (i.e. farm density) and $C_{f0} = \tau_{w0}/0.5\rho U_{F0}^2$ is the natural bottom friction coefficient.

The only empirical parameter employed in the Nishino model is γ in Eq. 1. This parameter describes how the bottom friction (shear stress) changes in response to the reduction of the average wind speed through the farm layer. γ is defined such that

$$\frac{<\tau_w>}{\tau_{w0}} = \beta^{\gamma} = \left(\frac{U_F}{U_{F0}}\right)^{\gamma} \tag{2}$$

where $<\tau_w>$ is the bottom shear stress averaged across the farm. Further details of the Nishino model can be found in [5] and [6].

3. RANS simulations

3.1. Geometry and mesh

An 'infinitely large' array of NREL 5MW rotors is considered in this paper. The 126m diameter (D) turbine is based on the initial design by NREL [13] and takes into account the modifications suggested by Sandia National Laboratories [14] (i.e. a smoother blade thickness distribution and a smoother root/blade transition compared to the initial design). This turbine concept has been widely used [15, 16, 17]. The tip shape, not included in the reference designs [13, 14], is defined as a simple 'rounded' tip and the hub diameter has been increased to 5.4m in this study (to simplify the generation of the structured mesh). The center of the hub is located 90m above the ground which gives a ground clearance of $0.21D$. The rotational axis is tilted by $5°$ with respect to the horizontal plane and a precone of $2.5°$ is used for the blades, as defined in [13].

All farm simulations in this study consider a fully developed boundary layer flow driven by a constant streamwise pressure gradient over an 'infinitely large' wind farm, similarly to [3] and [5]. This 'infinitely large' wind farm is modeled as a doubly periodic array of turbines. Only one turbine is placed in the middle of a relatively small ($6D \times 6D$), but periodic, computational domain (Fig. 1(a)). The height of the computational domain is $8D$ ($\sim$ 1km). The turbine is modeled as an Actuator Disk (AD) or as a Fully Resolved Rotor (FRR). Turbines are arrayed in a staggered configuration which means that the streamwise spacing between two turbines is $12D$ (Fig. 1(a)).

Multi-block structured meshes are created for the AD and FRR simulations. The computational domains have the same dimensions in both cases: $(L_x, L_y, L_z) = (6D, 6D, 8D)$. The AD mesh contains a 126m diameter porous disk whose center is located at $x = y = 0$m (center of the domain) and $z = 90$m. An 'O-grid' mesh topology is used inside and around the disk, as shown in Figure 2(a). 104 nodes are used on the disk edge. The first cell height on the bottom boundary (ground) is 1m. Around the disk, the streamwise node spacing starts from 0.5m and expands with a ratio of 1.15 to reach a maximum spacing of 11m near the inlet and outlet boundaries. The maximum lateral (y) spacing is 13m. In total, the AD mesh contains 8×10^5 cells.

IOP Conf. Series: Journal of Physics: Conf. Series **854** (2017) 012009 doi:10.1088/1742-6596/854/1/012009

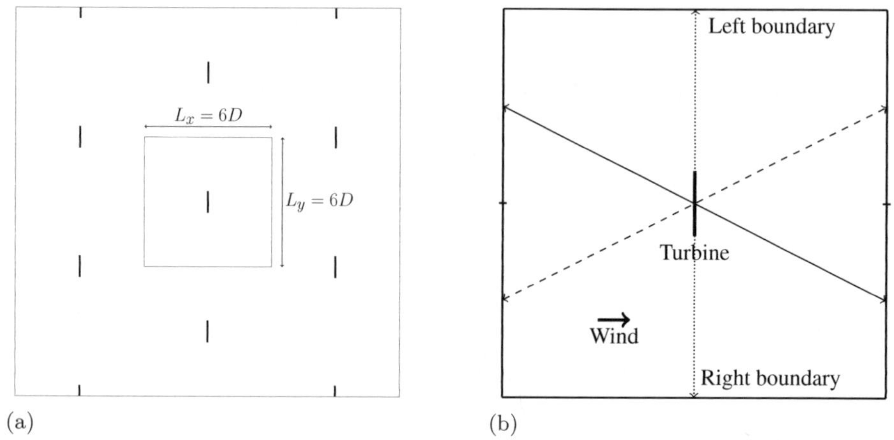

Figure 1: Horizontal section of the staggered array of rotors (a) and sketch showing the connections (arrows) between the periodic boundaries in a horizontal section of the computational domain (b).

The FRR mesh is divided into two sub-domains: a rotating part containing the three blades and a stationary outer part. The rotor part is meshed with an O-grid topology around the blades. The wall normal spacings ensure $y^+_{max} \sim 1$, except on the hub where wall functions are used to reduce the total number of cells. 186 nodes are used in the chordwise direction and 107 nodes in the spanwise direction for each blade (Fig. 2(b)). This gives 5.7×10^6 cells in the rotor sub-domain. The outer part contains the nacelle and the front part of the hub (i.e. the streamwise extent of the rotor sub-domain is small, Fig. 2(b)). The cell sizes immediately upstream and downstream of the rotor sub-domain are $\Delta x = 0.5$m and 0.1m, respectively, to match the cell sizes with those inside the rotor sub-domain. The maximum streamwise spacing is 6m and the maximum lateral spacing is 7m. The first cell height on the bottom boundary is 1m, as with the AD mesh. The outer sub-domain contains 6.9×10^6 cells. Therefore, the FRR mesh contains 12.6×10^6 cells in total.

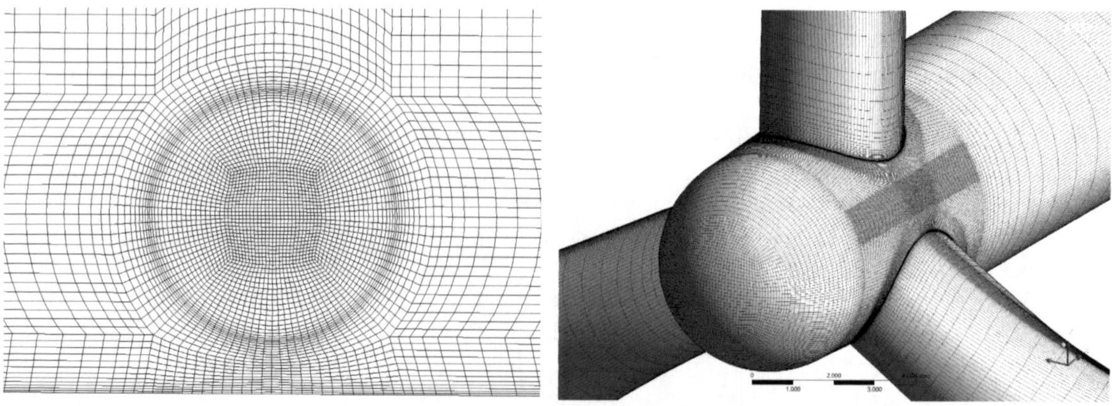

(a) (b)

Figure 2: Computational meshes: (a) cross sectional view around the disk (for AD simulations) and (b) surface mesh of the rotor hub (for FRR simulations).

3.2. Boundary conditions and simulation parameters

In all simulations presented in this paper, the bottom boundary is set as 'wall' with a nominal roughness height $K_s = 1$m or 5m. K_s is a sand-grain type roughness related to the aerodynamic roughness parameter z_0 as $K_s = (E/C)z_0$ [18], where $E = 9.793$ is an empirical constant and $C_s = 0.5$ is the roughness constant. The top boundary is set as 'symmetry'. To model the staggered configuration of the turbines, both inlet and outlet boundaries are vertically split into two equal parts (left and right). Periodic interfaces are then used to connect the right outlet boundary to the left inlet boundary and vice versa (Fig. 1(b)). The side boundaries are set as periodic interfaces. The doubly periodic computational domain obtained enables the simulation of one turbine operating inside an infinitely large wind farm with a staggered configuration. A constant pressure gradient is employed in the streamwise direction for all simulations using the same ground roughness. In the AD simulations, a 'porous jump' condition is used for the disk surface, where a disk resistance K is used to calculate the streamwise momentum sink following the equation $M_x = K(\frac{1}{2}\rho U_d^2)$, where U_d is the local streamwise velocity through the disk [19]. In the FRR simulations, blades are considered as 'wall'.

The Reynolds-averaged Navier-Stokes (RANS) equations are solved using the ANSYS Fluent software [20] with the $k - \omega$ SST turbulence model [21]. In the AD simulations (steady RANS), the SIMPLE algorithm is used for pressure-velocity coupling and second order schemes are used for pressure, momentum and turbulence spatial discretization. In the FRR simulations (unsteady RANS), the pressure-based coupled algorithm is used as it was found to be more stable. Only first order schemes are used for the first rotor revolution; then, 'Standard' scheme is used for the pressure equation and second order schemes are used for momentum and turbulence spatial discretization as well as for the transient formulation. The time step size always corresponds to a variation of the rotor azimuthal angle $\Delta\theta = 0.5°$ and 15 iterations are used per time step. For all simulations, the density and viscosity of air are assumed to be $\rho = 1.225\text{kg/m}^3$ and $\mu = 1.7894 \times 10^{-5}\text{kg/(ms)}$, respectively.

The FRR simulations require to specify an appropriate rotational speed of the rotor. This is not a straightforward task considering that the flow in the fully developed farm boundary layer is not known a priori. Therefore, we use the results of the AD simulations to estimate the wind speed at hub height and $3D$ upstream the turbine. This value ($U_\infty = 6$m/s with $K_s = 1$m and $U_\infty = 6.9$m/s with $K_s = 5$m) is used to calculate the rotational speed of the rotor that corresponds to the tip-speed-ratio (TSR) used in [13] (7.98rpm or TSR = 8.8 with $K_s = 1$m and 8.41rpm or TSR = 8.0 with $K_s = 5$m).

3.3. Single turbine simulation (for validation)

The NREL 5MW turbine has no experimental data available (concept design only). However, [13] gives performance predictions for an isolated turbine with a uniform upstream wind speed, using the FAST code. These data are used for validation of our resolved rotor model before using it for wind farm simulations. One additional mesh for the FRR simulation has been created based on the one described in section 3.1, with extended horizontal dimensions: $(L_x, L_y) = (40D, 40D)$. Unlike the farm simulations with periodic boundary conditions, the bottom boundary is set as 'symmetry' to simulate a uniform (not sheared) flow around the isolated turbine. A uniform wind speed $U_\infty = 11.4$m/s is defined on the upstream boundary with a turbulence intensity of 5% and a turbulent viscosity ratio of 10. The turbulence intensity decays while traveling downstream (as there is no shear) and reaches low values around the turbine ($\sim 0.03\%$). The rotational speed is 12.1rpm which corresponds to TSR = 7.0 [13]. The simulation has been run for 8 revolutions so that the difference of power coefficient between the last two revolutions is sufficiently small (0.77%).

Averaged results of power and thrust coefficients are presented in Table 1. The FRR simulation predicts a C_P value close to the reference value. The predicted C_P is supposed to be slightly higher than what would be obtained for a perfectly isolated turbine because of the local (partial) blockage effect due to the proximity of the bottom boundary [22]. It should also be noted that the result of the FRR simulation is not supposed to be exactly the same as the reference value since the reference value and the FRR simulation are based on two different numerical methods (BEM and CFD, respectively). However, the fact that the FRR simulation result is close to the reference value suggests that the rotor simulation (in particular the generation of rotor geometry) has been conducted appropriately.

3.4. Empty box simulations

Empty box simulations are run with the AD mesh described in section 3.1. The disk boundary condition is set as 'interior' instead of 'porous jump' and all other boundary conditions are the same as in section 3.2. Simulations are run with both $K_s = 1$m and $K_s = 5$m with a constant mass flow rate. The mass flow rate is calculated based on a power-law velocity profile (with an exponent of 0.1) with a reference wind speed $U_{ref} = 15$m/s at $z = 90$m (hub height). This gives a mass flow rate equal to 16.528×10^6kg/m^3. The empty box simulations are initialized with a uniform velocity field and they are run for about 10^6 iterations to obtain a fully converged flow field. Figure 3 shows the velocity profiles obtained with the two roughness heights. The wind speed is lower with $K_s = 5$m than with $K_s = 1$m in the region of the turbine. The empty box simulation is used to obtain the background pressure gradient (for a given K_s) that will be used in the AD and FRR farm simulations (Table 2). It is also used to obtain some parameters used in the theoretical farm model [5] such as the natural farm-layer wind speed (U_{F0}), the farm layer height (H_F) and the natural bottom shear stress (τ_{w0}). The farm layer height was found to be $H_F = 1.80D$ with $K_s = 1$m and $H_F = 1.76D$ with $K_s = 5$m.

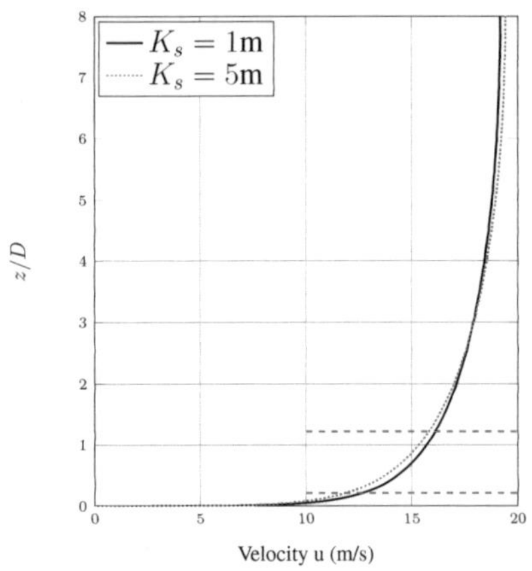

Figure 3: Streamwise velocity profiles of the non-disturbed atmospheric boundary layer with $K_s = 1$m and $K_s = 5$m. The rotor position is represented by the dashed gray lines.

Table 1: Power (P), power coefficient (C_P) and thrust coefficient (C_T) for an isolated turbine with $U_\infty = 11.4$m/s.

	P (MW)	C_P	C_T
FRR	5.60	0.49	0.75
FAST, [13]	5.29	0.47	0.81

Table 2: Summary of pressure gradients obtained from the 'empty box' simulations.

Roughness height, K_s (m)	Pressure gradient (Pa/m)
1	-7.49×10^{-4}
5	-10.59×10^{-4}

3.5. Wind farm simulations

Once the 'empty box' simulations detailed in section 3.4 have been conducted, it is possible to run the farm simulations using the background pressure gradients given in Table 2. The actuator disk simulations have been run for 10^6 iterations to obtain fully converged results (maximum residual $= 10^{-11}$). The fully resolved rotor simulations use the flow calculated in their corresponding AD simulation (with $K = 2$) as the initial flow condition (except for the rotor sub-domain). FRR simulations have been run for 92 revolutions (K_s = 1m) and 50 revolutions (K_s = 5m). After 92 revolutions with K_s = 1m, the turbine wake has traveled 26D downstream the turbine (assuming the average wind speed calculated between two rotors, along the horizontal line at hub height and going through the nacelle). This means that the wake created at the beginning of the simulation had time to reach the rotor twice (because of the periodic boundary conditions).

Figure 4 shows the variations of (spatially averaged) bottom shear stress with the number of revolutions for both ground roughness heights. The bottom shear seems to follow the same pattern in both cases and is fluctuating close to the value obtained in the AD simulations. The case K_s = 5m has been run for only 50 revolutions but the pattern observed for the K_s = 1m case suggests that the bottom shear would not vary significantly with additional revolutions.

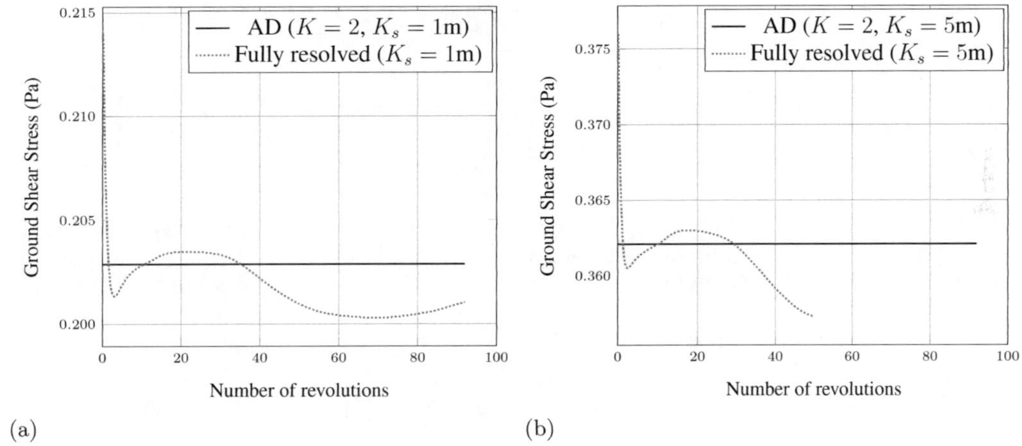

Figure 4: Average ground shear stress variations with the number of revolutions: (a) K_s = 1m and (b) K_s = 5m.

Figure 5 shows a comparison of the streamwise velocity profiles obtained from the AD (K = 2) and FRR simulations with K_s = 1m. The profiles are plotted 1.5D, 6D and 10.5D downstream the turbine. At x/D = 1.5, the difference between the AD and FRR simulations is significant in

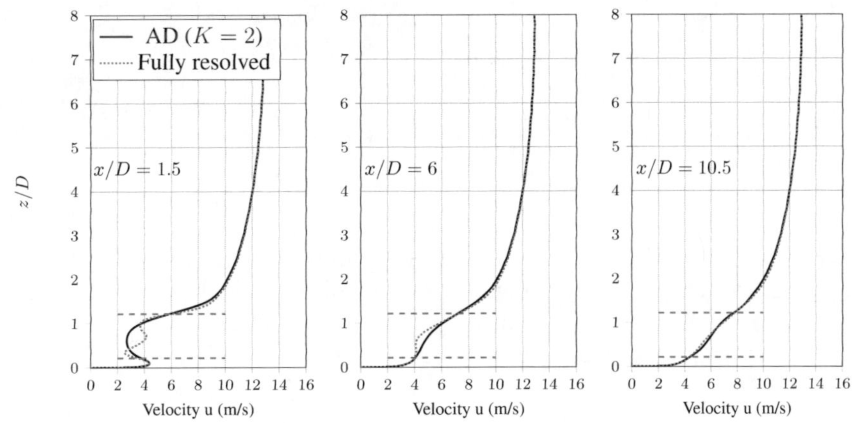

Figure 5: Velocity profiles at different locations in the wake of the rotor, K_s = 1m. The rotor position is represented by the dashed gray lines.

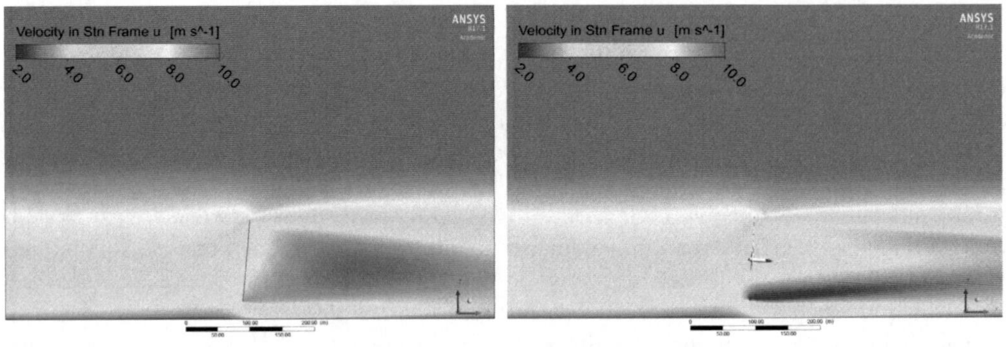

(a)

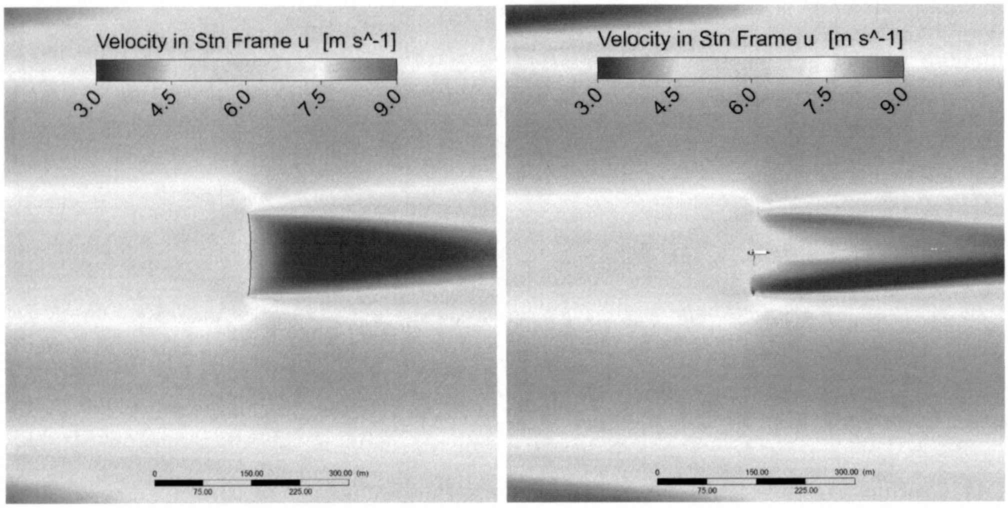

(b)

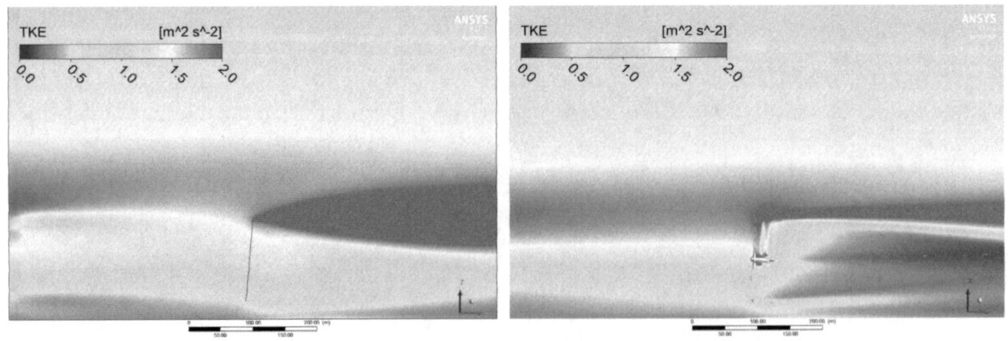

(c)

Figure 6: Flow field comparison between the AD simulation ($K = 2$, left) and the FRR simulation (right) in terms of (a): streamwise velocity in a vertical plane containing the rotor axis, (b): streamwise velocity in a horizontal plane located at hub height and (c): turbulent kinetic energy in a vertical plane containing the rotor axis. $K_s = 1$m.

the rotor area. This has already been observed in previous studies such as [8]. Away from the rotor area, both AD and FRR simulations give very similar results. When x/D increases, the difference between the AD and FRR simulations decreases. Although not shown in this paper, similar comments can be made for the results obtained with $K_s = 5$m.

Figure 6 shows a comparison of the flow field between the AD ($K = 2$, left) and FRR (right) simulations. Figures 6(a) and 6(b) show the streamwise velocity field in a vertical plane containing the rotational axis and in a horizontal plane at hub height, respectively. In the AD simulation, the wake downstream the turbine is relatively homogeneous, whereas in the FRR simulation, the velocity deficit is more significant in the lower part (Fig. 6(a)) and in the left part of the wake (Fig. 6(b)), when viewed from downwind). However, upstream the turbine and away from the rotor region, both simulations give similar velocity fields. Figure 6(c) shows the turbulent kinetic energy (TKE) field in the vertical plane containing the rotational axis. The TKE values are lower in the wake of the turbine in the FRR simulation compared to the AD simulation. Furthermore, in the FRR simulation, a thin layer of high TKE values can be observed at the top of the wake region, whereas this layer is much wider in the AD simulation. In the upstream region, the vertical distribution of TKE is also affected by the turbine model.

The results of all six wind farm simulations performed in this study are summarized in Table 3, including two additional AD simulations with $K = 1.6$ and 1.2 (for $K_s = 1$m). It should be noted that the results of FRR simulation for $K_s = 5$m are presented in brackets as this simulation may not have converged sufficiently. The coefficients of thrust ($C_T = T/(0.5\rho U_{F0}^2 A)$ and power ($C_P = P/(0.5\rho U_{F0}^3 A)$ are calculated using the undisturbed farm layer wind speed U_{F0}. The 'local' or 'effective' coefficients of thrust (C_T^*) and power (C_P^*) are also calculated using the farm layer wind speed U_F instead of U_{F0}. Similarly to [6], an increase in K_s (or decrease in λ/C_{f0}) results in an increase in C_P and C_T, regardless of the turbine model used. Of particular interest is the comparison between AD and FRR simulations for $K_s = 1$m. It can be seen that the AD simulation with $K = 1.2$ gives a higher C_T but also a lower C_T^* compared with the FRR simulation. This means that it is not possible to match both C_T and C_T^* at the same time between the AD and FRR simulations by adjusting the disk resistance K in the AD simulation. This is essentially because the present AD simulation tends to predict stronger wake mixing and hence a higher β value (i.e. faster average wind speed through the farm layer) than the FRR simulation for a given thrust.

Although the difference in the turbine model affects the average wind speed through the farm as well as the local flow pattern around the turbine, the γ value (defined in Eq. 2) is found to be around 1.9 to 2 in this study regardless of the turbine model used, while this value is around 1.5 to 1.6 in AD simulations in [6]. [6] uses a different turbine diameter ($D = 100$m) and turbine hub height (100m) which may explain the difference in γ observed. It can also be noted that $\gamma = 2.03$ is obtained with the FRR simulation at $K_s = 5$m while this parameter is expected to be smaller than 2 as the turbines usually increase the turbulence intensity within the farm layer [5]. However, as shown earlier in Fig. 6(c), the FRR simulations performed in this study predicted a large low TKE region behind the turbine, which may explain why the value of γ slightly exceeded 2 in this particular case. Note, however, that this may also be due to the RANS simulations not being able to predict the complex near-wake mixing accurately. To confirm this, we would need higher-fidelity simulations, such as detached-eddy simulations, resolving turbulent eddies in the turbine wake.

Figure 7 shows a comparison between the farm simulations and the Nishino model [5] for the power coefficient. As can be seen from the figure and similarly to [6], the agreement between the AD simulations and the theoretical model is very good. The FRR simulations show some consistent discrepancies from the theoretical model. However, the FRR results follow the same pattern as the theory, i.e. C_P increases as λ/C_{f0} decreases. The discrepancies between the theoretical farm model and FRR farm simulations are understandable since the theoretical model employs the classical actuator disk theory as its local (turbine-scale) flow model [5], i.e. the theoretical model considers 'ideal' turbines to predict an upper limit to the farm performance.

Table 3: Summary of computational results for K_s = 1m (λ/C_{f0} = 3.93, U_{F0} = 14.9m/s, $\tau_{\omega 0}$ = 0.7547Pa) and K_s = 5m (λ/C_{f0} = 2.60, U_{F0} = 14.4m/s, $\tau_{\omega 0}$ = 1.0671Pa).

Simulation	K_s	α	β	C_T	C_T^*	C_P	C_P^*	$\tau_\omega/\tau_{\omega 0}$	γ
AD ($K = 2$)	1	0.612	0.498	0.188	0.756	0.058	0.471	0.269	1.88
AD ($K = 1.6$)	1	0.647	0.520	0.183	0.675	0.063	0.444	0.288	1.91
AD ($K = 1.2$)	1	0.689	0.553	0.176	0.575	0.068	0.403	0.317	1.94
FRR	1	-	0.509	0.159	0.614	0.042	0.340	0.266	1.96
AD ($K = 2$)	5	0.620	0.574	0.256	0.778	0.093	0.492	0.339	1.95
FRR	5	-	(0.583)	(0.209)	(0.635)	(0.068)	(0.358)	(0.335)	(2.03)

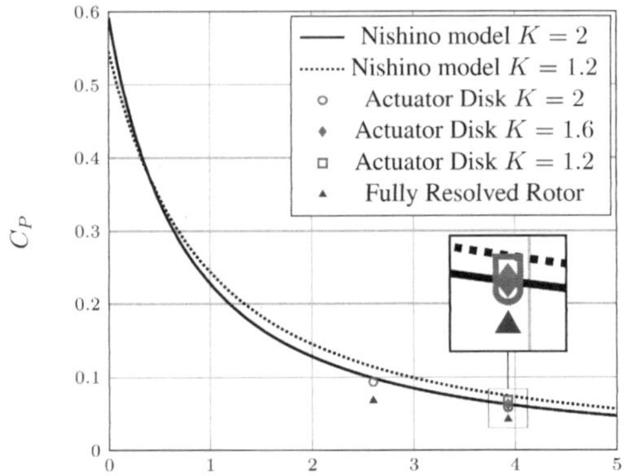

Figure 7: Comparison between the AD and FRR simulations and the Nishino model ($\gamma = 2$).

4. Conclusions

3D RANS simulations of an 'infinitely large' wind farm, consisting of a periodic staggered array of NREL 5MW turbines, have been conducted with two different turbine models: actuator disk and fully resolved rotor. These farm simulations employ an assumption that the atmospheric boundary layer is driven by a constant streamwise pressure gradient, the value of which was obtained (for a given bottom roughness) from the so-called 'empty box' simulation with a fixed mass flow rate. The results show that the difference in the turbine model affects the average wind speed through the farm as well as the local flow pattern around the turbine. In particular, the present actuator disk farm simulations tend to predict stronger wake mixing and hence a faster average wind speed through the farm than the resolved rotor farm simulations. The results have also been compared with the simple theoretical wind farm model proposed recently by Nishino [5]. The actuator disk farm simulation results show a very good agreement with the theoretical model, whereas the resolved rotor farm simulations show some consistent and expected discrepancies from the model. This suggests that the theoretical model could be further extended to predict the performance of real large wind farms in future studies.

5. Acknowledgments

This study was conducted as part of the MULTI-WIND (integrated multi-scale modeling for optimal wind energy extraction) project, supported by EPSRC through the Institutional

Sponsorship Fund allocated to Cranfield University (EP/P511134/1).

References

[1] RenewableUK 2015 Wind energy in the UK, State of the industry report 2015 *Tech Rep*

[2] EWEA 2015 Aiming high, Rewarding ambition in wind energy *Tech Rep*

[3] Calaf M, Meneveau C and Meyers J 2010 Large eddy simulation study of fully developped wind-turbine array boundary layers *Phys. Fluids* **22** 015110

[4] Meneveau C 2012 The top-down model of wind farm boundary layers and its applications *J. Turbul.* **13** N7

[5] Nishino T 2016 Two-scale momentum theory for very large wind farms *J. Phys.: Conf. Ser.* **753** (TORQUE 2016, Munich, Germany) 032054

[6] Zapata A, Nishino T and Delafin PL 2017 Theoretically optimal turbine resistance in very large wind farms *J. Phys.: Conf. Ser.* (Wake Conference 2017, Visby, Sweden) (under review)

[7] Troldborg N, Zahle F, Sorensen NN and Réthoré PE 2012 Comparison of wind turbine wake properties in non-uniform inflow predicted by different rotor models *J. Phys.: Conf. Ser.* **555** (TORQUE 2012, Oldenburg, Germany) 012100

[8] Kalvig S, Manger E and Hjertager B 2012 Comparing different CFD wind turbine modelling approaches with wind tunnel measurements *J. Phys.: Conf. Ser.* **555** (TORQUE 2012, Oldenburg, Germany) 012056

[9] Wilson JM, Davis CJ, Venayagamoorthy SK and Heyliger PR 2015 Comparisons of horizontal-axis wind turbine wake interaction models *J. Sol. Energ. Eng.* **137**

[10] Krogstad PA and Saetran L 2015 Wind turbine wake interactions; Results from blind tests *J. Phys.: Conf. Ser.* **625** (Wake Conference 2015, Visby, Sweden)

[11] Frandsen S 1992 On the wind speed reduction in the center of large clusters of wind turbines *J. Wind Eng. Ind. Aerodyn.* **39** 251-265

[12] Frandsen S, Barthelmie R, Pryor S, Rathmann O, Larsen S, Hjstrup J and Thgersen M 2006 Analytical modelling of wind speed deficit in large offshore wind farms *Wind Energ.* **9** 39-53

[13] Jonkman J, Butterfield S, Musial W and Scott G 2009 Definition of a 5-MW reference wind turbine for offshore system development *Tech Rep* NREL/TP-500-38060

[14] Resor BR 2013 Definition of a 5MW/61.5m wind turbine blade reference model *Tech Rep* SAND2013-2569, Sandia National Laboratories

[15] Bazilevs Y, Hsu AC, Akkerman I, Wright S, Takizawa K, Henicke B, Spielman T and Tezduyar TE 2011 3D simulation of wind turbine rotors at full scale. Part I: Geometry modeling and aerodynamics *Int. J. Numer. Meth. Fluids* **65**

[16] Chow R and van Dam CP 2012 Verification of computational simulations of the NREL 5MW rotor with a focus on inboard flow separation *Wind Energ.* **15**

[17] Tran T, Kim D and Song J 2014 Computational fluid dynamic analysis of a floating offshore wind turbine experiencing platform pitching motion *Energies* **7**

[18] Blocken B, Stathopoulos T and Carmeliet J 2007 CFD simulation of the atmospheric boundary layer: wall function problems *Atmospheric Environment* **41(2)** 238-252

[19] Nishino T and Willden RHJ 2012 Effects of 3-D channel blockage and turbulent wake mixing on the limit of power extraction by tidal turbines *Int. J. Heat Fluid Flow* **37** 123-135

[20] ANSYS Inc. 2016 ANSYS Fluent User's Guide Release 17.1

[21] Menter FR 1994 Two-equation eddy-viscosity turbulence models for engineering applications *AIAA J.* **32**

[22] Nishino T and Draper S 2015 Local blockage effect for wind turbines *J. Phys.: Conf. Ser.* **625** (Wake Conference 2015, Visby, Sweden) 012010

IOP Conf. Series: Journal of Physics: Conf. Series **854** (2017) 012010 doi:10.1088/1742-6596/854/1/012010

Effect of Wind Turbine Wakes on the Performance of a Real Case WRF-LES Simulation

P Doubrawa[1], A Montornès[2], R J Barthelmie[1], S C Pryor[3], G Giroux[4], P Casso[2]

[1]Sibley School of Mechanical and Aerospace Engineering, Cornell University, Ithaca, NY, USA
[2]VORTEX, Barcelona, Spain
[3]Earth and Atmospheric Sciences, Cornell University, Ithaca, NY, USA
[4]Wind Energy Institute of Canada, North Cape, PEI, Canada

E-mail: pd343@cornell.edu

Abstract. The main objective of this work is to estimate how much of the discrepancy between measured and modeled flow parameters can be attributed to wake effects. The real case simulations were performed for a period of 15 days with the Weather Research and Forecasting (WRF) model and nested down to a Large-Eddy Simulation (LES) scale of $\sim$ 100 m. Beyond the coastal escarpment, the site is flat and homogeneous and the study focuses on a meteorological mast and a northern turbine subjected to the wake of a southern turbine. The observational data set collected during the Prince Edward Island Wind Energy Experiment (PEIWEE) includes a sonic anemometer at 60 m mounted onto the mast, and measurements from the two turbines. Wake versus free stream conditions are distinguished based on measured wind direction while assuming constant expansion for the wake of the southern turbine. During the period considered the mast and northern turbine were under the southern turbine wake $\sim$ 16% and $\sim$ 11% of the time, respectively. Under these conditions, the model overestimates the wind speed and underestimates the turbulence intensity at the mast but not at the northern turbine, where the effect of wakes on the model error is unclear and other model limitations are likely more important. The wind direction difference between the southern and northern turbines is slightly underestimated by the model regardless of whether free stream or wake conditions are observed, indicating that it may be due to factors unrelated to the wake development such as surface forcings. Finally, coupling an inexpensive wake model to the high-fidelity simulation as a post-processing tool drives the simulated wind speeds at the mast significantly closer to the observed values, but the opposite is true at the coastal turbine which is in the far wake. This indicates that the application of a post-processing wake correction should be performed with caution and may increase the wind speed errors when other important sources of uncertainty in the model and data are not considered.

1. Introduction

Atmospheric phenomena across a wide range of spatial and temporal scales affect wind plant aerodynamics. Due to computational limitations, the majority of wind turbine wake analyses so far have relied on semi-idealized simulations where some physical processes are disregarded and a simplified version of a complex system is then simulated. For example, the widely used Dynamic Wake Meandering model (DWM, [1]) solves a steady-state approximation of the Navier-Stokes equations and then superimposes to the solution a turbulence field using stochastic methods and assuming a form for the energy spectrum [2]. On the other hand, the National Renewable

Energy Laboratory Simulator for Wind Farm Applications (SOWFA, [3]) simulates unsteady velocity fields at a high resolution, but its computational requirements preclude the simultaneous inclusion of some relevant atmospheric phenomena (e.g., cloud physics) and limit the achievable simulation time to minutes instead of hours or days.

Ongoing advances in computational resources have only recently started to allow real case (i.e., non-idealized) numerical simulations of the atmosphere to be run at spatial resolutions on the order of a typical wind turbine (WT) rotor diameter. As a result, it has become imperative to consider the effect that individual WTs have on the flow and it has also become possible to use these high resolution simulations for wake analyses. Research in this field may be conducted using the Weather Research and Forecasting (WRF) model [5] which considers the full physics of the atmosphere and can be nested down to the meso and micro scales, and even be run in Large-Eddy Simulation (LES) mode by switching off parameterizations for grid resolutions $\leq \mathcal{O}(10^2)$ m. Parameterizations are simplified mathematical descriptions of processes based on a set of assumptions, and used to model quantities that are not resolved under a given grid resolution. Unlike SOWFA, WRF does not include an explicit WT model and the effect of WTs on the flow is instead added in the form of a drag parameterization. In other words, individual wind turbine wakes are not modeled.

The idea of a wind farm parameterization first appeared in the context of global models to investigate the large-scale effect of wind farms on the Earth's climate [6]. Later, their impact on the regional climate was also considered [7] and more recently, shorter-term simulations using WRF are also seeking to include a description of wind farm effects [8, 9]. The wind farm parameterization currently distributed with WRF was developed for horizontal grid sizes higher than five rotor diameters [10] and real case WRF-LES simulations at a higher spatial resolution cannot benefit from it. Research is needed to investigate how the effect of WTs on the flow can be best modeled when performing WRF-LES simulations.

As a first step in this direction, the current work seeks to determine how the absence of such a parameterization affects the model performance. This is done by quantifying the role of wakes in explaining the discrepancy between simulated and observed flow conditions. The analysis consists in comparing WRF simulated wind and turbulence with observations from a meteorological mast and a northern wind turbine which are sometimes influenced by the presence of a wind turbine to the south. Finally, we investigate whether the application of an analytical wake model as a post-processing tool applied to the simulation output improves the degree of agreement between measured and simulated values. The use of this post-processing tool starts to allow for a quantification of wake error sources versus other errors sources in the model such as the level of accuracy and detail in the lower boundary conditions.

2. Data and Methods

2.1. Measurements

The measurements were collected at the Wind Energy Institute of Canada during the Prince Edward Island Wind Energy Experiment (PEIWEE) [11]. A period of 15 days is considered (May 11-25, 2015). Wind measurements used herein were made at a frequency of 10 Hz with a 3-D Gill Windmaster Pro sonic anemometer mounted to a meteorological mast (compliant with International Electrotechnical Commission standards) at 60 m above ground. Additionally, 10-minute mean wind speed, nacelle position and power production data from two 2-MW DeWind D9.2 turbines are used. The WT hub height is 80 m and the rotor diameter (D) 93 m. The location of the turbines and the mast is shown in Fig. 1.

2.2. Simulation

The simulation was performed by running the Advanced Research core of the WRF model for a dynamical downscaling consisting of two well differentiated phases: meso scale and micro

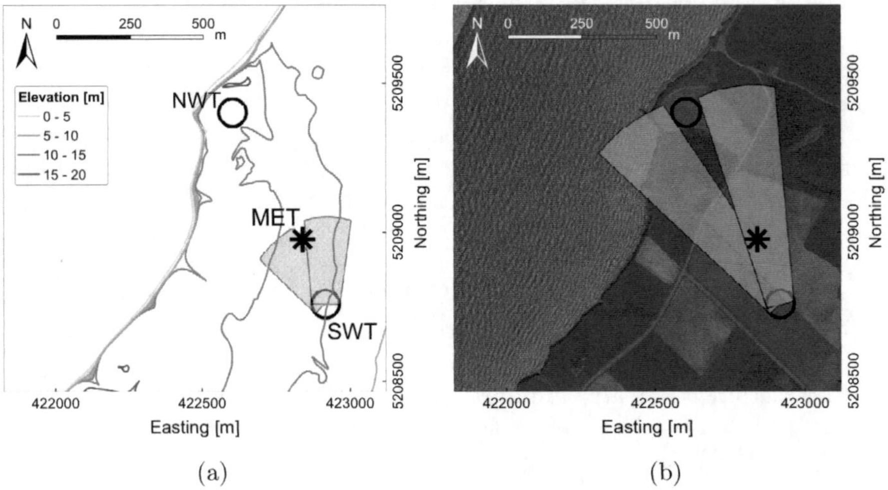

Figure 1: Instrument locations at the measurement site: wind turbines (circles of diameter 93 m) and meteorological mast (star). Elevation contours [m] (a) and satellite imagery from Environmental Systems Research Institute (b). Shaded areas represent wake from SWT impacting MET (a) and NWT (b).

scale. Five domains were defined, centered at the meteorological mast location (see Section 2.1) following a set of telescopic nests from 9 km to 111 m under a two-way nesting approach. The domains were configured to have 70 points in both directions and 50 vertical levels distributed every $\sim$10 m within the atmospheric boundary layer. The model source code was modified to enable meso-micro scale coupling based on the potential temperature perturbation method described in [12, 13]. Previous work has shown that this modification results in an improvement to simulations of wind speed and turbulence intensity under different meteorological regimes and terrain complexities [14, 15]. Details of the implementation constitute proprietary information and cannot be disclosed. The innermost domain was run on LES mode, switching off the physical parameterizations. The other domains (including the terra incognita [27] domain at a 333 m grid resolution) were run with the YSU [16] boundary layer scheme.

Initial and boundary conditions are taken from the Climate Forecast System Reanalysis [17] with a spatial resolution of $\sim$ 38 km. The static data for the high resolution domains are taken from the Shuttle Radar Topography Mission (at a 90 m resolution) [18] for topography and from the GlobCover [19] for land use. Other fields are taken from the default data sets provided by the model. The model was run for a 15-day period and time series were saved at three points: the northern wind turbine (NWT), the meteorological mast (MET), and the southern wind turbine (SWT), whose locations are shown in Fig. 1. The model saves historic output with a temporal frequency of 4 Hz. These values are then used to compute 10-minute averages and standard deviation, considering seven heights between $\sim$ 8 m and $\sim$ 182 m.

2.3. Wake Periods

To quantify the effect of WT wakes on model performance, the simulated values were compared to observations in the presence and absence of wakes. Hereinafter these periods will be referred to as "waked" and "free", respectively. The only wake considered is that of SWT which is located $\sim$ 230 m ($\sim$ 2.5 D) from MET and $\sim$ 715 m ($\sim$ 7.7 D) from NWT.

To determine the interval of nacelle positions γ in which the SWT wake impacts the downstream locations considered, the wake width w at any downstream distance x is

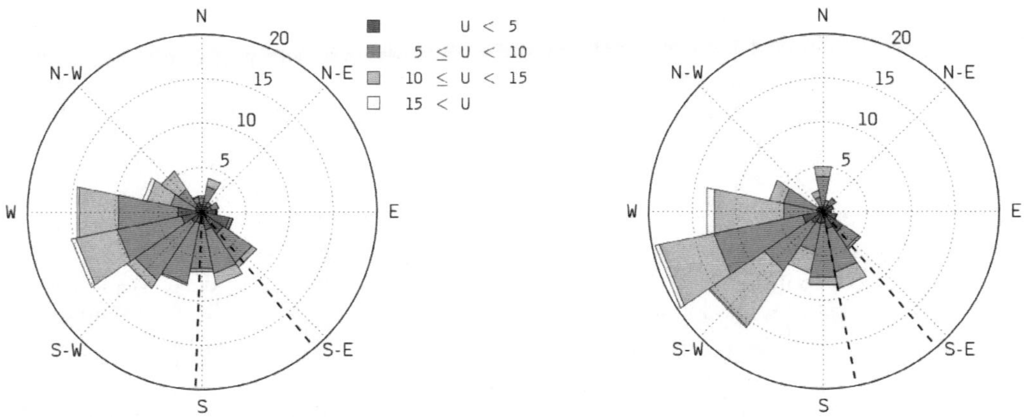

(a) Measured (b) Simulated

Figure 2: Wind roses at SWT during 15-day period considered (measurement height $z = 80$ m) using 10-minute mean measurements of wind speed and nacelle position in (a) and 10-minute mean simulated wind speed and direction at the grid cell containing SWT in (b). Color shading distinguishes between horizontal wind speed bins starting at 0 m s^{-1} in 5 m s^{-1} intervals. Radial axis gives normalized frequencies [%]. Dashed lines represent sectors in which MET (a) and NWT (b) are subjected to the SWT wake.

approximated as

$$w(x) = D + 2kx \tag{1}$$

following [20] where D is the rotor diameter of the wake-generating turbine and k the wake expansion coefficient taken here as 0.05 (i.e., 5% of wake expansion). This value was chosen based on previous recommendations for offshore sites [21] because of the predominantly offshore footprint seen in the measurements collected at the island [11].

The range of waked sectors is shown for both downstream sites in Fig. 1. For MET (Fig. 1a) this interval is $\gamma_{SWT} \in (140°, 182°)$ (where γ_{SWT} is the SWT 10-minute mean nacelle position) and represents $\sim$ 16% of the entire time series. For NWT (Fig. 1b) it is $\gamma_{SWT} \in (140°, 169°)$ representing $\sim$ 11% of the data. The simulation reproduced well the frequency of occurrence and the magnitude of wind speeds in these directional sectors, as can be seen in Fig. 2. Note that the wake of NWT on the other locations is not considered because northwesterly winds were less frequent during the experiment as shown in Fig. 2.

2.4. Post-simulation wake correction

We propose and evaluate a post-simulation wake correction method which seeks to drive the model results closer to the observations after the simulations have been conducted, for sites where wind turbines are present. The method is built on the Jensen wake formulation [20], widely used in industry and the basis of the well-known Park model [22]. Following this approach, the expected velocity deficit (VD) at a point can be estimated as

$$\mathrm{VD}(t) = \frac{1 - \sqrt{1 - C_T(t)}}{\left(1 + \frac{kx}{R}\right)^2} \tag{2}$$

where C_T [-] is the thrust coefficient of the WT generating the wake, k [-] the wake expansion factor (also known as decay coefficient), x [m] the downstream distance between the WT generating the wake and the location of interest, and R [m] the WT rotor radius. We consider unsteady C_T values calculated from the SWT wind speed measurements and thrust curve. A constant expansion factor is assumed as recommended for the first turbines in the row of a wind farm [23]. From Eq. (2) and a free stream velocity U_∞, a velocity corrected for the wake can be estimated as $U = U_\infty (1 - \mathrm{VD})$.

2.5. *Variables and Notation*

All of the variables considered throughout the analysis are 10-minute means. We focus on quantities of relevance to wind energy such as power P, horizontal wind speed U, wind direction β, nacelle position γ, and turbulence intensity $I = \sigma_U / U$ (where σ_U is the standard deviation of U over a 10-minute period). Wind direction measurements at the turbines were not provided. We assume negligible yaw misalignment at the two turbines so that nacelle position measurements can be treated as wind direction measurements (i.e., $\beta \sim \gamma$). The predicted power for the simulations is obtained from the simulated 10-minute mean wind speeds at hub height and the turbine power curve.

The analysis is predominantly based on calculated differences δ between the simulated and measured variables such that $\delta = x_{\mathtt{sim}} - x_{\mathtt{obs}}$, evaluated at each 10-minute mean time stamp for which observation were available. Absolute differences $|\delta|$ are used to determine the overall magnitude of the model error, and averages over a given time series are symbolized as $\overline{\delta}$ so that a mean absolute error is $\overline{|\delta|}$. The subscripts "w" and "f" (i.e., δ_w versus δ_f) differentiate between values in waked versus free directional sectors following the definition given in Section 2.3. Note that only coinciding time stamps are included in analyses that consider both observational and simulation data. To minimize the uncertainty, model values are not interpolated to different heights to conduct the analysis. Because there is a high density of values in the vertical (z) direction in the model, we simply consider the model level ($z_{\mathtt{sim}}$) that is closest to the observation height ($z_{\mathtt{obs}}$). At MET (NWT) the measuring height is $z_{\mathtt{obs}} = 60$ m (80 m) and the simulation height $z_{\mathtt{sim}} \sim 52$ m (~ 80 m).

3. Results

The results presented herein are divided into three sections. First, a general evaluation of the model performance during waked conditions (Section 3.1). Second, a comparison of model performance under waked versus free stream conditions (Section 3.2). Finally, an assessment of the post-simulation wake correction technique is given (Section 3.3).

3.1. *Model performance under "waked" conditions*

When operational, WTs convert a portion of the mean kinetic energy into electricity. The momentum deficit downstream leads some of this mean energy to cascade down the spectrum increasing the turbulent kinetic energy budget [24] and the ambient turbulence intensity. It is therefore expected that a simulation which fails to account for the effect of WTs on the flow will overestimate U and underestimate I. In terms of wind speed (Fig. 3, left) this expectation was only confirmed at the site MET. At NWT the model mostly underestimated the wind speed with a small mean error of $\overline{\delta} \sim -0.1$ m s^{-1}. For I (Fig. 3, center) the expectations are confirmed at both sites, with mean underestimations of $\overline{\delta} \sim -0.06$ at NWT and $\overline{\delta} \sim -0.15$ at MET.

The wind direction errors (Fig. 3, right) were similar at both sites, with the largest occurrences in the bin $\beta \in (0°, 30°)$ at NWT and $\beta \in (-30°, 0°)$ at MET. It is important to note that these direction errors can be brought on not only by the presence of wind turbine wakes, but also by the presence of other roughness elements at the site, which can lead to the development of internal boundary layers over roughness changes and to a modification in the displacement height [25]. More details on the effect of roughness and topography changes on near-surface flow

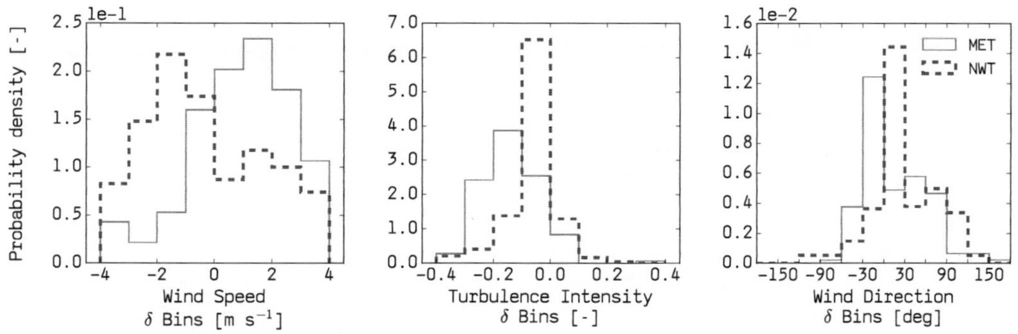

Figure 3: Probability density of model errors during waked conditions at MET (solid) and NWT (dashed) for time series of 10-minute means: wind speed (left), turbulence intensity (center) and wind direction (right). For wind direction, the standard 30° bin width is chosen.

Table 1: Temporal mean of model absolute errors at MET and NWT when considering waked ($\overline{|\delta_w|}$) and free ($\overline{|\delta_f|}$) conditions. Note that power error values are only available at the NWT site, and do not consider time stamps in which the NWT was not producing power.

		MET		NWT	
Variable	Unit	$\overline{\lvert\delta_w\rvert}$	$\overline{\lvert\delta_f\rvert}$	$\overline{\lvert\delta_w\rvert}$	$\overline{\lvert\delta_f\rvert}$
Wind Speed U	[m s^{-1}]	3.1	2.3	2.1	2.1
Wind Direction β	[deg]	35.5	34.7	40.2	18.6
Turbulence Intensity I	[-]	0.16	0.07	0.08	0.06
Power P	[kW]			495.4	385.8

can be found in [26]. More details for the present site can be found in [11]. In the next section, the role of wakes in explaining this discrepancy is estimated.

3.2. Model performance under waked versus free conditions

To estimate how much the presence of WT wakes affects the model performance, mean absolute errors $\overline{|\delta|}$ can be computed separately over the entire free and waked periods as given in Table 1. Note that in these averages, only absolute values are considered and therefore the information given in Table 1 differs from that in Fig. 3 where the sign of the error values was kept.

The results show that $\overline{|\delta_w|} > \overline{|\delta_f|}$ consistently. For wind speed, the error was ~ 35% higher at MET during the waked period but remained the same for both periods at the NWT site. These results along with those shown in Fig. 3 indicate that the model performance at MET was more affected by the presence of a wake than at NWT. It is important to note that although the wind speed errors under both flow scenarios are indistinguishable at NWT, the same is not true for power. Relatively small differences in wind speed may result in large differences in estimated power depending on the magnitude of the wind speeds. The power estimated from the simulation data presented larger errors under waked conditions, and consider only periods in which the NWT was operating.

The turbulence intensity error differences between waked and free conditions are especially large at MET, where the model failed to simulate high turbulence periods (Fig. 4a). These high turbulence periods in the measurements coincide with waked periods where sometimes the variation in wind speeds over 10 minutes is very large due to the intermittence and inhomogeneity of the flow under these conditions. At NWT (Fig. 4b) the model better reproduced the variation

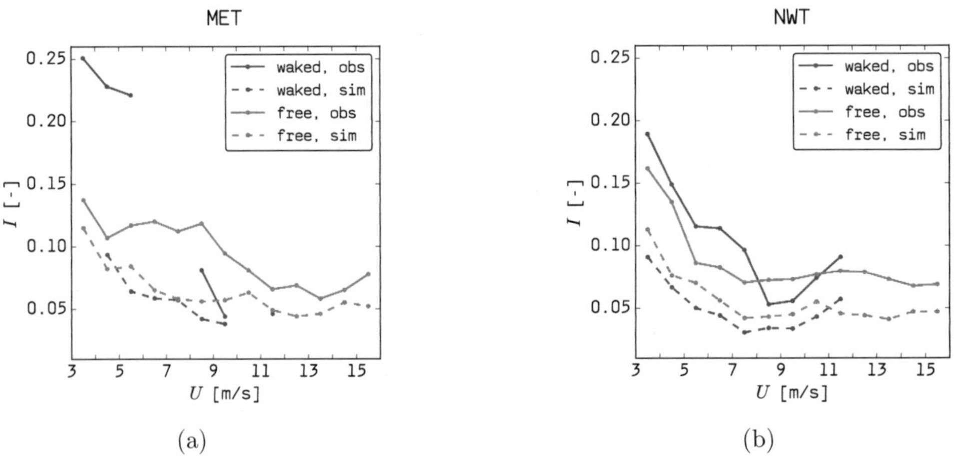

Figure 4: Median turbulence intensity I [-] as a function of horizontal wind speed U [m s^{-1}] at MET (a) and NWT (b) when subjected to free stream flow (gray) and to the wake of SWT (red) for measurements (solid) and simulations (dashed). Each median value corresponds to a discrete horizontal wind speed bin starting at 3 m s^{-1} in 1 m s^{-1} intervals.

of turbulence with wind speed, but still performed better during free conditions.

As mentioned in Section 3.1 and verified in Fig. 4, the model consistently underestimated the turbulence levels not only under waked but also under free conditions. The underestimation of turbulence by the model should be interpreted with caution because it can be brought on by data and model limitations that are not necessarily related to the absence of a wind turbine wake model. For example, the presence of an escarpment also contributes to enhanced turbulence at this site for the heights considered [11] and limitations in the spatial resolution of the lower boundary conditions could partly explain the I underestimation seen also during free stream periods. Moreover, a portion of this underestimation of turbulence intensity can likely be attributed to the numerical treatment of the transition from the meso to the micro scale, and the method chosen to simulate turbulence in the terra incognita [27] domain (see Section 2.2). Fig. 4 also shows that the simulated turbulence intensity was slightly higher for the free sectors which include not only an onshore but also an offshore footprint, adding to the flow complexity as it transitions over different roughness and terrain regimes. Note that error metrics given in Table 1 are not necessarily representative of the overall model skill, which should be evaluated with longer (e.g., one year) simulations. These investigations are beyond the scope of the present work, which seeks to differentiate the model performance under the two scenarios considered.

The wind direction differences between the SWT and NWT locations was also underestimated by the model as shown in Fig. 5. However, the model errors are very similar under free and waked periods, indicating that this was not an inability to simulate wake-induced meandering and rather a consequence of other model limitations likely relating to surface phenomena as such as the spatial resolution of the lower boundary conditions (i.e., roughness elements, terrain elevation, land-sea mask, sea surface temperatures) and the surface layer treatment to ensure the development of internal boundary layers both for on and offshore flow depending on the wind direction.

3.3. Post-simulation wake correction

In this section, the effect of the SWT wake is added to the simulated wind speed sampled at MET and NWT after the simulation has been completed. We seek to determine whether this

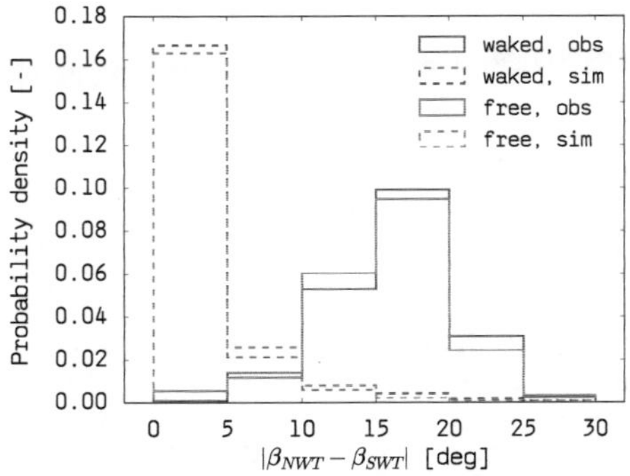

Figure 5: Histogram of wind direction differences between SWT and NWT during waked (red) and free (gray) periods from observations (solid) and simulation (dashed). Absolute differences are forced to the interval $[0^\circ, 180^\circ]$.

approach can improve the results with respect to local wind speed and turbine power prediction. Such an approach is of great value to the scientific community for being easy to implement and for eliminating the need for the costly alternative of fully coupling fluid and structural dynamics solvers, but it does not affect the turbulence and wind direction estimates. The method employed is described in Section 2.4.

We find that the wake correction method improves the error metrics by bringing the MET wind speed values closer to the observations during waked periods. Namely, the mean error goes from ~ 3.1 m s^{-1} (Table 1) at this site before the wake correction to ~ 2.0 m s^{-1} after the correction is applied (Fig. 6). The sensitivity of the results to different wake expansion factors was evaluated as shown in Fig. 6. All k values resulted in a reduction of the mean error at MET, but the largest improvement was seen for $k = 2\%$ where $\overline{|\delta_w|}$ was reduced by $\sim 38\%$ relative to the mean error for the non-corrected U values. This result agrees with expectations for the site during the period considered, where atmospheric conditions were mostly stable [11]. This expansion factor is close to the WAsP recommended value of $k = 5\%$ for offshore conditions [21] and is in agreement with the expectations for the measurements site (i.e., flat and narrow island with flow characteristics that often resemble those of offshore environments [11]).

At the NWT site, the post-processing wake correction had the opposite effect and increased the mean model error. This is expected following the analyses in Sections 3.1 and 3.2 which showed that the model already underestimated U, demonstrating that errors at this location are not necessarily related to the absence of a WT wake representation, but are rather a result of a combination of factors including limitations in the data and model, and in the post-processing technique itself which assumes a constant wind direction.

4. Conclusion

When wind turbines and their wakes are not represented in atmospheric models, the simulation performance is compromised. In this work, we quantify this effect by comparing model and observational data separately during free stream and wake conditions. Three unique data sets are considered which include a sonic anemometer mounted onto a meteorological mast compliant with the International Electrotechnical Commission standards [28], two utility-scale wind turbines, and cutting-edge WRF-LES simulations at a spatial resolution of 111 m. We focus on flow parameters of relevance to wind energy: wind speed, wind direction, and turbulence intensity. The wake of one turbine on two downstream sites is considered, namely a meteorological mast and another turbine.

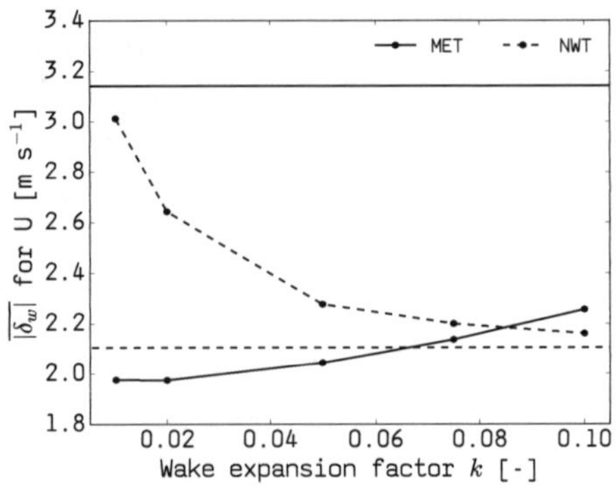

Figure 6: Mean absolute errors $\overline{|\delta_w|}$ for horizontal wind speed U [m s^{-1}] during waked periods without the wake correction (horizontal lines) and with the wake correction (lines with markers) at MET (solid) and NWT (dashed) sites, as a function of the expansion factor k [-] used in the Jensen wake expression (Eq. (2)).

We find that the model consistently underestimates turbulence intensity at both sites, but the relationship of this result with the absence of a wake description is only evident at one of the sites, where the model wind speed error in the presence of a wake is $\sim$ 35% larger than during free stream conditions. At the other site, the wind speed is underestimated indicating that these errors are related to other model and data limitations, such as the resolution and fidelity of the lower boundary conditions. The model also consistently underestimates the wind direction difference between the turbines regardless of whether a wake is present, evidence that this is not an inability of the model to simulate wake-induced meandering. Despite these shortcomings, it is important to keep in mind that the simulation considered is not long enough to undertake a thorough evaluation of the model skill which is beyond the scope of the work.

Finally, we find that a post-simulation wake correction reduces the error in wind speed estimates by $\sim$ 38% at the site that presented larger errors under wake conditions, but increases the errors at the other site which is in a coastal location, and in the far wake. The wake correction is done using the Jensen wake formulation and the best results are found using a wake expansion parameter of $k = 2\%$. This value is lower than has typically been found for offshore environments [29] but is consistent with the strongly stable conditions observed during the PEIWEE field campaign [11]. These preliminary results alert to the dangers of blindly applying a post-processing tool without consideration for other model error and uncertainty sources, which under certain conditions (e.g., complex terrain) may be larger in magnitude than the errors brought on by the absence of a wind turbine model. This analysis also reiterates the need for a wind turbine wake parameterization for high resolution WRF-LES which will improve simulations of wind speed, wind direction and turbulence intensity within and around wind power plants.

Acknowledgments

This work was partly funded by National Science Foundation 1565505, and the U. S. Department of Energy DE-EE0005379 and DE-SC0016438. The model data was provided by Vortex FDC, and the turbine data by WEICAN.

References

[1] Larsen G C, Madsen Aagaard H, Bingöl F, Mann J, Ott S, Sørensen J N, Okulov V, Troldborg N, Nielsen N M, Thomsen K and others 2007 Dynamic wake meandering modeling Tech. rep. Risø National Laboratory

[2] Keck R E, de Maré M, Churchfield M J, Lee S, Larsen G and Aagaard Madsen H 2014 *Wind Energy* **17** 1689–1710 ISSN 1099-1824 URL http://onlinelibrary.wiley.com/doi/10.1002/we.1662/abstract
[3] Churchfield M J, Lee S, Moriarty P J, Hao Y, Lackner M A, Barthelmie R, Lundquist J K and Oxley G 2014 *33rd Wind Energy Symposium* (Kissimmee, Florida: AIAA) URL http://arc.aiaa.org/doi/abs/10.2514/6.2015-0724
[4] Han Y, Stoellinger M and Naughton J 2016 *Journal of Physics: Conference Series* **753** 032044 ISSN 1742-6596 URL http://stacks.iop.org/1742-6596/753/i=3/a=032044
[5] Skamarock W C, Klemp J B, Dudhia J, Gill D O, Barker D M, Duda M G, Huang X Y, Wang W and Powers J G 2008 A Description of the Advanced Research WRF Version 3 Technical Note TN-475+STR National Center for Atmospheric Research
[6] Keith D W, DeCarolis J F, Denkenberger D C, Lenschow D H, Malyshev S L, Pacala S and Rasch P J 2004 *Proceedings of the National Academy of Sciences of the United States of America* **101** 16115–16120 ISSN 0027-8424, 1091-6490 URL http://www.pnas.org/content/101/46/16115
[7] Chatterjee F, Allaerts D, Blahak U, Meyers J and van Lipzig N 2016 *Quarterly Journal of the Royal Meteorological Society* **142** 3152–3161 ISSN 1477-870X URL http://onlinelibrary.wiley.com.proxy.library.cornell.edu/doi/10.1002/qj.2896/abstract
[8] Fitch A C, Olson J B, Lundquist J K, Dudhia J, Gupta A K, Michalakes J and Barstad I 2012 *Monthly Weather Review* **140** 3017–3038 ISSN 0027-0644 URL http://journals.ametsoc.org/doi/abs/10.1175/MWR-D-11-00352.1
[9] Volker P J H, Badger J, Hahmann A N and Ott S 2015 *Geosci. Model Dev.* **8** 3715–3731 ISSN 1991-9603 URL http://www.geosci-model-dev.net/8/3715/2015/
[10] Fitch A C 2016 *Wind Energy* **19** 1757–1758 ISSN 1099-1824 URL http://onlinelibrary.wiley.com/doi/10.1002/we.1945/abstract
[11] Barthelmie R J, Wang H, Doubrawa P, Giroux G and Pryor S C 2016 *Wind Energy* **19** 2271–2286 ISSN 1099-1824 URL http://onlinelibrary.wiley.com/doi/10.1002/we.1980/abstract
[12] Muñoz-Esparza D, Kosović B, García-Sánchez C and van Beeck J 2014 *Boundary-layer meteorology* **151** 453–478
[13] Muñoz-Esparza D, Kosović B, van Beeck J and Mirocha J 2015 *Physics of Fluids* **27** 035102
[14] Montornès A, Casso P, Kosovic B and Lizcano G 2016 *22nd Symposium on Boundary Layers and Turbulence (AMS)* URL https://ams.confex.com/ams/32AgF22BLT3BG/webprogram/Paper295742.html
[15] Montornès A, Casso P, Kosovic B and Lizcano G 2016 *17th Annual WRF Users' Workshop* URL http://www2.mmm.ucar.edu/wrf/users/workshops/WS2016/oral_presentations/3.3.pdf
[16] Hong S Y, Noh Y and Dudhia J 2006 *Monthly Weather Review* **134** 2318–2341 ISSN 0027-0644 URL http://journals.ametsoc.org/doi/abs/10.1175/MWR3199.1
[17] Saha S, Moorthi S, Pan H L, Wu X, Wang J, Nadiga S, Tripp P, Kistler R, Woollen J, Behringer D and others 2010 *Bulletin of the American Meteorological Society* **91** 1015–1057
[18] Jarvis A, Reuter H I, Nelson A, Guevara E and others 2008 *available from the CGIAR-CSI SRTM 90m Database (http://srtm. csi. cgiar. org)*
[19] Bontemps S, Defourny P, Bogaert E V, Arino O, Kalogirou V and Perez J R 2011
[20] Jensen N O 1983 A note on wind generator interaction Tech. Rep. 2411 Risø National Laboratory
[21] Barthelmie R J, Larsen G C, Frandsen S T, Folkerts L, Rados K, Pryor S C, Lange B and Schepers G 2006 *Journal of Atmospheric and Oceanic Technology* **23** 888–901 ISSN 0739-0572 URL http://journals.ametsoc.org/doi/abs/10.1175/JTECH1886.1
[22] Katic I, Højstrup J and Jensen N 1987 *EWEC'86. Proceedings* vol 1 ed Sesto E (Rome, Italy) pp 407–410
[23] Peña A, Réthoré P E and Rathmann O 2014 *Renewable Energy* **70** 164–171 ISSN 0960-1481 URL http://www.sciencedirect.com/science/article/pii/S0960148114001001
[24] Nathan J, Bautista M, Masson C and Dufresne L 2014 *Journal of Physics: Conference Series* vol 524 (IOP Publishing) p 012141
[25] Benson J 2005 *Boundary-layer response to a change in surface roughness.* Ph.D. thesis The University of Reading
[26] Hedegaard K and Larsen S E 1983 *Wind Speed and Direction Changes Due to Terrain Effects Revealed by Climatological Data from Two Sites in Jutland* vol 434 (Risø)
[27] Wyngaard J C 2004 *Journal of the Atmospheric Sciences* **61** 1816–1826 ISSN 0022-4928 URL http://journals.ametsoc.org/doi/abs/10.1175/1520-0469(2004)061%3C1816:TNMITT%3E2.0.CO;2
[28] International Electrotechnical Comission 2005 Wind Turbines - Part 12-1: Power performance measurements of electricity producing wind turbines Tech. Rep. IEC 61400-12-1:2005(E)
[29] Barthelmie R J and Jensen L E 2010 *Wind Energy* **13** 573–586 ISSN 1099-1824 URL http://onlinelibrary.wiley.com/doi/10.1002/we.408/abstract

IOP Conf. Series: Journal of Physics: Conf. Series **854** (2017) 012011 doi:10.1088/1742-6596/854/1/012011

Investigation of the effect of inflow turbulence on vertical axis wind turbine wakes

P Chatelain[1], M Duponcheel[1], S Zeoli[2], S Buffin[1], D-G Caprace[1], G Winckelmans[1] and L Bricteux[2]

[1] Institute of Mechanics, Materials and Civil Engineering, Université catholique de Louvain, 1348 Louvain-la-Neuve, Belgium

[2] Mechanical Engineering Department, Université de Mons, 7000 Mons, Belgium

E-mail: philippe.chatelain@uclouvain.be

Abstract. The aerodynamics of Vertical Axis Wind Turbines (VAWTs) is inherently unsteady, which leads to vorticity shedding mechanisms due to both the lift distribution along the blade and its time evolution. In this paper, we perform large-scale, fine-resolution Large Eddy Simulations of the flow past Vertical Axis Wind Turbines by means of a state-of-the-art Vortex Particle-Mesh (VPM) method combined with immersed lifting lines. Inflow turbulence with a prescribed turbulence intensity (TI) is injected at the inlet of the simulation either from a precomputed synthetic turbulence field obtained using the Mann algorithm [1] or generated on the-fly using time-correlated synthetic velocity planes. The wake of a standard, medium-solidity, H-shaped machine is simulated for several TI levels. The complex wake development is captured in details and over long distances: from the blades to the near wake coherent vortices, then through the transitional ones to the fully developed turbulent far wake. Mean flow and turbulence statistics are computed over more than 10 diameters downstream of the machine. The sensitivity of the wake topology and decay to the TI and to the operating conditions is then assessed.

1. Introduction

The operation of Vertical Axis Wind Turbines (VAWTs) essentially involves unsteady aerodynamics, and thus the shedding of complex vorticity structures in their wake. Indeed, VAWTs exhibit a wake topology that is governed by both the spanwise lift variations (like their horizontal axis counterparts (HAWTs)) and the temporal lift variations that arise over the blade rotation. This leads to a strongly three-dimensional vorticity field in the wake of VAWTs, an aspect that could translate into a more intense wake decay. Another interesting aspect of VAWTs is that they are naturally indifferent to variations in the wind direction; this could clearly translate into a lower sensitivity to ambient turbulence. These characteristics of VAWTs and their wakes hint at a possible lead over HAWTs once they will be installed in a farm; this has, in particular, led to predictions of higher land-specific power densities [2, 3].

Such promises, together with potential operational gains (maintenance costs, the disappearance of yawing actuation), have spurred some definite research momentum in VAWT aerodynamics, in the shape of experimental [4, 5] and numerical [6, 7] studies. However, because of their unsteady aerodynamics, VAWT simulation and modeling tools have not reached yet the maturity of those for HAWTs, e.g. the Blade Element Momentum (BEM) method. The wake phenomenon for VAWTs has only been approached recently, either from an experimental [5] or a numerical perspective [8, 9]. Among those, very few works have been dedicated to the investigation of the large scale development of the wake [9], and none have thus far considered the operation of the machine within a turbulent wind.

This paper addresses this very last point and follows directly the work of [9]. That work concerned a standard, medium-solidity, H-shaped machine; all simulations were performed without ambient turbulence, and rather considered the details of the vortex dynamics at work in the near wake and their influence on the wake decay. In the present work, we introduce a turbulent wind and perform a parametric study of its influence. In particular, this study highlights the behavior of the main flow structures identified in the absence of turbulence [9] and the relative contributions of the intrinsic vortex dynamics and the interplay between the wake vorticity and the ambient turbulence. We consider the same baseline VAWT geometry and keep it operating at its optimum Tip Speed Ratio (TSR); we vary the Turbulence Intensity (TI) and the nature of the turbulence: isotropic or anisotropic. The present work relies on the same numerical tools as in [9]: the large-scale and fine-resolution Large Eddy Simulations are performed by a state-of-the-art Vortex Particle-Mesh (VPM) method combined with immersed lifting lines [10]. We add the management of a turbulent inflow performed through the injection of a vorticity field in a Lagrangian fashion; this field is either pre-generated through the Mann algorithm [1] or is generated on-the-fly during the simulation.

This paper is structured as follows. We only briefly recall the Vortex Particle-Mesh (VPM) method with immersed lifting lines and turbulent inflows in Section 2. In Section 3, we study the effects of turbulence on the reference VAWT and also perform a verification of the synthetic turbulence approach. We close this paper in Section 4 with our conclusions and perspectives.

2. Methodology

2.1. The Vortex Particle-Mesh method with immersed lifting lines

The coarse scale aerodynamics and the wake of the VAWT are simulated using a massively parallel implementation of a Vortex Particle-Mesh flow solver. The present method relies on the Large Eddy Simulation in the vorticity-velocity formulation for incompressible flows ($\nabla \cdot \mathbf{u} = 0$):

$$\frac{D\boldsymbol{\omega}}{Dt} = (\boldsymbol{\omega} \cdot \nabla)\,\mathbf{u} + \nu \nabla^2 \boldsymbol{\omega} + \nabla \cdot \mathbf{T}^M \qquad (1)$$

where ν is the kinematic viscosity, and $\mathbf{T}^M$ is the Sub-Grid Scale (SGS) model, here a multiscale model solely acting on the small scales of the LES field [11]. The advection term of this equation is treated with particles while their right-hand sides are evaluated efficiently on a mesh. The Poisson solver for velocities operates in Fourier space and it simultaneously allows for unbounded directions and inlet/outlet boundaries [12]. The resulting hybrid method requires back-and-forth interpolations between the particles and the grid but does not detract from the good numerical accuracy (in terms of diffusion and dispersion errors) and the stability properties of a particle method. We refer to [13, 14, 11] for detailed descriptions.

The generation of vorticity along the blades is accounted for through an immersed lifting line approach [10]. The approach is very much akin to a Vortex Lattice method but only over the span of a time step. It relies on the Kutta-Joukowski theorem (see e.g. [15]) that relates the developed lift $\mathbf{L}$ to the relative flow $\mathbf{V}_{\rm rel}$ and the circulation bound around the local 2D airfoil $\mathbf{L} = \rho \mathbf{V}_{\rm rel} \times \boldsymbol{\Gamma}$ and it discretizes the bound circulation and the shed vortex sheets with particles, which are then added to the bulk flow. In this work, we account for unsteady effects through a Leishman-Beddoes dynamic stall model [16], following the indications of Dyachuk [17] and Scheurich [6] who present coefficients for various airfoils validated in the particular case of VAWT.

2.2. Turbulent inflow

In order to model a realistic incoming inflow, we generate synthetic turbulence at the domain inlet. A perturbation vorticity field $\boldsymbol{\omega}'$ is injected at the inflow as particles. This Lagrangian treatment, as opposed to a boundary condition, is consistent with the rest of the methodology and does not impose any additional time step constraint: a several grid points-thick slab of turbulent wind can enter the domain over a single time step. This technique, originally developed in [10, 18], is similar to that applied in other recent Vortex Particle methods works [19, 20, 21] but has been further developed here for more flexibility.

In a refinement of what was done in [18], the introduction of $\boldsymbol{\omega}'$ into the computational domain now relies on a smooth blending of $\boldsymbol{\omega}'$ with the domain vorticity, instead of a direct insertion of the $\boldsymbol{\omega}'$ particles into the domain. More specifically, a buffer computational domain co-exists with the main computational domain; this turbulent inflow contains particles which are progressively let free: their velocities and their stretching terms transition smoothly from a uniform flow at the prescribed inflow velocity $\mathbf{U}_\infty$ into the fully three-dimensional flow at the end of the buffer region. This change allows to handle arbitrary remeshing frequencies and non-uniform inflows.

In this paper, two $\boldsymbol{\omega}'$ generation techniques are used and compared. A first one relies on simply computing $\boldsymbol{\omega}' = \nabla \times \mathbf{u}'$ of a pre-generated database of velocities $\mathbf{u}'$, here using Mann's [22] algorithm. We introduce a second technique that avoids the pre-generation and storage of this database. At each time step, we generate a two dimensional slab of a turbulent velocity field using Mann's algorithm again; this velocity slab is correlated in time with the previous slabs using an asymmetric time filter as in [23] in order to produce realistic correlations in the streamwise direction.

Both methods produce fields that are periodic in the transverse directions; those are made compact through a smooth clipping, as we are working with unbounded transverse directions. Preliminary results for the second technique are provided below and are compared to those obtained using the pre-computed turbulence.

3. Results

3.1. Parameters

This whole section focuses on the same low solidity H-VAWT studied numerically in [24, 6], and then in [9] in configurations without any turbulence. For the sake of completeness, we here briefly recall its main parameters: an aspect ratio $AR = H/D = 1.5$, a solidity $\sigma = nc/D = 0.1725$, and constant-chord NACA0015 airfoils. We here only consider its optimum power operating point, $\mathrm{TSR} = \Omega R/U_\infty = 3.21$, as found by [25]. Unless otherwise specified, the numerical simulations involved $D/h = 96$ mesh points/particles per rotor diameter and extended up to 17 diameters downstream of the rotor axis. This discretization and the chosen turbulent inflow sizes lead to simulations with $3\,10^8$ to $7\,10^8$ particles/mesh points; those were run on 2560 CPU cores for wall-clock times between 30 (low TI cases) and 54 (high TI anisotropic case) hours.

Table 1. H-VAWT turbulent inflows: parameters of Mann's synthetic turbulence.

Run	TI	L	Γ	Box size
I2	2%	$D/3$	0	$24\,D$
I7.5	7.5%	$D/3$	0	$24\,D$
A7.5	7.5%	D	3.9	$24\,D$

Table 1 summarizes the inflow turbulence conditions considered. They allow to investigate the roles of the Turbulence Intensity $TI = (\frac{2}{3}\bar{k})^{1/2}/U_\infty$, where $\bar{k}$ is the mean Turbulent Kinetic

Energy (TKE), and of the anisotropy of this inflow, as characterized by the parameter Γ which is equal to zero for isotropic turbulence. The runs with isotropic turbulence I2 and I7.5 compare with wind tunnel conditions at low and intermediate grid-generated turbulence intensities; run A7.5, is a scenario with anisotropic turbulence at the same TI as I7.5, which is representative of a small-to-medium VAWT ($D \simeq 20 - 50$m) in atmospheric turbulence. All these turbulent inflows were pre-generated using Mann's algorithm [1].

3.2. Aerodynamics

We first consider the effect of these turbulence scenarios on the Angle-of-Attack (AoA) and on the normal and tangential force coefficients at mid-height, respectively $F_n/(c\,q_0)$ and $F_t/(c\,q_0)$, where q_0 is the dynamic pressure of the mean free stream. We will take, as reference, the behavior without ambient turbulence computed in [9]. We recall here the main features of these quantities over a blade rotation in the absence of turbulence: (1) the behavior is not symmetrical for the upstream and downstream legs because of the reduced velocity encountered downstream; (2) the downstream leg also exhibits oscillations in the AoA and force coefficients as the blade encounters vortex sheets shed during the upstream leg.

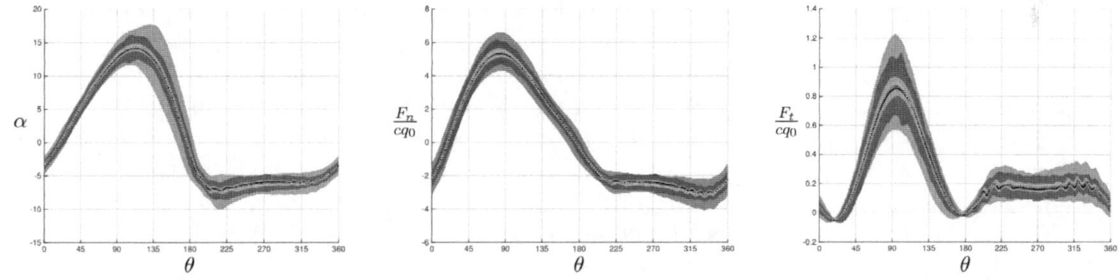

(a) Angle of attack (b) Normal force coefficient (c) Tangential force coefficient

Figure 1. H-type VAWT aerodynamics: profiles of the angle of attack and of the normal, $F_n/(c\,q_0)$, and tangential, $F_t/(c\,q_0)$, force coefficients at mid-height versus the blade angular position θ for the turbulent inflows: I2 (- - - -), I7.5 ($\cdot$), and A7.5 ($\cdots$) and the reference without turbulence [9] (——); the two-standard deviations ($\pm 2\sigma$) envelopes are indicated by the colored areas.

The average aerodynamic behaviors do not deviate markedly from the reference case without turbulence. The downstream oscillations decrease with the TI, as the blades interact with the upstream-shed structures at more randomized locations. The standard deviations globally exhibit orders of magnitude consistent with the turbulence level. They are however modulated by the blade position: the angle-of-attack is perturbed the most when the blade travels perpendicularly to the freestream. The angles-of-attack are practically zero around $\theta = 20°$ and $180°$; the zero slope of the drag curve at $\alpha = 0$ then translates such values into a collapse of the tangential forces perturbations, i.e. very small envelopes (Fig. 1(c)). Finally, we note the strong difference between the deviations achieved in the isotropic and anisotropic cases. For equal TIs, an anisotropic inflow redistributes more of the turbulent kinetic energy to the velocity components which impact the aerodynamics the most: those in the horizontal plane (u'_1 and u'_3).

The mean power coefficient $C_p = P/(\frac{1}{2}\rho A U_\infty^3)$ obtained without inflow turbulence was $C_p = 0.339$ [9]. When inflow turbulence is added, the obtained power coefficients are 0.346, 0.342 and 0.357 for the cases I2, I7.5 and A7.5, respectively. The increase compared to the case without turbulence is mainly due to the increased averaged energy flux in the turbulent cases: these cases maintain the averaged inflow velocity at $\bar{u} = U_\infty$ but the average $\overline{u(u^2 + v^2 + w^2)}$ differs from U_infty^3. Yet, the C_p of case I7.5 is slightly lower than that of case I2, denoting a small decrease of the aerodynamic efficiency due to the larger fluctuations.

3.3. Wakes

3.3.1. Vortex dynamics Instantaneous visualizations of the wakes are shown in Fig. 2 for the considered TIs. The turbulence-free baseline (Fig. 2(a)) provides a clear view of the fundamental vortex dynamics at work in the near wake. On the sides, the blade sheds vortex sheets due to the temporal variation of the circulation, $d\Gamma/dt$, and at the top and bottom, it sheds tip vortices with varying circulations. The near-wake is then governed by the instabilities and the merging phenomena of vortices of unequal circulations, all of which can be seen to originate in the corners of the wake. We refer to [9] for a more thorough discussion.

The ambient turbulence affects the wake at several scales. In the near-wake, one can observe the disturbances on the shed tip vortices (see the zooms in Fig. 2(b) to 2(d)) increasing with the TI. From very slight, they increase markedly for the high TI cases, and even lead to reconnections between successive tip vortices in the vicinity of the blades in the downwind part of their rotation. From a global point of view, if an absence of turbulence produces a quite periodic array of vortices and reconnection events in the corners of the wake, increasing the turbulence level leads to more irregularly spaced and distorted vortices and reconnections occurring on the top and bottom of the near-wake. One thus expects a more distributed production of turbulence in the near wake.

The wake behavior at large scales exhibits meandering with amplitudes increasing with the TI, and we note that it is already present in the baseline without turbulence. The wavelength of this meandering motion is of the order of $5D$ and thus much larger than the integral length scale of the synthetic turbulence boxes ($D/3$ and D). We leave the detailed investigation of this meandering behavior as a subject of future work.

3.3.2. Statistics We study the average behavior of the wakes by means of the mean axial velocity $\bar{u}$, and the mean turbulent kinetic energy $\bar{k}$; these statistics were collected over a period $T_{\text{avg}} = 24\, D/U_\infty$, corresponding to the flow-through time of the pre-computed inflow turbulence field. Figure 3 shows a horizontal slice of these statistics for the studied TIs and for the baseline. The previous study had identified a backflow region in the wake (Fig. 4(a)) at intermediate and high TSRs. This feature is here shown to persist for the isotropic turbulent inflows; it also moves closer to the rotor as the TI is increased, going from 6 D for no TI to 4 D and less for TI $= 2\%$ and 7.5%. This structure disappears for the anisotropic TI case; its large scale perturbations are then able to disrupt the near-wake. The TKE signatures translate and quantify the above-mentioned accelerated onset of vortical instabilities in the near-wake. Even at a low TI, TKE production increases on the sides of the wake and progresses upstream towards to the rotor (Figs. 4(b) and 3(c)). The TKE of the anisotropic case exhibits a strongly increased intensity; its proximity to the rotor and the angular spread of its signature are indicative of more pronounced large scale wind direction changes.

This higher dissipation translates in an increased smearing of the velocity deficit, with the most important gain for the anisotropic case; this is confirmed in the transverse slices of Fig. 4. We note the more even distribution of TKE in the periphery of the near wakes with a slight early bias for the blade-travelling-downwind side. The bean-shaped distributions for the deficit and decay are well preserved for the isotropic cases, up to 10 diameters. For the anisotropic case, the wake has already much decayed at $10D$ downstream.

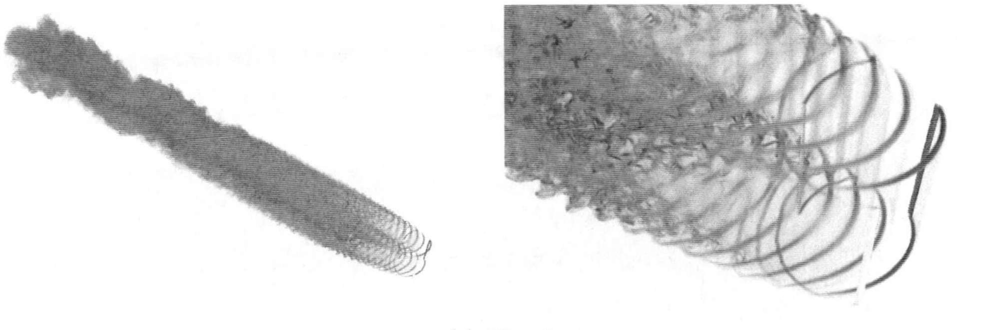

(a) TI = 0

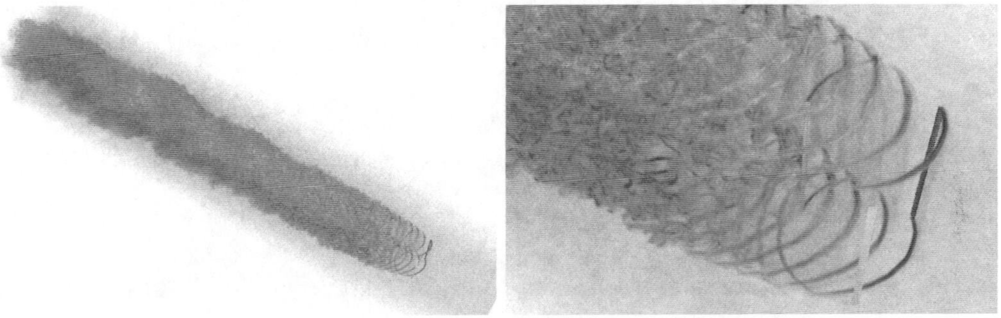

(b) TI = 2% isotropic

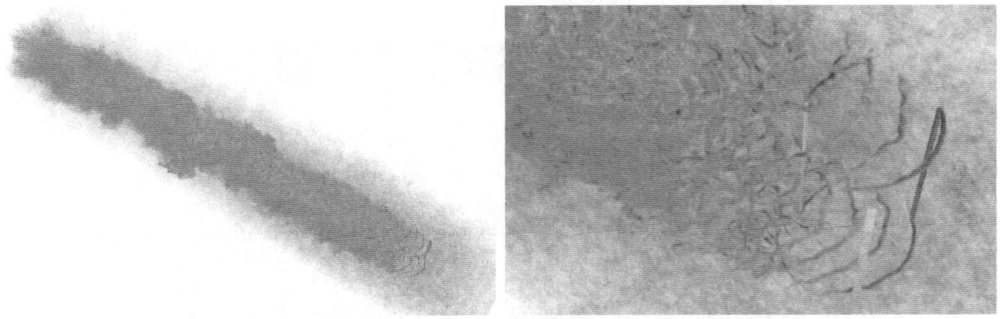

(c) TI = 7.5% isotropic

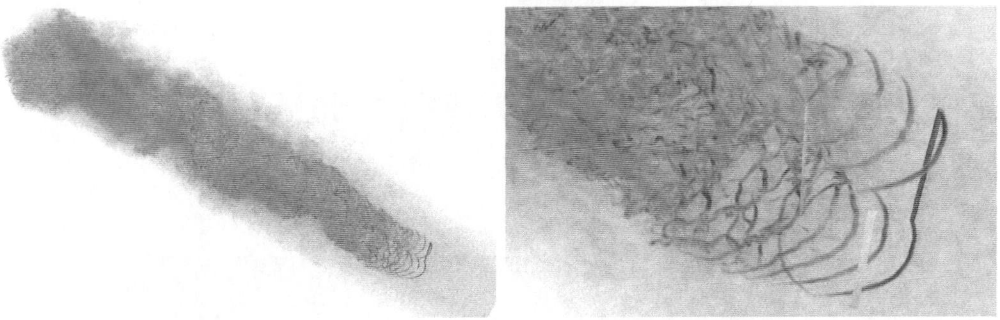

(d) TI = 7.5% anisotropic

Figure 2. H-type VAWT wake and very near wake: volume rendering of the vorticity magnitude $\|\boldsymbol{\omega}\|$; the lifting lines are also shown as 3-D blades.

IOP Conf. Series: Journal of Physics: Conf. Series **854** (2017) 012011 doi:10.1088/1742-6596/854/1/012011

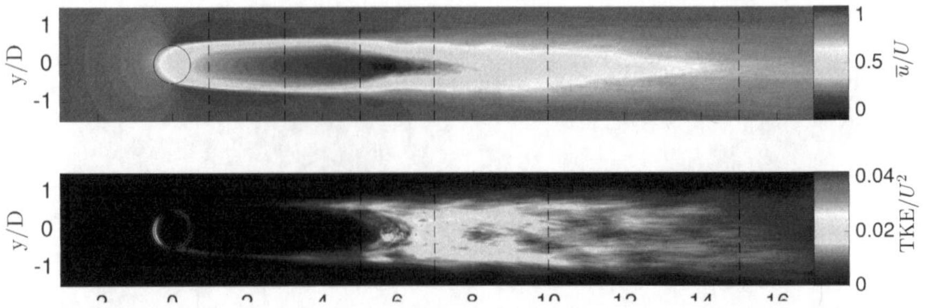

(a) TI = 0

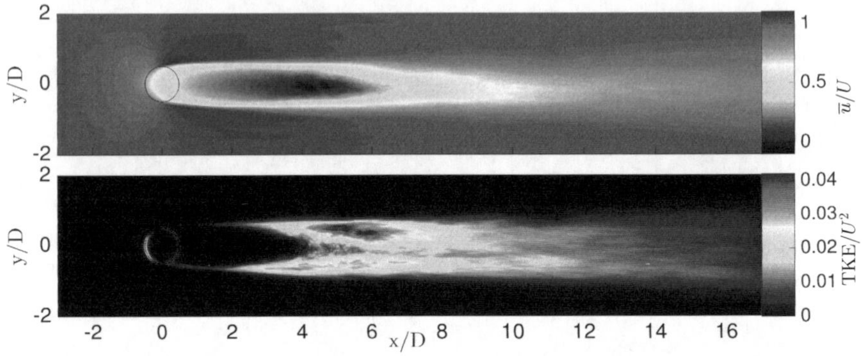

(b) TI = 2% isotropic

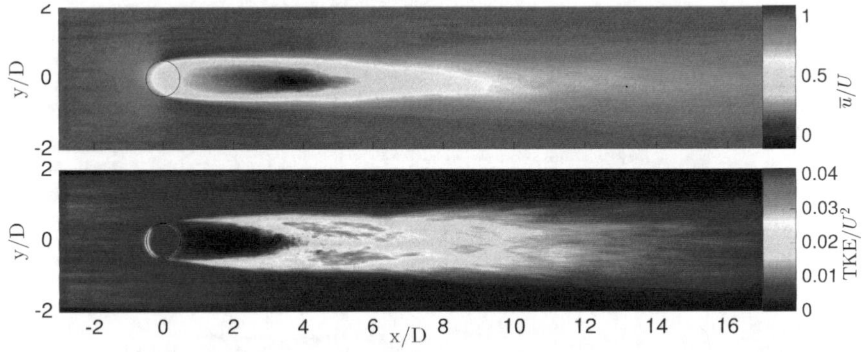

(c) TI = 7.5% isotropic

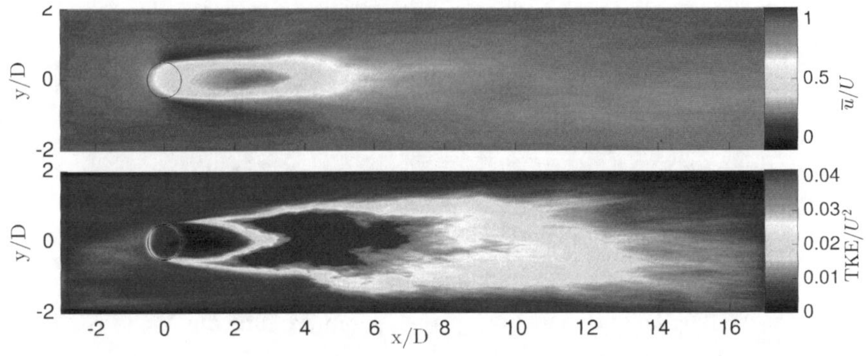

(d) TI = 7.5% anisotropic

Figure 3. H-type VAWT wake statistics: mean streamwise velocity $\bar{u}/U_\infty$ and mean turbulent kinetic energy TKE/U_∞^2 in the $y/D = 0$ plane.

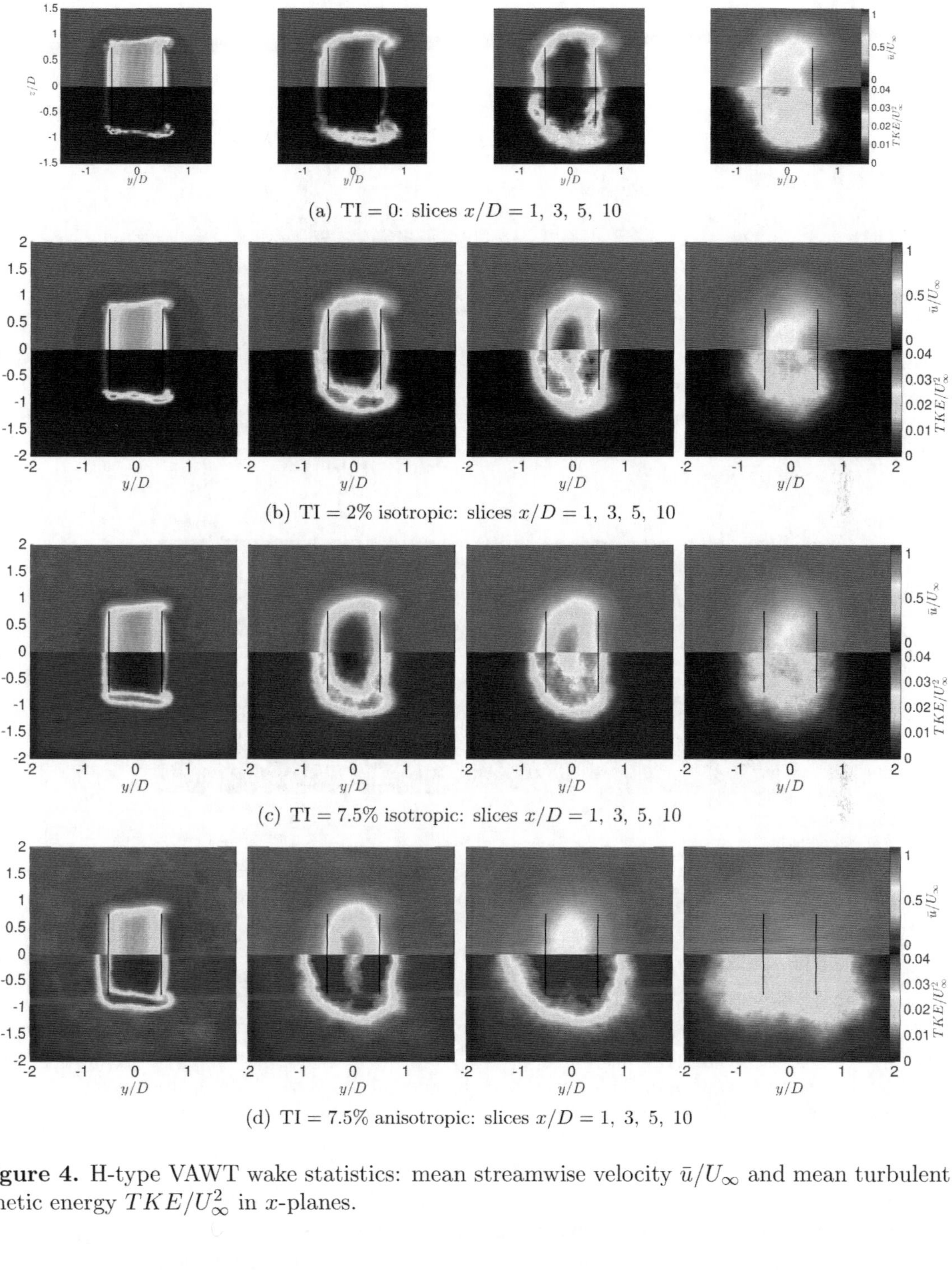

(a) TI = 0: slices $x/D = 1,\ 3,\ 5,\ 10$

(b) TI = 2% isotropic: slices $x/D = 1,\ 3,\ 5,\ 10$

(c) TI = 7.5% isotropic: slices $x/D = 1,\ 3,\ 5,\ 10$

(d) TI = 7.5% anisotropic: slices $x/D = 1,\ 3,\ 5,\ 10$

Figure 4. H-type VAWT wake statistics: mean streamwise velocity $\bar{u}/U_\infty$ and mean turbulent kinetic energy TKE/U_∞^2 in x-planes.

3.4. Preliminary results on turbulent inflow comparison

The generation of a turbulent inflow using a precursor has two drawbacks. First, it requires to generate a 3D turbulence database, store it and manage its input into the code. Second, as this database has a finite streamwise extension, the information entering the flow domain is repeated in time leading to periodic repetition of flow structures. With the alternative method

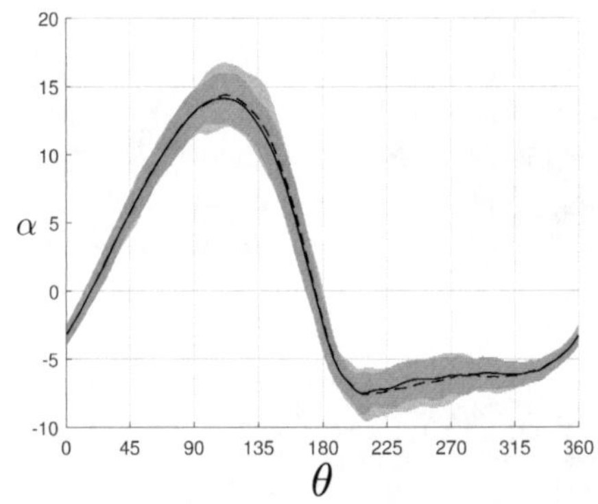

Figure 5. H-type VAWT aerodynamics ($TI = 7.5\%$): comparison of angle of attack mean profiles for Mann precursor (solid) and time-correlated 2D Mann planes (dash). The standard deviation enveloppes are also represented in grey shading (light and dark grey, respectively) and are almost superimposed.

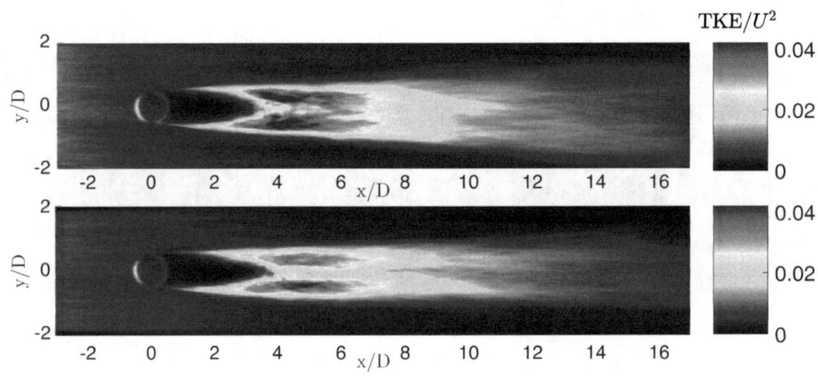

Figure 6. H-type VAWT aerodynamics ($TI = 7.5\%$): comparison of TKE fields for $TI = 7.5\%$, isotropic inflow turbulence generated using time-correlated 2D Mann planes (top) and Mann precursor (bottom).

investigated here, these two drawbacks are overcome as information is generated on the fly. Comparisons of flow diagnostics ($TI = 7.5\%$) obtained using the two methods are provided in Figs 5 and 6: a good agreement is found between both approaches. This comparison constitutes a first step towards validation of the new proposed approach.

4. Conclusions

A Vortex Particle-Mesh method with immersed lifting lines, here briefly presented, has been applied to large scale and high resolution LES of VAWT wakes with various turbulent inflows. The method is capable of tracking vortical structures over very long times and distances. Two TI levels (2.5% and 7.5%) have been considered for isotropic inflow turbulence whereas anisotropic turbulence has also been considered for the higher TI case. For each case, the mean flow and mean TKE fields have been obtained.

Increasing the turbulence level accelerates the transition of the wake and this tends to bring the low speed recirculating region closer to the turbine. For the higher TI considered (7.5%), the anisotropy of the turbulence has a strong effect since the decay is much faster in the anisotropic case than in the isotropic one. The recirculation zone is also suppressed in the case of the anisotropic turbulence and the TKE in the wake is also much higher. The wake meandering is also found to increase with the TI level, and to be further enhanced by the turbulence anisotropy. These large numerical databases contain rich information that will be further analyzed.

Acknowledgments

The development work benefited from the computational resources provided by the supercomputing facilities of the Université catholique de Louvain (CISM/UCL) and the Consortium des Équipements de Calcul Intensif (CÉCI) en Fédération Wallonie Bruxelles (FWB) funded by the Fond de la Recherche Scientifique de Belgique (F.R.S.-FNRS) under Convention No. 2.5020.11. The production simulations used computational resources made available on the Tier-1 supercomputer of the FWB, infrastructure funded by the Walloon Region under the Grant Agreement No. 1117545.

References

[1] Mann J 1994 *Journal of Fluid Mechanics* **273** 141–168
[2] Paquette J and Barone M F 2012 *EWEA 2012, April 16-19, 2012 in Copenhagen, Denmark.* (EWEA)
[3] Kinzel M, Mulligan Q and Dabiri J O 2012 *Journal of Turbulence* **13** N38
[4] Beaudet L 2014 *Etude expérimentale et numérique du décrochage dynamique sur une éolienne à axe vertical de forte solidité* Ph.D. thesis Université de Poitiers
[5] Tescione G, Ragni D, He C, Ferreira C S and van Bussel G 2014 *Renewable Energy* **70** 47 – 61 ISSN 0960-1481 special issue on aerodynamics of offshore wind energy systems and wakes
[6] Scheurich F 2011 *Modelling the aerodynamics of vertical-axis wind turbines* Ph.D. thesis University of Glasgow
[7] Ferreira C S, Madsen H A, Barone M, Roscher B, Deglaire P and Arduin I 2014 *Journal of Physics: Conference Series* **524** 012125
[8] Scheurich F and Brown R E 2013 *Wind Energy* **16** 91–107 ISSN 1099-1824
[9] Chatelain P, Duponcheel M, Caprace D G, Marichal Y and Winckelmans G 2016 *The Science of Making Torque from Wind (TORQUE 2016), Munich, Germany*
[10] Chatelain P, Backaert S, Winckelmans G and Kern S 2013 *Proceedings of The 9th International Symposium on Engineering Turbulence Modelling and Measurements (ETMM-9), June 6–8, 2012, Thessaloniki, Greece* (*Flow, Turbulence and Combustion* vol 91) ed Rodi W ERCOFTAC (Springer) pp 587–605
[11] Cocle R, Winckelmans G and Daeninck G 2008 *Journal of Computational Physics* **227** 9091–9120
[12] Chatelain P and Koumoutsakos P 2010 *Journal of Computational Physics* **229** 2425–2431
[13] Koumoutsakos P 2005 *Annual Review of Fluid Mechanics* **37** 457–487
[14] Chatelain P, Curioni A, Bergdorf M, Rossinelli D, Andreoni W and Koumoutsakos P 2008 *Computer Methods in Applied Mechanics and Engineering* **197** 1296–1304
[15] Prandtl L 1923 Application of modern hydrodynamics to aerodynamics Technical Report 116 NACA
[16] Leishman J G 2006 *Principles of helicopter aerodynamics* 2nd ed (Cambridge University Press)
[17] Dyachuk E, Goude A and Bernhoff H 2014 *AIAA journal* **52** 72–81
[18] Backaert S, Chatelain P, Winckelmans G and De Visscher I 2014 *EUROMECH Colloquium 528 "Wind energy and the impact of turbulence on the conversion process"* (*Research Topics in Wind Energy* vol 2) ed Hölling M, Peinke J and Ivanell S (Springer-Verlag, Berlin Heidelberg)
[19] Rasmussen J, Hejlesen M, Larsen A and Walther J 2010 *Journal of Wind Engineering and Industrial Aerodynamics* **98** 754 – 766 ISSN 0167-6105
[20] Hejlesen M, Rasmussen J, Larsen A and Walther J H 2015 *Journal of Wind Engineering and Industrial Aerodynamics* **146** 117 – 127 ISSN 0167-6105
[21] Branlard E, Papadakis G, Gaunaa M, Winckelmans G and Larsen T J 2015 *Journal of Physics: Conference Series* vol 625 (IOP Publishing) p 012019
[22] Mann J 1998 *Probabilistic Engineering Mechanics* **13** 269–282
[23] Davidson L 2007 *5th International Symposium on Turbulence and Shear Flow Phenomena, Munich, Germany* vol 2 pp 689–694
[24] Scheurich F, Fletcher T M and Brown R E 2010 *48th AIAA Aerospace Sciences Meeting Including the New Horizons Forum and Aerospace Exposition* (AIAA) pp AIAA 2010–1579
[25] Duponcheel M, Winckelmans G and Chatelain P 2016 *ECCOMAS Congress* (European Community on Computational Methods in Applied Sciences)

IOP Conf. Series: Journal of Physics: Conf. Series **854** (2017) 012012 doi:10.1088/1742-6596/854/1/012012

The Long distance wake behind Horns Rev I studied using large eddy simulations and a wind turbine parameterization in WRF

O Eriksson[1], M Baltscheffsky[2], S-P Breton[1], S Söderberg[2], S Ivanell[1]

[1]Uppsala University, Department of Earth Sciences, Wind Energy Campus Gotland, 621 67 Visby, Sweden

[2]WeatherTech Scandinavia AB, Uppsala Science Park, 751 83 Uppsala

E-mail: ola.eriksson@geo.uu.se

Abstract.

The aim of the present paper is to obtain a better understanding of long distance wakes generated by wind farms as a first step towards a better understanding of farm to farm interaction. The Horns Rev I (HR) wind farm is considered for this purpose, where comparisons are performed between microscale Large Eddy Simulations (LES) using an Actuator Disc model (ACD), mesoscale simulations in the Weather Research and Forecasting Model (WRF) using a wind turbine parameterization, production data as well as wind measurements in the wind farm wake. The LES is manually set up according to the wind conditions obtained from the mesoscale simulation as a first step towards a meso/microscale coupling.

The LES using an ACD are performed in the EllipSys3D code. A forced boundary layer (FBL) approach is used to introduce the desired wind shear and the atmospheric turbulence field from the Mann model. The WRF uses a wind turbine parameterization based on momentum sink. To make comparisons with the LESs and the site data possible an idealized setup of WRF is used in this study.

The case studied here considers a westerly wind direction sector (at hub height) of 270 $\pm$ 2.5 degrees and a wind speed of 8 $\pm$ 0.5 m/s. For both the simulations and the site data a neutral atmosphere is considered. The simulation results for the relative production as well as the wind speed 2 km and 6 km downstream from the wind farm are compared to site data. Further comparisons between LES and WRF are also performed regarding the wake recovery and expansion.

The results are also compared to an earlier study of HR using LES as well as an earlier comparison of LES and WRF. Overall the results in this study show a better agreement between LES and WRF as well as better agreement between simulations and site data.

The procedure of using the profile from WRF as inlet to LES can be seen as a simplified coupling of the models that could be developed further to combine the methods for cases of farm to farm interaction.

1. Introduction

As the number of wind farms offshore increases there will be more occasions where offshore wind farms will be situated relatively closely to each other due to the limited number of suitable sites and the desire to use the best sites first. Since the flow behind a wind farm is disturbed the question of how this might impact other wind farms, so-called farm to farm interaction, is therefore becoming more central. It is also worth noting that compared to onshore wakes the long distance wakes offshore are more persistent since the lower roughness and turbulence levels mean a slower recovery of the velocity behind the wind farm. An understanding of not only the wake inside wind farms but also the impact of farm to farm interaction due to the long distance wakes behind wind farms is needed to be able to perform accurate estimations of production for offshore wind farms in wind farm clusters.

The long distance wakes behind offshore wind farms can be in the order of 10 km [1] and can be studied with different wake and mesoscale models [2][3][4]. Large Eddy Simulations (LES) have been used in many studies of the near and far wake behind turbines and the wake interaction inside wind farms [5][6][7][8]. Compared to the mentioned models LES is much more computationally demanding but as computational power increases LES gradually becomes an alternative also for farm to farm interaction studies. The author has earlier performed studies of the long distance wakes behind a row of turbines and wind farms [9].

Mesoscale models are used for atmospheric simulations and generally include more atmospheric physics. Compared to LES the resolution in the grids used by mesoscale models are lower which allows for less demanding simulations. Wind turbines can be included using a parameterization [10] but the wake flow will include fewer details compared to LES. To combine LES and mesoscale models can be of interest and allows for the advantages of both models to be used. A first alternative in combining the two is to use a mesoscale simulation as an input to an LES (1-way nesting). A second alternative is to also feed back the LES into the mesoscale domain (2-way nesting) [11].

The aim of the current study is to obtain a better understanding of wakes generated by entire wind farms in order to improve the understanding of farm to farm interactions. In the study both the mesoscale model Weather Research and Forecasting Model (WRF) and LES using an actuator disc model (ACD) are used to study the same case.

The LES are set up to have a wind shear and turbulence level similar to those used in the simulation in WRF. The results are compared between the models and site data for the Horns Rev I wind farm (HR) for both energy production and recovery of the velocity behind the wind farm. For the simulations the wake expansion and the development of the atmospheric boundary layer (ABL) are further compared.

The current study is a continuation from earlier studies of the long distance wakes behind the wind farms of Horns Rev I [12] and Lillgrund [13]. The main novelties are that a) the whole wind farm is included in the LES (compared to the previous study of HR where only two rows were included with periodic boundary conditions at the lateral boundaries creating an infinitely wide wind farm) b) the wind veer is included in the LES (compared to a wind shear only in the streamwise direction in the earlier studies) c) compared to the former Lillgrund study the WRF simulations are performed with a higher resolution and an updated parameterization as in [14].

2. Study of wakes in and behind Horns Rev I wind farm

The simulations are performed for Horns Rev I, a wind farm situated in the North Sea west of Denmark. The 80 Vestas V80 wind turbines with a hub height of 70 m, a radius (R) of 40 m and a rated power of 2 MW are laid out in 8 rows and 10 columns (turned 7 degrees (deg) from North (N)-South (S)) with an internal spacing of 14 R, see Figure 1. In the figure the placement of the two met towers at 2 km respectively 6 km east of the wind farm can also be seen and the profiles along which the simulation results will be compared are marked with lines or an x. North (N) is marked with an arrow in the figure. A flow from the north corresponds to a 0 deg inflow angle.

The studied case considers a (at hub height) westerly wind direction sector of 270 ± 2.5 degrees (deg) with flow along the rows and a wind speed (U_0) of 8 ± 0.5 m/s. For the simulations and the site data a neutral atmosphere is considered. The vertical profile of the wind velocities (w) for streamwise direction (z) and (u) for the spanwise direction (x) can be seen in Figure 2

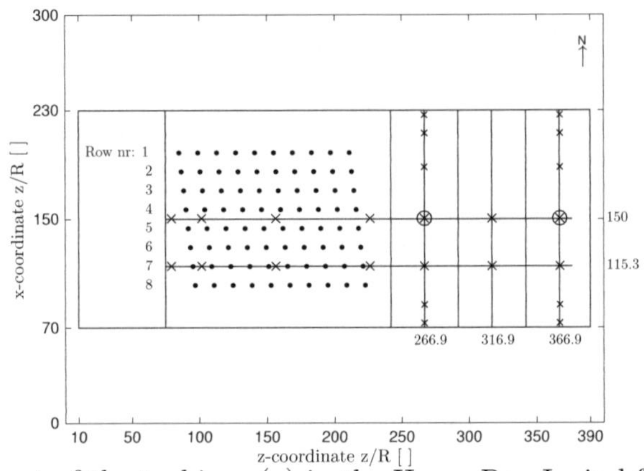

Figure 1. The placement of the turbines (•) in the Horns Rev I wind farm and the met towers (∘) in the LES domain covering 300 R ∗ 400 R with an equidistant region (marked with the rectangle) of 160 R ∗ 380 R. The flow is studied along the marked lines and for vertical profiles at the x's.

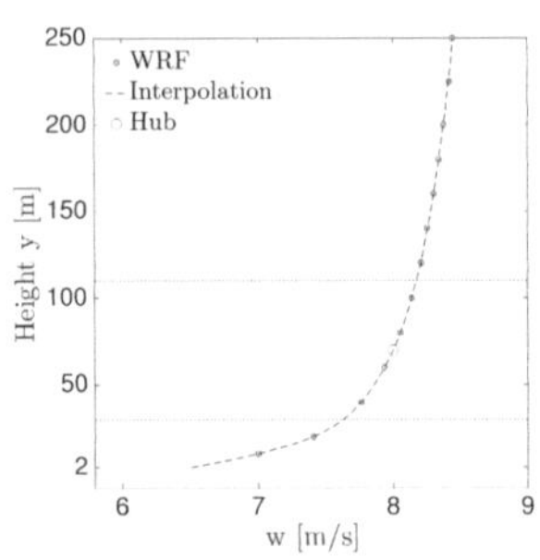

Figure 2. Wind shear with streamwise velocity (w) interpolated to fit the WRF-levels.

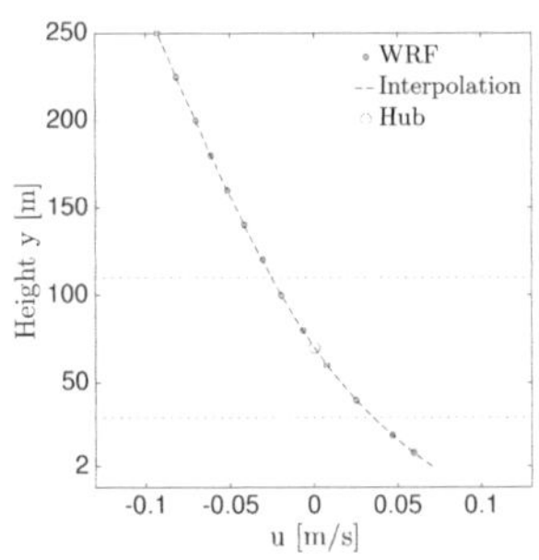

Figure 3. Wind shear with spanwise velocity (u) interpolated to fit the WRF-levels.

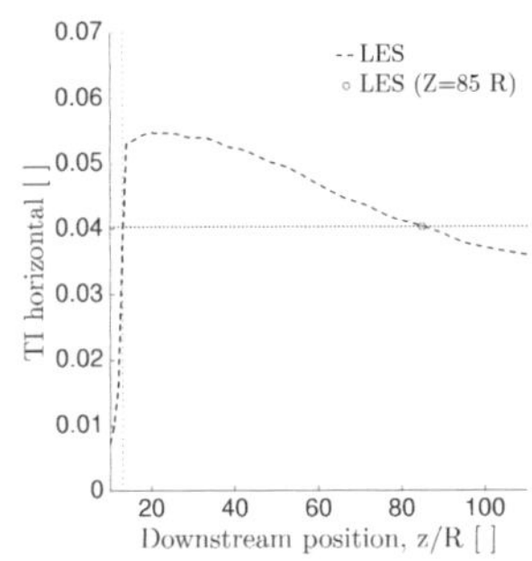

Figure 4. Downstream development of horizontal turbulence intensity (TI) at hub height in LES.

respectively in Figure 3. The vertical profile is based on the WRF simulation and is interpolated to be used in LES. From the two wind components the directional change with height can be calculated and over the rotor the wind veer in this case is 0.4 deg. At hub height the total turbulence intensity from WRF is 4.5 % and the horizontal turbulence for the LES is 4 % at the placement of the first turbine, see Figure 4. The simulation in LES using an actuator disc method uses the wind profile upstream of the wind farm from WRF as input. This is performed manually but has similarities to a 1-way nesting as information is transferred in one direction from WRF to LES (and no information is transferred back from LES to WRF).

The LES is further described in Section 3. The idealized WRF simulations using a wind farm parameterization are performed according to Section 4. Production data and met tower data are filtered to be comparable to the simulated wind direction sector and wind speed interval, see Section 5. The comparison between the simulations and the site data is presented in Section 6.

3. LES

The EllipSys3D code is used for the LES. The numerical model and the specific setup for the studied case is described below.

3.1. Numerical model LES
Large Eddy Simulations (LES) resolve the largest and most energetic eddies and only the smallest eddies are modelled, here using the subgrid-scale model from Ta-Phouc [15]. The turbines are introduced into the simulations using an actuator disc method based on airfoil data [16]. A generator-torque controller is used to adjust each turbine's rotational speed in the variable flow in a realistic way [17]. The incoming neutral atmospheric boundary layer is mimicked by using a forced boundary layer (FBL) that uses body forces [7] to introduce the wind shear as well as a realistic atmospheric turbulence from the Mann model [18][19].

3.2. Simulation setup LES
The simulations are performed in a domain of 637 M-cells with 300 R width and 400 R length with an inner equidistant region with the non dimensionalized (with R) resolution (Δx) of 0.1 (4 m) covering 160 R $*$ 380 R, see Figure 1. The height of the equidistant region is 8 R and the total boundary layer height is 25 R (1000 m). The turbulence that is introduced at 17 R is ideally allowed to develop with the flow in the LES domain until the turbulence levels out at a level dictated by the wind shear before the first turbine is introduced. Here the first turbine is at 85 R, see Figure 4, as a compromise considering the needed computational power. The domain has fixed values for the inlet according to the used wind shear, cyclic boundary conditions for the sides, convective for the outlet and farfield for the top. The ground uses farfield that is setup as a noslip condition. A non dimensionalized (with U_0 and R) timestep (Δt) of 0.05 is used to conservatively fullfill the Courant-Friedrichs-Lewy (CFL)-condition (Equation 1) based on the non dimensionalized (with U_0) wind speed at hubheight ($U_{0_{norm}}$):

$$\frac{U_{0_{norm}} * \Delta t}{\Delta x} = \frac{1 * 0.05}{0.1} = 0.5 < 1 \tag{1}$$

The simulations are first run for 12000 time steps (ts) to allow the turbine to adjust to the flow and the flow to pass the domain with some margin. After that the analysed averaged values are studied for 12000 ts which corresponds to 50 min in physical time.

The wind shear used is interpolated from the coarser vertical grid levels used in WRF, see Figures 2 and 3. The used Mann turbulence is created for the roughness length (z_0) of 0.0001 m (the same as the surface roughness in WRF) and a wind speed of 8 m/s at 70 m. The Mann box covers the equidistant region with a resolution of 1.56 Δx /1.56 Δx in width/height. The length is 10 min using the Taylor frozen hypothesis and the length resolution gives a non dimensionalized time step of 0.2344.

The airfoil data representing the turbine in the ACD is based on the NREL 5 MW turbine [20] and is downscaled to correspond to the Vestas 80 regarding power and thrust coefficient [8].

The simulations are performed for three inflow directions 270 deg, 267.5 deg and 272.5 deg. For the cases $\pm$ 2.5 deg the inlet wind shear and the fluctuations in the Mann box are turned (for the Taylor frozen hypothesis the flow along 270 deg is, however, still assumed). The turbines are also yawed to face the mean wind direction in each case.

4. Mesoscale simulations
The Weather Research and Forecasting Model (WRF) is used for the mesoscale simulations. The numerical model and the specific setup for the studied case is described below.

4.1. Numerical model WRF
WRF is an atmospheric model system developed by the National Center for Atmospheric Research (NCAR) and maintained as a community model by the Mesoscale and Microscale Meteorology Laboratory (MMM) of NCAR. A detailed description of WRF can be found in [21]. WRF v3.5.0, which was used here, includes a wind farm parameterization where wind turbines are represented as an elevated momentum sink and a source of turbulent kinetic energy (TKE) [10],[22]. The source of TKE compensates for any wind turbine induced turbulence that is not

accounted for by the sharper shear in the wake. Here a slightly modified parameterization with no source of TKE was used. Let us note that the used parametrization was orginally developed to study the impact of wind farms on the flow. For this reason the interaction between individual turbines inside the farm is expected to be less accurate.

4.2. Simulation setup WRF
WRF was setup for idealized simulations with 2 one-way nested domains over a sea surface. The parent domain was set up with dx = dy = 333 m model grid resolution and periodic boundary conditions. In the nested domain with dx = dy = 111 m boundary information was fed from the parent domain. The simulations were initialized with a uniform wind profile of w = 8.427 m/s and u = -0.462 m/s and uniform potential temperature below 1200 meters, increasing by 3 K/km above. The initial profile was chosen so that the wind speed and wind direction at hub height should match observed conditions after model spin-up. The coriolis parameter was set according to the location of the wind farm. The sea surface has a roughness length of $z_0 = 0.0001$.

5. Site data
Site data for the production and wind measurements at 2 km respectively 6 km east of the wind farm are used, see Figure 1 for the met tower positions. The data is filtered according to Hansen [23] to correspond to the studied case. The filtering includes suitable 10 min averages with correct wind direction and wind speed. The filtered data includes only periods with neutral conditions. It also excludes non suitable periods, for example when the turbines were not operational and during large changes of the weather conditions. The turbulence level according to site data is around 7 % in HR [24] for the studied case. The same site data was also used in the earlier study of HR assuming an infinitely wide wind farm [12].

6. Results
A comparison is performed between WRF, LES and available site data with an emphasis on increasing understanding of the modelling of flows behind wind farms. To highlight the level of improvements some results from the earlier study [12] are also included. For LES the mean of the results for the three directions (270 ± 2.5 deg) is presented (if nothing else is specified) and for WRF the results of 270 deg are presented (with an included spread of around ± 2 deg for the different time periods included in the simulation result). The presented data is non dimensionalized (with U_0 = 8 m/s and R = 40m). The results for the long distance wake are however also presented for the distance behind the wind farm in km.

Firstly the relative production (Section 6.1) of the turbines in the wind farm is studied. Secondly the development is studied for the normalized horizontal wind speed (U_{hor}) and the standard deviation of all velocity components (σ_{tot}) which describe the turbulence level (and divided with the local mean velocity correspond to the turbulence intensity). As illustrated in Figure 1 the downstream development of the flow is studied along the met towers and row 7 at hub height (Section 6.2). The vertical profile of the downstream development of the flow is also examined (Section 6.3). The downstream spanwise expansion of the wake is studied at hub height and for vertical profiles in Section 6.4.

6.1. Production
The relative production is the production normalized with the production of the first turbine in the row. In Figure 5 the relative production based on the mean value of the production for the turbines in row 2-7 is seen. Compared to the farm data the LES results show a lower production in the first portion and in contrast to the site data the level is relatively constant throughout the farm and ends up at the same level in the end of the rows. WRF shows a more gradual

decrease of relative production with down stream distance, especially for the first part of the wind farm, and goes towards a production lower than both LES and farm data.

The sensitivity to the correct inflow angle for LES is seen in Figure 6. A 2.5 deg change from the full wake direction (270 deg) gives a higher relative production. Due to the layout the direction of the change also has an impact.

Uncertainties in the actual direction of the site data gives a wider range of directions compared to the performed simulations and could explain the behaviour of the higher relative production of the site data compared to the simulations. The turbulence level in the simulations is also lower compared to the site data which has more impact on the mixing in the first part of the wind farm. The lower relative production in WRF could indicate that the recovery of the flow between the turbine is too slow which also is discussed further in next section.

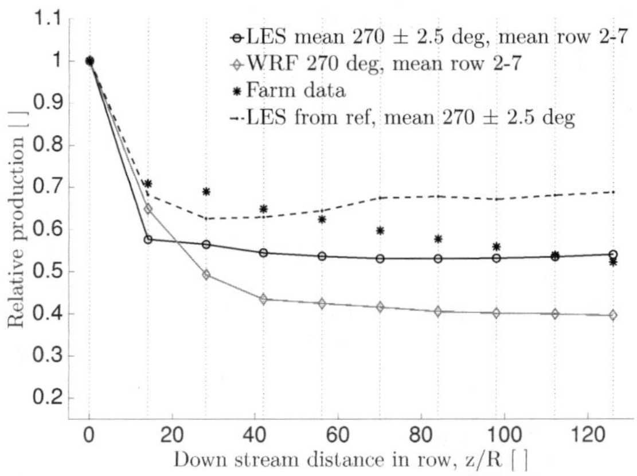

Figure 5. Mean of the relative production for the turbines in row 2-7 for LES, WRF and wind farm data. The older study [12] using LES with periodic boundary conditions is included for comparison.

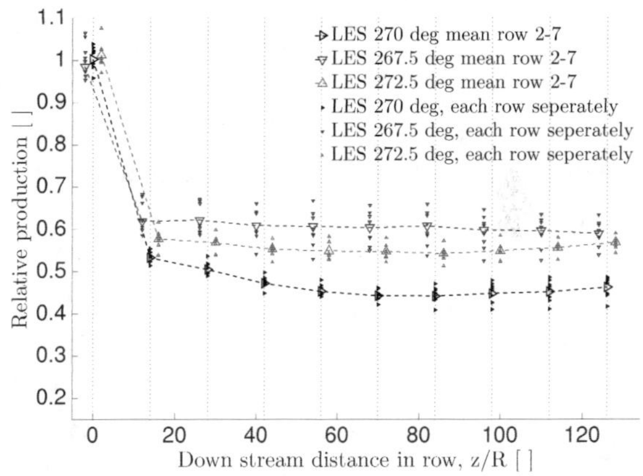

Figure 6. Dependency of inflow angle (270 ± 2.5 deg) on the relative production in LES. The relative production for each row shown separately as individual data points around each dotted line representing the average over the rows.

6.2. Flow recovery

The flow at hub height along the met towers respectively row 7 is here presented regarding horizontal velocity and turbulence level.

For the velocity in Figure 7 the general development down stream is similar in LES and WRF. As expected due to the more distinct wake in LES compared to WRF more fluctuations of the wind speed are seen for the LES along row 7 and a smaller decrease of velocity along the met tower inside the farm is seen.

The figure also shows the three different directions used in LES separately for the flow along the met towers. The velocity for the full wake direction (270 deg) in LES is relatively high at 2 km behind the wind farm possibly because the wakes of the rows have not expanded fully at the met tower position. Deviations from the full wake direction (270 deg) gives a decrease in the velocity along the met towers downstream in and behind the wind farm. The values for ± 2.5 deg have a trend closer to the met tower data.

In Table 1 the velocities at 2 km and 6 km behind the wind farm are seen. The WRF results show a faster recovery of the velocity compared to LES at 6 km. For the LES the results show an improvement when compared to the earlier study which had more than double as large overestimation of the velocities compare the new results.

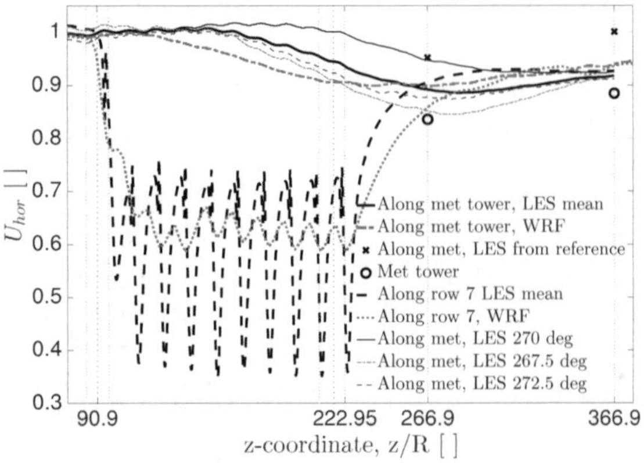

Figure 7. Normalized horizontal velocity (U_{hor}) in LES and WRF at hub height along the met towers and row 7. Site data and the older study [12] using LES with periodic boundary conditions are included for comparison.

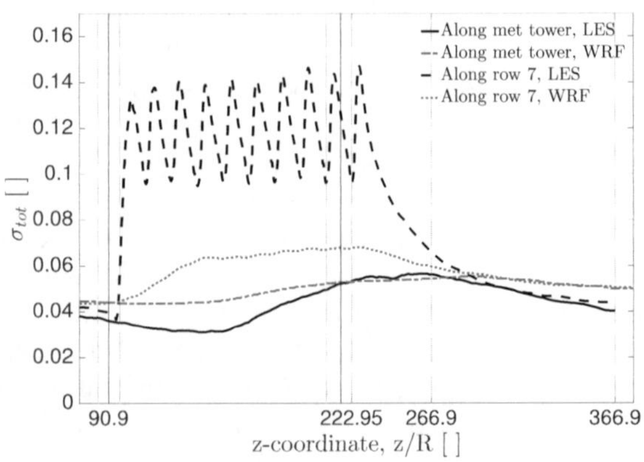

Figure 8. Total standard deviation of the velocity (σ_{tot}) in LES and WRF at hub height along the met towers and row 7.

Table 1. Normalized horizontal velocity (U_{hor}) at 2 km respectively 6 km behind the wind farm for LES, WRF, site data and results from the older study [12] using LES with periodic boundary conditions. In brackets the difference in percentage compared to the site data is shown.

Distance	LES	WRF	Site data	LES ref [12]
2 km	0.89 (+5.9%)	0.9 (+7.1%)	**0.84**	0.95 (+13.1%)
6 km	0.92 (+3.7%)	0.94 (+5.6%)	**0.89**	1.0 (+12.6%)

Along row 7 U_{hor} in LES the velocity is higher at 2 km behind the wind farm compared to the velocity along the met towers. This could be due to increased vertical mixing due to the higher turbulence level along this row which can be seen in Figure 8. The less distinct wake in WRF (due to the used resolution) can also be seen in the slightly higher levels of turbulence along the met towers. Along row 7 the turbulence level is lower in WRF compared to LES due to the used parameterization that does not add any extra turbine induced turbulence. The slightly slower recovery of the flow between the turbines can be a consequence of this.

6.3. Boundary layer

The development of the atmospheric boundary layer over and after the wind farm is studied along the met towers and row 7 for horizontal velocity and turbulence level. Profiles of horizontal velocity (Figure 9) and turbulence level (Figure 10) are studied for a vertical profile upstream from the farm, for three profiles in the wind farm as well as profiles at 2, 4 and 6 km behind it.

Looking first along the met towers it can be seen for the velocity before the wind farm that LES has higher velocities closest to the ground when compared to the inlet but at higher heights the profile is unchanged. Inside the wind farm a faster decrease can be seen using WRF compared to LES. However at heights above the farm the LES shows a larger reduction of the wind speed. WRF has higher turbulence level compared to LES with the exception of the last portion of the farm and WRF's vertical extension of increased turbulence is higher. After the wind farm in LES the velocity continues to decrease until 2 km before a recovery of the flow can be seen at the turbine height, however the impact above the turbine height in the first part of

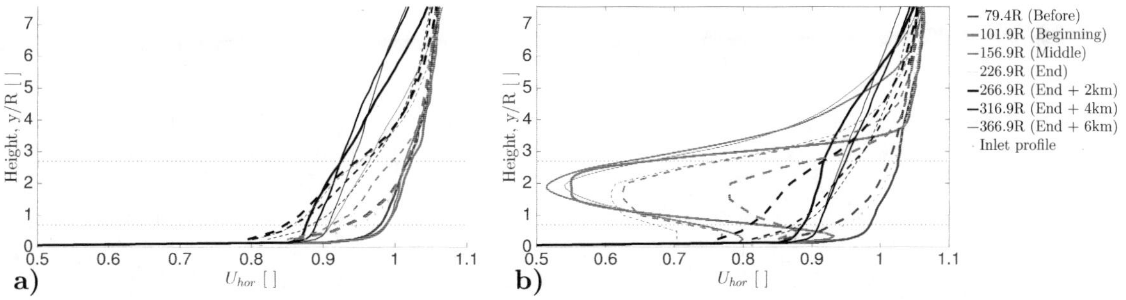

Figure 9. Normalized horizontal velocity (U_{hor}) in LES (solid lines) respectively WRF (dashed lines) for different downstream distances in the wind farm for: **a)** along the met towers **b)** along row 7.

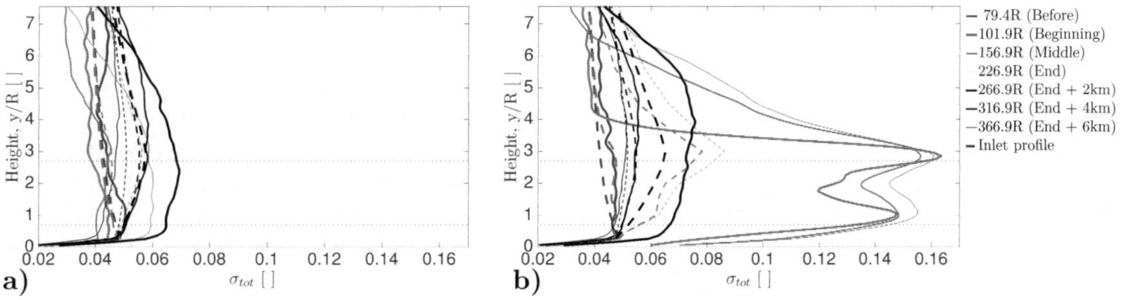

Figure 10. Total standard deviation of the velocity (σ_{tot}) in LES (solid lines) respectively WRF (dashed lines) for different downstream distances in the wind farm for: **a)** along the met towers **b)** along row 7.

the farm wake still increases. For WRF the velocity recovers slightly faster at the turbine height while at higher height no development is seen.

Along row 7 the more distinct wake in LES gives a larger reduction of the velocity inside the farm compared to WRF. At heights above the farm a larger reduction can be seen in LES as also was the case along the met towers. The level of turbulence is higher in LES inside the farm and in the first part of the wind farm wake while further down stream the level is similar in WRF. After the wind farm a recovery of the velocity in LES can be seen between 2 km and 4 km while the values at 6 km behind the wind farm are about the same level as at 4 km. A similar trend is seen for the turbulence but here the level decreases slightly after 4 km. For the turbulence it can also be seen that the level at higher heights decreases more slowly. For WRF the velocity reduction is higher at 2 km compared to LES. A faster recovery can be seen until 4 km after which a slower recovery can still be seen.

The vertical velocity profile has overall a sharper shear over the turbine height in the farm wake in the WRF results which could explain the slightly faster recovery of the velocity behind the wind farm.

6.4. Wake expansion

The expansion of the wake is studied for spanwise profiles at hub heigt for different downstream distances. The horizontal velocity is presented in Figure 11 and the turbulence level in Figure 12.

Looking 1 km behind the wind farm in LES the rows and the different distances from the last turbine in each row (due to the tilt of the layout) can be clearly seen. At 2 km the increased turbulence along the rows is still seen. For the velocities at 2km the reduction between the rows

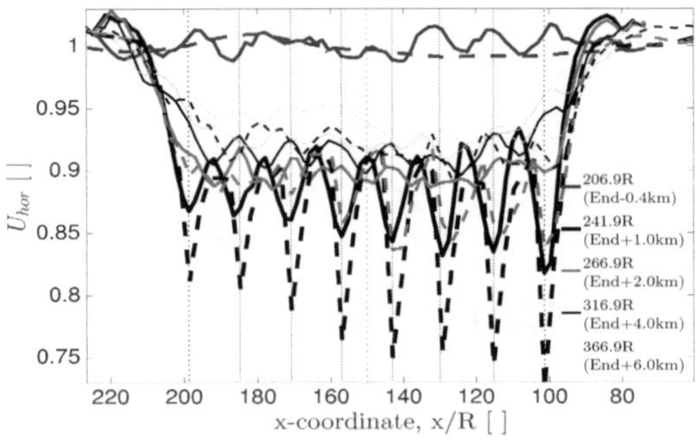

Figure 11. Normalized horizontal velocity (U_{hor}) in LES (solid lines) and WRF (dashed lines) along spanwise profiles at hub height for different downstream distances behind the wind farm.

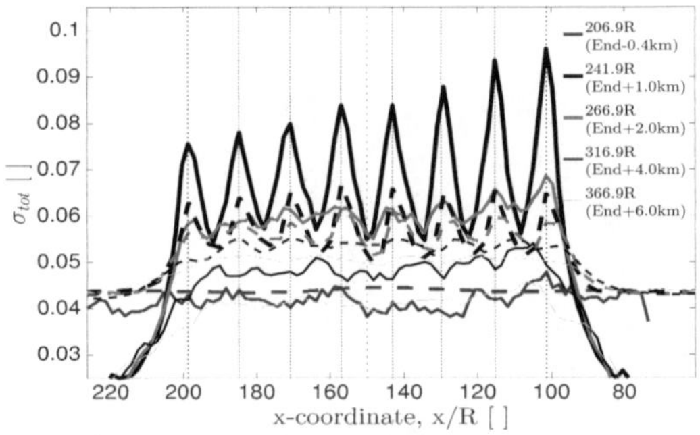

Figure 12. Total standard deviation of the velocity (σ_{tot}) in LES (solid lines) and WRF (dashed lines) along spanwise profiles at hub height for different downstream distances behind the wind farm.

is actually higher compared to along the rows, which could be due to the lower turbulence level. For 6 km the impact of the rows is hard to see. The expansion of the wake is about the same at 1 km and 2 km after which it increases downstream for 4 km and 6 km. Looking at the flow outside the farm wake an increase of the velocity can be seen in the first part of the farm wake. The turbulent fluctuations have at 6 km reached the same level as in front of the wind farm, however outside the farm the turbulence level is lower. The WRF results show that in comparison to LES along the rows of turbines there is a slower recovery for the first part of the farm wake while further down a slightly faster recovery can be seen in WRF compared to LES. The turbulence level starts at a lower level and decreases more slowly in the farm wake in WRF.

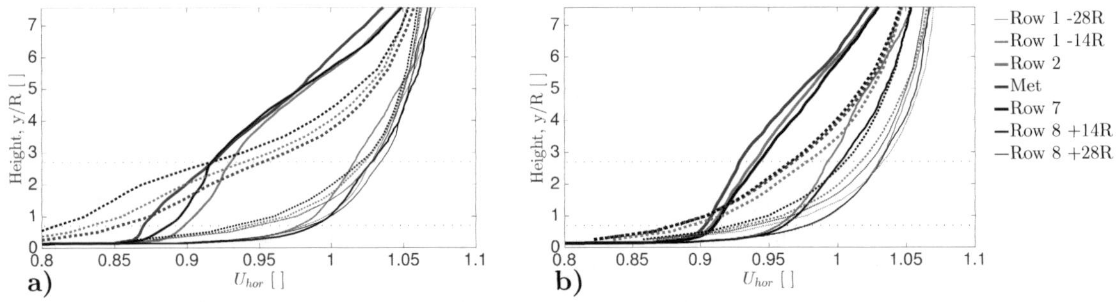

Figure 13. Normalized horizontal velocity (U_{hor}) in LES (solid lines) respectively WRF (dashed lines) for different spanwise positions in the wind farm for: **a)** 2 km behind the wind farm **b)** 6 km behind the wind farm.

The vertical profiles along a spanwise line at 2 km respectively 6 km behind the wind farm give further input to the wake expansion. In Figure 13 the horizontal velocity and in Figure 14 the vertical profiles of the turbulence level are seen. For LES 2 km behind the wind farm the wind speed is higher and the turbulence is lower along row 2 compared to row 7 as expected due to the distance to the last turbine. The velocity outside row 1 is slightly lower compared to outside row 8. The wake expansion seen at 6 km due to the differences between the profiles at 14 R respectively 28 R outside the outer rows should also be noted. The profile central along the met towers show at both 2 and 6 km a lower velocity compared to those along the rows also for higher heights above the wind farm. In comparison to LES the WRF results show at 2 km a higher velocity along the met towers in comparison to along the rows. It should be noted that

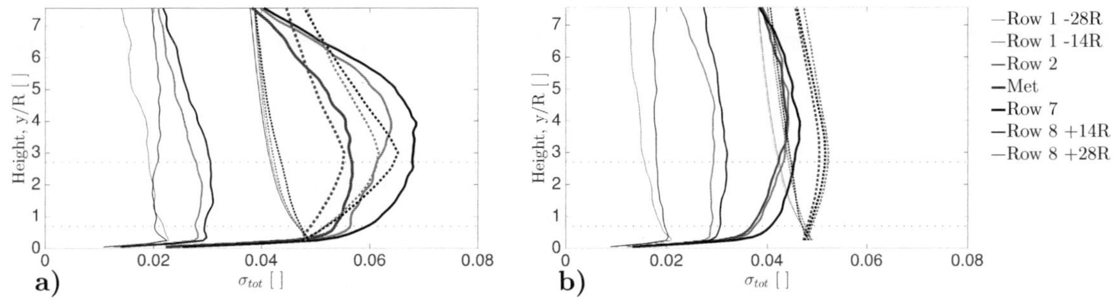

Figure 14. Total standard deviation of the velocity (σ_{tot}) in LES (solid lines)respectively WRF (dashed lines) for different spanwise positions in the wind farm for: **a)** 2 km behind the wind farm **b)** 6 km behind the wind farm.

there are also larger differences between the +14R and +28 R in LES compared to WRF.

7. Conclusion

To get an increased understanding of the modelling of long distance wakes behind whole wind farms the Horns Rev I (HR) wind farm has been considered for comparisons between microscale Large Eddy Simulations (LES) using an Actuator Disc model (ACD), mesoscale simulations (WRF) using a wind turbine parameterization, site data for production as well as wind measurements in the farm wake. The LES is manually set up according to the wind conditions obtained from the mesoscale simulation as a first step towards a meso/microscale coupling.

The relative production in LES is lower for the first portion of the farm compared to the wind farm data but by the end of the wind farm the level is about the same. The values for WRF show, like the site data, a more gradual decrease with down stream distance but the level of relative production is still underestimated. Inside the wind farm the horizontal velocity is in relatively close agreement between LES and WRF, but the more smoothed wake due to the lower resolution in WRF can still be noticed and WRF has a slightly lower velocity. The turbulence level is lower inside the wind farm in WRF with the used parameterization that adds no extra turbine induced turbulence. The lower turbulence could explain the slower recovery of velocity between the turbines and the resulting decreasing relative production. It should be noted that the turbulence level is distributed to greater heights in WRF and in LES a reduction in wind speed can be seen at greater heights. Behind the wind farm LES shows relatively good agreement with the met tower data for velocity, especially at 6 km. The long distance wake in WRF also shows relatively good agreement but a slightly faster recovery can be seen which could be related to a higher shear with resulting downward momentum transfer from higher heights.

The current study can be seen as a continuation from earlier studies of the long distance wakes behind wind farms of Horns Rev I [12] respectively Lillgrund [13].

In comparison to the earlier study of Horns Rev which assumed an infinitely wide farm [12] the down stream development of the relative production is closer to the site data. The velocity level in the farm wake is also closer to met data in the current study. The lower velocities in this study could partly be explained by the use of a less sharp shear which results in less turbulence downstream in the domain. The larger distance between the turbulence plane and the first turbine (letting the turbulence adjust to the LES) and lower blockage due to a larger domain cross section could also be part of the improvements. Given the other differences in the setup it is not possible to determine the relative impact of the increased realism in the simulation given by including the wind veer and the use of the real layout of the wind farm.

In comparison to the earlier study comparing LES and WRF [13] both the turbulence level and the velocity is here much closer between LES and WRF. The earlier study with WRF showed a much lower reduction of the velocity inside the wind farm and much higher added turbulence. The improved setup of WRF is the result of the increased resolution and the updated wind turbine parameterization which add less turbine induced turbulence that in the earlier study was found to be to large [14]. The turbulence is however lower than the LES in this study indicating that part of the turbine induced turbulence still needs to be added in the parameterization.

It can also be noted that there are a number of uncertainties in comparison between the site data and the simulations. Direction uncertainty is one of the difficulties in comparing measurements with simulations [25]. A change in the direction impacts not only the direction of the long distance wake but also the level and results of the relative production due to the resulting changes in wake interaction [24][26]. Another parameter of potential uncertainty is that the site data also might include data close to neutral on the unstable or stable side and this stability has an impact on the atmospheric boundary layer and the downstream development of the wake [27].

Overall an improvement could be seen comparing the LES results with the earlier LES results for the same case. The impact of the wind veer that was included in the LES to get as close as possible to the profile used in WRF (compared to a wind shear only in the streamwise direction in the earlier studies) and the change from an infinitely wide wind farm to the full layout could not be quantified. It would be of interest in a future study to more systematically quantify the relative importance of these parameters. Overall the results show a better agreement between LES and WRF compared to the earlier study. The results are a step towards bettering the study of long distance wakes by combining LES and WRF.

Acknowledgments
The simulations (LES) were performed on resources provided by the Swedish National Infrastructure for Computing (SNIC) at the National Supercomputer Centre in Sweden (NSC).

References
[1] Eriksson O and Ivanell S 2012 *A survey of available data and studies of farm-farm interaction.* (EAWE PhD Seminar on Wind Energy in Europe, 2012)
[2] Frandsen S, Barthelmie R, Rathmann O, Jørgensen H, Badger J, Hansen K, Ott S, Rethore P E, Larsen S and Jensen L 2007 *Summary report: The shadow effect of large wind farms: measurements,data analysis and modelling.* (Risø-R-1615(EN), Denmark)
[3] Brand A 2009 *Wind Power Plant North Sea – Wind farm interaction, The effect of wind farming on mesoscale flow.* (ECN, the Neatherlands)
[4] Hansen K S, Réthoré P E, Palma J, Hevia B G, Prospathopoulos J, Peña A, Ott S, Schepers G, Palomares A, van der Laan M P and Volker P 2015 *Simulation of wake effects between two wind farms* (Journal of Physics: Conference Series 625 012008)
[5] Ivanell S 2009 *Numerical Computations of Wind Turbine Wakes.* (PhD thesis, ISBN 978-91-7415-216, KTH Engineering Sciences, Sweden)
[6] Wu Y T and Porté-Agel F 2012 *Atmospheric turbulence effects on wind-turbine wakes: An LES study.* (Energies 2012, 5, 5340-5362)
[7] Troldborg N, Sørensen J N, Mikkelsen R and Sørensen N N 2014 *A simple atmospheric boundary layer model applied to large eddy simulations of wind turbine wakes.* (Wind Energy 2014; 17: 657-669.)
[8] Nilsson K, Ivanell S, Hansen K S, Mikkelsen R, Breton S P and Henningson D 2015 *Large-eddy simulations of the Lillgrund wind farm.* (Wind Energy 18, 449–467)
[9] Eriksson O 2015 *Numerical computations of wakes behind wind farms.* (Department of Earth Science Licentiate Thesis 2015, Uppsala University)
[10] Fitch A C, Olson J B, Lundquist J K, Dudhia J, Gupta A K, Michalakes J and Barstad I 2012 *Local and Mesoscale Impacts of Wind Farms as Parameterized in a Mesoscale NWP Model.* (Monthly Weather Review, Volume 140, Issue 9 .)

[11] Lundquist J K, Mirocha J D and Kosovic B 2010 *Nesting large-eddy simulations within mesoscale simulations in WRF for wind energy applications.* (Proceedings of the Fifth International Symposium on Computational Wind Engineering, Chapel Hill, NC)
[12] Eriksson O, Mikkelsen R, Nilsson K and Ivanell S 2012 *Analysis of long distance wakes of Horns rev 1 using actuator disc approach.* (J. Phys.: Conf. Ser. 555 012032)
[13] Eriksson O, Lindvall J, Breton S P and Ivanell S 2015 *Wake downstream of the Lillgrund wind farm - A Comparison between LES using the actuator disc method and a Wind farm Parametrization in WRF.* (J. Phys.: Conf. Ser. 625 012028)
[14] Baltscheffsky M, Söderberg S, Bergström H and Svensson N 2013 *Using a mesoscale atmospheric model to study wind turbine wakes and farm-farm interaction.* (EAWE Offshore 2013)
[15] TaPhuoc L 1994 *Modéles de sous maille appliqués aux écoulmements instationnaires décollés.* (Tech. rep. LIMSI 93074 LIMSI France.)
[16] Mikkelsen R 2003 *Actuator Disc Methods Applied to Wind Turbines.* (DTU, Denmark)
[17] Breton S P, Nilsson K, Ivanell S, Olivares-Espinosa H, Masson C and Dufresne L 2012 *Study of the effect of the presence of downstream turbines on upstream ones and use of a controller in CFD wind turbine simulation models.* (J. Phys.: Conf. Ser. 555 012014)
[18] Mann J 1998 *Wind field simulation.* (Risø, Denmark)
[19] Jakob M, Ott S, Jørgensen B H and Frank H P 2013 *WAsP Engineering 2000.* (Risø–R–1356(EN))
[20] Jonkman J, Butterfield S, Musial W and Scott G 2009 *Wind Turbine for Offshore System Development.* (NREL, the United states of America)
[21] Skamarock W C, Klemp J B, Dudhia J, Gill D O, Barker D M, Duda M G, Huang X, W W and Powers J G 2008 *A Description of the Advanced Research WRF Version 3.* (Technical Note NCAR/TN-475+STR)
[22] Fitch A C, Olson J B, Lundquist J K, Dudhia J, Gupta A K, Michalakes J, Barstad I and Archer C L 2013 *Corrigendum.* (Monthly Weather Review, 141, 1395–1395.)
[23] Hansen K S 2011 *Guideline to wind farm wake analysis, Guideline to wind farm wake analysis.* (Department of Mechanical Engineering, DTU, Denmark.)
[24] Hansen K S, Barthelmie R, Cabezon D and Politis E 2008 *Wp8: Flow Deliverable D8.1 Data Wake measurements used in the model evaluation .* (DTU.)
[25] Gaumond M, Réthoré P E, Ott S, Peña A, Bechmann A and Hansen K S 2013 *Evaluation of the wind direction uncertainty and its impact on wake modeling at the Horns Rev offshore wind farm.* (Wind Energy doi: 10.1002/we.1625)
[26] Ivanell S, Mikkelsen R, Sorensen J N, Hansen K S and Henningson D 2009 *The impact of wind direction in atmospheric BL on interacting wakes at Horns Rev wind farm.* (PhD thesis, ISBN 978-91-7415-216, KTH Engineering Sciences, Sweden)
[27] Hansen K S, Barthelmie R J, Jensen L E and Sommer A 2012 *The impact of turbulence intensity and atmospheric stability on power deficits due to wind turbine wakes at Horns Rev wind farm.* (Wind Energ., 15: 183–196.)

IOP Conf. Series: Journal of Physics: Conf. Series **854** (2017) 012013 doi:10.1088/1742-6596/854/1/012013

Full-Scale Field Test of Wake Steering

Paul Fleming[1], Jennifer Annoni[1], Andrew Scholbrock[1], Eliot Quon[1], Scott Dana[1], Scott Schreck[1], Steffen Raach[2], Florian Haizmann[2], David Schlipf[2]

[1]National Wind Technology Center, National Renewable Energy Laboratory, Golden, CO, 80401, USA
[2]University of Stuttgart, Stuttgart Wind Energy, Stuttgart, Germany

E-mail: paul.fleming@nrel.gov

Abstract. Wind farm control, in which turbine controllers are coordinated to improve farmwide performance, is an active field of research. One form of wind farm control is wake steering, in which a turbine is yawed to the inflow to redirect its wake away from downstream turbines. Wake steering has been studied in depth in simulations as well as in wind tunnels and scaled test facilities. This work performs a field test of wake steering on a full-scale turbine. In the campaign, the yaw controller of the turbine has been set to track different yaw misalignment set points while a nacelle-mounted lidar scans the wake at several ranges downwind. The lidar measurements are combined with turbine data, as well as measurements of the inflow made by a highly instrumented meteorological mast. These measurements are then compared to the predictions of a wind farm control-oriented model of wakes.

1. Introduction

Wind farm (or plant) control is a field of research in which the controller of individual wind turbines located within a farm are coordinated. The objective of this coordination is to improve global power production by accounting for wake interactions, or similarly improved loads, or the provision of grid frequency support services such as inertia response or active power control. We refer to [1] for an in-depth overview of the current activities and research questions in wind farm control. Among the implementations of wind farm control gaining interest, yaw-based wake steering has shown promise as a method of improving wind plant power output by reducing wake losses. In this method, wakes are steered away from downwind turbines through deliberate yaw misalignment of upwind turbines [2].

Simulation studies using computational fluid dynamics (CFD) simulations of wind farms demonstrated this wake steering effect when turbines are yawed ([3, 4]), and that although the upwind turbine is expected to lose power by being yawed, the net power can go up based on the increased power of downwind turbines now out of the wake [5]. Later work expanded on these findings by using engineering models of wakes [6] coupled to system-engineering tools [7], revealing that a meaningful impact on a wind farm's annual energy production could be achieved [8].

There is a need for experimental field testing of wake steering to determine if the observed and predicted benefits of simulation can be realized. There have been tests conducted in wind tunnels that have thus far aligned with simulation predictions [9, 10]. Also, tests have been conducted at scaled wind farm facilities [2]. Finally, there are some preliminary results from test campaigns

IOP Conf. Series: Journal of Physics: Conf. Series **854** (2017) 012013 doi:10.1088/1742-6596/854/1/012013

at commercial wind farms [11]. However, a critical need is filled by a detailed full-scale test in which a utility-scale turbine operates at various yaw offsets while its wake is measured. In this paper, we describe such a campaign and how it can be used to validate simulation results.

In this paper, lidar measurement results from the field-test campaign are compared with the predictions of a control-oriented engineering model of wind turbine wakes. The model is not tuned to the field-test data, but was instead tuned beforehand to CFD data [6]. A first contribution of the paper is a validation of important aspects of engineering models of wake control. We note here that, at this time, this is a preliminary validation of the larger-scale effects, and that detailed validation using rigorous practices, such as uncertainty quantification, is the subject of future work.

Additionally, recent research using CFD has suggested that the wake shape and relationship to atmospheric stability ([12, 13]) can be critical in ensuring the correct implementation of wake steering. A second contribution of this paper is a first comparison of wake shape versus atmospheric conditions and yaw setting.

Finally, it has been proposed that some potential difficulties in wake steering, such as model uncertainties and disturbances, can be overcome by making use of a lidar system and feedback controller. The lidar system provides wake position information and a closed-loop wake redirection controller sets the yaw for the turbines [14]. A final contribution of this paper is to demonstrate the ability of a rear-facing, nacelle-mounted lidar (as would be used in wind farm control) to detect and characterize wakes for use in a control system.

This paper is structured as follows. Section 2 briefly describes the wake steering models for comparison. The test setup is described in Section 3. Section 4 describes data processing. Finally, Section 5 covers the results and discussion and conclusions are provided in Section 6.

2. Models of wind farm control

In the literature, there are many models of wind farms used to understand turbine wake interactions. One category of these models is known as "control-oriented modeling," and these models are meant to be computationally inexpensive, such that they can be used inside a controller or optimization routine. These models contain descriptions of wakes, including the effect of changes of turbine control and atmospheric conditions on the wake behavior. Examples of models like this include that of [15], [16], and [6]. The latter model, known as FLOw Redirection and Induction Steady State (FLORIS), was developed by the National Renewable Energy Laboratory (NREL) and the Delft University of Technology. This model will be used in this paper to provide insights into the lidar data in yawed and nonyawed conditions.

The FLORIS model is an augmentation of the Jensen model [17], with the model of wake steering provided by [3] and the division of the wake into zones. It predicts the average steady-state behavior of wakes, and accounts for the impact in changes in yaw as well as pitch and torque control. More recently, the model was updated to include the impact of certain atmospheric stability [13], and improvements to the model are still ongoing, based on field campaigns discussed here, as well as being informed by recent results from other institutes (for example, [12] and [15].)

The FLORIS model is tuned to match the simulation results performed in CFD simulations of wind farms, particularly NREL's Simulator fOr Wind Farm Applications (SOWFA) wind farm simulation tool [18]. Later, comparisons will be made between field-test data and the FLORIS model, wherein the FLORIS model has not been retuned to match the experimental data).

3. Experimental setup

In this field-test campaign, a utility-scale turbine is run with a yaw misalignment while a nacelle-mounted lidar continually scans the wake. This campaign began in September 2016 and is ongoing. In this section, we describe the turbine, nearby meteorological mast, and the lidar system employed. Finally, we describe the testing procedure.

IOP Conf. Series: Journal of Physics: Conf. Series **854** (2017) 012013 doi:10.1088/1742-6596/854/1/012013

Figure 1: The U.S. Department of Energy (DOE) 1.5 turbine at the National Wind Technology Center is shown with the University of Stuttgart lidar being installed. Photo by Dennis Schroeder, NREL 38271

3.1. General setup of field testing

The test turbine and meteorological (met) tower are located at the National Wind Technology Center in Boulder, Colorado. The site is located near and to the east of the Rocky Mountains, where flow from the mountains to the west is the predominant wind direction.

The turbine used in this test campaign is the DOE 1.5, a GE 1.5 SLE owned by the U.S. Department of Energy (DOE) and operated by NREL (shown in Figure 1). Details of the turbine are provided in Table 1.

Rated Power (kW)	1500
Hub Height (m)	80
Nominal Rotor Diameter (m)	77
Rated Wind Speed (m/s)	14

Table 1: Test Turbine Details.

The met tower is located 161 m in the predominant upwind direction of the turbine at a bearing of 276° relative to true north. It was instrumented in accordance with International Electrotechnical Commission 61400-12-1. Table 2 lists some of the met tower instrumentation by tower elevation. All booms are pointing in the 278° direction. In addition to the met tower, the turbine nacelle wind speed and wind direction are time-synchronously measured and recorded with the met tower instruments.

In this work, test data are limited to a cone of directions in which the met tower is relatively upwind with respect to the turbine. We use the hub-height wind speed and direction measurements to describe mean wind speed, direction, and turbulence intensity, whereas veer will be described using the 38 m and 87 m wind directions.

IOP Conf. Series: Journal of Physics: Conf. Series **854** (2017) 012013 doi:10.1088/1742-6596/854/1/012013

Instrument	Elevations (m)
Precipitation	1
Wind Speed	38, 55, 80, 87, 90, 92
Wind Direction	38, 87
Humidity	90
Temperature	38, 90
Barometric Pressure	90

Table 2: Met Tower Instrumentation Details.

3.2. Lidar specification

Figure 2: The University of Stuttgart lidar installed on the DOE 1.5 turbine. Photo by Andrew Scholbrock, NREL

The Stuttgart scanning lidar system was developed in 2008 for nacelle measurement campaigns to redirect the laser beam of a standard Windcube lidar system [19]. The complete system consists of two parts: a Windcube V1 from Leosphere and a scanner unit developed at the University of Stuttgart. A picture of the lidar from the University of Stuttgart is shown in Figure 2. Because the original Windcube was designed for site assessment with its beam pointing upwards, a two-degree-of-freedom mirror for redirecting the beam in any position within the mirror's range was installed in a second casing. The accessible area is a 0.75 D-by-0.75 D square in 1 D distance. The modified software allows up to 49 measurement positions and 5 scan distances to be used. Further, the scan rate depends on the number of pulses used for each measurement position. The lidar system has been successfully used for several inflow measurements for lidar-assisted control and wake measurements.

The lidar performs a grid measurement pattern to record the wake with a 1 Hz sampling frequency. In five different distances (1 D to 2.8 D), the wind flow is measured (as depicted in Figure 3). At each measurement point, the lidar uses $10,000$ laser pulses to measure the line-of-sight wind speed, v_{los}, that is a projection of the three wind vector components $[u, v, w]$ onto the laser beam.

The complete scan pattern consists of 49 points in each scan plane, and the five planes are

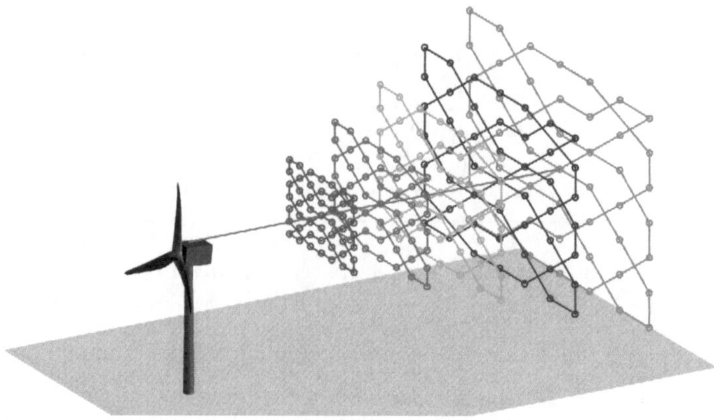

Figure 3: Visualization of the lidar scan pattern that is used to measure the wake behind the wind turbine. The lidar is measuring with a sampling frequency of 1 Hz simultaneously in five distances from 1 to 2.8 times the rotor diameter (D).

measured simultaneously. One scan takes an average of 48 s to complete. In this work, we refer to one scan, as one of these 48 s scans. In the upcoming analysis, scans of similar characteristics are aggregated to produce a mean or median scan for some conditions.

3.3. Field-testing procedure

The field test of wake steering was accomplished by implementing an outer control system above the built-in turbine control system. This control system would cause the turbine yaw control to track misaligned positions or offsets. These offset positions were changed every hour, and always included the baseline of zero offset in regular rotation. This practice helps ensure that for post-processing, each offset position can be compared with baseline operation in similar inflow conditions. Note that as Figure 4 shows, this target is tracked, but not perfectly, given wind direction variability and the limitation of the yaw controller. Therefore, analysis in this paper will be limited to periods when the target and actual yaw angles are close.

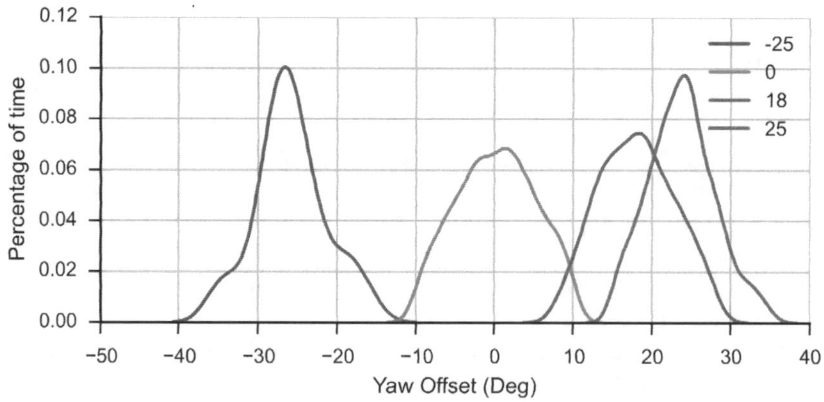

Figure 4: Percentage of actual time at a given yaw position grouped by the target yaw position.

Additionally, the lidar, located at the back of the GE1.5 SLE turbine facing downwind, is mounted on a rotational frame, which is motorized with a linear actuator. The same outer control system pivots the lidar opposite the yaw offset target. This setup allows the lidar to be directed downwind while the wind turbine is yawed.

4. Data processing

Processing of the recorded data is important to this work, which is briefly reviewed in this section. First, the recorded lidar measurement data are processed to filter out implausible data. Several methods are applied to check for hard target measurements, filter out lidar data with a bad carrier-to-noise ratio, and check for plausibility of the measurement data. All processed lidar data are grouped into scans, and combined with statistics of turbine and met mast sensor data, such as mean and standard deviation, over the period in which the scan is collected.

Finally, the data are filtered to include only certain conditions, including periods in which the met tower is nominally upwind of the turbine; periods in which the turbine is producing at least 100 kW, to eliminate faults and idling; and periods in which the target and realized offset are close.

At the time of this writing, there were approximately 15k scans completed, and the above filtering process reduces this to a set of approximately 1.5k scans to be used in this analysis, of which approximately half are aligned and half are yawed. For the analysis, all lidar scans belonging to a group (for example, scans of 8 m/s inflow while yawed) are aggregated, by finding the median velocity of each point. Median was used, and not mean, in order to diminish the impact of outliers.

5. Results

As described earlier, the results in this paper will be presented in comparison with predictions from the FLORIS model of wind farm control. The comparisons will be made with respect to the following key model properties: wake deflection, wake deficit and recovery, power loss from yaw offset, and atmospheric influence and wake shape.

5.1. Wake deflection

We first compare the wake deflection predicted by FLORIS with what is observed by the lidar. As stated earlier, the FLORIS model of deflection is based on [3], and predicts deflection at a given distance downstream as a function of coefficient of thrust and yaw angle.

In Figure 5, the median scan at 8 m/s wind for the first four ranges (with blue indicating low wind speed, and red faster wind speed) is compared with the deflection predicted by FLORIS (indicated by a green disk of the size of the turbine rotor). The model and test data show relatively good agreement. Note,for example, that FLORIS predicts nonzero deflection in the aligned case, which is observed in the lidar data. Also, the model predictions and observations are in agreement, showing that significant deflection has occurred by 1.5 D. Note that at the 1D range, the lidar scan is largely within the wake. However, a difference in deflection between aligned and offset operation can be observed on the left edge of wake.

5.2. Recovery and deficit

A second analysis can be made with predictions of the velocity deficit and its rate of recovery as the wake progresses downstream. Using the same aggregated scans shown in Figure 5, the average velocity within the disk area predicted by FLORIS can be computed. This average can then be compared with the same average velocity that would have been computed in the same way within FLORIS. This comparison is shown in Figure 6.

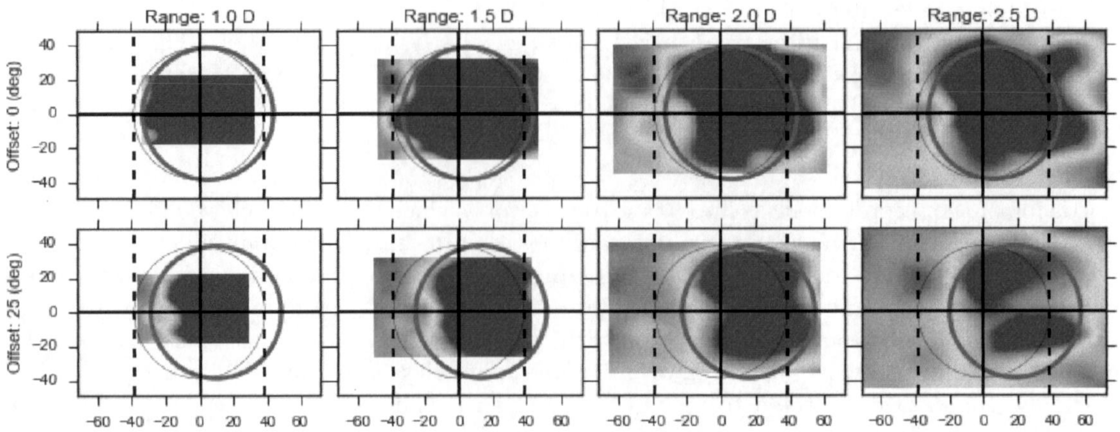

Figure 5: Median scans of the wake at 8 m/s for aligned and yawed conditions. The green circles indicate the position of the wake predicted by SOWFA/FLORIS.

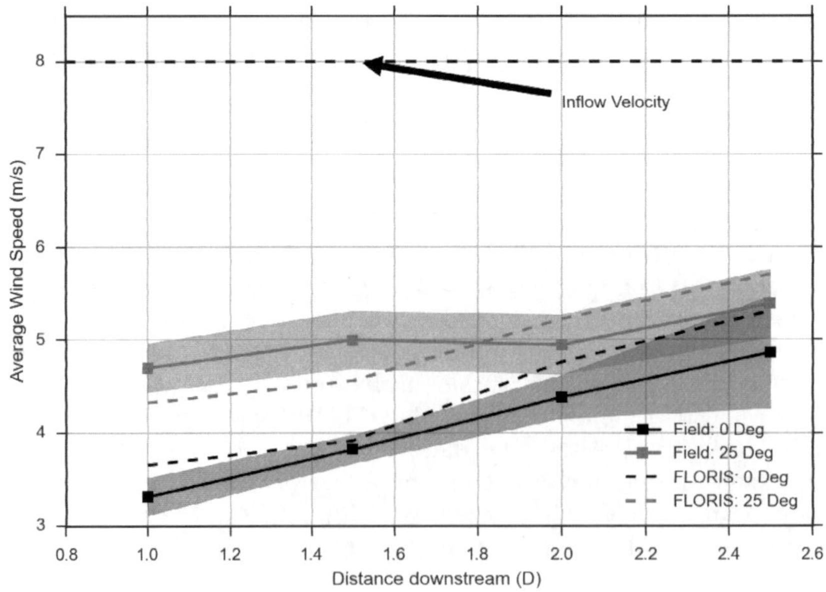

Figure 6: Comparison of the deficits predicted by FLORIS with those of observations. The deficits from field tests are computed by averaging over the circles shown in Figure 5. Additionally, to indicate some range of possibilities, for each range, 10 separate median scans are made from a random subset of all scans, and the standard deviation of the 10 aggregates provides the band radius. Finally, the values predicted from FLORIS, using the same approach of averaging over the rotor disk, are shown.

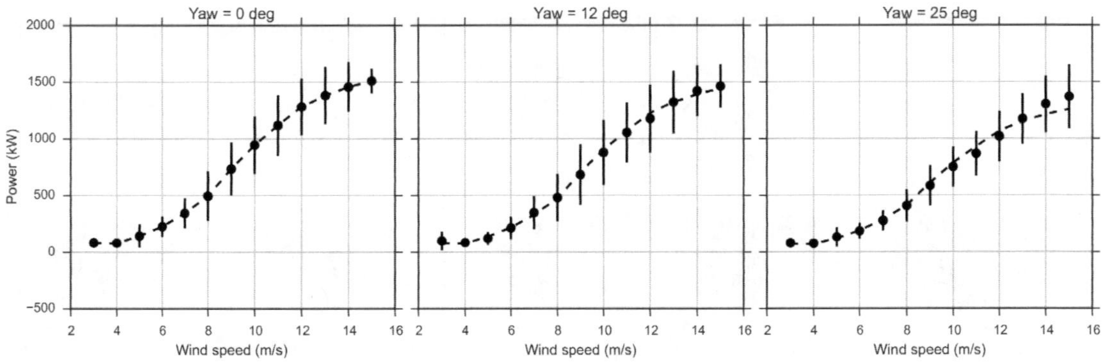

Figure 7: Comparison of the observed power curves for three yaw operating points to those assumed by FLORIS. The dashed lines are power curves drawn assuming the power loss of cosine of yaw with exponent 1.88. The points show the mean and standard deviation of observed powers.

Although the results shown in Figure 6 are not an exact match, the overall initial deficit level at 1D, the rate of recovery, and the difference in velocity between aligned and yawed conditions, are reasonably similar between model predictions and field-test data.

5.3. Power

Another important value predicted by FLORIS is the rate by which a turbine loses power because of yaw misalignment. FLORIS uses the standard rule that power is lost according to the cosine of the yaw misalignment raised to a certain exponent. Through comparisons with SOWFA, this exponent (called pP) has been fit to 1.88 [6]. In Figure 7, this function of power loss is compared with the data. The results show reasonable agreement.

5.4. Influence of the atmospheric stability and shape

A final comparison with FLORIS can be made with respect to the shape of the wake. In [13], the FLORIS model is extended to predict that the shape of the wake will go from circular to a skewed ellipsoid in the presence of veer.

In Figure 8, the scans of the wake at 7 m/s are divided by a simple function of veer, which compares the difference between the wind direction measured at 87 m and 38 m and divided the data into groups depending on the size of this difference.

Additionally, a contour algorithm is used to describe the shape of the wake in greater detail. Determining contours of wake data can be achieved through different methods and present an important method for quantifying wake behavior. Determining the "center" of a wake can be challenging when wake cut-throughs take shapes such as the bottom right of Figure 5 or in Figure 8, and contours provide a way to describe a wake shape within a given slice. Observing the contours in Figure 8, the skewing impact of veer can be observed.

In this analysis, contour detection algorithms for wake tracking are used mainly for identifying a wake shape visually, however, contours present a good opportunity for implementing systematic comparisons between wake models and test data and are the subject of ongoing research and development [20].

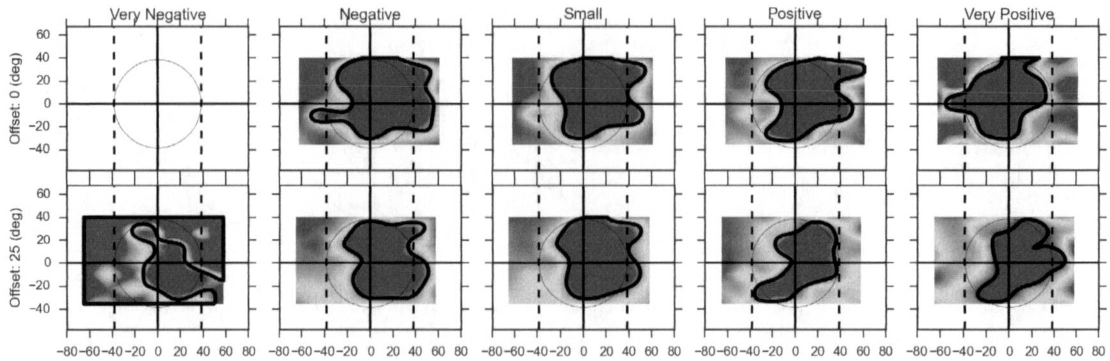

Figure 8: Using contour detection to demonstrate the change in wake shape under different veer conditions. The data is divided by the amount of difference in velocity between wind direction measurements at two heights.

6. Conclusions

In this work, the preliminary results of a field test of wake steering at full scale are reported. The wake of a full-scale turbine is measured by a lidar and compared to the predictions of FLORIS, a control-oriented wake model used in the design of wake steering controllers. Comparisons in terms of the key predictions of FLORIS, wake deflection, wake deficit and recovery, power loss, and wake skew, indicate good agreement between the model and observations. FLORIS simulations indicate wake steering can make a positive impact on wind farm annual energy production (AEP), and this work helps to validate the conclusions of AEP studies such as [8].

The results reported in this paper are based on the data that were available at the time of writing. However, the field test is ongoing and future results can be better converged when based on the larger data set. Additionally, this full data set will be made publicly available on the Atmosphere to Electrons Data Archive and Portal (https://a2e.energy.gov/about/dap).

An important aspect of future work will be to characterize the wake shape from field data. The relationship between veer and wake skew was demonstrated in this study, however, the tendency of the wake of a yawed turbine to form a kidney-bean shape (as discussed in [12]) was observed clearly in the full data, but is not currently modeled in FLORIS and was considered outside of the current scope of work.

Additionally, future work will employ a more rigorous approach to quantify the performance of wake models with respect to field measurements. Following the completion of data collection and quality control, a thorough quantitative analysis can be undertaken.

Finally, although this work has focused on validating reduced-order wake models to predict the wake behavior under wake steering control, it is possible that a wake control strategy will need direct wake measurement to operate fully. Therefore, this work can also be seen as aligning with other activities that focus on a closed-loop wake redirection concept based on direct lidar feedback of wake position. A first controller concept was introduced in [14]. Further, a $\mathcal{H}_\infty$ controller design approach for closed-loop wake redirection was described in [21]. To provide real-time wake tracking with lidar data, we refer to [22], wherein a model-based approach was presented and successfully tested to track the wake center. An application of this work demonstrates the ability of a scanning lidar to measure and estimate wake properties.

7. Acknowledgements
Matthew Churchfield's help (NREL) in the use and analysis of high-fidelity simulation of wind farms was, as it always is, instrumental to the completion of this work. The authors also thank Scott Lambert and Robb Wallen for the design and fabrication of the lidar mount and controller.

This work was supported by the U.S. Department of Energy under Contract No. DE-AC36-08GO28308 with the National Renewable Energy Laboratory. Funding for the work was provided by the DOE Office of Energy Efficiency and Renewable Energy, Wind Energy Technologies Office.

References

[1] Boersma S, Doekemeijer B, Gebraad P, Fleming Pand Annoni J, Scholbrock A, Frederik J and van Wingerden J W 2017 A tutorial on control-oriented modeling and control of wind farms *Proceedings of the American Control Conference (ACC)*

[2] Wagenaar J W, Machielse L and Schepers J 2012 Controlling wind in ECN's scaled wind farm *EWEA Annual Meeting* (Copenhagen, Denmark)

[3] Jiménez Á, Crespo A and Migoya E 2010 *Wind Energy* **13** 559–572

[4] Fleming P A, Gebraad P M, Lee S, van Wingerden J W, Johnson K, Churchfield M, Michalakes J, Spalart P and Moriarty P 2014 *Renewable Energy* **70** 211–218 ISSN 0960-1481

[5] Fleming P, Gebraad P M O, Lee S, van Wingerden J W, Johnson K, Churchfield M, Michalakes J, Spalart P and Moriarty P 2014 *Wind Energy*

[6] Gebraad P, Teeuwisse F, van Wingerden J, Fleming P, Ruben S, Marden J and Pao L 2014 *Wind Energy*

[7] Fleming P A, Ning A, Gebraad P M O and Dykes K 2015 *Wind Energy* ISSN 1099-1824

[8] Gebraad P, Thomas J J, Ning A, Fleming P and Dykes K 2016 *Wind Energy*

[9] Campagnolo F, Bottasso C L, Croce A *et al.* 2016 Wind tunnel testing of wake control strategies *American Control Conference (ACC), 2016* (American Automatic Control Council (AACC)) pp 513–518

[10] Schottler J, Hölling A, Peinke J and Hölling M 2016 Wind tunnel tests on controllable model wind turbines in yaw *34th Wind Energy Symposium* p 1523

[11] Fleming P, Annoni J, Shah J J, Wang L, Ananthan S, Zhang Z, Hutchings K, Wang P, Chen W and Chen L 2017 *Wind Energy Science Discussions* **2017** 1–17 URL `http://www.wind-energ-sci-discuss.net/wes-2017-4/`

[12] Vollmer L J, Steinfeld G, Heinemann D and Kühn M 2016 Estimating the wake deflection downstream of a wind turbine in different atmospheric stabilities: An LES study *Wind Energy Science Discussions*

[13] Gebraad P, Churchfield M and Fleming P 2016 Incorporating atmospheric stability effects into the floris engineering model of wakes in wind farms *Journal of Physics: Conference Series* vol 753 (IOP Publishing) p 052004

[14] Raach S, Schlipf D, Borisade F and Cheng P W 2016 Wake redirecting using feedback control to improve the power output of wind farms *Proceedings of the American Control Conference (ACC)*

[15] Bastankhah M and Porté-Agel F 2016 *Journal of Fluid Mechanics* **806** 506–541

[16] Boersma S, Gebraad P, Vali M, Doekemeijer B and van Wingerden J 2016 *Journal of Physics: Conference Series* **753** 032005 URL `http://stacks.iop.org/1742-6596/753/i=3/a=032005`

[17] Jensen N O 1984 A note on wind generator interaction Tech. Rep. Risø-M-2411 Risø National Laboratory

[18] Churchfield M, Lee S, Moriarty P, Martinez L, Leonardi S, Vijayakumar G and Brasseur J 2012 A large-eddy simulation of wind-plant aerodynamics *50th AIAA Aerospace Sciences Meeting Including the New Horizons Forum and Aerospace Exposition, Nashville, TN* (Nashville, Tennessee, USA)

[19] Rettenmeier A, Anger J, Bischoff O, Hofsäß M, Schlipf D and Würth I 2013 Nacelle-based lidar systems *Remote Sensing for Wind Energy, DTU Wind Energy-E-Report-0029(EN)* chap 8, pp 157–170

[20] Quon E W, Churchfield M J and Jonkman J 2017 *under Review: Wind Energy Science*

[21] Raach S, van Wingerden J W, Boersma S, Schlipf D and Cheng P W 2017 Hinf controller design for closed-loop wake redirection *Proceedings of the American Control Conference (ACC)*

[22] Raach S, Schlipf D and Cheng P W 2017 *Wind Energy Science Discussions* **2017** 1–17 URL `http://www.wind-energ-sci-discuss.net/wes-2017-3/`

Modelling lidar volume-averaging and its significance to wind turbine wake measurements

A R Meyer Forsting[1], N Troldborg[1], A Borraccino[1]

[1] DTU Wind Energy, Technical University of Denmark, Risø Campus, DK-4000 Roskilde, Denmark

E-mail: alrf@dtu.dk

Abstract. Lidar velocity measurements need to be interpreted differently than conventional in-situ readings. A commonly ignored factor is "volume-averaging", which refers to lidars not sampling in a single, distinct point but along its entire beam length. However, especially in regions with large velocity gradients, like the rotor wake, can it be detrimental. Hence, an efficient algorithm mimicking lidar flow sampling is presented, which considers both pulsed and continous-wave lidar weighting functions. The flow-field around a 2.3 MW turbine is simulated using Detached Eddy Simulation in combination with an actuator line to test the algorithm and investigate the potential impact of volume-averaging. Even with very few points discretising the lidar beam is volume-averaging captured accurately. The difference in a lidar compared to a point measurement is greatest at the wake edges and increases from 30% one rotor diameter (D) downstream of the rotor to 60% at $3D$.

1. Introduction

Lidars rely on the reflection of emitted light from aerosols back to the receiver to measure wind speed. To determine the velocity at the desired measurement location, the beam can be focused (continous-wave) or the backscattered signal split by the time of flight (pulsed). In both cases there are reflections from aerosols away from the desired probe location. Lidars therefore sample the flow velocity all along their beams, albeit with a spatial filter. This is also commonly referred to as "volume-averaging" or "range-weighting". Schematically this is shown in Figure 1 with the weighting function $W(r)$ acting in the local beam coordinate system with $r = |\mathbf{x} - \mathbf{x}_L|$ and peaking at the probe location F. The measured line-of-sight velocity v_{los} at a point in space $\mathbf{x}_F$ is thus given by the convolution of the radial velocities sampled along the beam and the weighting function

$$v_{los}(\mathbf{x}_F) = -\int_0^\infty \mathbf{e}(\mathbf{x}_F, \mathbf{x}_L) \cdot \mathbf{V}(r) W(r)\,\mathrm{d}r \tag{1}$$

Here $\mathbf{e}(\mathbf{x}_F, \mathbf{x}_L)$ represents the beam direction unit vector. The difference between the point and lidar measurement, $\Delta v_r = v_{los} - (-\mathbf{e}(\mathbf{x}_F, \mathbf{x}_L) \cdot \mathbf{V}(\mathbf{x}_F))$, therefore depends on the combination of flow-field and weighting function. This implies that large flow gradients and/or broad weighting functions can lead to significant Δv_r. In wind energy and the atmospheric sciences lidars are becoming ever more popular, especially in model validation, but usually their data are still treated as conventional point measurements. However, with the currently available lidar technology the effect of volume-averaging can be non-negligible, especially in areas of large

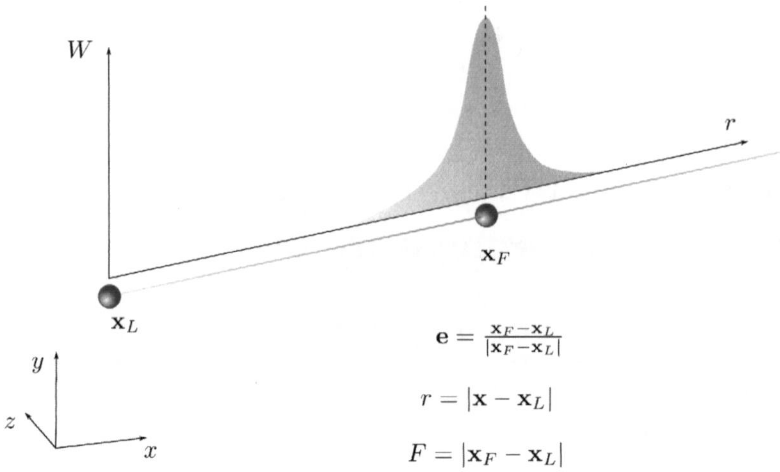

Figure 1. Lidar range weighting of the sampled velocities along the beam. The beam originates from $\mathbf{x}_L$ to measure at $\mathbf{x}_F$. The weighting function acts in the local beam coordinate system.

velocity gradients like the wind turbine wake. Furthermore, continous-wave (CW) and pulsed lidars possess very distinct weighting functions. The large domains of numerically generated flow-fields readily provide $\mathbf{V}(r)$, so only the weighting function remains to be implemented to arrive at a numerical estimate of $v_{los}(\mathbf{x}_F)$. These in turn are valuable to experimentalists using lidars, as they give insight into the magnitude of Δv_r given some measurement setup. This allows optimising the setup and estimating uncertainties, as done by Churchfield *et al.* [1] for a wake experiment with a CW lidar. Simley *et al.* [2] implemented a numerical version of the latter to study the errors from lidar measurements and their implications for feed-forward turbine control. Similarly, Mirocha *et al.* [3] used a virtual pulsed lidar to compare measurements and simulations. Despite presenting the respective weighting functions, there are no details given regarding the actual beam discretisation.

The aim of this paper is to increase the awareness of volume-averaging in lidar measurements, with a focus on wake measurements, and providing a guide for implementing numerical lidars in post-processing. For this purpose computational fluid dynamics (CFD) simulations of the wind turbine wake in flat terrain using an actuator line (AL) are sampled over large regions using numerical, nacelle-based CW and pulsed lidars.

2. Lidar modelling

The fundamental part underlying a numerical implementation of a lidar is the approximation of the weighting function and its efficient discretisation. The former is well established for each respective lidar technology (refer to the references for more detail):

- Continous-wave [4]

$$W_C(r) = \frac{1}{\pi}\frac{z_R}{z_R^2 + (r-F)^2} \quad \text{with} \quad z_R = \frac{\lambda F^2}{\pi \alpha_0^2} \tag{2}$$

- Pulsed [5, 6]

$$W_P(r) = \frac{1}{2\Delta p}\left\{\mathrm{Erf}\left[\frac{(r-F)+\Delta p/2}{r_p}\right] - \mathrm{Erf}\left[\frac{(r-F)-\Delta p/2}{r_p}\right]\right\}$$
$$\text{with} \quad \mathrm{Erf}(x) = \frac{2}{\sqrt{\pi}}\int_0^x \exp(-t^2)\,\mathrm{d}t \quad \text{and} \quad r_p = \frac{\Delta l}{2\sqrt{\ln(2)}} \tag{3}$$

IOP Conf. Series: Journal of Physics: Conf. Series **854** (2017) 012014 doi:10.1088/1742-6596/854/1/012014

Here $\lambda, \alpha_0, \Delta p, \Delta l$ are lidar specific constants, usually provided by the manufacturer, and denote laser wavelength, effective telescope radius, range-gate length and lidar beam full width at half maximum, respectively. In figure 2 the influence of these parameters on the lidars' weighting functions are shown. The baseline parameters, which are used throughout the paper, are taken from the ZephIR Dual Mode (CW) and Avent 5-beam Demonstrator (pulsed) (see table 1) and $F = 100$ m. Both functions are symmetric about the probe location F, independent of the choice of parameters. The pulsed weighting function is broader than the one for the continous-wave lidar at this probe location, but it is independent of F. The continous-wave lidar's weighting function on the other hand broadens rapidly with increasing F, as $z_R \propto F^2$. This has some important implications regarding the discretisation of each weighting function, as in a real measurement scenario the only parameter changing is F. Hence, the CW weighting function needs to be re-discretised for each F, whereas the one for a pulsed lidar can be determined once and shifted along r according to F.

Table 1. Lidar parameters used throughout this paper.

CW (ZephIR Dual Mode)	α_0	28×10^{-3} m
	λ	1565×10^{-9} m
Pulsed (Avent 5-beam Demo)	Δl	24.75 m
	Δp	38.4 m

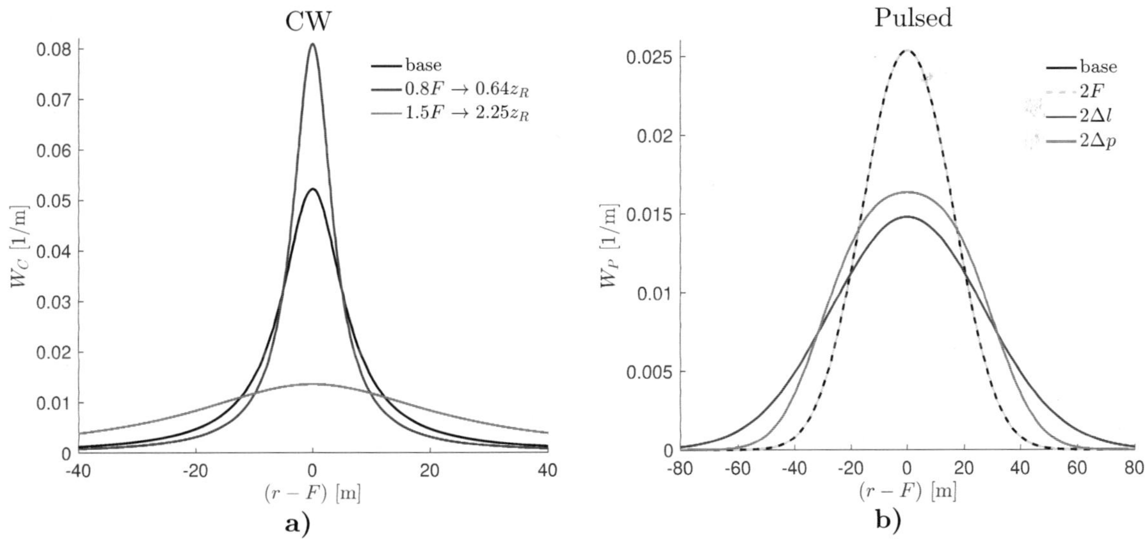

Figure 2. Idealised lidar weighting functions for $F = 100$ m and with baseline parameters given in table 1: a) Continous-wave, equation (2); b) Pulsed, equation (3).

2.1. Discretisation of weighting functions

In discrete space equation (1) becomes

$$v_{los}(\mathbf{x_F}) = \frac{\sum_{i=1}^{n_W} \mathbf{e}(\mathbf{x}_F, \mathbf{x}_L) \cdot \mathbf{V}(r_i) W(r_i)}{\sum_{i=1}^{n_W} W(r_i)} \tag{4}$$

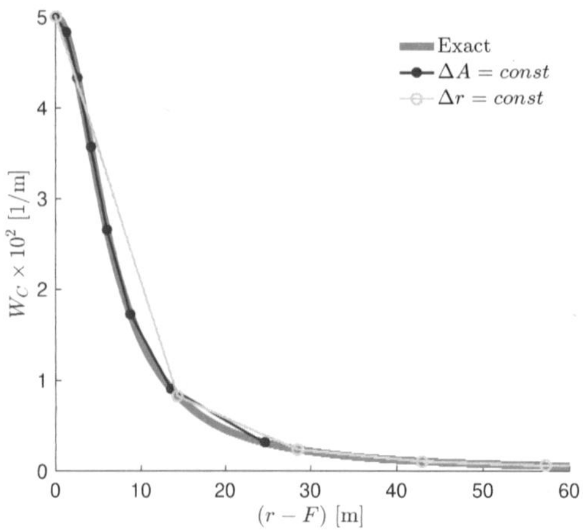

Figure 3. Continous-wave lidar weighting function as defined in equation (2) with $F = 100$ and discretised with either linear spacing in ΔA or Δr each with $n_W = 15$. Lidar parameters: $\lambda = 1565 \times 10^{-9}$ m; $\alpha_0 = 28 \times 10^{-3}$ m.

where n_W denotes the number of points used to discretise the lidar beam. It is directly linked to the numerical efficiency of estimating $v_{los}(\mathbf{x}_F)$, as it determines the number of locations at which $\mathbf{V}$ needs to be evaluated by some costly interpolation. Using a linear point distribution in r the convergence of v_{los} is poor, as the functions are not sampled sufficiently in regions where W is large. Instead, the the area under the curve should be conserved to preserve the shape of the function as shown in Figure 2.1 for $n_W = 15$. Adopting this strategy the integral of the respective weighting function needs to be determined:

$$\Delta A = \int_{r_i}^{r_{i+1}} W(r)\,\mathrm{d}r = const \tag{5}$$

with

$$\Delta A = \frac{2A_F}{(n_W - 1)} \quad \text{and} \quad A_F = \int_{-\infty}^{F} W(r)\,\mathrm{d}r \tag{6}$$

Here the symmetry of the weighting functions is used. The discretisation approach of each function differs, due to the nature $\int W(r)$.

2.1.1. Continous-wave Equation (2) has a definite integral

$$\Delta A = \int_{r_i}^{r_{i+1}} W_C(r)\,\mathrm{d}r = \frac{1}{\pi}\left[\tan^{-1}\left(\frac{r-F}{z_R}\right)\right]_{r_i}^{r_{i+1}} \tag{7}$$

It follows that

$$A_F = \int_{0}^{F} W_C(r)\,\mathrm{d}r = \frac{1}{\pi}\tan^{-1}\left(\frac{F}{z_R}\right) \tag{8}$$

Here the lower bound is set to zero instead, as $\int_{-\infty}^{0} W_C(r)\,dr \approx 0$. By rearranging equation (7) the point distribution can be determined

$$r_{i+1} = z_R \tan\left(\tan^{-1}\left(\frac{r_i - F}{z_R}\right) - \Delta A \pi\right) + F \tag{9}$$

The best starting location is $r_1 = F$ and the sampling points only need to be evaluated for one side and reflected in $r - F$.

2.1.2. Pulsed The error function has no definite integral, requiring a numerical approach. As W_P is independent of F, the discretisation should be performed only once for $s = r - F$. The shape of the weighting function requires choosing a certain cut-off value $w_{\min}$ until which the weighting function is discretised. The corresponding point $s_{\min}$ can be determined numerically down to a certain tolerance. Consequently s can be discretised linearly with a high resolution in the region $0 \le s \le s_{\min}$ to evaluate $\mathbf{w}^+ = W_P(\mathbf{s}^+)$. Some quadrature rule can then be used to determine $\mathbf{A}^+(\mathbf{s}^+)$, which is summed to determine A_F and consequently ΔA (see equation 6). Lowering the level of discretisation reduces computational time as less interpolation is performed in the flow domain. The final discretised weighting function should be sampled at each s_i corresponding to an area $i\Delta A$, which can be interpolated from $\mathbf{A}^+(\mathbf{s}^+)$. Reflecting $\mathbf{w}(\mathbf{s})$ in 0 gives the fully discretised W_P and to recover the beam coordinate system $\mathbf{r} = \mathbf{s} + F$.

2.2. Domain boundary exceeding beams

Depending on the probe location and the beam direction, some part of the lidar beam may lie outside the computational domain. Additionally, pulsed lidars give weight to velocities sampled at $r < 0$, when $F < 2\Delta l$. One solution to obtain an estimate of v_{los}, despite missing parts of the weighting function, is to scale the weights with regard to the total inside the domain

$$\tilde{W}(r) = \frac{W(r)}{\int_0^{r_\Omega} W(r)\,\mathrm{d}r} \tag{10}$$

Here r_Ω denotes the intersection of the beam with the domain boundaries. In a real measurement scenario it is not unusual that $\int_0^{r_\Omega} W(r)\,\mathrm{d}r \neq 1$, which corresponds to a beam intersecting hard targets like towers, turbines, ground etc. Usually these measurements are outright rejected by the lidar signal processing software, as hard targets show up as anomalies in the Doppler spectra. Similarly, a pulsed lidar cannot measure at around $F < 2\Delta l$ for technical reasons (internal reflections etc. [7]). Therefore numerical estimates of v_{los} for which $\int_0^{r_\Omega} W(r)\,\mathrm{d}r < 1$ should be treated with caution.

3. Computational method

3.1. Flow solver and modelling approach

The finite-volume solver, EllipSys3D, discretises the Navier-Stokes equations over a block-structured domain [8–10]. The SIMPLE algorithm [11] solves the pressure-linked terms of the Navier-Stokes equations and a modified Rhie-Chow algorithm the pressure equation [12]. Time-stepping is achieved through an iterative, second-order scheme. The turbulence is either modelled by a Reynolds-averaged Navier-Stokes (RANS) formulation with a Menter k–ω shear-stress transport closure [13] or by solving the filtered Navier-Stokes equations with a sub-grid scale (SGS) model by Ta Phouc [14]. Switching between models is determined by a limiter function as defined by Strelets [15]. This also determines whether the QUICK [16] (RANS) or a fourth-order CDS scheme (LES) discretises the convective terms. The rotor forces are introduced by an actuator line [17, 18].

3.2. Siemens SWT 2.3-93
The turbine is a commercial product by Siemens Wind Power with a rated power of 2.3 MW and a three-bladed rotor with 93 m diameter. The hub is located 80 m above. At a hub height wind speed of 8 m/s it performs 15 rotations per minute and $C_T = 0.81$.

3.3. Numerical domain
A box domain with 25 radii (R) side length reduces the blockage from the turbine ($\pi/25^2 = 0.5\%$). The bottom boundary is a no-slip wall and the top, side and rear boundary conditions follow the shear profile (Dirichlet). The AL is located centrally, at a hub height of 80 m, and is surrounded by a finely meshed box with 2.5 R side length and a grid spacing of $R/32$. This resolution is sufficient for accurate wake simulations using an AL [19, 20]. The wall cell of the structured mesh is set to 0.05 m and grows [8] hyperbolically in the vertical direction until reaching the equispaced finely meshed box. From there the mesh spacing increases hyperbolically towards the outer domain edges.

3.4. Numerical setup
The hub height velocity was 8 m/s and the inflow profile followed the power law with an exponent of 0.3. The kinematic viscosity and air density were set to 1.789×10^{-5} kg/m/s and 1.225 kg/m^3, respectively. Neither nacelle nor tower were modelled. The smearing factor of the AL is set to twice the grid spacing as suggested by Troldborg *et al.* [19].

The flow was sampled with a CW and pulsed lidar situated at the rotor centre. The parameters of the lidars were those given in table 1. The probe locations $\mathbf{x}_F$ lay in a vertical ($x - z$) and horizontal ($x - y$) plane passing through the rotor centre. The spacing between probe locations was $D/50$ in both directions.

4. Results
4.1. The mean flow-field
The developing mean flow-field is shown in Figure 4. The large C_T of 0.81 induces strong velocity gradients and peak deficit. The wake is asymmetric, due to the interaction of shear with wake rotation. The rotor blocks the incoming flow, leading to deceleration in the wall-bound regions upstream and re-acceleration downstream. This flow-field presents a challenging test case for lidars with its large velocity gradients.

4.2. Beam discretisation & v_{los} convergence
The convergence of the line-of-sight velocity Δv_{los} with respect to the number of points discretising the lidar beam is investigated over the region outlined in Section 3.4. Here the residual error is defined as

$$\epsilon^i = \frac{v^i_{los} - v^{i\text{-}1}_{los}}{v^{i\text{-}1}_{los}} \tag{11}$$

with i representing the discretisation level. For the vertical plane sampled by a CW lidar and $n_W = 19$ the residuals are shown in Figure 5. Only probe locations where the entire beam lies inside the domain ($\int_0^{r_\Omega} W(r)\,\mathrm{d}r = 1$) are presented. Close to the rotor v_{los} has converged further than downstream in the wake. This is linked to the CW lidar's weighting function: it is very acute close to the rotor but stretches quickly, necessitating more points for an accurate representation. Furthermore, close to large velocity gradients, as at the wake edge, the exact point distribution starts to play an important role. Figure 6 summarises the change of the v_{los} residual with n_W over the vertical plane for both lidars. Convergence is of second order for the CW and nearly third order for the pulsed lidar. Therefore even with few points discretising the beam the effect of volume-averaging can be captured. Note that beyond $n_W > 100$ the beam

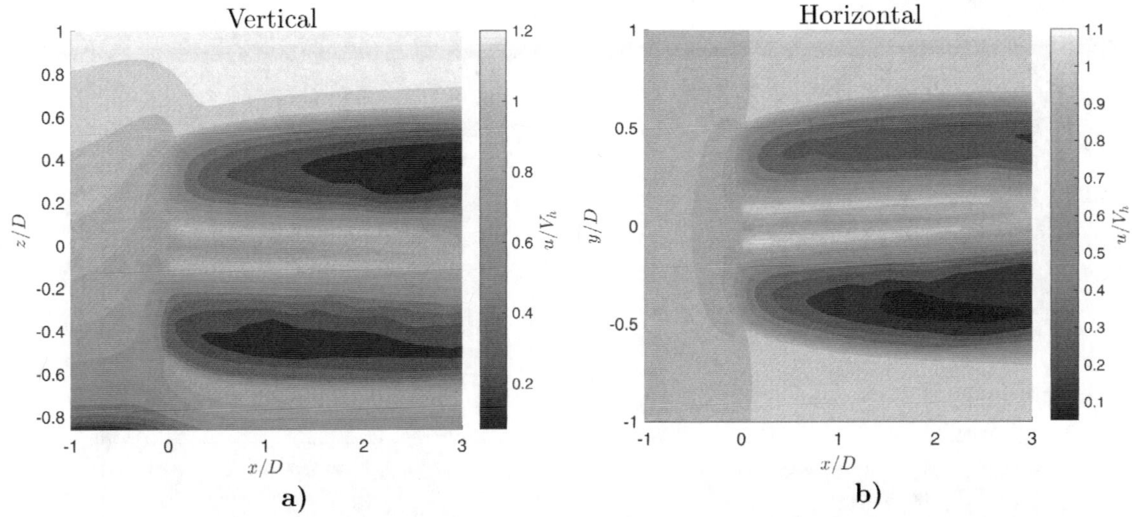

Figure 4. Contours of the time-averaged streamwise velocity component normalised by the hub height velocity $V_h = 8$ m/s in the a) vertical b) horizontal plane through the rotor centre.

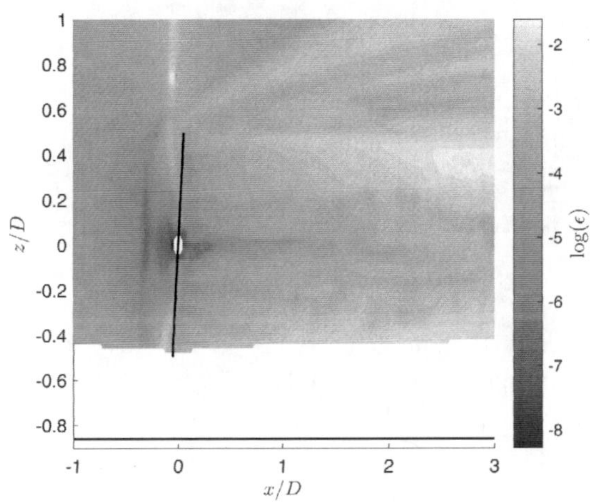

Figure 5. Contours of the v_{los} residual for the CW lidar with $n_W = 19$ discretising the beam. Regions where $\int_0^{r_\Omega} W(r)\,\mathrm{d}r \neq 1$ are removed.

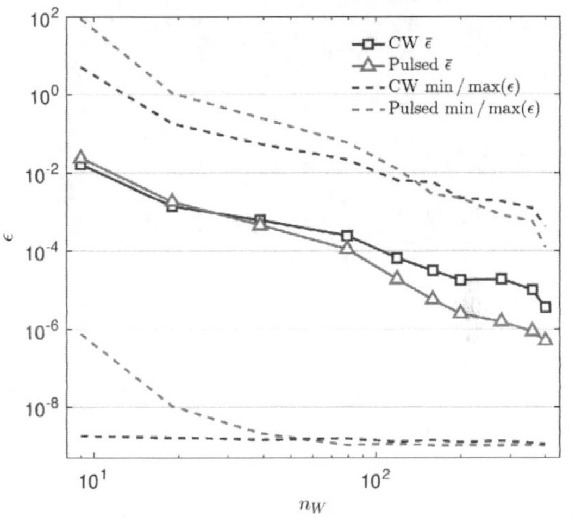

Figure 6. Evolution of the line-of-sight velocity v_{los} residual with number of beam points n_W. Mean and extreme values are determined over a vertical plane $(-3 \leq x/D \leq 3, y/D = 0, -0.86 \leq z/D \leq 1)$.

discretisation becomes finer than that of the flow-field and the residuals continue to fall due to interpolation in the flow domain.

4.3. The effect of volume-averaging

To investigate the effect of volume-averaging the flow-field sampled by lidars are compared to point-like measurements, such that

$$\Delta\tilde{v}_{los} = \frac{v_{los}^{\mathrm{lidar}} - v_{los}^{\mathrm{point}}}{v_{los}^{\mathrm{point}}} \tag{12}$$

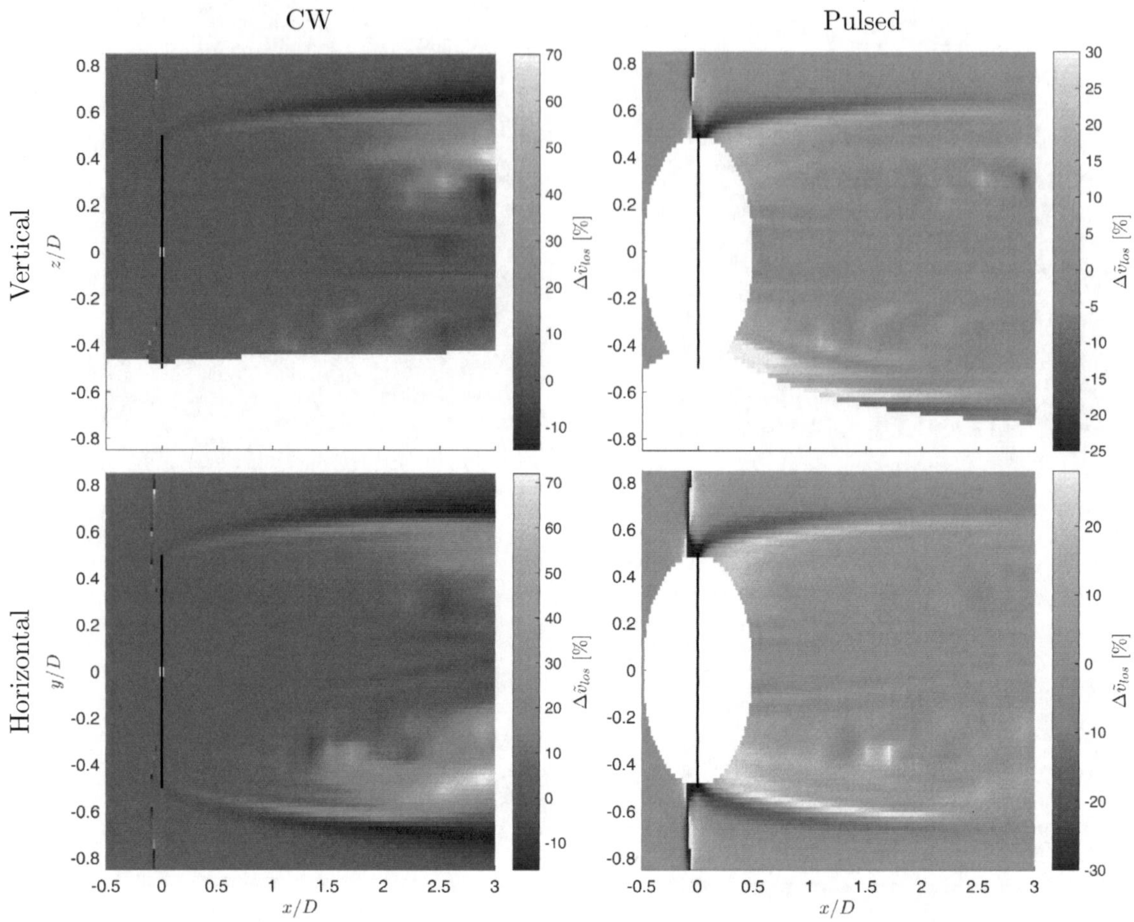

Figure 7. Time-averaged relative difference in the line-of-sight velocity $\Delta\tilde{v}_{los}$ measured by a CW (left) and pulsed (right) in a vertical (top) and horizontal plane (bottom) with respect to point-like velocities.

Time-averaged contours of this quantity are shown in Figure 7 for both planes and lidars. Note that here the beam was discretised with $n_W = 200$ and again probe locations where $\int_0^{r_\Omega} W(r)\,\mathrm{d}r \neq 1$ are excluded. Unsurprisingly, for both lidars the difference to point-like measurements are most prominent around the wake edges, due to the large velocity gradients. However, whereas the CW sampled velocities present nearly no volume-averaging close to the rotor, the pulsed system suffers greatest in this region. The inverse is true the further the probe locations move downstream. This is clearly related to the CW lidar's weighting function spreading quickly with F^2. The pulsed lidar's weights are unchanged with distance, but close to the rotor the beam cuts the wake edge nearly orthogonally.

This results in a relatively small v_{los} and in combination with sampling along a large velocity gradient introduces substantial amounts of volume-averaging. The large difference encountered in the rotor plane also stems from sampling the flow perpendicular to the mean flow direction. To give a more quantitative impression of volume-averaging in the wake, Figure 8 compares the difference in the line-of-sight velocities at different longitudinal stations between lidars. Upstream the velocity profile is nearly unchanged, except close to the ground, whereas in the wake there are significant alterations, mostly leading to larger measured velocities. This is most prominent around the wake edges.

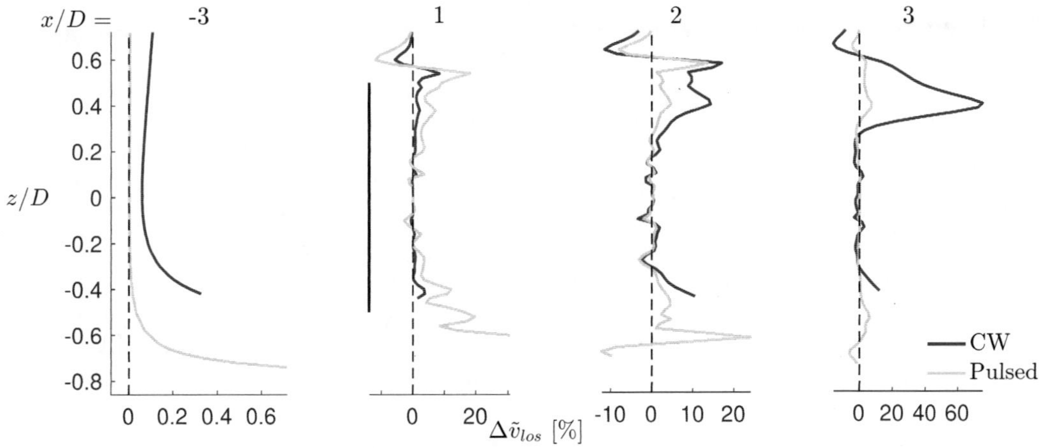

Figure 8. Time-averaged difference in the measured line-of-sight to point-like velocities along z at changing locations along the centreline.

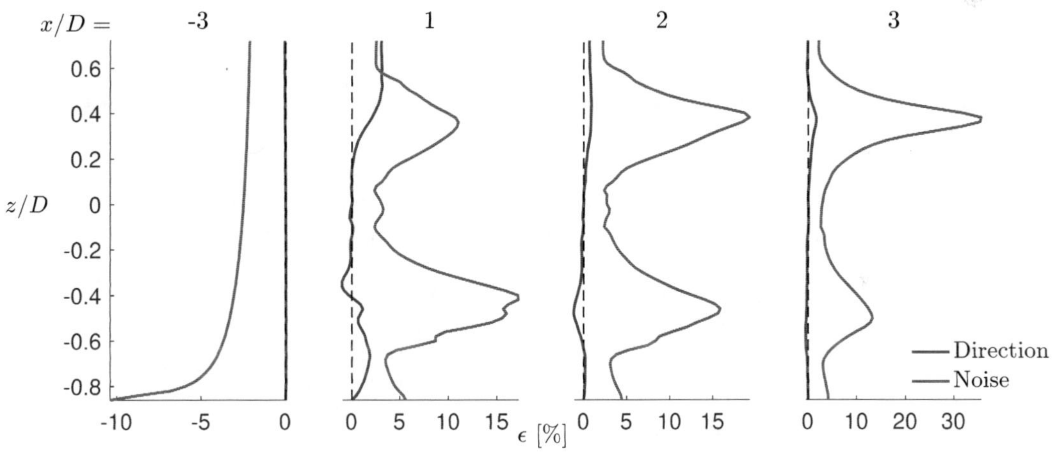

Figure 9. Directional and noise errors (only positive) along the same lines as Figure 8.

However, as discussed by Simley *et al.* [21], there are various other errors impacting lidar measurements, namely directional bias, signal noise and flow evolution. Here the directional bias refers to the error incurred by estimating the streamwise velocity from line-of-sight measurements ("cyclops dilemma"). Random measurement noise is unavoidable and introduces uncertainty in the measured velocities by disturbing the Doppler spectra. Both errors were computed for the transects already shown in Figure 8 following the definition of Simley *et al.*, which assumes a maximum noise error of $\pm$ 0.2 m/s. They are presented in Figure 9, which highlights that signal noise is the more significant of both. Referring to Figure 8, noise plays a detrimental role with respect to the total error, as it is of the same magnitude as the volume-averaging effect. Upstream it is in fact the only significant error source. Nevertheless, it should be kept in mind that the noise level needs to be reassessed for each measurement campaign, as it can vary significantly [21]. Therefore, the impact from noise presented here cannot be generalised.

5. Conclusion

Lidar volume-averaging can be readily implemented in existing flow simulation tools, as it is solely an additional post-processing step. An efficient and accurate method for discretising the lidar beam weighting function is presented, that allows lidar sampling at each time step at limited computational cost. Wake measurements are affected by volume-averaging especially at the wake edges, due to the large velocity gradients being smeared out. However the type of lidar, CW or pulsed, determines the region in which it is most prominent: close to the rotor the CW shows nearly none, whereas the pulsed lidar's velocity might reach 30% of a point-like measurement; the opposite is true beyond 200 m downstream. The lidar measurement location and type therefore determine whether volume-averaging should be incorporated into simulation data post-processing. As there are many variables determining the lidar line-of-sight velocity it is, however, hard to make a general statement regarding the significance of volume-averaging.

Acknowledgments

This work was performed inside the UniTTe project (unitte.dk), which is financed by The Innovation Fund Denmark, grant number 1305-00024B. Computational resources were provided by the Risø DTU central computing facility. Furthermore, we would like to thank Siemens Wind Power and the lidar manufacturers ZephIR and Avent.

References

[1] Churchfield M, Wang Q, Scholbrock A, Herges T, Mikkelsen T and Sjöholm M 2016 *Journal of Physics: Conference Series* **753**
[2] Simley E, Pao L Y, Frehlich R, Jonkman B and Kelley N 2012 Analysis of wind speed measurements using continous wave lidar for wind turbine control *49th AIAA Aerospace Sciences Meeting* AIAA 2011-263 (Orlando, Flordia, US)
[3] Mirocha J D, Rajewski D A, Marjanovic N, Lundquist J K, Kosovic B, Draxl C and Churchfield M J 2016 *Journal of Physics: Conference Series* **753**
[4] Sonnenschein C M and Horrigan F A 1971 *Applied Optics* **10**
[5] Banakh V A and Smalikho I N 1997 *Atmos. Oceanic Opt.* **10** 957–965
[6] Sonnenschein C M and Horrigan F A 1997 *J. Atmos. Oceanic Technol.* **14** 54–75
[7] Vasiljevic N, Courtney M and Mann J 2014 *A time-space synchronization of coherent Doppler scanning lidars for 3D measurements of wind fields* Ph.D. thesis DTU Wind Energy
[8] Sørensen N 1995 *General purpose flow solver applied to flow over hills* Ph.D. thesis Risø National Laboratory
[9] Michelsen J 1994 Basis3d - a platform for development of multiblock pde solvers Tech. rep. Dept. of Fluid Mechanics, Technical University of Denmark, DTU
[10] Michelsen J 1994 Block structured multigrid solution of 2d and 3d elliptic pdes Tech. rep. Dept. of Fluid Mechanics, Technical University of Denmark, DTU
[11] Patanker S and Spalding D 1972 *International Journal of Heat and Mass Transfer*
[12] Réthoré P E and Sørensen N 2012 *Wind Energy* **15** 915–926
[13] Menter F R 1993 *AIAA Journal*
[14] Phouc T 1994 *Modèles de sous maille appliqués aux écoulmements instationnaires décollés* (LIMSI)
[15] Strelets M 2001 Detached eddy simulation of massively separated flows 39th *AIAA Aerospace Sciences Meeting and Exhibit* AIAA Paper 2001-0879 (Reno,NV)
[16] Leonard B 1979 *Computer Methods in Applied Mechanics and Engineering* **19** 59–98
[17] Srensen J N and Shen W Z 2002 *Journal of Fluids Engineering* **124** 393–399
[18] Mikkelsen R 2003 *Actuator Disc Methods Applied to Wind Turbines* Ph.D. thesis Technical University of Denmark
[19] Troldborg N, Sørensen J and Mikkelsen R 2009 *Actuator Line Modeling of Wind Turbine Wakes* Ph.D. thesis Technical University of Denmark
[20] Meyer Forsting A and Troldborg N 2015 *Journal of Physics: Conference Series (Online)* **625** ISSN 1742-6596
[21] Simley E, Pao L Y, Frehlich R, Jonkman B and Kelley N 2014 *Wind Energy* **17** 413–433

Experiments in the wind turbine far wake for the evaluation of an analytical wake model

Luis García[1], Mari Vatn[1], Franz Mühle[2] and Lars Sætran[1]

[1] Department of Energy and Process Engineering, Norwegian University of Science and Technology, Trondheim, Norway

[2] Faculty of Environmental Sciences and Natural Resource Management, Norwegian University of Life Sciences, Ås, Norway

E-mail: franz.muhle@ntnu.no

Abstract. Nowadays, not only the size of single wind turbines but also the size of wind farms is increasing. Understanding the interaction between the turbines and especially the wakes formed behind them are getting more important to further improve such wind turbine arrays. Consequently, new issues in wind energy research arise. An experimental wind tunnel study was conducted, in order to analyze and understand the far wake of a wind turbine. The experimental results were used to test if an analytical wake model derived by H. Schilichting for blunt bodies can be used to describe the velocity and width development in the far wake of wind turbines. The results of the evaluation show that the wake of a wind turbine agrees fairly well with the model according to Schlichting. The velocity deficit as well as the width in the wake behind the turbine, are found to deviate with around only 2% from the results obtained applying the analytical model. Thus, it can be concluded that the analytical wake model by Schlichting is well suited to estimate the velocity deficit and the width in the far wake of a wind turbine.

1. Introduction

Over the last decades, not only the size of single turbines but also that of wind farms has increased [1]. Consequently, it is getting more and more important to understand the interactions between single turbines in a wind farm to further improve the overall efficiency. Here, the wake formed behind the wind turbine plays a major role. This wake is mainly characterized by a velocity deficit and higher turbulence. Even though a lot of research has been devoted to the understanding of wind turbine wakes, it is not fully understood yet and further research in this topic is needed.

Typically, the wake behind a wind turbine is divided into two sections; near wake and far wake, where the boundary between the two sections is difficult to exactly determine. However, the end of the near wake is considered to be where the shear layer reaches the wake axis [11]. Crespo has concluded in his previous work [8] that this normally occurs at around $2-5$ D. In the far wake, the velocity deficit gradually decays downstream of the turbine, and the wake is fully developed. Consequently, axis-symmetry and thus a self-similar wake structure can be assumed [11].

There are several types of models that describe the wake characteristics. Theoretical models, or kinematic models, are based on self-similar velocity deficit profiles. The momentum equation is used to model the velocity deficit of the wake and the wake growth rate, which is calculated by using information about the different turbulence involved. Examples of these models are, among others, proposed by Lissaman [2], Voutsinas et al. [3] and Vermeulen [4]. Numerical models, or field models, which calculate the characteristics of the flow field at every point, have been, among others, developed by Sforza [5] and Ainslie [10]. Other important models are analytical expressions that calculate the wake parameters, which are based on experimental results rather than theoretical formulation. These expressions are an alternative to the methods mentioned above for controlling a wind farm as they are fast methods and do not require high computational power. For a detailed review of the different wake models see Crespo et al. [8].

As the wake of a wind turbine influences the downstream wind turbines in a wind farm, it is of interest to find a good estimation of how the wake develop. In this paper, an analytical expression from Schlichting [7], which was derived in order to describe the wake behind by blunt bodies, is studied in the interest to evaluate if the model is also applicable to predict wind turbine wakes.

2. Theory

In the following the work by H. Schlichting, Boundary Layer Theory [7], which describes with simple analytical expressions the width and the velocity decay evolution of the wake behind blunt bodies is considered. A wake is created behind a solid body when it is introduced to a fluid flow. The body experiences drag, which generates a loss of momentum, and as a result the velocity of the flow in the wake is reduced.

The studied model is constrained to free turbulent flows, where the laminar friction can be neglected due to a much higher turbulent friction [7]. It is additionally assumed that the pressure across the wake remains constant. Consequently, this assumption is only valid from a certain distance away from the body, limiting the model to the far wake. The model is derived from the boundary layer equation for incompressible flow.

The results from the derivation are the power laws that describe the width and the velocity decay, which are presented in table 1.

Type of flow	Width	Velocity
Two-dimensional wake	$x^{1/2}$	$x^{-1/2}$
Circular wake	$x^{1/3}$	$x^{-2/3}$

Table 1: Power laws for the calculation of width and velocity decay in terms of downstream distance from the blunt body [7].

Two parameters need to be defined in order to comprehend the model derived by Schlichting. The first parameter is the *velocity decay* (u_1), which is defined as the difference between the free stream flow velocity (U_∞) and the velocity at the measured point (U); $u_1 = U_\infty - U$. The second parameter is the *half depth width* ($b_{1/2}$), defined as the width of where the wake has the half of the velocity decay. The reason for defining this parameter is due to the asymptotic behavior of the wake at the edges, which makes it difficult to determine the end of the wake.

To be able to apply the model for different geometries and compare the results, the expressions for $b_{1/2}$ and u_1 are normalized. Consequently, $b_{1/2}$ is normalized by the diameter of the blunt body, giving the expression

$$\frac{b_{1/2}}{D} = C_b \cdot \left(\frac{x}{D}\right)^{1/3} \tag{1}$$

and u_1 is normalized by the inlet velocity, giving

$$\frac{u_1}{U_\infty} = C_u \cdot \left(\frac{x}{D}\right)^{-2/3} \tag{2}$$

where C_b and C_u are proportional constants of $b_{1/2}$ and u_1 respectively, x is the downstream distance behind the turbine and D is the diameter of the blunt body. For the investigated case of a wind turbine wake, D refers to the diameter of the wind turbine rotor. The constant C_b is calculated by plotting experimental values of normalized $b_{1/2}$ against the normalized distance, both in a logarithmic scale. This creates a linear line, which equation is calculated. Concurrently, the power law that defines the width, showed in table 1, is used for calculating $b_{1/2}/D$. Rewritten as $ln(b_{1/2}/D) = ln(C_b) + \frac{1}{3}ln(x/D)$, which has a similar distribution of a linear equation. Equating both intercept terms, the line and the rearranged power law, the parameter C_b can be determined. The same procedure is used to calculate the parameter C_u.

The expression derived by Schlichting to calculate the velocity distribution is

$$\frac{u_1}{U_\infty} = \frac{\sqrt{10}}{18\beta} \cdot \left(\frac{x}{C_D d}\right)^{-1/2} \cdot \left\{1 - \left(\frac{z}{b}\right)^{3/2}\right\}^2 \tag{3}$$

where β is the mixing length factor that is experimentally determined, C_D is the drag coefficient, d is the diameter of the cylinder, z is transverse position in the wake, and b is the width of the wake.

Notice that equation 3 was originally derived for a two-dimensional wake. However, L. M. Swain [9] investigated the circular wake and concluded that the same expression can be applied if the coefficients of the power laws were modified. In this study, the wake of a model wind turbine is investigated and thus, a circular wake is explored, the modified power laws presented in table 1 are being used. Thus, the expression used for the velocity distribution is:

$$\frac{u_1}{U_\infty} = \frac{\sqrt{10}}{18\beta} \cdot \left(\frac{x}{C_T A}\right)^{-2/3} \cdot \left\{1 - \left(\frac{z}{b}\right)^{3/2}\right\}^2 \tag{4}$$

A in this formula refers to the rotor swept area and C_D is replaced by the thrust coefficient of the turbine (C_T).

3. Experimental setup

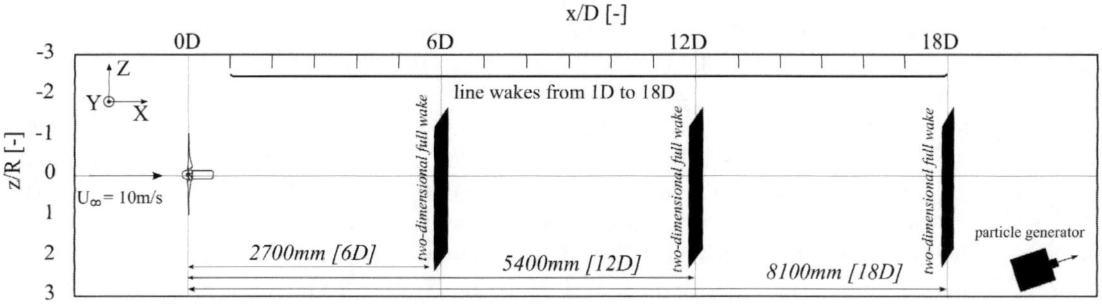

Figure 1: Non scaled sketch of the experimental set up in the wind tunnel.

The experiment in this study was carried out in the wind tunnel at the Department of Energy and Process Engineering at the Norwegian University of Science and Technology (NTNU). A

IOP Conf. Series: Journal of Physics: Conf. Series **854** (2017) 012015 doi:10.1088/1742-6596/854/1/012015

simple sketch of the experimental setup is presented in figure 1. The wind tunnel has a cross sectional area (zy-plane) of 2.7 m × 1.8 m and a length (x-axis) of 11 m. The turbine, which has a rotor diameter (D) of 0.45 m, was installed at a distance 2.71 D downstream of the inlet of the wind tunnel. The hub height was adjusted to 1.97 D (0.89 m) to have the rotor in the center of the cross section. The inlet velocity (U_∞) was set to 10.0 m/s and kept constant during the wake measurements. To ensure this, the velocity of the incoming flow was measured for every measurement point using the pressure difference at the contracted duct at the wind tunnel inlet. All the experiments were conducted at a low turbulence inflow with a turbulence intensity (TI) of 0.23%. In order to have uniform flow conditions, the roof of the test section was adjusted for a zero pressure gradient in the streamwise flow direction. Consequently, the mean velocity in the empty tunnel is found to be uniform within ± 0.6% [16]. A two-component Laser Doppler Velocimetry (LDV) probe, which was used to measure the velocity components, was mounted onto a traverse in order to move the LDV probe in the wind tunnel. In order for the LDV probe to detect the airflow, a small smoke machine generating particles with a size of 0.5-2.0 μm was placed in the far back of the wind tunnel. As the wind tunnel is of the type closed-loop, the particles were evenly distributed in the airflow during the returning process. 50 000 samples measured with the LDV probe were collected at every measuring point, resulting in an uncertainty of 7.3% in the measurements.

Figure 2: Picture of the wind turbine in the wind tunnel.

Three two-dimensional full wakes in the zy-plane were measured at distances in the x-direction of 6 D, 12 D and 18 D behind the turbine. The measured cross sectional area for the full wakes was ranging from -1.33 D to 1.33 D in z-direction, and from -1.07 D to 1.07 D in y-direction, with the center located at the rotor axis. With an increment of 0.13 D (0.06 m) in both directions, a measurement grid consisting of 357 points was investigated. In addition, line wake measurements were performed at every distance D from the turbine, ranging from 1 D to 17 D, at a center line 0.22 D (0.10 m) below the rotor axis. The z-axis remained the same, ranging from -1.33 D to 1.33 D, as well as the increment size, 0.13 D. The height for the line wake measurements was chosen at a lower y-value than the rotor axis due to the center of the wake shifted downward behind the turbine, which was observed from the full wake measurements which are presented in detail in chapter 4.

3.1. Turbine
For the presented study a new rotor was developed. The basis of the new rotor design is the rotor developed at the Department of Energy Process Engineering at NTNU Trondheim which

is described in detail by Krogstad and Lund [6]. The new rotor is also based on the NREL S826 airfoil, however, the blade parameters are modified to get a smaller rotor diameter. Here, the twist and the chord distribution were not changed, only the increment between the blade elements was halved to also get a rotor radius which has half the radius of the reference rotor. Thus, the new rotor has a diameter of 0.450 m. This smaller rotor was designed to be able to investigate the far wake and thus, a larger range behind the turbine. Furthermore, blockage effects and wake interference with the wind tunnel walls should be as small as possible such that the wake can develop without being contracted. Whereas the reference rotor has a wind tunnel blockage ratio of around 13% the new rotor has only a blockage ratio of around 3%. A higher limit of 10% blockage ratio is suggested to avoid the interference between the measurements and the walls of the wind tunnel [15]. Hence, it is expected that the wake can expand without being influence by the wind tunnel walls too much. The single rotor blades were produced with a 3D printer based on the multi-jet technology and connected at the hub with metal rings. The rotor was mounted on a newly designed turbine test rig, which was developed to have a small nacelle. The assembly in the wind tunnel is shown in figure 2. With this assembly the performance and the thrust for the new rotor were measured.

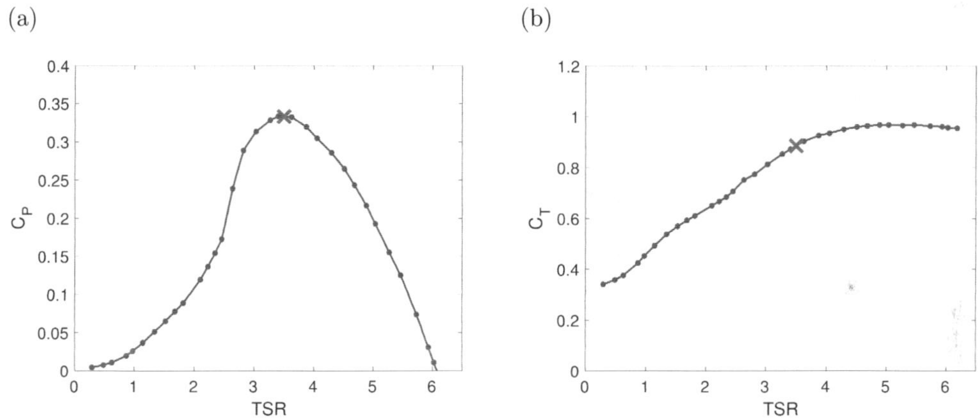

Figure 3: Experimental results for (a) C_p and (b) C_T over TSR. The red X marks the TSR where all the wake measurements were conducted.

In figure 3, the resulting power coefficient (C_P) and thrust coefficient (C_T) are presented. The optimum tip speed ratio (TSR) for the new rotor is 3.5, which is where the turbine reaches the highest power efficiency. This point is marked in figures 3a and 3b by a red cross. All the wake measurements were conducted at this optimum TSR of 3.5. The presented C_T represents the thrust of the rotor only. To separate the rotor thrust from the total thrust, the tower thrust was determined by exposing only the tower without the rotor to the flow. The resulting constant tower thrust was then subsequently subtracted from the thrust for the complete assembly for every TSR to get the rotor thrust. As all the wake measurements were done at the optimal TSR with an inlet velocity of 10.0 m/s, consequently, the resulting Reynolds number at the tip (Re_{tip} = 60 000) is rather low. However, Chamorro, Arndt and Sotiropolus showed in their study [17] that such low Reynolds numbers could effect the structure of the near wake as the turbulence characteristics in this region are influenced by the aerodynamic of the turbine blades. On the contrary, the flow characteristics of the far wake are rather dominated by the lateral transport and thus the mean velocity in the wake as well as higher order statistical moments show only a weak dependence to the Reynolds number. Consequently it can be concluded that the small rotor is suited well for far wake investigations.

4. Results

4.1. Velocity deficit in the yz-plane

In figures 4a, 4b and 4c, two-dimensional contour plots of the full wake in the zy-plane are presented at distances of 6 D, 12 D and 18 D, respectively. As expected, the velocity decay decreases as the wake moves downstream. The evident development of the wake can be seen, with an expansion of the wake as the downstream distance increases. This behaviour was expected, as previous work has already shown the same wake development [11]. In all three cases, a downshift of the center line of the wake can be observed. This is assumed to be due to interaction of the wake and the turbine tower. An uneven momentum induced by the presence of the tower pushes the wake downwards. The effect was investigated in detail by Pierella [12]. Consequently, the line measurements presented in the following sections were measured at a distance 0.22 D below the rotor axis. This distance was chosen after analyzing the wake and the wake center was defined where the lowest velocities were found. Its position is marked in figure 4 by a plus sign (+).

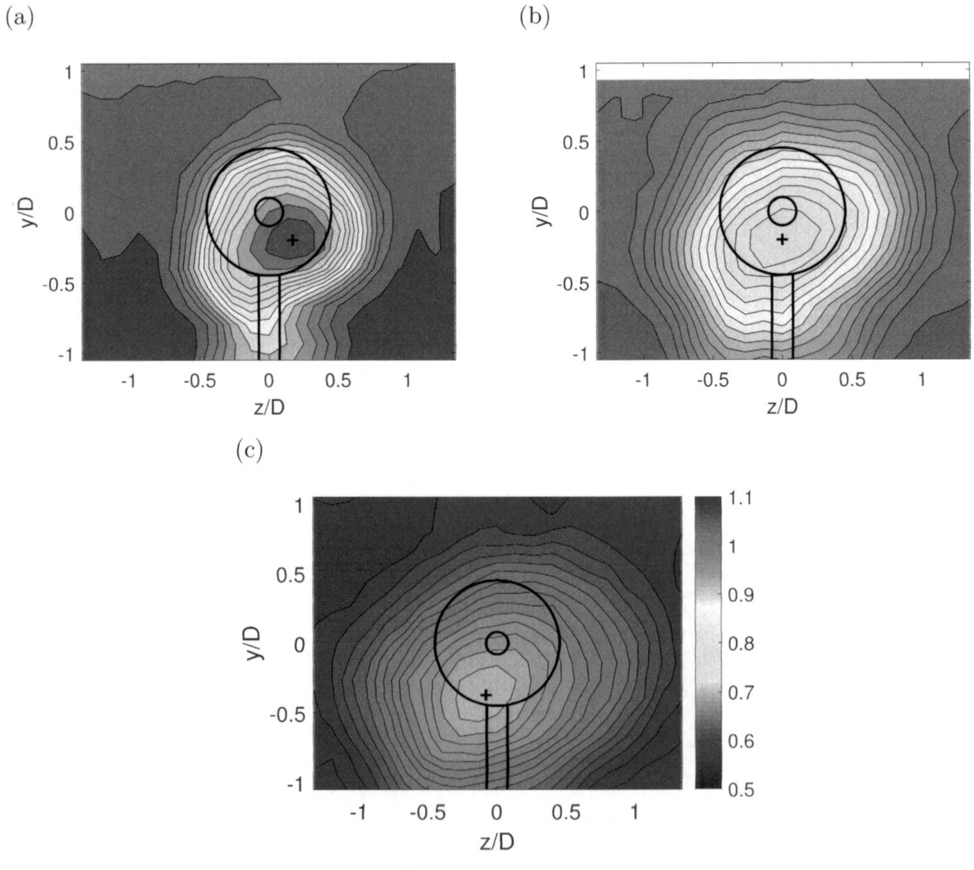

Figure 4: Two-dimensional velocity full wake plots in zy-plane at distances of (a) 6 D, (b) 12 D, and (c) 18 D behind the turbine with a rotor diameter of 0.450 m. The plus sign (+) in the plots marks the center of the wake.

At 6 D and 12 D, as expected from the observations by Pirealla [12], a slightly larger horizontal expansion can be seen, nevertheless the wakes are approximately circular whereas at 18 D a quite large expansion in horizontal direction can be observed and the wake has a rather oval shape.

This is expected to be due to the interference with the traverse in the back of the wind tunnel since, at 18 D, the LDV probe had to be placed beneath the framework supporting the traverse, instead of being placed in front of the framework, which was the case for the other downstream distances. This resulted in a contraction of the cross section and thus, a compression of the flow in y-direction. Hence, there is a clear influence of the framework of the traverse on the wake, and consequently, the measurements at 18 D are not considered in the further evaluation of the analytical Schlichting model.

4.2. Velocity development in the xz-plane

In order to see how the wake was developing with increasing distance downstream from the turbine, line wakes were measured from 1 D to 17 D. The measured points from the line wakes were interpolated to get a fine grid and thus counters for the velocity in flow direction could be determined. The resulting counter plot of the velocity development is shown in figure 5. The highest velocity deficit is found in the near wake region, whereas, as expected, the velocities increase with increasing distance due to turbulent mixing. In the near wake, which is where the wake is not yet fully developed and thus, does not has a Gaussian shape, the influence of the turbine geometry is still dominant and can clearly be observed by the two peaks in the velocity deficit. In the measured wake, the near wake can be defined as the region from right behind the turbine and downstream to 5 D. However, with increasing downstream distance the wake is getting more uni-modal and is considered to be fully developed in the far wake, which can be identified by a Gaussian shape. The outlined geometric curve in figure 5 represent the width development of the wake calculated according to the analytical model by Schlichting presented in section 2. In figure 5, it can be seen that the geometric curve is representing the outline of the far wake region, starting at around 6D, quite well.

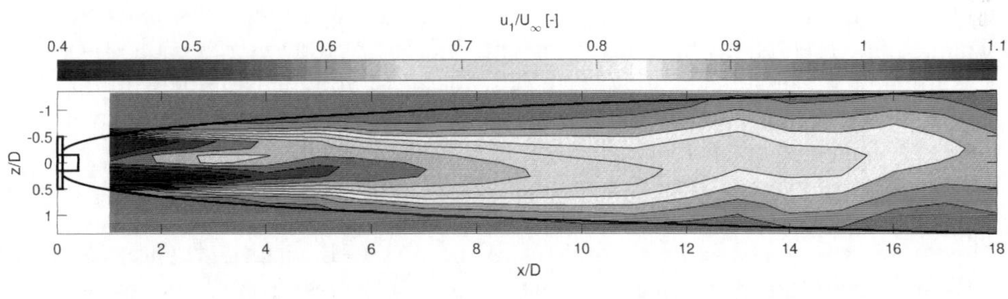

Figure 5: Velocity profiles of the wake in the xz-plane with a center line at y = -0.22 D. The geometric curve represents the wake outline applying the analytical wake model according to Schlichting.

5. Discussion

The velocity development of the experimental results analyzed in the previous chapter already suggests, that the analytical model derived by Schlichting seems to represent the far wake regime of a wind turbine very well. The velocity deficit depicted in figure 5 shows that the far wake begins approximately at a distance of 6 D, when the wake shows a Gaussian shape, meaning that the wake is fully developed. This is slightly later than suggested by Crespo et al. [8] who estimates the end of the near wake region between 2 and 5 diameters downstream of the turbine. A reason for this behaviour could be the low turbulent inflow, described in section 3, that results in less turbulent mixing of the wake and thus, a longer distances before the wake is fully developed [11]. In addition, Helmis et al [14] found that in wind tunnel experiments, the near wake region is overestimated. As the analyzed Schlichting model is asymthotic and thus has not such a good performance at low distances, the results at the distances 5 D and below are not considered in the comparison analysis.

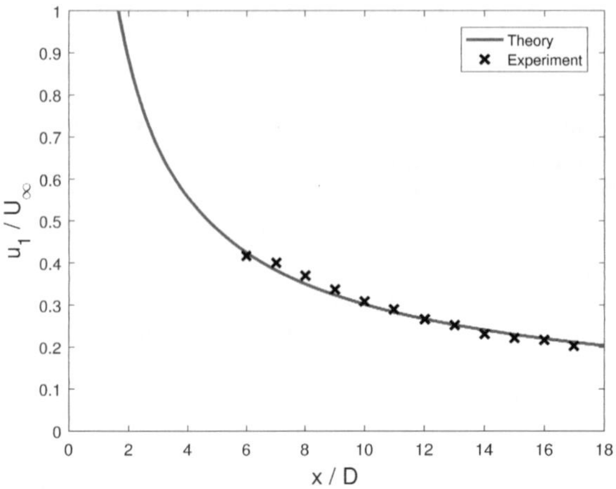

Figure 6: Comparison of the normalized velocity decay for the experimental results and the analytical model.

In figure 6, the experimental results of the normalized velocity decay of the line wakes from 6 D to 17 D are compared to the velocity deficit calculated using formula 2. The measurement results correspond very well with the theoretical line. As Schlichting's model behaves asymptotically, the performance is expected to be improved with increasing downstream distances. However, due to measurement uncertainty a small random spread of the measurement points around the line can be observed. Nevertheless the deviations are only around 1.9% in average and the maximum, which is found at 14D is with 3.9% also rather small.

The comparison between experiments and theory for the development of the width is presented in figure 7. Here, almost the same trend as for the velocity deficit can be observed. From 6 D and downstream the experimental results fit very well with the theoretical development of the width and a random distribution of the measured values around the graph can be seen. The average deviation is with 2.25% of the same order of magnitude as for the velocity deficit, with the higher deviation of up to -3.6% at 10 D and 3.4% at 15 D.

The velocity distribution along the z-axis at different downstream distances is shown in figure 8. To avoid an offset between the experimental and analytical results due to accelerated flow outside of the wake, the velocity in the wake was normalized with the velocity outside the wake for every measurement. Thus, U_∞ was selected to be around 10.3 m/s instead of the constant inlet flow velocity U_∞ of 10.0 m/s. That way, negative values in the outer area of the wake could be avoided, and the theoretical graph could be compared better to the values resulting from the experiments.

The wake profile resulting from the measurements show a similar shape at all different distances. When comparing them to the theoretical velocity profile it can be seen that the theory does not exactly describe the velocity distribution at the edges of the wake $|z/b_{wake}| > 0.70$ as the deviation is over 50% at all distances. However, it should be clarified that these derivations are not relevant, as the order of the magnitude is very small in that region and even very small differences create large deviation that are not significant. Thus, only the inner region of the

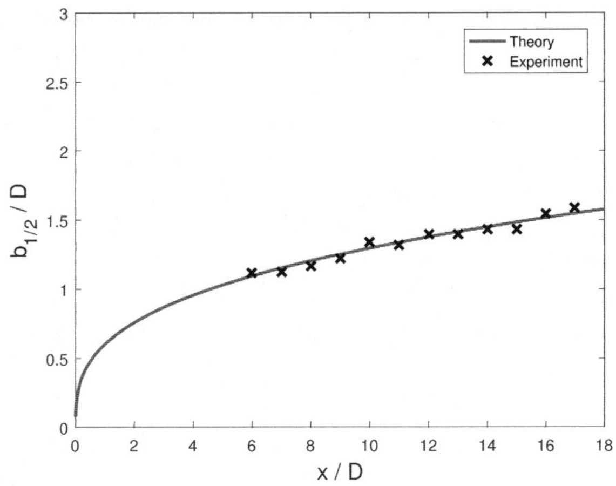

Figure 7: Comparison of the normalized width development for the experimental results and the analytical model.

distribution is considered. In the inside of the wake $|z/b_{wake}| < 0.70$, the deviation is reduced to an average deviation of around 11.3%. Whereas the clearest distinctions can be observed in the center of the wake, where the experimental results tend to form a shallower peak than the wake resulting from the theory. Here the velocity distribution at 6 D shows the highest differences from the theory, which was expected as the the model is asymptotic and thus has it worst performance at low distances.

The velocity distributions from 6 - 15 D show a similar profile, which has a Gaussian shape, meaning that the wake in the far wake regime has self-similarity behaviour [13], as the velocity is distance independent from the turbine. The analytical model is well suited to estimate the shape of the velocity profile in the far wake.

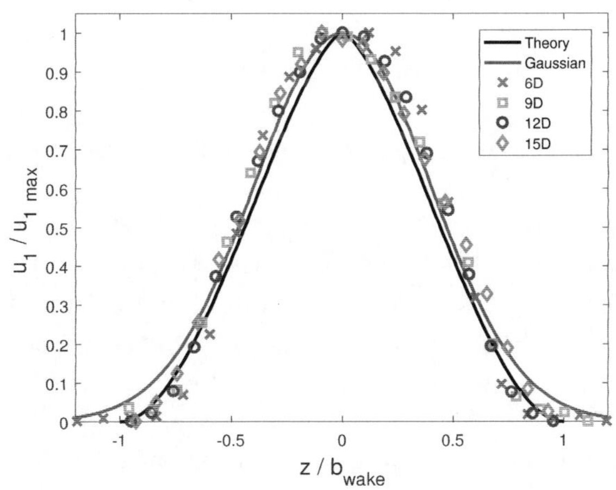

Figure 8: Comparison of the wake shape from the analytical model and velocity measurements at 6 D, 9 D, 12 D and 15 D.

6. Conclusion

An analytical wake model derived for blunt bodies was investigated for the application of the prediction of wind turbine wakes. Results of the analytical expressions were compared to experimental results of the wake formed behind a model wind turbine. The theoretical results for the velocity deficit and the wake width are in good agreement with the experimental results. For both investigated parameters the deviations between experiments and theory are in the range of only 2% in the analyzed far wake region. Also the Gaussian shape of the velocity deficit in the far wake regime of a wind turbine rotor can be predicted fairly well by the analytical model. Consequently, it can be concluded that the investigated theory derived by Schlichting, which was originally derived for blunt bodies, is applicable to predict the wake of a wind turbine rotor.

References

[1] Wiser R H and Bolinger M 2016 *2015 Wind technologies market report*
[2] Lissaman P B S 1979 Energy effectiveness of arbitrary arrays of wind turbines *J Energy* **3**(6) 323-8
[3] Voutsinas S G, Rados K G and Zervos A 1990 On the analysis of wake effects in wind parks *J Wind Eng Ind Aerod* **14** 204-19
[4] Vermeulen P E J. 1980 An experimental analysis of wind turbine wakes *3rd Int Symp on Wind Energy Systems* Lyngby 431-50
[5] Sheerin P M, Sforza and M. Smorto 1981 Three-dimensional wakes of simulated wind turbines *AIAA J* **19**(9) 1101-7
[6] Krogstad P Å and Lund J 2012 An experimental and numerical study of the performance of a model turbine. *Wind Energy* **15**(3) 443-57
[7] Schlichting H 1979 *Boundary Layer Theory* McGRAW-HILL series in Mechanical Engineering 7th
[8] Crespo A, Hernández J and Frandsen S 1999 Survey of modelling methods for wind turbine wakes and wind farms *Wind Energy* **2**(1) 1-24
[9] Swain L M 1929 On the turbulent wake behind a body of revolution *Proc. R. Soc. A* **125**(799) 647-59
[10] Ainslie J F 1988 Calculating the flow field in the wake of wind turbines *J Wind Eng Ind Aerod* **27**(1) 213-24
[11] Sanderse B 2009 Aerodynamics of wind turbine wakes - literature review Report ECN-E-09e016
[12] Pierella F and Sætran L 2017 Wind tunnel investigation on the effect of the turbine tower on wind turbines wake symmetry Paper accepted for publication in *Wind Energy*
[13] Gratton J 1991 Similarity and self similarity in fluid dynamics *FCPH* **15** 1-106
[14] Helmis CG, Papadopoulos KH, Asimakopoulos DN, Papageorgas PG, Soilemes AT. 1995 An experimental study of the near wake structure of a wind turbine operating over complex terrain *Solar Energy* **54**(6) 413–28.
[15] De Vries O. 1983 On the theory of the horizontal-axis wind turbine *Ann Rev Fluid Mech* **15** 77-96
[16] Bartl J and Sætran L 2017 Blind test comparison of the performance and wake flow between two in-line wind turbines exposed to different turbulent inflow conditions *Wind Energ Sci* **2** 55-76
[17] Chamorro L.P., Arndt R. and F. Sotiropoulos 2012 Reynolds number dependence of turbulence statistics in the wake of wind turbines *Wind Energy* **15**(5) 733-742

Account of ambient turbulence for turbine wakes using a Synthetic-Eddy-Method

Grégory Pinon[1,*], Clément Carlier[1,2], Arnaud Fur[1,2], Benoît Gaurier[2], Grégory Germain[2], Élie Rivoalen[1,3]

1 - Normandie Univ, UNIHAVRE, CNRS, LOMC, 76600 Le Havre.
2 - IFREMER, LCSM, 150 quai Gambetta, BP 699, 62321 Boulogne-sur-Mer, France.
3 - Normandie Univ, INSA Rouen, LMN, 76000 Rouen, France.

E-mail: `gregory.pinon@univ-lehavre.fr`

Abstract. The present paper aims at describing the use of a Synthetic-Eddy-Method (SEM), initially proposed by Jarrin *et al.* [12], in the 3D Lagrangian Vortex method framework. The SEM method is used here in order to generate a far-field incoming flow with a prescribed ambient turbulence intensity. However, for the account of the diffusive term in the Navier-Stokes equations, a classical Particle Strength Exchange model with a LES eddy viscosity is used.

Firstly, the general characteristics of the Synthetic-Eddy-Method will be presented together with its integration in the framework of the developed 3D unsteady Lagrangian Vortex software [27]. The capability of the ambient turbulence model to reproduce a perturbed flow that verifies any turbulence intensity I_∞ and any anisotropic ratio (σ_u:σ_v:σ_w) will be discussed and validated. Then, the capability of the presented ambient turbulence model to compute turbine wakes will also be presented together with first results. Finally, comparisons will be made between the obtained numerical results against experimental data [22, 23] for two levels of ambient turbulence, namely $I_\infty = 3\%$ and $I_\infty = 15\%$. Although the present study was initially performed in the framework of tidal energy, its application to wind energy is straightforward.

1. Introduction

The development of renewable energy has been rather intensive for some years, mainly using wind turbines. But tidal turbine are also arising. However, at the scientific level, many studies are still necessary in order to fully understand the behaviour of such arrays. One of these topics is the impact that ambient turbulence has on the behaviour of turbines in a farm. Indeed some studies demonstrated that ambient turbulence highly modifies the behaviour of turbines [26, 1, 6, 5, 4, 8]. In the tidal energy community, many recent studies have shown that ambient turbulence intensity I_∞ (see eq. (1)) highly modifies the behaviour of turbines [8, 2, 25, 22, 19].

$$I_\infty = 100\sqrt{\frac{\frac{1}{3}[\sigma^2(u_\infty)+\sigma^2(v_\infty)+\sigma^2(w_\infty)]}{\bar{u}_\infty^2+\bar{v}_\infty^2+\bar{w}_\infty^2}} \quad (1)$$

In the previous equation, $\boldsymbol{u} = (u, v, w)$ is the 3D velocity vector, $\bar{u}$ stands for the average-value of any velocity record u and $\sigma(u)$ represents the corresponding standard deviation. One of the most noticeable influence can be observed in the wake, downstream of the turbine. Indeed as shown in Ref. [22] for instance, the higher the turbulence intensity I_∞ is, the faster the wake effects decrease. Moreover the interactions between turbines are also highly modified by ambient turbulence as shown by Mycek *et al.* [23] in their study of the performances and wake of two interacting turbines. The relative spread in *in situ* ambient turbulence levels inhibits the definition of general guidelines on tidal arrays with respect to turbulence. For wind energy, although the fact that atmospheric boundary layer has a different behaviour than in sea, the situation may be similar. Therefore, the influence of each level of ambient turbulence needs to be characterised, either experimentally or numerically.

Consequently numerical simulations have to represent the ambient turbulence, or at least its effects on the performance and wake of the turbines. In the present paper, the objective is not to describe a turbulence model (like a RANS closure, LES or DES models) but how to generate accurate turbulent inflow conditions, that represent a given ambient turbulence. In the literature, some numerical works were carried out in that aim. One of the most famous initialisation of ambient turbulence boundary conditions is TurbSim [13] from NREL. But many others exist. For example, Chatelain *et al.* [6] used the Mann's algorithm [17] to synthesise a turbulent inflow in the case of wind turbines simulations. Branlard *et al.* [5, 4] used a similar Vortex method approach with frozen and un-frozen approach for turbulence and possibly shear. The present study considers the Synthetic-Eddy-Method (SEM), initially proposed by Jarrin *et al.* [12, 11], because an easier adaptation to the Vortex Method framework is foreseen. Togneri *et al.* [30, 31] already investigated a similar Synthetic-Eddy-Method in order to generate synthetic turbulent inflow conditions for their BEMT-CFD software.

This paper presents the latest numerical developments carried out to integrate a SEM module in the developed 3D Vortex Method software [27]. This software already possesses an LES turbulent model, its turbulent eddy viscosity being based on the work of Mansour *et al.* [18].

The first part of this paper is devoted to the presentation of the numerical method used to compute the flow around a wind or tidal turbine. Then in a second part, the Synthetic-Eddy-Method is presented as well as its integration in the present software. The last part of this paper focuses on preliminary results obtained using this method to represent the ambient turbulence I_∞. Eventually, conclusions are drawn on the presented work and an outlook on future development is proposed.

2. Numerical method

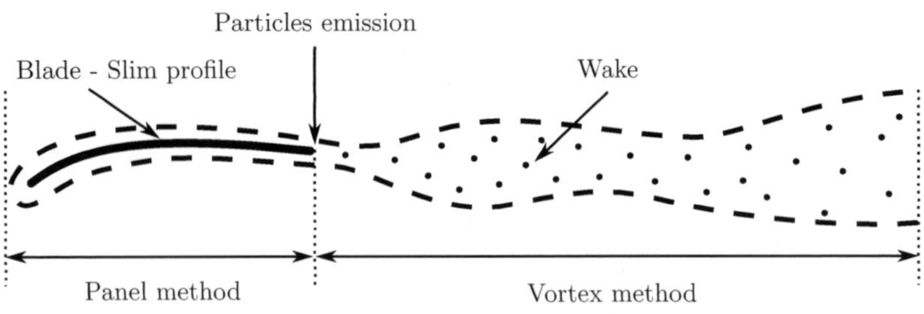

Figure 1. Scheme describing the different methods used in the software. The blade is supposed infinitely thin and is presented as cut-section.

The developed software to simulate turbines and wakes is based on the Vortex method [29, 14, 9]. This method is an unsteady Lagrangian method, where the flow is discretised using vorticity carrying particles (see Fig. 1) and the turbines are represented using a panel method [3]. In this numerical method, the Navier-Stokes equations for an unsteady and incompressible flow are used in their velocity/vorticity $(\boldsymbol{u}, \boldsymbol{\omega})$ formulation:

$$\nabla \cdot \boldsymbol{u} = 0, \tag{2}$$

$$\frac{D\boldsymbol{\omega}}{Dt} = (\boldsymbol{\omega} \cdot \nabla)\,\boldsymbol{u} + \nu\Delta\boldsymbol{\omega}, \tag{3}$$

where $\boldsymbol{u}$ is the velocity field, $\boldsymbol{\omega} = \nabla \wedge \boldsymbol{u}$ is the vorticity field and ν is the kinematic viscosity. Equation (3) represents the momentum equation in the velocity-vorticty formulation. $(\boldsymbol{\omega} \cdot \nabla)\,\boldsymbol{u}$ stands for the stretching term and $\nu\Delta\boldsymbol{\omega}$ for diffusion. By means of a viscous splitting (see Chap. 5 of Cottet & Koumoutsakos [9]), diffusion is handled via a Particle Strength Exchange model initially developed by Degond & Mas-Gallic [10] and Choquin & Huberson [7]. Additionally, an LES model with a turbulent eddy viscosity based on the work of Mansour *et al.* [18] completes the numerical model for diffusion. At this point of the document, it has to be reminded that the pre-cited methods to treat the viscous term of equation (3) (via a PSE scheme) and turbulent eddy viscosity (the LES of Mansour *et al.*) are rather well known in the scientific community. One can refer to [27] for more detailed information of the implementation. But the real novelty of the present paper is the use of the Synthetic Eddy Method (SEM) to generate far-field ambient turbulence as it will be discussed in Section 3.

The Helmholtz decomposition of the velocity field (eq. (4)) is used in order to decompose the velocity vector:

$$\boldsymbol{u} = \nabla \wedge \boldsymbol{\psi} + \nabla\phi + \boldsymbol{u}^{\infty} = \boldsymbol{u}^{\psi} + \boldsymbol{u}^{\phi} + \boldsymbol{u}^{\infty}, \tag{4}$$

Thus the velocity field is divided into three parts:

- a potential component $\boldsymbol{u}^{\phi}$ representing the influence of the turbines blades on the flow (see Fig. 1 and 2), a panel method (boundary integrals) is used to solve the Laplace equation for ϕ.
- a rotational component $\boldsymbol{u}^{\psi}$ representing the wake of the turbines (see Fig. 1 and 2), a 3D Vortex method is used to solve a Poisson equation for $\boldsymbol{\psi}$.
- a velocity vector $\boldsymbol{u}^{\infty}$ representing the upstream tidal current.

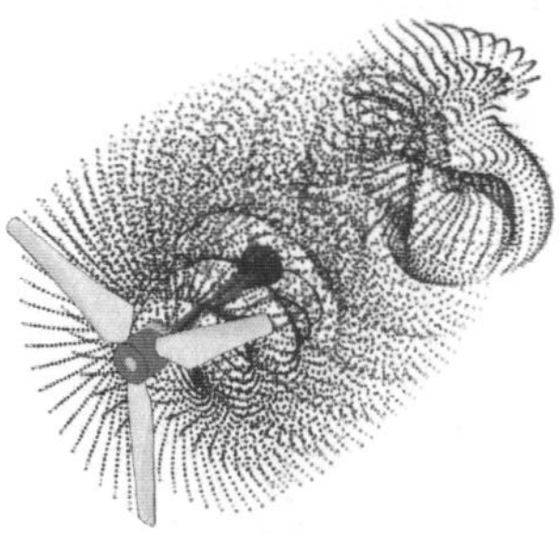

Figure 2. Rough results of the computation showing the blades coloured by the dipole-value and the free vorticity carrying particles.

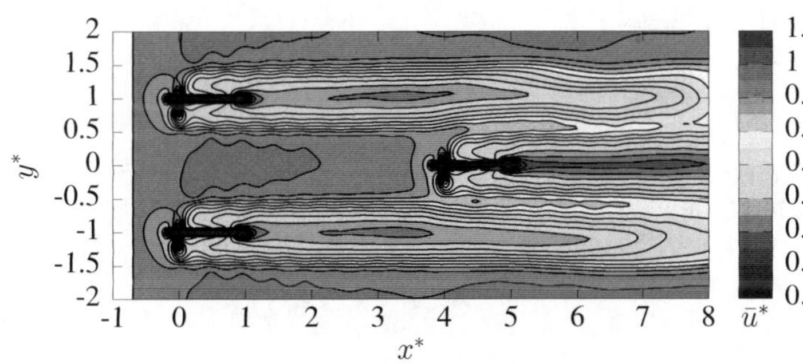

Figure 3. Post-treated velocity map of 3 interacting turbines immersed in incoming flow. Here, the ambient turbulence intensity is set to $I_{\infty} = 0\%$. The image is reproduced from Mycek *et al.* [24].

In order to make the link between the rotational and the potential parts of the velocity, particles are emitted at the blade's trailing edge using a Kutta condition [14]. The emitted particles are then advected in the flow as shown in Fig. 2. The rotational part of the velocity field is calculated as the sum of the influence of these vorticity carrying particles. The expression of the component $\boldsymbol{u}^{\psi}$ is obtained in every point of the fluid domain by solving the Biot-Savart law [32]. The calculation of this $\boldsymbol{u}^{\psi}$-component for each particles represents the major part of the computational cost. For that reason, a *Treecode* algorithm inspired from Lindsay and Krasny [15] is used to reduce the computational time. Then, the results are post-treated in order to obtain a map of the velocity field as presented in Fig. 3. This previous velocity map is reproduced from Mycek *et al.* [24], where a judicious pre-conditioner is presented for solving the potential component $\boldsymbol{u}^{\phi}$ in case of several turbines in interaction. In that respect, a pre-conditioned Bi-CGSTAB solver is presented in Ref. [24] that really speeds up the computation of this velocity component. For more details on the computations, including Biot-Savart equation desingularisation, remeshing or time-stepping, etc., one can refer to [27].

3. Synthetic-Eddy-Method

The Synthetic-Eddy-Method was initially developed to generate an inflow velocity boundary condition for a given turbulence intensity I_{∞} and a given anisotropic ratio $(\sigma_u{:}\sigma_v{:}\sigma_w)$ by adding a perturbation term, which is added to the far-field mean velocity of the flow $\boldsymbol{u}^{\infty} = (u^{\infty}, v^{\infty}, w^{\infty})$. The present paper aims at describing its implementation in the Lagrangian Vortex framework, and not only as inflow velocity condition but applied to the whole flow domain.

3.1. General overview of the method

Similarly to a Reynolds decomposition of the velocity, a fluctuating term $\boldsymbol{u}^{\sigma}$ is added to the upstream mean velocity $\overline{\boldsymbol{u}^{\infty}}$ and the upstream velocity $\boldsymbol{u}^{\infty}$ then becomes:

$$\boldsymbol{u}^{\infty}(\boldsymbol{x}) = \overline{\boldsymbol{u}^{\infty}} + \boldsymbol{u}^{\sigma}(\boldsymbol{x}). \tag{5}$$

On the contrary to the already presented method [12, 11], this perturbation term is not only added at the inlet of the fluid domain but everywhere in the studied space. This aspect will be treated later in sub-sec. 3.3.

The perturbation term $\boldsymbol{u}^{\sigma}$ is calculated as the influence of N generated "turbulent structures", also called "eddies" in [12, 11], randomly placed in the studied space. Then the perturbation term $\boldsymbol{u}^{\sigma}$ induced by those N turbulent structures is simply defined as the sum of the influences of each structure k:

$$\boldsymbol{u}^{\sigma}(\boldsymbol{x}) = \sum_{k=1}^{N} \boldsymbol{u}^{\sigma,k}(\boldsymbol{x}), \tag{6}$$

with $\boldsymbol{x}$ being any point of the flow and $\boldsymbol{u}^{\sigma,k}$ the velocity induced by a single turbulent structure k. This last term can be expressed as:

$$\boldsymbol{u}^{\sigma,k}(\boldsymbol{x}) = \sqrt{\frac{V_{\sigma}}{N}}\boldsymbol{c}^{k} F_{\boldsymbol{\lambda}}(\boldsymbol{x} - \boldsymbol{x}^{k}), \forall k \in [1, N], \tag{7}$$

where V_{σ} is the volume of the three-dimensional space containing all the N "turbulent structures". At this point, the turbulent structure k and more particularly its intensity $\boldsymbol{c}^{k}$ needs to be defined together with the function $F_{\boldsymbol{\lambda}}$.

3.2. Intensity and shape function

Each turbulent structure k is defined by its position $\boldsymbol{x}^{k}$ in the flow and its intensity $\boldsymbol{c}^{k}$. The intensity $\boldsymbol{c}^{k}$ is defined as:

$$c_i^k = \sum_{j=1}^{3} a_{i,j}\epsilon_{i,j}^k \quad \forall i \in \{1, 2, 3\}, \forall k \in [1, N]. \tag{8}$$

The different $\epsilon_{i,j}^k$ terms that intervene in equation (8) are independent random values that follow a normal or Gaussian distribution centred around 0 with a variance of 1. This $\epsilon_{i,j}^k$ term is supposed to represent the random aspect of turbulence. The elements $a_{i,j}$ represents any element of a matrix $\underline{\underline{A}}$, which is basically the Cholesky decomposition of the Reynolds Stress Tensor $\underline{\underline{R}}$ (see eq. (9)).

$$\underline{\underline{R}} = \begin{pmatrix} R_{1,1} & R_{1,2} & R_{1,3} \\ R_{2,1} & R_{2,2} & R_{2,3} \\ R_{3,1} & R_{3,2} & R_{3,3} \end{pmatrix} = \underline{\underline{A}}\underline{\underline{A}}^T \quad \text{with} \quad \underline{\underline{A}} = (a_{i,j}) \tag{9}$$

With these equations (8) and (9), the link between the intensities $\boldsymbol{c}^k$ of the turbulent structures and the Reynolds Stress Tensor $\underline{\underline{R}}$ ensures the generation of a velocity field that statistically replicates any given turbulence intensity I_∞ and any given anisotropic ratio $(\sigma_u{:}\sigma_v{:}\sigma_w)$ [16, 12]. Indeed the three components of the anisotropic ratio $(\sigma_u{:}\sigma_v{:}\sigma_w)$ are just the square root of the diagonal components over the component R_{11} of the Reynolds Stress Tensor $\underline{\underline{R}}$. Moreover the turbulence intensity I_∞ expression can be rewritten in function of the trace of the Reynolds Stress Tensor $\underline{\underline{R}}$ as shown in equation (10):

$$\begin{aligned} I_\infty &= 100\sqrt{\frac{\frac{1}{3}[\sigma^2(u_\infty)+\sigma^2(v_\infty)+\sigma^2(w_\infty)]}{\bar{u}_\infty^2+\bar{v}_\infty^2+\bar{w}_\infty^2}} \\ &= \frac{100}{|\boldsymbol{u}^\infty|}\sqrt{\frac{R_{1,1}+R_{2,2}+R_{3,3}}{3}} \\ &= \frac{100}{|\boldsymbol{u}^\infty|}\sqrt{\frac{\mathrm{tr}(\underline{\underline{R}})}{3}} \end{aligned} \tag{10}$$

This last equation (10) ensures that the generated turbulence will have the desired turbulence intensity I_∞ and the desired anisotropic ratio $(\sigma_u{:}\sigma_v{:}\sigma_w)$.

The second part of equation (7) is $F_{\boldsymbol{\lambda}}$ also called the shape function and it can be expressed as:

$$F_{\boldsymbol{\lambda}}(\boldsymbol{y}) = \prod_{i=1}^{3} f_{\lambda_i}(y_i). \tag{11}$$

$\boldsymbol{\lambda}$ defines the size of the zone of influence of each turbulent structure k. This size is probably linked to a turbulent length scale (Taylor or Kolmogorov length scales) but this will need further investigation. As it is defined by a vector $\boldsymbol{\lambda}$, and presented in equation (11), the zone of influence can ideally have different sizes regarding each coordinate (λ_i). And additionally, each turbulent structure could have its own size vector $\boldsymbol{\lambda}$ but this will be treated in the following. The sub-functions f_λ that intervene in the evaluation of the shape function $F_{\boldsymbol{\lambda}}$ have to meet some conditions to ensure that the ambient turbulence has the chosen characteristics:

$$argmax_y(f_\lambda(y)) = 0, \tag{12a}$$

$$f_\lambda(y) = f_\lambda(-y), \tag{12b}$$

$$\int_{-\lambda}^{\lambda} (f_\lambda^2(y)dy = 1. \tag{12c}$$

In the preliminary results presented in section 4, the basic shape function $F_{\boldsymbol{\lambda}}$ (eq. (11)) indicated in Jarrin *et al.* [12, 11] is used. Indeed its sub-function f_λ simply is a tent function, represented by a triangular function centred at zero with a base of 2λ.

$$f_\lambda(y) = \begin{cases} \sqrt{\frac{3}{2\lambda^3}}(\lambda - |y|) & \text{if} \quad |y| < \lambda \\ 0 & \text{otherwise.} \end{cases} \tag{13}$$

Figure 4 depicts tent functions for different values of λ. It is anticipated that this basic tent function (eq. 13) will not meet all the requirement of smoothness, especially if ones want to computed the velocity gradient, etc. However, in order to start with, this basic function issued from the bibliography will be used but other shape function were already tested and validated.

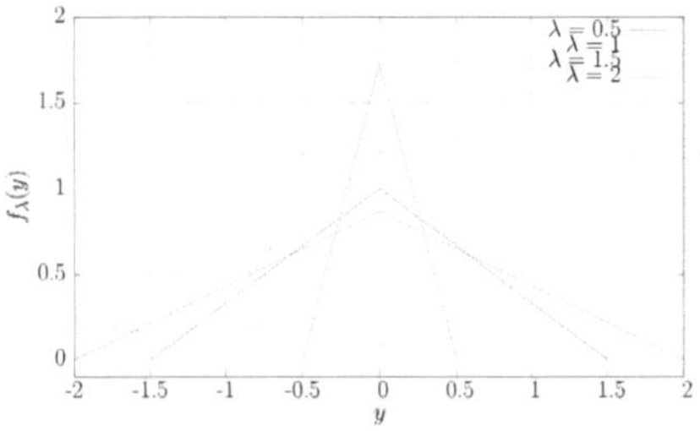

Figure 4. The tent function for different sizes λ of turbulent structure.

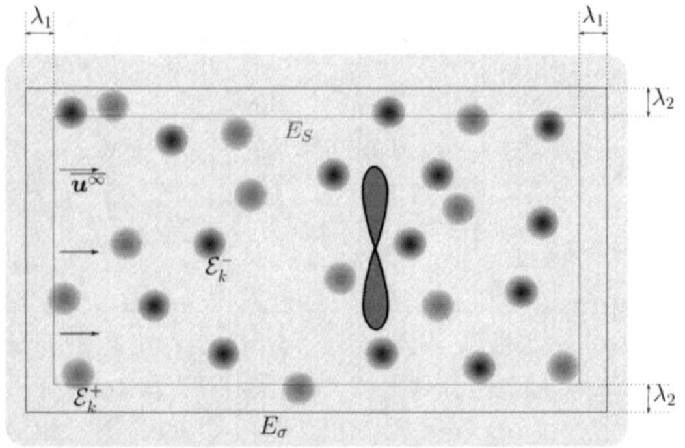

Figure 5. General representation scheme of the developed ambient turbulence model.

3.3. Integration of the Synthetic-Eddy-Method into the Vortex method

The numerical method presented in Section 2 is an unsteady Lagrangian method defined for an unbounded domain. Indeed, the fluid domain is supposed to be infinite but the zone where particles are present, grows with the particles emission and their advection in the flow. In fact, there is no domain boundaries as in Eulerian formulation. Unfortunately the Synthetic-Eddy-Method need the definition of a precise volume V_σ (see eq. (7)) containing all the turbulent structures, which is not required for classical Vortex method. Thus in order to combine the Synthetic-Eddy-Method with the Vortex method, a studied space E_S needed to be defined. The desired velocity fluctuations induced by the Synthetic-Eddy-Method will be applied inside this E_S domain.

Once the studied space E_S is defined, a turbulent space E_σ of volume V_σ can be created. This turbulent space E_σ contains all the turbulent structures used in the Synthetic-Eddy-Method. If one wants to have a statistically correct velocity fluctuations everywhere inside the studied space E_S, some conditions on the turbulent space E_σ are needed, especially at the limit. Therefore, the turbulent space E_σ needs to be a little larger than the studied space E_S:

$$\min(x_i \in E_\sigma) \leq \min(x_i \in E_S) - \lambda_i \quad \forall i \in \{1,2,3\}, \tag{14a}$$

$$\max(x_i \in E_\sigma) \geq \max(x_i \in E_S) + \lambda_i \quad \forall i \in \{1,2,3\} \tag{14b}$$

with $\boldsymbol{x}$ a position vector and $\boldsymbol{\lambda} = \lambda_i$, $(i = 1,3)$ the size vector of the turbulent structures. Those conditions simply ensure that all the study space is influenced by the turbulent structures.

Now that the turbulent space E_σ is defined, the generation of the turbulent structures can be performed. E_σ is initialised with the N turbulent structures k and their intensities $\boldsymbol{c}^k$ are calculated using equation (8) with $\epsilon_{i,j}^k$ taking the value 1 with a probability of 1/2 and the value -1 with an equal probability. Then each turbulent structure is randomly placed in the turbulent space E_σ.

During the simulation, the turbulent structures will be advected by the flow. Here, different strategies can be applied: either the structures can be advected by $\overline{\boldsymbol{u}^\infty}$, which is somehow too simplistic, or the structures can be advected by $\boldsymbol{u}$ as defined in equation (4). Additionally, these structures could also be used in order to compute the stretching term of the Navier-Stokes equation (first term of the right end side of equation (3) but not detailed here). In this preliminary study, the first option was chosen that is: the structures are advected by $\overline{\boldsymbol{u}^\infty}$ without influencing the stretching term. These structures work somehow as an under-layer, which means that they will influence the velocity distribution on the turbines (and hence the particles emission process) and also the particles in the wake. But they will not be influenced reciprocally.

Studied case	$R_{11} = \sigma_u^2$ $\times 10^{-2}$	$R_{22} = \sigma_v^2$ $\times 10^{-2}$	$R_{33} = \sigma_w^2$ $\times 10^{-2}$	Err/I_∞^{theo}
$I_\infty^{theo} = 15\%$	3.60	2.02	1.13	0%
$N = 100$	3.30	2.03	1.13	3.13%
$N = 500$	3.49	2.00	1.20	1.55%
$N = 1000$	3.51	1.96	1.09	1.18%
$I_\infty^{theo} = 15\%$	3.60	2.02	1.13	0%
$\boldsymbol{\lambda} = (0.5 : 0.5 : 0.5)$	3.59	2.04	1.17	0.09%
$\boldsymbol{\lambda} = (0.75 : 0.75 : 0.75)$	3.59	2.03	1.11	0.08%
$\boldsymbol{\lambda} = (1.0 : 1.0 : 1.0)$	3.51	1.96	1.09	1.18%
$I_\infty^{theo} = 15\%$	3.60	2.02	1.13	0%
$\boldsymbol{\lambda} = (0.5 : 0.4 : 0.5)$	3.68	2.08	1.16	1.29%
$\boldsymbol{\lambda} = (0.5 : 0.3 : 0.5)$	3.58	2.04	1.11	0.41%
$\boldsymbol{\lambda} = (0.5 : 0.2 : 0.5)$	3.57	2.01	1.11	0.05%
$I_\infty^{theo} = 15\%$	3.60	2.02	1.13	0%
$\sigma(\lambda) = 25\%$	3.56	1.93	1.08	1.9%
$\sigma(\lambda) = 50\%$	3.58	2.02	1.15	0.37%
$\sigma(\lambda) = 100\%$	3.66	2.08	1.10	0.49%
$I_\infty^{theo} = 15\%$	3.60	2.02	1.13	0%
Triangular function	3.51	1.96	1.09	1.18%
Polynomial function	3.68	2.04	1.22	0.15%
Cosine function	3.68	2.10	1.22	1.37%
Gaussian function	3.47	1.95	1.14	2.30%

Table 1. Error on the rebuild turbulent flow-field for $I_\infty = 15\%$. Unless indicated, $\boldsymbol{\lambda}_i = (1.0, 1.0, 1.0)$, $\sigma(\lambda) = 0\%$, 1000 structures and a triangular function were used. The value of $\sigma(\lambda)$ indicates the percentage of spread in the turbulent structure sizes.

Regarding the second option (the structures are advected by the actual fluid velocity $\boldsymbol{u}$), several questions arise. In fact, the validity of the method relies on the randomness of the turbulent structures locations; will this property still be valid when advected by $\boldsymbol{u}$? Thus, with the second approach, the Synthetic-Eddy-Method may not replicate the prescribed Reynolds stress tensor. Some more work needs to be performed on that prior to choose this second option.

Finally during the advection of the turbulent structures, each structure leaving the turbulent space E_σ will be suppressed and replaced by a newly generated one at the inlet of the turbulent space E_σ.

4. Numerical results

The numerical results presented in this section are issued from preliminary computations and will only be used as proof of concept. At first in sub-section 4.1, the ambient turbulence module will be run without any turbine to ensure that the model correctly reproduces the different input parameters. Then, computations will be presented in sub-section 4.2 with the presence of a single turbine for a moderate and a higher turbulence intensity I_∞ levels.

4.1. Results on the ambient turbulence model without turbine

The ambient turbulence module was firstly studied without turbine to ensure that the model correctly reproduces the different input parameters such as the turbulence intensity I_∞ and the anisotropic ratio (σ_u:σ_v:σ_w). This verification is quite straight forward as I_∞, σ_u, σ_v and σ_w can be deduced from the Reynolds Stress Tensor $\underline{\underline{R}}$ (see eq. (10)).

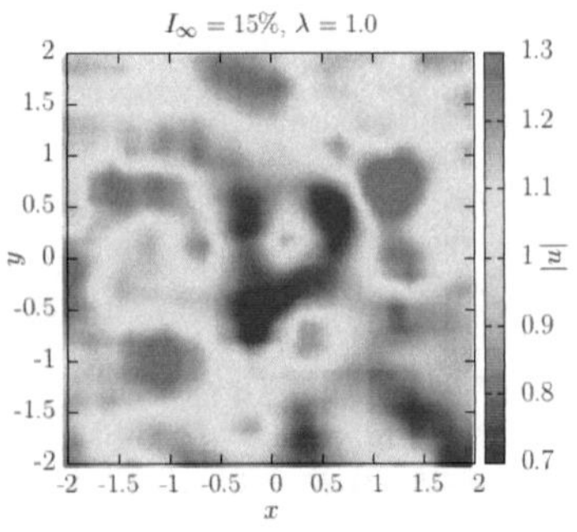

Figure 6. Rebuilt velocity field with $N = 1000$ structures of size $\boldsymbol{\lambda} = (1.0, 1.0, 1.0)$ and $I_\infty = 15\%$.

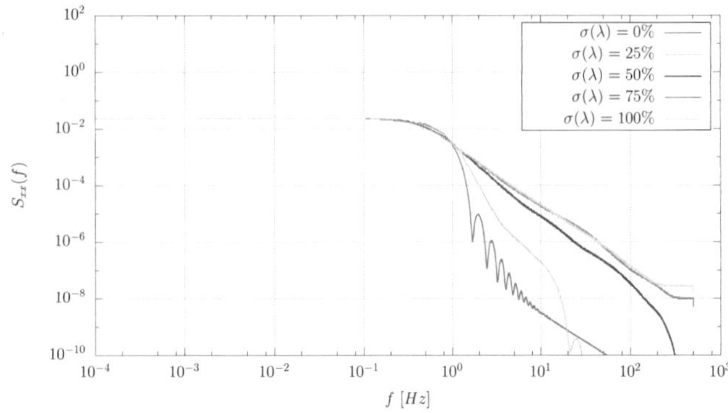

Figure 7. Influence of enabling a variation of the turbulent structure sizes (represented by $\sigma(\lambda)$) on the power spectral density with a Gaussian function, a mean $\boldsymbol{\lambda} = (0.5, 0.5, 0.5)$, a filling ratio of $R_f = 2$ and a given turbulence intensity of I_∞.

To this aim, 3D turbulent velocity fields were generated on a fluid domain E_S of dimensionless size $6 \times 6 \times 6$. In order to validate the model, the generated flow-field characteristics are analysed. Table 1 gives examples of comparison between the Reynolds Stress Tensor $\underline{\underline{R}}$ prescribed to the model and the recalculated ones from the generated flow. These examples are for diagonal Reynolds Stress Tensors $\underline{\underline{R}}$, which correspond to an ambient turbulent I_∞ of 15%. But the validation was also performed for other intensity values. The anisotropic ratio was arbitrarily set to the one observed in Milne *et al.* [21]. The recomputed values from the generated flow show an excellent agreement with the prescribed ones and thus regardless of the number of used turbulent structures, their size, functions, etc. This good agreement between the prescribed and recomputed values of the Reynolds Stress Tensor $\underline{\underline{R}}$ proves the model capacity to generate velocity fluctuations replicating a given turbulence intensity I_∞ and anisotropic ratio (σ_u:σ_v:σ_w).

Figure 6 displays the results obtained with the Synthetic-Eddy-Method for $I_\infty = 15\%$. It shows an example of "turbulent" flow-field generated with the model for the same anisotropic ratio (σ_u:σ_v:σ_w)=(1.0,0.75,0.56). One can observe that the generated flow seems to be very realistic, although the SEM model only is a mathematical tool that produces velocity fluctuations, without solving the fluid dynamics equations. Additionally, even though the turbulent structures sizes were equally set to $\boldsymbol{\lambda} = (1.0, 1.0, 1.0)$, this length cannot be clearly observed. At the present time, work is under progress to identify what is the link between this size vector $\boldsymbol{\lambda}$ and the turbulent length scales of the generated flow. This vector $\boldsymbol{\lambda}$ revealed to be an important parameter for this Synthetic-Eddy-Method and its influence needs to be further investigated.

Figure 7 displays the power spectral density function obtained with the use of a Gaussian shape function, with a mean value for $\boldsymbol{\lambda} = (0.5, 0.5, 0.5)$ and increasing the spread in the $\boldsymbol{\lambda}$ values, characterised by $\sigma(\lambda)$. $\sigma(\lambda)$ represents the standard deviation in the size of the turbulent structures sizes $\boldsymbol{\lambda}$. For instance, a $\sigma(\lambda) = 50\%$ for a mean $\boldsymbol{\lambda} = (1.0, 1.0, 1.0)$ means that each λ_i value may vary between 0 and 2 and that the spread is characterised by a standard deviation of $+/-0.5$ around 1.0. The differences between those curves are quite striking. Indeed, when a single size of structure is used, the obtained power spectral density is spiked for the higher frequencies. This spiking behaviour of single sized-structures was analysed theoretically and validated numerically. However, increasing the spread in the $\boldsymbol{\lambda}$-structures sizes qualitatively improve the results and quantitative comparisons with the experimental results of Medina *et al.* [20] are scheduled.

More generally speaking, a complete study of the different intervening parameters is under progress. The main parameters to be assessed are the turbulent structure size $\boldsymbol{\lambda}$, the number N of required turbulent structures for a given volume V_σ of E_σ, etc.

4.2. Results using the ambient turbulence model on a single turbine

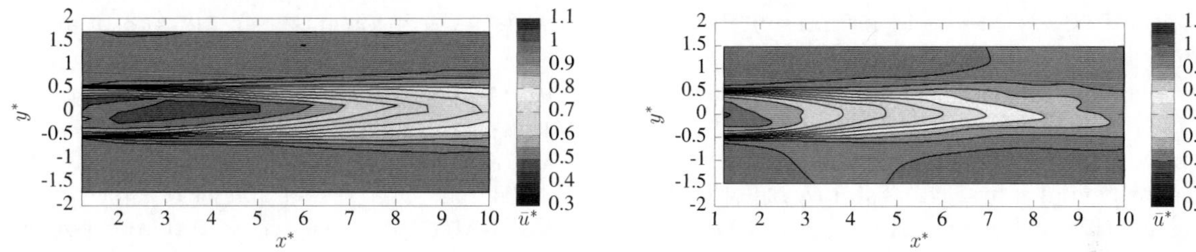

Figure 8. Wake velocity comparison with the experimental velocity maps of Mycek *et al.* [22]. Left is the experimental and right is the numerical velocity map for TSR = 3.67, $I_\infty = 3\%$ and $\boldsymbol{\lambda} = (0.1, 0.1, 0.1)$.

Further to the previous study of turbulent flow without any turbine, few computations were run with a turbine. These runs were performed for the two turbulent intensities I_∞ of 3% and 15% , namely the lowest and highest turbulent intensity values for which experimental results are accessible in the IFREMER wave and current flume tank. Numerical parameters were those from Ref. [27] and the computation were scaled on the Reynolds number of the experiments, that is to say $Re = \dfrac{u_\infty R}{\nu} = 0.28 \times 10^6$. The obtained results are very encouraging and a qualitative analysis can already be performed. Figures 8 and 9 depict the numerically computed wakes for $I_\infty = 3\%$ and 15% and compared to the experiments performed by Mycek *et al.* [22].

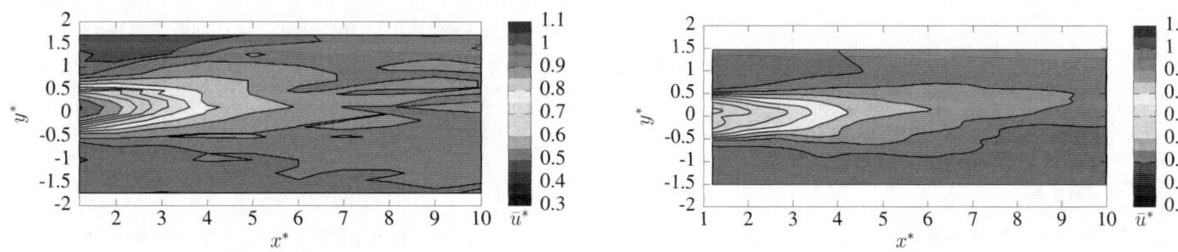

Figure 9. Wake velocity comparison with the experimental velocity maps of Mycek *et al.* [22]. Left is the experimental and right is the numerical velocity map for TSR = 3.67, $I_\infty = 15\%$ and $\boldsymbol{\lambda} = (0.1, 0.1, 0.1)$.

The numerical velocity map presented on Figure 8 seems to indicate that the velocity deficit is underestimated for an ambient turbulence set to $I_\infty = 3\%$. For the turbulent intensity of $I_\infty = 15\%$ an over-estimation seems to be observed. This discrepancies could partially be attributed to the fact that, since now, only rather coarse computations were tested. With finer discretisations, as those already used in [27], more accurate velocity deficits are expected. As already mentioned, these are only preliminary results, which need to be further validated with more computations (finer discretisations, a mesh independence analysis, other parameters validation, etc.). Hopefully, in a near future, more intense turbulent conditions are scheduled (several turbulence intensities and other anisotropic cases) in order to cover a larger range of ambient turbulence intensities.

5. Conclusions and Further works

The important role played by the ambient turbulence intensity on wind or tidal turbines' behaviour was addressed using a numerical investigation. A new module based on this Synthetic-Eddy-Method initially proposed by Jarrin *et al.* [12, 11] was implemented in the numerical code in order to represent the ambient turbulence around turbines. The general characteristics of the Synthetic-Eddy-Method were presented together with its integration in the framework of a 3D unsteady Lagrangian Vortex method.

The capability of the ambient turbulence model to reproduce a perturbed flow that verifies any turbulence intensity I_∞ and any anisotropic ratio $(\sigma_u{:}\sigma_v{:}\sigma_w)$ was validated. Moreover the compatibility of the presented ambient turbulence model with the use of a turbine on the software was proven and first results in terms of turbine wake were presented. As a matter of short-term perspective, a better qualification of the Synthetic-Eddy-Method is needed in order to better understand the input parameters of the model. For instance, the role of the shape function need to be identified and improvements are already planned.

Another development on the ambient turbulence model could be the modification of the Synthetic-Eddy-Method in order to generate a divergence-free flow. In fact, divergence-free flow is a compulsory hypothesis inherent to the Vortex method in order to apply the Biot & Savart law. Unfortunately, the present implementation is for sure not divergence-free. For low to moderate turbulence intensities, this should not introduce a tremendous error in the flow. But increasing the ambient turbulence intensity might become an issue. Such kind of development were already initiated by Poletto *et al.* [28] but need a lot of work in order to be introduced in the framework of Vortex method.

Acknowledgment

The authors would like to thank the Normandy Regional Council and IFREMER for their financial supports for C. Carlier and A. Fur Ph.D. Grants. The authors also would like to thank the CPER-ERDF program Neptune also financed by the Normandy Regional Council. This work received some support managed by the National Research Agency under the Investments for the Future programmes [ANR-10-IEED-0006-11]. Finally the authors wish to thank the CRIANN (Centre des Ressources Informatiques et Applications Numérique de Normandie) for their available numerical computation resources.

References

[1] S. Aubrun, S. Loyer, P. Hancock, and P. Hayden. Wind turbine wake properties: Comparison between a non-rotating simplified wind turbine model and a rotating model. *Journal of Wind Engineering and Industrial Aerodynamics*, 120:1 – 8, 2013.

[2] T. Blackmore, W. M. Batten, G. U. Műller, and A. S. Bahaj. Influence of turbulence on the drag of solid discs and turbine simulators in a water current. *Experiments in Fluids*, 55(1), 2013.

[3] J. Bousquet. *Méthode des singularités.* Cépaduès - Editions, 1990.

[4] E. Branlard, P. Mercier, E. Machefaux, M. Gaunaa, and S. Voutsinas. Impact of a wind turbine on turbulence: Un-freezing turbulence by means of a simple vortex particle approach. *Journal of Wind Engineering and Industrial Aerodynamics*, 151:37 – 47, 2016.

[5] E. Branlard, G. Papadakis, M. Gaunaa, G. Winckelmans, and T. Larsen. Aeroelastic large eddy simulations using vortex methods: unfrozen turbulent and sheared inflow. *J. Phys.: Conf. Ser.*, 625:1 – 12, 2015.

[6] P. Chatelain, S. Backaert, G. Winckelmans, and S. Kern. Large eddy simulation of wind turbine wakes. *Flow Turbulence and Combustion*, 91:587–605, 2013.

[7] J. Choquin and S. Huberson. Particles simulation of viscous flow. *Computers & Fluids*, 17(2):397 – 410, 1989.

[8] W. E. R. Cluster. A southern german wind energy test site in complex terrain. http://www.windfors.de/english/test-site.html, WindForS, last viewed April the 21st, 2017.
[9] G. Cottet and P. Koumoutsakos. *Vortex methods: theory and practice.* Cambridge University Press, 2000.
[10] P. Degond and S. Mas-Gallic. The weighted particle method for convection-diffusion equations. Part I: The case of an isotropic viscosity. *Math. Comp.*, 53(188):485–507, 1989.
[11] N. Jarrin. *Synthetic Inflow boundary conditions for the numerical simulation of turbulence.* PhD thesis, University of Manchester, 2008.
[12] N. Jarrin, S. Benhamadouche, D. Laurence, and R. Prosser. A synthetic-eddy-method for generating inflow conditions for large-eddy simulations. *International Journal of Heat and Fluid Flow*, 27:585–593, 2006.
[13] N. Kelley and B. Jonkman. A stochastic, full-field, turbulence simulator primarialy for use with inflowwind/aerodyn-based simulation tools. http://nwtc.nrel.gov/turbsim, NREL, last viewed March the 7th, 2017.
[14] R. Lewis. *Vortex element methods for fluid dynamic analysis of engineering systems.* Cambridge University Press, 1991.
[15] K. Lindsay and R. Krasny. A particle method and adaptive treecode for vortex sheet motion in three-dimensional flow. *J. Comput. Phys.*, 172:879–907, 2001.
[16] T. S. Lund, X. Wu, and K. D. Squires. Generation of turbulent inflow data for spatially-developing boundary layer simulations. *Journal of Computational Physics*, 140:233–258, 1998.
[17] J. Mann. The spatial structure of neutral atmosphere surface-layer turbulence. *Journal of Fluid Mechanics*, 273:141–168, 1994.
[18] N. Mansour, J. Ferziger, and W. Reynolds. Large-eddy simulation of a turbulent mixing layer. Technical report, Report TF-11, Thermosciences Div., Dept. of Mech. Eng., Stanford University, 1978.
[19] O. D. Medina, F. Schmitt, R. Calif, G. Germain, and B. Gaurier. Caractérisation expérimentale du sillage généré par une hydrolienne - influence du taux de turbulence ambiant. In *14mes Journes de l'Hydrodynamique*, Novembre 2014. Val de Reuil, France.
[20] O. D. Medina, F. Schmitt, R. Calif, G. Germain, and B. Gaurier. Correlation between synchronised power and flow measurements, a way to characterize turblence effecs on marine current turbine. In *11th European Wave and Tidal Energy Conference (EWTEC)*, September 2015. Nante, France.
[21] I. A. Milne, R. N. Sharma, R. G. J. Flay, and S. Bickerton. Characteristics of the turbulence in the flow at a tidal stream power site. *Philosophical Transactions of the Royal Society A: Mathematical, Physical and Engineering Sciences*, 371(1985):pp, February 2013.
[22] P. Mycek, B. Gaurier, G. Germain, G. Pinon, and E. Rivoalen. Experimental study of the turbulence intensity effects on marine current turbines behaviour. part I: One single turbine. *Renewable Energy*, 66(0):729 – 746, 2014.
[23] P. Mycek, B. Gaurier, G. Germain, G. Pinon, and E. Rivoalen. Experimental study of the turbulence intensity effects on marine current turbines behaviour. part II: Two interacting turbines. *Renewable Energy*, 68(0):876 – 892, 2014.
[24] P. Mycek, G. Pinon, C. Lothod, A. Dezotti, and C. Carlier. Iterative solver approach for turbine interactions: application to wind or marine current turbine farms. *Applied Mathematical Modelling*, 41:331 – 349, 2017.
[25] L. Myers, K. Shah, and P. Galloway. Design, commissioning and performance of a device to vary the turbulence in a recirculating flume. In *10th European Wave and Tidal Energy Conference*, September 2013.
[26] G. E. na, S. Aubrun, S. Loyer, and P. Devinant. Wind tunnel study of the wake meandering downstream of a modelled wind turbine as an effect of large scale turbulent eddies. *Journal of Wind Engineering and Industrial Aerodynamics*, 101:24 – 33, 2012.
[27] G. Pinon, P. Mycek, G. Germain, and E. Rivoalen. Numerical simulation of the wake of marine current turbines with a particle method. *Renewable Energy*, 46(0):111 – 126, 2012.
[28] R. Poletto, T. Craft, and A. Revell. A new divergence free synthetic eddy method for the reproduction of inlet flow conditions for les. *Flow, Turb. and Combustion*, 91:519–539, 2013.
[29] C. Rehbach. Calcul numrique d'coulements tridimensionnels instationnaires avec nappes tourbillonaires. *La Recherche Arospatiale*, 5:289–298, 1977.
[30] M. Togneri and I. Masters. Parametrising turbulent marine flows for a blade element momentum model of tidal stream turbines. In *9th European Wave and Tidal Energy Conference (EWTEC)*, September 2011. Southampton, UK.
[31] M. Togneri and I. Masters. Synthetic turbulence generation for turbine modelling with BEMT. In *3rd Oxford Tidal Energy Workshop*, Oxford, UK, 7-8 April 2014.
[32] G. S. Winckelmans and A. Leonard. Contributions to vortex particle methods for the computation of three-dimensional incompressible unsteady flows. *Journal of Computational Physics*, 109(2):247–273, 1993.

Statistical meandering wake model and its application to yaw-angle optimisation of wind farms

E Thøgersen[1], B Tranberg[1,2], J Herp[3] and M Greiner[1]

[1] Department of Engineering, Aarhus University, Inge Lehmanns Gade 10, 8000 Aarhus C, Denmark

[2] Danske Commodities A/S, Værkmestergade 3, 8000 Aarhus C, Denmark

[3] The Maersk Mc-Kinney Moller Institute, University of Southern Denmark, Campusvej 55, 5230 Odense M, Denmark

E-mail: greiner@eng.au.dk

Abstract. The wake produced by a wind turbine is dynamically meandering and of rather narrow nature. Only when looking at large time averages, the wake appears to be static and rather broad, and is then well described by simple engineering models like the Jensen wake model (JWM). We generalise the latter deterministic models to a statistical meandering wake model (SMWM), where a random directional deflection is assigned to a narrow wake in such a way that on average it resembles a broad Jensen wake. In a second step, the model is further generalised to wind-farm level, where the deflections of the multiple wakes are treated as independently and identically distributed random variables. When carefully calibrated to the Nysted wind farm, the ensemble average of the statistical model produces the same wind-direction dependence of the power efficiency as obtained from the standard Jensen model. Upon using the JWM to perform a yaw-angle optimisation of wind-farm power output, we find an optimisation gain of 6.7% for the Nysted wind farm when compared to zero yaw angles and averaged over all wind directions. When applying the obtained JWM-based optimised yaw angles to the SMWM, the ensemble-averaged gain is calculated to be 7.5%. This outcome indicates the possible operational robustness of an optimised yaw control for real-life wind farms.

1. Introduction

Because of their simplicity engineering wake models like the Jensen wake model (JWM) [1] have been used extensively for the optimisation of several wind farm objectives, like layout optimisation, power optimisation, load reduction and noise reduction. Originally, the engineering wake models have been developed for resource analysis, where it is standard to average wind-farm power over larger, 10min time intervals. The wakes have been treated as static and of rather broad nature. However, this is not the case for smaller time scales. The wakes appear to be rather narrow and reveal a meandering dynamics [2]. Consequently, the static and broad wake profiles described by the engineering wake models have to be interpreted as a time-averaged mean field resulting from the meandering dynamics.

This has inspired us to generalise the deterministic engineering wake models to a simple statistical meandering wake model (SMWM), where a random directional deflection is assigned to a narrow wake in such a way that on average it resembles a broad Jensen wake. See Figure 1 for an illustration. Furthermore, we want to find out what impact the statistical meandering dynamics has on the wind farm optimisation. Section 2 presents the description and calibration of the statistical meandering wake model. Its application to the yaw-angle optimisation of wind-farm power is presented in Section 3. An outlook is given in Section 4.

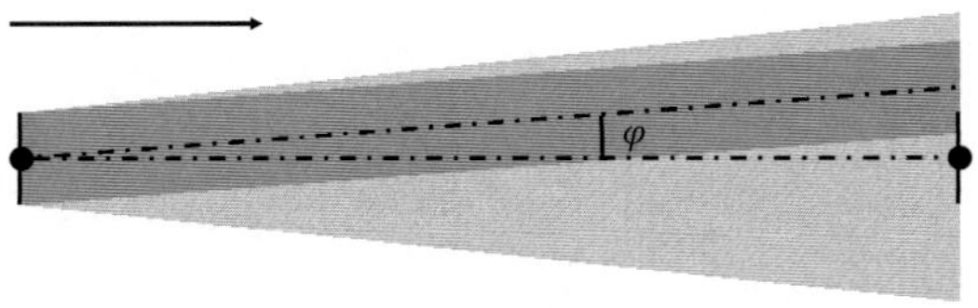

Figure 1. Narrow SMWM meandering wake (dark blue) and broad time-averaged JWM wake (light blue).

2. Statistical meandering wake model

The Jensen wake model [1]uses a linear expansion $R_{wake}(x) = R + kx$ of the wake produced by an upwind turbine with radius R. The wake parameter k represents the cone opening. We adopt the standard value $k = 0.04$ for the description of offshore wind farms. A downwind turbine, which is placed at the downwind distance x with a lateral displacement D from the wake centreline (see Figure 2), experiences an effective wind speed $v(x, D)$ corresponding to the relative deficit

$$\delta(x, D) = \frac{u - v(x, D)}{u} = \frac{2/3}{\left(1 + \frac{kx}{R}\right)^2} \frac{A_{overlap}}{\pi R^2} \tag{1}$$

with respect to the free wind speed u, where A_{overlap} represents the overlap area between the wake disc and the rotor disc. The downwind turbine generates the power

$$P_{downwind}(x, D) = \frac{\rho \pi R^2}{2} C_p v(x, D)^3 \,. \tag{2}$$

Its dependence on the downwind distance is illustrated as the grey curve in Figure 3; the Betz optimum $C_p = 16/27$ is used for the power coefficient.

Due to the observed wake meandering [2], the rather broad Jensen wake profile should be interpreted as an ensemble (or time) average over randomly displaced very narrow wakes. See again Figures 1 and 2. The meandering wakes are assumed to be very narrow and are described by a wake parameter $k_{meander} = 0.0001$, which is much smaller than the Jensen wake parameter. Larger, maybe more physical values of $k_{meander}$ are not discussed in this contribution. The displacement is described by a random azimutal angle ζ, which is drawn from a uniform distribution, and a random radial distance $r = x \tan \varphi$. The second angle is drawn from a Gaussian-mixture distribution $p(\varphi \geq 0) = N(\mu, \sigma) + N(-\mu, \sigma)$, with $\mu = \arctan k$ being identical to the opening angle of the broad Jensen wake and $\sigma = 1.65\mu$. These parameter settings have been found to best describe the average power generation of the downwind turbine. The downwind turbine's power generation is at first determined analogously to Equations (1) and (2) for each random displacement described by ζ and φ, and then averaged over all possible displacements. Figure 3 compares the outcome with the power generation calculated from the standard Jensen model. The agreement holds over an extended range of downwind distances.

So far the Statistical Meandering Wake Model has been tested and calibrated for a two-turbine configuration. A further and more stringent test is to examine its performance for a large wind farm without a parameter recalibration. The Nysted wind farm is chosen as example. Its layout is illustrated in Figure 4. The wind turbines are facing multiple wake interactions. These are described by the Katic wake superposition [3]

$$\delta_i^2(\theta) = \sum_{i \neq j} \delta_{ij}^2(\theta) \,, \tag{3}$$

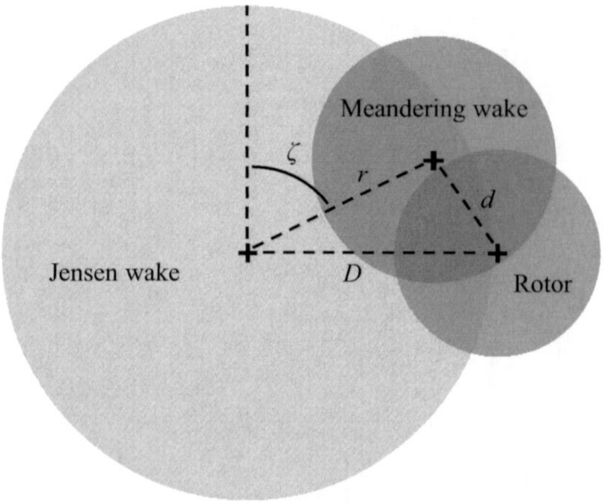

Figure 2. Overlaps between the rotor disc of the downwind turbine (red), the broad Jensen wake (light blue) and the narrow meandering wake (dark blue).

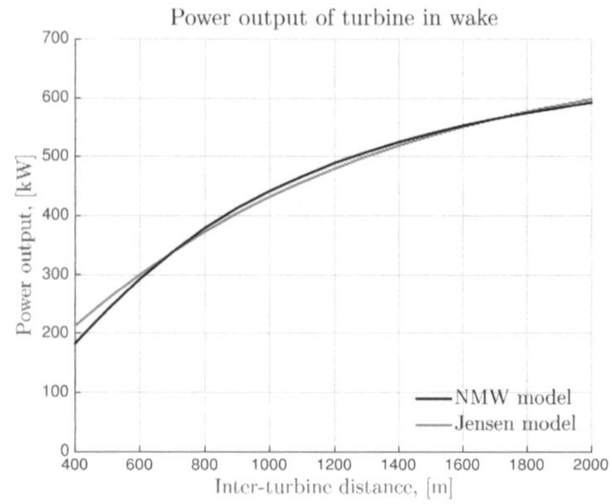

Figure 3. Dependence of the downwind turbine's power generation as a function of the downwind distance according to the Jensen wake model (grey) and the averaged Statistical Meandering Wake Model (black, denoted as NMW). The lateral displacement, the rotor-disc radius and the free upwind speed have been set to $D = 0$m, $R = 41.2$m and $u = 7$m/s.

where δ_i is the resulting overall velocity deficit at turbine i and δ_{ij} is the velocity deficit as if only turbine j is present in the wind stream upwind from turbine j. The velocity deficit δ_i depends on the wind direction and, in case of the Statistical Meandering Wake Model, also on the random displacement angles ζ_j and φ_j of the respective upwind turbines. Insertion of Equation (3) into (2) and summing over all turbines results in the total wind farm power $P_{farm}(u, \theta)$. In case of the SMWM the total wind farm power also needs to be averaged over all possible sets of random displacements, which are assumed to be statistically independent from each other. The wind farm efficiency

$$\eta_{farm}(\theta) = \frac{P_{farm}(u,\theta)}{N\frac{\rho\pi R^2}{2}\frac{16}{27}u^3} = \frac{1}{N}\sum_{i=1}^{N}(1-\delta_i(\theta))^3 \tag{4}$$

compares the total wind farm power to the total power of N single wind turbines standing in free wind. Figure 5 illustrates the wind farm efficiency obtained from the standard Jensen wake model and from the Statistical Meandering Wake Model. For all wind directions the agreement is good. This result is quite remarkable since the SMWM parameters have not been recalibrated.

3. Application to yaw-angle optimisation of wind farms

The yaw-angle optimisation of wind farms is a recent topic of interest [4]. With a variable non-zero yaw angle a wind turbine will on the one hand loose some power, but on the other hand it is then able to deflect its wake away from the downwind turbines [5], so that those will be able to face stronger intra-farm winds and to generate more power. The reduced power (2) of a yawed turbine is described by a modified power coefficient

$$C_p(\gamma) = \frac{16}{27}\cos^\eta\gamma\ , \tag{5}$$

where γ represents the yaw angle. We adopt the exponent $\eta = 1.88$ from [4], but other values like $\eta = 2$ and $\eta = 3$ have also been discussed in the literature. Further following [4], the lateral

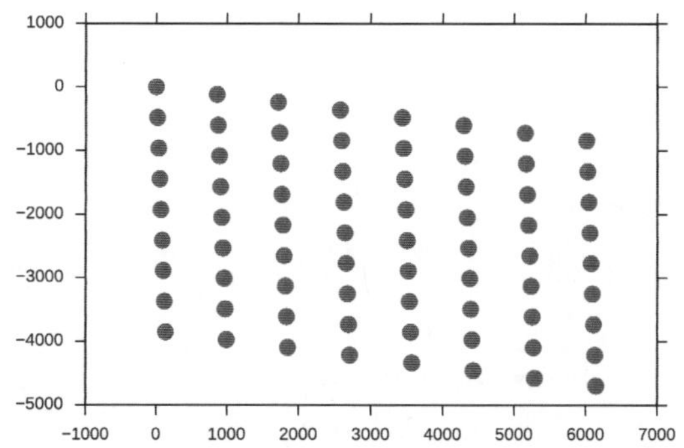

Figure 4. Layout of the Nysted wind farm. The spacing of the turbines with rotor-disc radius $R = 41.2$m is 867m in East-West direction ($\Theta = 98°$) and 482m in South-North direction ($\Theta = 178°$).

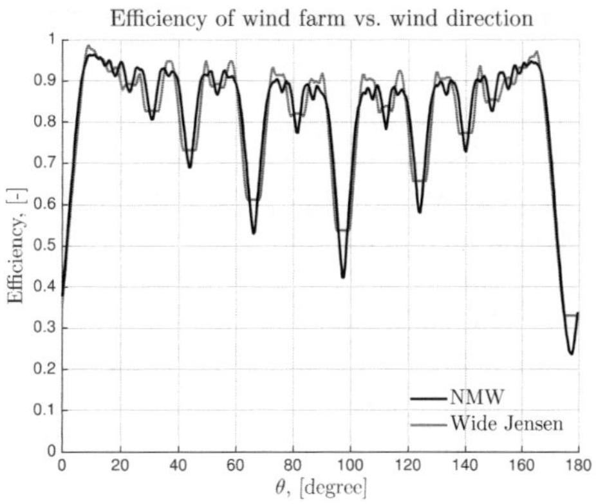

Figure 5. Efficiency of the Nysted wind farm calculated from the standard Jensen wake model (grey) and the Statistical Meandering Wake Model (black, denoted as NMW).

wake deflection

$$y(x,\gamma) = y_{rotation}(x) + y_{yaw}(x,\gamma) \tag{6}$$

consists of two parts, a rotation-induced wake deflection

$$y_{rotation}(x) = -4.5 - 0.01x \tag{7}$$

and a yaw-induced deflection

$$y_{yaw}(x,\gamma) = \frac{\xi_{init}(\gamma)\left(15\left(\frac{0.15x}{R}+1\right)^4 + \xi_{init}^2(\gamma)\right)}{\frac{2.25}{R}\left(\frac{0.15x}{R}+1\right)^5} - \frac{\xi_{init}(\gamma)R(15+\xi_{init}^2(\gamma))}{2.25} \tag{8}$$

with the initial wake deflection angle

$$\xi_{init}(\gamma) = \frac{C_T}{2}\cos^2\gamma\sin\gamma \tag{9}$$

at the rotor disc and the Betz-optimal thrust coefficient $C_T = 8/9$. For several yaw angles the resulting lateral wake deflection (6) is shown in Figure 6 as a function of the local downwind distance. The wake deflections are comparable to those described in [6].

This sets the stage for the yaw-angle optimisation of wind-farm power generation. In the first round we discuss the standard Jensen wake model. For a given wind direction, the optimal yaw angles are obtained from a two-step optimisation. The first part with large angular step sizes uses a sequential optimisation analogous to the algorithm described in [7]. The second part of the optimisation with smaller angular step sizes uses a conjugate gradient method.

Figure 7 illustrates the resulting optimised yaw angles for the Nysted wind turbines when the wind is exactly coming from the North. As expected, for the most downwind turbines in the bottom row the optimised yaw angles are zero. All other turbines come with an optimised yaw angle close to 25°. The resulting power gain factors for each turbine are shown in Figure 8. They are slighlty negative for the most upwind turbines in the upper row. For all other turbines the gain factors are quite significant, reaching a value close to 170% for the most downwind

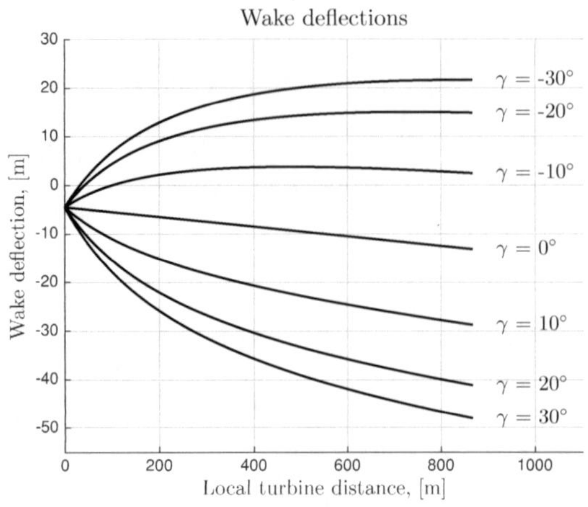

Figure 6. Lateral wake deflection (6) for several yaw angles of a Nysted wind turbine with rotor-disc radius $R = 41.2$m as a function of the local downwind distance.

turbines. For the North-South wind direction the gain factor for the total wind farm turns out to be 78%.

The wind-farm gain factors for the other wind directions are shown in Figure 9. Close to the main symmetry axes they are very pronounced. Also a double-hump structure appears. Once the wind direction is not perfectly aligned to a main axis, it is easier for the upwind turbines to yaw their wakes away from the downwind turbines. For a clockwise misalignment they yaw to the left and for a counterclockwise misalignment they yaw to the right. Inbetween main-axes wind directions the gain factors are close to zero. When uniformly averaged over all wind directions the overall wind-farm gain factor still turns out to be 6.7%.

For different wind directions different optimal sets of yaw angles have been derived. This can be used as input into a table-based control of the wind farm. Now we investigate the performance of the table-based control derived from the standard Jensen wake model once the wind farm faces the narrow meandering wakes. On top of the yaw-induced lateral wake deflections we introduce again the random displacements described by the random angles ζ and φ of the Statistical Meandering Wake Model. For each wind direction the wind-farm power then again

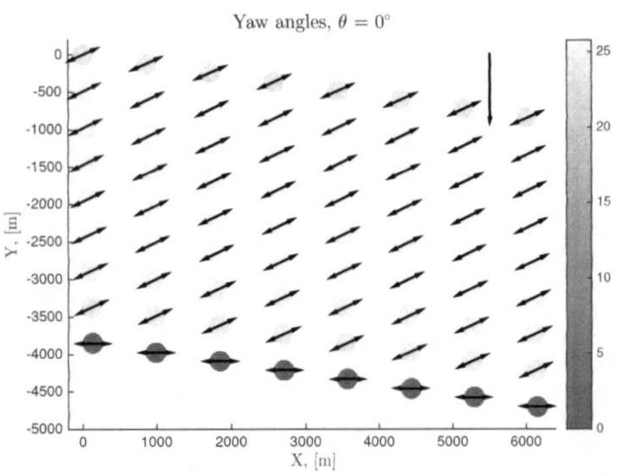

Figure 7. Optimised yaw angles of the Nysted wind turbines for a North-South wind direction. The standard Jensen wake model has been used.

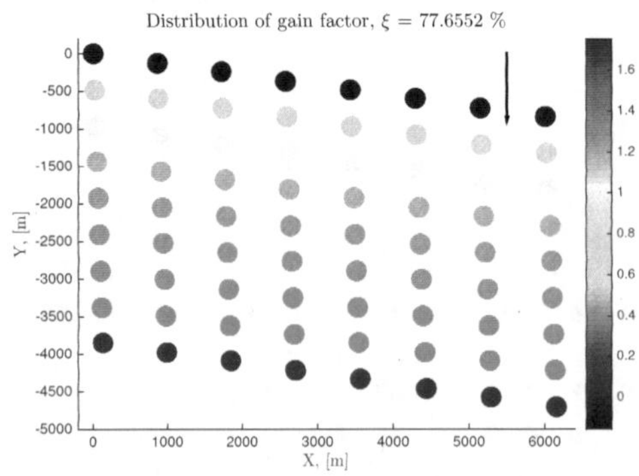

Figure 8. Single-turbine-power gain factors for the Nysted wind farm for a North-South wind direction. The standard Jensen wake model has been used.

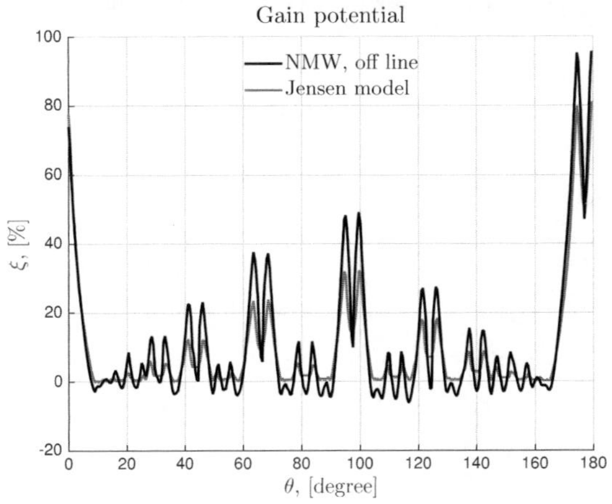

Figure 9. Power gain factors for the Nysted wind farm in dependence of the wind direction. The standard Jensen wake model has been used for the grey curve, and the Statistical Meandering Wake Model for the black curve (denoted as NMW).

needs to be first averaged over a large ensemble of random realisations before the gain factors are determined. The wind-farm gain factors obtained from the SMWM subject to the JWM-based optimised yaw angles are also shown in Figure 9. They are very similar to those obtained only from the standard Jensen model. However, there are a few small differences. Close to the main symmetry axes, the SMWM gain factors are a little bit larger than the JWM gain factors. For wind directions safely away from the symmetry axes the SMWM gain factors can become slightly negative. Nevertheless, when uniformly averaged over all wind directions the overall wind-farm gain factor results to be 7.5% and is even slightly larger than the 6.7% obtained from the standard JWM. This almost identical outcome indicates the possible operational robustness of an optimised yaw control for real-life wind farms.

The obtained absolute percentage gains should be considered as generic. A fine-tuning of the JWM and SMWM parameters to the Nysted wind farm has not been performed. Note however, that the order of magnitude of the absolute percentage gains is the same as obtained in [8] for a different wind farm.

4. Outlook

The optimised yaw angles can become quite large, of the order 25°. This will cause large turbine loads. In this respect and based again on the standard engineering wake models and the newly developed SMWM, it makes sense to study a conflicting two-objective optimisation between power maximisation and load reduction. Also a combined yaw angle and pitch control might be of interest, in particular when the wind direction coincides with the symmetry axes of the wind farm.

References

[1] Jensen N O 1983 A note on wind generator interaction *Risø-M-2411*

[2] Larsen G C, Madsen H A, Bingöl F, Mann J, Ott S, Sørensen J N, Okulov V, Troldborg N, Nielsen M, Thomsen K, Larsen T J and Mikkelsen R 2007 Dynamic wake meandering modeling *Risø-R-1607*

[3] Katic N J I and Højstrup 1987 A simple model for cluster efficiency *European Wind Energy Association Conference and Exhibition (Rome)* vol 1, ed W Palz and E Sesto, p 407

[4] Gebraad P M O, Teeuwisse F W, van Wingerden J W, Fleming P A, Ruben S D, Marden J R and Pao L Y 2016 Wind plant power optimization through yaw control using a parametric model for wake effects – a CFD simulation study *Wind Energy* **19** 95-114

[5] Jimenez A, Crespo A and Migoya E 2010 Application of a LES technique to characterize the wake deflection of a wind turbine in yaw *Wind Energy* **13** 559-72

[6] Guntur S, Troldborg N and Gaunaa M 2012 Application of engineering models to predict wake deflection due to a tilted wind turbine *Proceedings of EWEA 2012 – European Wind Energy Conference & Exhibition*

[7] Herp J, Poulsen U V and Greiner M 2015 Wind farm power optimization including flow variability *Renewable Energy* **81** 173-81
[8] Gebraad P, Thomas J J, Ning A, Fleming P and Dykes K 2017 Maximization of the annual energy production of wind power plants by optimization of layout and yaw-based wake control *Wind Energy* **20** 97-107

Simulation of a 7.7 MW onshore wind farm with the Actuator Line Model

A Guggeri, M Draper, G Usera

Universidad de la República, Facultad de Ingeniería, Julio Herrera y Reissig 565 11300, Montevideo, Uruguay

E-mail: aguggeri@fing.edu.uy, mdraper@fing.edu.uy, gusera@fing.edu.uy

Abstract. Recently, the Actuator Line Model (ALM) has been evaluated with coarser resolution and larger time steps than what is generally recommended, taking into account an atmospheric sheared and turbulent inflow condition. The aim of the present paper is to continue these studies, assessing the capability of the ALM to represent the wind turbines´ interactions in an onshore wind farm. The 'Libertad' wind farm, which consists of four 1.9MW Vestas V100 wind turbines, was simulated considering different wind directions, and the results were compared with the wind farm SCADA data, finding good agreement between them. A sensitivity analysis was performed to evaluate the influence of the spatial resolution, finding acceptable agreement, although some differences were found. It is believed that these differences are due to the characteristics of the different Atmospheric Boundary Layer (ABL) simulations taken as inflow condition (precursor simulations).

1. Introduction

Wind energy has expanded rapidly in recent decades all over the world. The increase in wind power in 2015 was close to half of global electricity generation growth and wind power capacity increased by 17% from 2014 to 2015 [1]. The growth rates have become lower in recent years, nevertheless the installed capacity is projected to increase in the following years [2]. This development of wind energy has been supported by technological improvements which are related to an increase in hub height, rotor diameter and unit power, which has led to larger capacity factors for the same wind speed [3]. This growth in size means, among other things, that the area of interest of the atmospheric boundary layer (ABL) swept by these wind turbine rotors, reaches higher altitudes and it is placed further away from the surface. It also means that the influence of different atmospheric conditions, such as turbulence intensity and vertical velocity gradient, in the wake of a wind turbine and in its performance, is larger [4][5][6][7].

Two approaches exist in order to represent the presence of a wind turbine in a simulation [8]: 1) actuator models, in which the blades are represented as body forces, 2) direct representation of the blade's geometry through the computational mesh. The Actuator Line Model (ALM) has been widely validated [9][10][11][12][13][14], showing to reproduce with reasonable accuracy the wind flow in the wake of a wind turbine with moderate computational cost. In general, it is recommended to use a spatial resolution of at least R/30, being 'R' the rotor radius, and to limit the movement of the rotor tip to a grid cell length in each time step [14]. Recently, the ALM has been evaluated with coarser resolutions and also larger time steps, considering a uniform inflow condition [15] as well as an Atmospheric Boundary Layer (ABL) like inflow condition. For the latter the 'Horns Rev' offshore wind farm and a well-known wind

tunnel campaign, were considered as validation cases [16][17], finding good agreement with the SCADA data in the power deficits in [17] and accuracy to reproduce the wake of the wind turbine in [16].
The aim of the present paper is to continue the research done in the past years, analyzing the power produced by the turbines and the influence of the wakes. For this, 'Libertad' onshore wind farm was simulated, considering different wind speeds and wind directions. The paper is organized as follows: Section 2 presents the solver and the ALM implementation, Section 3 describes the validation case used for the evaluation and the numerical setup, Section 4 presents the main results, discussing the effect of considering three different resolutions, and summary is given in Section 5.

2. Numerical method

2.1. Solver

caffa3d.MBRi [18] [19] is an open source, finite volume (FV) code, with second order accuracy in space and time, parallelized with MPI, in which the domain is divided in unstructured blocks of structured grids. Representation of complex geometries can be handled through a combination of body fitted grids and the immersed boundary method over both, Cartesian and body fitted grid blocks. Geometrical properties and flow properties, which are expressed in primitive variables, are always expressed in a Cartesian coordinate system, using a collocated arrangement. The large-eddy simulation (LES) technique is used in the simulations. For further information please see [18] [19].
The ALM has been implemented in the referred code [20]. In this model, the wind turbine rotor is represented as a body force field, where each blade is represented as a line that moves with the rotational speed of the rotor and it is discretized in radial sections where the aerodynamic forces are computed. To compute the force in each radial section, geometrical properties (chord length and twist angle) as well as aerodynamic properties (lift and drag coefficients) are determined. The latter are obtained from tabulated airfoil data. After computing the aerodynamic forces, it is required to project them from the representation of the rotor onto the computational domain. In order to compute the additional source term, a Gaussian smearing function with three coefficients, is used, taking into account the distance between each grid cell and radial section (d), for each direction (n-normal direction, r-radial direction, t-tangential direction):

$$f(d) = \frac{1}{E_n E_r E_t \pi^{1.5}} e^{-\left(\frac{d_n}{E_n}\right)^2} e^{-\left(\frac{d_r}{E_r}\right)^2} e^{-\left(\frac{d_t}{E_t}\right)^2} \tag{1}$$

3. Validation case

3.1. Libertad wind farm

Libertad is a 7.7MW onshore wind farm, located in the south of Uruguay. It consists of four Vestas V100 wind turbine generators (WTG), two with rated power of 1.9MW (WTG1 and WTG2) and two of 1.95MW (WTG3 and WTG4), all four with a hub height of 95m and a rotor diameter of 100m. It has a meteorological mast with anemometers and wind vanes that measure wind speed and wind direction at 93m height. The wind farm location and layout are shown in Figure 1.

Figure 1. Libertad wind farm location (left) and layout (right)

3.2. Numerical setup

The size of the computational domain is 3.00km in the streamwise direction, 1.50km in the spanwise direction and 0.75km in the vertical direction. The domain is uniformly divided into 144 x 128 grid cells in the streamwise and spanwise directions respectively, while a stretched grid with 80 grid cells, is used in the vertical direction. The spatial resolution implies a resolution of R/2.4, R/4.3 in the streamwise and spanwise directions respectively, while in the vertical direction, 19 grid nodes cover the rotor diameter. A zero velocity gradient is imposed at the outlet and a wall model based on the log law is used to compute the stress at the surface while periodic conditions are used in the lateral boundaries. We use the Crank-Nicolson scheme to advance in time and the scale dependent dynamic Smagorinsky model to compute the subgrid scale stress [21][22]. The inflow condition is obtained from a precursor simulation. Figure 2 shows vertical profiles at the inlet boundary of the mean wind speed and turbulence intensity, averaged in the spanwise direction. At hub height the mean wind speed is 8.85m/s and the turbulence intensity 7.5%. Turbulence intensity is defined as the ratio between the standard deviation and the mean of the streamwise velocity.

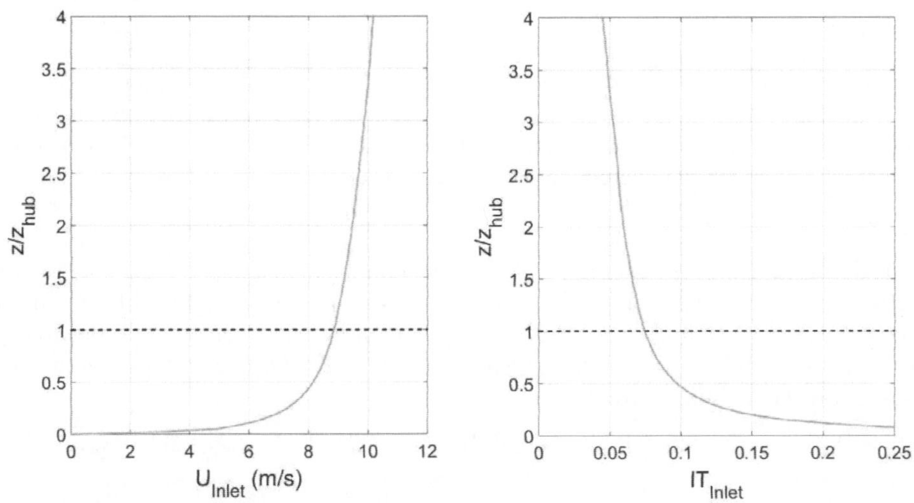

Figure 2. Mean wind speed (left) and turbulence intensity (right) vertical profiles at inlet boundary

To represent the wind turbine rotor, the ALM is used with 12 radial sections in each line. To find the chord length and twist angle in each radial section, the geometrical characteristics of the Vestas V80 wind turbine presented in [23], were modified according to the following methodology: We apply the Blade Element Momentum (BEM) model [24] with different wind speeds (U_{BEM}), all of them lower than the rated wind speed. For each U_{BEM} the SCADA data is sorted by $U_{MET} \in [U_{BEM} \pm 0.25\,m/s]$, where U_{MET} is the wind speed measured by the met mast anemometer. The blade rotating speed we use in BEM is the mean angular speed of the sorted dataset. The pitch angle is 0º in every case. Then, we compare the obtained power curve from BEM with the manufacturer´s one and with the power curve obtained from the filtered SCADA data. We modify the geometrical characteristics of the blade and repeat the procedure until an acceptable agreement is obtained between the power curves. The airfoil used is the NACA 63-415 for the entire blade. Lift and drag coefficients were also obtained from [23]. The fact that the chord and twist angle obtained along the blades with this procedure are not the exact values of a V100 blade, contributes to the uncertainties of the results presented below. Figure 3 shows the designed chord length and twist angle, the power curve obtained from BEM compared with the SCADA data and the manufacturer´s power curve (V100 PC), and the angular speed considered in BEM method.

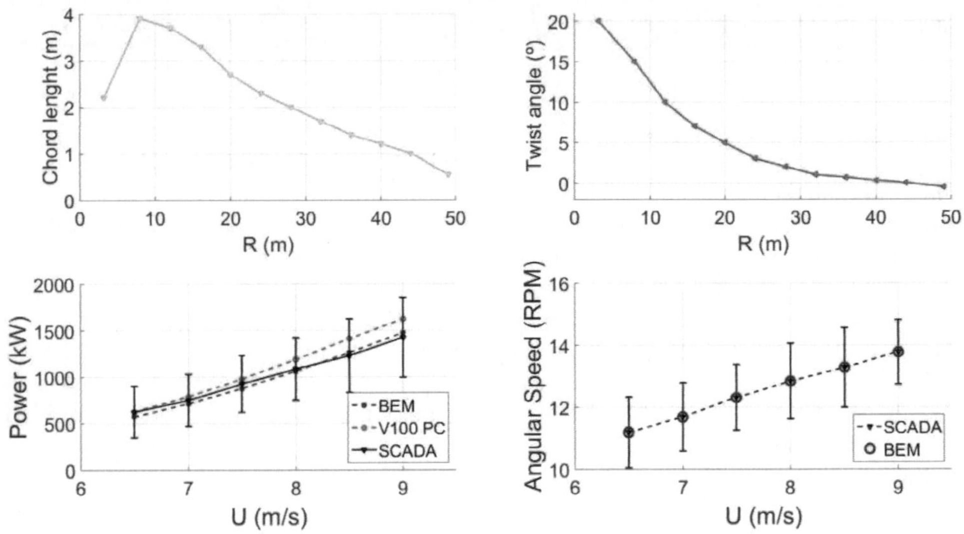

Figure 3. Designed chord length (top left), twist angle (top right), power curve (bottom left) and angular speed (bottom right) from BEM, SCADA data and manufacturer´s power curve.

In the ALM implementation a variable rotational speed is considered, in a similar approach as explained in [23] for the Actuator Disk Model with Rotation, obtaining the relationship between the rotational speed of the rotor and its torque from the wind farm SCADA data. An efficiency of 0.93 is considered to compute the generated power from aerodynamic power [23].

4. Results

Simulations considering different wind directions were performed by changing the layout while keeping the distance between wind turbines and the West boundary as inlet boundary [17]. Three wind sectors were considered, $150 \pm 5^{\underline{o}}$, $150 \pm 2.5^{\underline{o}}$ and $132.5 \pm 2.5^{\underline{o}}$, as shown in Figure 1. In order to compute the power associated to each wind sector, weighted averages were computed taking into account the results of the simulations of different directions:

$$P_{av\ 150\pm5^{\underline{o}}} = 12.5\%\ P_{145^{\underline{o}}} + 25\%\ P_{147.5^{\underline{o}}} + 25\%\ P_{150^{\underline{o}}} + 25\%\ P_{152.5^{\underline{o}}} + 12.5\%\ P_{155^{\underline{o}}} \quad (2)$$

$$P_{av\ 150\pm2.5^{\underline{o}}} = 25\%\ P_{147.5^{\underline{o}}} + 50\%\ P_{150^{\underline{o}}} + 25\%\ P_{152.5^{\underline{o}}} \quad (3)$$

$$P_{av\ 132.5\pm2.5^{\underline{o}}} = 25\%\ P_{130^{\underline{o}}} + 50\%\ P_{132.5^{\underline{o}}} + 25\%\ P_{135^{\underline{o}}} \quad (4)$$

The power obtained in the simulations and the weighted averages were compared with the power registered in the SCADA data, which was filtered as follows. Only 10-minute periods which met the following criteria were considered:

- $U_{MET} \in [U_{MET\,sim} \pm 0.2\ ^{m}/_{s}]$, where U_{MET} is the wind speed registered by the met mast anemometer and $U_{MET\,sim}$ is the mean wind speed obtained in the simulation at the met mast position
- $D_{MET} \in [D_{sim} \pm \Delta D\ ^{\underline{o}}]$, where D_{MET} is the wind direction measured by the met mast wind vane, D_{sim} is the wind direction which is being simulated and ΔD is the wind sector considered, 5° in the first case and 2.5° in the other two.
- The four wind turbines were available, their power was greater than zero and no alarms were reported.

No filtering was made in terms of turbulence intensity or atmospheric stability. The sorted data is represented as the mean power value ± 1 standard deviation.
When comparing the simulation results with SCADA datasets, the yaw misalignment of the turbines, the spatial variability of the wind direction within the wind farm and the variability of the wind direction within the averaging period, contribute to the uncertainties, as presented in [25].

The power simulation results and their average, are presented in Figure 4 and Figure 5, including also the SCADA data.

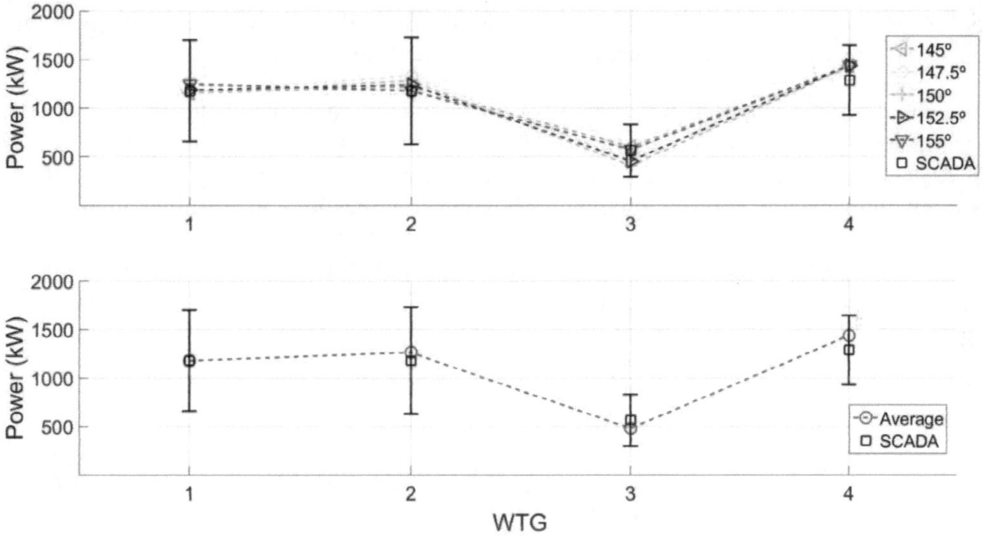

Figure 4. Power production of wind turbines for each direction (150+/-5°) (top) and power weighted average by direction (bottom).

IOP Conf. Series: Journal of Physics: Conf. Series **854** (2017) 012018 doi:10.1088/1742-6596/854/1/012018

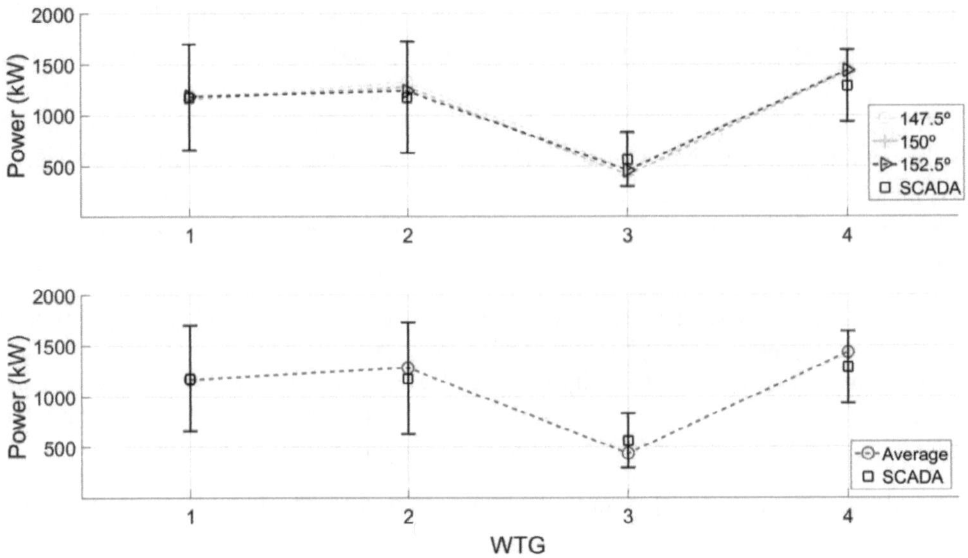

Figure 5. Power production of wind turbines for each direction (150+/-2.5°) (top) and power weighted average by direction (bottom)

The results show good agreement with the SCADA data in both cases, with no major differences between them, so a wind direction window of ±2.5º was used when taking into account another wind sector below. The power deficit of WTG3 due to the wake of WTG4 is clearly shown in both figures, and it is slightly higher at 150° than in other directions, because the two wind turbines are aligned in that direction.

Another case was simulated, considering direction 132.5°, where WTG3, WTG2 and WTG1 are aligned. The mean wind speed at the met mast position is 8.89m/s. The results are depicted in Figure 6.

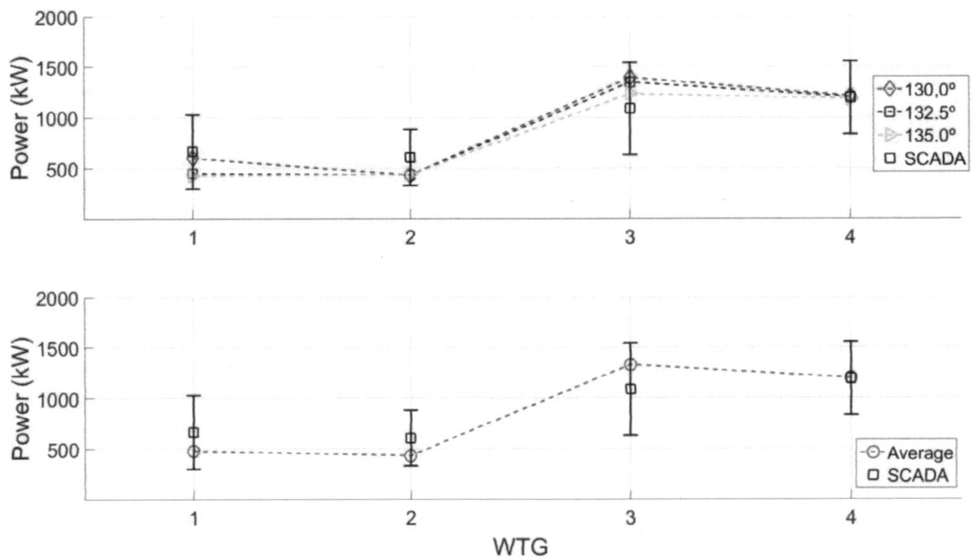

Figure 6. Power production of wind turbines for each direction (132.5+/-2.5°) (top), power weighted average by direction (bottom)

Results also show acceptable agreement with SCADA data, capturing WTG2 and WTG1 power deficits and overestimating WTG3 power production. An explanation of why this happens is presented below.

A simulation considering another inflow condition was performed, only for wind direction 150°. In this case, at the inlet and at hub height the mean wind speed is 7.60m/s and the turbulence intensity is 8.4%. In Figure 7, the SCADA data is represented as scattered points, showing mean wind speed measured at the met tower and power production of the wind turbines, including the results of the simulations and also the manufacturer's power curve (V100 PC). The SCADA data is filtered by wind direction (150º ± 2.5º). Good agreement is observed for the four WTG. Notice how the SCADA scattered points of WTG3 are 'deviated' from the manufacturer power curve, due to the WTG4 wake.

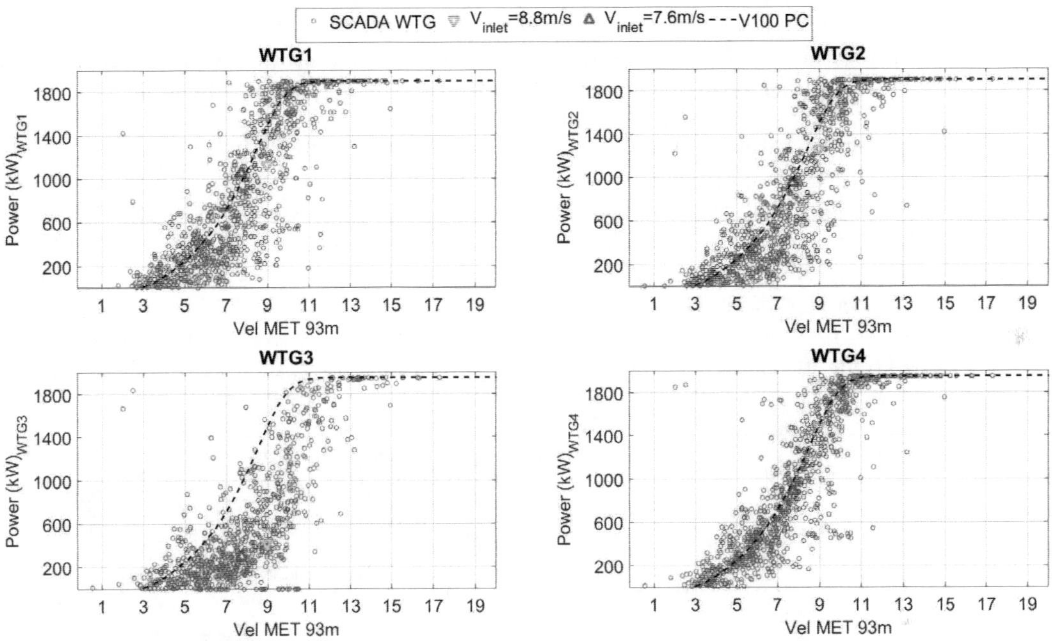

Figure 7. Simulated wind turbines mean power and scattered SCADA data filtered (wind direction 150°+/-2.5°), for two inflow conditions.

The influence of the spatial resolution in the results was analyzed, by considering two more resolutions, which are R00 and R02, being R01 the original resolution. Their characteristics are presented in Table 1. As the length in the vertical direction is not uniform, only the height of the first cell is shown. NzD represents the number of cells which cover the WTG rotor in the vertical direction.

Table 1. Characteristics of resolutions

	Nx	Ny	Nz	Δx(m)	Δy(m)	Δz_{min} (m)	R/Δx	R/Δy	NzR
R00	192	144	96	15.6	10.4	2.5	3.2	4.8	23
R01	144	128	80	20.8	11.7	3.0	2.4	4.3	19
R02	128	96	72	23.4	15.6	3.0	2.1	3.2	17

Simulations considering wind direction 150° and mean velocity at the inlet and at hub height of 8.8m/s, were performed. It is important to highlight that for each case a different precursor simulation was used, as the resolutions of the precursor simulations and the simulations with WTG must be the same. It is planned to implement the same precursor simulation for cases with different resolution in the near future. In Figure 8 the results for the three spatial resolutions are presented.

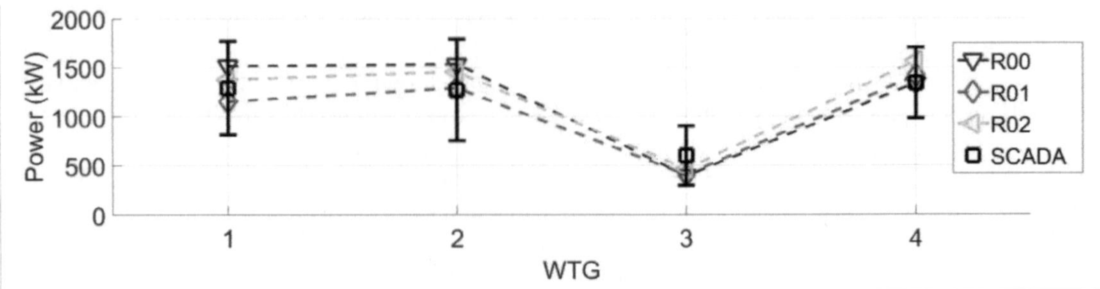

Figure 8. Power production of wind turbines considering three spatial resolutions, wind direction 150°. SCADA data filtered by wind direction: 150°+/-2.5°, wind speed 8.91+-0.2m/s.

Notice the dispersions in the results. For example, for R00 the power production of WTG4 is well predicted but the power production of WTG2 and WTG1 is overestimated, while in R01 the power production of WTG4 is slightly overpredicted and that of WTG1 is underpredicted. The reason for this is that as the precursor simulations are different, the velocity upstream the wind turbines is different in each case despite having almost the same velocity profile averaged in the spanwise direction. This is observed in Figure 9, where the mean streamwise velocity component (U) is plotted, as vertical profiles at the inlet and at the met mast position. This figure also shows a surface plot of the streamwise velocity at the inlet section, with the WTG positions represented by the dotted circles, and the met mast position, by the smaller circle.

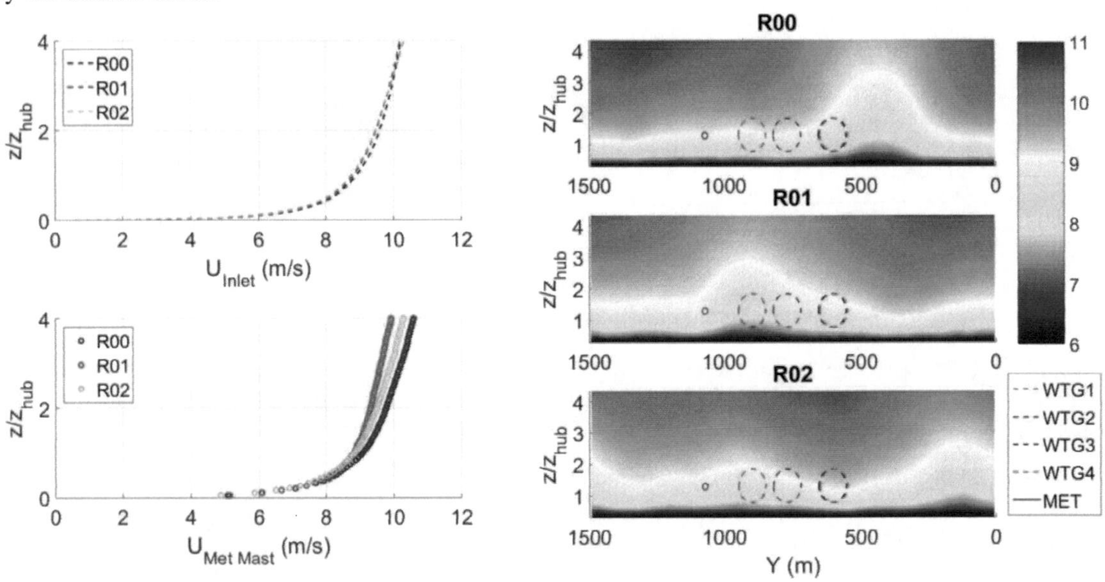

Figure 9. Vertical profiles of the mean streamwise velocity component (U), averaged in the spanwise direction at the inlet (left above) and at the position of the meteorological mast (left bottom), and at the inlet section (right), for the three spatial resolutions considered.

Although the mean velocity averaged in the spanwise direction at the inlet is very similar for the three cases, notice how there are zones where the velocity is lower or higher than this value. When this zone is upstream of a WTG, the power production will be associated to this low/high velocity. For the same reason, it is also important to observe that at the position of the met mast the velocity may not represent the velocity averaged at the inlet, so when comparing the SCADA data of the wind turbines and the met mast data with the results of the simulations, this has to be taken into account. This explains why the power production of WTG2 and WTG1 are overestimated in R00, and why the power production of WTG1 is underestimated in the case of R01. This issue can also be observed in Figure 10, which shows the mean streamwise velocity (U) in a horizontal plane at hub height, for the three resolutions.

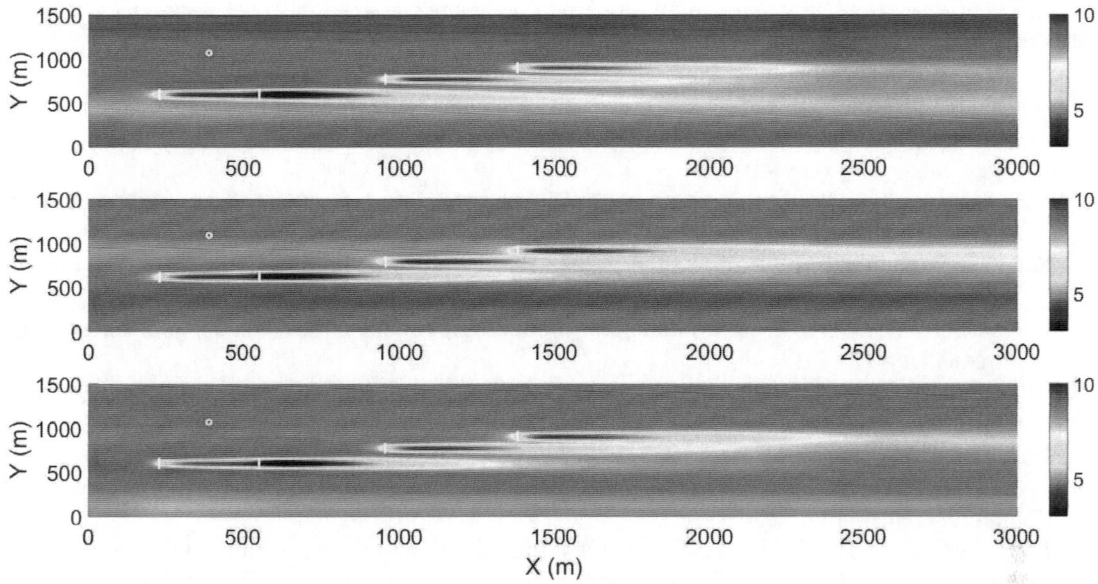

Figure 10. Mean streamwise velocity (U), in a horizontal plane at hub height, for R00 (top), R01 (center) and R02 (bottom). White bars represent wind turbines, and white circle shows the position of the meteorological tower.

5. Summary and conclusions

The ALM has been validated under different conditions and used in different numerical codes, being one of the chosen approaches in the academic community, to represent the presence of wind turbines in a simulation. Expanding the work previously presented by the authors and to continue validating the capability of the numerical solver caffa3d.MBRi to simulate the flow through wind farms, the 'Libertad' onshore wind farm was simulated, representing the wind turbines with the ALM and with an atmospheric boundary layer like inflow condition.

Two different wind sectors (150° and 132.5°) with the same precursor simulation, and two precursor simulations with the same wind direction (150°) and different wind speed at hub height, were considered, in order to analyse the interactions and wakes of the wind turbines. Also for wind direction 150°, three spatial resolutions were considered. Good agreement with the SCADA data was found when assessing the power deficits observed in the data under those different conditions. When analysing the results with the three spatial resolutions, some differences were obtained. The causes may be related to the fact that the velocity at the position of the met mast may not be an accurate representation of the velocity at the inlet, as the velocity of the precursor simulation is not uniform in the spanwise direction, which leads to different velocities upstream of the positions of the wind turbines and the met mast.

Future research will be focused on the implementation of different control strategies to adjust the angular velocity of the rotor and the individual pitch of each blade, enabling to estimate power production at higher wind speeds and also assessing the operation of down-regulated wind farms. In addition to this, we plan to implement the same precursor simulation for different spatial resolutions, including mesoscale forcing, and to develop and evaluate wind farm control strategies, aiming to maximize the global power production of a wind farm.

References

[1] Global Wind Energy Council 2016 Global wind Report Annual Market Update 2015 (Brussels).
[2] World Wind Energy Association 2014 Key statistics of world wind energy report 2013 (Shanghai).
[3] International Energy Association 2013 Technology roadmap Wind Energy 2013 (Beijing).
[4] Abkar M, Porté-Agel F 2013 The effect of free-atmosphere stratification on boundary-layer flow and power output from very large wind farms *Energies* **6** 2338-2361.

[5] Lee S, Churchfield M, Moriarty P, Jonkman J, Michalakes J 2012 Atmospheric and Wake Turbulence Impacts on Wind Turbine Fatigue Loading *50th AIAA Aerospace Sciences Meeting* (Nashville,Tennessee).

[6] Hansen K S, Barthelmie R J, Jensen L E, Sommer A 2012 The impact of turbulence intensity and atmospheric stability on power deficits due to wind turbine wakes at Horns Rev wind farm *Wind Energy* **15** 1 183-196.

[7] Vanderwende B J, Lundquist J K 2012 The modification of wind turbine performance by statistically distinct atmospheric regimes *Environmental Research Letters* **7** 3 034035.

[8] Sanderse B, van del Pijl S P and Koren B 2011 Review of computational fluid dynamics for wind turbine wake aerodynamics *Wind Energy* **14** 799-819.

[9] Sørensen J N, Shen W Z 2002 Numerical modelling of wind turbine wakes *Journal of Fluids Engineering* **124** 2 393-99.

[10] Troldborg N 2008 Actuator Line Modeling of Wind Turbine Wakes. PhD thesis, Technical University of Denmark.

[11] Churchfield M J, Lee S, Moriarty P, Martínez L A, Leonardi S, Vijayakumar G and Brasseur J G 2012 A large-eddy simulation of wind-plant aerodynamics *50th AIAA Aerospace Sciences Meeting* (Nashville, Tennessee).

[12] Porté-Agel F, Wu Y T, Lu H, Conzemius R J 2011 Large-eddy simulation of atmospheric boundary layer flow through wind turbines and wind farms *Journal of Wind Engineering and Industrial Aerodynamics* **99** (4) 154-168.

[13] Troldborg N, Zahle F, Réthoré P E and Sorensen N N 2014 Comparison of wind turbine wake properties in non-sheared inflow predicted by different computational fluid dynamics rotor models *Wind Energy* doi: 10.1002/we.1757.

[14] Pankjak K J, Churchfield M J, Moriarty P J and Schmitz S 2014 Guidelines for volume force distributions within actuator line modeling of wind turbines on large-eddy-simulation type grids *Journal of Solar Energy Engineering* **136** 031003-1-11

[15] Draper M, Usera G 2015 Evaluation of the actuator line model with coarse resolutions *Wake Conference* (Visby).

[16] Draper M, Guggeri A, Usera G 2016 Validation of the Actuator Line Model with coarse resolution in atmospheric sheared and turbulent inflow *Journal of Physics: Conference Series*, 753/8 (2016), 082007.

[17] Draper M, Guggeri A, Usera G 2016 Modelling one row of Horns Rev wind farm with the Actuator Line Model with coarse resolution *Journal of Physics: Conference Series*, 753/8 (2016), 082028.

[18] Usera G, Vernet A, Ferré J A 2008 A parallel block-structured finite volume method for flows in complex geometry with sliding interfaces *Flow, Turbulence and Combustion* **81** 471-495.

[19] Mendina M, Draper M, Kelm Soares A P, Narancio G and Usera G 2014 A general purpose parallel block structured open source incompressible flow solver *Cluster Computing* **17** 2 231-41.

[20] Draper M 2015 Simulación del campo de vientos y de la interacción entre aerogeneradores. PhD thesis (public defense July 2016), Universidad de la República.

[21] Porté-Agel F, Meneveau C, Parlange M B 2000 A scale-dependent dynamic model for large-eddy simulation: application to a neutral atmospheric boundary layer *Journal of Fluid Mechanics* 415 261-284.

[22] Stoll R, Porté-Agel F 2006 Dynamic subgrid-scale models for momentum and scalar fluxes in large-eddy simulations of neutrally stratified atmospheric boundary layers over heterogeneous terrain *Water Resources Research* 42 WO1409.

[23] Wu Y T, Porté-Agel F 2015 Modeling turbine wakes and power losses within a wind farm using LES: An application to the Horns Rev offshore wind farm *Renewable Energy* 75 945-55.

[24] Hansen M 2008, Aerodynamics of Wind Turbines Second Edition, *Earthscand.*

[25] Gaumond M, Réthore P-E, Ott S, Peña A, Bechmann A, Hansen K S 2014 Evaluation of the wind direction uncertainty and its impact on wake modeling at the Horns Rev offshore wind farm *Wind Energ.* **17** 1169-1178.

Comparison study between wind turbine and power kite wakes

T Haas and J Meyers

Department of Mechanical Engineering, University of Leuven, 3001 Heverlee, Belgium

E-mail: thomas.haas@kuleuven.be, johan.meyers@kuleuven.be

Abstract. Airborne Wind Energy (AWE) is an emerging technology in the field of renewable energy that uses kites to harvest wind energy. However, unlike for conventional wind turbines, the wind environment in AWE systems has not yet been studied in much detail. We propose a simulation framework using Large Eddy Simulation to model the wakes of such kite systems and offer a comparison with turbine-like wakes. In order to model the kite effects on the flow, a lifting line technique is used. We investigate different wake configurations related to the operation modes of wind turbines and airborne systems in drag mode. In the turbine mode, the aerodynamic torque of the blades is directly added to the flow. In the kite drag mode, the aerodynamic torque of the wings is directly balanced by an opposite torque induced by on-board generators; this results in a total torque on the flow that is zero. We present the main differences in wake characteristics, especially flow induction and vorticity fields, for the depicted operation modes both with laminar and turbulent inflows.

1. Introduction

In the past decade, the global cumulative installed wind power capacity has increased significantly. The size of the wind turbines has also been increasing constantly, with modern turbines reaching rotor diameters and hub heights of more than hundreds of meters; this is a logical trend, considering that wind speed increases with height. Already in 1980, Loyd [1] proposed an alternative technology that can tap into high-altitude wind by using tethered aircrafts. Airborne Wind Energy systems present many advantages upon conventional wind turbines. First, at operation altitudes up to 500m, AWE systems can tap into stronger and more consistent winds [2], thus surpassing the capacity factor of conventional wind power [3]. Second, this technology reduces material requirements and third, ground-based electricity generation facilitates the maintenance of such systems[4]. Despite this high potential, this new technology is still in an early development stage and thus not yet competitive with actual conventional wind turbines. Different operation modes are being investigated, among others the so called lift and drag modes[1], but no leading design has yet emerged, and many research efforts go towards insuring reliable operation and autonomous launching and landing.

The present study aims to address the interaction of such airborne wind energy systems with their wind environment. In the context of conventional wind turbines and large clusters of wind turbines, the interaction between the systems and the atmospheric boundary layer has been investigated intensively [5][6][7]. In particular, the wake effects on power generation of downstream turbine rows in wind farms are intensively being studied in current research activities. Thus, an extensive understanding of the wake characteristics is preferable and many studies have been carried out in this field, see [8] and [9] for instance. The complex dynamics of vortical structures accounting for the development of wind turbine wakes have been addressed in [10] and [11]. However in AWE, the interaction between airborne systems and their environment has not yet been thoroughly investigated in this context and thus this paper aims to give first insights on the dynamics of different operation modes.

In his pioneering work [1], Loyd introduced different operation modes for energy harvesting with tethered aircrafts. He first defined the so-called *lift mode*, for which the tethered system pulls a load on the ground while the tether reels out. Power is then generated on the ground via the tension in the tether. Second, he defined the *drag mode* for which additional on-board generators are mounted on the wing, producing electricity that is transferred to the ground via the tether. These small turbines induce additional drag on the wing. A review of the different implemented concepts is given in [12].

Wind turbine blades rotating at a constant angular speed experience a constant torque from the aerodynamic forces. In real systems, this torque is captured by the generator and turned into useful mechanical power. Hence, the blades induce a net torque addition onto the flow, which results in the rotation of the flow in the direction opposite to the blades. In this study, we only consider *blade portions* (approximately the top 30% of the blade length at the tip) that we will refer to as the wings of the system. This wing system rotates at constant angular speed and sweeps an annulus in a vertical plane orthogonal to the main flow direction. The force distribution along the wings span is similar to conventional turbine blades, thus we will refer to this turbine-like operation mode of our wing system as the *turbine mode*. In addition, we will also consider triple-kite systems in drag mode flying in similar configurations. We assume that the additional drag forces are solely acting in the system's rotation plane, in the direction opposite to the tangential component of the aerodynamic lift. The torque of this additional drag balances the aerodynamic torque and thus the resulting torque added to the flow is zero. The aerodynamic force distribution is computed according to the Blade-Element/Momentum theory applied to an optimal, drag-free, blade design [13].

This paper is organized as follows. In section 2, the methodology and the computational framework of the study are introduced. In section 3, numerical results for simulation of both turbine and power modes are presented and discussed. Finally, we conclude this paper with some outlook and motivation for future work in section 4.

2. Simulation framework and methodology

2.1. LES methodology

The simulations are performed using the large-eddy simulation code SP-Wind developed in-house at KU Leuven [6][14][15]. The governing equations are the filtered incompressible Navier-Stokes equations for neutral flows. The continuity and momentum equations are given by

$$\partial_i \tilde{u}_i = 0\,, \tag{1}$$

$$\partial_t \tilde{u}_i + \partial_j \left(\tilde{u}_i \tilde{u}_j\right) = -\partial_i \tilde{p}^* - \partial_j \tau_{ij}^d + f_i \tag{2}$$

where the filtered velocity is denoted by $\tilde{u}_i$, $\tilde{p}^*$ is the filtered modified pressure and τ_{ij}^d is the deviatoric part of the subgrid-scale stress tensor, which is modeled using an eddy-viscosity model. Its trace is included in the filtered modified pressure given by $\tilde{p}^* = \tilde{p}/\rho + \delta_{ij}\tau_{kk}/3$. The effects of the wing system on the flow are modelled using a body force f_i, as discussed further below. The governing equations are discretized using a Fourier pseudo-spectral method in the horizontal directions (x_1, x_2) while in the vertical direction (x_3), a fourth-order energy-conserving finite-difference scheme is used. The horizontal discretization requires periodic boundary conditions which can be circumvented in the streamwise direction x_1 by employing a fringe-region technique [16]. Time integration is performed with a classic four-stage fourth-order Runge Kutta scheme, and the time step is computed using a Courant-Friedrichs-Lewy number of 0.4.

2.2. Force distribution along the wing

The effect of the wing system on the flow is added to the momentum equations via the forcing term f_i. Using an actuator-line technique, as described in [8], we discretize the wings into a finite number of elements for which the aerodynamic forces are computed and subsequently filtered onto the LES grid using a Gaussian convolution filter.

The local force contributions are computed according to the optimality condition given by the Betz-Joukowsky limit [13]. The aerodynamic drag is neglected and so only the lift force on the wing is considered. The lift distribution can be derived from the axial and tangential induction factors, defined as $a = 1/3$ and $a' = a(1-a)/(\lambda^2\mu^2)$ respectively, where λ is the tip-speed ratio and $\mu = r/R$ the normalized radial distance from the virtual rotation center. The lift force L per unit span and the local flow angle ϕ are computed accordingly,

$$L = \rho 4\pi R \frac{U_\infty^2}{\lambda}\sqrt{(1-a)^2 + (\lambda\mu(1+a'))^2}, \qquad \tan\phi = \frac{1-a}{\lambda\mu(1+a')}, \tag{3}$$

where U_∞ is the inflow velocity. For the drag mode, the additional drag forces are computed such that the total torque on the system is zero, ie

$$\int_{r_i}^{r_o} (L\sin(\phi) \times r)dr + \int_{r_i}^{r_o} (D \times r)dr = 0\,. \tag{4}$$

The number of on-board generators on the wing is set to two and the individual drag forces are each time divided over approximately 5% of the wing span, respectively at locations $\mu = 0.76$ and $\mu = 0.88$.

3. Results and discussion

3.1. Simulation setup

The numerical framework was verified against reference simulations presented in [8] for the turbine mode using full-span turbines blades instead of wings. In the present study, we perform simulations for both operation modes (turbine mode and drag mode) in laminar and turbulent inflows. The wing size and the path radius are chosen according to [1], in which an aircraft of type C-5A is following an orbital trajectory (at an angle of 0.4rad from the orbit axis) at a fixed tether length ($l_t = 400\,m$). The span length of the aircraft is $s = 68\,m$ and the path radius is $R = 155.77\,m$.The angular speed is set to represent a tip speed ratio $\lambda = 7$ at an inflow velocity $U_\infty = 10\,ms^{-1}$. The turbulent inflow is generated using the *tugen library* [17]. The parameters for the spectral tensor model are taken from [18] for turbulence over sea, assuming a wind velocity of $10ms^{-1}$ at a height $z = 180m$ which corresponds to the lower altitude bound at which airborne wind energy systems are meant to fly. The eddy length scale is $L_T = 60\,m$, the eddy dissipation scale is $\alpha\epsilon^{2/3} = 0.022\,m^{4/3}s^{-2}$, and the eddy lifetime parameter is $\Gamma = 2.85$.

The size of the computational domain is chosen such that the blockage ratio of the swept wing-area to the frontal domain area is small, ie. equal to 8%. The height and width of the domain are equal to 5.85 radii and the length is 23.42R. The last 10% of the domain length is used as fringe-region to impose the required inflow conditions. The rotation plane of the kite is situated 2.36 radii downstream from the domain inflow. The resolution of the numerical grid is $N_x \times N_y \times N_z = 640 \times 160 \times 320$ which corresponds to a grid cell size of approximately $5.70 \times 5.70 \times 2.85[m^3]$ that enables to capture the vortices shedding from the wing tips. A sketch of the computational domain is given in figure 1.

3.2. Wake characteristics

In this section we present wake characteristics of turbine and drag modes.

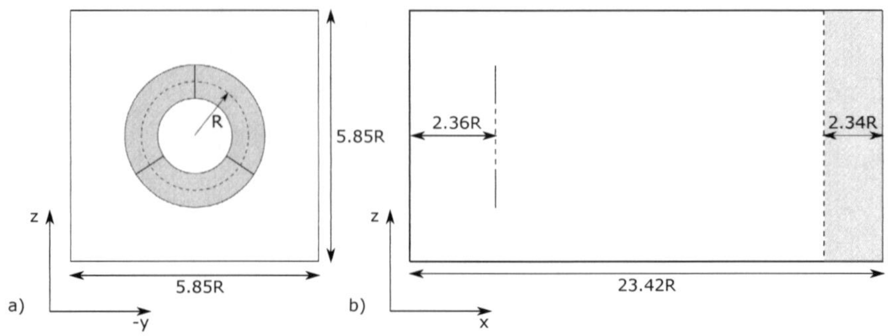

Figure 1. Computational domain used in this study; (a) cross-section; (b) side view.

The velocity induction in the wake is analysed by means of interference factors as shown in figure 2. The instantaneous axial interference factor is given by $a = 1 - U_x/U_\infty$, where U_x is the streamwise component of the velocity field. In figure 2.a, negative values of the axial induction factor indicate the location of tip vortices shedding from the wing in the annulus plane. Spreading of the wake is observable in positive and negative radial direction at downstream locations (fig.2.b-c). Due to the wake expansion, the surrounding flow as well as the flow in the core of the annulus is accelerated. In drag mode, no substantial difference in terms of axial induction is observed, thus induction of this mode is not shown here.

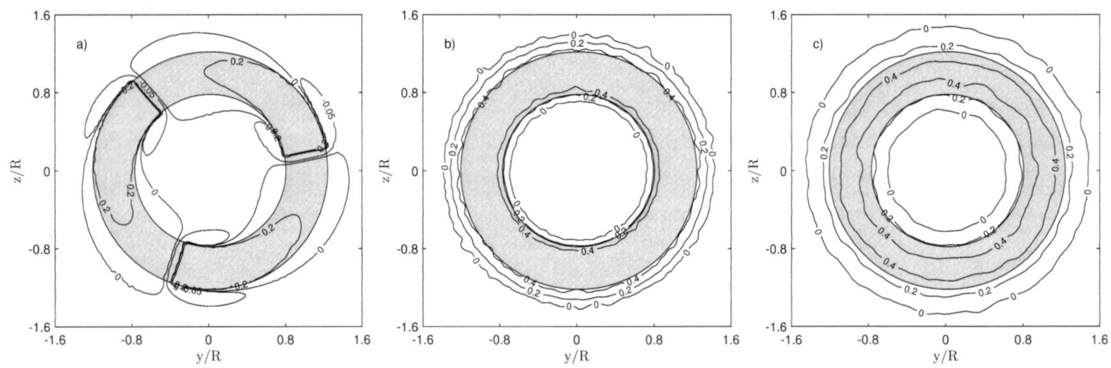

Figure 2. Axial induction factor of turbine mode for laminar inflow conditions; (a) Instantaneous distribution at annulus location; Time-averaged distribution downstream from annulus plane at (b) $x = 3R$ and (c) $x = 6R$.

The downstream wake development is shown in figure 3. In drag mode, we observe additional trailing vortices at the generator locations, associated to the modified lift force distribution. Although they rapidly break down between the annulus plane and $x/R \approx 1$, the wake mixing begins approximately at the same downstream location as in turbine mode, ie $x/R \approx 3$. The instantaneous location of the vortices with vorticity magnitude $|\omega| \geq 1.2s^{-1}$ is depicted in figure 4. In uniform inflow, the tip vortices are in both cases at the same position. From this we can see that the generator vortices have only little impact on the wake development. With turbulent inflow, the vorticity structure of the wake looses its coherence one radius downstream of the annulus plane, causing the wake to break down earlier. The additional perturbations due to drag mode do not seem to affect this process.

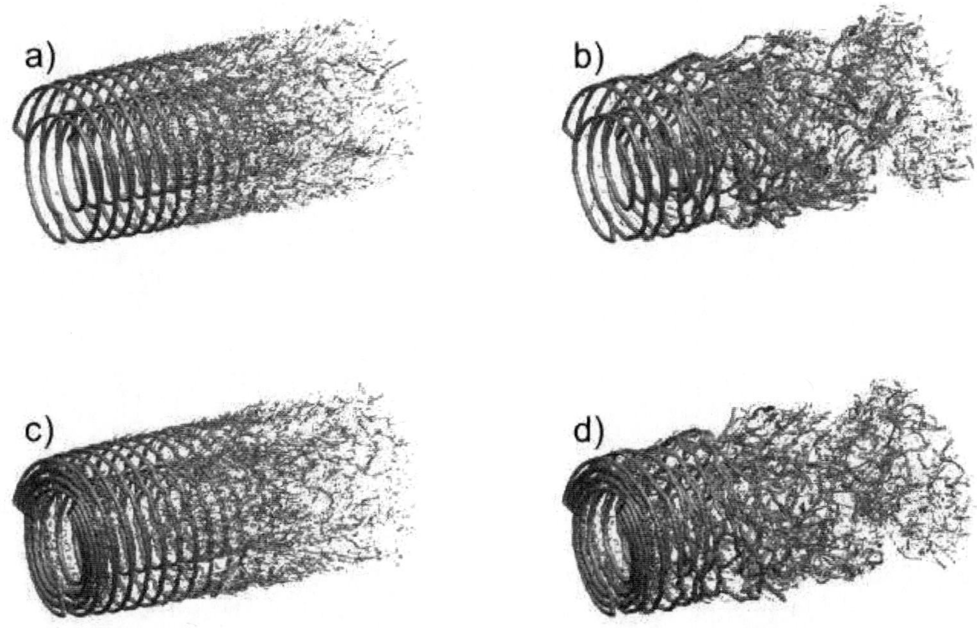

Figure 3. Downstream flow development visualized by vorticity isosurfaces; Turbine mode in (a) uniform and (b) turbulent inflow; Drag mode in (c) uniform and (d) turbulent inflow.

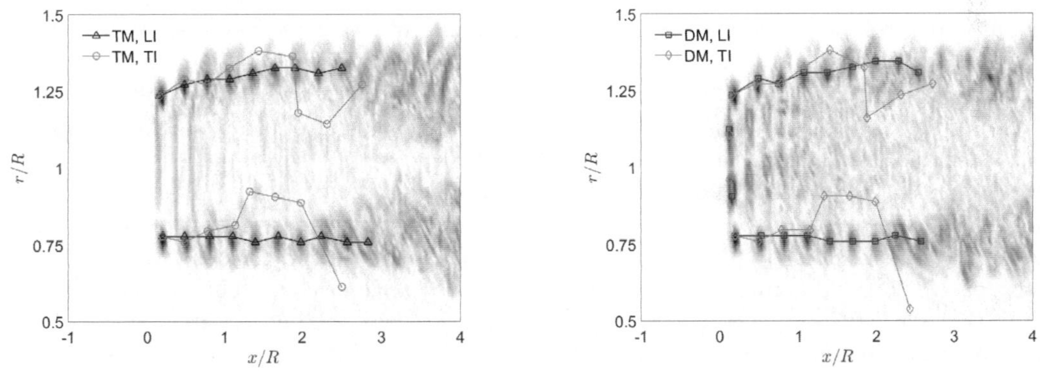

Figure 4. Instantaneous positions of vortices at $t = 360\,s$. (Left) Vorticity magnitude of turbine mode in laminar inflow is shown in background; instantaneous vortex positions in turbine mode for laminar (-▲-) and turbulent (-⊖-) inflow. (Right) Vorticity magnitude of drag mode in laminar inflow is shown in background; instantaneous vortex positions in drag mode for laminar (-⊟-) and turbulent (-◇-) inflow.

Radial profiles of time-averaged axial velocity component are shown at the annulus plane, in the near wake ($x = 3R$, $x = 6R$) and in the far wake ($x = 9R$, $x = 12R$) in figure 5. Again, no difference is observable between turbine and drag modes. In turbulent inflow, the wake mixing increases and lead to a faster wake recovery in the far wake.

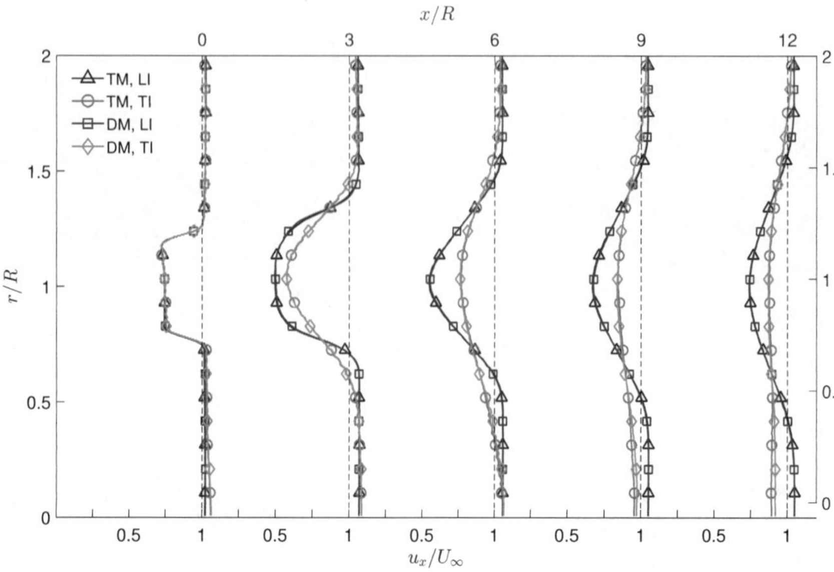

Figure 5. Radial profiles of time-averaged streamwise velocity component at different locations downstream of the annulus plane; Turbine mode in uniform (-▲-) and turbulent (-⊖-) inflows; Drag mode in uniform (-⊟-) and turbulent (-◇-) inflows.

4. Conclusions

In this paper, we have investigated the difference between two operation modes in the context of Airborne Wind Energy. Compared to a turbine-like mode, a torque-free operation mode of kite systems was presented. At the chosen operation and inflow conditions, this drag mode did not exhibit a different behaviour in terms of wake breakdown and recovery. The only feature specific to the drag mode was the shedding of additional vortices at the location of the on-board generators due to the modified force distribution along the wing. However, these vortices where not strong enough to interfere with the wing-tip vortices and had no substantial effect on the flow field. It is known from literature that the operation conditions of wind turbines, see for instance [9], affect the structure of the wake. Thus, simulations at different operation parameters (ie. tip-speed ratio) are required to further investigate the difference between the two modes introduced in this study. The addition of propellers onto the wing and the resulting modified near-rotor aerodynamics might also have an effect on the noise generation or the wind loading of the AWE system. These are interesting questions for further research. Next to that, an improved aerodynamic model also including the tether is of interest.

Acknowledgments

This project has received funding from the European Union's Horizon 2020 research and innovation programme under the Marie Skłodowska-Curie grant agreement No 642682. The computational resources and services used in this work were provided by the VSC (Flemish Supercomputer Center), funded by the Research Foundation Flanders (FWO) and the Flemish Government - department EWI.

References

[1] Loyd ML 1980 Crosswind kite power *J. Energy* **4** 106-11
[2] Diehl M 2013 Airborne Wind Energy: basic concepts and physical foundations *Airborne Wind Energy* ed U Ahrens *et al* (Berlin-Heidelberg: Springer) chapter 1 pp 3-22

[3] Fechner U 2016 Downscaling of airborne wind energy systems *J. Phys.: Conf Ser.* **753** 102002
[4] Coleman J, Ahmad H, Pican E and Toal D 2013 Non-reversing generators in a novel design for pumping mode airborne wind energy farm *Airborne Wind Energy* ed U Ahrens *et al* (Berlin-Heidelberg: Springer) chapter 34 pp 587-597
[5] Porté-Agel F, Wu Y T, Lu H and Conzemius R J 2011 Large-eddy simulation of atmospheric boundary layer flow through wind turbines and wind farms *J. Wind Eng. Ind. Aerodyn.* **99** 154-68
[6] Calaf M, Meneveau C and Meyers J 2010 Large eddy simulation study of fully developed wind-turbine array boundary layers *Phys. Fluids* **22** 015110
[7] Allaerts D and Meyers J 2017 Boundary-layer development and gravity waves in conventionally neutral wind farms *J. Fluid Mech.* **814** 95-130
[8] Sørensen J N and Shen W Z 2002 Numerical modeling of wind turbine wakes *J. Fluids Eng.* **124** 393-99
[9] Troldborg N, Sørensen J N and Mikkelsen R 2010 Numerical simulations of wake characteristics of a wind turbine in uniform inflow *Wind Energ.* **13** 86-99
[10] Iungo G V, Viola F, Camarri S, Porté-Agel F and Gallaire F 2013 Linear stability analysis of wind turbine wakes performed on wind tunnel measurements *J. Fluid Mech.* **737** 499-526
[11] Sørensen J N 2011 Instability of helical tip vortices in rotor wakes *J. Fluid Mech.* **682** 1-4
[12] Cherubini A, Papini A, Vertechy R and Fontana M 2015 Airborne wind energy systems: a review of the technologies *Renew. Sustainable Energy Rev.* **51** 1461-76
[13] Burton T, Jenkins N, Sharpe D and Bossanyi E 2011 *Wind Energy Handbook* John Wiley & Sons Ltd
[14] Goit J and Meyers J 2015 Optimal control of energy extraction in wind-farm boundary layers *J. Fluid Mech.***768** 5-50
[15] Munters W, Meneveau C and Meyers J 2016 Turbulent inflow precursor method with time-varying directions for large-eddy simulations and applications to wind farms *Boundary-Layer Meteorol.***159** 305-28
[16] Nordström J, Nordin N and Henningson D 1999 The fringe region technique and the fourier method used in direct numerical simulation of spatially evolving viscous flows *SIAM J. Sci. Comput.***20** 1365-93
[17] Gilling L 2009 Tugen - synthetic turbulence generator manual and user's guide. Technical report. Aalborg University. Aalborg
[18] Mann J 1998 Wind field simulation *Prob. Engng. Mech.* **13** 269-82

Detecting wind turbine wakes with nacelle lidars

D P Held[1,2], A Larvol[2] and J Mann[1]

[1] Department of Wind Energy, Technical University of Denmark (DTU), Frederiksborgvej 399, 4000 Roskilde, Denmark

[2] Windar Photonics A/S, Helgeshøj Alle 16-18, 2630 Taastrup, Denmark

E-mail: domhel@dtu.dk

Abstract.

Because the horizontal homogeneity assumption is violated in wakes flows, lidars face difficulties when reconstructing wind fields. Further, small-scale turbulence which is prevalent in wake flows causes Doppler spectrum widths to be broader than in the free stream. In this study the Doppler peak variance is used as a detection parameter for wakes. A one month long measurement campaign, where a continuous-wave lidar on a turbine has been exposed to multiple wake situations, is used to test the detection capabilities. The results show that it is possible to identify situation where a downstream turbine is in wake by comparing the peak widths. The used lidar is inexpensive and brings instalments on every turbine within economical reach. Thus, the information gathered by the lidars can be used for improved control at wind farm level.

1. Introduction

Within a wind farm wakes from neighbouring turbines cannot be avoided; optimal wind farm design can minimize these effects but never completely avoid them. Much effort has been put into wake models as wind farm design tools (for an overview see [1]) and their validation, while significant interest in measuring wake with the use of remote sensing devices originates from optimal wind farm operation using wake redirection.

In this paper we want to present a novel method of detecting wakes by using a forward-looking nacelle lidar. Information of the detection can be applied as input to wind farm based yaw steering.

A very straight-forward approach to detect wakes was introduced by [2] of merely using wind farm SCADA data. Aim was to reduce turbine fault activations that are caused by wakes. Based on turbine data a wake pattern for each turbine was measured and used to decide in real time whether a power reduction was due to a wake or an actual fault.

An alternative method to detect wakes was presented in [3]. The local wind speed and turbulence intensity are derived from strain gauges or optical fibres. It was shown that this method is capable of detecting differences of these quantities on the two sides of the rotor and thus detect wakes. These theoretical methods have more recently been studied on scaled wind turbines in a wind tunnel showing similar success of detecting the lateral position of upstream wakes [4]. One drawback of this method is that to estimate wind speeds from blade bending moments requires an aerodynamic model of the turbine. Wind farm operators often do not have access to this kind of information.

IOP Conf. Series: Journal of Physics: Conf. Series **854** (2017) 012020 doi:10.1088/1742-6596/854/1/012020

Remote sensing devices have shown potential at measuring wakes. Studies have shown application to detecting aircraft tip vortex using ground-based [5] and on-board devices [6]. Further lidar ship wake detection has been studied in [7].

Most recently, interest has grown in the application of lidar systems to wind energy. [8, 9] have used a scanning lidars to track the wake behind a turbine and measured wake deficit, width and centre. In [10] two lidar systems were used to measure a wake immediately behind a turbine. The study was extended in [11] to investigate the influence of the stability of the atmosphere on wake recovery. However, these lidar systems are very complex and expensive and are not realizable as commercial products for employment of several turbines within a wind farm.

Here we are presenting wake detection capabilities of an inexpensive, commercial lidar. It is mounted on the nacelle and measures the incoming wind field and thus our interest lies in detecting wakes in the inflow.

2. Experimental Setup and Motivation

The data analysed here have been gather during a one month long period in late 2015. A WindEYE lidar by Windar Photonics A/S was mounted on a Vestas V52 at the Risø test site. A schematic of the setup is shown in figure 1. The Vestas turbine, where the WindEYE was mounted, was exposed to a wake of a smaller Nordtank 500kW turbine, see figure 2.

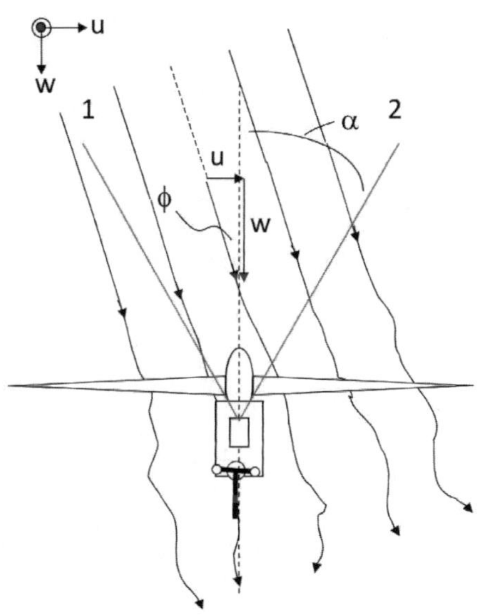

Figure 1. Setup schematic of the WindEYE lidar on a wind turbine ($\alpha = 30$ deg).

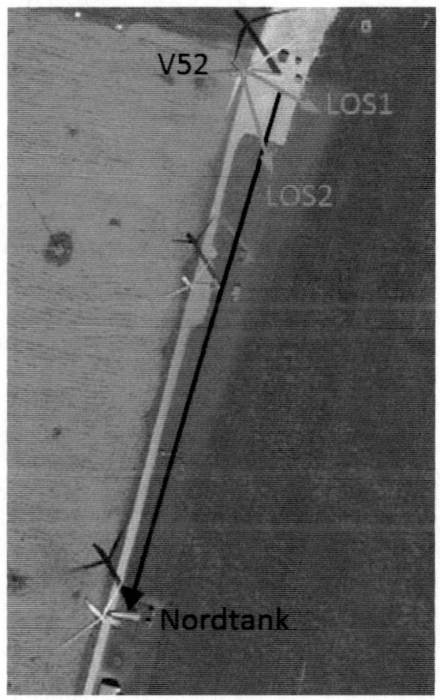

Figure 2. Overview of the Ris test site.

The lidar is mounted on the nacelle of a wind turbine and is looking upwind to measure the incoming wind field. From the two measured line-of-sight velocities ($v_{\mathrm{LOS},1}$ and $v_{\mathrm{LOS},2}$) the horizontal components (U and W) of the wind can be estimated. In the process it is necessary to make the following assumptions about the incoming wind field:

(i) The wind field is horizontally homogeneous and

(ii) no vertical components.

In situations where the turbine is exposed to a wake of an upstream turbine, assumption (i) is heavily violated and thus the derived wind field information will be invalid. Thus a wake detection algorithm is necessary to identify conditions where the applied assumptions are not valid.

3. Approach

The approach is based on the assumption that the lidar can detect enhanced turbulence, and that this enables the system to determine whether one or more beams of the system are measuring inside wakes. However, turbulence measured with a lidar is affected by the relatively long effective measurement volume. The lidar effectively averages the wind speeds within the volume [12].

Small-scale turbulence manifests itself in the width of the Doppler peak of the lidar, so that the width is a measure of the small-scale turbulence [13, 14]. Specifically, it has been shown that the width of the Doppler spectrum of a continuous-wave lidar is proportional to the turbulence intensity within the probe volume of the instrument [15]. At the same time, it is known that small-scale turbulence, which would be particularly responsible for increasing the Doppler spectrum width, is prevalent in wakes [16, 15].

As a measure of the Doppler spectrum width we use the 2nd central statistical moment of the peak. The first moment is defined as the centroid. In order to reduce noise in the width estimation a running average filter with a size of 600s has been used.

4. Results

In this section two periods are presented: one with a wake situation and one without a wake situation.

The results for the first period are shown in figure 3. The top panel shows the difference in peak variance between beam 1 and 2. Two vertical lines at 20 and -20 are also plotted. Large differences indicate enhanced turbulence levels, and thus a wake, in one of the beams. The centre panel shows the yaw position of the two turbines including an indication of the wake sector. The bottom panel shows the power production of the wake-emitting turbine.

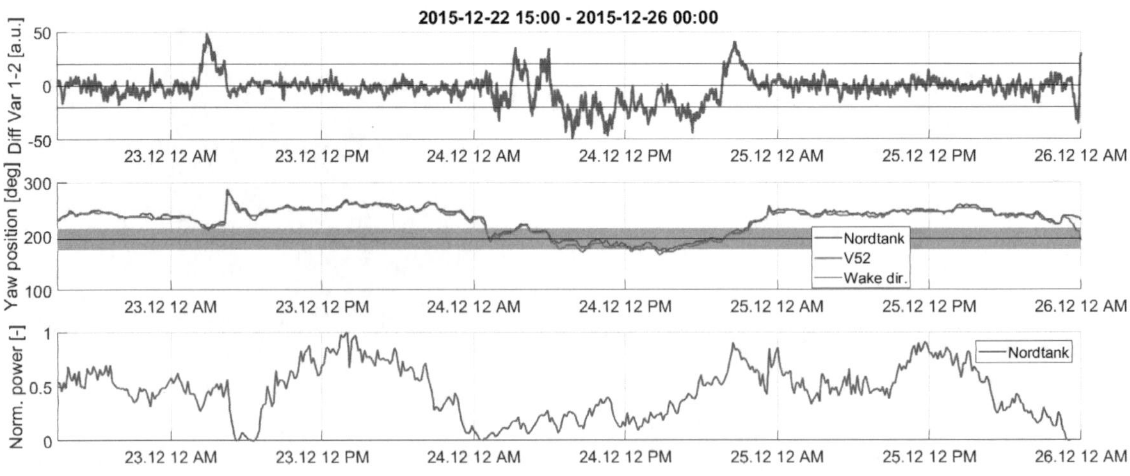

Figure 3. Top: difference of peak variance between beam 1 and 2. Centre: yaw position of both turbines. Bottom: Power production of Nordtank turbine.

It can be seen that as soon as the upstream turbine is expected to emit a wake on the downstream turbine (based on their yaw positions), there is a sudden jump in the difference of the peak variances. This is shown especially after Dec 23rd 12AM, where a negative spike in variance difference can be seen as soon as a wake position is entered. This means that a wake on the left half of the rotor is detected, which agrees with the yaw position showing higher values than the wake direction (195 deg). Similarly, the period around the Dec 24th 12AM shows a wake on the right half of the rotor (positive difference in peak variance), which is also confirmed by the turbine yaw position. Hence, the sign of the peak variance difference can detect a wake situation and show which half of the rotor is affected.

Figure 4 shows the results for a period where according to the yaw positions no wake situation is expected. When looking at the difference in peak variance, it can be seen that at all time the values fluctuate around zero. Contrarily to the wake situation no sudden deviation in peak variance difference can be seen and so no wakes are predicted by the lidar.

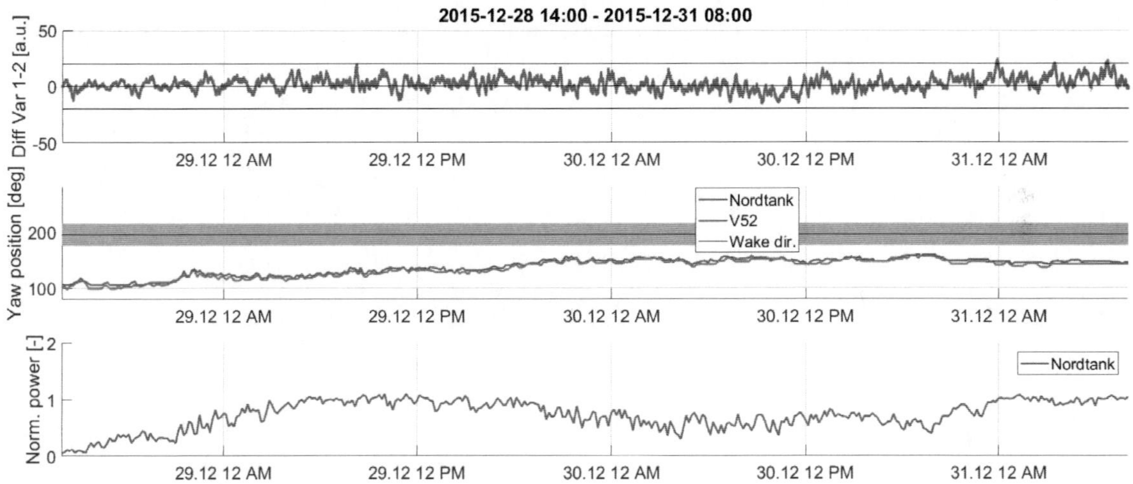

Figure 4. Top: difference of peak variance between beam 1 and 2. Centre: yaw position of both turbines. Bottom: Power production of Nordtank turbine.

5. Conclusion

This study investigated the wake detection possibilities of a nacelle lidar. It was shown that by comparing the Doppler peak width of the two line-of-sight velocities wake situations can be predicted. A suitable measure of the peak width is the 2nd central statistical moment. The sign of the peak width difference indicates which side of the turbine is exposed to a wake. A challenge remains in setting appropriate thresholds for the difference of peak variances. Such a limit is probably dependent on ambient turbulence levels and future experiments will be conducted to test the algorithm in different turbulence levels.

Wake detection is necessary to identify where the assumption used for wind field reconstruction are violated. Further, information about the position of wakes is very valuable information and can be exploited for wind farm control.

Acknowledgments

This study was supported by Innovationsfonden Danmark in form of an industrial PhD stipend (project number: 5016-00182).

References

[1] Vermeer L J, Sørensen J N and Crespo A 2003 *Prog. Aerosp. Sci.* **39** 467–510 ISSN 03760421
[2] Yan Y, Zhang J Z and Karayaka H B 2015 *Int. J. Progn. Heal. Manag.* 1–12
[3] Bottasso C L, Cacciola S and Schreiber J 2015 *J. Phys. Conf. Ser.* **625** 012007 ISSN 1742-6588 URL `http://dx.doi.org/10.1088/1742-6596/625/1/012007`

[4] Schreiber J, Cacciola S, Campagnolo F, Petrović V, Mourembles D and Bottasso C L 2016 *J. Phys. Conf. Ser.* **753** 032027 ISSN 1742-6588
[5] Harris M, Young R I, Koepp F, Dolfi A and Cariou J P 2002 *Aerosp. Sci. Technol.* **6** 325–331
[6] Douxchamps D, Lugan S, Verschueren Y, Mutuel L, Macq B and Chihara K 2008 *IEEE Trans. Aerosp. Electron. Syst.* **44** 1276–1290 ISSN 00189251
[7] Bunkin A F, Klinkov V K, Lukyanchenko V a and Pershin S M 2011 *Appl. Opt.* **50** A86–A89 ISSN 0003-6935
[8] Wang H and Barthelmie R J 2015 *J. Phys. Conf. Ser.* **625** 012017 ISSN 1742-6588
[9] Aitken M L, Rhodes M E and Lundquist J K 2012 *J. Atmos. Ocean. Technol.* **29** 347–355 ISSN 07390572
[10] Iungo G V, Wu Y T and Porté-Agel F 2013 *J. Atmos. Ocean. Technol.* **30** 274–287 ISSN 0739-0572
[11] Iungo G V and Porté-Agel F 2014 *J. Atmos. Ocean. Technol.* **31** 2035–2048 ISSN 0739-0572
[12] Sathe A and Mann J 2013 *Atmos. Meas. Tech.* **6** 3147–3167 ISSN 18671381
[13] Angelou N, Mann J, Sjöholm M and Courtney M 2012 *Rev. Sci. Instrum.* **83** ISSN 00346748
[14] Branlard E, Pedersen A T, Mann J, Angelou N, Fischer A, Mikkelsen T, Harris M, Slinger C and Montes B F 2013 *Atmos. Meas. Tech.* **6** 1673–1683 ISSN 18671381
[15] Larsen G C, Hansen K S, Mann J, Bingöl F and Enevoldsen K 2010 *Sci. Mak. Torque from Wind*
[16] Larsen G C, Madsen H A, Thomsen K and Larsen T J 2008 *Wind Energy* **11** 377–395 ISSN 10954244

IOP Conf. Series: Journal of Physics: Conf. Series **854** (2017) 012021 doi:10.1088/1742-6596/854/1/012021

High resolution wind turbine wake measurements with a scanning lidar

T G Herges[1], D C Maniaci[1], B T Naughton[1], T Mikkelsen[2] and M Sjöholm[2]

[1]Sandia National Laboratories, Albuquerque, NM, 87185, United States of America
[2]DTU Wind Energy, Risø Campus, Frederiksborgvej 399, 4000 Roskilde, Denmark.

Abstract. High-resolution lidar wake measurements are part of an ongoing field campaign being conducted at the Scaled Wind Farm Technology facility by Sandia National Laboratories and the National Renewable Energy Laboratory using a customized scanning lidar from the Technical University of Denmark. One of the primary objectives is to collect experimental data to improve the predictive capability of wind plant computational models to represent the response of the turbine wake to varying inflow conditions and turbine operating states. The present work summarizes the experimental setup and illustrates several wake measurement example cases. The cases focus on demonstrating the impact of the atmospheric conditions on the wake shape and position, and exhibit a sample of the data that has been made public through the Department of Energy Atmosphere to Electrons Data Archive and Portal.

1. Introduction

Sandia National Laboratories (SNL) and the National Renewable Energy Laboratory (NREL) are conducting an experiment at the SNL Scaled Wind Farm Technology (SWiFT) facility located in Lubbock, Texas [1, 2] to investigate the use of wind turbine yaw control to direct wakes for increased wind plant performance [3-5]. One of the primary objectives of the experiment is to collect experimental data to improve the predictive capability of wind plant computational models to represent the response of the turbine wake to varying inflow conditions and turbine operating states. The baseline SWiFT facility has well-instrumented meteorological towers and research turbines, but in order to adequately measure the turbine wakes, the team partnered with the Technical University of Denmark (DTU) Wind Energy Department to leverage their wake measurement expertise and their custom-built SpinnerLidar instrument, which is uniquely capable of measuring wakes at the temporal and spatial resolution required for the experiment [6-8].

Data collected as part of the multi-month field campaign will be used to improve both high-fidelity wind plant models as well as demonstrate control concepts to facilitate future wind industry technology development for reducing wake losses. To confirm model predictions, long-term continuous measurements of the wake velocity profile downwind of the turbine are needed to obtain a statistical representation. The methodology for this effort is based on a formal Verification and Validation (V&V) framework developed for the U.S. Department of Energy (DOE) Atmosphere to Electrons (A2e) initiative. This V&V framework is being used for the development and execution of a coordinated simulation and experimental program to assess the predictive capability of computational models of complex systems through focused, well structured, and formal processes [9].

The present work summarizes the experimental setup and illustrates several wake measurement example cases from the current SWiFT facility measurement campaign. The focus of this paper is to provide a few sample cases that demonstrate the impact of the atmospheric conditions on the wake shape and position, and exhibit some of the data that is available through the DOE Atmosphere to electron (A2e) Data Archive and Portal (DAP) as part of the current experimental campaign [10].

2. Experimental Setup

The primary quantities of interest for the experiment include the trajectory of the wake center, the wake deficit strength, the wake shape, and the wind turbine rotor loads and power for a range of fixed yaw offsets subjected to a range of atmospheric inflow conditions [3, 5, 11]. To meet this objective, a variety of synchronized instrumentation were integrated at the SWiFT test site, including an upstream meteorological tower, turbine structural and performance sensors and the DTU SpinnerLidar to characterize the wake center location and deficit magnitude [8, 12]. The following sections describe the experimental setup in detail.

2.1. SWiFT Site Configuration

Sandia National Laboratories operates the Scaled Wind Farm Technology (SWiFT) facility located in Lubbock, Texas. The baseline site instrumentation includes three research wind turbines (WTG) and two meteorological towers (MET) as shown in figure 1 with the relative as-built survey locations listed in table 1.

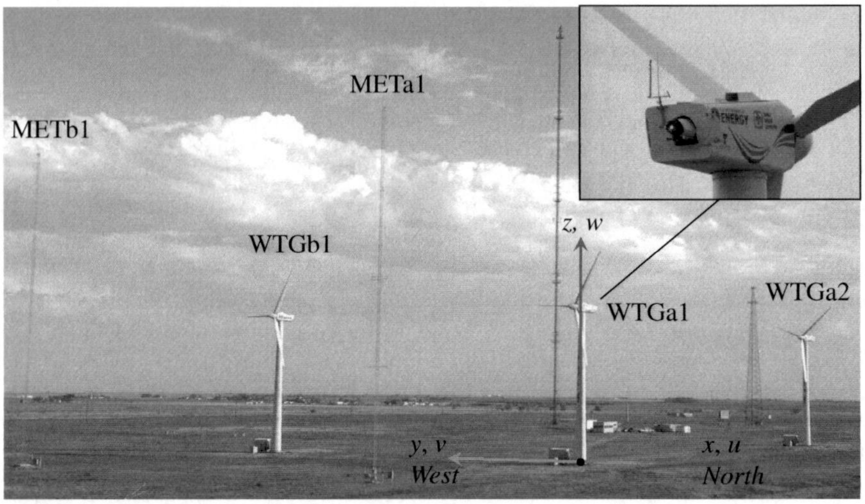

Figure 1. SWiFT site layout and coordinate system with the DTU SpinnerLidar installed in WTGa1 [1].

Table 1. As-built survey locations of the baseline SWiFT site instruments

Instrument	*x* (m), North	*y* (m), West	*z* (m), Base Height
WTGa1	0.00	0.00	0.00
WTGa2	134.97	0.17	0.09
WTGb1	-5.41	80.95	-0.02
METa1	-69.50	-3.29	0.22
METb1	-74.95	77.53	0.20

The layout of the SWiFT facility is seen from an overhead view in figure 2 along with the wind rose at a height of 75 m (north is 0 degrees). The met tower and the two turbines used in this campaign (METa1, WTGa1, and WTGa2) are all aligned with the predominant wind direction at the site. This configuration allows measurement of the atmospheric inflow with the met tower and measurement of the wake of the WTGa1 turbine using the nacelle-mounted DTU SpinnerLidar.

IOP Conf. Series: Journal of Physics: Conf. Series **854** (2017) 012021 doi:10.1088/1742-6596/854/1/012021

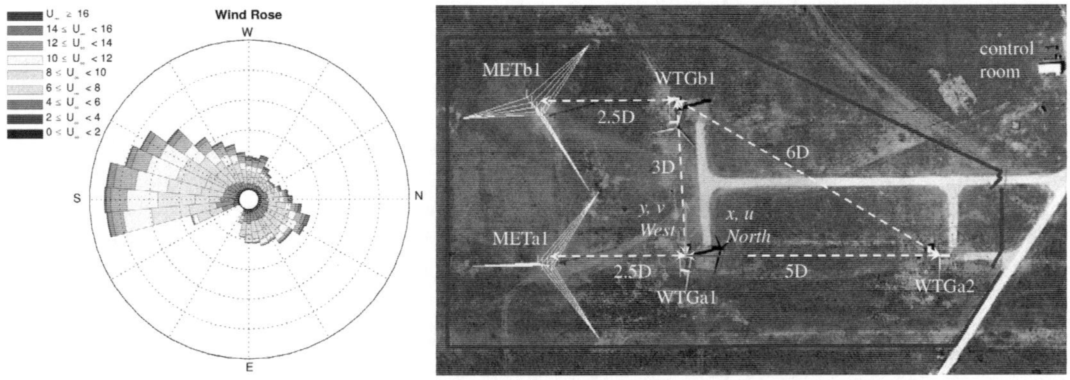

Figure 2. Top view of the SWiFT facility layout and wind rose at 75 m height (D = 27m) [2].

The SWiFT turbines are highly-modified Vestas V27 machines with a hub height of 32.1 meters, a rotor diameter of 27 meters, and a power rating of 192 kW. The meteorological towers are 60 m tall with a suite of atmospheric sensors located at heights as indicated in figure 3. The entire site is on a fiber optic network facilitating the logging of over 400 data channels at a frequency between 20 to 100 Hz, synched to a GPS time-signal and stored centrally in the control building shown in the upper right corner of figure 2.

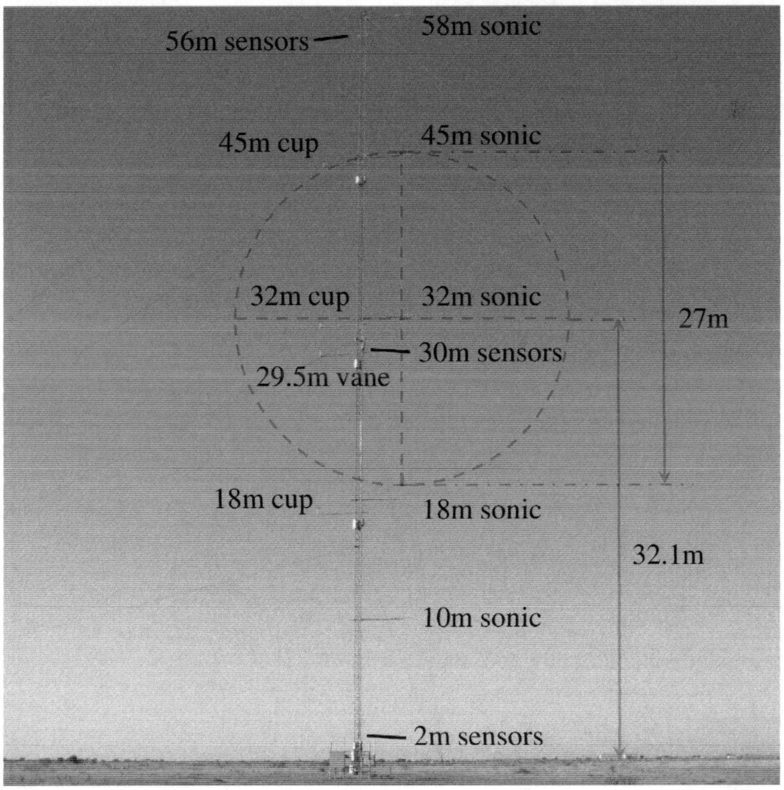

Figure 3. Photo of the METa1 with locations of the cups, vanes, sonic anemometers and atmospheric sensors (temperature, relative humidity, barometric pressure). The image was acquired looking in the Southerly direction and includes an approximate projection of the WTGa1 rotor diameter on the met tower.

2.2. SpinnerLidar Configuration

The DTU SpinnerLidar (figure 4a) is uniquely capable of measuring the wind turbine wake at the temporal and spatial resolution required for the experimental campaign. DTU developed the SpinnerLidar to be a turbine mounted lidar for rapid scanning of the wind field in a two-dimensional plane. For this experiment, the SpinnerLidar was mounted as shown in figure 1 to point out of the rear of the upwind SWiFT turbine nacelle (WTGa1), optimally positioned to capture the full wind turbine wake up to five rotor diameters downstream (D = 27 m). The SpinnerLidar has a scan head consisting of two co-rotating ~15° wedge-shaped prisms integrated on a ZephIR 300 continuous-wave coherent Doppler lidar. The lidar produces laser light at a wavelength of 1565 nm and was configured to stream averaged Doppler spectra at a rate of about 492 measurements per second. The prisms have a fixed gear ratio with adjustable motor settings to change the duration and number of measurements per scan (motor speed fixed for each scan) [8, 13]. The resulting rosette scan pattern is displayed in figure 4b. At each focus distance, the SpinnerLidar scans the two-dimensional surface of a sphere with an approximately 30° half angle. The lidar also has the ability to cycle through focus distances with a change in focus distance occurring in the same amount of time as a full scan (figure 4b). The position of the measurement relative to the symmetry axis is calculated from the instantaneous position of the two wedge-shaped prisms. Changes in the orientation of the rotation axis are accounted for using an integrated three-axis accelerometer.

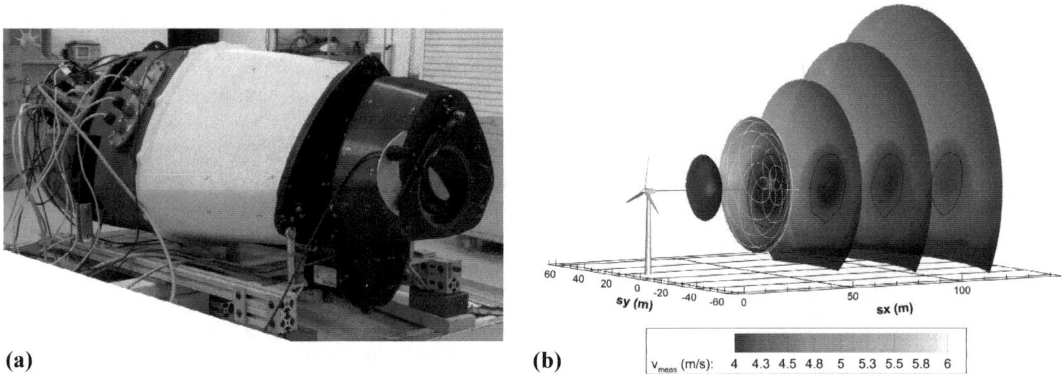

(a) **(b)**

Figure 4. (a) A photograph of the DTU SpinnerLidar in the lab, and (b) the SpinnerLidar scanning pattern at the SWiFT site, overlaid on line-of-sight (LOS) speed profiles extracted form virtual lidar simulations [4, 14].

Figure 4b shows one of the scanning configurations that has been used during the wake steering experiment with rosette scans at each of 5 rotor diameter distances downstream. Each rosette pattern of approximately 984 data points takes 2 seconds to complete and another 2 seconds to refocus to each sequential measurement plane downstream. In this configuration, the lidar captures five slices of the wake about every 18 seconds. So far the lidar has captured this wake data as the turbine operated under a variety of intentional yaw offsets, or misalignments, between ± 10° with respect to the predominant inflow direction and under various atmospheric conditions characterized by wind speed, temperature profile, turbulence intensity, veer, and shear. Future yaw offsets will include a range from -18° to +25°. The lidar is mounted to a manually adjustable platform that has fine resolution (~1°) yaw shift increments to ensure the lidar field of view captures the wake at these larger yaw offsets.

2.3. Data Collection and Processing

The SpinnerLidar calculates the line-of-sight velocity from the returned Doppler spectra by calculating the median of frequencies observed within the probe volume above a noise threshold. The quality assurance and quality control (QA/QC) method removes data points that have a low level of laser signal return relative to the noise threshold as well as points contaminated with laser signal returns from stationary objects due to the introduction of slight object movement from the motion of the laser scan pattern. The removal of points with laser returns from stationary objects primarily addresses points contaminated with ground returns and boresight returns. The boresight returns are due to laser reflections from the outer window and appear at the center of the scan pattern. Sometimes this removal includes points due to returns from the met towers, neighboring SWiFT wind turbines and other objects.

The lidar measurement location is aligned and calibrated using the method provided in Ref. [12]. The measurements are transformed into the coordinate system in figure 5 (referred to as the lidar coordinate system), where the *sx* direction is aligned with the centerline of the nacelle axial direction, *sz* is vertical, and *sy* is oriented to create a right-handed coordinate system. Figure 5 also shows the relationship between the overall SWiFT site coordinate system and the lidar coordinate system. The location of the measurements in the lidar coordinate system are scaled from an initial normalized frame (figure 6a, which was calibrated in Ref. [12]) using the average focus distance over the scan. The actual focus distance at each measurement point can fluctuate as high as ± 2 m. This fluctuation is relatively small when compared to the probe volume length of the lidar as shown in figure 6b, resulting in a minimal change in the corresponding velocity measurement [15, 16]. The orientation and position of the lidar relative to the ground and nacelle was determined using total station theodolite (TST) measurements [12]. The lidar roll and pitch angles are measured by a calibrated 3-axis accelerometer [12], and applied to the data to create the lidar coordinate system (*sx*, *sy*, *sz*).

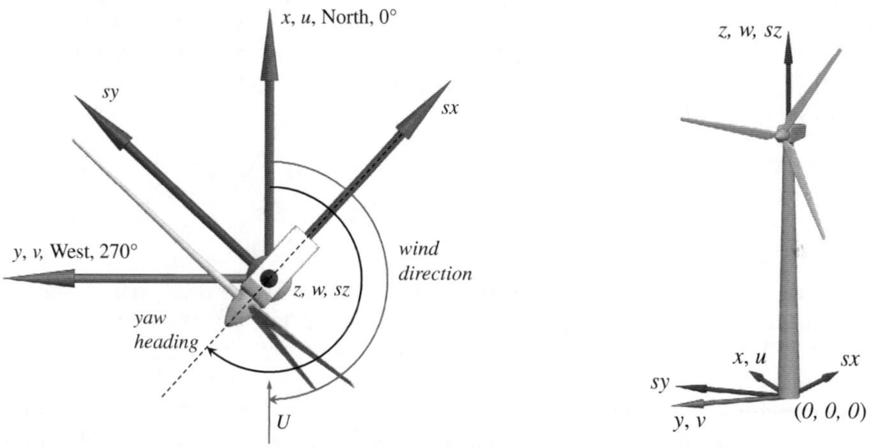

Figure 5. Schematic of the SWiFT site coordinate system relative to the lidar coordinate system (*sx*, *sy*, *sz*).

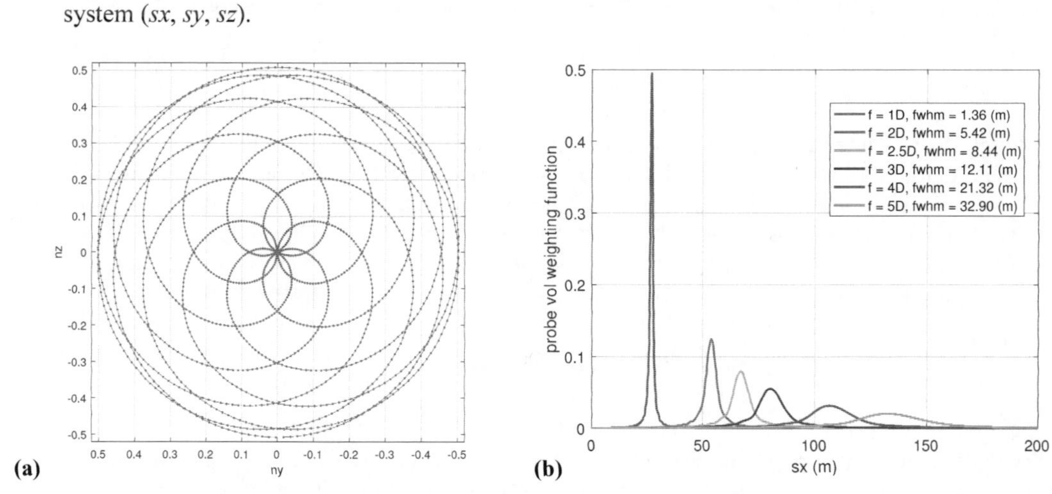

Figure 6. (a) Calibrated DTU SpinnerLidar scan points in normalized lidar coordinates and (b) lidar probe volume weighting function versus measurement distance.

The origin of the coordinate system is centered along the nacelle yaw rotation axis and located at the base of the turbine foundation. The vertical coordinate in the lidar coordinate frame matches the vertical coordinate of the SWiFT site coordinate frame, and the SWiFT site *x-y* plane matches the lidar *sx-sy* plane. Thus, the only difference between the lidar coordinate system and cardinal direction coordinate systems is due to the yaw heading of the turbine. The lidar coordinate system can be transformed into the SWiFT site coordinate system or wind direction coordinate system by applying a rotation around the *sz* axis with a magnitude corresponding to the yaw heading or yaw offset, respectively.

3. Results

Several wake measurement datasets are presented here, all of which strongly indicate the effect of atmospheric conditions on the wake shape and position. Figures 7 – 12 display sample contour surfaces from the measured lidar data between 1 – 5 D (D = 27 m) downstream of the turbine as the line-of-sight speed (v_{LOS}). The irregular scan pattern was interpolated to a regular grid using a smoothing surface fit, while the wake position was calculated using an image processing object detection method of the wake velocity deficit. The calculated wake boundary and center are displayed as a white line and black dot, respectively. The wake location in time and space is the primary quantity of interest to assess and improve the wake steering control model throughout the current experimental campaign [3, 5]. The atmospheric conditions have been calculated as defined in Ref. [2]. Specifically, the atmospheric boundary layer power-law velocity profile exponent, α, is defined in equation (1), where α was approximated using a least-squares fit of the wind speed measured by the met tower sonic anemometers at a height of 10 m, 18 m, 32 m, 45 m, and 58 m (figure 3). The power-law exponent approximation has an average uncertainty of ±0.01 in the presented cases using 10 min. bins of sonic anemometer data.

$$\frac{U(z)}{U_{\infty,\text{32m Sonic}}}=\left(\frac{z}{z_{\text{32m sonic}}}\right)^{\alpha} \tag{1}$$

Positive veer of the atmospheric inflow is expressed in equation (2) as a clockwise rotation of the wind direction (from an overhead view, as in figure 5) with increasing height measured across the height of the rotor. The SWiFT site atmospheric characterization reveals a stable atmospheric boundary layer is frequently required for high veer and shear profiles, positive veer is more common than negative veer, and turbulence reduces the presence of veer [2]. Only stable atmospheric cases with veer are presented because unstable and neutral cases with large amounts of veer are rare. Observation of the impact of veer on the wind turbine wake is also more difficult under neutral and unstable atmospheric inflows due to increased levels of atmospheric turbulence as compared to the stable atmosphere condition.

$$veer=\theta_{\text{45m sonic}}-\theta_{\text{18m sonic}} \tag{2}$$

Although the presented cases are at a range of wind speeds, the rotor is operating in region 2 or 2.5 for all cases [17]. The yaw offset of the wind turbine is described as the wind direction of the met tower hub-height sonic anemometer minus the wind turbine yaw heading (equation (3)). The wind turbine coordinate system in figure 5 defines both the wind direction and the yaw heading relative to north. Figure 5 illustrates a wind direction and wind turbine yaw heading that corresponds to a negative yaw offset.

$$\gamma_{\text{yaw offset}}=\theta_{\infty,\text{32m sonic}}-\psi_{\text{turbine yaw heading}} \tag{3}$$

3.1. Neutral Atmospheric Boundary Layer Case

The first case (figure 7) displays lidar data acquired during a neutrally stable atmospheric inflow as defined by the normalized Obukhov length, *z/L*. The turbine was operating in region 2.5, with moderate turbulence and shear in addition to low veer. The measurements in figure 7 were acquired over a duration of 22 s with a 2 s scan rate at each focus distance (1, 2, 2.5, 3, 4, and 5 D). The wake position and shape appears to fluctuate between measurement surfaces, relative to the inflow direction.

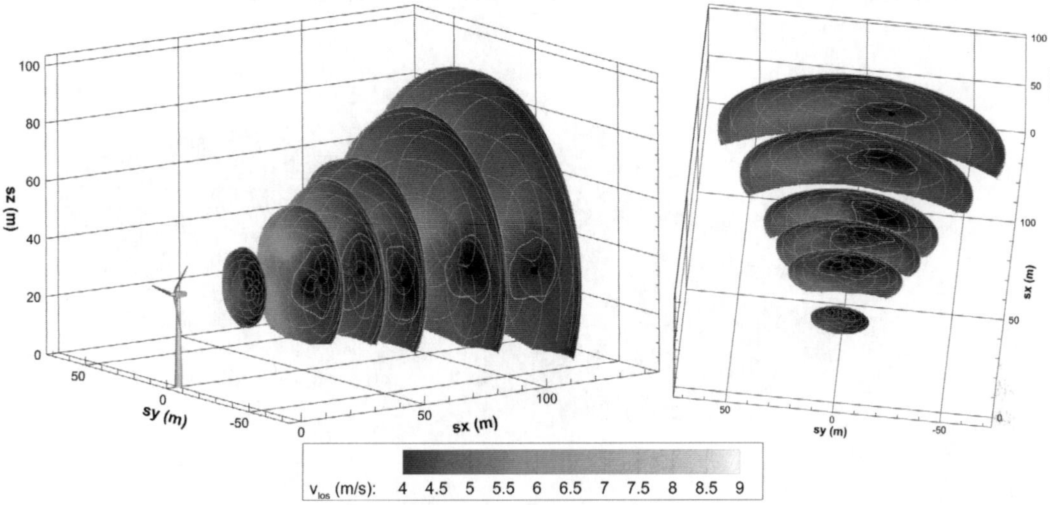

Figure 7. Example lidar measurements from 1 – 5 D over 22 s with a neutrally stable inflow: normalized Obukhov length $z/L = 0.0$, $\alpha = 0.12$, wind speed = 8.2 m/s, TI = 0.11, veer = 1.3°, yaw offset = 5.9°, and yaw heading = 243.8 degN.

3.2. Stable Atmospheric Boundary Layer with No Veer Case

The next case displays the wake under stable atmospheric conditions with low veer, shown in figure 8. The rotor is operating in region 2 with minimal yaw offset. The turbulence was very low and the shear was stronger than in the neutral case. The measurements in figure 8 were acquired over a duration of 22 s with a 2 s scan rate at each focus distance (1, 2, 2.5, 3, 4, and 5 D). Notice that the shape of the wake deficit appears round and follows the wind direction, with little fluctuation in position or shape between measurement surfaces.

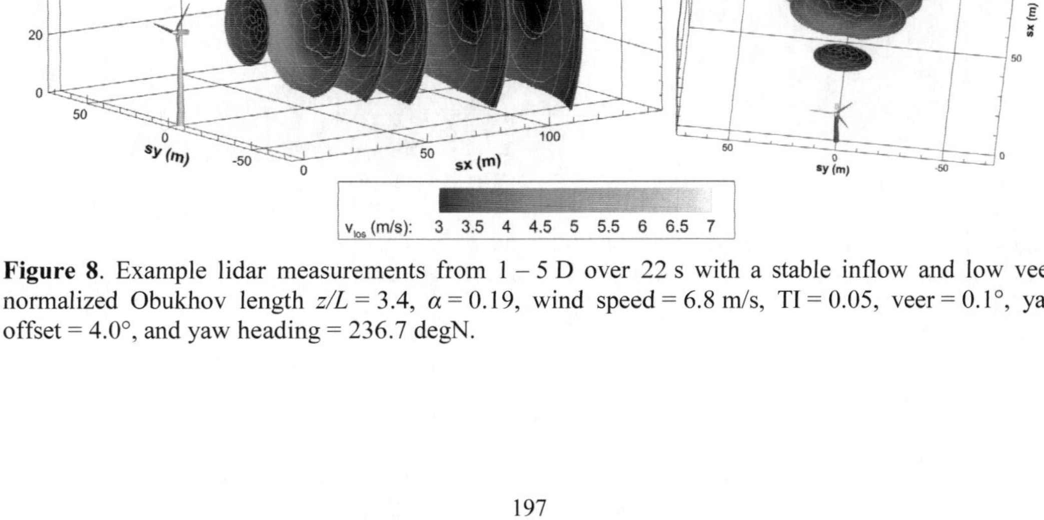

Figure 8. Example lidar measurements from 1 – 5 D over 22 s with a stable inflow and low veer: normalized Obukhov length $z/L = 3.4$, $\alpha = 0.19$, wind speed = 6.8 m/s, TI = 0.05, veer = 0.1°, yaw offset = 4.0°, and yaw heading = 236.7 degN.

3.3. Stable Atmospheric Boundary Layer with Positive Veer Case

The case shown in figure 9 is also stable, but has stronger shear and substantial positive veer. The rotor was also operating in region 2, with similar wind speed, turbulence intensity, and yaw offset as the previous stable case with minimal veer (figure 8).

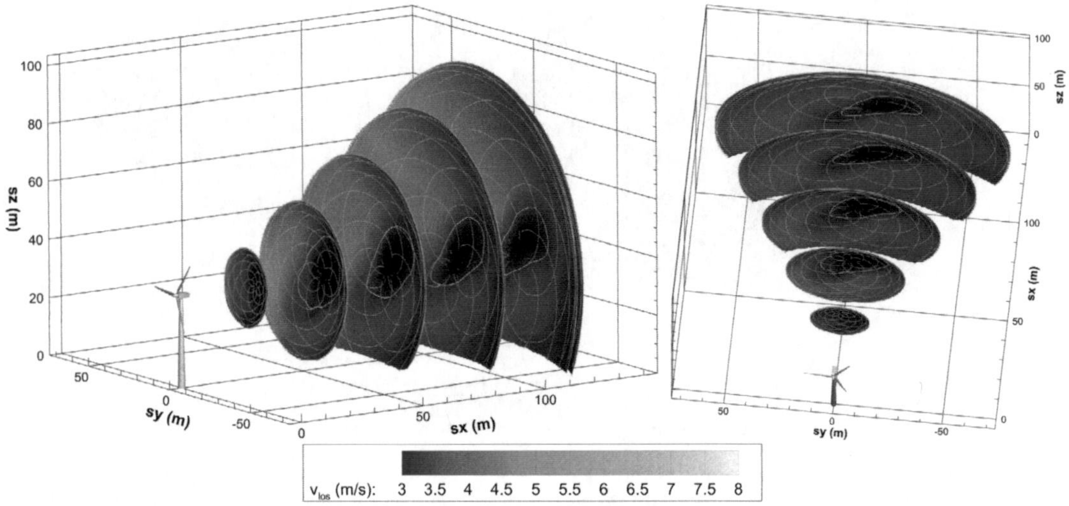

Figure 9. Example lidar measurements from 1 – 5 D over 18 s with a stable inflow and positive veer: normalized Obukhov length $z/L = 2.3$, $\alpha = 0.37$, wind speed = 6.9 m/s, TI = 0.04, veer = 14.6°, yaw offset = -0.12°, and yaw heading = 195.3 degN.

As expected, the strong positive veer causes the wake velocity deficit shape (wake shape) to skew in the direction of the changing wind direction as a function of height. The skew in the wake shape is displayed more distinctly in the flattened wake cross sections shown in figure 10 at 2 and 5 rotor diameters downstream of the rotor. The measurements in figure 9 were acquired over a duration of 18 s with a 2 s scan rate at each focus distance (1, 2, 3, 4, and 5 D). Observe that the inflow is highly stable and that even though the wake shape changes dramatically as it moves downstream, the overall wake position aligns with the flow direction, without fluctuations.

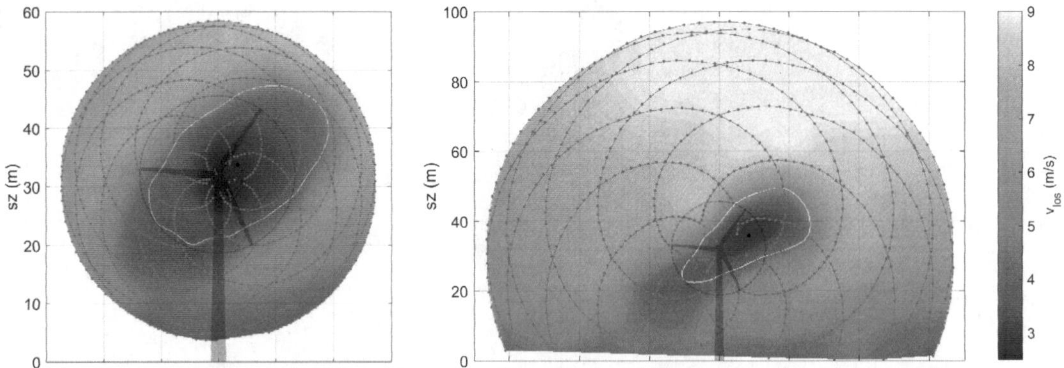

Figure 10. SpinnerLidar measurements viewed in the *sy-sz* plane to help visualize the impact of a highly stable, large positive veer inflow on wake shape at (a) 2 D and (b) 5 D downstream, silhouette of WTGa1 included for reference and scale.

3.4. Stable Atmospheric Boundary Layer with Negative Veer Case

The effect of negative veer is illustrated in figure 11, which includes the combined influence of moderate negative veer and moderate positive yaw offset on the wake shape and position. The measurements in figures 11 were acquired over a duration of 18 s with a 2 s scan rate at each focus distance (1, 2, 3, 4, and 5 D) with the turbine operating in region 2. As with the previous case, the stable atmosphere results in a wake trajectory with few fluctuations.

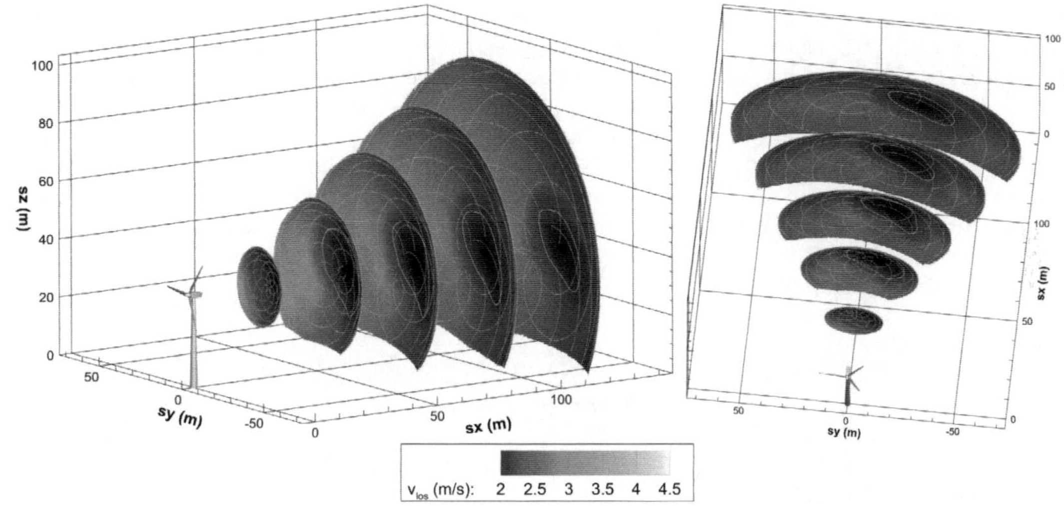

Figure 11. Example lidar measurements from 1 – 5 D over 18 s with a stable inflow and negative veer: normalized Obukhov length $z/L = 0.9$, $\alpha = 0.15$, wind speed = 4.8 m/s, TI = 0.08, veer = –5.0°, yaw offset = 10.9°, and yaw heading = 29.5 degN.

3.5. Unstable Atmospheric Boundary Layer Case

The final case (figure 12) displays the wake velocity deficit under unstable inflow conditions, commonly found during daytime (acquired at 11:51 AM CST). The wind speed places the rotor near the beginning

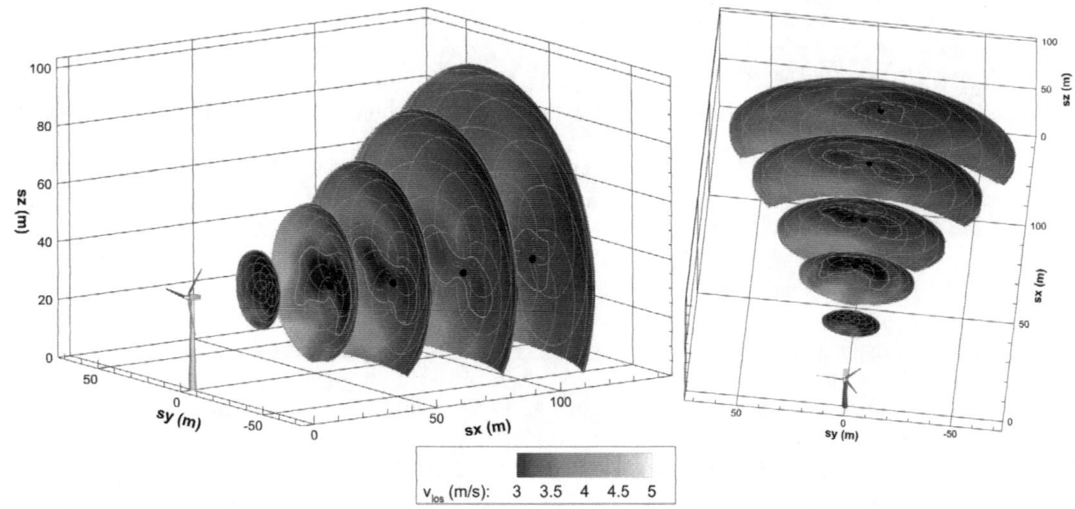

Figure 12. Example lidar measurements from 1 – 5 D over 18 s with an unstable inflow: normalized Obukhov length $z/L = -0.8$, $\alpha = 0.05$, wind speed = 5.8 m/s, TI = 0.25, veer = -2.0°, yaw offset = 5.7°, and yaw heading = 47.4 degN.

of region 2 turbine operation. Shear and veer were relatively low and turbulence relatively high, as expected during the increased mixing in an unstable atmospheric boundary layer. The wake shape is irregular since it is heavily influenced by the turbulent structures in the atmospheric boundary layer.

4. Conclusion

Several wake measurement example cases with varying inflow conditions have been presented in the current work along with a thorough description of the experimental setup. One of the primary objectives of the experiment is to collect validation-quality experimental data to improve the predictive capability of wind plant computational models to represent the response of the turbine wake to varying inflow conditions and turbine operating states. Considerable attention has been paid to the experimental setup and design in order to produce high quality data sets for this purpose. These data sets quantify the effect of a range of inflow conditions (including highly stable atmospheric boundary layers) and turbine operating states on the detailed, high-resolution development of the wake velocity deficit. These measurements are being used in a series of validation studies for both high fidelity wind plant models and lower fidelity controls models. The data sets produced during this test campaign are publicly available through the DOE Atmosphere to electron (A2e) Data Archive and Portal (DAP) at https://a2e.energy.gov/projects/wake [10].

5. References

[1] Berg J, Bryant J, LeBlanc B, Maniaci D C, Naughton B, Paquette J A, Resor B R, White J and Kroeker D 2014 *32nd ASME Wind Energy Symposium*: American Institute of Aeronautics and Astronautics

[2] Kelley C L and Ennis B L 2016 SWiFT Site Atmospheric Characterization. In: *SAND-2016-2016*: (Sandia National Laboratories)

[3] Fleming P, Gebraad P M O, Lee S, Wingerden J-W v, Johnson K, Churchfield M, Michalakes J, Spalart P and Moriarty P 2015 Simulation comparison of wake mitigation control strategies for a two-turbine case *Wind Energy* **18** 2135-43

[4] Churchfield M, Wang Q, Scholbrock A, Herges T, Mikkelsen T and Sjöholm M 2016 Using High-Fidelity Computational Fluid Dynamics to Help Design a Wind Turbine Wake Measurement Experiment *Journal of Physics: Conference Series* **753** 032009

[5] Fleming P, Churchfield M, Scholbrock A, Clifton A, Schreck S, Johnson K, Wright A, Gebraad P, Annoni J, Naughton B, Berg J, Herges T, White J, Mikkelsen T, Sjöholm M and Angelou N 2016 Detailed field test of yaw-based wake steering *Journal of Physics: Conference Series* **753** 052003

[6] Mikkelsen T, Angelou N, Hansen K, Sjöholm M, Harris M, Slinger C, Hadley P, Scullion R, Ellis G and Vives G 2013 A spinner-integrated wind lidar for enhanced wind turbine control *Wind Energy* **16** 625-43

[7] Machefaux E, Larsen G C, Troldborg N, Hansen K S, Angelou N, Mikkelsen T and Mann J 2016 Investigation of wake interaction using full-scale lidar measurements and large eddy simulation *Wind Energy* **19** 1535-51

[8] Sjöholm M, Pedersen A T, Angelou N, Abari F F, Mikkelsen T K, Harris M, Slinger C and Kapp S 2013 Full two-dimensional rotor plane inflow measurements by a spinner-integrated wind lidar. In: *European Wind Energy Association Conference*

[9] Hills R G, Maniaci D C and Naughton J W 2015 V&V Framework. In: *SAND-2015-7455,* (Sandia National Laboratories)

[10] Naughton B 2017 Wake Steering Experiment A2e Data Archive Portal. (https://a2e.energy.gov/projects/wake)

[11] Churchfield M J, Lee S, Michalakes J and Moriarty P J 2012 A numerical study of the effects of atmospheric and wake turbulence on wind turbine dynamics *Journal of Turbulence* **13** N14

[12] Herges T G, Maniaci D C, Naughton B, Hansen K, Sjöholm M, Angelou N and Mikkelsen T 2017 *35th Wind Energy Symposium*: American Institute of Aeronautics and Astronautics

[13] Angelou N and Sjöholm M 2015 UniTTe WP3/MC1: Measuring the inflow towards a Nordtank 500kW turbine using three short-range WindScanners and one SpinnerLidar. (DTU Wind Energy)
[14] Herges T 2016 Wake Steering Experimental Campaign Animation. In: *Sandia Energy & Climate,* (https://vimeo.com/162440208)
[15] Angelou N, Mann J, Sjöholm M and Courtney M 2012 Direct measurement of the spectral transfer function of a laser based anemometer *Review of Scientific Instruments* **83** 033111
[16] Horváth Z L and Bor Z 2003 Focusing of truncated Gaussian beams *Optics Communications* **222** 51-68
[17] Resor B R and LeBlanc B 2014 An Aeroelastic Reference Model for the SWIFT Turbines. (Albuquerque, NM: Sandia National Laboratories)

Acknowledgments
This work was supported by the U.S. Department of Energy under Contract No. DE-AC04-94AL85000 with Sandia National Laboratories, a multi-mission laboratory managed and operated by Sandia Corporation, a wholly owned subsidiary of Lockheed Martin Corporation, for the U.S. Department of Energy's National Nuclear Security Administration and contract No. DE-AC36-08GO28308 with the National Renewable Energy Laboratory. Funding for the work was provided by the DOE Office of Energy Efficiency and Renewable Energy, Wind Energy Technologies Office. In addition to our DOE sponsors, the authors would like to acknowledge Jonathan Berg, Joshua Bryant, Andrew Scholbrock, David Jager, and Jeroen van Dam for their help with supporting and running the SWiFT wind turbine. The authors also thank the SWiFT support staff, David Mitchell, Miguel Hernandez, Kolton Foster, and David Martinez, as well as Andrew Scholbrock and Donald Baker from the NREL for their support installing the SpinnerLidar at SWiFT and the NREL colleagues working on the Wake Steering Experiment.

Influence of the conservative rotor loads on the near wake of a wind turbine

I. Herráez[1], D. Micallef [2] and G.A.M. van Kuik[3]

[1] Dept. of Technology, University of Applied Sciences Emden/Leer, Constantiaplatz 4, 26723 Emden, Germany

[2] Dept. of Environmental Design, Faculty for the Build Environment, University of Malta, Msida MSD 2080, Malta

[3] Faculty of Aerospace Engineering, Technical University Delft, Kluyverweg 1, 2629HS, Delft, The Netherlands

E-mail: ivan.herraez@hs-emden-leer.de

Abstract. The presence of conservative forces on rotor blades is neglected in the blade element theory and all the numerical methods derived from it (like e.g. the blade element momentum theory and the actuator line technique). This might seem a reasonable simplification of the real flow of rotor blades, since conservative loads, by definition, do not contribute to the power conversion. However, conservative loads originating from the chordwise bound vorticity might affect the tip vortex trajectory, as we discussed in a previous work. In that work we also hypothesized that this effect, in turn, could influence the wake induction and correspondingly the rotor performance.

In the current work we extend a standard actuator line model in order to account for the conservative loads at the blade tip. This allows to isolate the influence of conservative forces from other effects. The comparison of numerical results with and without conservative loads enables to confirm qualitatively their relevance for the near wake and the rotor performance. However, an accurate quantitative assessment of the effect still remains out of reach due to the inherent uncertainty of the numerical model.

1. Introduction

The study of the fluid mechanics of propellers dates back to the 19th century. Rankine [1] and Froude [2] settled the basis of the momentum theory explaining the origin of propeller thrust and torque from the momentum change in the fluid. This theory defines a limit to the efficiency of an ideal propeller but, due to the fact that this theory is one-dimensional, it does not provide any information about the required propeller geometry. On the contrary, the blade element theory, which is attributed to Drzewiecki [3], takes the rotor geometry into account by dividing each blade into a finite number of independent lifting surfaces (blade elements), although it does not define any limit to the propeller efficiency. This limitation is owed to the lack of consideration of the velocity induction of the propeller itself. In order to overcome the above mentioned drawbacks of both theories, Glauert [4] introduced the Blade Element Momentum (BEM) theory combining the momentum and the blade element theories. This theory has been successively improved over the years with many corrections that make it an essential tool for rotor design [5]. It is worth remarking that only fluid loads contributing to the power conversion (i.e. non-conservative loads) are considered in BEM codes [6, 7]. The same applies to all other methods relying on the blade element theory. For example, the actuator line method (introduced by Sørensen and Shen [8]), replaces the momentum theory from the BEM method with the 3D Navier-Stokes equations for obtaining the induced velocities in the rotor plane. However, it also uses the blade element theory for computing the blade loads. Consequently, the actuator line (AL) method, which is gaining much attention in the field of wind farm simulations [9, 10, 11], also disregards the conservative loads. Conservative loads do not contribute to the power conversion and do not generate vorticity. Gravitational loads are a well known example of that type of load. However, aerodynamic loads on lifting surfaces, as rotor blades, might have a conservative component, too [12]. The origin of that load is the same as for the lift force: the bound circulation. At the root and tip of the blade, the bound circulation is deflected from the radial towards the chordwise direction before becoming free vorticity. The Kutta-Joukowsky load acting on the radial bound circulation gives rise to the lift force. In the same manner, the Kutta-Joukowsky load acting on the chordwise bound circulation gives rise to a force contained in a plane normal to the chord. Therefore, the load originating from the chordwise bound circulation can not contribute to the power conversion [12]. This article discusses the adequacy of neglecting this type of load in numerical models making use of the blade element theory.

Currently, rotor models that do not rely on the blade element theory are also widely-available (although they are more computationally expensive). For instance, Navier-Stokes models based on body-fitted grids have demonstrated to be very useful for the study of rotor aerodynamics [13, 14, 15, 16, 17]. Also vortex lattice models have proven to be reliable under attached flow conditions [18]. It is worth to recall that these types of models obtain the loads from the pressure field, which makes them capable of automatically considering the conservative loads (in opposition to the AL technique).

In Ref. [12] we used an AL, a vortex lattice and a body-fitted grid Navier-Stokes model for simulating the same wind turbine like in the present article. The models not relying on the blade element theory predicted a short inboard motion of the tip vortex just after release and before the wake expansion drove it outboard. This was in agreement with our experimental observations. Further experimental evidence of the same tip vortex behaviour is also available for other wind turbines [19, 20]. However, the AL model, which as mentioned above disregards the conservative load, was unable to predict that behaviour. This suggests a clear relationship between the tip vortex trajactory and the conservative load, what supports the hypothesis from previous analytical efforts. Furthermore, as we also hypothesized, the change in the tip vortex trajectory might influence the rotor performance. In order to analyse the validity of these hypothesis, in the current work we have extended an AL model for computing and applying automatically the conservative load at the tip. Hence, we aim at answering the following questions by comparing the baseline case (which only considers the non-conservative loads) with the extended model (which also considers the conservative load):

(i) Is it possible to modify the tip vortex trajectory of an actuator line model by considering the conservative load? In other words, is it possible to corroborate numerically by means of an enhanced actuator line model that the inboard motion of the tip vortex is a direct consequence of the conservative load at the blade tip?

(ii) If the previous question is answered affirmatively, is the influence of the mentioned effect on the rotor performance appreciable?

Section 2 introduces the numerical models that we use in our study. In §3 we first verify the reliability of the models and then compare the results of the simulations with and without conservative loads. This allows us to clarify what is the effect of the conservative loads. Finally, in §4 we present the main conclusions of this work in relation to our initial research questions.

2. Methods

2.1. Modelled wind turbine rotor

In this work we use the same experimental results as in our previous work [12] for verification purposes. Correspondingly, our study is based on the so called TUDelft-B wind turbine. The two-bladed rotor has a diameter of 2 meters. The blades are twisted and tapered. They make use of the DU96-W-180 airfoil type along the whole span except at the connection to the hub, where the blade section becomes cylindrical. The blade tip is not pointed but rectangular. The wind turbine is operated at its nominal conditions (tip speed ratio $\lambda = 7$). In the experiment the tip vortex trajectory was tracked by means of particle image velocimetry (PIV). Further details about the wind turbine and the experimental set-up can be found in Ref. [20, 21, 22, 23].

2.2. Numerical model

2.2.1. Navier-Stokes solver The simulations have been carried out with the open source toolbox OpenFOAM, which consists of a set of C++ libraries for solving partial differential equations [24]. All the computations are of the type Large Eddy Simulation (LES) and make use of a standard Smagorinsky sub-grid scale eddy viscosity model [25] for modelling the eddies of smaller size than the grid elements. The standard Smagorinky constant was set to $C_s = 0.168$, in accordance to the recommendations for isotropic turbulence [26, 27]. The discretization of the governing equations was based on the finite volume method. The time was discretized with a second order backward scheme. For the convective and diffusive terms we used second order linear upwind and linear limited differencing schemes, respectively. The pressure-velocity coupling was accomplished by means of the so called PIMPLE algorithm [24], which merges the PISO [28] and SIMPLE [29] algorithms.

2.2.2. Baseline actuator line model Our implementation of an actuator line (AL) model in OpenFOAM is based on the AL from the SOWFA package [30], which in turn is based on the original formulation proposed by Sørensen and Shen [8]. In this kind of model the actual geometry of each blade is substituted by a body force distributed along a line. In order to compute the body force, the actuator line is discretized into a finite number of blade elements. The user must predefine the blade geometry (spanwise distributions of chord and twist) as well as the aerodynamic characteristics of each blade element as a function of the angle of attack (AoA). Then, the AoA and the local lift $\boldsymbol{F_l}$ and drag $\boldsymbol{F_d}$ forces can be obtained iteratively from the above mentioned predefined data and the velocity field computed by the Navier-Stokes solver. In order to avoid numerical instabilities when applying the actuator line loading to the fluid, a gaussian smearing function is used for converting the point loads of each blade element into volumetric loads. The load applied to the fluid $\boldsymbol{f}$ was computed as:

$$\boldsymbol{f} = \frac{(\boldsymbol{F_l} + \boldsymbol{F_d})}{\epsilon^3 \pi^{3/2}} exp[-(r/\epsilon)^2] \tag{1}$$

where r is the distance between grid points and blade elements and ϵ represents a regularization parameter. Previous studies have shown a high dependency of the predicted power on the ϵ parameter. Therefore, in §3.1 we assess the sensitivity of our model to this parameter.

The hub is usually neglected in simulations based on the AL. As a consequence, an unrealistic jet appears along the centerline of the rotor. In order to overcome this issue, we modelled the nacelle as a permeable disk area, as described e.g. by Wu and Porte-Agél [31]. The force applied by the hub to the fluid is:

$$\boldsymbol{F_{hub}} = -\frac{1}{2}\rho \boldsymbol{u}_\infty^2 A_{hub} C_{D_{hub}} \tag{2}$$

where u_∞ is the undisturbed velocity upstream of the hub, A_{hub} is the hub area and $C_{D_{hub}}$ is the hub drag coefficient. In our simulations we assumed $C_{D_{hub}} = 0.85$, as Ref. [31] suggests.

The wake induction on the rotor plane is automatically taken into account by the AL and there is no need for considering any correction for the finite number of blades (like e.g. the one proposed by Prandtl [4]). However, the AL disregards the pressure equalization between the pressure and suction sides of the blade. In order to overcome this limitation, we used the correction model developed by Shen *et al.* [32] for accounting for this effect on actuator models (in accordance to the recommendations of Sørensen [33], Chapt. 8.7). The correction model applies the following function to the 2D airfoil data:

$$F_1 = \frac{2}{\pi} cos^{-1}\left[exp\left(-g\frac{B(R-r)}{2r sin\phi}\right)\right] \tag{3}$$

where B is the number of blades, R is the blade radius, r is the local radial position, ϕ is the flow angle and g is the function

$$g = exp\left(-0.125\left(\frac{B\Omega R}{U_\infty} - 21\right)\right) + 0.1 \tag{4}$$

where Ω is the rotational speed and U_∞ is the freestream velocity.

The numerical domain consisted of a square cylinder with a length of 22R and a width of 20R, in accordance to the recommendations of Troldborg [9]. The rotor was placed at a distance of 10R from the inlet. The greatest cell density was concentrated around the rotor. In that region, which has the width 1.6R and the length 4R, the cell size is R/50. Outside that region the grid is progressively stretched towards the outer boundaries. It is worth to recall that a grid resolution with the cell size R/30 in the rotor region is usually considered to be enough for AL simulations [34, 8]. The reason why a finer resolution was used in our case was to capture the tip vortex trajectory as accurately as possible. For the same reason, 60 blade elements were used for modelling the blade in spite of the fact that 20 elements are usually considered to be enough [35]. Ivanell *et al.* [36] documented that in actuator line simulations the tip vortex is released slightly inboard of the tip. In our experience, this problem is related to the size of the blade elements in that region. Jin [37] also showed that increasing the blade element density at the tip is a computationally efficient manner of improving the simulation accuracy. Therefore, apart from having a relatively large number of elements, the spacing between elements in our simulation follows a stretching function that allows the element density to increase from the hub towards the blade tip. As a consequence, the blade elements at the tip are smaller than the cells themselves. This is advantageous for azimuthal positions at which only a tiny fraction of a cell is covered by the very tip of the blade.

A Dirichlet boundary condition was used for the wind speed at the inlet and for the pressure at the outlet. A Neumann condition was set for the pressure at the inlet and for the wind speed at the outlet. At the lateral boundaries of the domain, both the wind speed and the pressure were set to Neumann conditions.

2.2.3. Enhanced actuator line model The enhanced AL model only differs from the baseline model in the consideration of the conservative load at the blade tip. As we described in our previous work [12], the origin of the conservative force is the Kutta-Joukowski load acting on the chordwise bound circulation Γ_{chord}, which is defined as the circulation around a contour normal to the blade chord (see e.g. figure 5 from Ref. [12]). Hence, the normal $\mathscr{N}$ and radial $\mathscr{R}$ components of the conservative load are computed as:

$$\mathscr{N} = \rho \cdot V_{radial} \cdot \Gamma_{chord} \cdot c \tag{5}$$

$$\mathscr{R} = \rho \cdot V_{axial} \cdot \Gamma_{chord} \cdot c \tag{6}$$

where ρ is the fluid density, V_{radial} and V_{axial} are the radial and axial components of the relative velocity, respectively, and c is the chord length. Γ_{chord} is concentrated at the tip (see figures 7 and 9 from Ref. [12] , where the chordwise vorticity and circulation are shown), so $\mathscr{N}$ and $\mathscr{R}$ are applied as body forces only at the last blade element.

In order to obtain Γ_{chord} at the tip for applying the conservative load at the last blade element, we need to compute first the radial bound circulation along the span as

$$\Gamma_{radial}(r) = \frac{F_l(r)}{\rho \cdot V_{rel}(r)} \tag{7}$$

where F_l and V_{rel} are the sectional lift force and the relative wind speed, respectively. Γ_{radial} is computed for every single blade element. Once the magnitude of the maximum Γ_{radial} along the span is known, it can be used for estimating Γ_{chord} at the tip after assuming conservation of circulation. This assumption implies that Γ_{radial} is transformed into Γ_{chord} at the tip just before being converted into the tip vortex itself:

$$\Gamma_{chord} = z \cdot max(\Gamma_{radial}(r)) \tag{8}$$

where z is an engineering factor between 0 and 1 accounting for the fact, that a fraction of the maximum circulation is released as trailing circulation along the blade span before reaching the tip. We assume $z = 0.75$, since as figure 9 from Ref. [12] shows, that is approximately the normalized strength of the chordwise circulation (with respect to the maximum circulation) at the tip (assumed to be the last 2% of the blade length). In other words, 0.25 of the maximum bound circulation is released approximately from r/R = 0.8 (the radial position of maximum bound circulation) to 0.98 of the blade span. It should be however remarked that this assumption is just a best guess and it should be tuned in the future with more experiments and/or simulations. Furthermore, we expect this factor to be highly dependent on the blade geometry and the operating tip speed ratio.

3. Results

3.1. Verification of the baseline numerical model

It is well known that the accuracy of AL models highly depends on the regularization parameter ϵ. As Ivanell *et al.* [36] reported, ϵ must be as small as possible in order to avoid influencing the wake structure with the Gaussian smearing of the body force. However, the ratio $\epsilon/\Delta x$ (where Δx is the cell size in the rotor plane) can not be too small if numerical instabilities are to be avoided. Therefore, in order to achieve a small ϵ and resolve accurately the tip vortex, we chose in the first place a mesh that clearly exceeds the usual requirements of actuator line simulations (see §2.2.2). The power obtained from the simulations is used for analysing the sensitivity of the numerical model to the the ratio $\epsilon/\Delta x$ (see figure 1). The best consistency with the experiment (power overprediction of just 3%) is achieved with $\epsilon/\Delta x = 2$ (i.e. $\epsilon = R/25$). Troldborg [9] also suggested the same ratio as the best compromise between simulation accuracy and stability. For smaller values of $\epsilon/\Delta x$ the simulation was too unstable. Therefore, we chose to use $\epsilon/\Delta x = 2$ for both the baseline and enhanced models of our study.

As we explained in §2.2.3, the axial and radial velocity components are required for computing the conservative load that acts on the chordwise vorticity. Figure 2 shows those velocity components at the rotor plane as predicted by two different models: the current baseline AL model and the RANS model with body-fitted mesh that was validated in Ref. [12].

In spite of the fact that modelling a 3D blade by means of a 1D line is a strong simplification, the agreement between the AL and the body-fitted RANS model is in general good for both the axial and radial velocity components. However, a shift can be seen in the azimuthal position of the region with high radial velocity close to the blade tip. This implies that the radial velocity

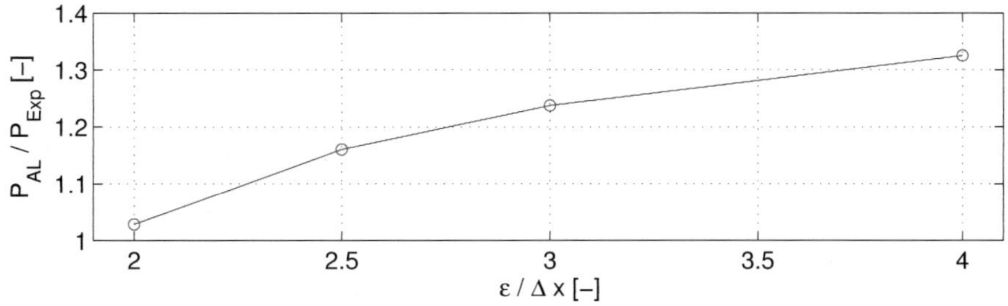

Figure 1. Sensitivity of the computed power from the baseline actuator model (normalized with the experimental results) to the ratio between the regularization parameter ϵ and the cell length Δx.

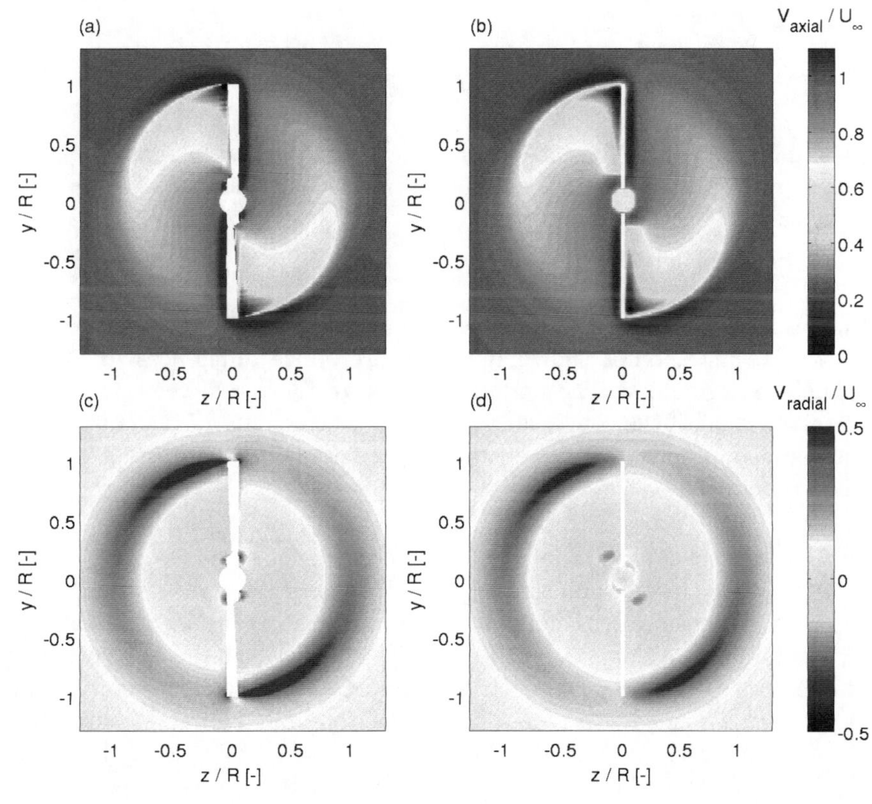

Figure 2. Velocity at the rotor plane: (a) V_{axial} from the body-fitted RANS simulation, (b) V_{axial} from the baseline AL-LES simulation, c) V_{radial} from the body-fitted RANS simulation and (d) V_{radial} from the baseline AL-LES simulation. The white surface in the results from the body-fitted RANS simulation represents the intersection of the rotor plane with the blades and nacelle. The white lines in the results from the AL-LES simulation represent the actuator lines.

	Body-fitted RANS	$AL_{baseline}$	$AL_{enhanced}$
$\mathscr{N}$	1.2	0	0.03
$\mathscr{R}$	1.2	0	1.2

Table 1. Normal and radial loads ($\mathscr{N}$ and $\mathscr{R}$, respectively) at the tip, as a percentage of the blade thrust. The body-fitted RANS results were obtained from the model presented in our previous work [12] and they were calculated by surface pressure integration. The enhanced AL model differs from the baseline model just in the consideration of the conservative loads attributed to chordwise vorticity.

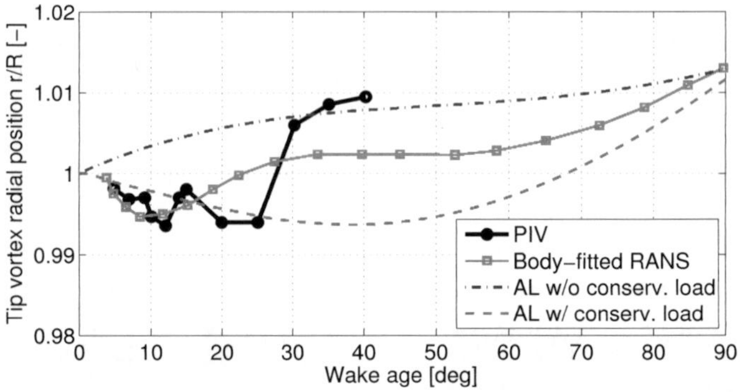

Figure 3. Radial position of the tip vortex as a function of the wake age. Experimental (Ref. [22]) and numerical results from different simulation models are compared.

at the tip of the baseline actuator line is strongly underpredicted. The same happens with the enhanced actuator line model (not shown in the figure, since differences in this regard with respect to baseline are negligible). The underprediction of the radial velocity component leads to a severe underestimation of the conservative load $\mathscr{N}$ (see equation 5), as represented in table 1. However, the radial component $\mathscr{R}$ is accurately predicted thanks to the good estimation of the axial velocity component.

3.2. Influence of the conservative load on the tip vortex trajectory

The λ_2 criteria [38] has been used for identifying the location of the vortex core at each azimuthal position. Figure 3 shows the vortex trajectory obtained from the experiment and different numerical models. Our analysis focuses on the AL-LES simulations, since the other results were already analysed in Ref. [12]. The AL-LES simulations without conservative load (labelled as *AL w/o conserv. load*) lead to the wake beginning to expand directly after the tip vortex release. On the contrary, the AL-LES simulations with conservative load (labelled as *AL w/ conserv. load*) present an inboard motion of the tip vortex prior to the wake expansion (as it also occurs in the experiment, the body-fitted RANS and panel lattice simulations). Furthermore, the computed radial position of the tip vortex at its innermost location agrees well with the experiment and the body-fitted RANS simulations. This behaviour confirms that the radial conservative load $\mathscr{R}$ is responsible for the inboard motion of the tip vortex, as we estimated in Ref. [12]. However, the azimuthal angle at which the wake expansion begins is clearly delayed in the AL-LES simulation. The reason for this seems to be related to the above described strong underestimation of $\mathscr{N}$ (see table 1).

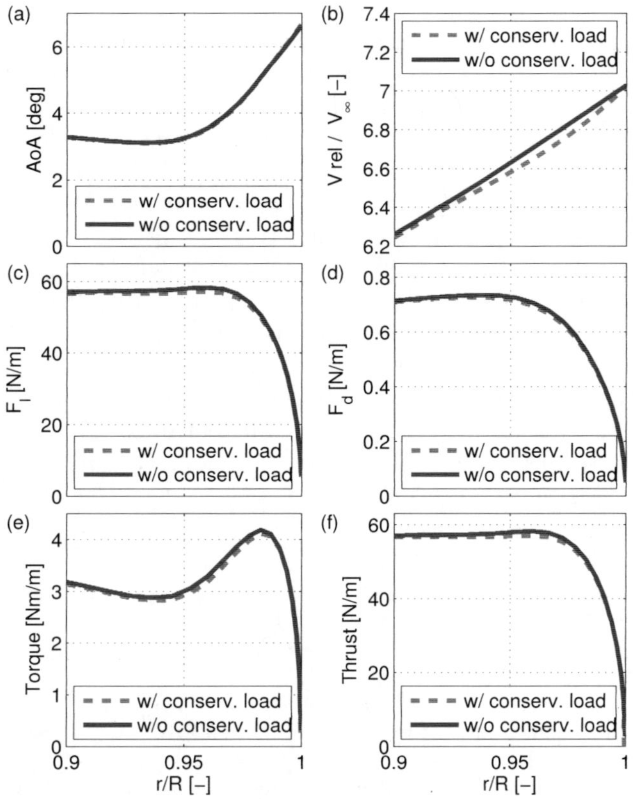

Figure 4. Influence of the conservative force on the spanwise distribution of (a) angle of attack AoA, (b) relative velocity V_{rel}, (c) sectional lift force F_l, (d) sectional drag force F_d, (e) sectional torque and (f) sectional thrust.

3.3. Influence of the conservative load on the rotor performance

In this section we aim at analysing if the influence of the conservative force on the tip vortex trajectory plays a role on the rotor performance. Figure 4-a shows the AoA in the tip region. No significant difference between the results with and without conservative force can be seen. This implies that the lift and drag force coefficients (C_l and C_d) do not change. However, the relative wind speed (figure 4-b), which is defined as $V_{rel} = \sqrt{V_x^2 + V_\theta^2}$, is slightly affected by the change in the vortex trajectory. Correspondingly, the lift and drag forces F_l and F_d are also influenced (figures 4-c and 4-d, respectively). This, in turn has a small but distinguishable effect on the driving torque and the thrust (figures 4-e and 4-f, respectively). Integrating the torque over the last 10% of the blade gives a difference of the torque at the tip with and without conservative load of approx. 2%. The total power of the turbine is reduced by approx. 1%. Hence, considering the conservative load contributes to mitigate the slight power overprediction of the simulation as compared to the experiment (see figure 1). This indicates that the conservative loads play an indirect, modest role on the rotor performance.

4. Conclusions

The use of an enhanced actuator line model capable of turning on and off the conservative load at the blade tip enabled studying the influence of the mentioned load on the aerodynamic performance and the near wake of a wind turbine. The model relates the maximum circulation along the span to the chordwise circulation at the tip by means of a factor z, which accounts for the trailing circulation. In order to estimate the value of that factor body-fitted CFD simulations (or experimental results with high spatial resolution) can be used. We expect our estimation of the mentioned parameter not to be valid for other turbines because of its high dependency on the blade geometry and the tip speed ratio. Once the chordwise vorticity at the tip is known, computing the conservative load is straightforward by means of the Kutta-Joukowski theorem.

The model allowed to corroborate that the conservative load can influence the tip vortex trajectory. Furthermore, it has been shown that this change might have a very small but non negligible impact on the rotor performance. However, quantifying this high order effect with sufficient accuracy is still not feasible because of the inherent uncertainty of the numerical model. Further research in this direction might be relevant for the development of physically-sound tip losses corrections models.

References

[1] W. J. M. Rankine. On the mechanical principles of the action of propellers. *Transactions Institute of Naval Architects*, 6:1339, 1865.

[2] R.E. Froude. On the part played in propulsion by differences of fluid pressure. *Transactions Institute of Naval Architects*, 30:390405, 1889.

[3] S. Drzewiecki. *Bulletin de L'Association Technique Maritime, Paris*, 1892.

[4] H. Glauert. The general momentum theory. In W.F. Durand, editor, *Aerodynamic Theory*, volume IV. Division L. Springer, 1935. Reprinted 1963 Dover.

[5] M. L. Hansen and H. Aagaard Madsen. Review paper on wind turbine aerodynamics. *Journal of Fluids Engineering*, 133(11):114001–114001–12, 2011.

[6] G. A. M. van Kuik, J. N. Sørensen, and V. L. Okulov. Rotor theories by Professor Joukowsky: Momentum theories. *Progress in Aerospace Sciences*, 73:1–18, 2015.

[7] V L Okulov, J N Sørensen, and D H Wood. The rotor theories by Professor Joukowsky: Vortex theories. *Progress in Aerospace Sciences*, pages 19–46, 2015.

[8] J. N. Sørensen and W. Z. Shen. Numerical modelling of wind turbine wakes. *Journal of Fluids Engineering*, 124(2):393–399, 2002.

[9] N. Troldborg. *Actuator Line Modeling of Wind Turbine Wakes.* PhD thesis, Technical University of Denmark, 2008.

[10] M. Shives and C. Crawford. Mesh and load distribution requirements for actuator line CFD simulations. *Wind Energy*, 16(8):1183–1196, 2013.

[11] K. Nilsson, W. Z. Shen, J. N. Sørensen, S. P. Breton, and S. Ivanell. Validation of the actuator line method using near wake measurements of the MEXICO rotor. *Wind Energy*, 18(3):499–514, 2015.

[12] G. A. M. van Kuik, D. Micallef, I. Herráez, A. H. van Zuijlen, and D. Ragni. The role of conservative forces in rotor aerodynamics. *Journal of Fluid Mechanics*, 750:284–315, 7 2014.

[13] N. N. Sørensen, J. A. Michelsen, and S. Schreck. Navier-Stokes predictions of the NREL phase vi rotor in the NASA ames 80 ft 120 ft wind tunnel. *Wind Energy*, 5(2-3):151–169, 2002.

[14] A. Bechmann, N. N. Sørensen, and F. Zahle. CFD simulations of the MEXICO rotor. *Wind Energy*, 14(5):677–689, 2011.

[15] I. Herráez, B. Stoevesandt, and J. Peinke. Insight into rotational effects on a wind turbine blade using Navier-Stokes computations. *Energies*, 7(10):6798, 2014.

[16] I. Herráez, B. Akay, G. J. W. van Bussel, J. Peinke, and B. Stoevesandt. Detailed analysis of the blade root flow of a horizontal axis wind turbine. *Wind Energy Science*, 1(2):89–100, 2016.

[17] N. N. Sørensen, A. Bechmann, P. E. Réthoré, and F. Zahle. Near wake Reynolds-averaged Navier-Stokes predictions of the wake behind the MEXICO rotor in axial and yawed flow conditions. *Wind Energy*, 17(1):75–86, 2014.

[18] D. Micallef, G. van Bussel, C. S. Ferreira, and T. Sant. An investigation of radial velocities for a horizontal axis wind turbine in axial and yawed flows. *Wind Energy*, 16(4):529–544, 2013.

[19] J. Xiao, J. Wu, L. Chen, and Z. Shi. Particle image velocimetry (PIV) measurements of tip vortex wake structure of wind turbine. *Appl. Math. Mech.*, 32(6):729–738, 2011.

[20] D. Micallef. *3D flows near a HAWT rotor: A dissection of blade and wake contributions.* PhD thesis, Delft University of Technology, 2012.

[21] D. Micallef, C.S. Ferreira, T. Sant, and G. J. W. van Bussel. Experimental and numerical investigation of tip vortex generation and evolution on horizontal axis wind turbines. *Wind Energy*, 19(8):1485–1501, 2016. we.1932.

[22] D. Micallef, B. Akay, C. S. Ferreira, T. Sant, and G. van Bussel. The origins of a wind turbine tip vortex. In *The Science of Making Torque from Wind 2012*, Oldenburg, Germany, October 2012.

[23] V. del Campo, D. Ragni, D. Micallef, B. Akay, F. J. Diez, and C. S. Ferreira. 3D load estimation on a horizontal axis wind turbine using SPIV. *Wind Energy*, 17(11):1645–1657, 2014.

[24] OpenFOAM. OpenFOAM: the open source CFD toolbox, 2015. Accessed: October 2015.

[25] J. Smagorinsky. General circulation experiments with the primitive equations. *Monthly Weather Review*, 91:99–164, 1963.

[26] D. K. Lilly. The representation of small–scale turbulence in numerical simulation experiments. In *Proc. IBM Scientific Computing Symposium on Environmental Sciences*, 1967.

[27] S. B. Pope. *Turbulent flows.* Cambridge University Press, 2000.

[28] R. I. Issa. Solution of the implicitly discretized fluid flow equations by operator-splitting. *Journal of Computational Physics*, 62:40–65, 1985.

[29] S. V. Patankar. *Numerical Heat Transfer and Fluid Flow.* Taylor & Francis, 1980.

[30] SOWFA. SOWFA: Simulator for wind farm applications, 2015. Accessed: October 2015.

[31] Y. T. Wu and F. Porte-Agél. Large-Eddy simulation of wind-turbine wakes: Evaluation of turbine parametrisations. *Boundary-Layer Meteorol*, 138:345–366, 12 2010.

[32] W. Z. Shen, J. N. Sørensen, and R. Mikkelsen. Tip loss correction for actuator/Navier-Stokes computations. *Journal of Solar Engineering*, 127:209–213, 2005.

[33] J. N. Sørensen. *General Momentum Theory for Horizontal Axis Wind Turbines.* Springer, 2015.

[34] W. Z. Shen, W. J. Zhu, and J. N. Sørensen. Actuator line/NavierStokes computations for the MEXICO rotor: comparison with detailed measurements. *Wind Energy*, 15(5):811–825, 2012.

[35] P. K. Jha, M. J. Churchfield, P. J. Moriarty, and S. Schmitz. Guidelines for volume force distributions within actuator line modeling of wind turbines on Large-Eddy Simulation-type grids. *J. Sol. Energy Eng.*, 136(3):031003–031003–11, 2014.

[36] S. Ivanell, J. N. Sørensen, R. Mikkelsen, and D. Henningson. Analysis of numerically generated wake structures. *Wind Energy*, 12(1):63–80, 2009.

[37] W. Jin. Numerical simulation of wind turbine wakes based on actuator line method in NEK5000. Master's thesis, KTH, Sweden, 2013.

[38] J Jeong and F Hussain. On the identification of a vortex. *Journal of Fluid Mechanics*, 285:69–94, 2 1995.

Qualitative analysis of wind-turbine wakes over hilly terrain

A Hyvärinen and A Segalini

Linné Flow Centre, KTH Mechanics, Royal Institute of Technology, SE-100 44 Stockholm, Sweden

E-mail: annhy@mech.kth.se

Abstract. In this work, wind-turbine wakes are studied over flat and hilly terrains. Measurements made by using stereoscopic PIV are compared to data obtained from numerical simulations using RANS equations and an actuator-disc method. The numerical and experimental data show similar qualitative trends, indicating that the wind-turbine wake is perturbed by the presence of the hills. Additionally, a faster flow recovery at hub height is seen with the hilly terrain, indicating that the hills presence is beneficial for downstream turbines exposed to wake-interaction effects. The Jensen wake model is implemented over the hilly terrain and it is shown that this model cannot accurately capture the wake modulations induced by the hills. However, by superimposing a wind-turbine wake simulated over flat terrain on the hilly-terrain flow field, it is illustrated that the commonly-used wake-superposition technique can yield reasonable results if the used wake model has sufficient accuracy.

1. Introduction

Wind turbines operating at onshore sites are often found over hilly or mountainous landscapes. Terrain variations at such sites influence the wind-turbine performance and the downstream wake evolution. At specific wind-farm sites, wind-turbine data can be collected and compared with numerical simulations including site-specific terrain models [1]. In such simulations, it has been observed that complex terrain can enhance the turbulent activity in the flow, resulting in a faster wake recovery and higher power outputs from the turbines [2]. While studies of this kind give a good overview of the site-specific flow fields, the terrain complexity makes it difficult to draw conclusions applicable for a wider range of topographical cases. In this respect, it is advantageous to study more generalised terrain models covering fundamental flow situations. This was for instance done through wind-tunnel measurements by Tian *et al.* [3], who observed higher power performance for wind turbines placed on the crest and on the leeward side of an isolated hill. It has also been found that an upstream hill can increase the turbulence intensity in the inflow seen by a turbine, and quicken the wake recovery [4]. This effect is reduced with increased distance between the hill and the downstream turbine. Terrain features far upstream can nonetheless influence the flow field over onshore sites, as observed by Walmsley *et al.* [5].

The flow interaction over multiple hills is largely dependent of the shape of the hill and its slope, where steep hill slopes with sharp crests often result in flow separations that significantly reduce the available wind power over downstream hills [6]. Improvements in the available power that a wind turbine can potentially extract have been observed when the distance between two adjacent hills is small, due to an upward deflection of the flow [7]. For a larger number of periodic hills, regions of correlated flow have been found to extend up to distances equal to the hill spacing [8], as expected. This makes the wake development behind a wind turbine placed over such terrain features intriguing to study, and that has been the focus of the current work.

In the present paper, the wake evolution behind a single wind turbine, operating over flat and hilly terrain, is studied under homogeneous inflow conditions. The used terrain model consisted of a series of streamwise periodic sinusoidal hills. Wind-tunnel measurements made using stereoscopic particle image velocimetry (PIV) were compared with numerical simulations performed using an actuator-disc approach and Reynolds averaged Navier-Stokes (RANS) equations in ANSYS CFX. The purpose of this was initially to investigate if the trends seen in the measurement data could be understood with a fairly simple numerical model. Additionally, since the PIV measurements had a limited field of view, the simulations provided additional insight concerning the studied flow conditions. Moreover, the simulations allowed to do extended comparisons with engineering wake-model approaches applied over hilly terrain.

The present paper is structured as follows: the wind-tunnel measurements and the numerical procedure are described in sections 2 and 3. Thereafter, the simulation results are compared with some of the experimental data in section 4. Finally, conclusions regarding the results are made in section 5.

2. Experimental procedure

The experimental measurements were performed in the closed loop Boundary Layer (BL) wind tunnel at KTH, which is a high flow quality facility with a 4.2 m long, 0.5 m wide and 0.75 m high test section [9]. Hilly terrain was introduced in the second half of the test section, where the hills (which had a sinusoidal shape) were made of XPS styrofoam. Five hills were used and the hill periodicity was four rotor diameters (based on the size of the used wind-turbine model). The hill height, H, was 7 cm (0.625 rotor diameters). The wind-turbine model had a rotor diameter, D, of 112 mm, with thin slightly-chambered fixed-pitch rotor blades made of carbon fibre reinforced plastic. The rotor Reynolds number was about $Re \approx 6 \cdot 10^4$ based on the rotor diameter and the inflow velocity. Although this is considerably lower than that of a full-scale wind turbine, a study by Chamorro *et al.* [10] indicated that the mean velocity statistics in the wake become independent of the Reynolds number at $Re \approx 4.8 \cdot 10^4$. Under this assumption, the flow characteristics seen with the present model will be similar to the conditions obtained in a full-scale situation. The turbine rotor was attached to a Faulhaber DC micro motor (working as a generator), and the angular velocity of the turbine was adjusted by altering its electrical load. The turbine DC generator was fastened to a 16 cm tall tower made of aluminium, and the wind turbine was placed on the top of the first hill. The tunnel free-stream velocity U_∞, which was kept constant at 7.5 m/s during the measurements, was measured by means of a Prandtl tube mounted upstream of the terrain model. An illustration of the experimental setup, including cameras and a laser used during the PIV measurements, can be seen in Figure 1.

To observe the wind-turbine wake development with and without the hilly terrain, measurements were made using stereoscopic PIV. The PIV measurements were made by utilising two high-speed 10-bit CMOS cameras of type Photron Fastcam APX RS (that captured image pairs at 50 Hz for a sampling period of 61.44 seconds per plane), a 600 mJ Litron Nano-PIV pulsed Nd:YAG laser that emitted light at a wavelength of 532 nm, and a LaVision PTU X high-speed controller unit. A spherical and a cylindrical lens were used to generate a 2 mm thick light sheet that illuminated particles of DXS oil that were injected into the tunnel free stream. The camera lenses had a focal length of 105 mm that, together with the camera resolution, made it possible to capture planes with a 10x12 cm^2 field of view during the measurements (corresponding to 0.9x1.1D^2). The cameras were angled so that they captured planes perpendicular to the free-stream direction, and an Isel traversing system was used to move both the cameras and the laser system (which were rigidly joined) in the streamwise direction.

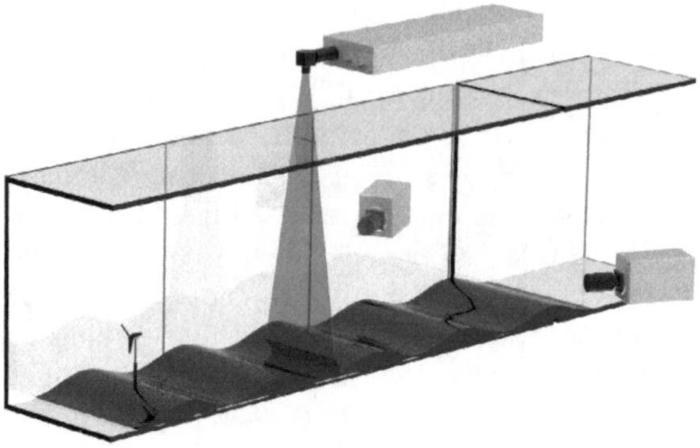

Figure 1. Illustration of experimental setup including PIV equipments.

The LaVision Davis 8.3 software was used for data acquisition and post processing. From the acquired flow images, three-component instantaneous flow fields were re-created through stereoscopic cross-correlations (using multiple passes with decreasing size). These were then averaged to view the mean flow fields at different downstream locations behind the turbine.

3. Numerical procedure

Numerical simulations were made using steady-state RANS equations in ANSYS CFX, where a geometry with the same cross-section and hill dimensions as in the wind-tunnel experiments was used. Only half of the domain was simulated to improve the computational efficiency: a symmetry boundary condition was used at the vertical symmetry plane of the domain, while a free-slip condition was used at the sidewall far from the rotor (namely the wall shear stress and the wall-normal velocity was set to zero). The ceiling and floor surfaces were modelled with free-slip and no-slip boundary conditions, respectively. To study the wake diffusion behind the turbine over a longer distance, the numerical terrain model was extended (compared to the experimental terrain model) by introducing six periodic hills instead of five. The simulation domain was elongated 1 m upstream and downstream of the terrain model to avoid undesirable effects from the inlet and outlet boundaries. Thereby, the total length of the simulation domain was 4.7 m ($42D$). The inlet boundary condition of the simulation domain was set to a homogeneous inflow velocity $U_\infty = 7.5$ m/s (corresponding to the free-stream velocity of the experiments) and a turbulence intensity of 5%. The turbulence intensity at the inlet was chosen by considering the resulting wake diffusion rate (over flat terrain) in comparison with empirical wake-model predictions. The outlet of the simulation domain was modelled with a constant pressure boundary condition. The Menter SST $K-\omega$ model [11] was used for the turbulence modelling in the flow domain. This method, which combines the standard $K-\epsilon$ and $K-\omega$ models, has been observed to be more accurate in capturing the wake development behind a single rotor than the $K-\epsilon$ model [12]. A first-order upwind advection scheme was used for the turbulence modelling, ensuring that sufficient numerical stability was achieved during the simulations with hilly terrain.

The wind-turbine rotor was described by using a simple actuator-disc approach, where a momentum loss, corresponding to a certain turbine thrust, was prescribed on a disc-shaped region projected by the rotor area (using the same rotor dimensions as in the wind-tunnel experiments). This was done by introducing a momentum source term, S_M, in the RANS equations, expressed as

$$S_{M,i} = \alpha \frac{\rho}{2} |\mathbf{U}| U_i, \qquad (1)$$

where $\alpha = C_T/[\delta(1-a)^2]$ is a quadratic loss coefficient, δ is the thickness of the actuator-disc domain (set to $0.04D$ during the simulations) and a is the axial induction factor defined as

$$a = \frac{1-\sqrt{1-C_T}}{2}, \tag{2}$$

using actuator-disc theory. The thrust coefficient was set to $C_T = 0.8$, which is common for wind turbines operating close to their optimal performance conditions. For simplicity, an isotropic loss was imposed over the rotor area and rotational effects were not accounted for.

In the performed simulations, a swept mesh with hexahedral elements was used for the rotor disc, whereas an unstructured grid was used in the rest of the domain. The element size in the rotor-disc domain was $0.018D$, while elements with a size of $0.036 - 0.089D$ were used in the vicinity of the actuator disc and over the hills. Further away (near the domain boundaries) the element size was gradually increased to $0.54D$, since these regions were of lesser importance. A mesh-convergence study was made to ensure that the major flow features in the domain were properly captured, regarding both the velocity gradients over the hills and the wake diffusion downstream of the actuator disc. To resolve the wall boundary layer at the surface of the hills, the mesh was refined so that the normalised wall distance was below $y^+ = 1$. This was done using 10 prismatic inflation layers with a growth rate of 1.8 in a region extending to $0.07H$ above the hill surfaces (where H is the hill height).

To ensure that the wake behind the actuator disc was properly resolved, and that the wake diffusion in the far-wake region was reasonably captured, comparisons were made with the Jensen and Ainslie wake models [13, 14] and the model by Porté-Agel *et al.* [15, 16]. Additionally, the flow upstream of the actuator disc was compared with the theoretical expression

$$\frac{U}{U_\infty} = 1 - a\left(1 + \frac{X}{\sqrt{D^2/4 + X^2}}\right), \tag{3}$$

derived along the rotor centreline using the Biot-Savart law [17], where X is the streamwise coordinate. The centreline velocity upstream and downstream of the actuator disc, estimated by the numerical simulations and theoretical wake models, can be seen in Figure 2 (in this case, the simulations were made over flat terrain). With the Jensen model, the wake constant k, which determines the slope of the wake expansion and influences the wake diffusion rate, was set to 0.04. As seen in the figure, the results predicted by the RANS simulations are in fairly good agreement with the empirical models (considering the approximations made with the different approaches).

4. Results

Experimental results obtained through the PIV measurements can be seen in Figure 3, where the mean-velocity values measured in planes behind the turbine (with and without hills) are shown. The results have been normalised with the tunnel free-stream velocity and the coordinate system has its origin at the location of the rotor hub. As visible from the results, the wake recovery is initially similar for the two cases. However, as the wake propagates over the second hill, there is a stronger downward wake deflection compared to the case with flat terrain. This is probably due to the pressure gradients induced by the hilly terrain.

Further insights, regarding how the hills influence the wake development downstream of the turbine, can be gained from the numerical simulations. Results from simulations with flat and hilly terrain can be seen in Figure 4: it is evident that the numerical results show the same qualitative trends as the experimental data, namely that the wake developing over the hilly terrain is deformed due to speed-up effects as it passes over the hill tops, and it generally follows the curvature of the hills. This is particularly clear in the lower part of the wake, while the top

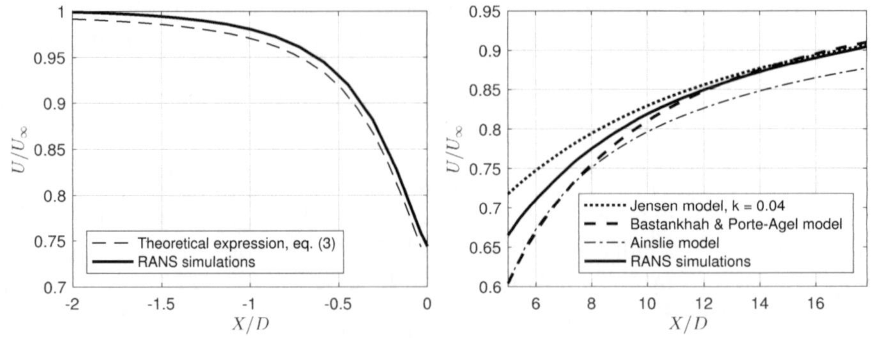

Figure 2. Normalised streamwise velocity, U/U_∞, along the rotor axis estimated by the RANS simulations, the wake models and the theoretical relationship (3). (*Left*) upstream deceleration, (*right*) downstream wake recovery.

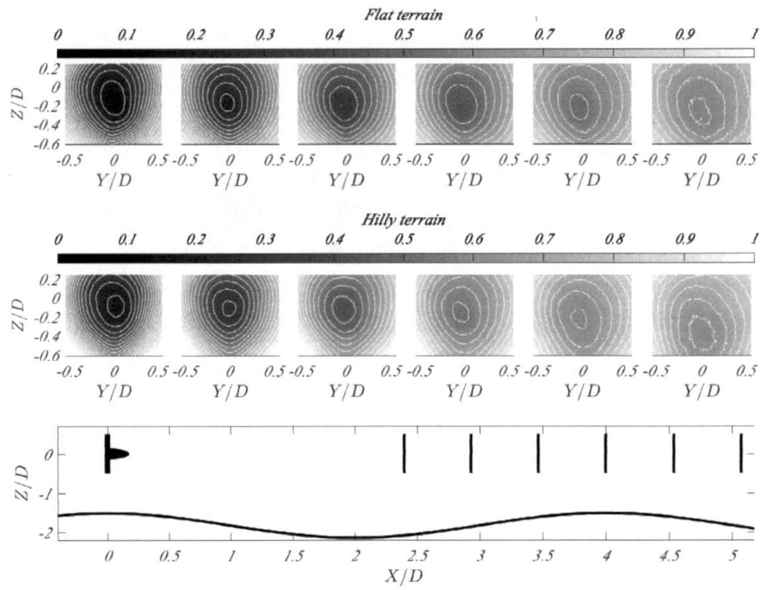

Figure 3. Normalised mean streamwise velocity, U/U_∞, measured in wake of one turbine without hills (*top*) and with hills (*middle*). The position of the planes with respect to the hills and turbine is also indicated (*bottom*). Velocity contours have been included to better visualise the wake-centre movement.

of the wake becomes flatter in comparison. Interestingly, the wake asymmetry introduced by the hills is also seen in the results viewed at the horizontal plane located at hub height, albeit to a lesser extent.

If an industrial wake model is used to study the wake propagation over complex terrain, the wake velocity deficit is usually superimposed on the velocity field computed over the terrain. To study the implications of using this methodology, simulations were made where the Jensen wake model [13] was superimposed on the flow field calculated over the hilly terrain (with no actuator disc). Figure 5 shows the resulting flow fields obtained by using this approach. The wake coefficient k in the Jensen model was set to 0.04 and 0.075, to study the result sensitivity to that

IOP Conf. Series: Journal of Physics: Conf. Series **854** (2017) 012023 doi:10.1088/1742-6596/854/1/012023

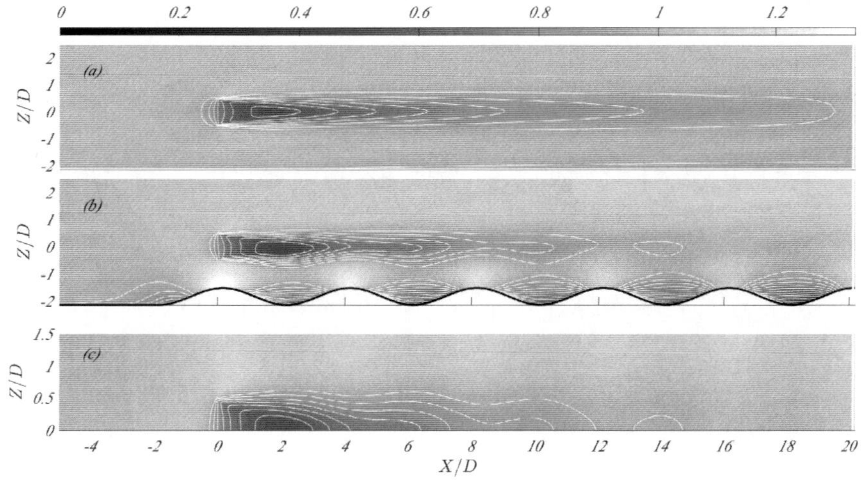

Figure 4. Normalised streamwise velocity, U/U_∞, estimated by the RANS simulations. Results are shown at the vertical symmetry plane with flat terrain (a) and with hilly terrain (b). Additionally, for the flow case with hills, the wake is viewed at a horizontal plane located at the rotor hub height (c). The results are viewed in planes parallel to the free-stream direction and velocity contours in the range of $U/U_\infty \in [0, 0.9]$ are included to better visualise the downstream wake evolution.

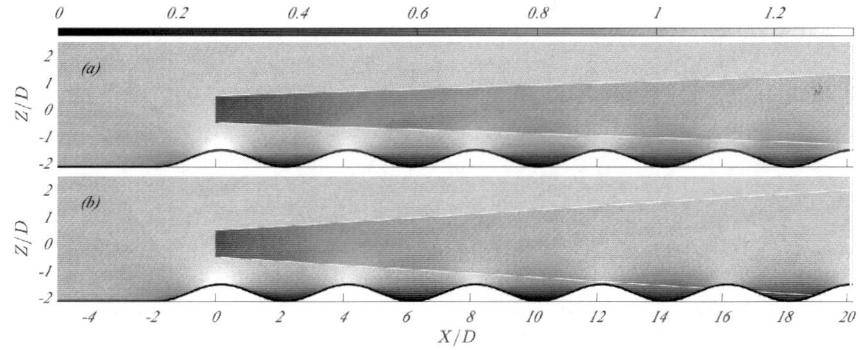

Figure 5. Normalised streamwise velocity, U/U_∞, obtained at the vertical symmetry plane when applying the Jensen wake model over hilly terrain. The Jensen wake constant k was set to 0.04 (a) and 0.075 (b). White lines are included along the wake edges for better visualisation.

constant (these k values are typically used for offshore and onshore wind farms, respectively). As seen when comparing the wakes in this figure to the results in Figure 4b, the wake expansion differs significantly when using this method. While the wake interior in Figure 5 is modulated by the surrounding velocity field, the effects from the hills on the wake expansion is completely absent. When the wake coefficient is $k = 0.075$ (Figure 5b), the wake expansion is exaggerated compared to the actuator-disc RANS simulations. This was expected since the inlet turbulence intensity level during the RANS simulations was fairly low.

Returning to the RANS simulations with an actuator-disc model, a closer view of the velocity along the rotor centreline is visible in Figure 6a. Results from the PIV measurements are also shown. The simulation results over flat and hilly terrains show a reasonable agreement with the experimental data (considering the modelling assumptions made). Over flat terrain, the

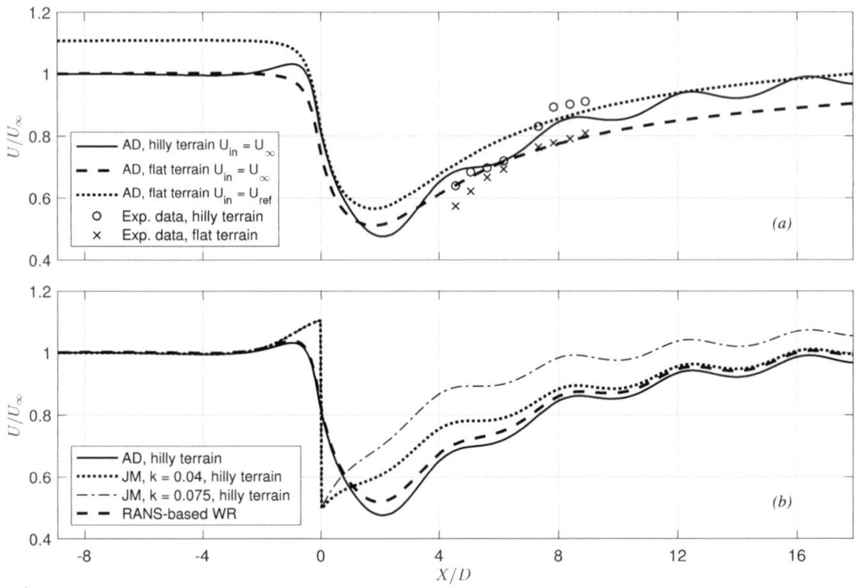

Figure 6. (a) Normalised streamwise velocity, U/U_∞, along the rotor centreline, estimated by RANS simulations with and without hills. Measurement results over flat and hilly terrains are also included. (b) Normalised streamwise velocity over hilly terrain, along the rotor centreline, estimated from RANS simulations, the Jensen wake model and the RANS-based wake-reconstruction technique. In the figure legends the actuator disc method is denoted by AD while the Jensen model is denoted by JM (with k as the wake constant of the Jensen model) and WR denotes wake reconstruction.

experimentally-measured velocity is seen to asymptotically approach the free-stream value in a similar manner as predicted by the RANS simulations. Meanwhile, the experimental data recorded with hills are seen to agree with the corresponding actuator-disc simulations in a qualitative sense: evidently perturbations induced by the hills occur at similar downstream distances in both cases. Quantitatively, deviations are noticed when comparing the simulation results with the experimental data. This is particularly true closer to the rotor, since the used actuator-disc model could not capture the tip-vortex characteristics and the rotational effects that are significant in this region, and the variation in the load distribution over the rotor blades was not introduced in the simplified model. Despite these approximations, the qualitative agreement seen further downstream of the rotor indicates that the simulations can be used to gain additional insight about the wake development in the far-wake region with the studied flow cases.

A comparison between the simulations made over flat and hilly-terrain (with an inlet velocity $U_{in} = U_\infty = 7.5$ m/s), shows that the velocity increment over the first hill leads to a higher velocity at the rotor disc located at $X/D = 0$. When comparing the two cases where $U_{in} = U_\infty$, the results in Figure 6a indicate that the flow recovery at hub height is enhanced when introducing hills, in agreement with the experimental results. The higher flow recovery is beneficial when introducing multiple turbines exposed to wake-interaction effects. Over flat terrain, a velocity speed-up naturally leads to a quicker wake recovery. To isolate the effect induced by the hills from this, a simulation was made over flat terrain, where U_{in} was set to the inflow velocity, U_{ref}, seen by the rotor over the hilly terrain. The turbine inflow velocity,

$U_{ref} = 8.3$ m/s, was taken from a simulation that included hills but no turbine. As seen from the results (shown together with the other data in Figure 6a), the disc velocity at $X/D = 0$ was approximately the same with this inlet velocity as in the case with hilly terrain. With hilly terrain, the wake velocity is perturbed by local gradients, introducing wiggles in the results (footprints of these are also seen in the experimental results). Moreover, the wake velocity deficit is initially larger than in the flat-terrain cases. Further downstream, the hilly-terrain velocity deficit gradually approaches the flat-terrain results with $U_{in} = U_{ref}$, but the wiggles caused by the presence of the hills remain.

In Figure 6b, the centreline velocity from the actuator-disc simulation with hills, is compared to results obtained with the Jensen wake model applied over hilly terrain. As seen in Figure 6b, the velocity deficit is underestimated by the Jensen model, particularly close to the rotor disc. When the Jensen wake coefficient $k = 0.075$, the results are generally in poor agreement with the actuator-disc simulations over hilly terrain, while the agreement is somewhat better with $k = 0.04$, as expected from the low turbulence intensity in the simulations.

From the results in Figure 5, it was clear that the hill-induced perturbations at the edges of the wake could not be captured with the linear wake expansion introduced in the Jensen model. However, the complexity of the wake-development description differs between different engineering wake models. To investigate the feasibility of using the wake-superposition technique over hilly terrain (with less errors introduced by the wake-expansion model), a RANS-based wake reconstruction was made by combining the simulation data over flat terrain with the flow field computed over the hills. In this case, the reconstructed velocity field (denoted $\widetilde{U}_*$) was obtained from

$$\widetilde{U}_* = \widetilde{U} - U_{ref} + \bar{U}_{AD}\Big|_{(U_{in}=U_{ref})}, \tag{4}$$

where $\widetilde{U}$ is the velocity field obtained from RANS simulations over hilly terrain (without an actuator disc), and $\bar{U}_{AD}$ is the velocity field resulting from the actuator-disc RANS simulations over flat terrain with an inlet velocity $U_{in} = U_{ref}$. The reference velocity U_{ref} is the velocity at the position of the rotor-disc centre, taken from the velocity field over the hills, $\widetilde{U}$. The method reconstructs the wake over the hilly terrain by superimposing the wake deficit obtained over flat terrain (with an adjusted inflow velocity) on the velocity field obtained over the hilly terrain, in agreement with what done by typical engineering wake models. The resulting flow field was compared with the flow field obtained from the actuator-disc RANS simulations over hilly terrain (denoted $\widetilde{U}_{AD}$). The results using the RANS-based wake reconstruction are shown together with the other data in Figure 6b. Close to the rotor disc, the RANS-based reconstruction significantly improves the results compared with the Jensen model: the results come closer to the velocities observed in the actuator disc simulation over hilly terrain. This becomes even clearer when looking at the entire velocity field in the vertical symmetry plane of the simulation domain. As seen in Figure 7b, the wake modulations induced by the hilly terrain are captured reasonably well with the RANS-based reconstruction, especially when compared to the actuator-disc simulations over hilly terrain (Figure 7a). In Figure 7c the difference between the velocity fields in Figures 7a and 7b, defined as

$$\Delta\widetilde{U}/U_\infty = (\widetilde{U}_* - \widetilde{U}_{AD})/U_\infty, \tag{5}$$

is shown. As seen in Figure 7c, some differences between the two velocity fields remain in the lower part of the wake (particularly over the valley behind the first hill at $X/D \approx 2$). However, the differences between the velocity fields decrease with downstream distance from the rotor disc. The results from the RANS-based wake reconstruction indicate that the wake-superposition technique can generate reasonable predictions of the wake development over hilly terrain, given that the wake-expansion model has sufficient accuracy. While some modelling simplifications were made during the simulations, the comparative analysis (with and without

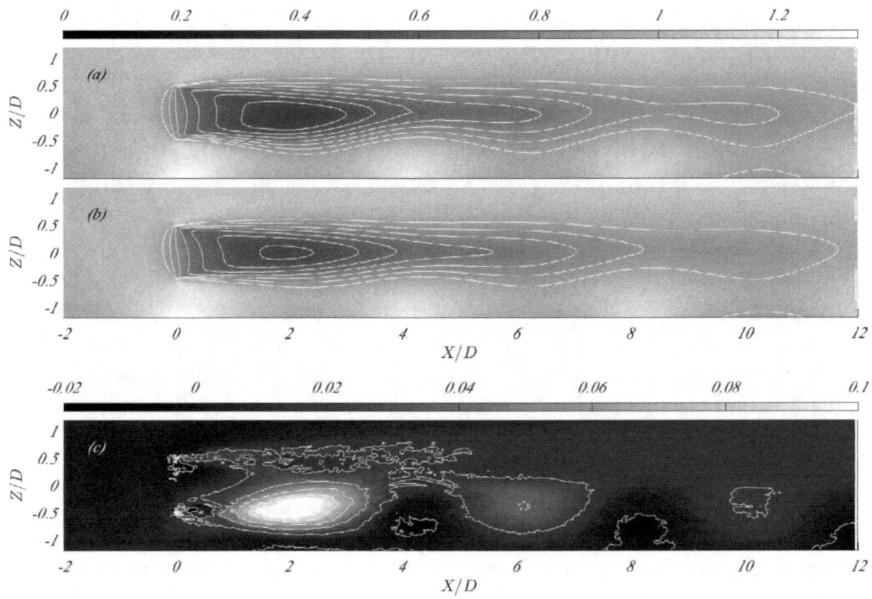

Figure 7. Normalised streamwise velocity at the vertical symmetry plane: (a) actuator-disc simulations over hilly terrain, $\widetilde{U}_{AD}/U_\infty$, ($b$) velocity field obtained using the RANS-based wake-reconstruction method, $\widetilde{U}_*/U_\infty$, (c) difference between the two velocity fields, $\Delta\widetilde{U}/U_\infty$. The velocity fields are zoomed in over the wake region and velocity contours are added for better wake visualisation.

the wake-superposition technique) ensured that the same errors were introduced for all simulated flow cases.

5. Conclusions

In the present paper, the wake development behind a single wind turbine placed over flat and hilly terrain has been studied. This was done both experimentally using stereoscopic PIV, and numerically by means of steady-state RANS simulations and an actuator-disc technique. The results clearly showed that the wind-turbine wake development can be significantly influenced by local terrain features, which are of great importance when planning for wind farms over complex terrain. The wake was deflected downward on the leeward side of a downstream hill, and generally followed the curvature over the hills. With hilly terrain, a faster flow recovery was seen at hub height, indicating that wind farms built over complex terrain (of similar appearance) could potentially produce more power than their flat-terrain counterparts.

The numerical model used in this work was somewhat simplified, but similar qualitative trends as in the experimental data could nevertheless be seen. Some engineering models used in industry to model wind-turbine wakes cannot account for perturbations induced by local pressure gradients. This was illustrated by using the Jensen wake model, which could not capture modulations of the wake induced by the hilly terrain. To perform a wake superposition with reduced errors due to the prescribed wake expansion, the wake-velocity deficit from an actuator-disc RANS simulation over flat terrain (with an adjusted inflow velocity) was superimposed on a flow field simulated over hilly terrain. The RANS-based wake reconstruction was seen to improve the results compared to the Jensen wake model, indicating that the wake-superposition technique can be used over hilly terrain if the wake expansion is properly modelled. The main advantage with engineering wake models (such as the Jensen model) is their computational efficiency, which is why they are often used in industrial applications. However, the present work shows that additional care must be taken when studying wind turbines over complex terrain to ensure that the hill-induced wake modulations are captured. Additional analyses are required to further validate the wake-superposition technique over hilly terrains, including studies with different terrain features and multiple turbines. A more accurate modelling of the wind-turbine rotor is recommended to improve the results accuracy and the performance of different engineering wake models should be investigated. Nevertheless, the performed wake reconstruction showed promising results regarding the wake development behind a single wind turbine operating over hilly terrain.

6. Acknowledgments

The work presented in the present paper has been funded by The Swedish Research Council (VR) under a framework grant for strategic energy research. The authors wish to thank Jan-Åke Dahlberg at Vattenfall Vindkraft AB for manufacturing the wind-turbine rotors and for lending equipment to the experimental measurements. Rune Lindfors and Jonas Vikström at KTH are acknowledged for manufacturing items used in the wind-tunnel setup. Finally, Prof. P. Henrik Alfredsson at KTH and Jari Hyvärinen at ANKER-ZEMER Engineering AB are acknowledged for giving useful advices during the work.

References

[1] Castellani F, Astolfi D, Burlando M and Terzi L 2015 *Journal of Wind Energy and Industrial Aerodynamics* **147** 320–329
[2] Schulz C, Klein L, Weihing P, Lutz T and Krämer E 2014 *Journal of Physics: Conference Series* **524**
[3] Tian W, Ozbay A, Yuan W, Sarakar P and Hu H 2013 *51st AIAA Aerospace Sciences Meeting including the New Horizons Forum and Aerospace Exposition* (Grapevine, Texas: AIAA 2013-0612)
[4] Yang X,, Howard K B, Guala M and Sotiropoulos F 2015 *Physics of Fluids*
[5] Walmsley J L, Troen I, P L D and Mason P J 1990 *Boundary-Layer Meteorology* **52** 259–281
[6] Røkenes K and Krogstad P Å 2009 *Wind Energy* **12** 391–410
[7] Ruck B, Boes R and Gruber M 2013 *International Journal of Energy* **7**
[8] Kähler C J, Scharnowski S and Cierpa C 2016 *Journal of Fluid Mechanics* **796** 257–284
[9] Lindgren B 2002 *Flow facility design and experimental studies of wall-bounded turbulent shear-flows* Ph.D. thesis KTH Royal Institute of Technology
[10] Chamorro L P, Arndt R E A and Sotiropoulos F 2011 *Wind Energy* **15** 733–742
[11] Menter F R 1994 *AIAA Journal* **32** 1598–1605
[12] Shives M and Crawford C 2016 *Renewable Energy* **92** 273–292
[13] Katic I, Højstrup J and Jensen N O 1986 *European Wind Energy Association Conference and Exhibition, Rome - Italy*
[14] Ainslie J F 1988 *Journal of Wind Engineering and Industrial Aerodynamics* **27** 231–224
[15] Bastankhah M and Porté-Agel F 2014 *Renewable Energy* **70** 116–123
[16] Niayifar A and Porté-Agel F 215 *Journal of Physics: Conference Series* **625**
[17] Koning C 1935 *Influence of the propeller on other parts of the airplane structure* (Springer Berlin Heidelberg) pp 361–430

Stability Impact on Wake Development in Moderately Complex Terrain

D Infield and G Zorzi

Wind Energy and Control Centre, University of Strathclyde, Royal College Building, 204 George Street, Glasgow G11XW, UK

david.infield@strath.ac.uk

Abstract. This paper uses a year of SCADA data from Whitelee Wind Farm near Glasgow to investigate wind turbine wake development in moderately complex terrain. Atmospheric stability measurements in terms of Richardson number from a met mast at an adjoining site have been obtained and used to assess the impact of stability on wake development. Considerable filtering of these data has been undertaken to ensure that all turbines are working normally and are well aligned with the wind direction. A group of six wind turbines, more or less in a line, have been selected for analysis, and winds within a 2 degree direction sector about this line are used to ensure, as far as possible, that all the turbines investigated are fully immersed in the wake/s of the upstream turbine/s. Results show how the terrain effects combine with the wake effects, with both being of comparable importance for the site in question. Comparison has been made with results from two commercial CFD codes for neutral stability, and reasonable agreement is demonstrated. Richardson number has been plotted against wind shear and turbulence intensity at a met mast on the wind farm that for the selected wind direction is not in the wake of any turbines. Good correlations are found indicating that the Richardson numbers obtained are reliable. The filtered data used for wake analysis were split according to Richardson number into two groups representing slightly stable to neutral, and unstable conditions. Very little difference in wake development is apparent. A greater difference can be observed when the data are separated simply by turbulence intensity, suggesting that, although turbulence intensity is correlated with stability, of the two it is the parameter that most directly impacts on wake development through mixing of ambient and wake flows.

1. Introduction

Wake models are an area of active research due to their importance in wind farm design, and also their potential to be used in real-time control for operational optimisation of wind farms taking into account power output and fatigue loading. They range in complexity from detailed computational flow models such as based on RANS and LES, to empirical models motivated by physics of flow considerations such as [1], and also highly simplified models, for example [2]. Model validation is essential if users are to have confidence in these models, and also to indicate the degree of accuracy that may be expected. The IEA Wakebench, [3], is an important international collaboration concerned with promoting wake model validation and providing guidance for the procedures that should be followed for reliable results. Much of the model validation work to date has concentrated on offshore wind farms where rectangular arrays lend themselves to wake analysis, and where terrain does not complicate the flow. These studies have highlighted the impact of atmospheric stability on wake development. There has been relatively limited

IOP Conf. Series: Journal of Physics: Conf. Series **854** (2017) 012024 doi:10.1088/1742-6596/854/1/012024

validation against wind farms located in complex terrain. This paper aims to help fill this gap through study of wind turbines in moderately complex terrain. Efforts have been made to identify the effect of atmospheric stability on the wake data, but also as will be seen, these effects turn out to be relatively modest in this case.

2. Site characteristics and turbine selection

Whitelee wind farm is located 15 km south of Glasgow on an area of partially forested upland. The site area is 55 km^2 and accommodates a total of 215 turbines with a total installed capacity of 539 MW. Of these 140 are Siemens 2.3 MW rated turbines comprising a mix of SWT-2.3-82 and SWT-2.3-93 turbines having rotor diameters of 82 m and 93 m respectively. The terrain is moderately complex with changes in elevation that have a distinct impact on wind speeds. The prevailing wind direction is south-westerly as shown clearly in the wind rose shown in Figure 1 depicting data measured at turbine hub height on mast 81, which is located to the west of the wind farm.

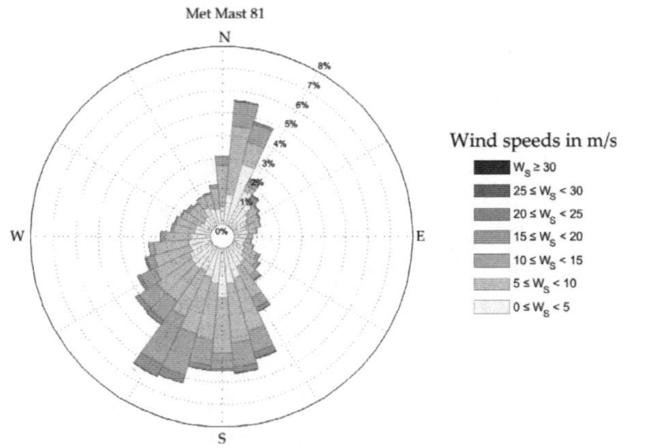

Figure 1: Wind rose from met mast 81

Because of the complexity of the terrain, the wind farm does not follow a regular grid structure. This makes the selection of turbines for wake analysis purposes more difficult. A group of six SWT-2.3-93 wind turbines, all with a hub height of 63.5 m, which lie more or less along a line at 237° have been selected due to their open and consistent exposure to the prevailing wind direction. These turbines are highlighted in Figure 2 showing a map of this part of the wind farm, taken from digimap, [4].

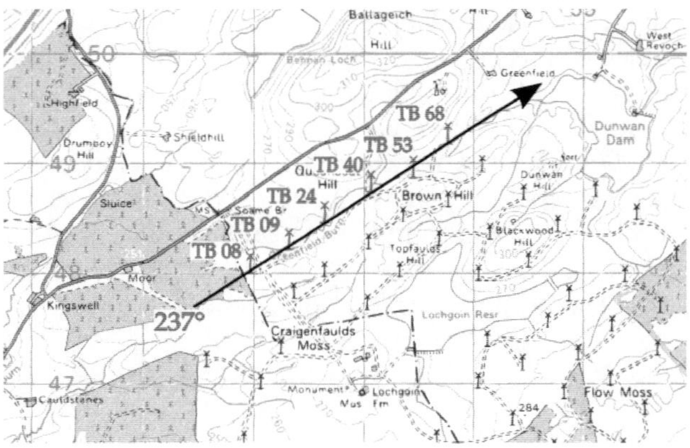

Figure 2: Selected wind turbines for wake analysis

Table 1 gives the distances between adjacent wind turbines and also the exact angle between the pairs of turbines. It is apparent that the turbines are not equally spaced, and that there are minor, but perhaps, in terms of wake impact, significant departures from the 237° line.

Turbine pair	08 - 09	09 - 24	24 - 40	40 - 53	53 - 68
Distance between turbines (m)	377	423	465	421	457
Distance between turbines (rotor diameters)	4.05	4.55	5.00	4.53	4.91
Angular alignment (degrees)	240.8	227.1	238.4	249.4	226.1

Table 1: Separation and angular alignment of the selected turbines

Along the wind direction of 237° the turbines lie on a steady upward slope. The ground elevation for the six turbines is given in Table 1 below.

Turbine number	08	09	24	40	53	68
Terrain elevation (m asl)	263	264	267	275	279	272

Table 2: Elevations of the selected turbines

3. Data processing and direction re-calibration

SCADA data for the complete wind farm has been made available by the site operator, Scottish Power Renewables, for the period 1st January 2012 to 1st January 2013. This includes all 140 turbines plus wind speed data from the two site met masts at a range of measurement heights together with wind direction, air temperature and pressure. Amongst other parameters, the individual turbine SCADA measurements include 10 minute averages of: nacelle wind speed; rotor angular velocity; blade pitch angle; active power output; and yaw position, together with maximum, minimum and standard deviation. These data has been imported into MATLAB for analysis. Since the met masts are some distance from the selected turbines, it was decided, considering the irregularity of the intervening terrain, not to rely on wind direction measurements made at the masts, but rather to identify wind direction from the yaw position sensor for the first turbine in the line, namely turbine 08.

3.1. Yaw angle re-calibration

It is well known that wind turbine yaw sensors are commonly very poorly calibrated; this is a result of poor sensor set up and also the fact that calibration is often lost when turbines are maintained. The yaw angle sensor data for all selected turbines have been corrected following the procedure outlined here. Each turbine is surrounded by a number of other turbines; the angular arrangement of which is precisely known from the turbine locations. Figure 3a shows the example of turbine TB 09. Depending on the wind direction, turbine TB 09 will exhibit a lower wind speed and a reduced output due to operating in the wake of one of the other turbines, as is clear from the direction bin averaged values shown in Figure 3b. To further assist in identification on the turbine angular positions, only data with wind speed in the range 6.7 to 7.6 m/s is shown. This results in an absence of power production from TB 09 when operating directly in a wake. The exact angular location can be identified from the gap centres in Figure 3c.

IOP Conf. Series: Journal of Physics: Conf. Series **854** (2017) 012024 doi:10.1088/1742-6596/854/1/012024

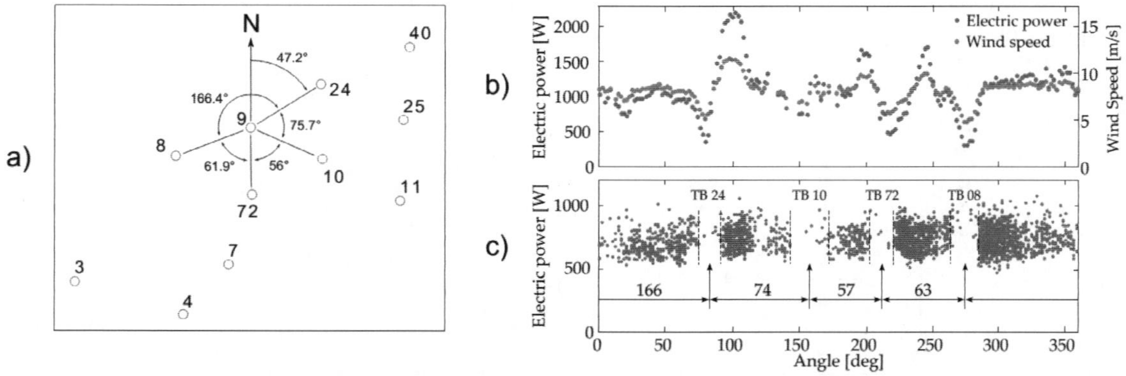

Figure 3: a) Relative angular position of TB 09 with surrounding turbines. b) Power and wind speed as a function of yaw angle position for TB 09. c) Power from TB 09 for wind speeds 6.7 to 7.6 m/s

3.2. Data selection

Three nacelle wind speed signals are available in Siemens SCADA data sets with different designated sensor names (in fact reflecting different ways in which the raw anemometer signals have been processed by the SCADA software): active; primary; and secondary wind sensors. The first of these has been chosen because it showed the highest correlation with the hub height wind speeds measured at met mast 81. For reliable wake assessment, it is essential that all the turbines selected are working properly, and not for example cutting in or out within the ten minute averaging period, or operating in a curtailed mode. Such unwanted data have been removed by applying a sequence of filters. Wind speeds below 4 m/s and power values below 100 kW have been filtered out as these are of no interest for wake analysis. In addition the known pitch characteristics of the Siemens turbine (ie the dependence of blade pitch angle on average wind speed) have been used to identify any curtailment and to remove such data.

In addition to the data reduction outlined above, data have been further filtered to ensure that all six turbines are operating, and that they are all reasonably aligned with the wind direction (defined as being within $\pm 10^{o}$). The result of this data processing was to eliminate 67% of the total number of data points. Finally, a very narrow wind directing sector of 237 $\pm 1^{o}$ has been adopted for the wake analysis.

4. Experimental results

The data has been analysed with several key objectives in mind: to explore the wake spread and confirm the yaw angle re-calibration procedure; to investigate the cumulative wake superposition effect and its interaction with terrain effects; and to explore the impact of atmospheric stability.

4.1. Assessment of wake spread

The simplified models generally assume a Gaussian shape for the wake as a function of distance from the wake centreline. Figure 4 shows the results calculated for the shape of the wake for turbine TB 08 as measured by the relative power output at turbine TB 09, which is 377 meters downstream (Table 1), as a function of the wind speed direction. Wind speeds between 7.5 and 8.5 m/s as measured by the nacelle anemometer of turbine TB 08 were used to give an approximately constant thrust coefficient at turbine TB 08 as indicated by Figure 5 since thrust has a major impact on the wake, and the data was split into high and low turbulence intensity (as assessed from the hub height mast data and defined as above and below the mean turbulence intensity of 0.15 respectively). As anticipated the wake shapes are broadly Gaussian but with some asymmetry caused by the wake rotation that is ignored in the simple models, and indeed most CFD models. High ambient turbulence intensity promotes greater mixing between the wake and ambient flows, giving rise to faster wake dissipation. This can be seen in the lower centreline power deficit apparent from Figure 4. However, the broader wake profile for the high turbulence case is unexpected. At first site, the fact that the centreline is displaced from zero might cause concern regarding the wind direction calibration, but actually this reflects the fact that turbines TB 08 and TB 09 do not lie exactly along the 237° line, but are displaced by 4°, in accord with Figure 4.

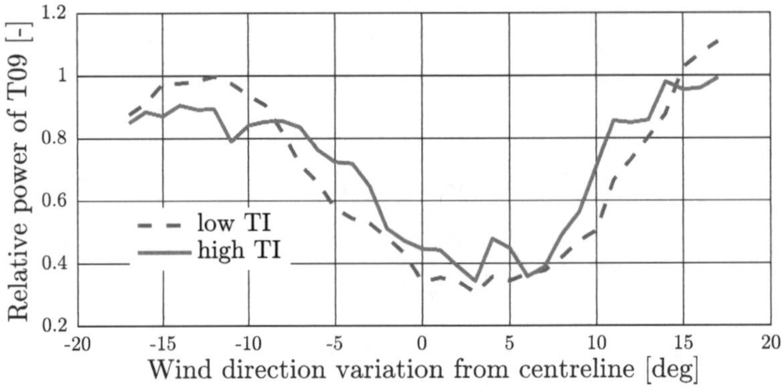

Figure 4: Wake spread for turbine TB 08 for wind speed interval [7.5 8.5] m/s

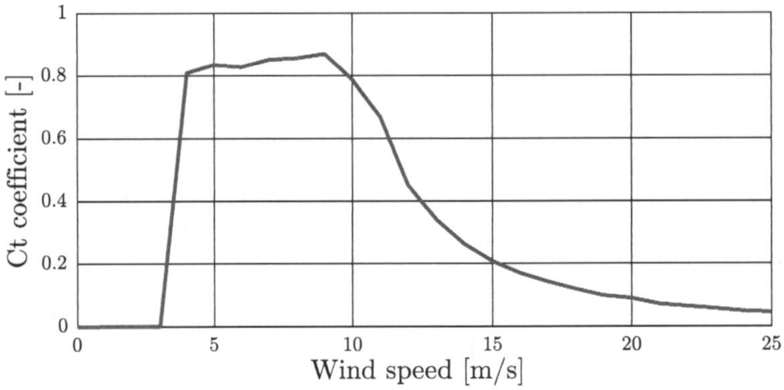

Figure 5: Thrust coefficient as a function of wind speed

4.2. Cumulative wake superposition and interaction with terrain effects

Two approaches to data selection have been used to assess the wakes of the group of turbines: wind speed filtering; and power filtering. In both cases the range has been selected to give an approximately constant thrust coefficient at turbine TB 08. As in Section 4.1 a wind speed range of 7.5 to 8.5 m/s was used, and a power range 800 to 1200 kW was used. Figure 6 shows the development of the wind speed and the power along the line of turbines, together with the standard deviation of the results. The wake impact at TB 09 is large and probably little influenced by the terrain. However, further down the line of turbines, with reference to Tables 1 and 2, it is clear that the results show some influence of the terrain, in particular the higher wind speed at the 5th turbine located on the terrain highpoint, falling off to the final turbine in the line. The results also reflect the exact alignment of the turbines that is expected to result in partial wake immersion in some cases.

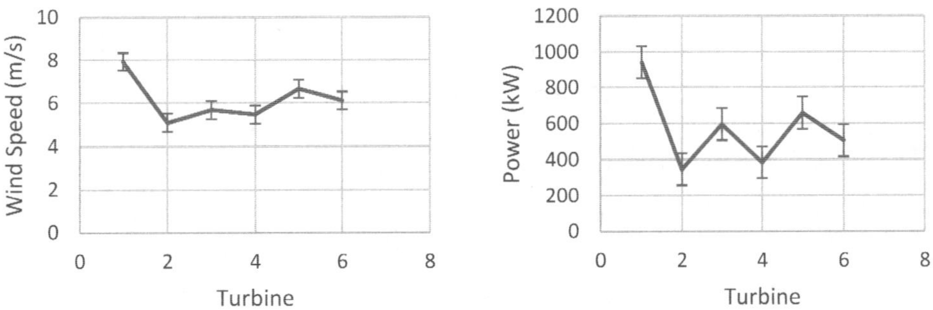

Figure 6: Wake development from TB 08 to TB 68 (left for wind speed, right for power)

4.3. Impact of atmospheric stability

Concurrent estimates of the bulk Richardson number (Ri) have been made available based on data collected by NEL at an adjoining site. The raw data was not supplied but it is known that Ri was calculated according to Equation 1, where ΔT, z and U are the differences in temperature (Kelvin), height (m), and wind speed (m/s) respectively, and that measurements were made at 1 m and 76 m height for temperature, and 10 m and 80 m for wind speed (the fact that the heights used for temperature and wind speed did not coincide exactly will result in some minor error in the estimation). Although Richardson number is a convenient measure of atmospheric stability, it is difficult to estimate accurately due to expected errors in measuring small temperature difference. To give some confidence to the estimates, correlations with wind shear and turbulence intensity at met mast 81 have been calculated derived from average values at each hour of the day that reflect the expected consistent changes in stability through the 24 hour period. These are presented in Figure 7 and show well defined and expected relationships; ie turbulence intensity increases with atmospheric instability (negative Ri), and wind shear (given by the shear exponent) increases with stability (positive Ri). However, if the Ri is plotted directly against shear or turbulence intensity for the individual time points, the result is widely scattered with no correlation apparent, indicating poor accuracy for individual estimates of stability. To contrast with Figure 7, and provide a more balanced assessment of the relationships, the data has been binned by shear exponent and turbulence intensity to create Figures 8 and 9 respectively. The results are not nearly as linear and very wide confidence limits are apparent reflecting the scatter mentioned; despite this the expected tendencies in the relationships are generally apparent.

$$Ri = \frac{g.(\frac{\Delta T}{\Delta z})}{T.(\frac{\Delta U}{\Delta z})} \qquad (1)$$

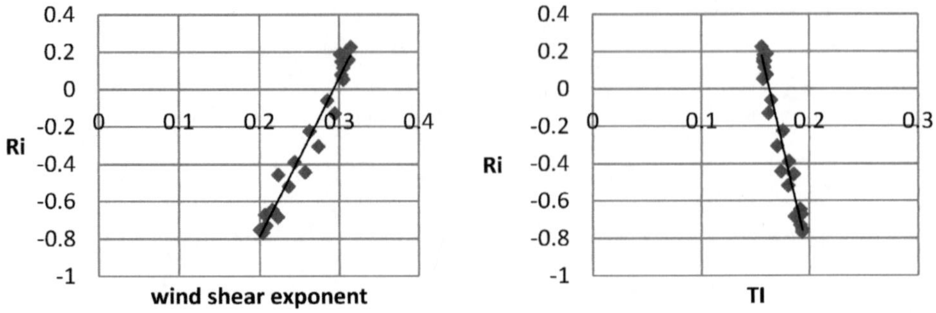

Figure 7: Correlations between Ri and shear exponent (left) and turbulence intensity (right)

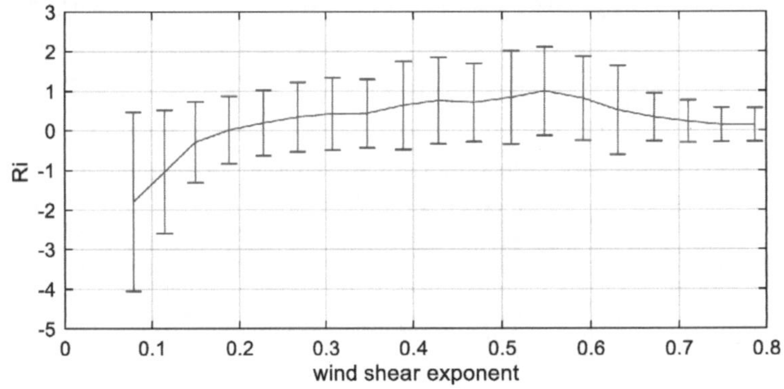

Figure 8: Binned relationship between Ri and shear exponent

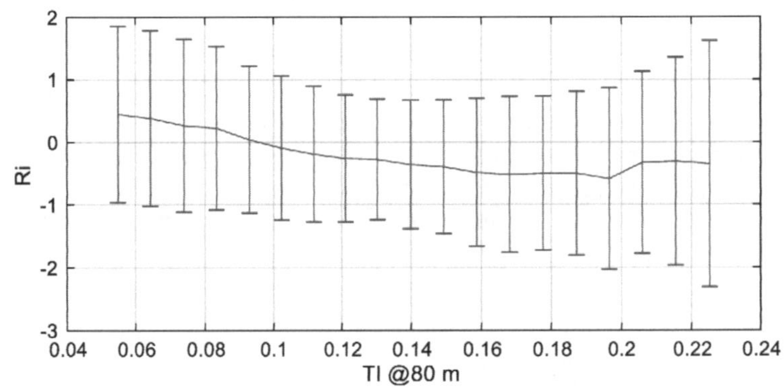

Figure 9: Binned relationship between Ri and turbulence intensity

It can be seen from Figures 7, 8 and 9 that atmospheric conditions at the windfarm range from unstable to neutral to slightly stable as might be expected for an onshore site near to the west coast of the UK. To explore whether changes in stability impact significantly on wake development at this site, the data filtered by power in the range 400 to 1700 kW to give the maximum amount of data consistent with near constant thrust coefficient has been split by positive and negative Ri, and the normalized power with respect to the power of TB 08, plotted for each of the turbines as shown in Figure 10. Very little difference can be identified; this is assumed to be due to the scatter between the raw hourly values of the parameters that is apparent in Figures 8 and 9. Regrettably, it has not been possible to present the errors bars for Figure 10, but these will be comparable to those shown in the corresponding plot of Figure 6.

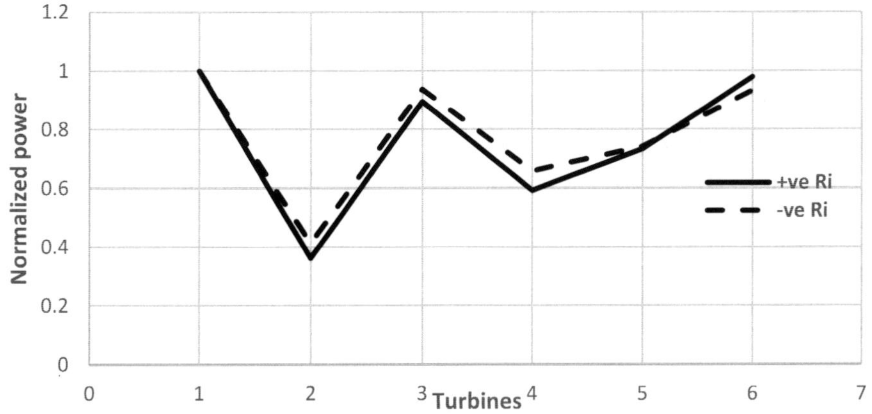

Figure 10: Wake development for positive and negative Ri

As an alternative to analysis by Richardson number, the same data has been plotted but this time separated into high and low turbulence intensity as measured at met mast 81. The result is shown in Figure 11, and shows a much clearer dependency. This suggests that it is the ambient turbulent level that is known to determine wake mixing and decay that is more important, although of course changes in turbulence intensity are related to changes in stability as shown above. The turbulence intensity was measured closer to the turbines than the mast used to calculate Ri but whether this might in part explain the result is unclear.

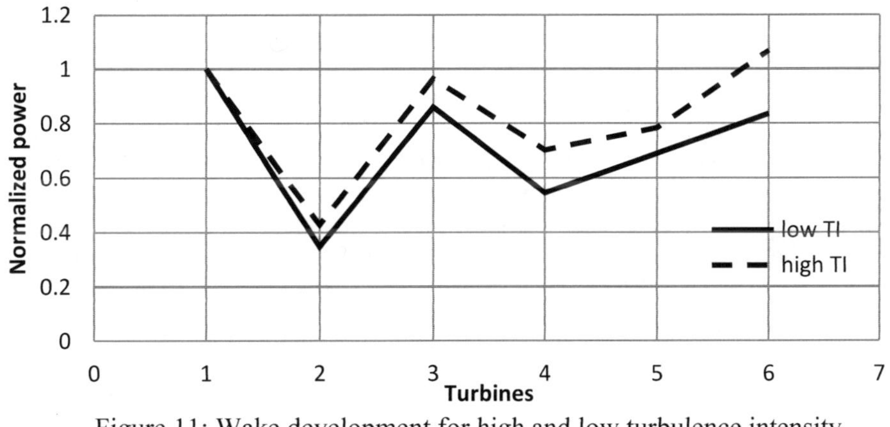

Figure 11: Wake development for high and low turbulence intensity

5. Provisional wake model comparisons

Two commercial CFD codes have been used to model the wake development of the selected turbines: WindModeller (based on ANSYS/CFX), [5], and STAR-CCM+, [6]. Only a single wind direction has been used in the results presented here and so variations in wind speed within the 10 minutes averaging period of the SCADA data are not accounted for. Both CFD codes employ an actuator disk representation for the wind turbine rotor, and both have represented the atmospheric boundary layer as neutral. There are however differences in the way the codes as applied in this instance have dealt with stratification above the boundary layer. STAR-CCM+ accounts for the stable stratification that typically occurs in a

IOP Conf. Series: Journal of Physics: Conf. Series **854** (2017) 012024 doi:10.1088/1742-6596/854/1/012024

free atmosphere above the boundary layer, from around 1 km above the ground, whilst although CFX has this capability, this option has not been invoked in this particular application. This is not considered to have a significant impact on the results presented. As mentioned, no attempt has been made in this provisional application of the CFD codes to take account of expected wind direction changes within the 10 minute averaged data (the standard deviation of wind directions was calculated as 8.83° from the wind vane on met mast 81). The results (Figure 12) show reasonable agreement between the models and data, and reflect, but to different extents and in different ways, the impact of the terrain and the differences in turbine alignment.

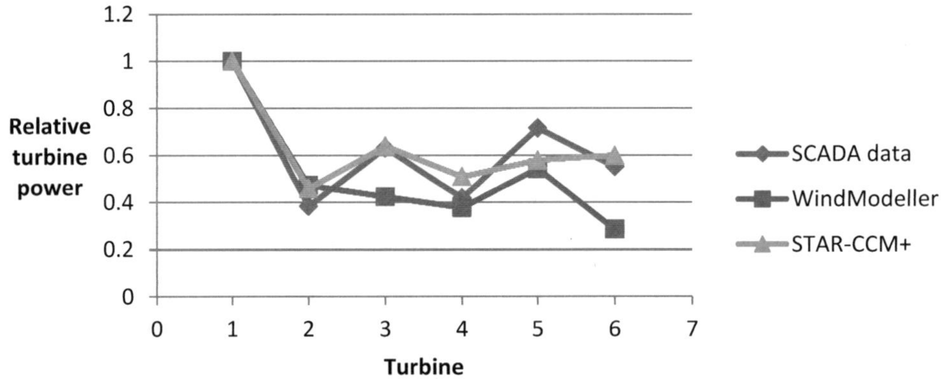

Figure 12: Comparison of wake models with SCADA data results

6. Conclusions

A year of SCADA data from Whitelee wind farm has been carefully processed to exclude potentially misleading data points and has been examined for a very narrow wind direction sector to examine the wind speeds and power production of 6 turbines, and thus the wakes of five turbines. The data has been split into two sets according to bulk Richardson number, and also ambient turbulence intensity, so as to examine the impact of atmospheric stability on wake development. Although this effect has been examined for offshore sites, very little corresponding work has been undertaken for onshore wind farms. The results indicate very little impact of atmospheric stability on wake strength which may be due to the limited range from unstable to slightly stable for this particular site. Splitting the data into high and low turbulence intensity however showed a marked impact, with high turbulence resulting in diminished wakes as would be expected. Some provisional wake calculations undertaken using ANSYS WindModeller and STAR-CCM+ and assuming a neutral boundary layer showed reasonable agreement with the data.

IOP Conf. Series: Journal of Physics: Conf. Series **854** (2017) 012024 doi:10.1088/1742-6596/854/1/012024

References

[1] Larsen, G. C. A simple stationary semi-analytical wake model; Risø-R-1713(EN), 2009.
[2] Katic, I., Hojstrup, J. and Jensen, N.O. (1986) A Simple Model for Cluster Efficiency; Proc. EWEC, October, 1986, Rome.
[3] https://www.ieawind.org/task_31.html
[4] Ordnance Survey, "OS MasterMap Topography Layer."
[5] http://resource.ansys.com/Resource+Library/Webinars/Wind+Farm+Flow+Modeling:+Including+Atmospheric+Stability
[6] http://mdx.plm.automation.siemens.com/star-ccm-plus

Acknowledgements

The authors would like to acknowledge the support of Christiane Montavon of ANSYS for her assistance in using WindModeller, Lindsey Amos of Strathclyde University for producing the results from WindModeller, James Bleeg of DNV-GL for STAR-CCM+ results, and Laura Maury, a visitor from INSA-Lyon, who made useful contributions to the data analysis. Funding from EPSRC under grant EP/N508536/1 as part of the InnovateUK project SWEPT2 is gratefully acknowledged. We thank Scottish Power Renewables for making available the data from Whitlee wind farm, this is subject to a confidentiality agreement and so is not available to other researchers.

High-Order Numerical Simulations of Wind Turbine Wakes

E. Kleusberg[1], R. F. Mikkelsen[2], P. Schlatter[1], S. Ivanell[1,3] and D. S. Henningson[1]

[1] Linné FLOW Centre and Swedish e-Science Research Centre (SeRC), KTH Mechanics, Royal Institute of Technology, SE-100 44 Stockholm, Sweden
[2] Technical University of Denmark, DTU Wind Energy, 2800 Kgs. Lyngby, Denmark
[3] Uppsala University, Wind Energy Section, Campus Gotland, SE-621 67 Visby, Sweden

E-mail: kleu@mech.kth.se

Abstract.
Previous attempts to describe the structure of wind turbine wakes and their mutual interaction were mostly limited to large-eddy and Reynolds-averaged Navier–Stokes simulations using finite-volume solvers. We employ the higher-order spectral-element code Nek5000 to study the influence of numerical aspects on the prediction of the wind turbine wake structure and the wake interaction between two turbines. The spectral-element method enables an accurate representation of the vortical structures, with lower numerical dissipation than the more commonly used finite-volume codes. The wind-turbine blades are modeled as body forces using the actuator-line method (ACL) in the incompressible Navier–Stokes equations. Both tower and nacelle are represented with appropriate body forces. An inflow boundary condition is used which emulates homogeneous isotropic turbulence of wind-tunnel flows. We validate the implementation with results from experimental campaigns undertaken at the Norwegian University of Science and Technology (NTNU Blind Tests), investigate parametric influences and compare computational aspects with existing numerical simulations. In general the results show good agreement between the experiments and the numerical simulations both for a single-turbine setup as well as a two-turbine setup where the turbines are offset in the spanwise direction. A shift in the wake center caused by the tower wake is detected similar to experiments. The additional velocity deficit caused by the tower agrees well with the experimental data. The wake is captured well by Nek5000 in comparison with experiments both for the single wind turbine and in the two-turbine setup. The blade loading however shows large discrepancies for the high-turbulence, two-turbine case. While the experiments predicted higher thrust for the downstream turbine than for the upstream turbine, the opposite case was observed in Nek5000.

1. Introduction

The wind-energy sector has been subject to continuous progress throughout the last decades due to the attempt of many countries to emancipate themselves from fossil fuels. Landscape restrictions and judicial regulations lead to wind turbines being placed closely together in small clusters or wind farms. Thus, apart from being subjected to the changing characteristics of the atmospheric boundary layer, individual turbines are also exposed to the wakes of other turbines. The wake from an upstream turbine usually leads to an increase in turbulent intensity and fatigue loading of downstream turbines. Due to the decreased energy content of the flow in the wake of a turbine the downstream turbines extract less energy from the atmosphere than the first row of turbines. Using large-eddy simulations Nilsson *et al.* [1] show that the power production of downstream turbines in the Lillgrund wind farm drops to less than 50% of the upstream turbines. If the flow approaching the upstream turbines is already turbulent, then the added momentum exchange between the wake and the freestream causes an increase in the mean streamwise velocity of the flow approaching the downstream turbines [2].

To construct and operate closely clustered wind turbines it is imperative to be able to estimate the influence of wake interaction on wind turbine loading. When using numerical simulations the limiting factor is the Reynolds number $Re_c = W_\infty c/\nu$, where W_∞ is the free-stream velocity, c is the chord, and ν the viscosity. The length scales of a fully-resolved wind-turbine simulation range from a fraction of the boundary-layer thickness of the rotor blades to the size of the largest eddies in the atmospheric boundary layer. To reduce the range of length scales to be resolved, modeling assumptions need to be introduced. Crespo *et al.* [3] provide an overview of wake models ranging from the assumption of turbines in wind farms as distributed roughness elements to individual wake models with superposition of interacting wakes. At the other end of the complexity spectrum lies the possibility to resolve the blades, as done in Zahle *et al.* [4] with a single wind turbine based on the Reynolds-averaged Navier–Stokes equations. Widely used modeling approaches of intermediate complexity are the actuator methods. The actuator-disk method (ACD) *e.g.* uses the blade element method to compute the blade forces on a disk, coupled with the Navier–Stokes equations [5]. The ACD simplification has been shown to provide accurate predictions of the blade loading and the averaged wake properties. However, the spiral tip and root vortices that are in reality released by wind turbines are replaced by a continuous vortex sheet due to the fact that the blade forcing is applied along the entire rotor disk plane. As a more advanced technique, the actuator-line method (ACL) was proposed in Sørensen and Shen [6] which models each rotor blade as a line force, thus enabling the release of distinct tip and root vortices. A main source of uncertainty in the actuator methods is the airfoil data with which, using two-dimensional airfoil theory, the sectional lift and drag forces are computed. Both ACL and ACD are currently being used in the context of large-eddy simulations (LES) to analyze single wind turbines and wind-turbine interaction. The solvers mostly employ finite-volume techniques of varying orders of accuracy for spatial discretization, *e.g.* EllipSys3D [7] or OpenFOAM [8]. In Ivanell *et al.* [9] and Sarmast *et al.* [10] the dominant modes responsible for the onset of instability of a single wind turbine wake are investigated while in Nilsson *et al.* [1] the ADM method is used to model the complete Lillgrund wind farm and analyze its performance. Other approaches such as free vortex wake methods have also been employed to investigate wind turbine wakes.

This paper studies the wake of a single wind-turbine setup and the interaction between two consecutive wind turbines by means of numerical simulation and compares them to experimental results ("Blind Tests") by Krogstad and Eriksen [11] and Krogstad *et al.* [12] (refered to as NTNU in the figures) and numerical simulations using the finite-volume solver EllipSys3D [7].

2. Numerical Setup

We integrate the incompressible three-dimensional Navier–Stokes equations with the spectral-element code Nek5000 [13], which has been shown to provide accurate results of wind turbine wakes while requiring lower resolution than comparable finite-volume codes [14]. This is due to the use of high-order Legendre polynomials on Gauss–Lobatto–Legendre (GLL) quadrature points for spatial discretization. However, due to non-equidistant meshing the required time step in Nek5000 was smaller than in the finite volume code.

2.1. Domain and boundary conditions

In the following we non-dimensionalize all quantities with the wind-turbine radius R and the free-stream velocity W_∞. The numerical domain is constructed such that it matches the wind-

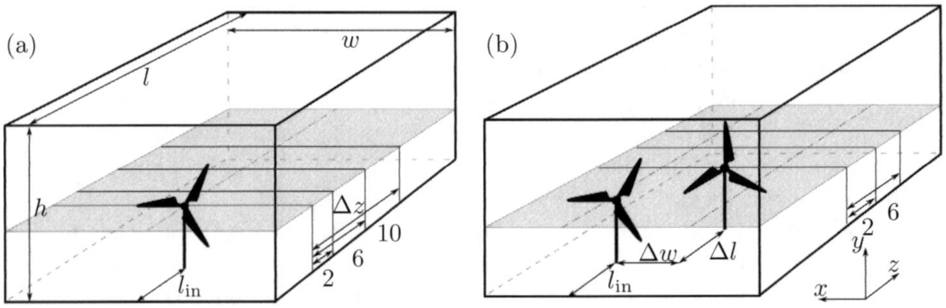

Figure 1. (a) Case 1: single wind turbine and (b) Case 2: two turbines with a total spanwise offset of $\Delta w = 0.89$ and distance $\Delta l = 6$; the horizontal red lines indicate where the time-averaged streamwise velocity and streamwise turbulent stress are extracted. The domain dimensions are: $l = 25, h = 4.03, w = 6.04$.

tunnel dimensions adjusted for the displacement boundary-layer thickness of the tunnel walls. This enables the use of a slip condition on the lateral boundaries without the need to resolve the boundary layers forming on the wind-tunnel walls. The distance between the inlet and the single turbine is $l_{\text{in}} = 7$. In the two-turbine cases with low turbulence the distance of the upstream turbine is also $l_{\text{in}} = 7$ while in the high-turbulence case the upstream turbine is placed at $l_{\text{in}} = 4$. We employ a convective outflow boundary condition at the outlet.

When comparing with wind-tunnel experiments, the turbulence in the freestream must be taken into account. This study includes the turbulence as random sinusoidal perturbations superposed with uniform plug flow as an inlet boundary condition [15] as opposed to the EllipSys3D simulations [16, 17], where the Mann turbulence model [18] is used to generate turbulence which is then included as a force field upstream of the turbines. The turbulence properties in the domain are set at the inlet to match the decay of the turbulence intensity throughout the experimental wind tunnel [11, 12] and the turbulence intensity at each wind turbine. In the high-turbulence case the experiments report a turbulence intensity of $Ti_{\text{up}} = 10\%$ at the upstream turbine and $Ti_{\text{down}} \approx 5\%$ at the downstream turbine. The grid is Cartesian and contains approximately 80000 equidistantly spaced spectral elements with 9th order polynomials. The actuator lines are discretized with $N_{\text{ACL}} = 70$. The radius-based Reynolds number used in the simulations is $Re_R = 50000$. The blade data is extracted using the chord-based blade tip Reynolds number is $Re_{\text{c,tip}} \approx 100000$ which is the Reynolds number reported in the experiments. Sørensen and Shen [6] noted that the Reynolds number does not affect the overall wake behaviour above a certain minimum therefore justifying the chosen Reynolds number in Nek5000.

2.2. Wind turbine model

The rotor blades are modeled as actuator lines which compute the blade forces at each time step based on the inflow velocity, the angle of attack α and the tabulated, chord-based Reynolds number dependent lift and drag coefficients $C_{L,D}$ obtained by Sarmast and Mikkelsen [19] and corrected for solid, wake and streamline blockage. After the lift and drag forces $\mathbf{F}_{2D} = (F_L, F_D)$ have been computed at the ACLs using two-dimensional airfoil theory the blade forces are spread out using the convolution of the force with a three-dimensional Gaussian kernel [6].

The wind turbine towers are also modeled using a body-force approach similar to the ACL method. The experimental data taken from Sumer and Fredsøe [20] is dependent on the Reynolds number and the incoming turbulence intensity. The tower forces are computed

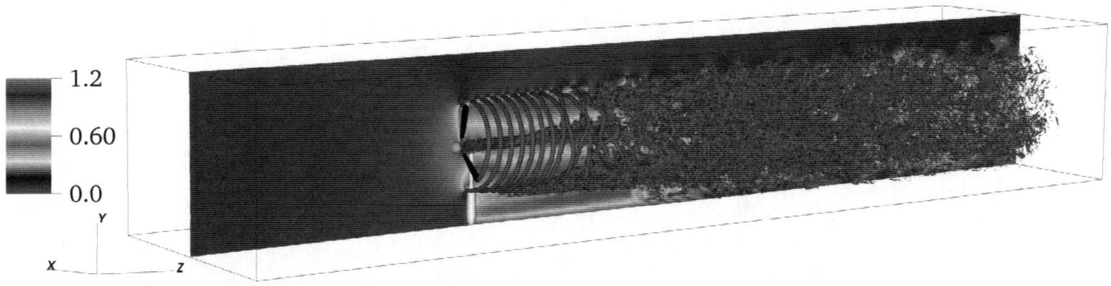

Figure 2. Slice of the streamwise velocity w and isosurfaces of vorticity magnitude (blue) $|\omega|$. Isosurfaces of the streamwise force show the blades (black), tower (white) and nacelle (grey).

such that $F^T_{L,D} = 0.5\rho u_T^2 d C^T_{L,D}$. Here, u_T is the local velocity magnitude of the flow around the tower along the line representing the tower, d is the tower diameter and $C^T_{L,D}$ are the tabulated lift and drag coefficients of the cylinder. The tower lift force F^T_L is composed of only a fluctuating component, while the tower drag force F^T_D consists of a constant drag with a superposed disturbance, oscillating at twice the frequency of the lift force. The frequency of the tower lift-force oscillation f is defined by the Strouhal number $Sr = fD/u_T$. The nacelle is represented as an actuator disk with a constant drag force $F^N_D = 1/2\rho C^N_D W^2_\infty \pi r^2_{\mathrm{nac}}$ with the drag coefficient of $C^N_D = 0.5$. Both nacelle and tower forces are distributed on to the GLL points using the three-dimensional Gaussian kernel.

3. Results

Figure 2 shows the wake development for the optimal tip-speed ratio (TSR) $\lambda = 6$ of a single wind turbine. The blade–tower interaction creates an additional velocity deficit.

3.1. Tower influence

The tower introduces a significant asymmetry in the wake shown in figure 3(a) while the nacelle model leads only to a slight decrease in the streamwise velocity in the wake center and a small increase in turbulence intensity. At approximately $x = -0.3$, in figure 3(a) the tower wake is observed in the averaged streamwise velocity at $\Delta z = 2$ (position shown in figure 1(a)). Further downstream a distinct tower wake cannot be observed in the averaged streamwise profiles. However, the velocity at the centerline is decreased. Due to the fact that the nacelle model only consists of a constant drag force, the unsteady three-dimensional turbulent nature of the flow downstream of the nacelle is not captured and the high streamwise turbulent stress at the rotor center of the experiments in figure 3(b) at $\Delta z = 2$ is not replicated in the simulation. The interaction between the tower and blade wake contributes to an early breakdown of the helical tip and root vortices. The turbulent stress peaks generated by the tip vortices and the wake shear layer in Nek5000 is five times smaller than in the experiment due to the larger vortex core size of the ACL method which is defined by the width of the Gaussian and the finite resolution of the simulation. Additionally, as the blade boundary layer is not resolved in the ACL method any boundary layer turbulence is not represented in the ACL simulation. A simulation conducted at a higher Reynolds number shows that the overall wake behaviour is similar to that of the low Reynolds number. The tower in EllipSys3D, though modeled similarly to Nek5000, influences the wake differently, creating a much more centered “tower dent” as seen in figure 3(a). The wake deficit generated by the blades is well predicted by EllipSys3D.

To the authors knowledge the influence of this type of tower model on the wake center has not

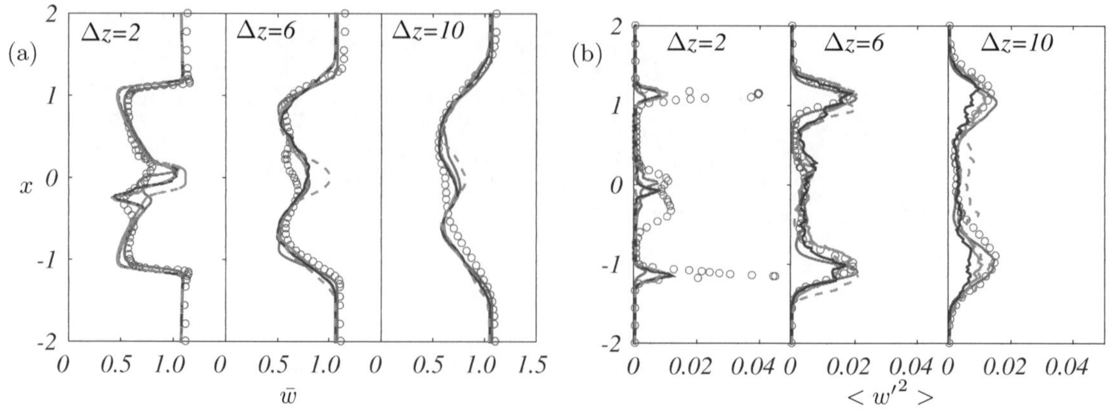

Figure 3. (a) Time-averaged streamwise velocity and (b) streamwise turbulent stress, TSR $\lambda = 6$, (blue circles) NTNU, (green solid line) Nek5000 (with tower and nacelle), (green dashed line) Nek5000 (no tower and nacelle), (purple) Nek5000 at $Re_R \approx 300000$ (with tower and nacelle) and (orange) EllipSys3D (with tower).

yet been investigated numerically. By computing the minimum of the time-averaged azimuthal velocity the wake center position along the streamwise axis is determined. Experimental data provided by Blomhoff [21] and Pierella [22] show that the wake center defined by the minimum azimuthal velocity is shifted downwards and spanwise due to the tower wake: At $\Delta z = 6$ ($z = 13$ from the inlet) the wake center is shifted to $(x_c, y_c) = (-0.1, -0.25)$ for $\lambda = 5.5$ [21] and $(x_c, y_c) = (-0.157, -0.26)$ for $\lambda = 6$ [22]. A similar effect may also be shown through numerical simulation, though at a smaller scale. Figure 4 shows the deviation of the wake center from the rotor centerline at TSR $\lambda = 6$. Further upstream the azimuthal velocity minimum is difficult to determine exactly due to blade–tower interaction. Without the tower the wake remains centered. The simple tower body force used in this study leads to a wake center which is shifted downwards to $(x_c, y_c) = (-0.06, -0.05)$ at $\Delta z = 6$. A reason for the discrepancy of the spanwise shift of the wake center may be that the tower model does not take into account the variation in the angle of attack with which the flow approaches the tower due to the tangential induction of the blades. If the variation of the tower angle of attack is taken into account, then the deviation in the spanwise direction x at $\Delta z = 6$ is closer to the experiments. The vertical displacement of the wake center is probably underestimated due to the fact that the tower model is based on two-dimensional theory and does not take the radial velocity component of the wind turbine wake into account, when computing the tower lift and drag forces.

4. Tip-speed ratio variation

This section discusses the TSR influence on the wake structure and loading. Figure 5 shows a comparison of the power and thrust coefficient

$$C_P = \frac{2P}{\pi \rho W_\infty^3 R^2}, \qquad C_T = \frac{2T}{\pi \rho W_\infty^2 R^2}, \tag{1}$$

(P is the power and T the thrust) with the experimental curves [11] and EllipSys3D [17]. The power coefficient is in good agreement with the experimental data at off-design conditions. However, at the optimal TSR $\lambda = 6$ the power coefficient of Nek5000 differs significantly ($\Delta C_P \approx 9\%$), while EllipSys3D agrees well. The numerical thrust coefficients are generally lower than the experimental data, which is in accordance with most simulation results comparing with

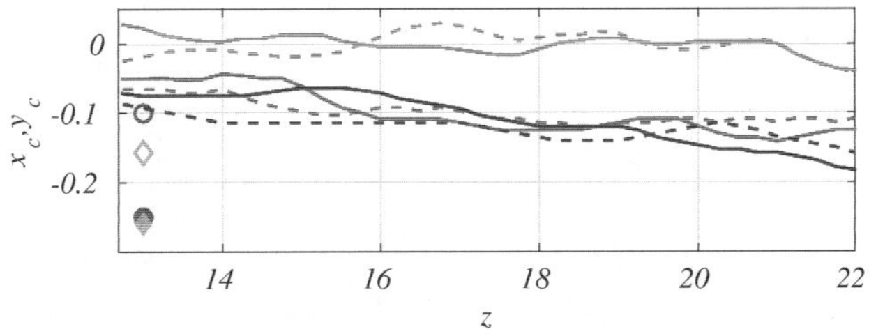

Figure 4. Wake center based on minimum time-averaged azimuthal velocity along the streamwise axis; NTNU: (empty symbols) x_c, (filled symbols) y_c, (blue) $\lambda = 5.5$, (yellow) $\lambda = 6$; (green) Nek5000 without tower, (orange) Nek5000 with tower, and (purple) Nek5000 with tower force at variable angle of attack; dashed lines: x_c, solid lines: y_c .

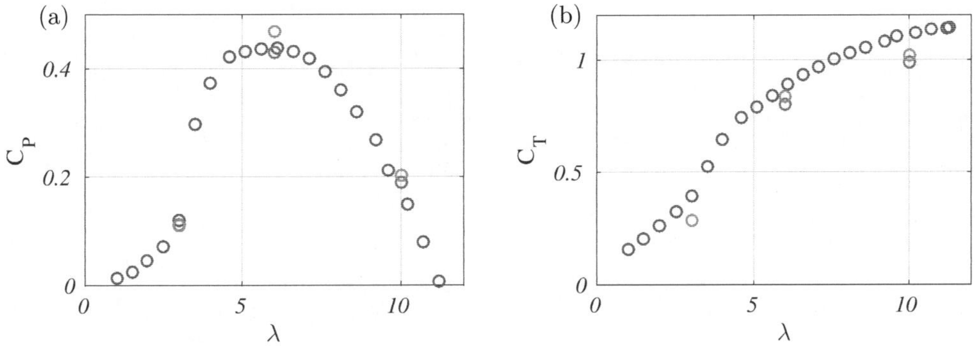

Figure 5. Case 1:(a) Power and (b) thrust coefficients, (blue) NTNU, (green) Nek5000 and (orange) EllipSys3D.

this set of experimental data. In Krogstad and Eriksen [11] it is shown that a fully resolved turbine simulation achieves a better agreement of the thrust coefficients with the experiments. Figure 6 shows the averaged streamwise velocity and turbulent stress profiles at off-design conditions. At $\lambda = 3$, the blades are fully stalled. Due to the larger distance between the vortices the tip and root vortices remain stable for a long distance downstream, causing the two distinct turbulent stress peaks at the tip vortex position to persist in the numerical simulations. In Nek5000 additionally a dip in the turbulent stress peaks are visible at $\Delta z = 2$, where the vortex center is located and the streamwise velocity fluctuations are smaller. Discrepancies with the experiments may be caused by the fact that the turbulence produced by the nacelle geometry, which is a main source of turbulence at $\lambda = 3$ is included in neither the Nek5000 nor the EllipSys3D simulations. Additionally, the heightened level of turbulence due to the flow separation along the entire blade is not represented in the simulations. For $\lambda = 10$, close to the runaway TSR, the rotational velocity of the turbine is very large. The distance between consecutive tip vortices is decreased, which leads to increased interaction between the vortices and a quicker breakdown to turbulence. Figure 6(c) shows the streamwise time-averaged velocity and 6(d) the streamwise turbulent stress. The averaged streamwise velocity compares very well between experiments and Nek5000. EllipSys3D predicts slightly lower velocities at the blade tip. This case gives larger turbulent stresses immediately downstream of the rotor which are captured well in Nek5000 in comparison with the experiments. It is not known why EllipSys3D predicts lower turbulent stress peaks at the tip vortex position at $\Delta z = 2$. In Nek5000 the tip

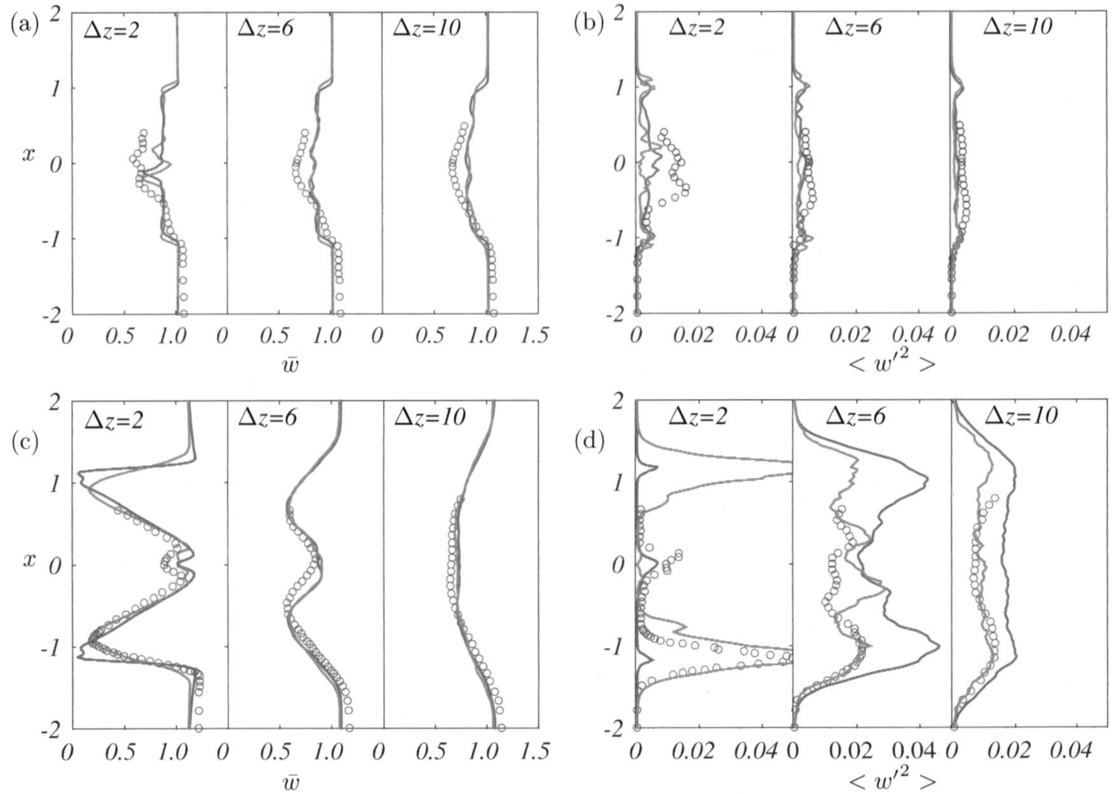

Figure 6. (a,c) Time-averaged streamwise velocity and (b,d) streamwise turbulent stress, TSR $\lambda = 3$ (a,b) and $\lambda = 10$ (c,d); (blue circles) NTNU, (green line) Nek5000 and (orange) EllipSys3D.

vortices are unstable and experience pairing and merging already before $\Delta z = 2$ giving rise to the high turbulent stress at the tip vortices. Where the experiments and Nek5000 predict similar turbulence intensity levels at $\Delta z = 10$ the streamwise turbulent stress in EllipSys3D increases and is significantly higher than the Nek5000 results.

5. Two turbines with spanwise offset

This section discusses the influence of partial wake interaction on the loading and the wake development of the two-turbine setup shown in figure 1(b). The experimental data is obtained from Krogstad *et al.* [12] and also compared to numerical data from EllipSys3D extracted from the same paper. The upstream turbine operates at the design TSR of $\lambda_1 = 6$ while the downstream turbine operates in stall $\lambda_2 = 3.5$, at optimum condition $\lambda_2 = 4.75$ and at a runaway TSR $\lambda_2 = 8$.

The power and thrust coefficients of both low and high turbulence cases are presented in figure 7. The power coefficient for the low turbulence case agrees well at all operating conditions. The thrust coefficient curves of the upstream and downstream turbines are on top of each other in the experiments. In Nek5000 the thrust coefficient of the upstream turbine T_1 is overpredicted by 11%. However, in the so-called "Blind Test 4" published later by Bartl and Sætran [23] a similar case is presented with $Ti = 0.23\%$ with aligned and centered turbines. There the thrust of the upstream turbine at $\lambda_1 = 6$ is significantly higher at $C_T \approx 0.81$ and matches the numerical results better. The thrust coefficient of the downstream turbine T_2 is underpredicted by up to

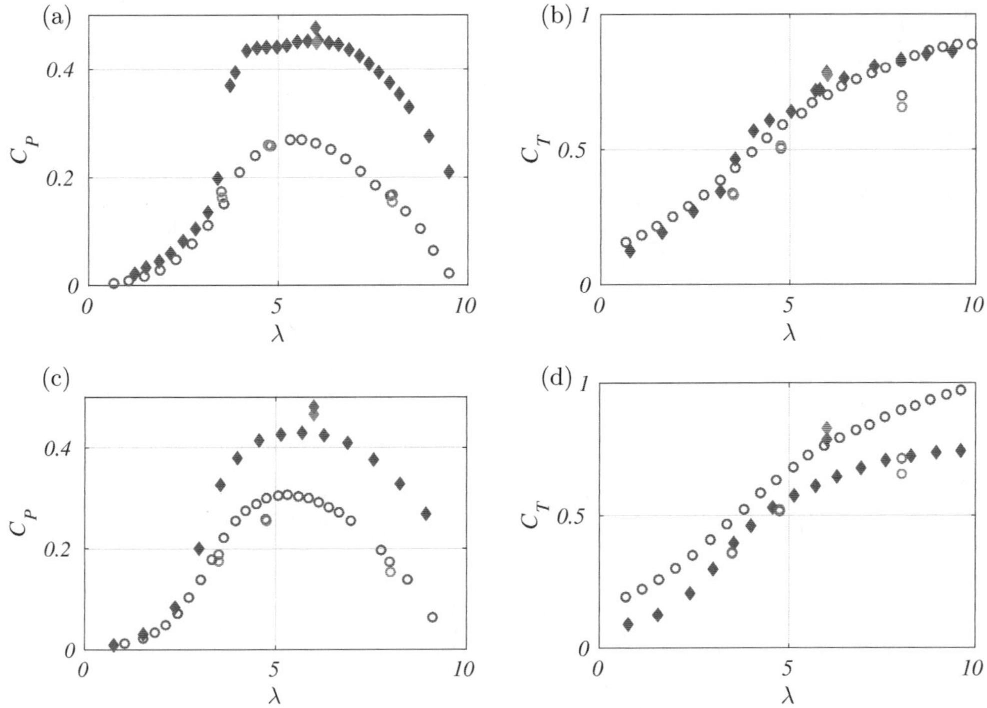

Figure 7. (a,c) Power and (b,d) thrust coefficients of Case 2; (a-b) $Ti = 0.3\%$, (c-d) $Ti = 10\%$, (blue) NTNU, (green) Nek5000 and (orange) EllipSys3D; Symbols: T_2: circles, T_1: diamonds.

25%, which may be due to the underestimation of the streamwise velocity speedup around the wake of the upstream turbine in the numerical simulations (visible in the wake profiles in figure 8(a-c)). In the high turbulence case, the experimental power coefficient curve is smoother in the stall region thus suggesting significant influence of the turbulence on the stall behaviour [12]. The thrust coefficient of the downstream turbine is in this case higher than that of the upstream turbine. For both solvers, Nek5000 and EllipSys3D, the influence of the turbulence on the thrust and the power coefficients is much smaller than in the experiments; the thrust coefficient of the upstream turbine is higher than that of the downstream turbine in both Nek5000 and EllipSys3D. This trend for the upstream turbine at high incoming turbulence is again consistent with the results published by Bartl and Sætran [23]. At similar inflow conditions ($Ti = 10\%$) the same upstream turbine (though centered in this case) has a thrust coefficient of $C_T \approx 0.83$ and a power coefficient of $C_P \approx 0.47$, which is very similar to the Nek5000 results.

Figure 8(a-c) shows the wake development at two streamwise positions behind the downstream turbine. The second turbine wake is superposed on the upstream wake between $-0.5 \leq x \leq 1.2$. In both codes the averaged velocity profiles for $\lambda_2 = 3.5, 4.75$ transition quicker than the experiments to a Gaussian-like profile. For $\lambda_2 = 8$ the wake deficit is overpredicted, which is counter-intuitive as the thrust coefficient of the downstream turbine was underpredicted by 25% in Nek5000. The streamwise turbulent stress is depicted in figure 8(d-f). The location of the peaks from the tip vortices of T_2 is well predicted by the numerical simulations. Apart from the turbulence peak in figure 8(e) at $\Delta z = 2$ at $x = 1.5$ the magnitude of the peaks is also captured well by Nek5000. At $\Delta z = 6$ the tip vortices of T_2 are either broken down or in the process of breaking down, resulting in a much smoother distribution of turbulent stress with the peak at $x = 1 - 1.5$ being the result of the tip vortex of T_2, which is not subjected to the

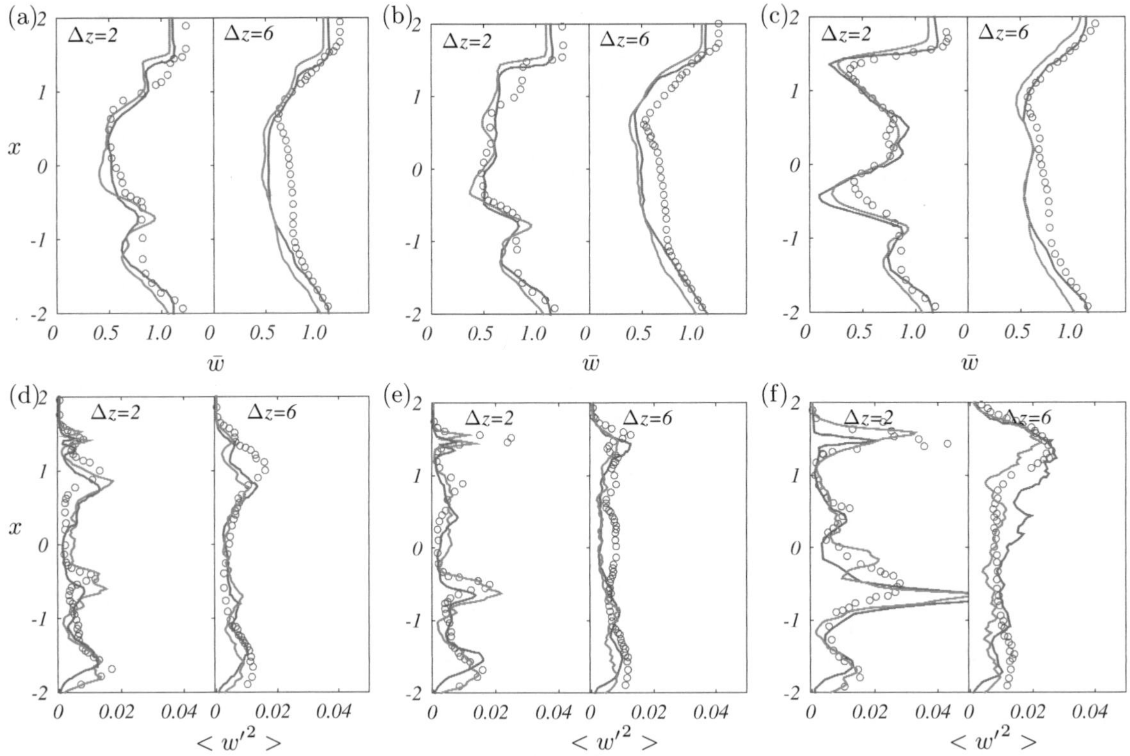

Figure 8. Case 2. Low turbulence level $Ti = 0.3\%$. (a-c) time-averaged streamwise velocity and (d-f) streamwise turbulent stress at downstream turbine TSR: (a,d) $\lambda_2 = 3.5$, (b,e) $\lambda_2 = 4.75$, (c,f) $\lambda_2 = 8$; (blue circles) NTNU, (green) Nek5000 and (orange) EllipSys3D.

upstream wake.

At high incoming turbulence intensity the averaged wake profiles compare well with the experimental profiles (not shown). At $\Delta z = 6$ the experimental velocity profiles are Gaussian shaped and match the numerical results. The turbulent stress profiles are well predicted.

6. Conclusions

The aim of this study was i) to validate the actuator line method in the spectral-element code Nek5000 against an extensive experimental dataset, ii) to test the turbulent inflow boundary condition and iii) to analyze the influence of different additional models (tower, nacelle). The ability of Nek5000 to capture the wake interaction is evaluated in comparison to experiments and additional data from the finite-volume code EllipSys3D. The paper first investigates the wake and loading of a single-turbine setup. The power coefficient is in good agreement with the experimental data, while the thrust coefficient is lower for both numerical solvers. The horizontal velocity and turbulent stress profiles show the main features of the experimental dataset for all tip-speed ratios. Difficulties are only observed where the turbulence generated by the nacelle and blade separation is dominant (*i.e.* for tip-speed ratio $\lambda = 3$). The wake center is identified based on the minimum azimuthal velocity and compared to the wake center identified in experiments [21]. While the wake center is shifted in the same direction as in the experimental case [21] the magnitude of the shift is much smaller. Changing the angle of attack of the tower lift and drag forces to take into account the tangential induction of the blades leads to an improved comparison of the spanwise displacement of the wake center with the experiments.

The second investigation reproduces the two-turbine experimental setup by Krogstad *et al.* [12], where the turbines are offset in the spanwise direction and subjected to different turbulence levels. The downstream turbine tip-speed ratio is varied to include stall, optimal and runaway conditions. For low turbulence intensity the power coefficients agree well with the experiments. The thrust coefficient of the downstream turbine is up to 25% different from the experimental results due to the underestimation of the speedup around the wake of the upstream turbine. The thrust coefficient of the upstream turbine differs both in the high and low turbulence case significantly from the experiments but is found to match data from comparable cases by Bartl and Sætran [23]. The wake profiles agree well for both low and high turbulence cases with both wake deficit and turbulent stresses being well predicted. No large differences are observed between the spectral-element code Nek5000 and the finite-volume code EllipSys3D. The only difference that may be observed is that the streamwise turbulent stress in Nek5000 sometimes showed a better agreement with the experimental results (*e.g* at tip-speed ratio $\lambda = 10$). Using a comparable grid size the time-averaged velocity and turbulent stress do not differ significantly.

The simulations were performed on resources provided by the Swedish National Infrastructure for Computing (SNIC). The first author acknowledges Dr. Sasan Sarmast for providing the numerical data from EllipSys3D.

7. References

[1] Nilsson K, Ivanell S, Hansen K S, Mikkelsen R, Sørensen J N, Breton S P and Henningson D 2015 *Wind Energy* **18** 449–467
[2] Troldborg N 2009 *Actuator Line Modeling of Wind Turbine Wakes* Ph.D. thesis Technical University of Denmark
[3] Crespo A, Hernandez J and Frandsen S 1999 *Wind Energy* **2** 1–24
[4] Zahle F, Sørensen N N and Johansen J 2009 *Wind Energy* **12** 594–619
[5] Mikkelsen R 2003 *Actuator Disc Methods Applied to Wind Turbines* Ph.D. thesis Technical University of Denmark
[6] Sørensen J and Shen W Z 2002 *J. Fluids Eng.* **124** 393–399
[7] Sørensen N N 1995 *General Purpose Flow Solver Applied to Flow over Hills* Ph.D. thesis Technical University of Denmark
[8] OpenFOAM 2016 Web page www.openfoam.org
[9] Ivanell S, Mikkelsen R, Sørensen J N and Henningson D 2010 *Wind Energy* **13** 705–715
[10] Sarmast S, Dadfar R, Mikkelsen R F, Schlatter P, Ivanell S, Sørensen J N and Henningson D S 2014 *J. Fluid Mech.* **755** 705–731
[11] Krogstad P Å and Eriksen P E 2013 *Renew. Energy* **50** 325–333
[12] Krogstad P Å, Sætran L and Adaramola M S 2015 *J. Fluids Struct.* **52** 65–80
[13] Fischer P F, Lottes J W and Kerkemeier S G 2008 Nek5000 Web page http://nek5000.mcs.anl.gov
[14] Kleusberg E, Sarmast S, Schlatter P, Ivanell S and Henningson D 2016 Actuator line simulations of a Joukowsky and Tjæreborg rotor using spectral element and finite volume methods *J. Phys. Conf. Ser.* vol 753 (IOP Publishing) p 082011
[15] Schlatter P 2001 *Direct Numerical Simulation of Laminar-Turbulent Transition in Boundary Layer Subject to Free-Stream Turbulence* Master's thesis KTH Mechanics, Royal Institute of Technology
[16] Sarmast S, Chivaee H S, Ivanell S and Mikkelsen R F 2014 Numerical investigation of the wake interaction between two model wind turbines with span-wise offset *J. Phys. Conf. Ser.* vol 524 (IOP Publishing) p 012137
[17] Sarmast S 2014 *Numerical Study on Instability and Interaction of Wind Turbine Wakes* Ph.D. thesis KTH Royal Institute of Technology
[18] Mann J 1994 *J Fluid Mech.* **273** 141–168
[19] Sarmast S and Mikkelsen R F 2012 The experimental results of the NREL S826 airfoil at low Reynolds numbers Technical report
[20] Sumer B M and Fredsøe J 2006 *Hydrodynamics around Cylindrical Strucures* vol 26 (World scientific)
[21] Blomhoff H P 2012 *An Experimental Investigation of Wind Turbine Wakes* Master's thesis Norwegian University of Science and Technology
[22] Pierella F 2014 *Experimental Investigation of Wind Turbine Wakes and their Interaction* Ph.D. thesis Norwegian University of Science and Technology
[23] Bartl J and Sætran L 2017 *Wind Energy Science* **2** 55–76

Medium fidelity modelling of loads in wind farms under non-neutral ABL stability conditions – a full-scale validation study

G.C. Larsen[1], T.J. Larsen[1] and A. Chougule[2]

[1] Technical University of Denmark, Department of Wind Energy
[2] University of Agder, Department of Engineering Sciences

E-mail: gula@dtu.dk

Abstract. The aim of the present paper is to demonstrate the capability of medium fidelity modelling of wind turbine component fatigue loading, when the wind turbines are subjected to wake affected *non-stationary* flow fields under *non-neutral* atmospheric stability conditions. To accomplish this we combine the classical Dynamic Wake Meandering model with a fundamental conjecture stating: *Atmospheric boundary layer stability affects primary wake meandering dynamics driven by large turbulent scales, whereas wake expansion in the meandering frame of reference is hardly affected.* Inclusion of stability (i.e. buoyancy) in description of both large- and small scale atmospheric boundary layer turbulence is facilitated by a generalization of the classical Mann spectral tensor, which consistently includes buoyancy effects. With non-stationary wind turbine inflow fields modelled as described above, fatigue loads are obtained using the state-of-the art aeroelastic model HAWC2.

The Lillgrund offshore wind farm (WF) constitute an interesting case study for wind farm model validation, because the WT interspacing is small, which in turn means that wake effects are significant. A huge data set, comprising 5 years of blade and tower load recordings, is available for model validation. For a multitude of wake situations this data set displays a considerable scatter, which to a large degree seems to be caused by atmospheric boundary layer stability effects. Notable is also that rotating wind turbine components predominantly experience high fatigue loading for stable stratification with significant shear, whereas high fatigue loading of non-rotating wind turbine components are associated with unstable atmospheric boundary layer stratification.

1. Introduction

For wind farm (WF) *production* estimation *stationary* WF flow field modeling as provided by e.g. full CFD RANS models or fast linearized CFD RANS models [1] may suffice. However, for *load estimation* of wind turbines (WT's) exposed to wake affected inflow conditions, a *non-stationary* WF flow field description is inevitable. Insisting on models reflecting the *basic physics* of the problem, and considering direct numerical simulation (DNS) precluded due to its excessive computational demand, we are thus left with either high-fidelity CFD LES actuator disc or line a type of approaches or, alternatively, the medium-fidelity Dynamic Wake Meandering (DWM) [2], [3] type of model approach. CFD LES models must – like the DWM model – be linked to an aeroelastic model of each and every WT in a WF to provide a complete picture of the load conditions inside a WF for all design load cases. This is extremely CPU demanding and considered unrealistic for design purposes even with the capacity of nowadays very powerful state-of-the-ark super computer clusters. This challenge is further enhanced when including atmospheric boundary layer (ABL) stability as an additional design load case dimension and/or considering optimization of WF layout.

Being a medium-fidelity approach the DWM model offers significant savings in CPU-demands compared to the above mentioned non-stationary CFD alternatives, and it thereby facilitates detailed WF production/load design simulations to be conducted even when including the extra design load case dimension constituted by ABL stability. The "classical" (i.e. associated with neutral ABL conditions) DWM model is well established and about to be included in the revised IEC standard as a recommended practice. The model has been extensively validated both *directly* against full-scale wake affected flow measurements [5], [6], [7] and *indirectly* [8], [9] by using DWM together with the aeroelastic code HAWC2 [10] to compare predictions of WT production and loading with full-scale data. In [9] a huge scatter in the measured load data was observed, and the motivation for the present study is thus to investigate whether at least part of this scatter can be attributed to ABL stability effects as well as to get qualitative and quantitative insight in ABL stability effects on WT loading in WF's.

The core of the DWM model is a *split of scales* in the wake affected flow field, with large turbulence scales being responsible for stochastic *wake meandering* and small scales being responsible for wake *attenuation* and *expansion* in the meandering frame of reference as caused by turbulent mixing. Thus, essentially the DWM model assumes that the transport of wakes in the ABL can be modeled by considering the wakes to act as passive tracers driven by a combination of large-scale turbulence structures and a mean downstream advection velocity, adopting the Taylor hypotheses [4]. This basic split in scales further facilitates generalization of the DWM model to non-neutral ABL stability conditions in a simplistic way, as ABL stability is known primary to affect the low wave number turbulence regime.

The paper is structured as follows. First the generalization of the DWM model to non-neutral ABL conditions is briefly described in Section 2; then follows in Section 3 a description of the Lillgrunden case study, which previously have been used in an extensive validation study but, however, limited to neutral ABL conditions [9]. In Section 4 results from comparing model simulations with the Lillgrunden full-scale measurements are presented. Finally, conclusions are drawn in Section 5.

2. DWM for non-neutral conditions

The large scale turbulence structures, used to describe wake dynamics in the DWM model, are traditionally provided by a fast Navier-Stokes (NS) consistent turbulence model [11] which, however, assumes neutral ABL stability and thereby neglect the effects of buoyancy on turbulence production and thus turbulence characteristics. As ABL stability mainly affects the large scale turbulence structures [4], the effect of buoyancy on wake meandering cannot be neglected. In addition, the significance of buoyancy effects on small turbulence structures relevant for wake deficit expansion and attenuation in the meandering frame of reference must be clarified. The crucial question in this respect is whether the effect of buoyancy can be "decomposed" in analogy with the description of wake dynamics.

The starting point is the following fundamental conjecture launched in [12]: *In a wake context, ABL stability affects primary wake meandering driven by large (lateral and vertical) turbulent scales, whereas wake expansion in the meandering frame of reference is a second order effect only.* This conjecture has been investigated for a variety of stability conditions in [13] and [14], respectively. Based on full-scale LiDAR and sonic measurements as well as detailed numerical simulations the wake affected flow field behind a 500kW turbine was examined with focus on both organized wake deficit flow structures (resolved in the meandering frame of reference) and wake dynamics. These extensive investigations have justified the above stated conjecture, and we will therefore generalize the applicability of the DWM model to non-neutral ABL stability regimes by adopting this conjecture.

For consistency, we need a generalization of the classical spectral tensor to the non-neutral regime. Such a model has been developed and validated against full-scale data in [15] and [16]. Being a generalization of the classical Mann spectral tensor, this model share some of the basic features of the

Mann spectral tensor (e.g. linear mean wind speed shear and eddy life time model), but includes a linear vertical temperature profile as well as buoyancy terms in the linearized NS equations used in the Rapid Distorsion approach. The inclusion of buoyancy comes with a cost of two additional input parameters. The generalized spectral tensor is used to model both the large- and small scale ABL turbulence used with the DWM model.

3. The Lillgrunden case

The Lillgrund WF consists of 48 Siemens SWT-2.3-93 WT's, and one of these (C-8) is instrumented with strain gauges resolving blade, main shaft and tower loads, respectively. These measurements have been made available by Siemens Wind Power, and they constitute probably one of the most comprehensive sets of wake affected wind turbine load measurements ever recorded. The measurement period extends from 2008-06-03 to 2013-03-19 – i.e. over a period of almost 5 years.

The layout of the WF is shown in figure 1.

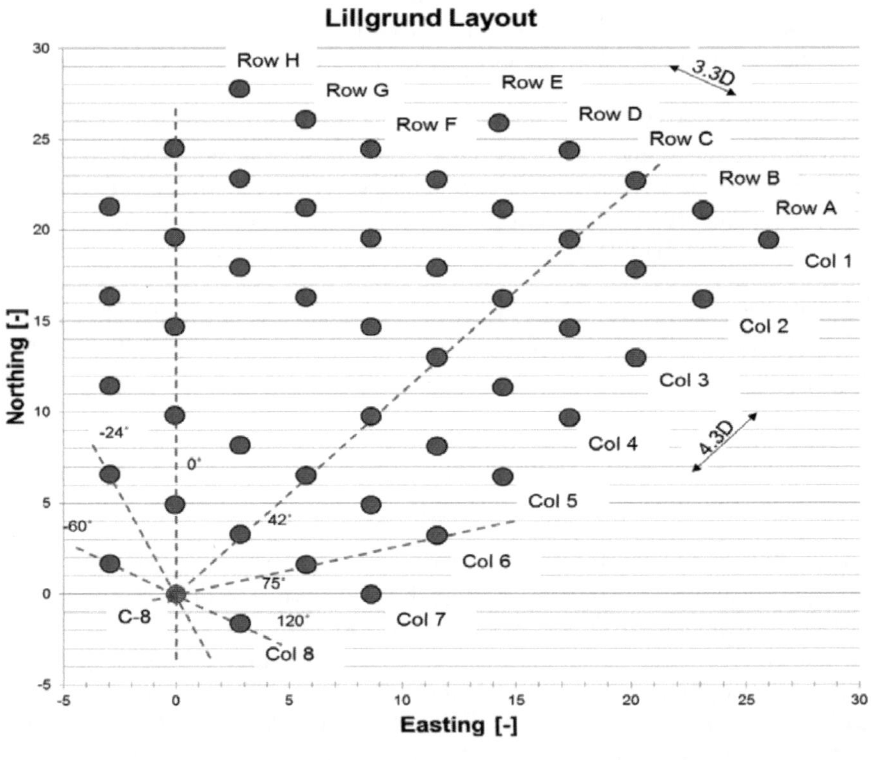

Figure 1. Layout of the Lillgrunden WF with the instrumented WT C-8 marked. Distances are non-dimensioned with the rotor diameter.

Whereas the Egmond aan Zee WF, previously investigated in [8], is characterized by a "conventional" WT inter spacing, the layout of the Lillgrunden WF is characterized by very small WT inter spacing's – i.e. down to 3.3 D. This makes the Lillgrunden WF especially interesting and challenging as a wake-load validation case, because the close spacing magnifies wake generated load effects compared to more traditional spaced WF's.

3.1. Measurements

WT C-8 is instrumented with strain gauges resolving respectively the blade root flap bending moment and the tower base for-aft bending moment. In addition to these high-frequency data, WT SCADA data (pitch setting, rotational speed and electrical power) were available for the WF WT's during the measuring period. Unfortunately no meteorological mast data was available within the recording period, but as the WT power and pitch angle are directly correlated with the inflow wind speed, the ambient undisturbed wind speed has been determined as based on power and pitch angle recordings for corner placed WT's located in free inflow conditions. A similar philosophy was used for estimating the ambient undisturbed wind direction, which was determined from nacelle orientations of corner placed WT's. These nacelle orientations were initially calibrated against power deficit polar's constrained to directions, where wake losses were dominated by the closest neighboring WT's.

Full polar load cases, associated with normal WT operation, are available for mean wind speeds ranging from 6m/s to 16m/s, and the data have been classified into 2m/s velocity bins. The blade root flap moments and the tower bottom for-aft moments have been post processed to fatigue equivalent moments using the Palmgren-Miner [17] approach and subsequently normalized with the respective fatigue equivalent moments associated with a inflow wind speed of 9m/s – i.e. here represented by mean equivalent moments associated with the velocity bin [8;10]m/s. Wöhler exponents of 5 and 10 were assumed for the tower and blade composite structures, respectively.

3.2. Simulations

For the DWM/HAWC2 validation study the load response of WT C-8 is simulated for mean wind speeds reflecting the median of the defined velocity bins. Measured wind speed dependent turbulence intensities (TI's) are used, reflecting the offshore wind speed dependent "surface" roughness. However, no attempt is done to resolve TI as function of upstream fetch (i.e. direction). Thus, in the mean wind speed regime 6m/s-14m/s a TI of 5.8% is used – gradually increasing to 6.2% at 16m/s. Both the *deterministic* and the *stochastic* part of the inflow are affected by ABL stability conditions. The deterministic mean wind shear and the stochastic turbulence input are described in the following.

3.2.1. Mean wind shear

The mean wind shear is stability dependent through the buoyancy impact on mixing in the ABL layer – in stable conditions with little mixing the shear profile gradient tend to increase, whereas in unstable conditions with increased mixing the shear profile tend to become more uniform.

In the IEC-61400 code [18] the *neutral* wind shear is specified in terms of a power law approach as

$$U(z) = U_{hub}(z/z_{hub})^{\alpha} \tag{1}$$

with U being the mean wind speed, z being the height above terrain, and z_{hub} being the hub height. The power exponent α is specified to 0.14 for an offshore site [19]. However, the new (not yet approved) revised version of the IEC code opens alternatively also for use of a log-profile defined as

$$U(z) = U_{hub}\frac{Ln(z/z_0)}{Ln(z_{hub}/z_0)} \tag{2}$$

with z_0 being the roughness length. The log-profile is physical consistent and used in Monin-Obukhov (MO) similarity theory [4] and, as seen, it takes the value 0 at $z = z_0$ and the value U_{hub} at $z = z_{hub}$. However, a recommended roughness length for offshore sites, consistent with the specifications of the power law exponent in equation (1), is not yet specified. We will therefore use equation (2) to introduce the effect of atmospheric stability in the power law description in an approximate manner.

As a first step we rewrite equation (2) as

$$U(z) = \frac{U_{hub}}{Ln(z_{hub}/z_0)}Ln(z/z_0) = \frac{u_*}{\kappa}Ln(z/z_0) \tag{3}$$

where the last identity provides the link to the conventional description of the log-profile, with u_* being the friction velocity, and κ being von Karman's constant (= 0.4), viz.

$$\frac{u_*}{\kappa} = \frac{U_{hub}}{Ln(z_{hub}/z_0)} \tag{4}$$

To establish a connection between the IEC specified power coefficient, α, and the roughness length, z_0, we will require the best possible agreement of respectively the power law profile and log profile over the vertical extent of the rotor, and we will define this best fit in terms of minimum least square deviation. For this purpose we define the functional Π as

$$\Pi(z_0) = U_{hub}{}^2 \int_{z_{hub}-R}^{z_{hub}+R} \left[\frac{Ln(z/z_0)}{Ln(z_{hub}/z_0)} - \left(\frac{z}{z_{hub}}\right)^{\alpha}\right]^2 dz \tag{5}$$

and define z_0 accordingly as the particular value, $\hat{z}_0$ that minimize the above functional, viz.

$$\hat{z}_0 = \left\{\hat{z}_0 \middle| \Pi(\hat{z}_0) = \min_{z_0}\{\Pi(z_0)\}\right\} \tag{6}$$

Adopting M-O similarity theory, the mean wind shear in *stable conditions* is traditionally formulated as [4]

$$\begin{aligned} U(z) &= \frac{u_*}{\kappa}\left[Ln\left(\frac{z}{z_0}\right) - \psi_m^{(s)}\left(\frac{z}{L_M}\right)\left(1 - \frac{z}{2z_i}\right)\right] \\ &\approx \frac{u_*}{\kappa}\left[Ln\left(\frac{z}{z_0}\right) - \psi_m^{(s)}\left(\frac{z}{L_M}\right)\right] \\ &= \frac{u_*}{\kappa}\left[Ln\left(\frac{z}{z_0}\right) + 5\frac{z}{L_M}\right] \; ; 0 < z/L_M \le 1 \end{aligned} \tag{7}$$

where L_M is the M-O length, z_i is the height of the atmospheric boundary layer, and $\psi_m^{(s)}$ is the stability function associated with stable stratification. Based on the reflections concerning the neutral shear profile, we will now "transform" relation (7) into a power law formulation for the stable ABL condition using eq. (6). The resulting expression is

$$U(z) = U_{hub}(z/z_{hub})^{\alpha} + \frac{5U_{hub}}{Ln(z_{hub}/z_0)}\frac{z}{L_M} \; ; 0 \le z/L_M \le 1 \tag{8}$$

where the value of the roughness length, z_0, will relate to the power law exponent as described in equations (5) and (6).

Again adopting M-O similarity theory, the mean wind shear under *unstable conditions* may be formulated as [4], [20]

$$\begin{aligned} U(z) &= \frac{u_*}{\kappa}\left[Ln\left(\frac{z}{z_0}\right) - \psi_m^{(us)}\left(\frac{z}{L_M}\right)\right] \\ &= \frac{u_*}{\kappa}\left[Ln\left(\frac{z}{z_0}\right) - Ln\left[\frac{1+x^2}{2}\left(\frac{1+x}{2}\right)^2\right] + 2Atan(x) - \frac{\pi}{2}\right] \; ; -2 \le z/L_M < 0; \\ x &= \left(1 - 16\frac{z}{L_M}\right)^{1/4} \end{aligned} \tag{9}$$

Taking the same approach as for the stable stratification, we arrive at the following approximate expression when taking a power law approach

IOP Conf. Series: Journal of Physics: Conf. Series **854** (2017) 012026 doi:10.1088/1742-6596/854/1/012026

$$U(z) = U_{hub}\left(\frac{z}{z_{hub}}\right)^{\alpha} + \frac{U_{hub}}{Ln(z_{hub}/z_0)}\left[-Ln\left[\frac{1+x^2}{2}\left(\frac{1+x}{2}\right)^2\right] + 2Atan(x) - \frac{\pi}{2}\right]; \quad -2 \le z/L_M < 0 \tag{10}$$

where, as in the stable case, the value of the roughness length, z_0, relates to the power law exponent as described in equations (5) and (6).

3.2.2. Turbulence

Using the coupled DWM-HAWC2 platform [3] requires simulation of three different turbulence fields – a large scale turbulence field dictating the wake meandering, a traditional turbulence field accounting for conventional WT turbulence loading, and an *isotropic* and *inhomogeneous* small scale turbulence field accounting for WT loading caused by wake self generated turbulence. The wake self generated turbulence field is assumed to be invariant with respect to the ABL stability condition, and it is simulated using the classic Mann turbulence simulator [11]. The two remaining turbulence fields, which are highly dependent on ABL stability conditions, are generated based on the generalized buoyancy dependent spectral tensor [15], [16], which degenerates to the classic spectral tensor for neutral stability conditions. In all cases the turbulence fields are simulated as three dimensional fields resolved in suitable Cartesian grid configurations.

The parameters of the buoyancy dependent spectral tensor are, for each stability class, obtained from fitting model auto- and cross spectra to respective spectra obtained from full-scale sonic measurements from the Høvsøre site in Denmark. Fits are performed based on data recorded at 60m and 40m altitudes, respectively. For both altitudes, data associated with the [8;9]m/s mean wind speed bin were used for the parameter fitting. However, the fits at 40m seem to fit slightly better the full-scale data compared to the 60m data in especially the neutral and the unstable regimes. As 40m altitude moreover is more likely to be within the surface layer, where L_M is defined, than the 60m altitude (especially for stable stratification), it was decided to base input parameters to the generalized spectral tensor on the 40m data fits.

Except for the turbulence intensity (described by the $\alpha\varepsilon^{2/3}$ parameter of the spectral tensor *when* the remaining parameters have been fixed), it was further decided to adopt the (neutral) turbulence input specifications from the IEC code to [18] mimic the turbulence conditions at the Lillgrunden site, where no high-frequency meteorological measurements, as mentioned, are available. The IEC spectral tensor input parameters were originally obtained by fitting the spectral tensor spectra to a target Kaimal spectra [18].

Since we have decided to base the neutral turbulence generation partly on IEC specifications of turbulence tensor input (i.e. turbulence length scale L and eddy life time parameter Γ), the spectral tensor input parameters for neutral conditions do not match the fitted parameter values from the Høvsøre full-scale data. As a consequence, it was decided to scale (L, Γ) for the non-neutral (i.e. diabatic) conditions accordingly, although this approximation might not be completely true. However, it is believed to be a fair approximation – alternatively the neutral case could defined by the directly fitted values for L and Γ at the Høvsøre site, which is just another approximation to the Lillgrunden conditions.

4. Results

The present analysis deviates from the analysis presented in [9] by the inclusion of ABL stability affected turbulence and shear in the load simulations. Two non-neutral stability cases, corresponding to Richardson number Ri = -0,015 (unstable ABL stratification) and Ri = 0,01 (stable ABL stratification) have been simulated. Based on such numerical predictions, conducted for various ABL stability conditions, it is possible to get insight in how ABL stability affects the WT load levels, as well as to what extend ABL stability can explain the rather high scatter level in the measurements.

Results are shown in figure 2 for below rated wind speed WT operation and in figure 3 for above rated wind speed WT operation.

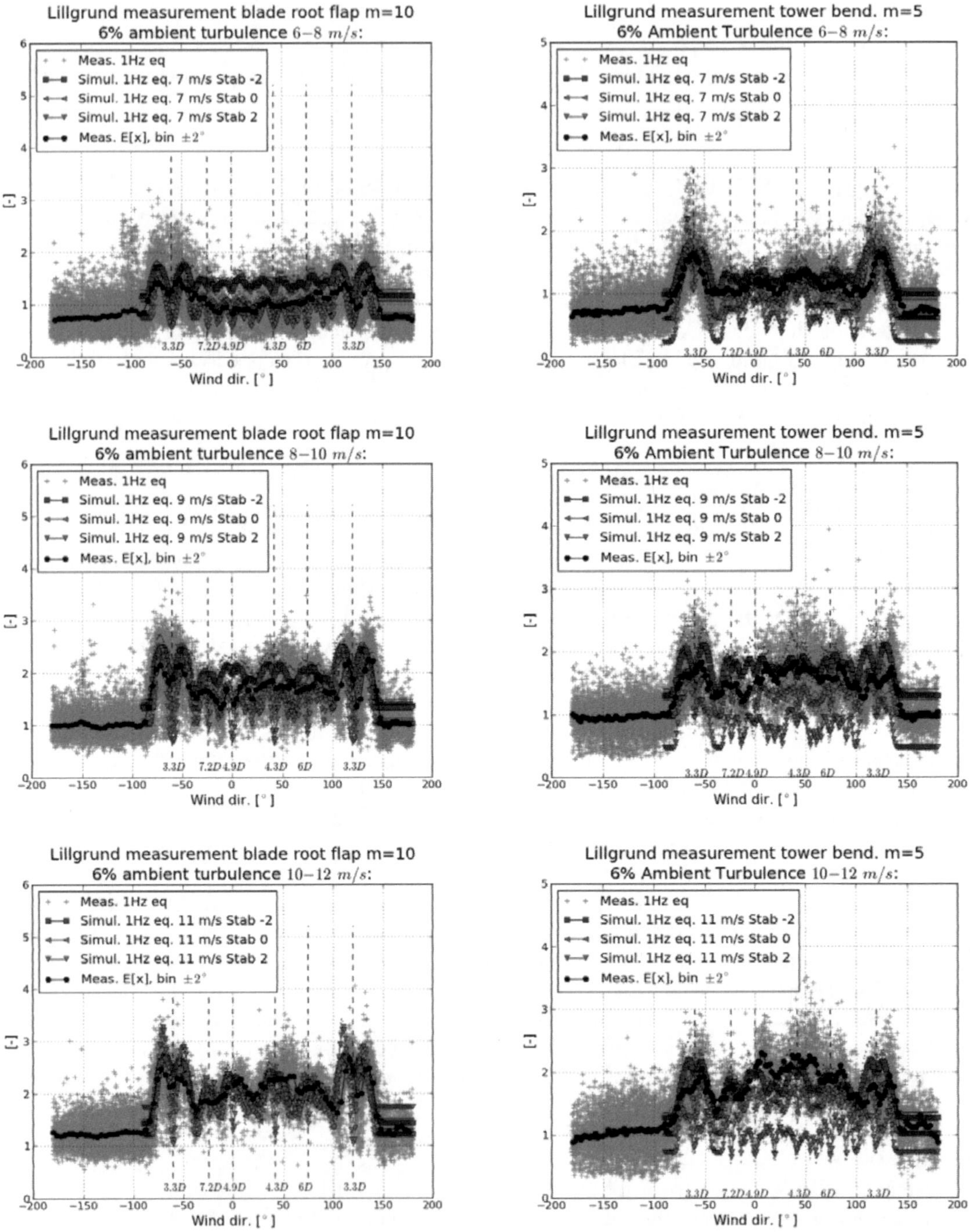

Figure 2: Comparison of measured and simulated fatigue loads for the C-8 turbine below rated wind speed. With reference to Table 1, simulations are conducted for unstable (-2), neutral (0) and stable (2) ABL stability conditions using the maximum deficit wake merging operator.

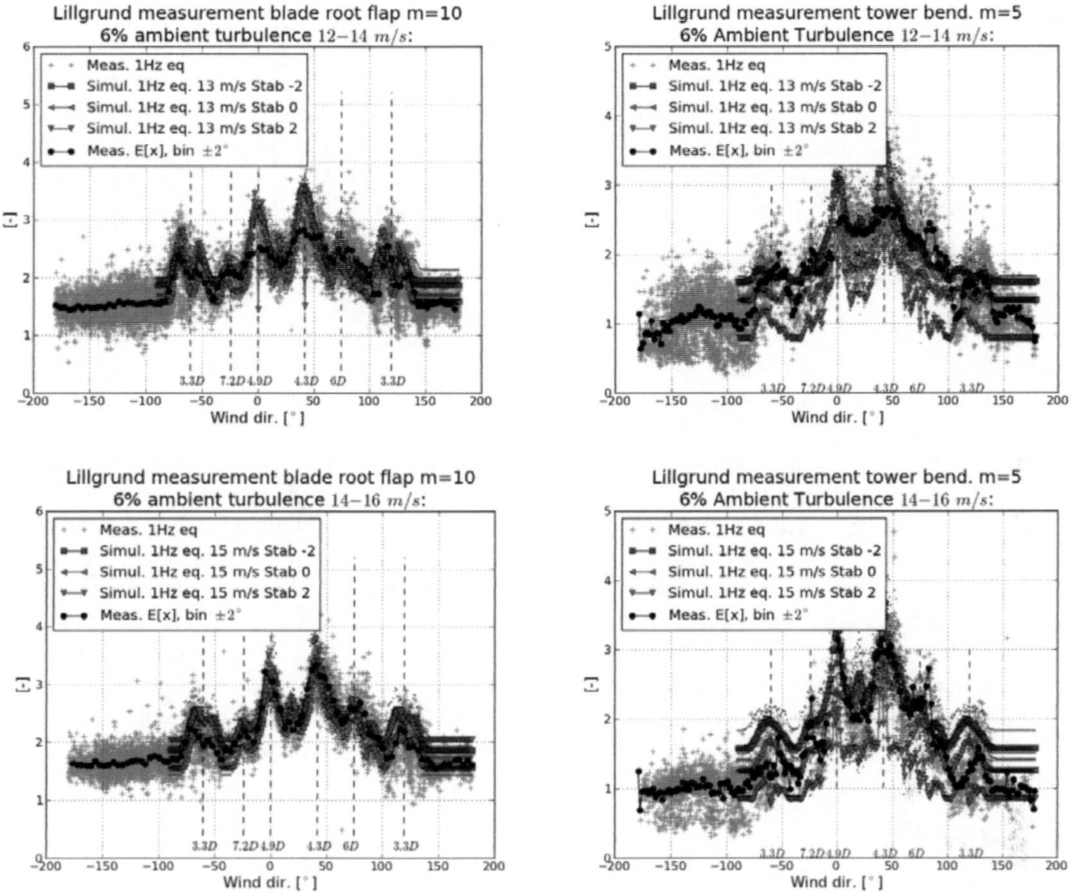

Figure 3: Comparison of measured and simulated fatigue loads above rated wind speed. With reference to Table 1, simulations are conducted for unstable (-2), neutral (0) and stable (2) ABL stability conditions using the linear perturbation deficit wake merging operator.

For all wind speeds it can be seen that the *tower bottom fatigue bending moments* are largest in unstable and smallest in stable ABL conditions. This is easily explained by the difference in turbulence intensity associated with the stability conditions, where unstable conditions causes a turbulence level of 9.8%, neutral conditions corresponds to a turbulence level of 6.0%, and stable conditions corresponds to a turbulence level of 1.99%.

In the inflow sector 150deg.-180deg., the investigated WT is not affected by upstream wakes, and it is thus possible to evaluate the impact from ABL stability on solitary WT loading. When observing the blade *flapwise fatigue loads* in this free sector, the results are significantly different than the tower fatigue loading. Contrary to tower loading, the blade loads are sensitive to both shear and turbulence levels, as both cause varying wind speeds over the rotor area. For rotating WT components these are *counter acting effects*, because stable conditions result in large shear and low turbulence intensity and vice versa for unstable conditions. For most wind speeds, it is seen that *both* stable and unstable conditions result in a larger fatigue loads than the neutral case. The only exception is at very low wind speed, where neutral and stable conditions result in approximately identical fatigue load levels, whereas larger fatigue load levels are observed for unstable conditions. Similar results were obtained in [21].

In the wake affected conditions the same trend, with increased tower fatigue loads in unstable conditions and decreased tower fatigue loads in stable conditions, are seen. Also for wake affected inflow conditions the neutral situation seems to cause the smallest blade fatigue loads. However, for *single wake* situations it is interesting to notice, that stable ABL conditions result in both the largest fatigue load level (i.e. in a half-wake situations) and the smallest load level (i.e. in the full wake situation) compared to unstable and neutral ABL conditions. Finally, it is noted that a significant part of the observed measurement scatter potentially can be explained by ABL stability effects, especially remembering that *very stable* and *very unstable* stability conditions are not included in the present analysis.

5. Conclusion

A medium-fidelity DWM/HAWC2 framework for modelling of wind turbine component fatigue loading, when the wind turbines are subjected to wake affected *non-stationary* flow fields under *non-neutral* atmospheric stability conditions, has been developed and demonstrated on full-scale load data from the Lillgrunden WF.

A comparative analysis of measured and simulated blade and tower fatigue loads has revealed that ABL stability effects can explain a significant part of the large measurement scatter observed in full-scale data. The study has further gained insight in important load mechanism, and shown that mean wind speed shear is the dominating fatigue load driver for the rotating WT components in stable ABL conditions, whereas turbulence (including its dictating wake meandering mechanism) is the dominating load driver for tower fatigue loading in general.

After conclusion of this study wind speed and temperature recordings, covering the major part of the measurement period, have been made available from the nearby Drogden lighthouse. In a future study, it is planned to use these measurements to classify the Lillgrunden measurements into stability classes, thus facilitating a one-to-one comparison of ABL stability affected measurements and full-scale data.

Acknowledgements

Siemens Wind Power has provided the unique set of Lillgrunden load measurements used for this study, which is gratefully acknowledged. Further, the EUDP project “Impact of atmospheric stability conditions on wind farm loading and production” under contract 64010-0462 is acknowledged for financial support to the study.

References

[1] Ott S, Berg J and Nielsen M (2011). Linearised CFD Models for Wakes, Risoe-R-1772(EN).

[2] Larsen GC, Madsen HAa, Thomsen K and Larsen TJ (2008). Wake meandering - a pragmatic approach. Wind Energy, **11**, pp. 377–395.

[3] Madsen HAa, Larsen GC, Larsen TJ and Troldborg N (2010). Calibration and validation of the dynamic wake meandering model for implementation in an aeroelastic code. Journal of Solar Energy Engineering, **132(4)**: 041014-1–041014-14. DOI: 10.1115/1.4002555.

[4] Kaimal JC and Finnigan JJ (1994). Atmospheric Boundary layer Flows – Their structure and Measurement. Oxford University Press.

[5] Bingöl F, Mann J and Larsen GC (2010). Light detection and ranging measurements of wake dynamics. Part I: one-dimensional scanning. Wind Energy, **13(1)**, pp. 51–61.

[6] Trujillo J, Bingöl F, Larsen GC, Mann J and Kühn, M (2011). Light detection and ranging measurements of wake dynamics. Part II: two-dimensional scanning. Wind Energy, **14** pp. 61–75.

[7] Machefaux E, Larsen GC, Troldborg N, Gaunaa M and Rettenmeier A (2015). Empirical modeling of single-wake advection and expansion using full-scale pulsed lidar-based measurements. Wind Energy, **18**, pp. 2085–2103.

IOP Conf. Series: Journal of Physics: Conf. Series **854** (2017) 012026 doi:10.1088/1742-6596/854/1/012026

[8] Larsen TJ, Madsen HAa, Larsen GC and Hansen KS (2012). Verification of the Dynamic Wake Meander Model for Loads and Power Production in the Egmond aan Zee Wind Farm. Wind Energy, **16**, pp. 605–624.

[9] Larsen TJ, Larsen GC, Madsen HAa, Thomsen K and Pedersen SM (2015). Wake effects above rated wind speed - An overlooked contributor to high loads in wind farms. Scientific Proceedings, EWEA Annual Conference and Exhibition 2015, pp. 95-99.

[10] Larsen TJ and Hansen AM (2007). How to HAWC2, the Users Manual, Risø-R-1597(EN), Risø National Laboratory - Technical University of Denmark.

[11] Mann J (1998). Wind field simulation**.** Probabilistic Engineering Mechanics, **13**, pp. 269-282.

[12] Larsen GC, Larsen TJ, Mann J, Peña A, Hansen KS and Madsen HAa (2009). The dependence of wake losses on atmospheric stability characteristics. EUROMECH Colloquium 508 on Wind Turbine Wakes 2009; pp. 35–37.

[13] Larsen GC, Machefaux E and Chougule A (2015). Wake meandering under non-neutral atmospheric stability conditions – theory and facts. Journal of Physics: Conference Series (Online), **625**, 012036.

[14] Machefaux E, Larsen GC, Tilman K, Troldborg N, Kelly MC, Chougul A, Hansen KS and Rodrigo JS (2016). An experimental and numerical study of the atmospheric stability impact on wind turbine wakes**.** Wind Energy, **19**, pp. 1785–1805.

[15] Chougule A, Mann J, Kelly MC and Larsen GC (2016). Modeling atmospheric turbulence via rapid-distortion theory: spectral tensor of velocity and buoyancy. Journal of the Atmospheric Sciences, Vol. **74**, Issue 4, pp. 949–974.

[16] Chougule A, Mann J, Kelly MC and Larsen GC (2016). Validation of buoyancy driven spectral tensor model using HATS data. Journal of Physics: Conference Series (Online), Vol. **753**, No. 3, 032038.

[17] Miner (1945) MA. Cumulative damage in fatigue. Journal of Applied Mechanics, 12, pp. 159–164.

[18] IEC 61400-1. Wind turbines—part 1: Design requirements, 2005.

[19] IEC 61400-3. Wind turbines – part 1: Design requirements for offshore wind turbines, 2009.

[20] Paulson CA (1970). The mathematical representation of wind speed and temperature profiles in the unstable surface layer. J. Appl. Meteor., 9, 857-861.

[21] Sathe A, Mann J, Barlas T, Bierbooms WAAM and van Bussel GJW (2013). Influence of atmospheric stability on wind turbine loads. Wind Energy, Vol. **16**, Issue 7, pp. 1013–1032.

IOP Conf. Series: Journal of Physics: Conf. Series **854** (2017) 012027 doi:10.1088/1742-6596/854/1/012027

Validation of the Dynamic Wake Meander model with focus on tower loads

T.J. Larsen[1], G.C. Larsen[1], M.M. Pedersen[1], K. Enevoldsen[1], H.A. Madsen[1]

[1]Technical University of Denmark, Department of Wind Energy, tjul@dtu.dk

Abstract. This paper presents a comparison between measured and simulated tower loads for the Danish offshore wind farm Nysted 2. Previously, only limited full scale experimental data containing tower load measurements have been published, and in many cases the measurements include only a limited range of wind speeds. In general, tower loads in wake conditions are very challenging to predict correctly in simulations. The Nysted project offers an improved insight to this field as six wind turbines located in the Nysted II wind farm have been instrumented to measure tower top and tower bottom moments. All recorded structural data have been organized in a database, which in addition contains relevant wind turbine SCADA data as well as relevant meteorological data – e.g. wind speed and wind direction – from an offshore mast located in the immediate vicinity of the wind farm. The database contains data from a period extending over a time span of more than 3 years. Based on the recorded data basic mechanisms driving the increased loading experienced by wind turbines operating in offshore wind farm conditions have been identified, characterized and modeled. The modeling is based on the Dynamic Wake Meandering (DWM) approach in combination with the state-of-the-art aeroelastic model HAWC2, and has previously as well as in this study shown good agreement with the measurements. The conclusions from the study have several parts. In general the tower bending and yaw loads show a good agreement between measurements and simulations. However, there are situations that are still difficult to match. One is tower loads of single-wake operation near rated ambient wind speed for single wake situations for spacing's around 7-8D. A specific target of the study was to investigate whether the largest tower fatigue loads are associated with a certain downstream distance. This has been identified in both simulations and measurements, though a rather flat optimum is seen in the measurements.

1. Introduction

Fatigue loads on wind turbines are mainly driven by the varying inflow condition, where especially turbulence and shear are the most dominating contributors. Especially for offshore turbines, the ambient turbulence levels are typically small with average turbulence intensity around 6%. As offshore turbines are typically placed in wind farms, the turbulence driven fatigue contribution is, depending on the wind farm layout, dominated by wake effects of the neighboring wind turbines. It is therefore important to be able to model these effects with high accuracy to ensure the right load level and to be able to utilize new advanced passive or active load reducing strategies to alleviate the wake induced loading. Usually these effects are included in simulations using the method from Frandsen [1],[2], which basically adjust the turbulence level based on the distance to the nearest wind turbine. This may be a sufficient method for many applications and site evaluations, especially when the turbine spacing is close to 8 diameters (D), which was the distance the model was originally derived for. The method used in this paper is however based on the Dynamic Wake Meander (DWM) approach [3],[4],[5],[6] where the wake is modeled as a wind speed deficit that is transported from the upstream turbine in a meandering way prescribed by the large turbulence structures. This model has been validated based on full scale measurements in [5],[6]. An interesting consequence of the wake meandering approach is that for very short distances, the meandering is limited, but the deficit is large, i.e. the downwind turbine to have a low thrust and power production while the blade load variations are high in half-wake situations. The tower load variations are not expected to increase, as this is mainly sensitive to varying thrust of the full rotor – typically not seen at short distances (<3D). For very large spacing (>15D) the deficit is reduced to a degree where the load impact on the downstream turbine is minimal.

Consequently, the tower load level reach its maximum level between 3D and 15D, which is the typical spacing in most wind farms today. To see if this can be observed experimentally, the Nysted II experiment was setup with tower load measurement on several turbines with different spacing, so that a broad range of distances could be covered. The result of this study, together with at general load comparison, is presented in this paper.

2. Experimental setup

The Nysted II wind farm consist of 90 Siemens SWT2.3 PRVS turbines installed in the southern part of Denmark, see Figure 1. A total of 6 turbines were instrumented with strain gauges in top and bottom of the tower. The available measurements can be divided into 5 data classes, each with their own characteristics, data coverage and time stamp reference.

1) High frequency sampled meteorological data.
2) High frequency sampled WT structural strain-gauge data.
3) Time series data from WT Supervisory Control And Data Acquisition (SCADA) systems.
4) Statistical 10-minute meteorological data.
5) Statistical data from the WT SCADA systems.

The high frequency sampled (i.e. 20 Hz) meteorological data consist of recordings from 3 Sonics mounted on an off-shore meteorological mast nearby the WF in altitudes 17m, 40m and 57m above mean sea level (AMSL), respectively. The high sampled (i.e. 40 Hz) structural data consist of strain-gauge measurements from the 6 instrumented turbines (L1, L2, L3, M1, M2, K18) – cf. Figure 1. For each WT, the strain-gauge setup resolves two WT tower bottom bending moments and two WT tower top bending moments in mutual perpendicular directions as well as the tower top torsion (i.e. yaw) moment. It was however problematic to use the tower top bending moment as local stress concentrations affecting the strain gauges removed the usable signal from several sectors. The analyses in this paper are therefore based on the tower bottom bending moment and the tower top yaw moment.

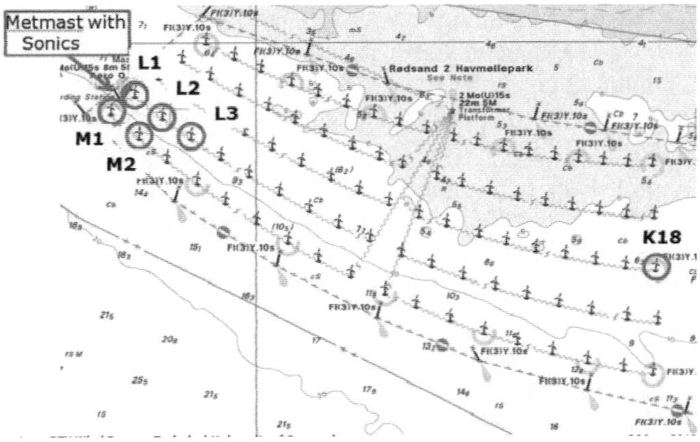

Figure 1 Site overview with instrumented WTs indicated.

3. Turbine and ambient inflow conditions

The baseline is a model of a 2.3MW pitch regulated turbine equipped with the DTU controller [14] in order to mimic as close as possible the Siemens 2.3MW WTs at the Nysted II site. Simulations have been performed for mean wind speeds ranging between 6m/s and 24m/s with increments of 2m/s. The ambient turbulence level has been chosen to 6%, corresponding approximately to the mean turbulence level at different previously investigated offshore WFs (Egmond aan Zee, Lillgrund, Horns Rev I) as well as to the mean of the undisturbed ambient turbulence intensity level at the present Nysted II site. In Figure 3 the measured turbulence intensity has been given for the full polar (i.e. [0°; 360°]), and as seen significant systematic variations are observed. These are partly caused by upstream (turbulence generating) WTs (i.e. L1, M1, M2, K1 etc.), partly to variations in the upstream fetch conditions e.g. the upstream roughness element constituted by summer cottage area in direction 340°.

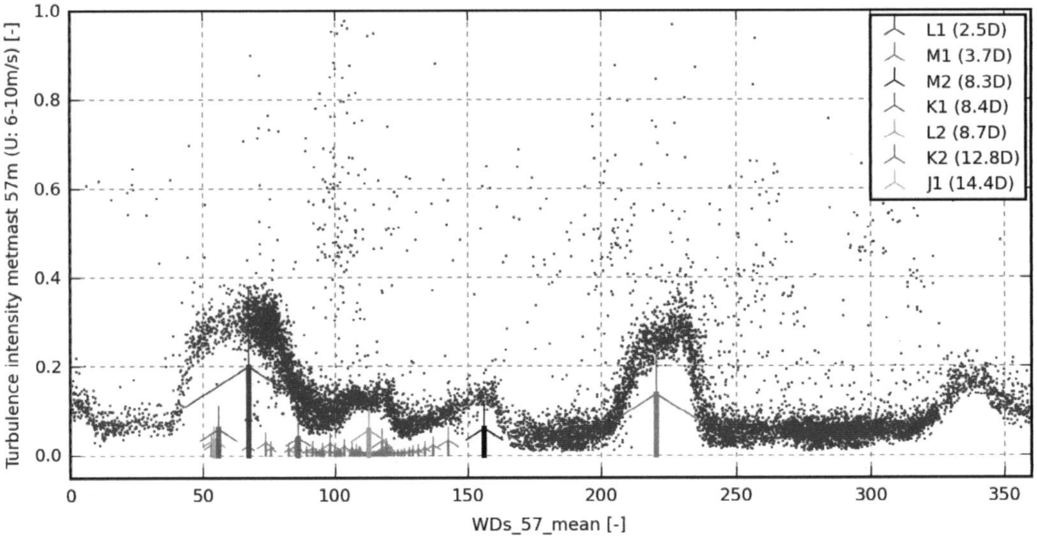

Figure 2. Measured turbulence intensity levels at the Nysted II site.

4. The numerical model

The DWM model complex [3] is based on the combination of three cornerstones: (1) Modeling of quasi-steady wake deficits; (2) a stochastic model of the downwind wake meandering; and (3) added or self-generated wake turbulence, see Figure 3.

The quasi-steady wake deficit is the wake deficit formulated in the moving (i.e. meandering) frame of reference and includes the wake expansion as a function of downstream transportation time caused partly by turbulence diffusion and partly by recovery of the rotor pressure field. The modeling of this deficit is based on a thin shear layer approximation of the Navier–Stokes equations in their rotational symmetric form combined with an eddy viscosity closure. The initial condition is constituted by the induced wind field in the rotor plane determined from a BEM approach. In the present formulation, the aerodynamic module of HAWC2 is used for this purpose. Further details on the implementation can be found in Madsen et al. [4].

The wake meandering part is based on a fundamental presumption stating that the transport of wakes in the atmospheric boundary layer can be modeled by considering the wakes to act as passive tracers driven by the large-scale turbulence structures in lateral and vertical directions [3]. Modeling of the meandering process consequently includes considerations of a suitable description of the 'carrier' stochastic transport media as well as a suitable definition of the cutoff frequency defining large-scale turbulence structures in this context. For the stochastic modeling of wake meandering, we imagine the wake as being constituted by a cascade of wake deficits, each 'emitted' at consecutive time instants in agreement with the passive tracer analogy [3], [4]. We then subsequently describe the propagation of each of the 'emitted' wake deficits, and the collective description of these thus constitutes the wake meandering model.

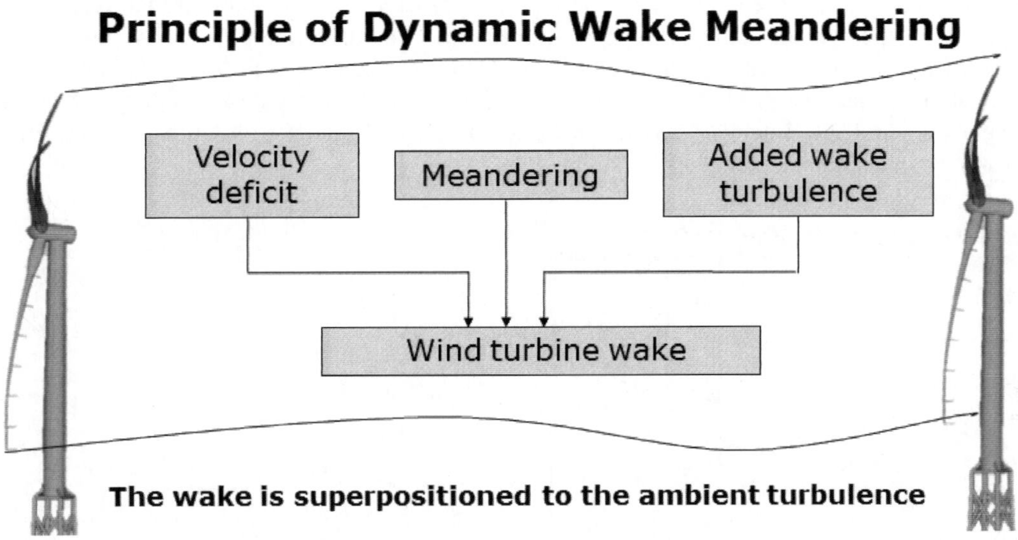

Figure 3. Overview of the three fundamental parts of the DWM model.

Adopting Taylor's hypothesis [7], the downstream advection of these is assumed to be controlled by the mean wind speed of the ambient wind field. With this formulation, the wake momentum in the direction of the mean flow is invariant with respect to downstream displacement. This is a considerable simplification allowing for a straight forward decoupling of the wake along the wind deficit profile (and its expansion) and the wake transportation process. As for the dynamics in the lateral and vertical directions, each considered wake cascade element is displaced according to the large-scale lateral and vertical turbulence velocities at the position of the particular wake cascade element at each time instant. The choice of a suitable stochastic turbulence field, that in turn defines the stochastic wake transport process, is not mandatory, but may be guided by the characteristics of the atmospheric turbulence at the site of relevance. These characteristics encompass in principle not only turbulence standard parameters such as turbulence intensity, turbulence length scale and coherence properties, but also features such as degree of isotropy, homogeneity of the turbulence, Gaussianity of the turbulence, etc.

The DWM model has been implemented into the aeroelastic code HAWC2. This code is based on a multibody formulation as described by Shabana [8] with Timoshenko beam elements as described in [9]. The aerodynamic model is based on an extended BEM model as described in [10]. Validation can be found in [5],[6],[11],[12].

5. Validation:

Measurement are compared to simulations with respect to 1Hz equivalent fatigue loads, see Figure 4-13. Note that all loads are non-dimensionalised with the fatigue load level in the free sector at 9m/s. When comparing the tower bottom bending loads (Figure 4-7) at 8m/s a small offset can be seen in the free sector between the measurement and simulations, however when comparing the relative load level in the wake sectors to the free sector there is an excellent agreement between measurement and simulation. This is yet another indication that the DWM model captures the wake effect very well. The tower yaw loads also have a very fine agreement, which can be seen in Figure 8-9. When comparing loads levels at rated wind speed at 12m/s (Figure 10-11) there is a fine agreement in the multiwake

IOP Conf. Series: Journal of Physics: Conf. Series **854** (2017) 012027 doi:10.1088/1742-6596/854/1/012027

sector, which is mainly caused by a change in the approach the individual wakes are superimposed, see more about this in [6]. However, the simulated load levels in the single wake situation at 12m/s are significantly underestimated. The reason for this may very well be caused by the simplified way the deficit is evaluated. In the present approach, the deficit is based on the assumption that the upstream turbine operates at the average setting of pitch and rotor speed corresponding to the average ambient wind speed. As the controller has a highly nonlinear influence on the thrust of the turbine, it seems that this leads to slightly underestimated load levels. One solution could simply be to subdivide the results into more narrow wind speed bins, or perhaps to expand the DWM model with a time varying deficit. The latter will be invested in future work.

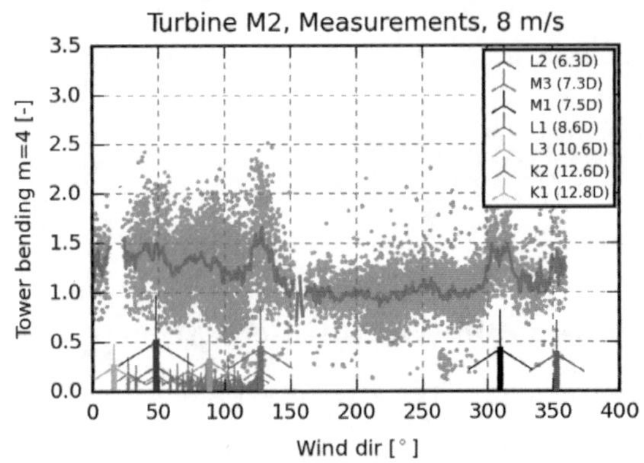

Figure 4. Measured tower bottom bending equivalent moment; WT M2; m = 4; U = 8m/s.

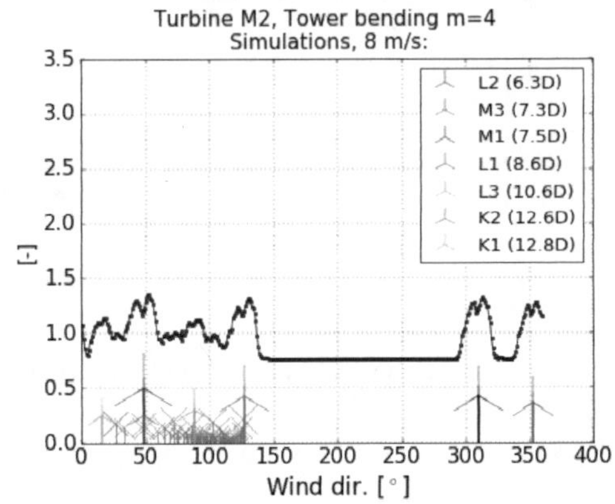

Figure 5. Simulated tower bottom bending equivalent moment; WT M2; m = 4; U = 8m/s.

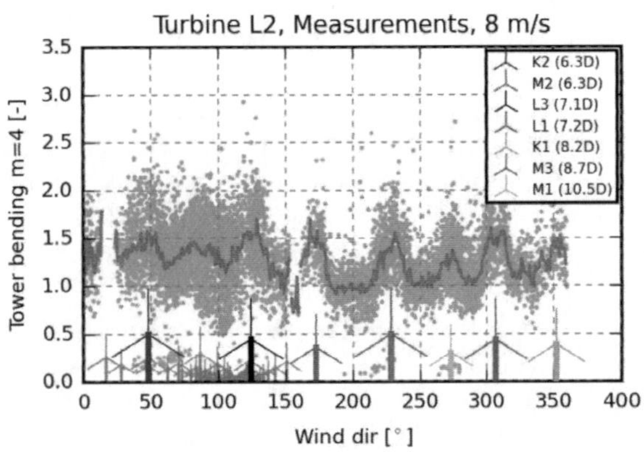

Figure 6. Measured tower bottom bending equivalent moment; WT L2; m = 4; U = 8m/s.

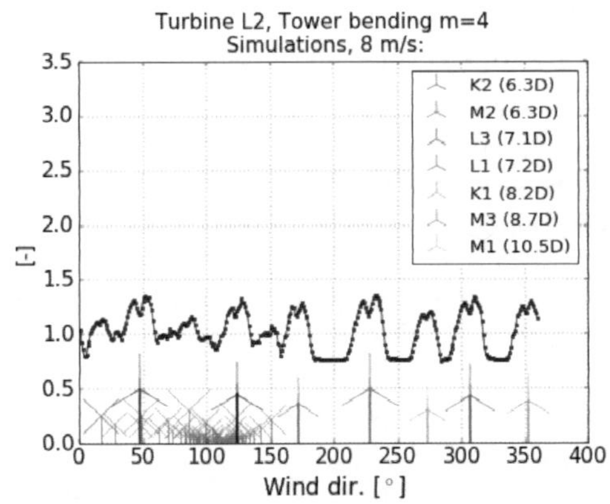

Figure 7. Simulated tower bottom bending equivalent moment; WT L2; m = 4; U = 8m/s.

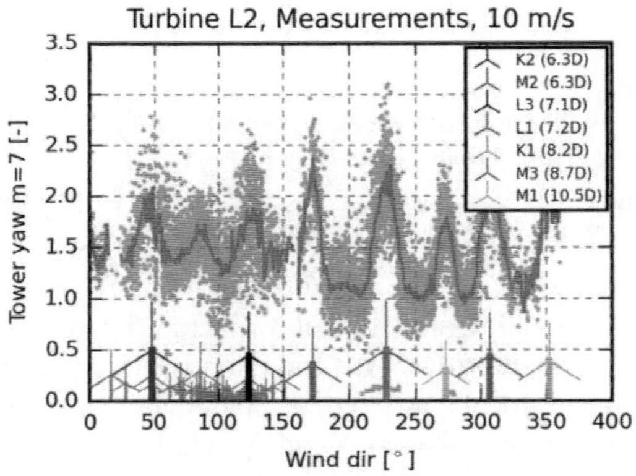

Figure 8. Measured tower top yaw equivalent moment; WT L2; m = 7; U = 10m/s.

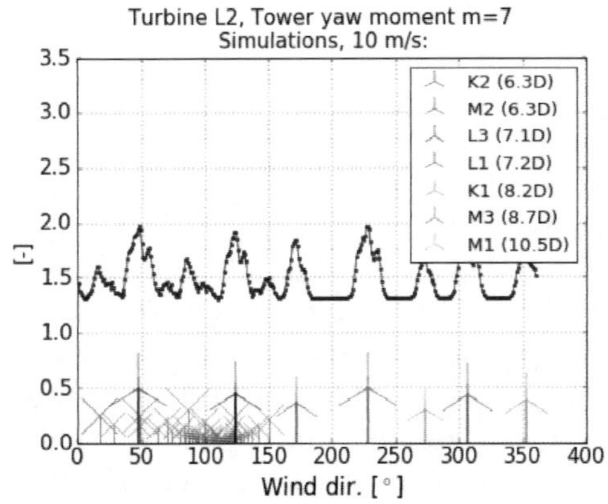

Figure 9. Simulated tower top yaw equivalent moment; WT L2; m = 7; U = 10m/s.

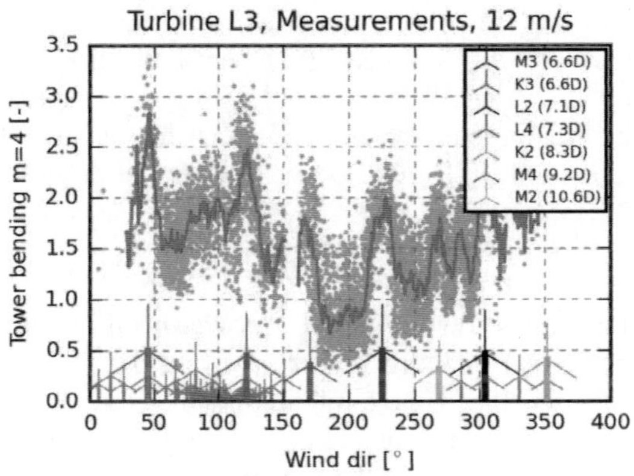

Figure 10. Measured tower bottom bending equivalent moment; WT L3; m = 4; U = 12m/s.

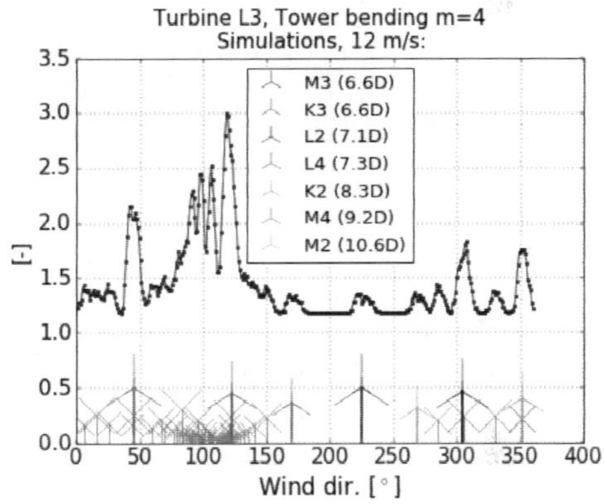

Figure 11. Simulated tower bottom bending equivalent moment; WT L3; m = 4; U = 12m/s.

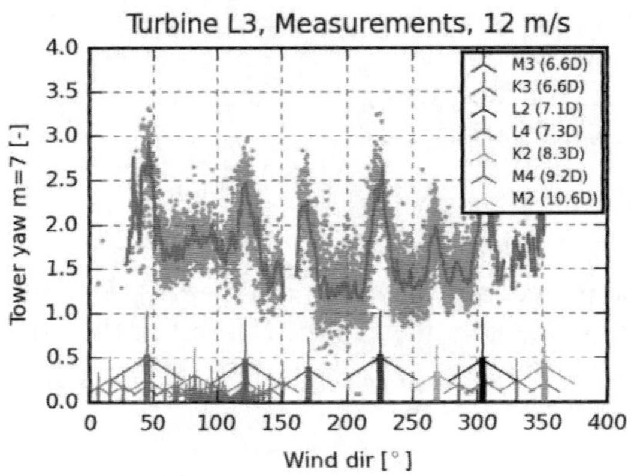

Figure 12. Measured tower top yaw equivalent moment; WT L3; m = 7; U = 12m/s.

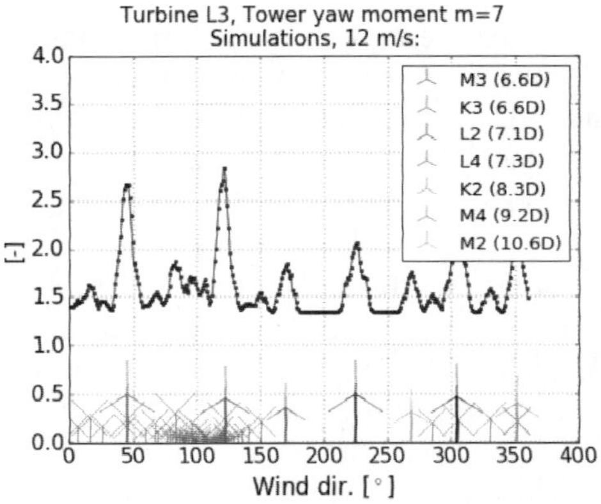

Figure 13. Simulated tower top yaw equivalent moment; WT L3; m = 7; U = 12m/s.

6. Tower loads as function of turbine interspacing

To investigate whether the load levels decrease for increased turbine spacing and find the distance causing the highest load levels, a simulation study was first carried out. It is important to realize that the largest loads may not occur in full wake situations, but in half wake conditions [13]. In Figure 14-16 the load levels of a turbine in a single wake situation were mapped with respect to fatigue load levels as function of wind direction and turbine distance.

At 8m/s the full wake situation shows a rather constant load plateau until 8D spacing for which load decrease beyond this spacing. This, however, only covers a narrow wind direction interval. For wind directions resulting in half wake situations, the loads levels are higher and also continuously decreasing for increased spacing. Therefore, one should expect that for wind speeds below rated, the tower loads are indeed decreasing for increased distance (or the maximum load level is for distances below 4D). This is also seen clearly at 10m/s, Figure 16, where a load peak is identified around 8D for the full wake case. The yaw loads at 8m/s, Figure 15, reveal a distance where the loads have a maximum close to 5D spacing.

Based on the measurements, presented in Figure 17 and 18, it was investigated whether this showed a similar trend. It is not possible from the measurements to identify a certain downstream distance where the loads have a maximum. It is, however, clear to see that between 6.2D until 9D the bending moment loads levels remain on a constant plateau, where a small however clearly visible lower load level is seen for 9-10D spacing's. A similar conclusion can be drawn from the yaw moment measurements. Loads remain at a rather constant plateau until 7D and decrease or further downstream distances.

7. Conclusion

In this study the Nysted II wind turbine park has been analyzed with respect to tower bending load and tower yaw loads. A total of 6 turbines, with different inter spacing's, were fully instrumented with strain gauges. First a load validation between the measured loads and simulated loads using the DWM approach in combination with the aeroelastic code HAWC2 was carried out. A very fine agreement was seen between measurements and simulations in single and multiple wake situations until just below rated wind speed at 12m/s. At 12m/s it seems as the DWM approach under predicts the load levels. It is expected that this is caused by the complex interaction between the nonlinear turbine control characteristics, which may cause a highly varying deficit depth not yet accounted for in the DWM approach.

The measured and simulated cases have also been used to study how load levels vary as function of downstream distance. It is found that the tower bottom bending moments are on same level between 6 and 9D spacing and decreasing for further distances. For the yaw moment, the load levels are on a constant level from 5 to 7D spacing and decreasing for further distances.

8. Acknowledgement

The work was conducted as part of the Energinet.dk project no. 10546 Funded by PSO ForskEL. The writing of the paper has taken place as part of ForskEL under the project Concert (2016-1-12396), which is gratefully acknowledged.

IOP Conf. Series: Journal of Physics: Conf. Series **854** (2017) 012027 doi:10.1088/1742-6596/854/1/012027

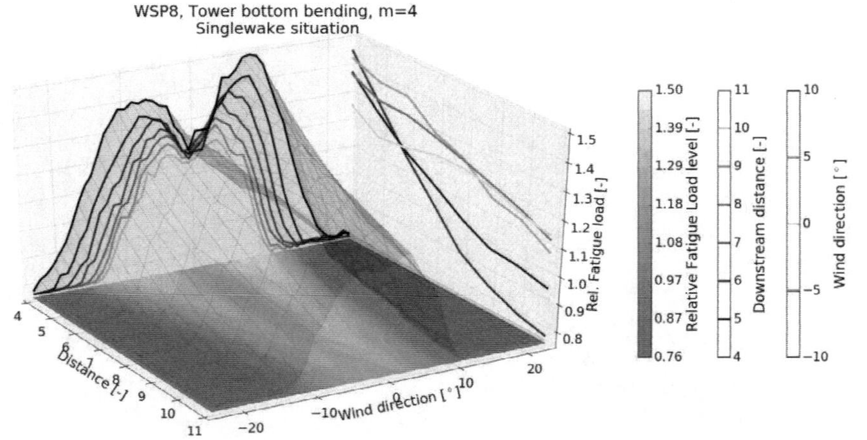

Figure 14. Tower bottom bending moments at 8m/s and 6% turbulence intensity.

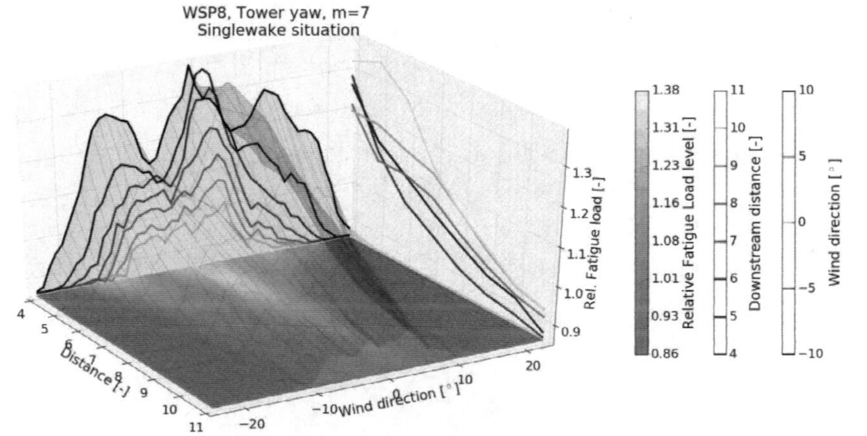

Figure 15: Tower top yaw moments at 8m/s and 6% turbulence intensity.

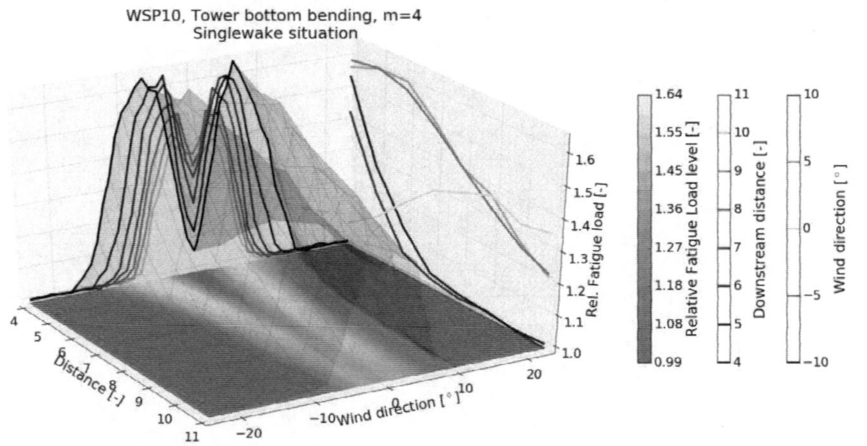

Figure 16: Tower bottom bending equivalent moment; U = 10m/s; TI = 6%.

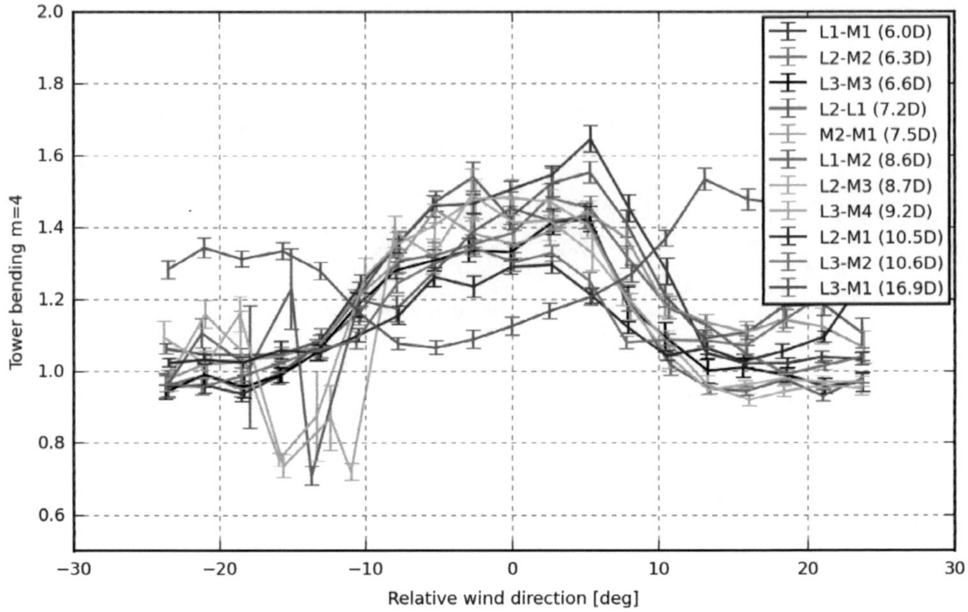

Figure 17: Measured tower bottom bending equivalent moment as function of the relative wind direction between wind and full wake direction; m = 4; U = 8m/s.

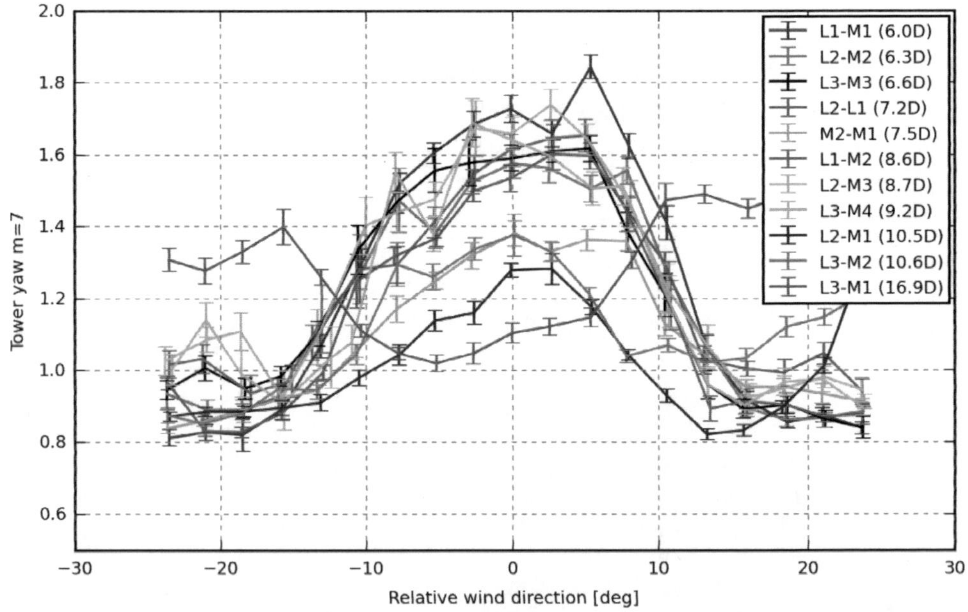

Figure 18: Measured tower top yaw equivalent moment as function of the relative wind direction between wind and full wake direction;
m = 7; U = 8m/s.

IOP Conf. Series: Journal of Physics: Conf. Series **854** (2017) 012027 doi:10.1088/1742-6596/854/1/012027

References

[1] Frandsen S. Turbulence and turbulence-generated structural loading in wind turbine clusters. Technical Report. Risø-R-1188(EN), Risø-DTU, 2005

[2] IEC 61400-1 ed. 4. Wind turbines - part 1: design requirements.

[3] Larsen, G.C.; Madsen, H.Aa.; Thomsen, K. and Larsen, T.J. (2008). Wake meandering - a pragmatic approach. Wind Energy, 11, pp. 377–395.

[4] Madsen, H.Aa.; Larsen, G.C.; Larsen, T.J.; and Troldborg, N. (2010). Calibration and Validation of the Dynamic Wake Meandering Model for Implementation in an Aeroelastic Code. J. Sol. Energy Eng., 132(4).

[5] Larsen, T.J.; Madsen, H.Aa.; Larsen, G.C. and Hansen, K.S. (2013). Validation of the Dynamic Wake Meander Model for Loads and Power Production in the Egmond aan Zee Wind Farm. Wind Energy, Volume 16, pp. 605–624.

[6] Larsen, T. J., Larsen, G. C., Aagaard Madsen , H., & Petersen, S. M. (2015). Wake effects above rated wind speed. An overlooked contributor to high loads in wind farms. In *Scientific Proceedings. EWEA Annual Conference and Exhibition 2015* (pp. 95-99). European Wind Energy Association (EWEA).

[7] Taylor G. The spectrum of turbulence. Proceedings of the Royal Society London A 1938 1937;164:476-490. DOI:10.1098/rspa.1938.0032

[8] Shabana, A. (1998). *Dynamics of Multibody Systems*. Cambridge University Press: New York.

[9] Kim, T., Hansen, A. M., & Branner, K. (2013). Development of an anisotropic beam finite element for composite wind turbine blades in multibody system. Renewable Energy, 59, 172-183. DOI: 10.1016/j.renene.2013.03.033

[10] Madsen, H.Aa.; Riziotis, V.; Zahle, F.; Hansen, M.; Snel, H.; Grasso. L.T.F.; Politis, E. and Rasmussen, F. (2011). BEM Blade element momentum modeling of inflow with shear in comparison with advanced model results. *Wind Energy*; **15**(1): 63–81.

[11] Vorpahl, F., Strobel, M., Jonkman, J. M., Larsen, T. J., & Passon, P. (2014). Verification of aero-elastic offshore wind turbine design codes under IEA Wind Task XXIII. *Wind Energy*, *17*(4), 519-547. DOI: 10.1002/we.1588

[12] Passon, P., Kühn, M., Butterfield, S., Jonkman, J., Camp, T., & Larsen, T. J. (2007). OC3 - Benchmark exercise of aero-elastic offshore wind turbine codes. *Journal of Physics: Conference Series (Online)*, *75*, 12. DOI: 10.1088/1742-6596/75/1/012071

[13] Steudel, D. Personal communication. 2013.

[14] Hansen, M. H., & Henriksen, L. C. (2013). Basic DTU Wind Energy controller. DTU Wind Energy. (DTU Wind Energy E; No. 0028).

CFD and Experimental Studies on Wind Turbines in Complex Terrain by Improved Actuator Disk Method

Xin Liu[1], Shu Yan[1], Yanfei Mu[1], Xinming Chen[1], Shaoping Shi[1]

[1]Huaneng Clean Energy Research Institute, Changping, Beijing 102209, China.
E-mail: liuxin@hnceri.com

Abstract. In this paper, an onshore wind farm in mountainous area of southwest China was investigated through numerical and experimental methods. An improved actuator disk method, taking rotor data (i.e. blade geometry information, attack angle, blade pitch angle) into account, was carried out to investigate the flow characteristic of the wind farm, especially the wake developing behind the wind turbines. Comparing to the classic AD method and the situ measurements, the improved AD shows better agreements with the measurements. The turbine power was automatically predicted in CFD by blade element method, which agreed well with the measurement results. The study proved that the steady CFD simulation with improved actuator disk method was able to evaluate wind resource well and give good balance between computing efficiency and accuracy, in contrary to much more expensive computation methods such as actuator-line/actuator-surface transient model, or less accurate methods such as linear velocity reduction wake model.

1. Introduction

Today, an improved estimation of the wake effects between wind turbines can benefit the design of a wind farm plan and siting. Industry sector gets used to and still continue to employ a number of engineering wake models for assessing the velocity deficits and power production behind single and multiple wind turbines[1]. These engineering models (e.g. Jensen and Frandsen model) are based on the analytical solution of the simplified momentum or mass conservation equations. They require the least computational time of all kinds of wake prediction models, and are occasionally capable of giving good agreement on overall efficiency of a flat terrain field.

It is extremely challenging to assess the power performance of a wind farm in complex terrain. Engineering wake models cannot be available any more in such cases. Therefore, direct rotor modelling (DRM) by CFD RANS based models or large-eddy simulation based models) is adopted to figure out complex terrain problems. It needs the largest computational time cost and exact geometry information of blades which are not acceptable for wind farm level assessment (neither flat terrain nor complex terrain), but are able to produce a complete and detailed representation of rotor dynamic flow for just a single or quite few units. Currently, DRM is regarded as the most accurate approach but also the least used in practice.

Simplifying the rotor modelling, variations of the actuator methods (i.e. actuator disk[2, 3], actuator line[4, 5] and actuator surface[6-8]) could be alternative options to DRM. There is no exact geometry of the rotor modeled in actuator methods. These geometrical parameters and aerodynamic performance are implicitly stored in tabulated airfoil data by establishing the relationships between power and thrust coefficients, or chord lengths, twist angles, attack angles, lift coefficients and drag coefficients. The specific parameters in airfoil data depend upon the selected actuator method. Through inputting this prepared tabulated airfoil data, actuator methods can calculate relative flow velocities and the body forces in a NS solver. The advantage of actuator methods compared to DRM is twofold: First, it enables to save mesh points because boundary layers need not to be resolved. Next, it enables to represent moving boundaries by body forces, hence intrinsic dynamically problems are easily represented without having to use moving meshes[6]. Actuator line (AL) and actuator surface (AS) methods can consider the unsteady characteristic of the flow, by means of modelling individual blades and their movement over time. Although some versions of the AD model can do unsteady simulation as well, it just simplifies the disk rotating over time, and cannot distinguish the individual blade motion. Therefore, in most cases the AD methods are adopted in steady simulations. AL and AS are able to predict the intricate details of the wake dynamics, such as tip vortices. AD is a more efficient approach than AL and AS by permitting lower spatial resolutions (i.e. mesh density) and solution as a steady-state problem [9].Depending on the available airfoil data, rotor loading may be either approximated as a constant thrust or calculated using the blade element theory[10].The former one is the classic AD method referred to as the Froude's [11] momentum theory, which loses the detail of flow over the individual blades. The later one is the improved AD method using the blade element theory evolved by Glauert[12], which takes the tabulated airfoil data as input parameters. The AD, AL and AS methods have been compared by various researchers under different conditions and looking at various quantities, e.g. velocities at different positions in the wake, power, etc., and the results of the comparison were dependent on the quantities compared[10, 13-16]. The turbine-induced forces (e.g., thrust, lift and drag) by means of LES [17] were compared with wind tunnel measurements and demonstrated that the improved AD and AL/AS methods gave very similar results that were in good agreement with measured data whereas the results of the classical AD method deviated significantly from the measurements. When all of these problems are considered together, improved actuator disk method would give a relatively good balance between the rotor physical characteristic fidelity, the prediction precision, and calculation efficiency. Improved AD method is a powerful tool to estimate wind resources and power output efficiency in a complex terrain farm[9].

In this paper, a real wind farm with complex terrain in mountainous area of southwest China was evaluated by RANS solver along with the k-ω turbulence model. The improved actuator disk method, taking rotor data (i.e. blade geometry information, attack angle, blade pitch angle) into account, was carried out to investigate each turbine power output and wake interaction. The rotor data is a tabulated format including chord length, twist angle, relationship between attack angles and lift/drag coefficients, pitch angles and wind speeds, in each blade section, provided by the wind turbine manufacturer. The numerical results were compared with the classic AD method and the situ measurements for two turbines with Lidar.

2. Numerical modelling

The numerical simulations in the wind farm with/without actuator disks have been performed with ANSYS Fluent Program, by solving the RANS equations with a Shear Stress Transport (SST k-ω) turbulence model.

2.1. The wind farm

The mesoscale terrain including four wind farms, as shown in Figure 1 (a), is located on a complex terrain in mountainous area of southwest China and operated by the largest energy generating company: China Huaneng Group. The wind farm has an altitude difference of about 800 m, from the lowest valley to the highest peak, and 25 km by 25 km in length and width. There are four small wind farms operated in this region. The layout of the test wind farm in this paper is in the northwest of this region, sketched in Figure 1 (b), whose the altitude range is 2400~2749 m above the sea. Thirty-three wind turbines are installed on site, with 1.5 MW of rated power at around 10.5 m/s. The rotor diameter is 82 m and the hub height is 65 m.

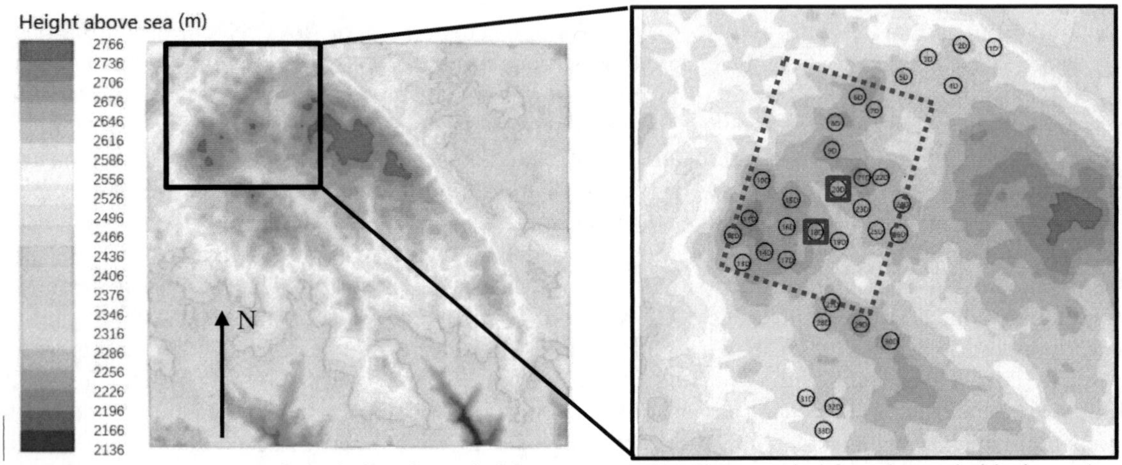

(a) The mesoscale terrain including four wind farms (b) The test wind farm (zoom in black zone)

Figure 1. Wind farm layout, plotted in altitude height.

2.2. Classic actuator disk method

The classic AD method is to simply the turbine rotor as an actuator disk with uniform thrust. The rotor is modelled by porous cells instead of the swept area of the rotor. Porous cells act as a momentum absorber upon which a uniform distribution of axial force is applied [10]. The pressure drop Δp is calculated from the thrust coefficient curve which is given by the wind turbine manufacturer.

The pressure drop through the classic AD is calculated from equation (1)[18],

$$\Delta p = \frac{T}{A} = C_T \frac{1}{2} \rho u_\infty^2 \tag{1}$$

The axial induction factor a, is defined by,

$$a = \frac{u_\infty - u_1}{u_\infty}, \text{ and } u_\infty = \frac{1}{1-a} u_1 \tag{2}$$

u_1 is the velocity at the hub of the rotor. From Betz's theory, if a fixed value of the thrust coefficient is assumed, the axial induction factor can be calculated as follow,

$$a = \frac{1}{2}\left(1 - \sqrt{1 - C_T}\right) \tag{3}$$

Final, the pressure drop is obtained from

$$\Delta p = C_T \frac{1}{2} \rho \left(\frac{1}{1-a} u_1 \right)^2 \tag{4}$$

2.3. Improved actuator disk method

The turbine rotors are modelled as improved ADs on which body forces are imposed, described in [19-22]. Improved AD defines the blades aerodynamically as two-dimensional airfoil data. The local angles of attack and the lift and drag coefficients, C_L and C_D, are used to compute the forces acting on the blade according to,

$$\mathrm{f} = \frac{1}{2}\rho U_{rel}^{2} cB\left(C_L e_L, C_D e_D\right) \tag{5}$$

Where U_{rel} is the local relative velocity, c is the blade chord length, B indicates the number of blades, e_L and e_D are the direction of the unit vector for the lift and drag forces.

The local angle of attack is determined by the local flow angle, the twist and the pitch of the blades. The U_{rel} is obtained from the components of absolute velocity in axial U_{axial} and tangential directions U_θ,

$$U_{rel} = \left(U_{axial}, r\Omega - U_\theta\right) \tag{6}$$

where Ω is the angular velocity of the rotor. The rotating speed of the turbines are determined from the inflow velocity.

If looking at a small area of the disc of differential size using cylindrical coordinates $dA = rdrd\theta$, the load applied in this area is given by

$$\mathrm{f}' = \frac{\mathrm{f}dr}{rdrd\theta} \tag{7}$$

The rotor sink terms are then subtracted from the momentum sources computed from the sum of the lift and drag forces in X, Y and Z directions by ANSYS Fluent UDF (User Defined Function).

2.4. The wind measurement by lidar

Two vertically profiling, ground-based Doppler light detection and ranging (lidar) instruments was deployed in front and at back of the turbine #18, respectively. The front lidar is used to measure the velocity profile of the free stream at 224 m distance in front of the rotor. The rear lidar is used to capture the wake of the rotor at 330 m distance at the back of the rotor. The lidar locations are plotted in **Figure 2**.The lidars continuously measured the wind flow from June 5 to August 31, 2015, in vertical directions at 12 measurement heights from 45 to 155 m for the front lidar and 80 to 190 m for the rear lidar above the ground level, insteps of 10 m and sampling frequency of 1 Hz. The slopes of the ridgeline are about 10°, hence the lidars had to be placed at lower elevations than the base of the wind turbine.

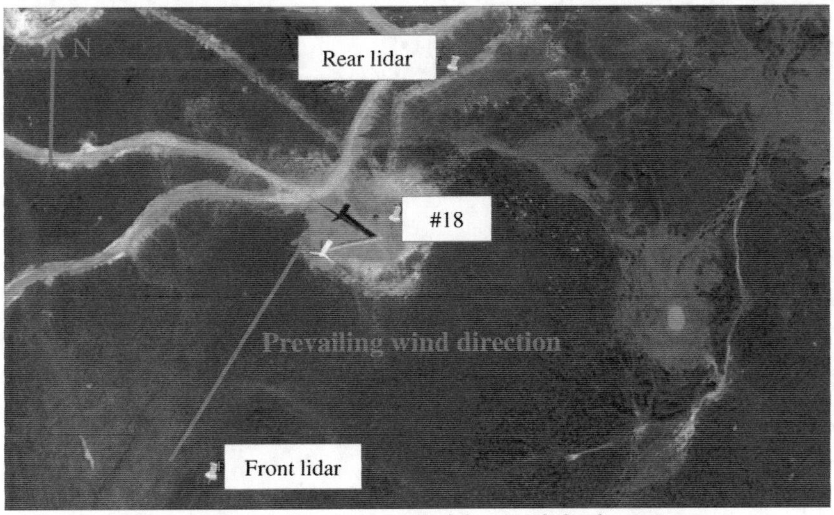

Figure 2. Satellite imagery of the two lidar locations.

2.5. Computational domain and boundary conditions

The evaluated case in this paper was limited to a small rectangle region which is 3 kilometers long, 2 kilometers wide and 20 kilometers high, as shown in dashed frame in Figure 1 (b). The main goal of this work was to study and validate the generic behavior of the wake effect of a wind turbine in complex terrain in qualitative way rather than providing a detailed expression of the actual wind farm. The final goal of the research is to introduce the wake evaluation and AEP estimation wind turbines. Hence only 2 turbines (No. #18 and#20, see purple frame in Figure 1b) are included in the computational domain to decrease the cost of calculation. Even there are other turbines exist in this computational domain. The selection of #18 and #20 is enough to explore the impact of the wake effect. The mesh of the terrain and turbine simplified models are plotted in Figure 3. The quadrilateral mesh of the ground surface is 5 m by 5 m. The boundary layer at the ground has hexahedral grids, meshed by the first layer height of 5 meters and 10 layers in height direction with 1.2 growth ratio.

Velocity-inlet condition in ANSYS-Fluent is imposed at the inlet for pressure and velocities, pressure-outlet condition is given at the outlet, symmetry conditions are applied at the sides and top, and no slip condition with the roughness height of 0.2 m is imposed at the bottom. The rotating speed of the turbines are determined from the inflow velocities before the rotors which had been given by the turbine manufacture. The rotating speed is a function of the rotor inflow velocity, which has been given in the airfoil data and provided by the wind turbine manufacture. Fluent UDF is used to get the averaged wind speed in front the actuator disk, then to set the rotor rotating speed by checking the airfoil data.It is noted that the determination of the inlet velocity of the whole wind farm is quite tricky. Because of the complexity of the terrain, it is difficult to choose an accurate inlet velocity directly. The approach we used was as follow: first, an average wind speed in the period of one hour during a relatively steady wind flow was selected from the massive measurement database of the front lidar. Then the average speed at the elevation of 100 m was set as the target speed. The inlet velocity, which is assumed to be a logarithmic profile, was iteratively adjusted to match this target speed, as shown in Figure 4.

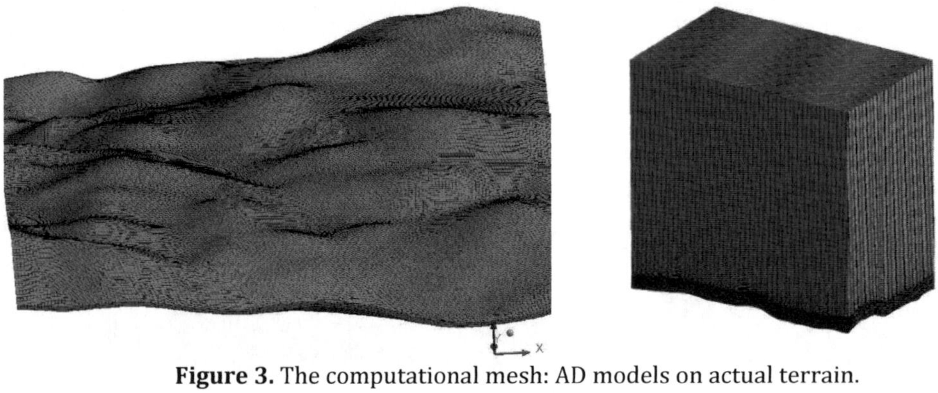

Figure 3. The computational mesh: AD models on actual terrain.

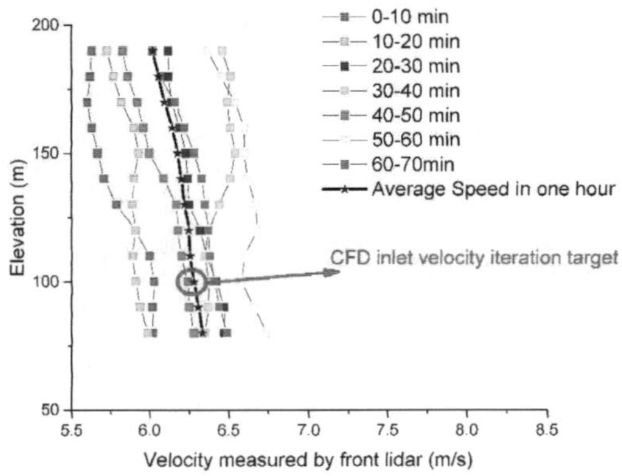

Figure 4. The reference wind speed selection.

3. Results

One of the major differences between complex and flat terrains is that the local velocity and pressure distribution in complex terrain highly depend upon the terrain features. The up and down of the complex terrain dramatically change wind shear, turbulence intensity, inflow angle, velocity magnitude, and so on, which in turn have important impacts on the turbine operating stability. The velocity distributions at the elevation of the turbine hub with both the classic and improved AD methods are shown in Figure 5. Wakes after the turbines (actuator disks) are clearly recognized. Unlike that in a flat terrain, the wake in complex terrain is not always straight.it changed alone with the flow direction affected by mountains downstream. The wake tended to go around the mountain and pass through the valley between two mountains. The basic wake flow tracks of these two methods are almost the same, except that the improved AD predicted the wake length much longer than the classic AD.

Moreover, for cases with complex terrain, the terrain may be a more important factor for a wind farm, rather than the wake interactions. On the one hand, majority of turbines are located at the upwind sides of mountains around the peaks in prevailing wind direction. The elevations of the turbines are commonly different from each other, as well as wakes after turbines. On the other hand, because of wake flow direction change along the terrain, even if two turbines coincide in the same line of prevailing wind direction and have the same elevations, there is still the chance that the wake of the front turbine bypasses the rear one.

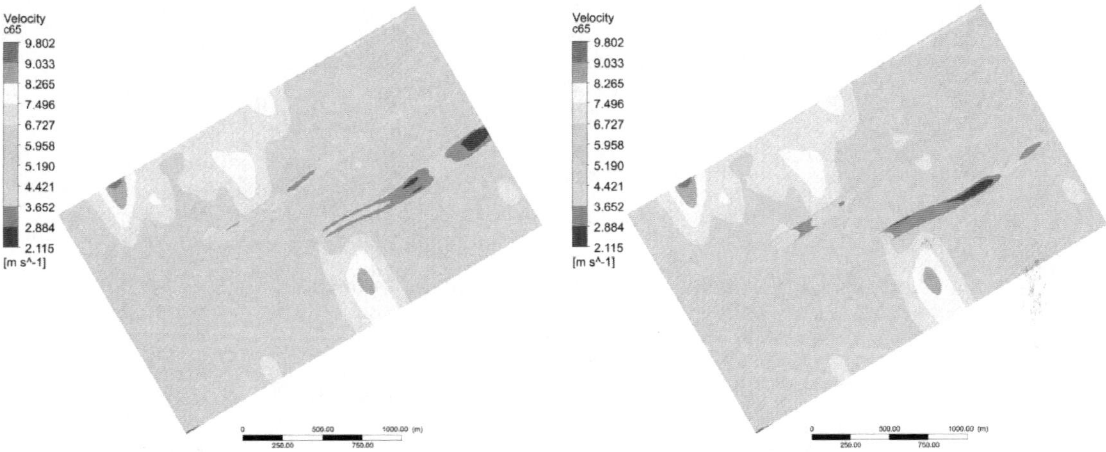

Figure 5. The velocity distributions at the elevation of the turbine hub. (left is the improved AD; right is the classic AD)

Figure 6 are the rose diagrams of the wind directions of two lidars, which were measured in one day. As it shown, the wind direction changed about 20° when the wind passed through the turbine rotor. At the same location of the rear lidar, the simulation result shown the wind direction changed about 12° (see Figure 7). This wake deflection angle was calculated as follows: the inflow direction is the free stream of the wind before the rotor, and the deflection direction is got due to the averaged velocity vector at the rear lidar location. This result indicated that CFD method was able to well predict the wake flow direction along the terrain change. The possible reason of such difference between calculation and measurement might be that the terrain model in CFD was not enough accurate and lost some details of the ground.

IOP Conf. Series: Journal of Physics: Conf. Series **854** (2017) 012028 doi:10.1088/1742-6596/854/1/012028

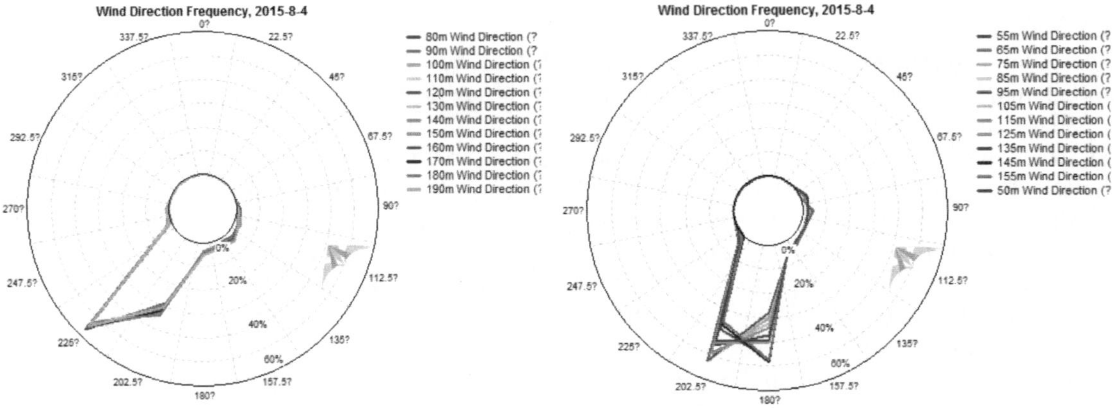

(a) Front lidar (b) Rear lidar

Figure 6. Rose diagrams of the wind direction frequency.

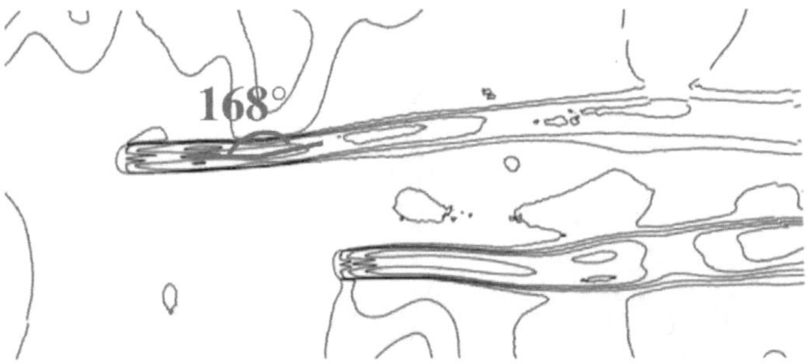

Figure 7. The wake deflection angle at the rear lidar location for#18.

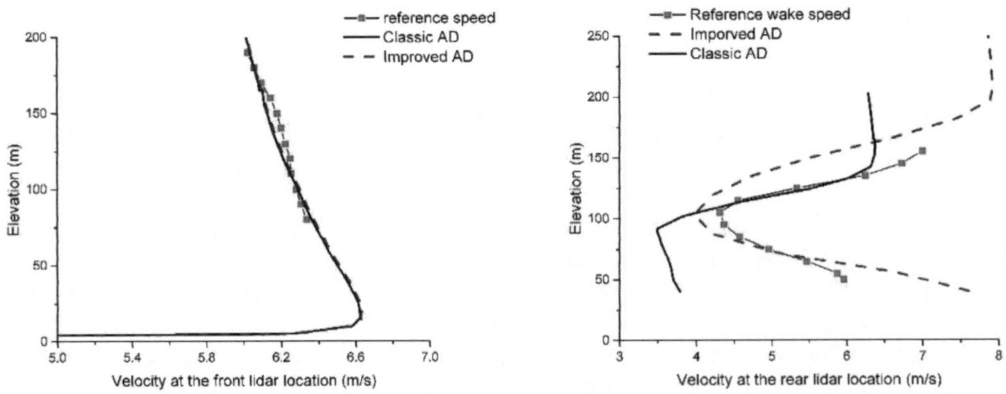

Figure 8. Comparison of the velocity profiles between lidars and CFD.

Before comparing velocity profiles between CFD calculations and measurements, it is noted that it is impossible to model the exact conditions of the measurements. For instance, for complex terrains it is very difficult to highlight climate effects, because multiple factors interact: orography, wakes, turbulence, wind speed, atmospheric stability [23, 24]. Meanwhile, even for measurement, it is nearly impossible to capture the totally same results, because the wind flow is extremely turbulent and results are recorded by means of time averaging method. As discussed before in Figure 4, the measurement results are carefully selected from a relatively steady wind flow. The average velocity of this period was set as the reference condition of CFD. Three different atmospheric stability, convective, near-neutral and stable, will induce totally different velocity profiles of the free stream and wind turbine wake, as indicated by lidar data. However atmospheric stability model was not taken into account in CFD at this stage. The predicted velocity profiles of the free stream in front of the rotor were nearly the same for both improved AD and classic AD, and agreed well with measurement results (see Figure 8 left). This results were easy to understand because the AD didn't play any role in area of the rotor upstream. Simulation result of improved AD slightly over-predicted the velocity deficit after the rotor, but the shape of the profile was very similar with the measurement results. On the contrary, the results of classic AD had much larger deviation from the measurement results (see 8 right).

The negative slope of the velocity profile at the front lidar location was different from the commonly known logarithmic profile. This is due to the location of the lidar and specific terrain shape. In this case, the front lidar was placed at the upwind side of the mountain, where the up-coming wind will be accelerated near the ground.

Another advantage of the improved AD method compared with the classic AD method is that the former one is able to directly predict the power output of the turbine at any operating condition in CFD (see Figure 9). This is because the improved AD method provides blade geometry information and blade pitch angle, and the Fluent provides wind flow characteristics; Based on blade element theory, combining these factors will obtain the turbine power output. Considering the large simplification in simulation, this calculated result by improved AD seemed satisfied and be able to assess the wind farm well with turbines included.

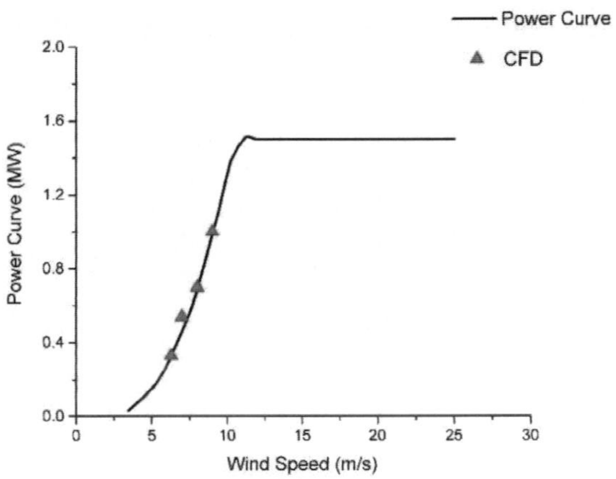

Figure 9. Comparison power prediction for #18 by improved AD with power curve by manufacturer.

4. Conclusion

In this work, the flow characteristics of wind turbines on an onshore wind farm in mountainous area of southwest China were investigated through numerical and experimental methods. Both the classic and improved AD models have applied to simulate the presence of the rotors. Two lidars were placed in front and at back of the turbine #18, respectively to measure the inflow and wake flow.The improved AD slightly over-predicted the velocity deficit after the rotor but captured the velocity profile much better comparing to the classic AD. Moreover the terrain will change the wake developing direction which makes the behaviour of wake more complex.Generally, the improved AD model is a powerful and high-efficient tool for investigating rotor aerodynamic behavior, as well as predicting the power output of the wind turbine.

Acknowledgments
This work was supported by China Huaneng Group Science and Technology Fund (HNKJ15-H17).

References

[1] Andersen S J, Sørensen J N, Ivanell S and Mikkelsen R F 2014 Comparison of Engineering Wake Models with CFD Simulations *Journal of Physics: Conference Series* **524** 012161

[2] Mahmoodi E, Jafari A and Keyhani A 2015 Wind Turbine Rotor Simulation via CFD Based Actuator Disc Technique Compared to Detailed Measurement **4**

[3] Sørensen J N, Shen W Z and Munduate X 1998 Analysis of wake states by a full - field actuator disc model *Wind Energy* **1** 73-88

[4] Porte-Agel F, Wu Y T, Iungo V and Lu H 2011 Large-eddy simulation of atmospheric boundary layer flow through wind farms *Journal of Wind Engineering and Industrial Aerodynamics* **99** 154-68

[5] Shen W Z, Sørensen J N and Mikkelsen R 2005 Tip Loss Correction for Actuator/Navier–Stokes Computations *Journal of Solar Energy Engineering* **127** 209-13

[6] Shen W Z and Zhang J H 2009 The Actuator Surface Model: A New Navier–Stokes Based Model for Rotor Computations *Journal of Solar Energy Engineering* **131** 284-9

[7] Shen W Z, Sørensen J N and Zhang J H 2007 Actuator Surface Model for Wind Turbine Flow Computations

[8] Zhang J 2004 Numerical Modeling of Vertical Axis Wind Turbine (VAWT) *Technical University of Denmark*

[9] Gomes V M M G C, Palma J M L M and Lopes A S 2014 Improving actuator disk wake model *Journal of Physics: Conference Series* **524** 012170

[10] Tzimas M and Prospathopoulos J 2016 Wind turbine rotor simulation using the actuator disk and actuator line methods **753**

[11] Froude R E 1889 On the part played in propulsion by differences of fluid pressure *Transactions Of The Royal Institution Of Naval Architects* **30**

[12] Glauert H 1935 *Airplane Propellers*: Springer Berlin Heidelberg)

[13] Martinez L, Leonardi S, Churchfield M and Moriarty P 2013 A Comparison of Actuator Disk and Actuator Line Wind Turbine Models and Best Practices for Their Use. In: *Aiaa Aerospace Sciences Meeting Including the New Horizons Forum and Aerospace Exposition,*

[14] Martínez - Tossas L A, Churchfield M J and Leonardi S 2015 Large eddy simulations of the flow past wind turbines: actuator line and disk modeling *Wind Energy* **18** 1047-60

[15] Panjwani B 2014 OffWindSolver: Wind Farm Design Tool Based on Actuator Line/Actuator Disk Concept in OpenFoam architecture. In: *Symposium on Openfoam in Wind Energy,*

[16] Watters C S, Breton S P and Masson C 2010 Application of the actuator surface concept to wind turbine rotor aerodynamics *Wind Energy* **13** 433-47

[17] Porteagel F, Wu Y T, Iungo V and Lu H 2011 Large-eddy simulation of atmospheric boundary layer flow through wind farms *Journal of Wind Engineering & Industrial Aerodynamics* **99** 154-68

[18] Crasto G and Gravdahl A R 2008 CFD wake modeling using a porous disc. In: *WindSim User Meeting,* (Tønsberg, Norway: WindSim AS)

[19] Mozafari J and Teymour A 2014 Numerical investigation of Marine Hydrokinetic Turbines: methodology development for single turbine and small array simulation, and application to flume and full-scale reference models *Software for Computer Control* **52** 129-34

[20] Nilsson K, Ivanell S, Hansen K S, Mikkelsen R, Sørensen J N, Breton S P and Henningson D 2014 Large - eddy simulations of the Lillgrund wind farm *Wind Energy* **18** 449-67

[21] Lavaroni L, Watson S J, Cook M J and Dubal M R 2014 A comparison of actuator disc and BEM models in CFD simulations for the prediction of offshore wake losses *Journal of Physics: Conference Series* **524** 012148

[22] Nilsson K 2012 Numerical computations of wind turbine wakes and wake interaction : Optimization and control. (Stockholm, Sweden: Royal Institute of Technology) p 62

[23] Castellani F, Astolfi D, Terzi L, Hansen K S and Rodrigo J S 2014 Analysing wind farm efficiency on complex terrains *Journal of Physics: Conference Series* **524** 012142

[24] Wharton S, Newman J F, Qualley G and Miller W O 2015 Measuring turbine inflow with vertically-profiling lidar in complex terrain *Journal of Wind Engineering and Industrial Aerodynamics* **142** 217-31

Impact of Complex Orography on Wake Development: Simulation Results for the Planned WindForS Test Site

‡Thorsten Lutz, ‡Christoph Schulz, ‡Patrick Letzgus and ♮Andreas Rettenmeier

‡Institute of Aerodynamics and Gas Dynamics (IAG),
University of Stuttgart, Pfaffenwaldring 21, 70569 Stuttgart, Germany
♮WindForS, c/o University of Stuttgart, Allmandring 5B, 70569 Stuttgart, Germany

E-mail: ‡{lutz, schulz, letzgus}@iag.uni-stuttgart.de, ♮rettenmeier@windfors.de

Abstract. In Southern Germany a test site will be erected in complex terrain. The purpose is to enable detailed scientific studies of terrain impact on the characteristics of two research wind turbines and to demonstrate new technologies. Within preparatory studies an appropriate site was identified and examined by field tests and numerical studies in more detail. The present paper summarizes CFD analyses on the impact of the local test site orography on the wake development of a virtual wind turbine. The effects of the orography are identified by comparative simulations for the same turbine using comparative wind situation in flat terrain.

1. Introduction

To further increase the share of onshore wind energy, the erection of wind farms in complex terrain and at exposed locations will be necessary. Due to the impact of orography and vegetation the inflow experienced by the turbines can feature noticeable inclination angles, differing turbulence and strong deviations in the shear profile compared to flat terain. This has an impact on turbine loads and load variations compared to operation in flat terrain citeschulz2014, [1], [2]. The changed load characteristics along with the altered local flow field in complex sites are expected to impact the development of the turbine wake [3], although detailed studies are still lacking.

Scientific studies on the use of wind energy in complex terrain are one focus of *WindForS*, the Southern German Wind Energy Research Cluster [4]. WindForS is comprised of 24 institutes from 7 institutions of higher education and research institutes of the German states Baden-Württemberg and Bavaria. One goal pursued since long is the erection and scientific utilization of a wind energy test site in complex terrain including research wind turbines and extensive measurement instrumentation [5]. The objective is to conduct combined field tests and numerical studies on the impact of complex orography on aerodynamics, loads, aeroelastic behaviour and aeroacoustics of wind turbines. Further, the research turbines shall enable development and prototype testing of new technologies and control strategies. A demand for such a test site is seen because all existing test sites, except one (Aláiz, Spain), are on flat terrain as they are used for certification measurements [1].

Studies to characterise the wind conditions in complex terrain, definition of requirements for the reseach turbines and the measuring techniques as well as the selection of an appropriate test site location was advanced in the two WindForS collaborative projects *LiDAR Complex* and *KonTest*, both funded by the German Federal Ministry for Economic Affairs and Energy (BMWi). While *LiDAR Complex* was aiming on enhancing and qualifying different field measuring techniques and numerical methods for application to complex terrain studies, preparatory plannings of the test site and preparation of the numerical tools were conducted in *KonTest*. Meanwhile, the funding of two research wind turbines of the 800kW class, four meteorological masts, remote sensing devices and other measurement technique was approved by the BMWi. The funding also inlcudes accompanying research activities and establishing detailed numerical models of the research turbines. The activities are pooled in the *WINSENT* collaborative project. Supplementary funding by the state of Baden-Württemberg will support the realization of the test site.

In the present paper, scale-resolving DES simulations (Detached Eddy Simulation) on the impact of the test site orography on the wake development of a virtual wind turbine are discussed. The results are contrasted with simulations of the same turbine in flat terrain to identify the impact of the orography. In the subsequent section, information on the test site, the research turbines and the intended techniques for flow field measurements are presented. Sec. 3 gives an overview of the flow solver, the numerical setup and the wake evaluation techniques applied in the present study. In Sec. 4 the impact of the complex orography on the unsteady wake development will be discussed for a virtual wind turbine.

2. Planned Test Site in Complex Terrain

The above mentioned *KonTest* project served for the preparation and planning of the test site in complex terrain. Firstly, the scientific requirements with respect to the topology of the site, the specification of the requirements for the research turbines, the instrumentation and supplementary field measuring techniques were defined. It was searched for an escarpment with an un–forested plateau as site for the turbines but with orography–caused disturbances of the inflow. In a second step, these criteria formed the basis for an evaluation of potential locations in Southern Germany. Among the candidate sites an area close to the village of Stötten in Baden-Württemberg was selected for further detailed local studies. The investigated location has a mean wind speed of 6 m/s at a height of 75 m above ground. Meteorological data, such as wind speed, turbulence and wind profile were collected by means of met mast, RASS and LiDAR measurements as well as by UAV-borne flow field surveys [3], [6], [7]. The measurements showed that the wind speed is accelerated close to escarpment by up 60%. It could be seen, that there is a wide range of different flow characteristics for different wind directions. The in-field measurements were supplemented by numerical simulations to further characterise the local flow field [3], [8], [9], [10].

The field tests and the numerical studies confirmed the suitability of the location and were used for specification of the position of the research turbines and met masts to be erected. The choice of the positions, however, was constrained by infrastructural aspects and building conditions. According to current planning two turbines equipped with dedicated measuring technique will be installed by the end of 2018. Each of the turbines will generate around 750 kW nominal output and feature a hub height of 75 m and 50 m rotor diameter [3], [5]. Consciously, a smaller size compared to current state-of-the-art commercial turbines was chosen to enable future turbine modifications at affordable costs. But this turbine size still allows transferring knowledge of parameters such as loads, controller settings and turbine behaviour to larger wind turbines because different detailed simulation models exist. The turbines will be erected side by side to have the possibility to directly compare the behaviour of one modified turbine to the baseline turbine. One important requirement was that the involved researchers have full access to all turbine and controller data.

It is planned to position one meteorological mast in front of and one behind each turbine whose maximum height should compare to the height of the highest blade tip position of the turbines. The topview of the orography of the selected test site is shown in Fig. 1 along with the planned locations of the research turbines and the met masts. Further, the simulation domain (compare Sec. 3) of the present CFD studies is illustrated. The plan is to equip the met masts with various meteorological sensors including ultrasonic anemometers in addition to the ground- and nacelle-based Lidars that will be installed. Like during the preparatory studies, UAVs will characterise the local flow field from the valley to the vicinity of the turbines including the wake behind the turbines. Lidar, met mast, eddy covariance and ceilometer measurements will serve to examine the wake of the turbines in detail, too.

Figure 1: Topview of the orography of the planned test site area [source: USTUTT-IAG, USTUTT-SWE, GoogleEarth].

3. Numerical Model and Evaluation

3.1. Flow Solver

In the present study scale-resolving Detached-Eddy-Simulations (DES) were performed to examine the impact of complex orography and inflow turbulence on the development of the unsteady wake characteristics. The simulations were performed using the flow solver *FLOWer* that was mainly developed by the German Aerospace Center (DLR) within the past *MEGAFLOW* project [11]. FLOWer is a compressible finite–volume flow solver that uses block–structured grids and features a wide range of turbulence models [11]. In addition, it can handle rotating structures and the user can apply the (*Chimera*) overset grid technique [12] to embed separately meshed components into one computational grid. During the last years the code was systematically extended for wind turbine and helicopter application at the institute of the present authors. Amongst other features, this involves time-accurate structural coupling for aeroelastic analyses of wind turbines, inclusion of a generalized grid deformation algorithm, implementation of a higher-order numerical scheme, of actuator line approaches as well as options for imposing inflow turbulence and shear.

To enable the consideration of turbulent, atmospheric inflow conditions during the simulation two options were implemented. A *Dirichlet* boundary condition allows to feed in a stationary or unsteady velocity field at the inflow boundary of the computational domain [13]. The velocity field can be obtained from precursor LES simulations or from generic shear profiles, optionally superimposed by synthetically generated turbulence. To reduce the impact of disturbances from the inflow boundary on the injected turbulence, the turbulence field can also be imposed by means of volume forces in a plane further downstream as proposed by Troldborg [14]. In the present study this approach was used to impose synthetic turbulence generated by means of the *Mann* model [15]. More information about these options can be found in Refs. [13], [16], [17]. In the past this flow solver was already applied and qualified for detailed wake studies, e.g. on the

Mexico rotor including yaw effects [18], [19], [20] or the *AVATAR* rotor in ambient turbulence [16].

3.2. Wind Turbine and Meshes

At the beginning of the present study the size and the type of the research turbines to be installed was not yet decided. Therefore, available data of a 2.4 MW low wind speed onshore wind turbine were used for this wake study. The turbine features a hub height of 95 m and a rotor diameter of 109 m. The actual tilt and cone angles were considered in the present setup. Making use of the *Chimera* overset grid approach dedicated meshes were generated for the different components of the turbine, overlapped and embedded into a background mesh that covers the complete flow domain. The rotor blades were meshed by using the *automesh* script developed at the institute of the present authors. This script enables automated generation of high-quality structured meshes around single rotor blades in combination with the commercial grid generator *Pointwise*. It offers several parameters to control the composition and the resolution of the grid. The present blade grid is of C-topology and features 141 grid cells in radial, 101 in wall normal and 221 in circumferential direction. Together with additional grid blocks in the blade wake and outwards of the blade tip this sums up to about 7 Mio. grid cells per blade. In wall normal direction the blade boundary layer is resolved by about 30 ~ 40 grid layers and the thickness of the first layer relates to a normalized wall distance of $y^+ \approx 1$. The grid is adapted in radial direction to ensure these characteristics along the complete blade span.

A single mesh consisting of 4.3 Mio. grid cells was introduced for nacelle and tower to minimise the interpolation effort needed for the grid overlap. For the hub a rotating separate grid (2.2 Mio. cells) was introduced that overlaps with the fixed nacelle and tower mesh during the simulation. The hub mesh is further connected with a joint mesh that was introduced to build a connection between blades and hub and features 3×0.3 Mio. grid cells [9]. This approach was chosen since a better mesh quality and a lower total amount of cells could be realised than meshing the joint together with the hub.

The objective of the present study was to examine the impact of complex orography on the development of the turbine wake. To have a reference, a background grid for a flat terrain scenario was created first. The background grid extends about 1900 m×1200 m×1000 m in streamwise, lateral and vertical direction, respectively. The value for the vertical direction has to be interpreted as mean value as, for obvious reasons, it is dependent on the position inside the terrain. Hanging grid node technique was utilised in the Cartesian domain to enable a refinement in the areas of interest while keeping the total number of grid cells within an acceptable limit. The refined mesh is cube-shaped and shows an extent of 300 m by 300 m in vertical and lateral direction [3], [17]. The resolution of the refined domain is 1 m in each direction of space and it ranges from the inlet towards the outlet to enable a resonable high resolution of the inflow turbulence and the wake region (compare Fig. 2). Outside of the refined domain the background mesh is successively coarsened, still capturing the general flow features but not the turbulence statistics. For the background mesh a total number of about 150 million cells were introduced.

A second backround grid was generated that features the real orography of the site described in Sec. 2 at the lower boundary of the computational domain. The orography data were taken from a digital terrain model provided by the LGL (State Agency for Spatial Information & Rural Development Baden-Württemberg). A script was applied that maps the flat terrain mesh onto the surface of the complex terrain by means of mesh deformation. Further away from the surface, the deformation fades out and cubic cells are created. The resulting complex terrain background mesh, therefore, features the same mesh topology and number of grid cells as the baseline mesh for flat terrain [3], [17]. Figure 2 shows a sketch of the complex terrain background grid along with the embedded meshes around the turbine components.

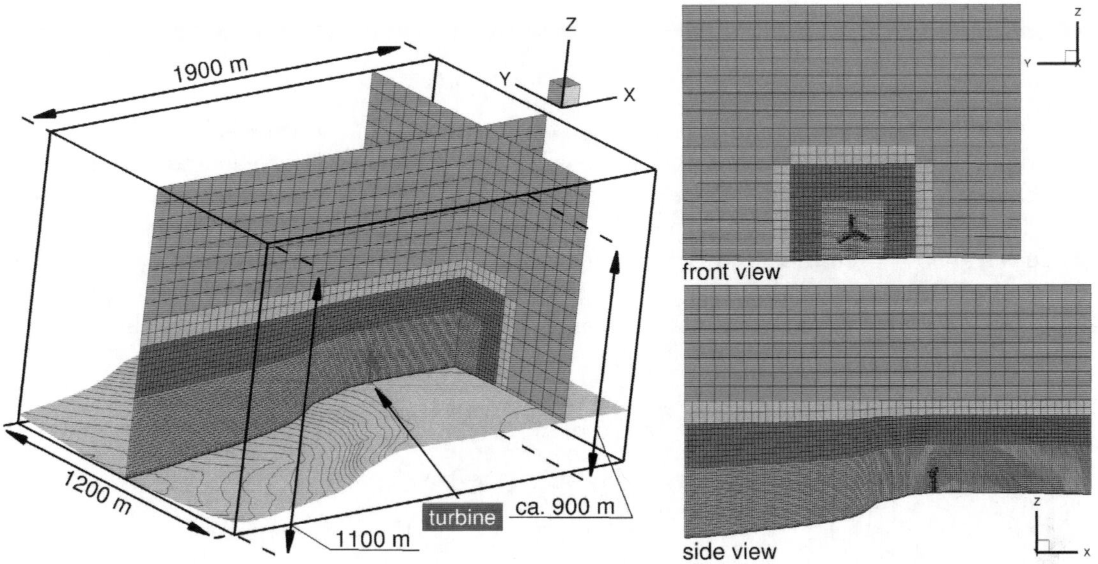

Figure 2: CFD meshes of the virtual turbine and background mesh in complex terrain [17].

3.3. Numerical Setup and Procedure of the Simulations

All simulations of the present study were conducted as zonal DES simulations utilising the *Spalart–Allmaras* turbulence model with *Edwards modification* [21]. For each simulation, a number of 40 inner iterations per physical time step was chosen with a dual–time stepping method for temporal discretisation according to Jameson [22]. A fifth order *WENO (Weighted Essentially Non–Oscillatory)* scheme [23], [24] was used to determine the convective fluxes in all regions of interest and a second order *Jameson–Schmidt–Turkel method* was utilised in regions of minor interest [25]. The boundary layers around the components of the wind turbine were calculated fully turbulent. Aeroelastic effects on rotor and tower were completely neglected in the present study.

In this paper results of four simulations will be discussed. Two simulations without wind turbine, one for flat terrain and one for complex terrain situation. Thereafter, two further simulations on the same background grids but with embedded wind turbine were performed. All simulations were started with a coarse time step of 0.26 s (corresponding to 20° rotor rotation) and were conducted over 36 rotor revolutions. This served to initialise the flow field and to propagate the inflow turbulence throughout the whole computational domain. Thereafter, another 36 revolutions were simulated with a refined time step of 0.026 s (2° azimuth) to develop smaller scale turbulence. Finally, the simulations were continued with the same time step for further 12 revolutions and flow field data were stored for the subsequent evaluation and wake analysis.

3.4. Wake Evaluation Techniques

To enable flow physics studies of the wake development the time-averaged velocity profiles and the spectra of the velocity fluctuations were evaluated. Further, the trajectory of the wake centerline and its spatial motion were analyzed. The centerline was determined by means of mass centers on cross axes of the wake. These cross axes are arranged parallel to the rotor plane, compare Fig. 3. The mass centers are calculated in analogy to the following equation:

$$y_{centerpoint} = \frac{\int y\,(u - u_\infty)dy}{\int (u - u_\infty)dy} \tag{1}$$

Here, u is the local velocity and u_∞ represents the anticipated mean velocity of 11 m/s at hub height (compare Sec. 4.1). This evaluation provides the mass center for each cross axis. The cross axes are placed one after the other at a certain distance downstream of the rotor plane, see Fig. 3. This figure visualizes the flow field in a horizontal plane extracted at hub height. The centerline can then be determined by connecting the resulting positions obtained by Eq. (1) for each cross axis. Care must be taken to ensure that the cross axes completely include the wake in y-direction (see Figure 3). However, the cross axes should not project too far into the ambient flow domain in order to avoid sources of error.

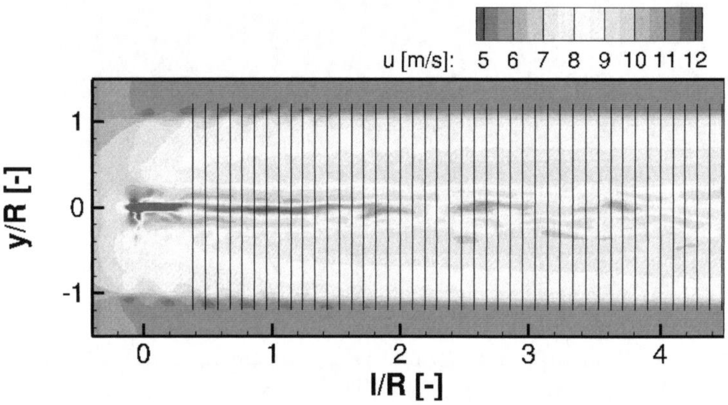

Figure 3: Procedure for the determination of the centerline via cross axes.

This method can also be used to display a whole array of centerlines for a sequence of time steps to examine the temporal variation of the wake geometry and the meandering process. The amplitudes and the standard deviation of the centerline can then be evaluated over an arbitrary period of time. In addition, the temporal mean value of the centerline is determined. This approach allows to characterise the unsteadiness of the wake development and to visualise the meandering.

4. Results

The objective of the present study was to examine the impact of complex orography on the development of the wake of a wind turbine. For this purpose, scale-resolving DES simulations with the numerical setup as described in Sec. 3.3 were conducted. One baseline simulation for the 2.4 MW turbine presented in Sec. 3.2 in flat terrain and one simulation considering the real orography of the planned test site near Stötten (compare Sec. 2) were performed. It was intended to mimic the real atmospheric inflow conditions (shear profile, turbulence) of the terrain at the prevailing wind direction based on in-field measurements. Two challenges are associated with this objective. Firstly, to define realistic unsteady time series of the inflow velocity vector for the complete inflow plane of the simulations and, secondly, to ensure consistent wind speed at the turbine location for the flat and the complex terrain simulations.

4.1. Inflow Conditions

The inflow plane of the complex terrain simulation is located in the flat valley usptream of the escarpment at a distance of 1200 m (about 11 rotor diameter) upstream of the virtual turbine. Near this location neither a met mast was installed, nor were LiDAR or UAV-borne measurements available. Therefore, data of a met mast that is located about 1 km downstream of the escarpment were utilised . As the test site and this met mast are located on a flat plateau it was assumed that the disturbances caused by the escarpment are largely abated at the position of the met mast and therefore represent a realistic basis for the definition of the inflow conditions in the present simulations. This mast has a height of 100 m and is equipped with four cup anemometers, three 3d sonic anemometers, vanes as well as pressure and temperature sensors. All data were recorded with a sampling rate of 20 Hz which enables to derive information about the turbulence level. Further information can be found in Ref. [3].

Met mast data with the above measuring technique have been taken since February 2015. To derive the inflow properties for the present simulations one bin of measurement data containing data of one year was filtered for neutral conditions and evaluated for the prevailing wind direction of 295° featuring a wind speed of 11 m/s at 95 m height, which represents the hub height of the virtual turbine [7]. With the chosen wind direction the flow in the mid-plane of the simulation is almost orthogonal to the shape of the escarpment. As the current version of flow solver FLOWer does not consider stratification effects, a neutral atmospheric boundary layer with an exponent of $\alpha = 0.14$ was used to define the inflow condition for the simulations. This velocity profile was prescribed at the inflow plane of the computational domain by means of a *Dirichlet* boundary condition (compare Sec. 3.1).

Preparatory simulations with the setup described above but without wind turbine were conducted to ensure comparative conditions for the flat and complex terrain simulations at the location of the virtual turbine. Following the met mast data for the selected wind situation a wind speed of 11 m/s at hub height was anticipated for both terrains. This velocity was realised in the definition of the power-law profile prescribed at the inflow plane of the flat terrain simulation. For the complex terrain the flow is accelerated by the displacement of the escarpment. Thus, a smaller velocity has to be defined at the inflow plane which is located in the flat valley, compare Fig. 2. Pre-simulations showed that an inflow speed of 8 m/s provides the intended hub height velocity of 11 m/s at the position of the virtual turbine [9], [17].

Based on the anemometer data different modelling parameters of the *Mann* model [15] for synthetic generation of the inflow turbulence were derived. As input parameters the length scale L, the stretching factor Γ and the dissipation $\alpha\varepsilon^{2/3}$ are needed. The distribution of the energy in the spectrum was derived from L and $\alpha\varepsilon^{2/3}$, whereas Γ controls the shear. For the present simulations L was set to 40 m, $\alpha\varepsilon^{2/3} = 0.035$ and $\Gamma = 3.9$. A turbulence intensity of $TI = 10\%$ (related to the respective mean inflow velocity of the two terrain simulations) was specified based on the anemometer data. The turbulence generated by the *Mann*–box is fed into the flow domain via source terms as described in Sec. 3.1.

4.2. Pure Terrain Simulations

First, simulations without consideration of the wind turbine were conducted to characterise the local flow field, to confirm that the intended mean velocity at hub height was achieved and to support the interpretation of the wake development. The resulting time-averaged boundary layer profiles of the pure terrain simulations are depicted in Fig. 4 for different streamwise positions. While the blue profiles represent the results of the flat terrain the red curves give the results for the complex terrain simulation. The dashed lines display the velocity profiles prescribed at the inflow plane. Comparisions between measurements and simulations have been published by [3] and [17] and will not be discussed here.

At 700 m, which is at the beginning of the slope, the predicted profile for the complex terrain already shows some flow acceleration at medium height (50 ~ 150 m) above ground. At 950 m the complex terrain simulation provides a significant acceleration particularly near ground, i.e. the boundary layer profile is enriched. The effect is even stronger at 1200 m which is the location of the virtual turbine. Note, that the hub and the rotor diameter are sketched in the graph. It is obvious that the average velocities at hub height (95 m) coincide for both terrain simulations. However, below hub height the complex terrain simulation shows significantly higher wind speed while above hub height the velocity is lower compared to the flat terrain result. Near tip height even negative shear can be observed. Across the rotor plane the velocity appears to be much more uniform compared to the flat terrain. In consequence, the azimuthal load variations caused by shear are reduced for the present complex terrain simulation. On the other side, the inflow shows some inclination at the position of the turbine which increases the load variations, compare Ref. [9]. It should be noted, however, that this result in first instance holds for the present orography (turbine located on a flat plateau downstream of an escarpment, no impacting vegetation) and wind condition (wind direction perpendicular to the ridge, neutral stratification) considered. The results, therefore, cannot be generalized for other complex terrain or wind conditions.

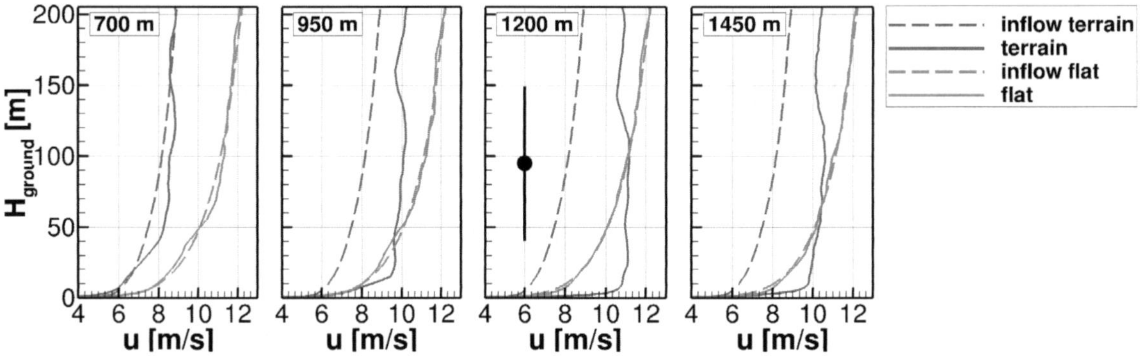

Figure 4: Boundary layer profiles at different positions [17].

4.3. Simulations Including a Virtual Wind Turbine

In a next step the meshes of the virtual turbine were embedded into the terrain background meshes to examine the impact of the complex terrain on the wake development. The same setup as for the pure terrain simulations was used. Fig. 5 visualises snap shots of the wake development for flat terrain (upper picture) and complex terrain (lower picture). The figures show the axial velocity in a vertical plane parallel to the inflow direction with Δz being the height above the local ground level.

On first glance, the higher speed close to the ground downstream of the tower in complex terrain is prompting. As a result of the terrain-caused flow enrichment near ground (compare Sec. 4.2), the tower wake deficit is fed with more energy from the surroundings than in the flat terrain case. In general the wake in complex terrain looks less disordered and its propagation seems much more straight than in flat terrain. Close to the turbine an upward deflection of the wake can be observed. This is the result of an upward flow inclination caused by the displacement effect of the upstream escarpment. The time-averaged deflection appears to be stronger than the average inclination angle across the rotor plane, i.e. comparable skew effects for turbines under yaw condition can be observed. Further downstream the inclination quickly decays and the upwards deflection of the wake is reduced accordingly [17]. Around $l/R = 6$ the wake seems to break up loosing its structure for both terrains. The mixing in the complex terrain, however, appears to be weaker than in flat terrain. This is attributed to a reduced turbulence level (related to the local mean velocity) caused by the flow acceleration near the ridge [9], [17].

Looking at some cuts parallel to the rotor plane the radial extent of the wake can be investigated in more detail. Figs. 6 and 7 show the mean velocity fields for both cases at

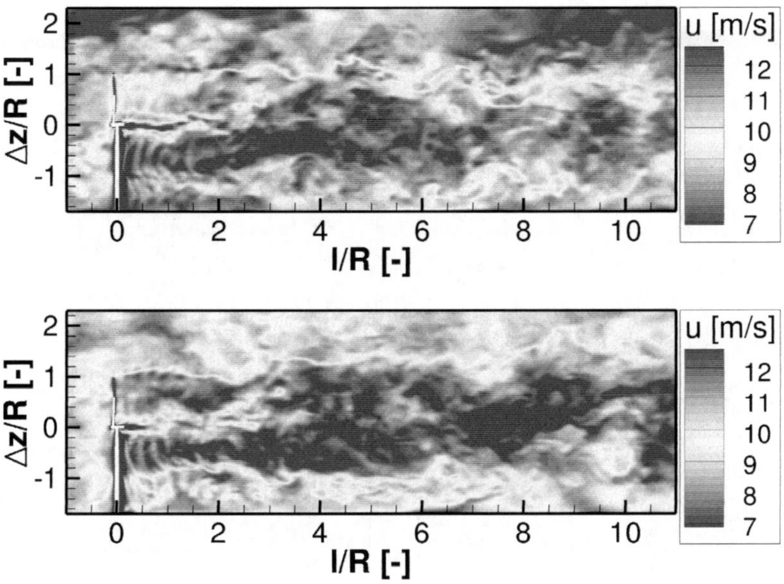

Figure 5: Snap shot of the wake flow field in flat terrain (upper picture) and complex terrain (lower picture) [17].

various downstream locations. In flat terrain (Fig. 6) the foot print of the rotor is nicely visible in a cricular area of reduced velocity at a relative disctance $l/R = 1$ downstream of the turbine position. The foot print of the tower diminishes in the low speed region close to the ground. Moving downstream the velocity deficit smears as fluid of higher energy is transported into the wake domain. As a result of the mixing process, the deficit gets more bell–shaped with increasing downstream distance and the wake edge begins to fade. In complex terrain (Fig. 7) the incoming velocity field is more homogeneous with less wind shear, compare Sec. 4.2. In consequence, the surrounding of the rotor and nacelle footprint features a more homogeneous field than in flat terrain. As the ambient velocity is higher close to the ground the tower foot print is much more prompting inside the flow field at $l/R = 1$. As already seen in the side view (Fig. 5) the tower deficit recovers quickly as the flow velocity close to the ground is higher than in flat terrain. In the downstream development the deficit starts to fade similar to the flat terrain case. The shape is hard to compare as the incoming flow fields are pretty different. Overall, a bell–shaped deficit can not be seen as clearly as in flat terrain.

Using the method described in Sec. 3.4 an investigation on the mean wake centerline and its motion over 12 revolutions has been performed. The time averaged evolution of the wake centerline is depicted in Fig. 8 along with the amplitudes of the lateral centerline motions. In flat terrain the wake centerline is virtually oriented in wind direction and shows no significant deflection in lateral direction [17]. The fluctuation amplitudes of the wake position is almost constant up to about $l/R = 6$ and starts to increase substantially further downstream. The wake breaks up at this position (compare Fig. 5) and larger scale motions are observed as documented in Ref. [17]. This larger scale movement is usually described as meandering. Larsen [26] explains it the following way. Each time a certain deficit sheet is created at the rotor and starts to propagate downstream. As the inflow varies temporarily these sheets propagte non–equally and may run up to each other. As a result, the wake structure is non–homogeneous over time at a certain position in space. The propagating deficit sheets may create a larger scale spatial movement in the wake depending on the turbulent length scale, ambient turbulence and flow homogenity, amongst others. Usually, meandering is related to length scales larger than the rotor area. This is not true for the investigated case suggesting that other effects like tip vortex interactions and instability are involved in the generation of large scale wake motions [17].

In contrast, for the present complex terrain case, the amplitudes are much smaller and the

IOP Conf. Series: Journal of Physics: Conf. Series **854** (2017) 012029 doi:10.1088/1742-6596/854/1/012029

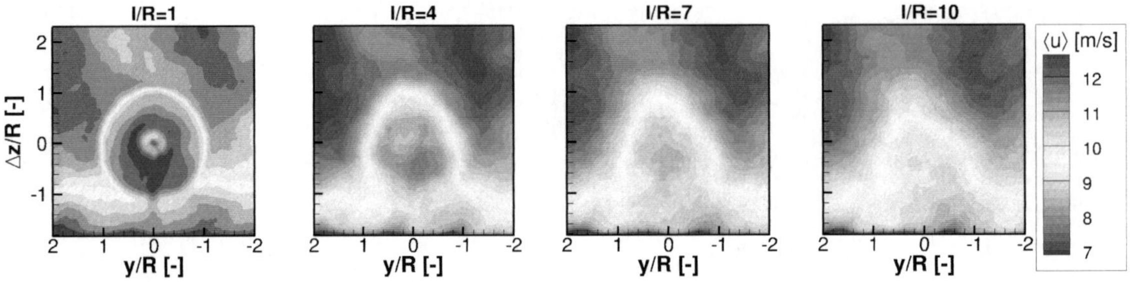

Figure 6: Time averaged wake deficit for the turbine in flat terrain [17].

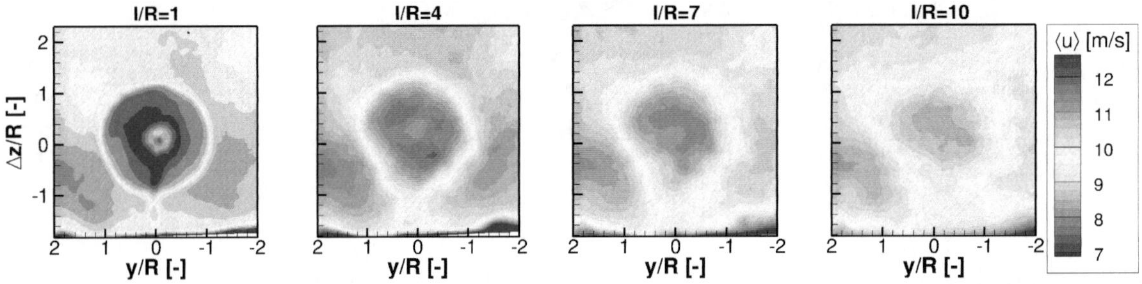

Figure 7: Time averaged wake deficit for the turbine in complex terrain [17].

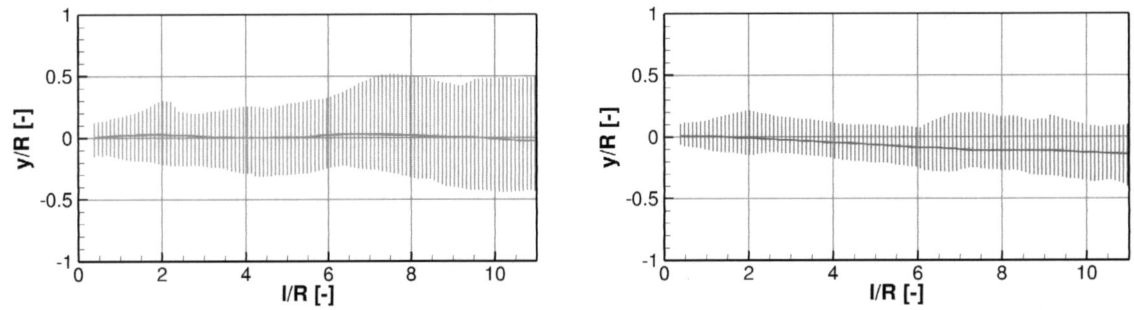

Figure 8: Development of wake center location and the lateral amplitude of the wake center location at hub height in flat (left) and complex terrain (right) [17].

wake propagates downstream more uniformly. A strong increase of the wake motion beyond $l/R = 6$ can not be observed [17]. The more homogeneous flow field and the reduced fluctuations seem to suppress the meandering tendency in the wake. Visually, the lateral deflection of the wake centerline is stronger and oriented towards negative y/R values for the present complex terrain case (compare Fig. 8). It has to be mentioned, however, that the axes are not to scale and the time-averaged deflection amounts to less than 1°. This lateral deflection can be attributed to a very small orography-caused local yaw angle but also to a combination of rotor cone and tilt yielding to non-symmetric loading and vortex shedding of the rotor.

Fig. 9 displays the spatially averaged power spectral density S_{uu} at various downstream locations in the wake. The spatial average has been performed along the y axis over the complete rotor diameter. In this way the spectra are much smoother and better comparable to each other than just one point spectra which may be dominated by single and strongly local events. The blue lines in the pictures depict data from the terrain simulations without turbine. The results were evaluated at hub height of the later turbine position. They can be interpreted as an undisturbed reference. The yellow and the red curves represent the results of the simulations including the virtual turbine and are given for two locations dowstream of the rotor. Prominent peaks in the signal directly downstream of the turbine (l/R = 0.25) can be noticed. These peaks appear at the harmonics of the blade passing frequencies and have also be shown in Refs. [17], [13]. For a small distance downstream of the turbine (l/R = 0.25) increased amplitudes result in the high frequency domain if compared to the inflow spectra of the pure terrain simulations. With increasing downstream distance the peaks diminish quickly and the spectra recover almost to the state of the inflow. The trends in the spectra are similar in complex and flat terrain.

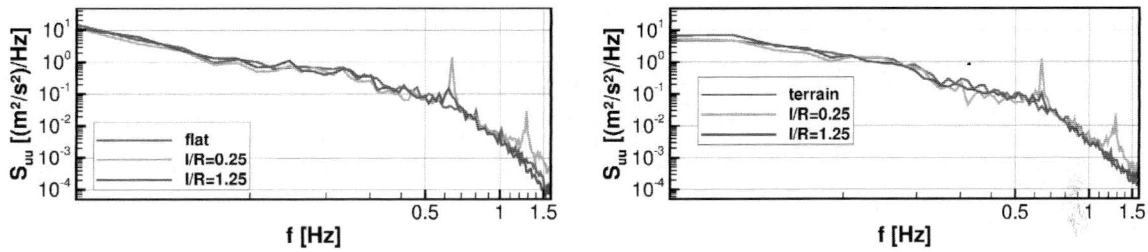

Figure 9: Wake spectra at two positions downstream of the turbine for flat (left) and complex terrain (right) compared to the spectra of the pure terrain simulations [17].

5. Conclusion and Outlook

In the present study numerical DES simulations were performed to investigate the impact of a complex terrain on the wake development of a virtual 2.4 MW wind turbine. For this purpose the orography at the planned WindForS test site near Stötten in Baden–Württemberg (Southern Germany) was considered. The test site is located at a flat plateau rearwards of an escarpment. To define the atmospheric inflow conditions for the simulations met mast data for the prevailing wind direction and neutral stratification were evaluated. The mean boundary–layer profile was approximated by a power–law profile and prescribed at the inflow boundary of the computational domain. The measured turbulence intensity of $TI \approx 10\%$ was generated synthetically by means of a *Mann* box and fed into the simulations by a volume force approach. Firstly, pure terrain simulations were conducted to characterise the local inflow conditions for the virtual turbine. The results for the complex terrain were compared to flat terrain results to identify the impact of the orography. It turned out that due to the flow accelaration near the ridge of the escarpment the velocity near ground was increased at the position of the virtual turbine, i.e. the boundary–layer profile was enriched and the turbine is exposed to a much more uniform inflow compared to the situation in flat terrain where significant shear effects appear.

In a next step the virtual wind turbine was embedded in the flat and in the complex terrain simulations and the resulting wake characteristics were evaluated. Analysis of the wake deficit showed that for the present complex terrain case the enrichment of the wake is slightly delayed. This was attributed to the reduced local ambient turbulence caused by the flow acceleration near the ridge of the escarpment. Further, the time-averaged evolution of the wake centerline was evaluated along with the amplitudes of their lateral motion. It was observed, that due to the higher local turbulence the wake meandering was more pronounced in flat terrain.

It shall be noted that these findings in first place hold for the present special orography, turbine location and wind condition. For differing wind directions stronger disturbances caused by forest and flow separation near the ridge are to be expected. Further detailed studies will be performed within the collaborative project *WINSENT*. Within this project two research turbines will be erected along with four met masts. Besides comprehensive field measurements detailed numerical models will be established and applied for flow field and aerodynamic studies.

Acknowledgements

The authors acknowledge the *German Federal Ministry for Economic Affairs and Energy* for funding this study in the framework of the German joint research project *Lidar complex* and *KonTest*, the *State Agency for Spatial Information & Rural Development Baden-Württemberg* for providing terrain data and the *High Perfomance Computing Center Stuttgart* for the contribution of the computational resources.

References

[1] Anger J *et al.* 2016 Abschlussbericht des forschungsprojektes: Erstellung einer konzeption eines windenergie–testgeländes in bergig komplexem terrain Tech. rep. WindForS
[2] Bergmann D, Bischoff O, Denzel J, Cheng P W, Emeis S, Hofsäß M, Lutz T, Peters B, Rauch J and Schulz C 2017 Abschlussbericht des forschungsprojektes: Entwicklung von lidar-technologien zur erfassung von windfeldstrukturen hinsichtlich der optimierung der windenergienutzung im bergigen, komplexen gelnde Tech. rep. WindForS
[3] Schulz C, Hofsäß M, Anger J, Rautenberg A, Lutz T, Cheng P W and Bange J 2016 *Journal of Physics: Conference Series* vol 753 (IOP Publishing) p 082017
[4] 17th February 2017 URL `http://www.windfors.de/english/index.html`
[5] Rettenmeier A, Anger J and Cheng P 2016 *Euromech Colloquium 576* (Euromech)
[6] Wildmann N, Bernard S and Bange J 2016 *Renewable Energy*
[7] Hofsäß M, Bischoff O and Cheng P W *IOP Conference Series: Earth and Environmental Science; ISARS 2016*
[8] Schulz C, Klein L, Weihing P, Lutz T *et al.* 2014 *Journal of Physics: Conference Series* vol 524
[9] Schulz C, Klein L, Weihing P and Lutz T 2016 *Journal of Physics: Conference Series* vol 753 (IOP Publishing) p 032016
[10] Weihing P, Schulz C, Lutz T and Krämer E 2015 *High Performance Computing in Science and Engineering 14* (Springer)
[11] Kroll N, Rossow C C, Becker K and Thiele F 2000 *Aerospace Science and Technology* **4**
[12] Benek J A, Steger J L, Dougherty F C and Buning P G 1986 *Chimera. A Grid-Embedding Technique.*
[13] Meister K 2015 *Numerische Untersuchung zum aerodynamischen und aeroelastischen Verhalten einer Windenergieanlage bei turbulenter atmosphärischer Zustömung* (Shaker)
[14] Troldborg N, Sørensen J N, Mikkelsen R and Sørensen N N 2014 *Wind Energy* **17** 657–669
[15] Mann J 1998 *Probabilistic engineering mechanics* **13** 269–282
[16] Kim Y, Jost E, Bangga G, Weihing P and Lutz T 2016 *Journal of Physics: Conference Series* vol 753 (IOP Publishing) p 032047
[17] Schulz C to be submitted 2017 *Numerische Untersuchungen des Verhaltens von Windenergieanlagen im komplexen Gelände unter turbulenter atmosphärischer Zuströmung* Ph.D. thesis Institute of Aerodynamics and Gas Dynamics, University of Stuttgart
[18] Lutz T, Meister K and Krämer E 2011 *EWEA 2011* (EWEA)
[19] Schulz C, Meister K, Lutz T and Krämer E 2016 *Notes on Numerical Fluid Mechanics and Multidisciplinary Design* (Springer)
[20] Schulz C, Meister K, Lutz T and Krämer E 2016 *New Results in Numerical and Experimental Fluid Mechanics X* (Springer) pp 871–882
[21] Spalart P R *et al.* 2006 *Theoretical and Computational Fluid Dynamics* **20**
[22] Jameson A 1991 *AIAA paper*
[23] Liu X D, Osher S and Chan T 1994 *Journal of computational physics* **115** 200–212
[24] Jiang G S and Shu C W 1995 Efficient implementation of weighted eno schemes. Tech. rep. DTIC Document
[25] Jameson A, Schmidt W, Turkel E *et al.* 1981 *AIAA paper*
[26] Larsen G C, Madsen H A, Thomsen K and Larsen T J 2008 *Wind energy* **11** 377–395

IOP Conf. Series: Journal of Physics: Conf. Series **854** (2017) 012030 doi:10.1088/1742-6596/854/1/012030

Towards physics-based operational modeling of the unsteady wind turbine response to atmospheric and wake-induced turbulence

Y Marichal[1], I De Visscher[1], P Chatelain[1,2] and G Winckelmans[1,2]

[1] WaPT - Wake Prediction Technologies, Rue Louis de Geer 6, 1348 Louvain-la-Neuve, Belgium

[2] Institute of Mechanics, Materials and Civil Engineering, Université catholique de Louvain, 1348 Louvain-la-Neuve, Belgium

E-mail: yves.marichal@wapt.be

Abstract. The objective of the present work is to develop a tool able to predict, in a computationally affordable way, the unsteady wind turbine power production and loads as well as its wake dynamics, as a function of the turbine dynamics and incoming wind conditions. Based on the lessons learned from a previous study about the characterization of the unsteady wake dynamics, the framework for an operational wake model is presented. The approach relies on an underlying vorticity-based skeleton consisting of different components, such as a regularized Vortex Sheet Tube (VST) and Vortex Dipole Line (VDL). Physically based evolution equations, accounting for the various flow phenomena occurring in the wake (such as advection, turbulent diffusion/core spreading, source/sink terms, etc.), are then derived. Once calibrated, the wake model is shown to be in good agreement with results of high-fidelity Large Eddy Simulations (LES) obtained using an Immersed Lifting Line-enabled Vortex Particle-Mesh method.

1. Introduction

The challenge for wind energy is to lower the levelized cost of energy (LCOE) in order to make it competitive compared to other types of energies. From a mechanics/aerodynamics perspective, this can be achieved by reducing as much as possible the material fatigue generating loads and/or by increasing the total power production. To that end, there is growing interest in global wind farm control, which would provide a superior global efficiency with respect to that resulting from current control strategies mostly focusing on individual turbines. At this level, one of the key aspects is the improvement of the response of wind turbine operations in wind farms to highly unsteady phenomena (wind gusts, meandering wakes impinging downstream rotors, emergency shutdown, etc.). In order to anticipate the incoming unsteady flow in the context of model-predictive control, very accurate and computationally efficient wake modeling is required. This goes beyond common steady state approaches mainly designed for the a priori optimization of wind farm layouts [1]. Many research efforts have been made in order to address this shortcoming, e.g. the Dynamic Wake Meandering (DWM) model [2], FLORIDyn [3], QBlade [4], etc. Yet, to the authors' opinion, these models still lack strong physical grounds, i.e. by assuming passive advection of the wake for the former or by lacking the transition from distinct tip vortices to far wake turbulence for free vortex wake models such as QBlade.

The present work aims to provide a wake modeling framework more in line with the underlying wake physics and vortex dynamics. Hence a vorticity-based skeleton is likely the best suited candidate for such a model. Based on a previous study about the wake characterization using such a vortex skeleton [5], the present paper focuses on the development of dynamical equations for the time evolution of the parameters describing the wake model components. The target is to develop an operational model capable of accurately predicting in real- or fast-time the behavior of a wind turbine and its wake with respect to upstream conditions; including atmospheric turbulence as well as turbulent wakes from upstream turbines.

First a description of the model and its components is provided in Section 2. Section 3 depicts the Immersed Lifting Line-enabled Vortex Particle-Mesh method that is used to calibrate and validate the wake model. The results are then presented in Section 4. Finally, conclusions are drawn in Section 5 along with perspectives of future developments.

2. Wake model formulation

We here consider the flow past a wind turbine of diameter D (radius $R = D/2$) with the hub centered at $\mathbf{x} = 0$. The upstream flow U is aligned with the rotor axis $\hat{\mathbf{e}}_z$, and it is assumed statistically uniform. The turbulence intensity is defined by $\mathrm{I} \triangleq w_{\mathrm{RMS}}/U$, where w_{RMS} is the root mean square of the axial velocity. The usual way to characterize the operational point of the wind turbine is given by the tip speed ratio $\lambda \triangleq \Omega R/U$, with Ω being the rotor rotational speed.

2.1. General description

Based on the wake characterization results obtained in [5], we use a vortex skeleton consisting of the following components (see Fig. 1) in order to model the wake of the wind turbine:

- The rotor operation is modeled using Blade Element Momentum (BEM) theory [6, 7].
- The near wake is represented by a regularized Vortex Sheet Tube (VST) of radius R^{VST}, azimuthal circulation per unit length γ^{VST} and mollification parameter σ^{VST} (here using a high order algebraic regularization, see [8, 9, 10]). This element can be related to the "tangential vorticity cylinder" from Branlard and Gaunaa [11].
- For the far wake, we use a regularized Vortex Dipole Line (VDL) of impulse per unit length i^{VDL} and mollification parameter σ^{VDL}.

Note that only the velocity deficit is modeled here, i.e. swirl effects related to axial vorticity, such as the root vortex emanating from the hub, are ignored. Therefore, in the current model formulation no vorticity is present at the rotor disc. For more details about this vortex skeleton, refer to [5]. In this study, it was shown through fitting of the model parameters that this skeleton is particularly well suited to represent the velocity deficit profiles measured around the centerline of a wind turbine wake (considering long and short time averages). Several evolution trends have also been identified for these parameters as a function of λ and I, e.g. low variability of the radius R^{VST}, increase of the core size σ^{VST}, etc. Based on these observations, we here focus on the development of a physical temporal evolution model for these parameters, which can be expressed in the form of a set of ODE's. The objective is to capture unsteady phenomena arising from dynamic inflow conditions (atmospheric and wake-induced turbulence from upstream turbines). This is the subject of the following section.

2.2. Rotor model: Blade Element Momentum theory (BEM)

The BEM inputs are the blade geometry and polar characterization, as well as the wind speed U, the collective blade pitch β and the rotor rotation speed Ω. The latter two are obtained from the static response curve of the wind turbine as a function of the wind speed (a realistic wind

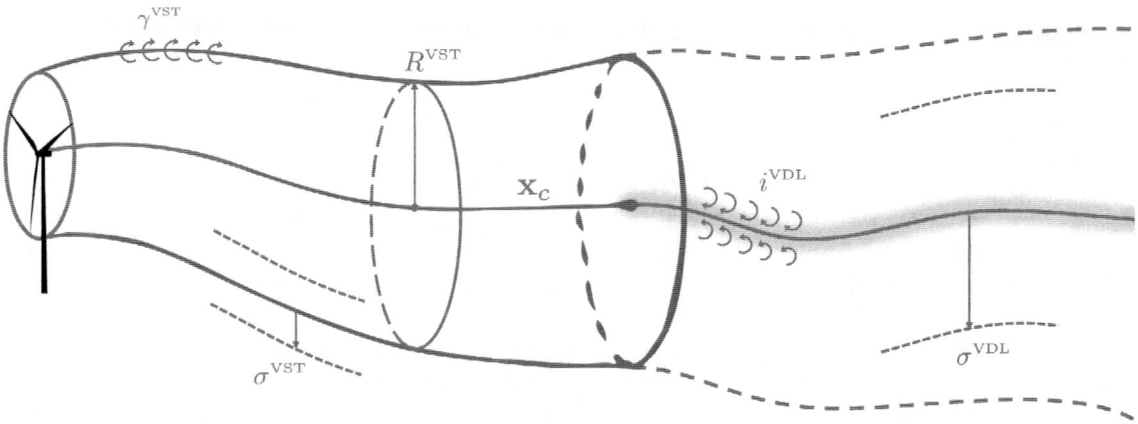

Figure 1: Sketch of the wake model components along with the component parameters: Vortex Sheet Tube (VST) and Vortex Dipole Line (VDL).

turbine controller could be used instead). Note that the wind speed required by the BEM is a "freestream" velocity which precludes the contribution of the wake. Since the vortex skeleton framework makes a clear distinction between the wakes coming from different wind turbines, one can easily evaluate the wind speed "as seen" by the designated turbine (i.e. rotor-averaged velocity) by simply excluding the own wake vorticity contribution.

One can then also obtain the circulation distribution resulting from the BEM

$$\Gamma(r) = \frac{1}{2}\, U_{\text{rel}}(r)\; c(r)\; C_l(r)\;,$$

with U_{rel} the relative velocity norm observed at the blade (in the plane normal to the blade), c the local chord and C_l the lift coefficient.

2.3. Near wake model: Vortex Sheet Tube (VST)

The evolution equations that are derived here rely on the conservation of a quantity $\phi(\mathbf{x}, t)$ integrated over a material line

$$\frac{D(\phi dl)}{Dt} \triangleq \frac{\partial(\phi dl)}{\partial t} + (\mathbf{u}\cdot\nabla)(\phi dl) = 0\;,$$

where $\mathbf{u}$ is the velocity field and $d\mathbf{l}(\mathbf{x}, t) = dl\,\hat{\mathbf{e}}_l$ is an infinitesimal material line element ($\hat{\mathbf{e}}_l$ is the unit orientation vector of the line element and dl its length). Applying this to a particular material line $\mathcal{L}(t)$ described by $\mathbf{X}_n(\lambda, t)$ and neglecting the line curvature, the previous equation becomes

$$\frac{\partial f}{\partial t} + \left(\frac{\partial l}{\partial \lambda}\right)^{-1} \frac{\partial}{\partial \lambda}\,(u_l\, f) = 0\;, \tag{1}$$

with $f(\lambda, t) \triangleq \phi(\mathbf{X}_n(\lambda, t), t)$, $l = l(\lambda, t)$ an arc-length coordinate and u_l the projection of $\mathbf{u}$ on $\hat{\mathbf{e}}_l$. λ is a "normal" Lagrangian coordinate in the sense that it characterizes the path $\mathbf{X}_n(\lambda, t)$ resulting from the advection using the velocity normal to the line $\mathbf{u}_n \triangleq \mathbf{u} - u_l\hat{\mathbf{e}}_l$. From a numerical perspective, the advection is hence decomposed into a normal and a tangential contribution, where the first part is handled in a Lagrangian way, and the second part of the advection is treated in a Eulerian fashion by solving Eq. (1) over the control points $\mathbf{X}_n(\lambda_j, t)$ $(j = 1 \ldots N)$.

The application of this framework to the wake model requires us to assume an axisymmetric wake with a straight centerline along $\hat{\mathbf{e}}_z$. The VST is considered to be a material surface. In the polar frame of reference (r, z), Eq. (1) can be applied over the tube surface described by $\mathbf{X}_n = R^{\text{VST}}\hat{\mathbf{e}}_r + z_c^{\text{VST}}\hat{\mathbf{e}}_z$.

The VST can be seen as a collection of vortex rings. For an inviscid flow, their circulation $\gamma^{\text{VST}}dl$ must be conserved in time. Therefore, Eq. (1) is applicable with $f = \gamma^{\text{VST}}$. Similarly, in order to model the vortex stretching of the vortex rings in an inviscid flow, Eq. (1) must be satisfied with $f = \sigma^{\text{VST}} R^{\text{VST}}$. Turbulent mixing is taken into account by adding source and sink terms based on an eddy viscosity model. The circulation equation is thus extended with a turbulent stress term that is assumed to be proportional to $\gamma^{\text{VST}}/(\sigma^{\text{VST}})^2$, and the mollification parameter equation is complemented using a core-spreading technique [12], similarly to what is done for free vortex wake models [13]. In summary, the following ODE's are obtained

$$
\begin{aligned}
&\frac{\partial R^{\text{VST}}}{\partial t} = \mathbf{u}_n \cdot \hat{\mathbf{e}}_r \\
&\frac{\partial z_c^{\text{VST}}}{\partial t} = \mathbf{u}_n \cdot \hat{\mathbf{e}}_z \\
&\frac{\partial \gamma^{\text{VST}}}{\partial t} + \left(\frac{\partial l}{\partial \lambda}\right)^{-1} \frac{\partial}{\partial \lambda}\left(u_l\, \gamma^{\text{VST}}\right) = -\nu_{\text{eff},\gamma}^{\text{VST}} \frac{\gamma^{\text{VST}}}{(\sigma^{\text{VST}})^2} \\
&\frac{\partial(\sigma^{\text{VST}} R^{\text{VST}})}{\partial t} + \left(\frac{\partial l}{\partial \lambda}\right)^{-1} \frac{\partial}{\partial \lambda}\left(u_l\, (\sigma^{\text{VST}} R^{\text{VST}})\right) = 2\, \nu_{\text{eff},\sigma}^{\text{VST}} \left(\frac{R^{\text{VST}}}{\sigma^{\text{VST}}}\right) .
\end{aligned}
\tag{2}
$$

The eddy viscosity is modeled in the spirit of Ainslie [14] by separating the contributions from the wake turbulence and from the ambient turbulence (i.e. atmosphere and impinging wakes from upstream turbines)

$$
\begin{aligned}
\nu_{\text{eff},\gamma}^{\text{VST}} &= \varepsilon_{\gamma,\text{w}}^{\text{VST}} \nu_{\text{w}}^{\text{VST}} + \varepsilon_{\gamma,\text{a}}^{\text{VST}} \nu_{\text{a}}^{\text{VST}} \\
\nu_{\text{eff},\sigma}^{\text{VST}} &= \varepsilon_{\sigma,\text{w}}^{\text{VST}} \nu_{\text{w}}^{\text{VST}} + \varepsilon_{\sigma,\text{a}}^{\text{VST}} \nu_{\text{a}}^{\text{VST}} ,
\end{aligned}
$$

where $\varepsilon_{(\cdot)}^{\text{VST}}$ are calibrated model coefficients. The wake contribution is modeled as

$$\nu_{\text{w}}^{\text{VST}} = F(z/l_{\text{BD}})\, \sigma^{\text{VST}} \Delta U_c ,$$

with ΔU_c the local wake deficit and $F(\cdot)$ a filter function accounting for the lack of equilibrium between the mean velocity field and the wake generated turbulent field [14]. The filter function is here adapted compared to its original definition [14]: a S-shaped function, going from 0.1 to 1 in the far wake, is used, and the original filter length scale D is replaced by the tip vortex breakdown length l_{BD} obtained by Sorensen *et al* [15] through stability analysis of the tip vortex system. The ambient contribution is modeled as $\nu_{\text{a}}^{\text{VST}} = \text{I}\, \nu_{\text{w}}^{\text{VST}}$, which also slightly differs from the original definition [14].

The boundary conditions for Eqs. (2) are obtained by extracting information from the BEM rotor model. The tip vortex core size emanating from the rotor is modeled through

$$\sigma^{\text{VST}}(0, t) = \sigma_T^{\text{VST}} \triangleq \alpha_\sigma^{\text{VST}} \frac{\Gamma_b}{\left|\frac{d\Gamma}{dr}\right|_{r=R}} ,$$

where $\Gamma_b = \max_r(\Gamma)$ and $\alpha_\sigma^{\text{VST}}$ is a calibrated coefficient. The VST radius and circulation per unit length are obtained through a momentum balance around the rotor actuator disc

$$R^{\text{VST}}(0, t) = R_T^{\text{VST}} \triangleq \sqrt{\frac{2T}{\rho N_b \Gamma_b \Omega}} \quad \text{and} \quad \gamma^{\text{VST}}(0, t) = \gamma_T^{\text{VST}} \triangleq \frac{N_b \Gamma_b \Omega}{2\pi U_{\text{TV}}} ,$$

with T the rotor thrust obtained by the BEM, ρ the air density, N_b the number of blades and U_{TV} the tip vortex advection velocity induced by the vortex skeleton at the rotor ($r = R^{\mathrm{VST}}(0,t)$). This result can be compared to [16], where the link between the circulation per unit length and the thrust is obtained using the pressure jump across an actuator disc. Note that the equation for $\gamma^{\mathrm{VST}}(0,t)$ is implicit, but no iteration is performed here thanks to the unsteady framework (the γ^{VST} value from previous time step can be used to evaluate U_{TV}).

Preliminary tests have shown that a few additional adjustments of the model are required. First, since we do not explicitly model the hub jet and its breakdown, as was done in [5], $\gamma^{\mathrm{VST}} = \gamma_T^{\mathrm{VST}}$ is here fixed for $z \leq z_0 = C_{\mathrm{BD}}\, l_{\mathrm{BD}}$ with C_{BD} a calibrated coefficient. Secondly, the material surface hypothesis for the VST is well suited to describe the initial wake expansion. As soon as $\sigma^{\mathrm{VST}}/R^{\mathrm{VST}}$ becomes significant, this hypothesis becomes inappropriate. Hence, the slope of R^{VST} is prescribed for $z > z^*$ (z^* is such that $\frac{\partial R^{\mathrm{VST}}}{\partial z}(z^*) = 0$):

$$\frac{\partial R^{\mathrm{VST}}}{\partial z} = k_{\mathrm{R}}^{\mathrm{VST}} \max(0, \mathrm{I} - \mathrm{I}_{\mathrm{th}}) \; ,$$

with $k_{\mathrm{R}}^{\mathrm{VST}}$ and I_{th} calibrated coefficients (note that the discontinuity in the R^{VST} derivative at z^* is not problematic since the radius slope is small in practice).

2.4. Far wake model: Vortex Dipole Line (VDL)
The ODE's describing the time of evolution of i^{VDL} and σ^{VDL} are similar to those of the VST. This is due to the fact that the VST and VDL models become equivalent when the ratio $\sigma^{\mathrm{VST}}/R^{\mathrm{VST}}$ becomes large. One then obtains that $i^{\mathrm{VDL}} = \pi(R^{\mathrm{VST}})^2\gamma^{\mathrm{VST}}$. From an operational perspective, the VDL is far less computationally expensive to evaluate compared to the VST. Yet, it provides a satisfactory approximation of the velocity profile in the far wake and can therefore replace the VST. As the scope of the present paper solely consists in the calibration of the wake model, the VDL component is left out here.

3. Wake flow simulation using a Vortex Particle-Mesh method (VPM)
The wake flow is obtained through simulations of a wind turbine evolving in a turbulent wind field using an Immersed Lifting Line-enabled Vortex Particle-Mesh (VPM) method, as presented in Chatelain *et al* [17]. The VPM flow solver relies on the vorticity formulation of the Navier-Stokes equations. Advection is handled in a Lagrangian fashion using vorticity-carrying particles, and all the remaining spatial differential operations, such as the solution of the Poisson equation to obtain the velocity, the vortex stretching term, as well as the subgrid-scale modeling term are efficiently computed on an underlying grid [18], thus in a Eulerian manner (information being interpolated back and forth between the particles and the grid using high order interpolation schemes). The presence of the blades is modeled using an Immersed Lifting Line approach.

The approach is very well suited to efficiently capture detailed flow physics of wind turbines, from the near wake to the far wake. This is enabled by the relaxation of usual linear stability constraints (CFL criterion for advection problems) and by low numerical dispersion and numerical diffusion errors. Moreover, the computational domain is very compact, thanks to the Hockney-Eastwood algorithm [19] that allows unbounded outer conditions on the lateral domain faces.

Adopting the same procedure as described in [5], the inflow turbulence is accounted for using vortex particles carrying the vorticity deduced from Mann's synthetic atmospheric turbulence model [20].

4. Results
The wake model described in Section 2 is calibrated using simulation results of the flow past the NREL-5MW offshore wind turbine [21].

4.1. Wake flow characterization

The flow characterization procedure is fully described in [5]. Basically, in order to quantify the behavior of the wake deficit, the characterization starts by identifying the unsteady wake centerline. Indeed, as the ambient turbulence intensity increases, the wake meandering phenomenon plays a more and more significant role in the wake dynamics, see Fig. 2, and the centerline is subject to a transversal fluctuation (lateral and vertical). The centerline computation can be applied to short and long time-averaged velocity fields. Here, the target is to obtain an unsteady description of the wake deficit with an averaging period ΔT that filters out the fine grain turbulence but preserves the fluctuations related to the wake meandering and large atmospheric gusts. Next, the azimuthally averaged axial velocity profiles are computed around the centerline at several z/D stations, similarly to [22]. The envelope of these profiles, as well as their time average are shown in Fig. 3 at $z/D = 4$ and $z/D = 7$ at I = 6.6% and I = 15%. The calibration of the wake model is performed based on this information.

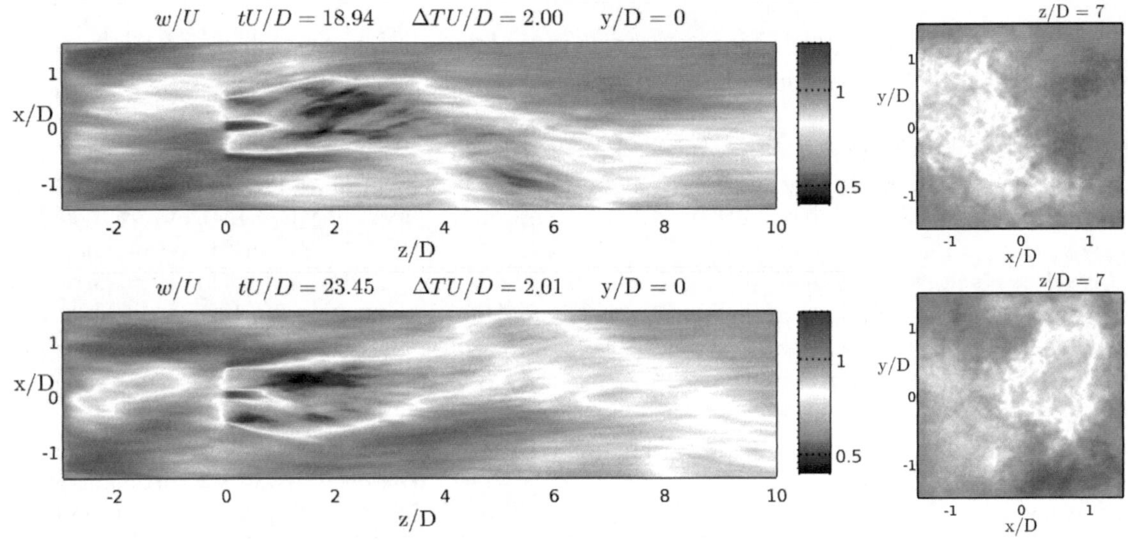

Figure 2: Axial velocity maps for the case with $\lambda = 7.55$ and I = 15%.

4.2. Wake model calibration and validation

Given the importance of accounting for the unsteady meandering of the wake centerline, the present model has been designed such that the centerline can handle a transversal displacement. This will be performed by introducing atmospheric turbulence into the computational domain, following an approach similar to that of the DWM model. However, as a first validation step, the model performance is here assessed with respect to time-averaged wake velocity profiles and diagnostics (yet, computed around the time dependent centerline as measured in the VPM simulations).

We here investigate the influence of the turbulence intensity I at optimal tip speed ratio $\lambda = 7.55$. Two cases are considered: I = 6.6% (typical offshore value) with $\Delta TU/D = 2.1$ and I = 15% (typical onshore value) with $\Delta TU/D = 2.0$. The model coefficients have been calibrated such that the velocity deficit profiles from the wake model match those of the VPM simulations. In the present study, we consider the time average of the time-filtered axial velocity profiles computed around the unsteady centerline from the VPM simulation results. The inflow velocity for the wake model is constant and the turbulence is simply characterized by I.

The advection terms in Eqs. (2) are discretized using first order upwind finite differences and the ODE's are integrated using the Euler time integration scheme. The time integration stops when a steady state is reached, based on a residual threshold criterion. The resulting axial velocity profiles are shown in Fig. 3 at two different axial positions downstream of the rotor. Figure 4 provides an overview of the spatial evolution of some wake characteristics, i.e. the wake radius (corresponding to the radial location associated to a deficit recovery of 25%, 50%, 75% and 90%) and the axial velocity at specific radial stations ($r/D = 0$, 0.5, 0.75 and 1).

There is a good agreement between the model and the VPM results, especially in the far wake. Despite the significantly higher variability of the case with I = 15%, the calibrated model provides a very accurate representation of the wake deficit profiles, even in the far wake. As to the axial velocity in the near wake, the accuracy of the model is lower. This is due to the fact that the hub jet is not intended to be explicitly captured by the model. Yet, the model is able to predict the wake flow in a wide range of atmospheric turbulence conditions.

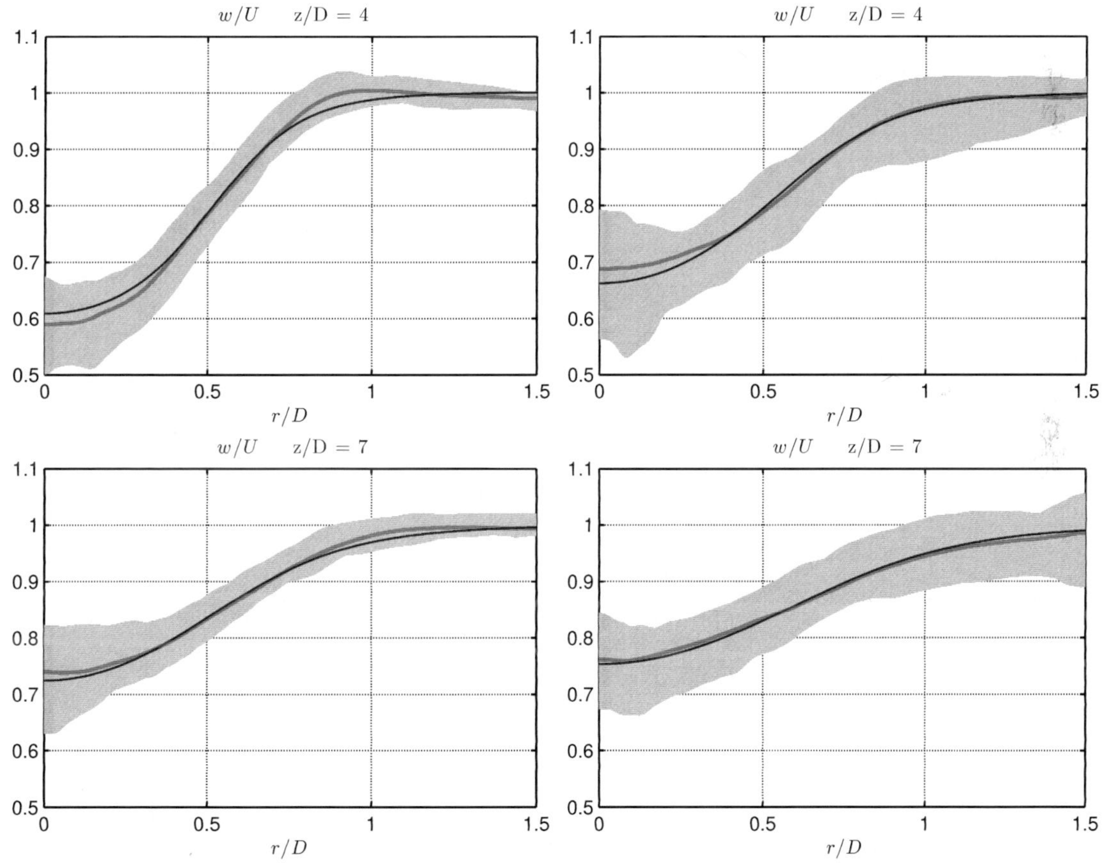

Figure 3: Axial velocity profiles: I = 6.6% (left) and I = 15% (right); VPM profile time average (dark green), VPM profile envelope (light green) and wake model (blue).

The computational time related to the evaluation of the model is not addressed in detail here since the current implementation of the model stands as a proof of concept Matlab tool and is thus non-optimized as such. The Biot-Savart velocity evaluation is quite efficient, though, since it is performed in a Fortran subroutine. Nevertheless, the CPU time and complexity related to the method lies between a simple BEM algorithm and a typical free vortex wake method. As an indication, a typical run takes a couple of minutes on a modern desktop PC (single process) to

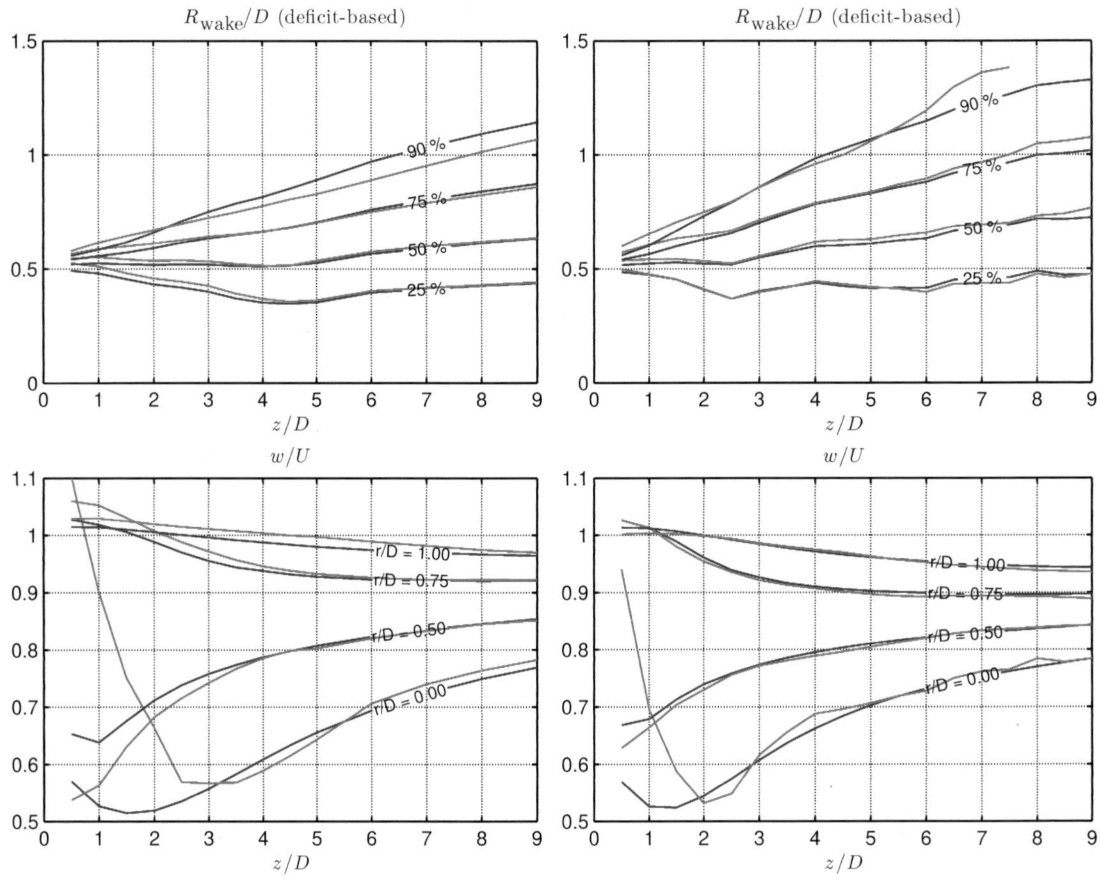

Figure 4: Wake radius based on the deficit recovery (top) and axial velocity at several radial positions (bottom): I = 6.6% (left) and I = 15% (right); VPM profile time average (dark green) and wake model (blue).

reach a steady state (i.e. 15 to $25D/U$ of physical time), depending on the considered operating point, flow condition and initial condition.

5. Conclusion and perspectives

An operational wake model has been presented for the prediction of the unsteady wake dynamics induced by the flow past a wind turbine subjected to different turbulent inflow conditions. The model is able to accurately reproduce the time-averaged wake behavior for different turbulence intensities, as obtained by high-fidelity Large Eddy Simulations using an Immersed Lifting Line-enabled Vortex Particle-Mesh method. The next calibration step will consist in comparing the model predictions with VPM simulations by also varying the tip speed ratio. Thanks to the unsteady nature of the model based on a set of ODE's, it is expected that the model will reproduce reasonably well low frequency wake deficit variations induced by an unsteady freestream velocity signal. This shall be confirmed by future validation studies, also using field measurement data.

As to perspectives of future developments, in order to further improve the unsteady response, the equilibrium state hypothesis in the momentum balance underlying the BEM theory could be relaxed by accounting for unsteady effects and turbulent fluctuations. Moreover, improved

BEM methods taking into account non uniform inflow conditions, as well as skewed/yawed inflow [6, 16] might be used. Another important feature will be the relaxation of the axisymmetric wake hypothesis by allowing a lateral displacement of the centerline in order to reproduce wake meandering and wake misalignment. One could also consider additional wake model components accounting for swirl, e.g. using elements similar to those of [11].

Acknowledgments

The present research was supported by the Walloon Region under the grant agreement n°1420169 and it benefited from computational resources made available on the Tier-1 supercomputer of the Fédération Wallonie-Bruxelles, infrastructure funded by the Walloon Region under the grant agreement n°1117545.

References

[1] Katic I, Hojstrup J and Jensen N O 1986 *European Wind Energy Association (Rome)* pp 407–409
[2] Larsen G C, Madsen H A, Thomsen K and Larsen T J 2008 *Wind Energy* **11** 377–395 ISSN 10954244
[3] Gebraad P M O and van Wingerden J W 2014 *Journal of Physics: Conference Series* vol 524 p 012186 ISBN 9781634393058 ISSN 1742-6596
[4] Marten D, Wendler J, Pechlivanoglou G, Nayeri C N and Paschereit C O 2013 *International Journal of Emerging Technology and Advanced Engineering* **3** 264–269
[5] Marichal Y, De Visscher I, Chatelain P and Winckelmans G 2016 *Journal of Physics: Conference Series* **753** 032050 ISSN 1742-6588
[6] Hansen M 2008 *Aerodynamics of Windturbines* (Earthscan) ISBN 9781844074389
[7] Burton T, Jenkins N, Sharpe D and Bossanyi E 2011 *Wind Energy Handbook Second Edition* (John Wiley & Sons, Ltd) ISBN 9780470699751
[8] Winckelmans G 1989 *Topics in Vortex Methods for the Computation of Three- and Two-Dimensional Incompressible Unsteady Flows* Ph.D. thesis Caltech Pasadena
[9] Winckelmans G S and Leonard A 1993 *J. Comput. Phys.* **109** 247–273
[10] Winckelmans G 2004 *Encyclopedia of computational mechanics* vol 3 (Fluids) ed Stein E, de Borst R and Hughes T (John Wiley & sons) pp 129–153
[11] Branlard E and Gaunaa M 2016 *Wind Energy* **19** 345358 ISSN 1095-4244
[12] Leonard A 1980 *J. Comput. Phys.* **37** 289–335
[13] Sant T 2007 *Improving BEM-based Aerodynamic Models in Wind Turbine Design Codes* Ph.D. thesis TUDelf
[14] Ainslie J 1988 *Journal of Wind Engineering and Industrial Aerodynamics* **27** 213–224 ISSN 0167-6105
[15] Sorensen J N, Mikkelsen R F, Henningson D S, Ivanell S, Sarmast S and Andersen S r J 2015 *Philosophical Transactions of the Royal Society A: Mathematical, Physical and Engineering Sciences* **373** 20140071–20140071 ISSN 1364-503X
[16] Yu W, Ferreira C and van Kuik G 2016 *34th Wind Energy Symposium* January pp 1–12 ISBN 978-1-62410-395-7
[17] Chatelain P, Backaert S, Winckelmans G and Kern S 2013 *Proceedings of The 9th International Symposium on Engineering Turbulence Modelling and Measurements (ETMM-9), June 6–8, 2012, Thessaloniki, Greece* (*Flow, Turbulence and Combustion* vol 91) ed Rodi W ERCOFTAC (Springer) pp 587–605
[18] Chatelain P, Curioni A, Bergdorf M, Rossinelli D, Andreoni W and Koumoutsakos P 2008 **197** 1296–1304
[19] Chatelain P and Koumoutsakos P 2010 *J. Comput. Phys.* 2425 – 2431 ISSN 0021-9991
[20] Mann J 1994 *J. Fluid. Mech.* **273** 141–168
[21] Jonkman J, Butterfield S, Musial W and Scott G 2009 Definition of a 5-mw reference wind turbine for offshore system development Tech. Rep. TP-500-38060 NREL
[22] Madsen H A, Larsen G C, Larsen T J, Troldborg N, Mikkelsen R, Box P O and Roskilde D 2010 *Journal of Solar Energy Engineering* **132** 2–32 ISSN 0199-6231

IOP Conf. Series: Journal of Physics: Conf. Series **854** (2017) 012031 doi:10.1088/1742-6596/854/1/012031

Wake Flow Simulation of a Vertical Axis Wind Turbine Under the Influence of Wind Shear

Victor Mendoza and Anders Goude

Department of Engineering Sciences, Division of Electricity, Uppsala University, Uppsala 751 21, Sweden

E-mail: victor.mendoza@angstrom.uu.se

Abstract. The current trend of the wind energy industry aims for large scale turbines installed in wind farms. This brings a renewed interest in vertical axis wind turbines (VAWTs) since they have several advantages over the traditional Horizontal Axis Wind Tubines (HAWTs) for mitigating the new challenges. However, operating VAWTs are characterized by complex aerodynamics phenomena, presenting considerable challenges for modeling tools. An accurate and reliable simulation tool for predicting the interaction between the obtained wake of an operating VAWT and the flow in atmospheric open sites is fundamental for optimizing the design and location of wind energy facility projects. The present work studies the wake produced by a VAWT and how it is affected by the surface roughness of the terrain, without considering the effects of the ambient turbulence intensity. This study was carried out using an actuator line model (ALM), and it was implemented using the open-source CFD library OpenFOAM to solve the governing equations and to compute the resulting flow fields. An operational H-shaped VAWT model was tested, for which experimental activity has been performed at an open site north of Uppsala-Sweden. Different terrains with similar inflow velocities have been evaluated. Simulated velocity and vorticity of representative sections have been analyzed. Numerical results were validated using normal forces measurements, showing reasonable agreement.

1. Introduction

The amount of wind farms that are being built is increasing considerably, and it produces a relevant interest in the scientific community for the understanding of the interaction between the produced wake from the turbines and the atmospheric turbulent flow [1–5]. Even more, today there is a renewed interest for vertical axis wind turbines (VAWTs) for offshore floating arrays and wind farms with closer spacing due to their faster wake recovery [6, 7]. Aditionally, the omni-directionality of VAWTs makes them insensitive to the wind direction, resulting in a simpler mechanical design with a few moving parts, which excludes the yawing equipment and (often) the pitching mechanism. This characteristic is considerably appreciated in offshore farm facilities, because the yawing mechanism is one of the biggest sources of failures in the horizontal wind turbines (HAWT) [8–10]: the operation and maintenance take big part of the total energy cost. Another relevant advantage in VAWTs, is the fact that the generator can be placed at sea level, which stabilizes the structure due to its low center of gravity and therefore could reduce the size and cost of the floating base. Consequently, generators with larger dimensions and weight can be used, which allows for heavy direct drive generators with permanent magnets [11]. Compared to HAWTs, VAWTs have higher potential for scalability [12, 13].

VAWTs are characterized by complex and unsteady three-dimensional aerodynamics which present considerable challenges for both simulations and measurements to describe it. The near-wake structure of a VAWT has been revealed to be mainly dominated by the effects of the blade tip vortex, which produces levels of recovery (due to the vertical advection) considerably higher than those from turbulent variations [14]. This effect is not present in HAWT wakes. The energy conversion process in VAWTs is based on the variation of the blade's circulation along the rotation axis of the turbine, producing a different creation of the wake compared to HAWTs, in which the wake is only generated by the tip vortices. In order to understand the performance of VAWTs better, it is necessary to have a fundamental knowledge about the flow phenomena involved. Moreover, to maximize the efficiency of a singular VAWT within a wind farm facility, an accurate prediction of the resulting flow from an operating VAWT is essentially required. For this goal, numerical simulations can contribute useful qualitative and quantitative understanding of VAWTs wakes in a relatively fast and economical way.

The numerical works that have been carried out to investigate wakes behind VAWTs can be separated into two principal approaches. The first one consists in the full resolution of the boundary layers around the blades [15–18]. Through this method it is possible to obtain detailed information about the load distribution and the flow fields within the rotor and near wake regions. However, there is a considerable computational cost inherently involved, which restricts its utilization for wake studies in realistic large scale environments. Additionally, the accuracy of the results is directly related to the numerical method and the applied turbulence model [19, 20]. In the second approach, the blades are simulated by using the actuator type technique, which excludes the need of the full resolution of the boundary layer flow around the blade. Therefore, these simplifications makes it feasible to run studies of the far wake of VAWTs and VAWT wind farms [4, 5, 21, 22].

The present work aims to simulate the produced wake from a VAWT applying an actuator line model (ALM), and also to identify the effects of the interaction between the wake structure and the incident flow. Different terrains with their own aerodynamic roughness length have been considered. A turbulent inflow generator was not used, in order to isolate the effect of wind shear from that of turbulence intensity. The model has been implemented using the open-source CFD toolbox library OpenFOAM to solve the governing equations and to compute the resulting wake [23]. An operational H-shaped VAWT model was tested, for which experimental activity has been performed at an open site north of Uppsala-Sweden [24]. Numerical results were compared to normal forces from experimental data.

2. Methodology

An actuator line model (ALM) coupled to a dynamic stall model (DSM) has been implemented to solve the governing equation of the involved phenomena. The ALM samples the velocity from the Navier-Stokes solver, and then it calculates the angle of attack and relative velocity in each blade element. The DSM computes blade force coefficients, which the ALM uses to impart back the body force into the flow field solver. The main characteristics of the model were described in this section, it is suggested to read [25, 26] for more detailed information.

2.1. Actuator Line Model

The applied ALM is a three-dimensional and unsteady aerodynamics model developed by Sørensen and Shen [27], and created to study the surrounded flow of wind turbines. It is a coupling between a Navier-Stokes solver for the flow field with the actuator line technique, which distributes body force along lines representing the blades of the turbine. The used forces are determined through the DSM, based on empirical data.

2.2. Dynamic Stall Modeling

The DSM is represented by the Leishman and Beddoes model [28], with the variation of Sheng et al. [29] and Dyachuk [26]. It is able to compute the unsteady lift, pithing moment and drag, giving a physical description of the aerodynamics. It has been validated with experimental results in [30]. A division of three subsystems is considered in the model: an attached flow model for unsteady linear airloads, a separated flow model for non-linear airloads, and a dynamic stall model for the airloads induced by the leading edge vortex. In [25], the results have shown small improvements in the accuracy and agreement with experimental data compared to the results where the attached flow correction of the DSM is not included.

3. Simulation Parameters: Validation case

Experimental data from a 12 kW VAWT with diameter of 6.48 m were used to validate the model (Figure 1). This device is operating in a open site in the north of Uppsala. The measurement campaign and the obtained forces are published in [31] and [24], respectively; four load cells were used to measure forces on one blade and its support arms. Due to the lack of experimental results for the far wake of VAWTs in open sites, normal forces for a TSR of $\lambda = 3.44$ are used as the validation parameter. The chosen TSR value is close to the optimal one. Specifications of the turbine are shown in Table 1. In order to represent the results in an easier way, a rectangular coordinate system has been used with the origin placed at the base of the turbine, such that the x-axis is pointing positively in the downwind direction. Central shaft and supporting struts were included in the computation.

Figure 1. The 12 kW turbine, designed and built by the Division of Electricity at Uppsala University. The turbine is equipped with load cells used for the force measurements [31]

Table 1. Specification of the 12 kW VAWT used for validation

Number of blades	3
Turbine diameter	6.48 m
Hub height	6.0 m
Blade length	5.0 m
Airfoil profile	NACA0021
Chord length	25 cm
Blade pitch angle	2°
TSR	3.44°

The turbine described above was operating in different terrains with different surface roughness such that the velocity at the equator of the turbine ($z_\infty = 5.75$ m) is $U_\infty = 6.4$

m/s. The friction velocity U_{x*} and the equator-height mean wind speed are used on the log law in order to do this,

$$U_x = \frac{U_{x*}}{K} ln \left(\frac{z + z_0}{z_0} \right) \tag{1}$$

where z_0 is the roughness length and $K = 0.40$ is the von Kármán constant.

A uniform flow with a non-slip boundary condition for the ground (common configuration used for simulations), and logarithmic inlet velocity profiles in snow, long grass and bushes have been tested. These parameters are summarized in Table 2. Simulations were carried out with a domain considering open site boundary conditions.

Table 2. Aerodynamic roughness lengths used for the different cases

Case	Inlet velocity profile	z_0 [m]
1	uniform	-
2	logarithmic	0.0025 (snow)
3	logarithmic	0.055 (long grass)
4	logarithmic	0.55 (bushes)

Figure 2 displays the main characteristics of the calculated mean wind shears over the different surface types. These fields were used as incoming flows for the simulations of the resulting wake.

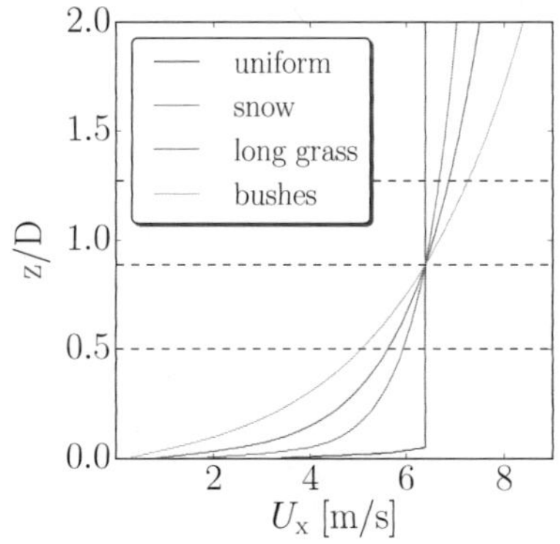

Figure 2. Vertical profiles of the streamwise velocity used at the inlet of the domain. Dashed lines represent the extent and center of the turbine

4. Results and discussion

In this section, streamwise velocity for representative planes are analyzed in order to study the wake behind the operating VAWT over the flat surfaces with different roughness lengths. The obtained normal forces were compared to experimental data. A LES Smagorinsky model has been used to predict the turbulence effects.

4.1. Mesh sensitivity

A reference mesh topology with local refinement around the region close to the turbine and in the downwind direction (in order to capture details of the obtained wake) has been used to

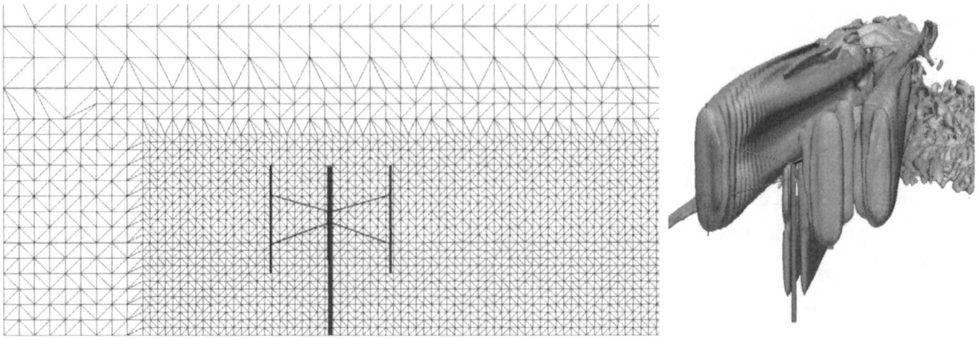

Figure 3. Illustration of the reference mesh section used with 0.43×10^6 cells (left) and vorticity isosurfaces generated by the operating turbine using 3.30×10^6 cells (right)

study the mesh sensitivity of the model, see Figure 3. This topology was kept constant and proportionally scaled in every direction for the refinements.

Figure 4 shows the normal force response during one revolution for both experimental and simulated values, varying the number of mesh points of the domain. All simulations were performed for 40 turbine revolutions to ensure convergence of the results. The surface roughness length used is $z_0 = 0.025$ m, which corresponds to the terrain of the open site where the turbine is placed. It should be noticed that there is no considerable improvement in the force predictions when increasing the mesh resolution beyond 3.3×10^6 mesh cells. Generally, the simulated results are in good agreement with the experimental values.

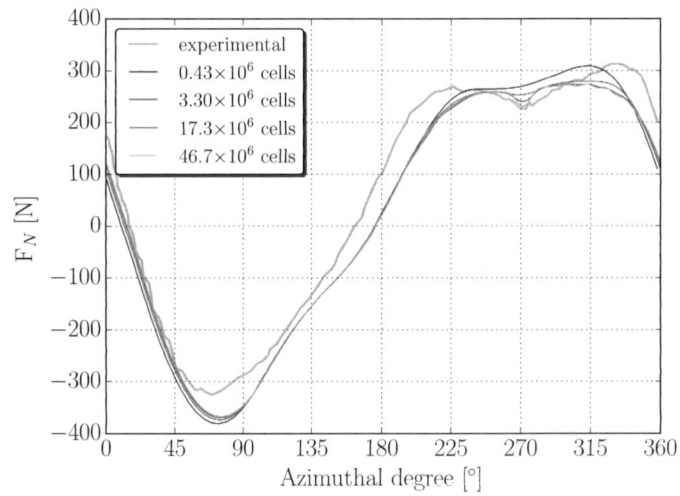

Figure 4. The normal force response for different mesh sizes

Figure 5 depicts the simulated instantaneous streamwise velocity components in the vertical plane for four different mesh sizes: 0.43, 3.3, 17.3 and 46.7×10^6 cells, which correspond to a cell size/blade-span ratio of 0.075, 0.375, 0.0213 and 0.0152, respectively. The two coarsest meshes do not have enough resolution to capture the vortical structures produced by the blade tips and the turbulent regime is not fully developed until the end of the domain. Figure 6 shows for all the cases,that the resulting wake has similar shape and position; hence the model is able to represent the global structure of the wake independently of the mesh size. However, the refinement is necessary to capture the details of the generated vortex which has an essential contribution to the wake recovery. The wake transition to the turbulent regime is closer to the upwind direction when the mesh is finer: at x/D $\simeq$ 12 and 8 for the meshes of 17.3 and 46.7$\times10^6$ cells, respectively.

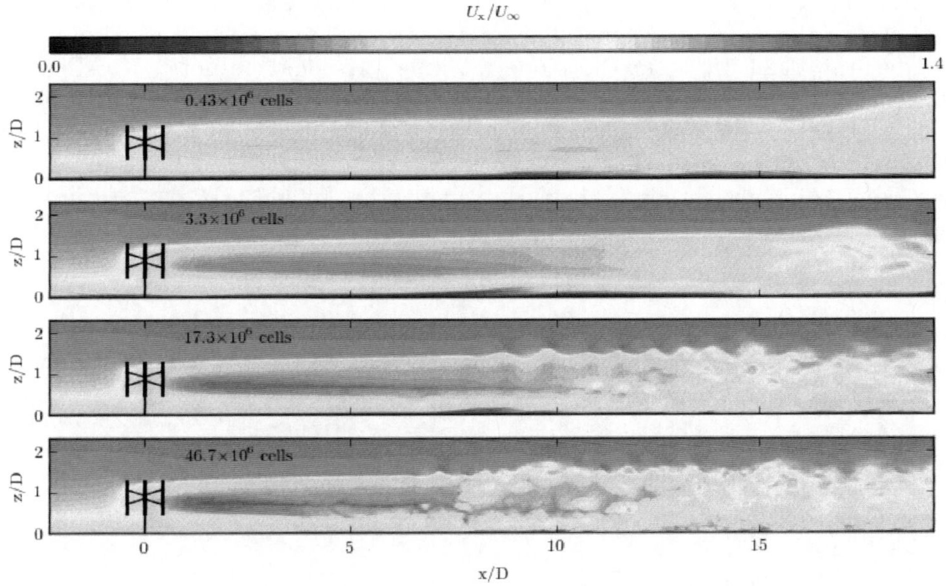

Figure 5. Normalized instantaneous streamwise velocity in the vertical plane at the centre of the turbine for different mesh sizes: 0.43, 3.3, 17.3 and 46.7$\times10^6$ cells

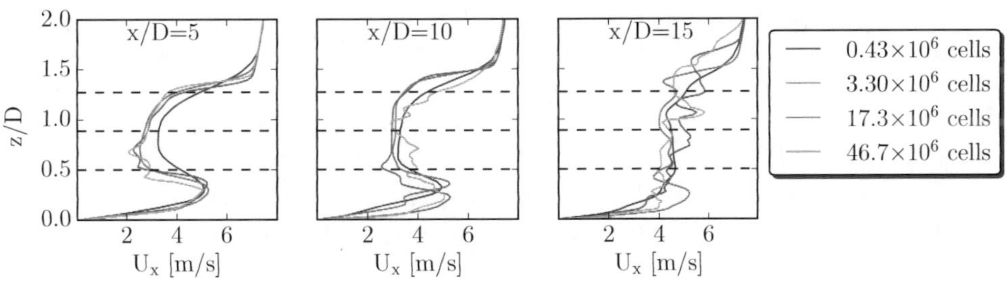

Figure 6. Vertical profiles of the instantaneous streamwise velocity component for different mesh sizes: 0.43, 3.3, 17.3 and 46.7$\times10^6$ cells. Dashed lines represent the extent and center of the turbine

Regarding to the temporal discretization, previous studies have shown that the model is less sensitive to temporal than spacial discretization [25, 32]. Nonetheless, the Courant-Friedrichs-Lewy (CFL) condition must be taken in account for the time discretization, to ensure convergence.

4.2. Terrain roughness variation

Figures 7 and 8 reveal the resulting wake produced by the turbine. These results were obtained using a mesh resolution of 17.3×10^6 cells. In all the cases the wake shape (geometry) and position are similar. Furthermore, expansion and recovery of the wake in the downwind direction is clearly identified, as well as the uniform and logarithmic profiles at the inlet of the domain. In the vertical plane of velocity (Figure 7), it is shown that the flow is accelerated in the upper part of the wake after facing the turbine. For the uniform flow, an acceleration is also observed in the lower part of the wake. The general characteristics are mainly dominated by the vortical structures from the blade tips and the blade pitching. It could be noticed in Figure 8 that there is strong creation of vortical structures in the interface between the resulting wake and the ground and, as expected, the rougher the surface the larger turbulence. Inside the wake there is a region with no vorticity, and this becomes smaller when the surface roughness increases, which means that the wake breaks earlier to the turbulent regime. In general, larger turbulence levels lead to faster wake recovery because of the improvement of the mixing process and momentum transfer.

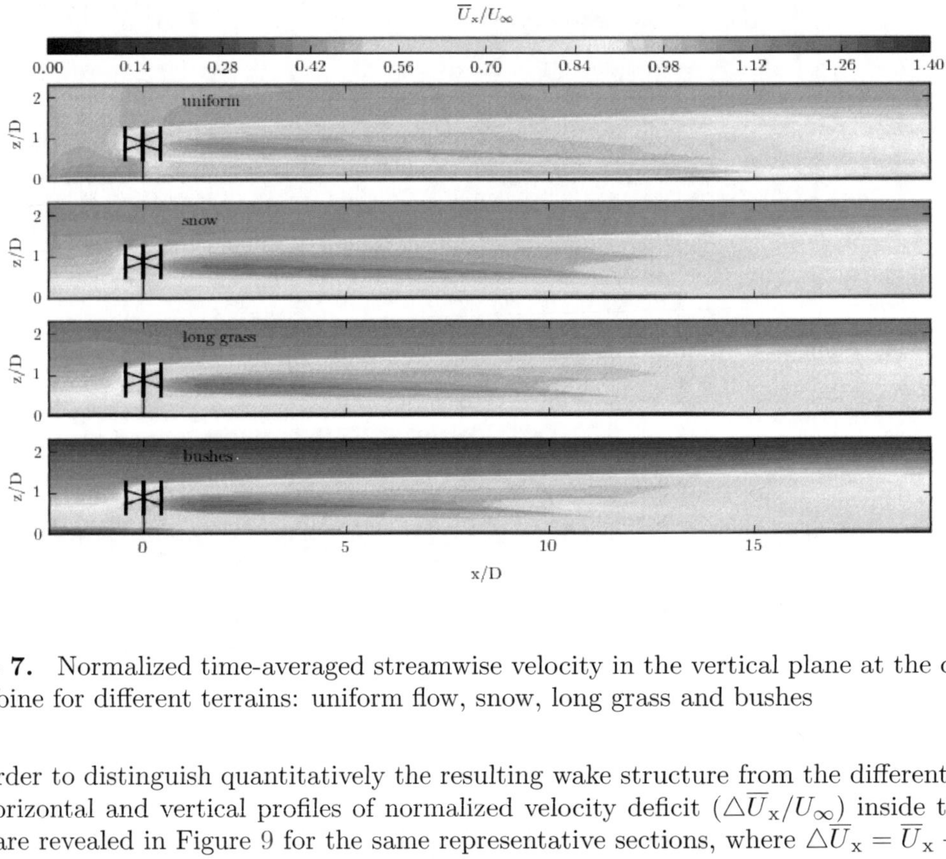

Figure 7. Normalized time-averaged streamwise velocity in the vertical plane at the centre of the turbine for different terrains: uniform flow, snow, long grass and bushes

In order to distinguish quantitatively the resulting wake structure from the different results, both horizontal and vertical profiles of normalized velocity deficit ($\triangle\overline{U}_{\mathrm{x}}/U_\infty$) inside the wake region are revealed in Figure 9 for the same representative sections, where $\triangle\overline{U}_{\mathrm{x}} = \overline{U}_{\mathrm{x}} - U_{\mathrm{inflow}}$ is the time-averaged velocity deficit, and U_{inflow} is the mean streamwise inflow velocity shown in Figure 2. As can be observed, the wake has a slight asymmetry in the horizontal direction, and a strong influence from the ground in the vertical structure. No significant changes are noticed until the section x/D = 10. After this, the velocity deficit decreases approximately from 50% to 30% in the equatorial area at x/D = 15. Effects leaded from the wind shears are mainly in the perpendicular direction to the main flow.

Figure 10 displays the vertical contour of the streamwise turbulence intensity for the different terrains. An evident production of turbulence intensity is noticed after the section x/D = 10,

IOP Conf. Series: Journal of Physics: Conf. Series **854** (2017) 012031 doi:10.1088/1742-6596/854/1/012031

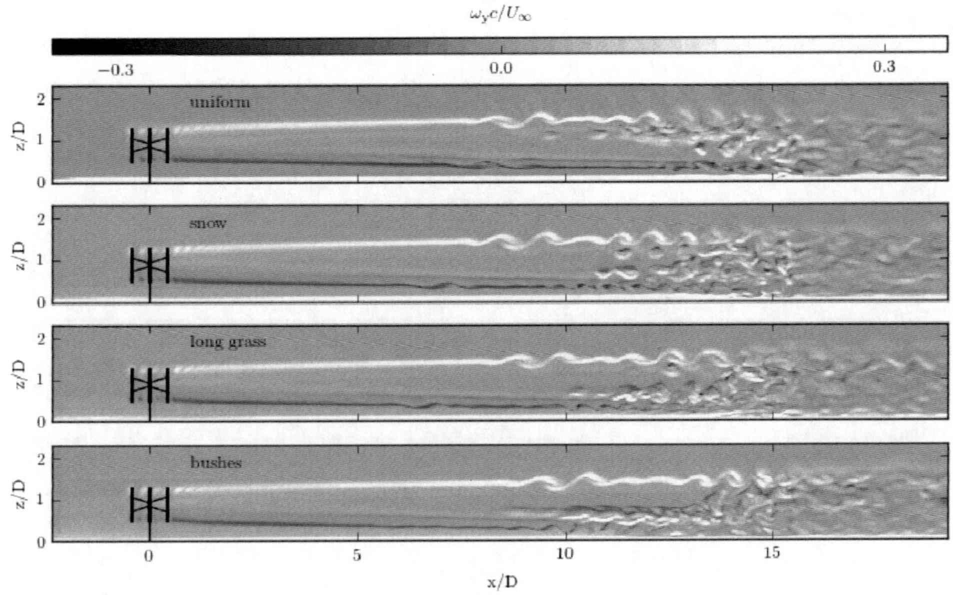

Figure 8. Normalized instantaneous cross-stream vorticity in the vertical plane at the centre of the turbine for different terrains: uniform flow, snow, long grass and bushes

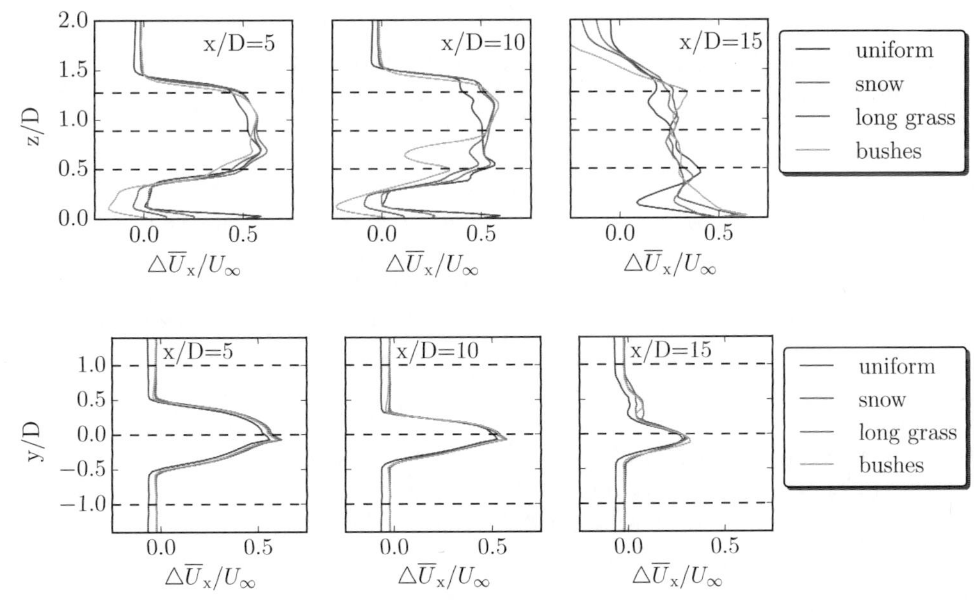

Figure 9. Vertical (up) and horizontal (down) profiles of the normalized velocity deficit for different terrains: uniform flow, snow, long grass and bushes. Dashed lines represent the extent and center of the turbine

with a higher intensity in the lower part of the wake where the shear is stronger (except for the uniform flow case). Turbulence intensity generated from the tip vortices can also be observed.

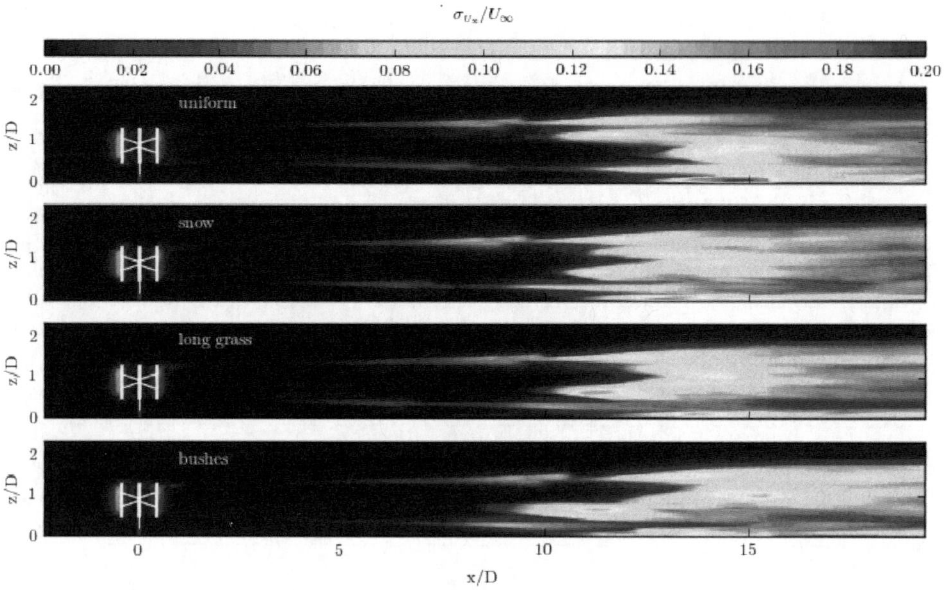

Figure 10. Streamwise turbulence intensity in the vertical plane for different terrains: uniform flow, snow, long grass and bushes

5. Conclusions

The presented ALM is able to reproduce the main pattern of a wake produced by an operating VAWT in an open site, revealing its shape and position. The numerical results showed reasonable agreement with experimental results of normal forces. The model allows to evaluate qualitatively and quantitatively the most relevant parameters and phenomenas involved in the flow pattern: velocity field with its zones of acceleration and deceleration, regions with high and low concentration of vorticity and turbulence production. Interaction between the wake and the ground is captured. Effects on the contribution from the wind shears in the mixing process and the transition to turbulent regime are clearly identified for the terrain roughness variation. The model shows stability for the tested cases, which makes it suitable for application in VAWTs simulations.

Acknowledgment

This work was conducted within StandUp for Wind, a part of the StandUp for Energy strategic research framework. The computational works were performed on resources provided by the Swedish National Infrastructure for Computing (SNIC) at NSC.

References

[1] Erik L P, Niels G M, Lars L, Jørgen H, and Helmut P F. Wind power meteorology. part i: climate and turbulence. *Wind Energy*, 1(S1):25–45, 1998.

[2] Vermeer L J, Sørensen J N, and Crespo A. Wind turbine wake aerodynamics. *Progress in aerospace sciences*, 39(6):467–510, 2003.

[3] Baidya Roy S, Pacala S W, and Walko R L. Can large wind farms affect local meteorology? *Journal of Geophysical Research: Atmospheres*, 109(D19), 2004.

[4] Shamsoddin S and Porté-Agel F. A large-eddy simulation study of vertical axis wind turbine wakes in the atmospheric boundary layer. *Energies*, 9(5):366, 2016.

[5] Abkar M and Dabiri J O. Self-similarity and flow characteristics of vertical-axis wind turbine wakes: an les study. *Journal of Turbulence*, 18(4):373–389, 2017.
[6] Dabiri J O. Potential order-of-magnitude enhancement of wind farm power density via counter-rotating vertical-axis wind turbine arrays. *Journal of renewable and sustainable energy*, 3(4):043104, 2011.
[7] Kinzel M, Mulligan Q, and Dabiri J O. Energy exchange in an array of vertical-axis wind turbines. *Journal of Turbulence*, 13(1):N38, 2012.
[8] Ribrant J and Bertling L. Survey of failures in wind power systems with focus on swedish wind power plants during 1997-2005. In *Power Engineering Society General Meeting, 2007. IEEE*, pages 1–8. IEEE, 2007.
[9] Tavner P J, Xiang J, and Spinato F. Reliability analysis for wind turbines. *Wind Energy*, 10(1):1–18, 2007.
[10] Arabian-Hoseynabadi H, Oraee H, and Tavner P J. Failure modes and effects analysis (fmea) for wind turbines. *International Journal of Electrical Power & Energy Systems*, 32(7):817–824, 2010.
[11] Eriksson S, Solum A, Leijon M, and Bernhoff H. Simulations and experiments on a 12kw direct driven pm synchronous generator for wind power. *Renewable Energy*, 33(4):674–681, 2008.
[12] Musgrove P J. Wind energy conversion: recent progress and future prospects. *Solar & wind technology*, 4(1):37–49, 1987.
[13] Peace S. Another approach to wind: vertical-axis turbines may avoid the limitations of today's standard propeller-like machines. *Mechanical Engineering-CIME*, 126(6):28–32, 2004.
[14] Bachant P and Wosnik M. Characterising the near-wake of a cross-flow turbine. *Journal of Turbulence*, 16(4):392–410, 2015.
[15] Castelli M R, Englaro A, and Benini E. The darrieus wind turbine: Proposal for a new performance prediction model based on cfd. *Energy*, 36(8):4919–4934, 2011.
[16] Bremseth J and Duraisamy K. Computational analysis of vertical axis wind turbine arrays. *Theoretical and Computational Fluid Dynamics*, 30(5):387–401, 2016.
[17] Marsh P, Ranmuthugala D, Penesis I, and Thomas G. Three-dimensional numerical simulations of straight-bladed vertical axis tidal turbines investigating power output, torque ripple and mounting forces. *Renewable Energy*, 83:67–77, 2015.
[18] Posa A, Parker C M, Leftwich M C, and Balaras E. Wake structure of a single vertical axis wind turbine. *International Journal of Heat and Fluid Flow*, 61:75–84, 2016.
[19] Lam H F and Peng H Y. Study of wake characteristics of a vertical axis wind turbine by two-and three-dimensional computational fluid dynamics simulations. *Renewable Energy*, 90:386–398, 2016.
[20] Chowdhury A M, Akimoto H, and Hara Y. Comparative cfd analysis of vertical axis wind turbine in upright and tilted configuration. *Renewable Energy*, 85:327–337, 2016.
[21] Pierce B, Moin P, and Dabiri J. Evaluation of drag forcing models for vertical axis wind turbine farms. In *APS Meeting Abstracts*, November 2013.
[22] Shamsoddin S and Porté-Agel F. Large eddy simulation of vertical axis wind turbine wakes. *Energies*, 7(2):890–912, 2014.
[23] Bachant P, Goude A, and W Martin. turbinesFoam: v0.0.7. Zenodo. `http://dx.doi.org/10.5281/zenodo.49422`, 2016.
[24] Dyachuk E, Rossander M, Goude A, and Bernhoff H. Measurements of the aerodynamic normal forces on a 12-kw straight-bladed vertical axis wind turbine. *Energies*, 8(8):8482–8496, 2015.
[25] Mendoza V, Bachant P, Wosnik M, and Goude A. Validation of an actuator line model coupled to a dynamic stall model for pitching motions characteristic to vertical axis turbines. In *Journal of Physics: Conference Series*, volume 753, page 022043. IOP Publishing, 2016.
[26] Dyachuk E. *Aerodynamics of vertical axis wind turbines: Development of simulation tools and experiments.* PhD thesis, Acta Universitatis Upsaliensis, 2015.
[27] *Computation of Wind Turbine Wakes using Combined Navier-Stokes/Actuator-line Methodology.*
[28] Leishman J G and Beddoes T S. A generalised model for airfoil unsteady aerodynamic behaviour and dynamic stall using the indicial method. In *Proceedings of the 42nd Annual forum of the American Helicopter Society*, pages 243–265. Washington DC, 1986.
[29] Sheng W, Galbraith R A, and Coton F N. A modified dynamic stall model for low mach numbers. *Journal of Solar Energy Engineering*, 130(3):031013, 2008.
[30] Leishman J G and Beddoes T S. A semi-empirical model for dynamic stall. *Journal of the American Helicopter society*, 34(3):3–17, 1989.
[31] Rossander M, Dyachuk E, Apelfröjd S, Trolin K, Goude A, Bernhoff H, and Eriksson S. Evaluation of a blade force measurement system for a vertical axis wind turbine using load cells. *Energies*, 8(6):5973–5996, 2015.
[32] Bachant P, Goude A, and Wosnik M. Actuator line modeling of vertical-axis turbines. *arXiv preprint arXiv:1605.01449, Submitted to Wind Energy*, 2016.

Comparative study on the wake deflection behind yawed wind turbine models

Jannik Schottler[1], Franz Mühle[2], Jan Bartl[3], Joachim Peinke[1], Muyiwa S Adaramola[2], Lars Sætran[3], Michael Hölling[1]

[1] ForWind, University of Oldenburg, Institute of Physics, Oldenburg, Germany
[2] Faculty of Environmental Sciences and Natural Resource Management, Norwegian University of Life Sciences, Ås, Norway
[3] Department of Energy and Process Engineering, Norwegian University of Science and Technology, Trondheim, Norway

E-mail: jannik.schottler@forwind.de

Abstract. In this wind tunnel campaign, detailed wake measurements behind two different model wind turbines in yawed conditions were performed. The wake deflections were quantified by estimating the rotor-averaged available power within the wake. By using two different model wind turbines, the influence of the rotor design and turbine geometry on the wake deflection caused by a yaw misalignment of 30° could be judged. It was found that the wake deflections three rotor diameters downstream were equal while at six rotor diameters downstream insignificant differences were observed. The results compare well with previous experimental and numerical studies.

1. Introduction

In order to reduce power losses in wind farms caused by wake effects, different active wake control strategies were investigated in the recent years. One promising approach is an intentional yaw misalignment which is able to deflect the wake deficit laterally and thus reduce its impact on downstream turbines. However, a detailed understanding of a yawed wind turbine's wake trajectory is needed before potentially applying the concept.

As full-scale field studies are limited and numeric simulations face modeling uncertainties, wind tunnel experiments are a useful tool for validation purposes. Wakes of yawed model wind turbines have been addressed in multiple wind tunnel studies using different model turbines of varying geometries and rotor designs. For a small model of $D = 0.12\,\mathrm{m}$ and a yaw angle of $\gamma = 30\,°$, Medici and Alfredsson [1] found a wake deflection of about $z/D \approx 0.5$ at a downstream distance of $x/D = 4$. For the same yaw angle, Krogstad and Adaramola [2] measured a wake deflection of $z/D \approx 0.2$ rotor diameters already at $x/D = 1$. The model turbine, however, had a rotor diameter of $D{=}0.90\,\mathrm{m}$ and thus was significantly larger. A slightly different wake deflection angle for $\gamma = 30°$ was found in an experimental PIV study by Bastankhah and Porté-Agel [3], who localized the wake center at $z/D \approx 0.16$ for $x/D{=}2$ and at $z/D \approx 0.34$ for $x/D{=}6$ downstream distance for a smaller wind turbine model of $D{=}0.15\,\mathrm{m}$. Howland et al. [4] used a yawed drag disc of $D{=}0.03\,\mathrm{m}$ and found a deflection between $z/D \approx 0.4$ and $z/D \approx 0.5$ at a distance of $x/D = 6$, dependent on the measurement technique and fitting method. All of these studies qualitatively showed consistent wake deflection effects, that can be generalized. However,

certain quantitative differences in the deflection can be observed. Several factors could influence the variation in the quantification of the wake deflection. First of all, the airfoil dependent force coefficients are different for the rotors, giving a different initial forcing on the air deflected around the turbines' blades. Secondly, the tower and nacelle geometries of the different model turbines are deemed to have a certain influence to interact with the flow. A third factor could be inflow and boundary conditions in different wind tunnel environments. Furthermore, Howland et al. [4] show a significant influence of the 2- or 3-dimensional fitting method on the wake deflection.
In order to judge effects of boundary conditions such as turbine geometry, blade design or wind tunnel blockage, comparative wake measurements behind two different model turbines developed at ForWind in Oldenburg and at the NTNU in Trondheim were performed. The models vary in size, geometry and blade design. Other dimensionless parameters of the setup such as tip speed ration (TSR) are equal, allowing for an isolation of effects caused by the turbine type. This paper describes the experimental setup and methods used in detail. Further, the wake deflection during a yaw misalignment of $\gamma = -30°$ is quantified for both turbines at two downstream distances using an approach estimating the available power of a potential downstream turbine.

2. Experimental setup

2.1. Wind tunnel facility & inflow conditions

The experiments were conducted at the closed-loop wind tunnel at the Norwegian University of Science and Technology (NTNU) in Trondheim, Norway. The test section is 11.15 m long with an inlet cross-section of 2.71 m × 1.81 m (width × height). During the experiments a shear-flow-generating turbulence grid was installed at the inlet to the test section, which is described by Bartl & Sætran in detail [5]. The installed turbulence grid generates a vertical shear in the inflow velocity, being non-uniform over the rotor area. A common way to describe wind shear is the power law [6]

$$\frac{u}{u_{ref}} = \left(\frac{y}{y_{ref}}\right)^{\alpha}, \tag{1}$$

which expresses the wind speed u as a function of height y, given that the wind speed at an arbitrary reference height y_{ref} is known. The coefficient α describes the strength of shear in the profile. The shear grid used here generates a profile that can be approximated by a shear coefficient of $\alpha = 0.11$, which is a realistic value for offshore boundary layers in stable atmospheric conditions [7]. Table 1 gives an overview of the inflow conditions for each turbine. The turbulence intensity,

$$TI = \frac{\sigma_u}{\langle u \rangle}, \tag{2}$$

and the free stream velocity u_∞ refer to the respective turbine's hub height.

Table 1. Overview of the inflow conditions at each turbine's location for the empty wind tunnel. α corresponds to Eq. (1), u_∞ and TI refer to the data at the respective hub height.

Turbine	u_∞ [ms^{-1}]	TI [%]	α [−]	Hub height [m]
ForWind	7.5	5.2	0.11	0.820
NTNU	10	10.1	0.11	0.890

IOP Conf. Series: Journal of Physics: Conf. Series **854** (2017) 012032 doi:10.1088/1742-6596/854/1/012032

2.2. Model wind turbines

Two different model wind turbines were used that vary in size, blade design and geometry. The turbines will be denoted *NTNU* and *ForWind*, respectively. The ForWind turbine has a rotor diameter of $D_{\text{ForWind}} = 0.580\,\text{m}$ and a hub diameter of 0.077 m. In order to adjust the hub height of the turbine to the center of the NTNU wind tunnel, the turbine tower was extended with four rods of a diameter of 0.02 m attached to the base plate. The rotor blades are based on an SD-7003 airfoil profile, and were manufactured by a vacuum casting method using a MG804 synthetic (isocyanate-polyol) compound. Further details about the blade design and airfoil characteristics can be found in Section 2.3. The rotor is designed for turning in clockwise direction when observed from upstream. The ForWind turbine features an automatic load control, which allows an operation at constant TSR during the experiments. The pitch angle β can be adjusted using a stepper motor (see [8] for details) and was set to have maximum performance in perfect yaw alignment, which resulted in a value of $\beta = 1.8°$ for this experiment. The pitch angle and tip speed ratio were then kept constant when the turbine was yawed to an angle of $\gamma = -30°$. Further details about the load control and the turbine design are described by Schottler et al. [8].

The NTNU turbine has a larger rotor of $D_{\text{NTNU}} = 0.894\,\text{m}$ with a hub diameter of 0.090 m. It is a re-design of the turbines used in [5], having a thinner tower and a smaller nacelle in order to minimize their effects on the wake flow behind the turbine. The aluminum blades are based on a NREL S826 airfoil, which was designed by the National Renewable Energy Laboratory (NREL). The NTNU rotor turns in counter-clockwise direction and is driven at by a 400W Panasonic LIQI electric servo motor. The servo motor ensures a rotation at constant tip speed ratio, while the excessive power is burned off in an external resistor. The pitch angle of the turbine has to be adjusted manually and was fixed to $\beta = 0°$ for this experiment.

Both model turbines were operated at the same TSR of $\lambda = 6$. Table 2 gives an overview of the main characteristics of both turbines used, Figure 1 shows scaled sketches of both turbine models.

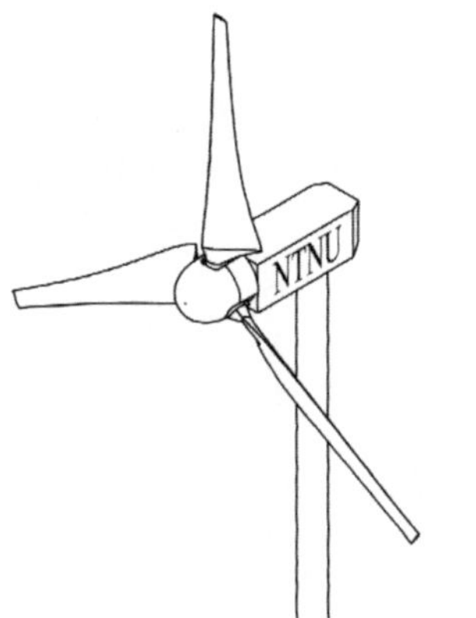

Figure 1. Scaled technical drawings of both model wind turbines. Left: NTNU turbine, right: ForWind turbine.

IOP Conf. Series: Journal of Physics: Conf. Series **854** (2017) 012032 doi:10.1088/1742-6596/854/1/012032

Table 2. Summary of main turbine characteristics. The TSR is based on the free stream velocity u_∞ at hub height. The Reynolds number at the blade tip, Re_{tip}, is based on the chord length at the blade tip and the effective velocity during turbine operation. The blockage corresponds to the ratio of the rotor's swept area to the wind tunnel's cross sectional area. The direction of rotation refers to observing the rotor from upstream.

Turbine	Rotor diameter	Hub diameter	Blockage	TSR	Re_{tip}	Rotation
ForWind	0.580 m	0.077 m	5.4 %	6	$\approx 6.4 \times 10^4$	clockwise
NTNU	0.894 m	0.090 m	13 %	6	$\approx 1.1 \times 10^5$	counter-clockwise

2.3. Rotor and airfoil comparison

The ForWind rotor has a total twist of about 22° and a chord length distribution varying from about 70 mm at the root to about 20 mm at the tip. The rotor is an upscaled version of the rotor described in Odemark and Fransson [9]. It is build from elements based on the SD-7003 airfoil, which is specifically designed for low Reynolds number operation. A detailed documentation is given by Selig et al. [10] and Counsil et al. [11].

The total twist distribution on the bigger NTNU rotor spans about 38° from the blade root to the tip, while the chord length varies from 82 mm to 26 mm. The rotor geometry is defined in detail by Sætran & Bartl [12] and is based on the NREL S826 airfoil from the root to the tip. The airfoil was originally designed for full-scale wind turbines operating at Reynolds numbers in the order of 10^6 and is documented by Somers [13]. As the Reynolds number for this model experiment is about one order of magnitude lower, experimental performance data were recorded for low to moderate Reynolds numbers. One comprehensive dataset is provided by Sarmast and Mikkelsen [14], another by Ostovan et al. [15].

The airfoil shapes, on which the two rotors are based, as well as the performance at a Reynolds number of $Re = 1 \times 10^5$ are compared in Figure 2. The drag distribution is observed to be very similar for the two airfoils, while the lift force is seen to be around 20% higher for the NREL S826 airfoil for a wide range of angles of attack. This noteworthy difference in lift is the reason for a lower power output of the ForWind rotor, but also implicates a difference in thrust force between the rotors, having an impact of the wake deflections.

2.4. Measurement techniques

Flow velocities in the wake were measured using a DANTEC FiberFlow two-component Laser Doppler Anemometer (LDA) system, recording the u- and v-component of the flow, cf. Figure 4. 5×10^4 samples were recorded at each point of measurement, resulting in time series of varying lengths of approximately 30 s. For one wake measurement, the LDA was traversed in the YZ-plane, cf. Figure 4. Each measured plane consists of 357 points, 21 in z-direction ranging from -D to +D and 17 in y-direction, ranging from -0.8 D to 0.8 D. The resulting distance separating two points of measurement is thus 0.1 D. Figure 3 illustrates the measurement grid behind the rotors.

Based on this grid of physically measured values, a much finer grid of a total of $401 \times 321 \approx 129000$ grid point is interpolated. The distance between the interpolated grid points is thus reduced to 0.005 D. Natural neighbor interpolation is used, which is resulting in a smoother approximation of the distribution of data points [16]. The interpolated values make it possible to calculate a rotor-averaged velocity of approximately 3.16×10^4 data points per rotor area for the wake center detection method described in Section 3.1.

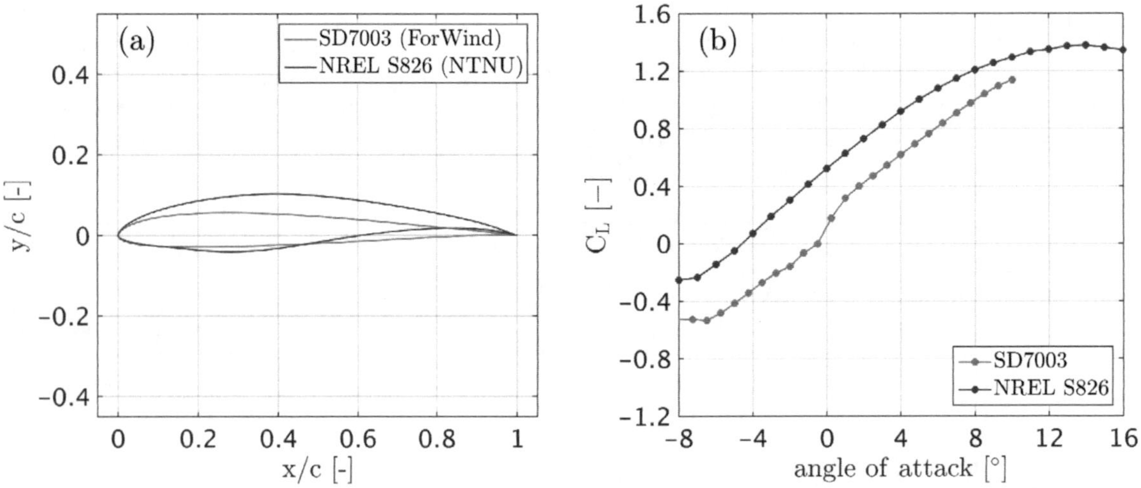

Figure 2. Comparison of the airfoils used at the ForWind and NTNU rotor. **(a)** airfoil shape, **(b)** lift and drag coefficients calculated for $Re = 1 \times 10^5$ using XFoil.

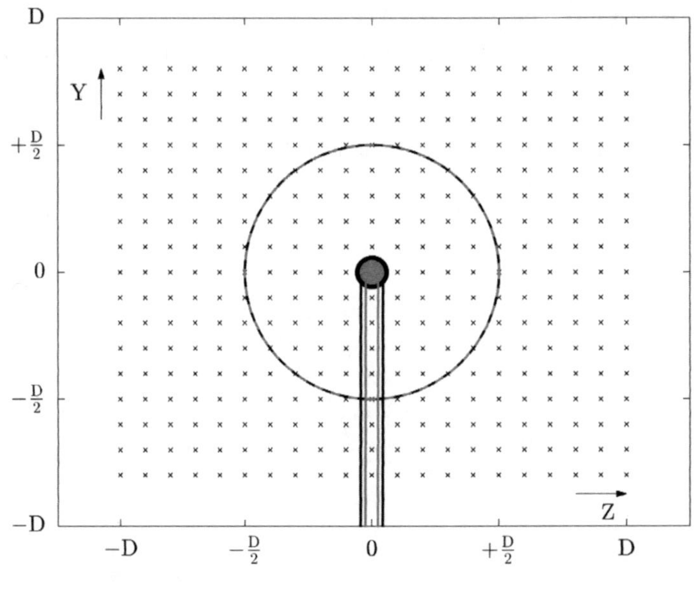

Figure 3. Illustration of the measurement grid behind the model turbines. The tower and hub of the NTNU turbine are marked in red, black corresponds to the ForWind turbine. Measurement positions are marked by ×.

2.5. Setup

The respective model turbines were positioned on a turning table allowing for a yaw misalignment. The orientation of the yaw angle γ is based on the tunnel geometry, which is sketched in Figure 4 along with the schematic setup.

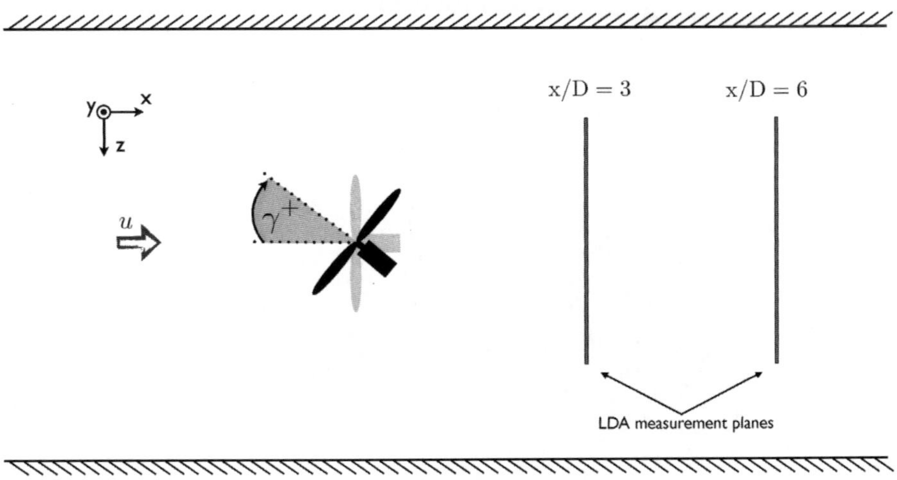

Figure 4. Sketch of the wind tunnel setup, top view. Scales do not match.

3. Methods

3.1. Wake location detection

As a major motivation for studying wind turbine wakes is the effect on downstream turbines, we use the power of a potential downstream turbine for estimating the wake center position in z-direction. A similar approach was shown by Vollmer et al. [17]. The potential power of a downstream turbine is

$$P^* = \sum_{i=1}^{10} \rho A_i \left\langle u_i(t) \right\rangle^3_{A_i,t} \,. \tag{3}$$

The rotor area is divided in ten ring segments. A_i is the area of the ith ring segment and $\langle u_i(t) \rangle_{A_i,t}$ denotes the temporally and spatially averaged velocity in main flow direction within the area A_i. For simplicity, the power coefficient is assumed to be constant for each ring segment and is not included here. P^* is estimated for 50 different hub locations in the range $-0.5\,\mathrm{D} \leq \mathrm{z} \leq 0.5\,\mathrm{D}$, at the original hub height. We define the wake center in z-direction by the z position resulting in the minimum of P^*. The procedure is illustrated in Figure 5.

3.2. Operating conditions

Both turbines were operated close to their optimal operation conditions at a yaw angle of $\gamma = 0\,°$. For $\gamma = -30\,°$, the TSR was kept constant at $\lambda = 6$ based on the u-component of the free stream velocity, resulting in a power decrease of approximately 21 % for the ForWind, and 29 % for the NTNU turbine. The average of both values is 25 %, which corresponds precisely to a cosine-squared relation between the relative power and angle of yaw misalignment as $\cos^2(30°) = 0.75$.

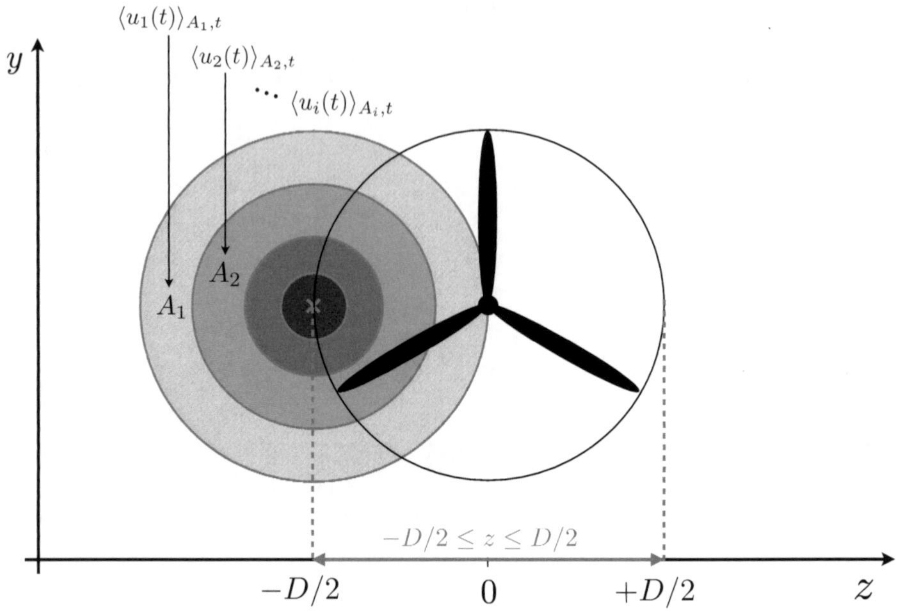

Figure 5. Illustration of the wake center detection method. A_i is the area of the ith ring segment of the potential downstream turbine's rotor area, whose hub is located at the red $\times$. $\langle u_i(t)\rangle_{A_i,t}$ is the spatially and temporarily averaged u-component of the velocity. The potential power P^* is calculated for each ring segment and then added up. This procedure is repeated for various hub locations $\times$, while the position resulting in the lowest value of P^* is interpreted as wake center.

4. Results

For the wakes of both turbines during a yaw misalignment of $\gamma = -30°$, the deflections in z-direction at x/D = 3 and x/D = 6 were quantified using the approach described in Section 3.1. Figure 6 shows the normalized potential power P^* for varying z-positions and both downwind distances. As expected, for both turbines the wake is deflected in z-direction, whereas the magnitude of deflection increases with the downstream distance behind the respective rotor. Table 3 summarizes the results derived from Figure 6 and will be used for further quantification of the wakes' deflections. At x/D=3, the wake deflection magnitude is 0.19 D for both turbines,

Table 3. Wake center location and corresponding skew angles based on Figure 6. The wake location corresponds to the value of z resulting in the lowest value of P^*, cf. Section 4.

Turbine	x/D	Wake location	Skew angle
NTNU	3	0.19 D	$\approx 3.7°$
ForWind	3	0.19 D	$\approx 3.7°$
NTNU	6	0.32 D	$\approx 3.0°$
ForWind	6	0.40 D	$\approx 3.8°$

which results in a wake skew angle of approximately 3.7°. The ForWind turbine shows a steeper graph, corresponding to a more pronounced deficit in the wake center, possibly due to the larger

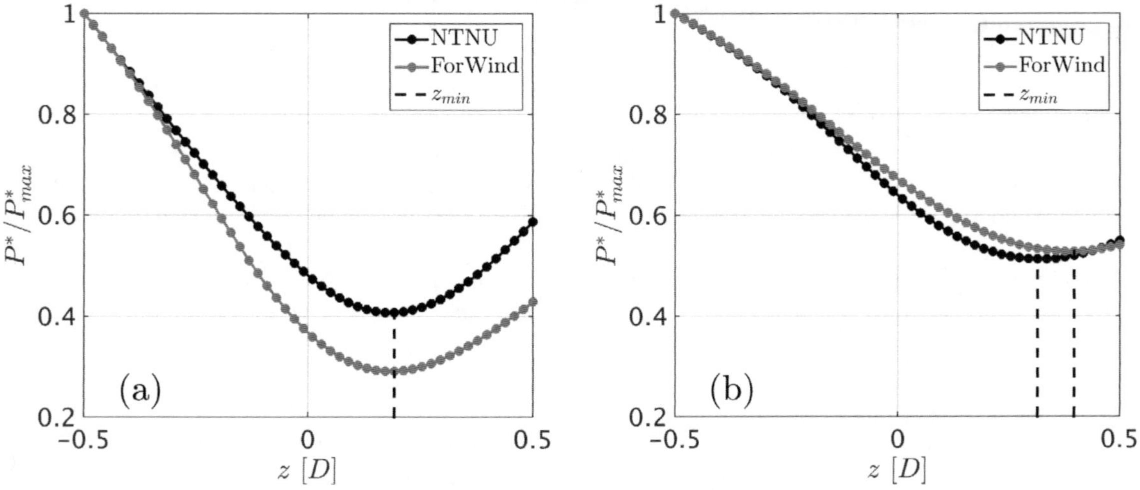

Figure 6. Normalized potential power P^* for various hub locations z of a potential downstream turbine. The respective minimum is marked by the dashed black line. (**a**): $x/D = 3$, (**b**): $x/D = 6$. In all cases the yaw angle was $\gamma = -30°$.

nacelle as compared to the NTNU turbine. Further downstream at x/D=6, the deflections are larger, as expected due to the wakes' expansion. However, differences between both turbines become apparent. The wake of the NTNU turbine is located at z/D=0.32, resulting in a slightly smaller skew of approximately 3.0°. The wake behind the ForWind turbine, on the other hand, is deflected slightly further to z/D=0.40, which corresponds to a skew of nearly 3.8° and is therewith nearly constant as compared to the distance x/D = 3.

5. Discussion

The fact that at x/D=3 the same wake deflection in z-direction is observed for both turbines having a different blade design and turbine geometry suggests that those aspects do not influence the wake trajectory at this distance. Further, it is assumed that wind tunnel effects such as blockage or wall-effects are insignificant for both turbines.

At x/D=6, however, differences in the wake deflections of both turbines can be observed, the wake behind the ForWind turbine is found to be deflected slightly further. A possible reason for this is a bigger influence of the increased wind tunnel blockage ratio of the 1.5 times bigger NTNU rotor. While the interaction of the flow behind the ForWind rotor with the wind tunnel side walls (blockage ratio 5.4%) is deemed to be insignificant, the considerably higher blockage ratio of 13% for the the NTNU rotor might influence the wake expansion and defection. Chen et al. [18] quantified the influence of blockage effects in a wind tunnel study, suggesting that those can be neglected for a blockage ratio below 10 %. This finding supports the assumption that blockage effects are of greater relevance for the NTNU turbine and might cause the differences in wake deflection at x/D=6 as the blockage ratios of both turbines are on either side of the suggested 10% limit.

In general, the observed deflections at x/D=3 and x/D=6 are in very good agreement with large eddy simulations (LES) by Vollmer et al. [17] of a full-scale scenario investigating the wake deflection behind a NREL 5MW turbine in stable atmospheric conditions. There, wake deflections of approximately 0.3 − 0.38 D were observed, depending on the tracking method. Further, an experimental study based on particle image velocimetry (PIV) measurements behind a significantly smaller model wind turbine of 0.15 m rotor diameter showed a very similar wake deflection

at $\gamma = 30\,^\circ$ and x/D=6, being $z/D \approx 0.34$ as reported by Bastankhah and Porté-Agel [3], which is in very good accordance with the current measurements. Another LES study performed by Fleming et al. [19] finds a slightly higher wake deflection of about 0.47 D at a downstream distance of x/D = 6. A recent study by Howland et al. [4] report a wake deflection of $0.40 - 0.55$ D behind a drag disc showing a somewhat bigger deflection. Summarized, the current experiments show that the setup and wake tracking method applied deliver very similar results as for much smaller model wind turbines on the one hand, and simulations of a full-scale turbine on the other hand.
It should be noted that the thrust force of both turbines is a parameter influencing the wake deflections and trajectories, which is, however, not included in this study.

6. Conclusion & future work

A comparative study of the wake deflection behind two different model wind turbines operated in $\gamma = -30°$ yaw misalignment was realized. A very similar defection behavior of the mean wake flow behind the two turbines was observed, resulting in skew angles of $3.0 - 3.8°$. Only insignificant differences in wake deflection were found at x/D=6, although model turbines of different sizes, rotor designs, and geometries were used.
As the deflection magnitudes and skew angles are furthermore observed to fit well with previous experiments on even smaller scale turbines [3] as well as simulations on full-scale turbines [19, 17], general estimations for the deflection can be deducted. Further, it can be concluded that the presented experimental setup and methods allow for a systematic investigation of wake details including their trajectories.
In future analyses of the acquired data, the wake shapes and trajectories will be further analyzed in detail. Besides effects of the turbine geometry, the influence of turbulence and shear in the inflow will be investigated. Moreover, a blind test comparison of the wake data with computational wake simulations of the wind tunnel setup is initiated.

Acknowledgments

The Authors thank Stefan Ivanell for proving the rotor blade design of the ForWind model turbine used in this study as well as the Reiner Lemoine Stiftung (RLS) for funding parts of this work.

References

[1] Medici D and Alfredsson P H 2006 *Wind Energy* **9** 219–236 ISSN 10954244
[2] Krogstad P Å and Adaramola M S 2012 *Wind Energy* **15** 743–756 ISSN 10954244 URL `http://doi.wiley.com/10.1002/we.502`
[3] Bastankhah M and Porté-Agel F 2016 *Journal of Fluid Mechanics* **806** 506–541 ISSN 0022-1120 URL `http://stacks.iop.org/1742-6596/625/i=1/a=012014`
[4] Howland M F, Bossuyt J, Martínez-Tossas L A, Meyers J and Meneveau C 2016 *Journal of Renewable and Sustainable Energy* **8** ISSN 19417012 URL `http://dx.doi.org/10.1063/1.4955091`
[5] Bartl J and Sætran L 2017 *Wind Energy Science* **2** 55–76 URL `http://www.wind-energ-sci.net/2/55/2017/`
[6] Tong W 2010 *Wind power generation and wind turbine design* (WIT press)
[7] Hsu S, Meindl E A and Gilhousen D B 1994 *Journal of Applied Meteorology* **33** 757–765
[8] Schottler J, Hölling A, Peinke J and Hölling M *Journal of Physics: Conference Series* 072030 ISSN 1742-6588
[9] Odemark Y and Fransson J H M 2013 *Experiments in Fluids* **54** ISSN 07234864
[10] Selig M, Guglielmo J, Broeren A and Giguere P 1995 *SoarTech Publications*
[11] Counsil J N N and Goni Boulama K 2013 *Journal of Aircraft* **50** 204–216 ISSN 0021-8669 URL `http://arc.aiaa.org/doi/abs/10.2514/1.C031856`
[12] Sætran L and Bartl J 2015 *Technical report NTNU*
[13] Somers D M 2005 THe S825 and S826 Airfoils, Technical Report NREL/SR-500-36344 `https://wind.nrel.gov/airfoils/Documents/S825,S826_Design.pdf` (last access: 9 February 2017)

IOP Conf. Series: Journal of Physics: Conf. Series **854** (2017) 012032 doi:10.1088/1742-6596/854/1/012032

[14] Sarmast S and Mikkelsen R F 2012 The experimental results of the NREL S826 airfoil at low Reynolds numbers http://www.diva-portal.org/smash/record.jsf?pid=diva2%3A615785&dswid=-528 (last access: 9 February 2017)

[15] Ostovan Y, Amiri H and Uzol O 2013 *Conference on Wind Energy Science and Technology-RUZGEM*

[16] Sukumar N 1997 *Theoretical and Applied Mechanics, Northwestern University, IL 60208*

[17] Vollmer L, Steinfeld G, Heinemann D and Kühn M 2016 *Wind Energy Science* **1** 129–141 ISSN 2366-7451 URL http://www.wind-energ-sci.net/1/129/2016/

[18] Chen T Y and Liou L R 2011 *Experimental Thermal and Fluid Science* **35** 565–569 ISSN 08941777 URL http://dx.doi.org/10.1016/j.expthermflusci.2010.12.005

[19] Fleming P, Gebraad P M, Lee S, van Wingerden J W, Johnson K, Churchfield M, Michalakes J, Spalart P and Moriarty P 2015 *Wind Energy* **18** 2135–2143 ISSN 10954244 URL http://doi.wiley.com/10.1002/we.1810

Comparison of OpenFOAM and EllipSys3D actuator line methods with (NEW) MEXICO results

J Nathan[1], A R Meyer Forsting[2], N Troldborg[2] and C Masson[1]

[1]ÉTS Montral, Mechanical Engineering, Montréal H3C 1K3, Canada
[2]Technical University of Denmark, Department of Wind Energy, 4000 Roskilde, Denmark

E-mail: joern.nathan.1@ens.etsmtl.ca

Abstract. The Actuator Line Method exists for more than a decade and has become a well established choice for simulating wind rotors in computational fluid dynamics. Numerous implementations exist and are used in the wind energy research community. These codes were verified by experimental data such as the MEXICO experiment. Often the verification against other codes were made on a very broad scale. Therefore this study attempts first a validation by comparing two different implementations, namely an adapted version of SOWFA/OpenFOAM and EllipSys3D and also a verification by comparing against experimental results from the MEXICO and NEW MEXICO experiments.

1. Introduction

In order to circumvent the modeling of the boundary layers attached to the wind rotor blades by representing their full geometry numerically, the Actuator Line Method[1] (ALM) has become a well established alternative for more than a decade. Numerous implementations exist and are used in the wind energy research community. Often those codes were compared to experimental results such as the MEXICO experiment, which is a wind tunnel experiment with a three-bladed rotor (with a radius of $R = 4.5\,m$) conducted at the German-Dutch wind tunnels (DNW). By placing pressure sensors on the blades and using particle image velocimetry (PIV) pertinent information about the near-wake could be obtained. After some amelioration of the setup a second round of the experiment was conducted, called NEW MEXICO. For more technical details see [2] and [3].

This work picks two of the most widely used implementatons, namely EllipSys3D[4] and the SOWFA[1] project[5] in OpenFOAM. Despite the fact that some adaptions have been made to the original SOWFA code by the authors, it will be referenced as SOWFA througout the article. This work tries to presents the ability to well reproduce near wake phenomena without modeling the boundary layer on the blades and at the same time show the models limitations due to its underlying assumptions.

First there will be brief comparison of similarities and differences between the two frameworks in section 2 alongside with a description of the numerical setup. Then there will be a discussion of the results by comparing the two numerical methods against each other and against the experimental data from the MEXICO and NEW MEXICO experiment in section 3. The

[1] This implementation is part of SOWFA (Simulator fOr Wind Farm Applications) by National Renewable Energy Laboratory.

examined cases include the cases for the reference velocities $U_\infty = \{10, 15, 24\}\, m/s$ and for the comparison with the experiment the PIV sheets and force measurements are used. Finally some concluding remarks will be given in section 4.

2. Numerical method

2.1. Numerical framework

Both implementations have been done within a CFD framework based on the control volume method with variables located at cell center and a Rhie-Chow[6] like correction. For a more in-depth description of the underlying CFD methods see [7] and [8].

In both cases the rotor is modeled by a force inserted as a momentum sink in the Navier-Stokes equations and distributed by a Gaussian distribution in order to avoid non-physical spikes in the velocity field around the affected cells. While SOWFA specifies a cut-off length for the 3D Gaussian curve in order to recover approximately 99.9% when integrated, the cut-off length of EllipSys3D is significantly larger to contain an even higher percentage. These two different cut-off lengths do not seem to have a significant impact on the simulation results.

Velocities are sampled in a very similar manner. While EllipSys3D uses explicitly trilinear interpolation to obtain the sampled velocity at the actuator point, SOWFA relies on correcting the cell center value by the velocity gradient. In the present case this gradient is obtained by linear interpolation and hence both sampling methods behave the same way.

While there are efforts to modify the orignal 2D airfoil coefficient data such as[9], the present work wants to use the ALM as an a priori tool. Hence the ALM is relying on the original airfoil data obtained from wind tunnel experiments of an infinite wing. As the centrifugal forces of the rotating blades keep the boundary layer attached longer than on a non-rotating wing [10], the angle of attack (AOA) for the maximum c_L is expected to be shifted towards higher values. Therefore it is expected that this ALM will lead to discrepancies between simulation and experiment for AOAs beyond the angle where stalling occurs in the airfoil data.

A tip correction is applied on the calculated forces. Without the tip correction, the forces where much higher than predicted by experiment. This is most probably due to the under-resolved tip vortex resulting in a weaker down-wash than in the experiment. By conducting the same simulation at different resolutions around the rotor, a higher induction in the tip region can be noticed. In order to compensate for this effect a Glauert tip correction is applied. This resulted also in a better agreement with the experimental data.

2.2. Numerical setup

The computational domain is cubic with an edge length of $20R$ with R as the rotor radius and the rotor positioned at the domain center. In both cases the cells in the vicinity of the rotor are cubic with the size $\Delta x = R/32$ and are stretched towards the domain boundaries in the case of EllipSys3D. Within SOWFA several refinement zones are applied each time halfing the cell edge length. Therefore the mesh of EllipSys3D consists of 7.1e6 cells while the mesh of the SOWFA case consists of $1.9e6$ cells.

For both cases the velocity boundary condition are given by a uniform inflow velocity of $\mathbf{U} = (U_\infty, 0, 0)$ and a zero gradient at the outlet. The lateral boundaries are set as symmetric.

As both simulations are Large Eddy Simulations, the subgrid-scale models are the dynamic Lagrangian method based on[11] for the SOWFA case and a DES model using a limiter to switch between $k - \omega$ SST and LES[12] for the EllipSys3D case. But as there is no inflow turbulence and the helical vortex structure does not break up within the examined region, not a lot of turbulence modeling has to be done. When comparing to a very coarse DNS by deactivating the sub-grid scale model, no significant difference is found.

The SOWFA case uses a discretisation scheme for the convection term that blends 75% of second order central differencing with 25% linear upwind differencing, which is a second order upwinding scheme, where the face value is corrected by the gradient in the upwind cell. EllipSys3D applies QUICK in RANS regions and central differencing of fourth order in LES regions using the same limiter as for DES model.

For parametrisation of the ALM the Gaussian distribution parameter is set at $\epsilon = 2\Delta x$ and 40 actuator points are used to represent one blade.

3. Results

The first interesting observation is the similarity in sampled velocities and AOAs as shown in figure 1 and figure 2. The directions in figure 2 are based on a local coordinate system on the blade. Despite the existence of steep velocity gradients due to the bound vorticity both frameworks obtain very similar values. This even holds for the relative small value of the tangential velocity component.

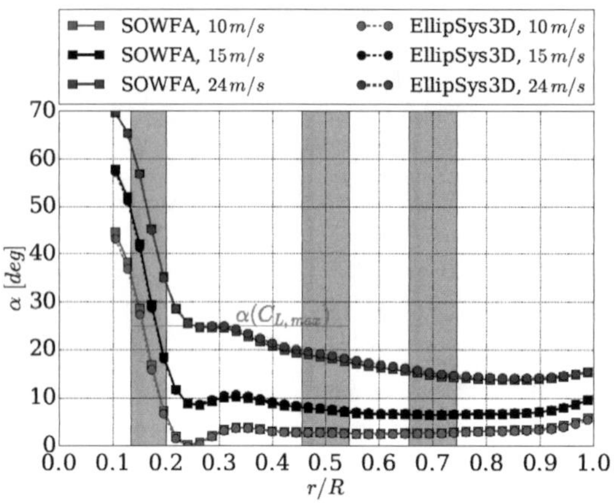

Figure 1. Angle of attack α for each actuator point along the blade. The grey zones represent transition between different airfoil types and the red line (—) the AOA for which the inbound airfoil data indicates stalling.

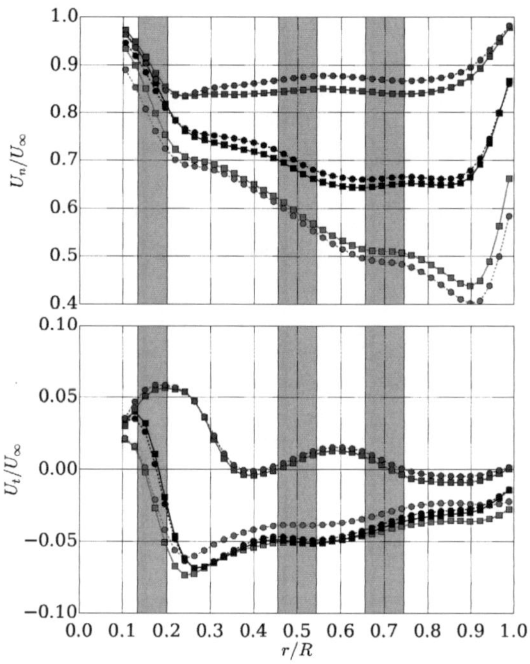

Figure 2. Normalized normal velocity and tangential velocity component U_n/U_∞ and U_t/U_∞ for each actuator point along the blade. The grey zones represent transition between different airfoil types. (see legend in figure 1)

When looking at the AOAs in figure 1 it can be seen that for the radial position $r/R < 0.3$ at $U_\infty = 24\,m/s$ it exceeds the angle after which stalling occurs according to the 2D airfoil data. For the other cases and airfoil sections the AOA always remains below the critical angle. Hence the ALM with the unaltered airfoil data breaks for the aforementioned case and the calculated forces will not match the ones obtained experimentally. In figure 3 the body forces associated with rotating blades can be seen. Again the directions are based on a local coordinate system on the blade. For $U_\infty = 10\,m/s$ and $15\,m/s$ exists a very good agreement, while the forces are not correctly evaluated for the high velocity case. A sudden drop in both forces can be seen. This is stemming from the fact that beyond a certain AOA no experimental data is available

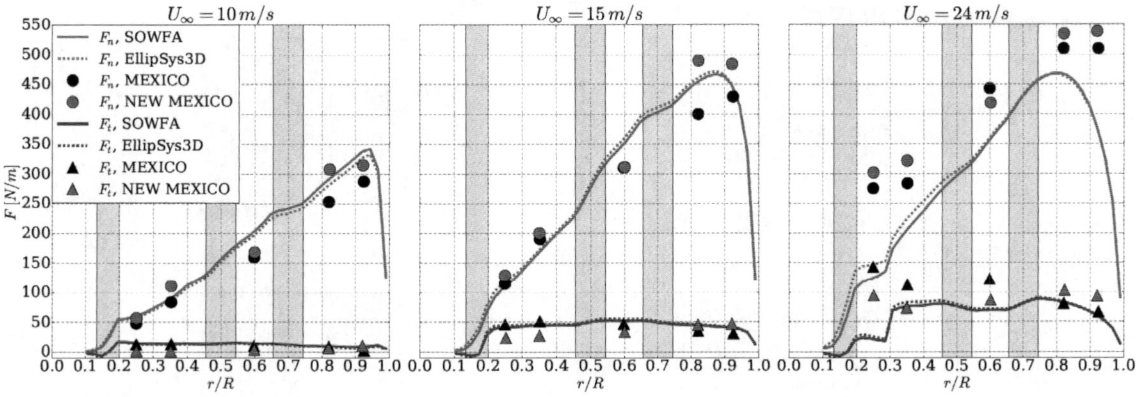

Figure 3. Comparion between the evaluated normal and tangential blade forces F_n and F_t by SOWFA and EllipSys3D against the experimental results of MEXICO and NEW MEXICO over the radial position.

and the void was filled by using the flat plate assumption. Therefore the airfoil data has a sharp drop in the lift coefficient at this point. Another interesting observation is that the forces for the MEXICO experiment are lower than for the NEW MEXICO experiment. This is probably due to the fact that the experiments were conducted with a slightly lower inlet velocity. Again the SOWFA and the EllipSys3D case are very similar even in the case where the models break down.

Due to the slight difference in the blade forces both frameworks produce a relatively similar flow field as shown in figure 4. In the following visualisations the flow field data is based on the domain coordinate system with x in flow direction, r parallel to the rotor plane and lying in the PIV sheet and t as the tangential direction normal to the PIV sheet. The blade position is expressed by the angle Ψ whereas $\Psi = 0$ means that the first blade is pointing upwards. It can be seen that neither SOWFA nor EllipSys3D can simulate distinct vortices shed in the case of turbulent wake state ($U_\infty = 10\,m/s$), but instead a continuous vortex is shed from the rotor. This is due to the rather coarse resolution of $R/32$. With higher inlet velocities a vortical structure can be noticed.

When looking at the radial profiles of the axial, normal and tangential velocity components U_x, U_r and U_t in figure 5 and figure 6, it can be seen that in general both frameworks are underestimating the velocity deficit in the ultimate rotor vicinity ($x/R = \pm 0.13$) compared to the experimental results but the overall trend is very well maintained. An exception is the high velocity case ($U_\infty = 24\,m/s$) where the models deficit becomes appearant. The high gradient of the radial velocity component in the tip region seems to be smeared out by the numerical simulation as can be seen in figure 6. This is most probably due to the numerical schemes and the mesh discretisation, which is expected to ameliorate when using a finer resolved mesh around the rotor. This would not only lead to a better resolution of the vortical structures, but due to the fixed force distribution parameter $\epsilon = 2\Delta x$ also to a more realistic distribution of the force in the tip region. The same applies to the tangential velocity component in the root region.

The axial profiles of the velocity components can be seen figure 7 and figure 8 and again both codes reproduce very similar results in the near wake further away from the rotor. While for the inbound position only data from NEW MEXICO experiments for $U_\infty = 15\,m/s$ are available, we can look at a bigger picture for the outbound position shown in figure 8. Again simulation results are very close to experimental data from the NEW MEXICO experiment, while overestimating with respect to MEXICO results due to reasons already mentioned. For

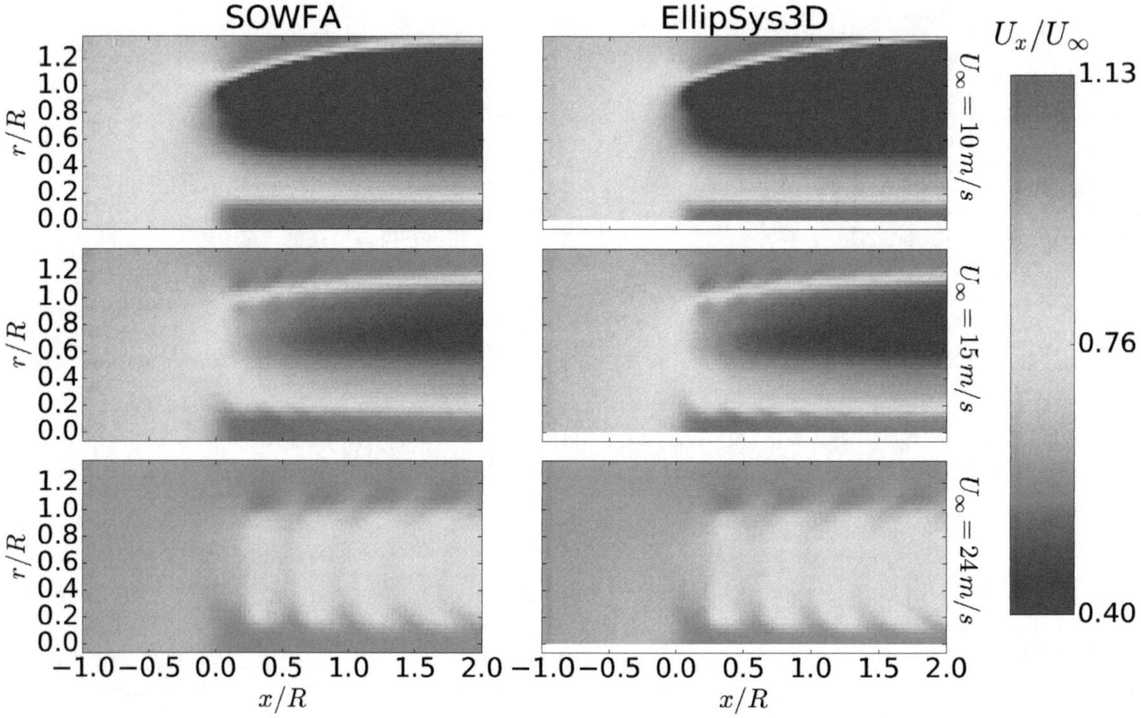

Figure 4. Planes of normalized axial velocity component U_x/U_∞ for all three cases for SOWFA and EllipSys3D. The rotor is situated at $x/R = 0$ and the velocity field is phase averaged for the rotor position $\Psi = 0^o$.

the high velocity case ($U_\infty = 24\,m/s$) the vortex sheets shed from the blades become visible by the oscillations in the axial velocity component U_x.

When looking at the vortex properties as examined in [13] in figure 9 and figure 10. it can be seen that there is a good general agreement despite the coarse resolution of the grid around the rotor ($\Delta x = R/32$). In figure 9 the vortex locations were calculated based on a rotor position of $\Psi = 0^o$ while the experimental was taken for $\Psi = 30^o$, which is the moment when the blade crosses the PIV sheet. Despite the fact that the origins of the vortices might not coincide as stated in [13] and the wake deficit was lower than in the experiments, the simulations appear to represent well the vortex propagation. This can also be seen by looking at figure 9 and comparing the axial locations of the vortices compared to the experimental results. Differences between the two simulations as seen in figure 9 stem from discrete representation of the vortices. In a more refined mesh both curves are expected to be closer.

In order to evaluate the vortex strength the circulation for each vortex is calculated. For obtaining the circulation of the vortices the vorticity magnitude ω in the plane at $\Psi = 0^o$ is integrated over a square with the edge length R_S weighted by the area. When looking at the evolution of vortex circulations in figure 10 it can be seen that the circulation remains almost constant througout the examined region, despite the fact that a constant integration radius R_S was used in order to evaluate the total circulation Γ. In figure 10 the total circulation can be seen in dependence of the square edge size R_S. At around $R_S/R = 0.113$ the saddle point can be perceived. Upto this point the integration only includes vorticity due to the examined vortex. Beyond that R_S also the vorticity of the neighbouring vortices is taken into account hence the steeper increase in circulation.

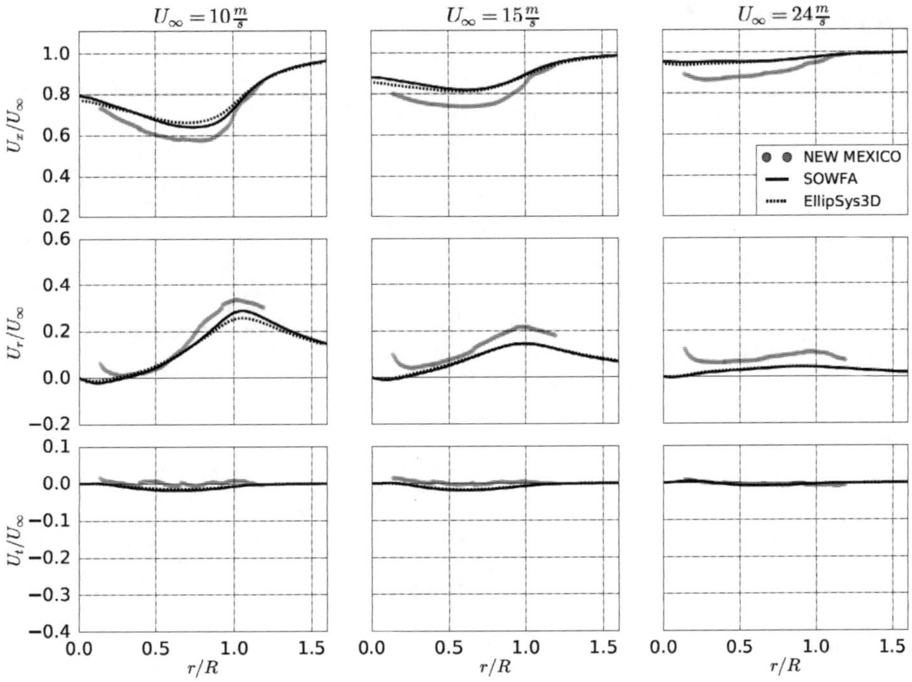

Figure 5. Radial profile of phase averaged ($\Psi = 0^o$) velocity components for different flow cases at upstream ($x/R = -0.13$) position.

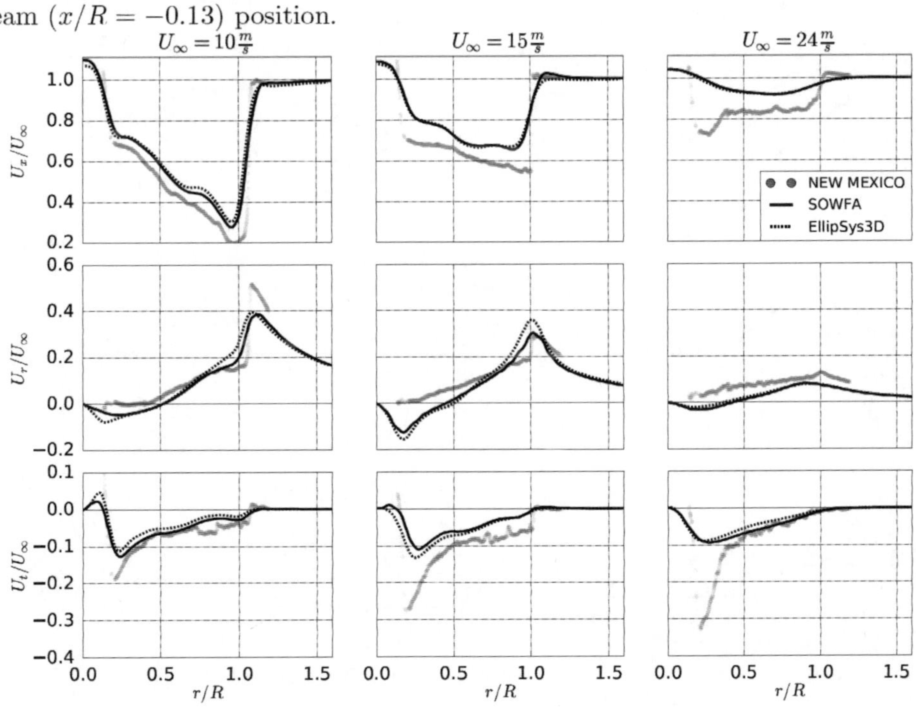

Figure 6. Radial profile of phase averaged ($\Psi = 0^o$) velocity components for different flow cases at downstream ($x/R = 0.13$) position.

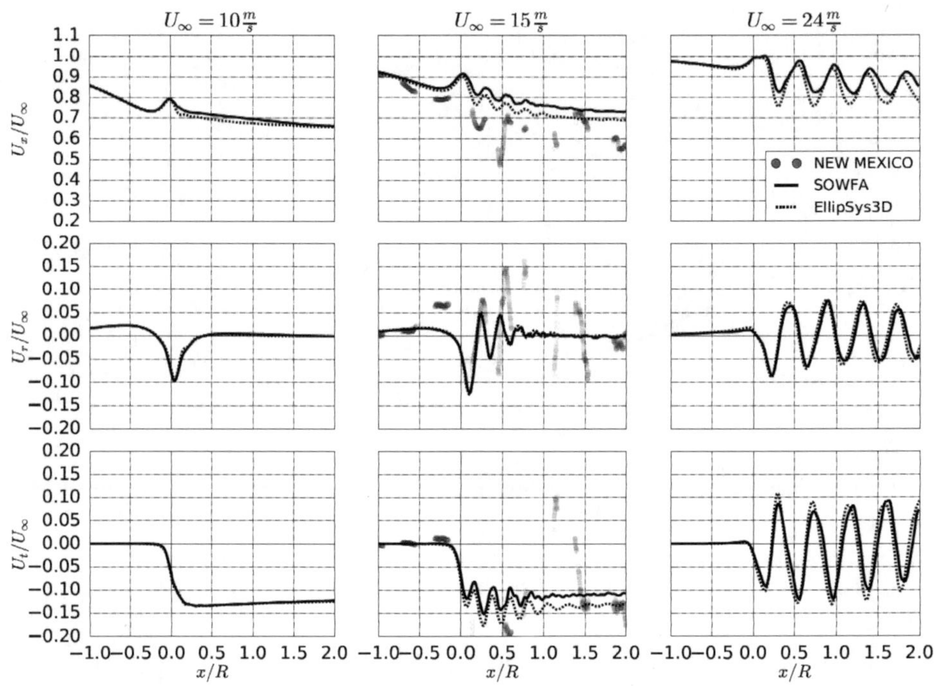

Figure 7. Axial profile of phase averaged ($\Psi = 0^o$) velocity components for different flow cases at inbound ($r/R = 0.22$) blade position.

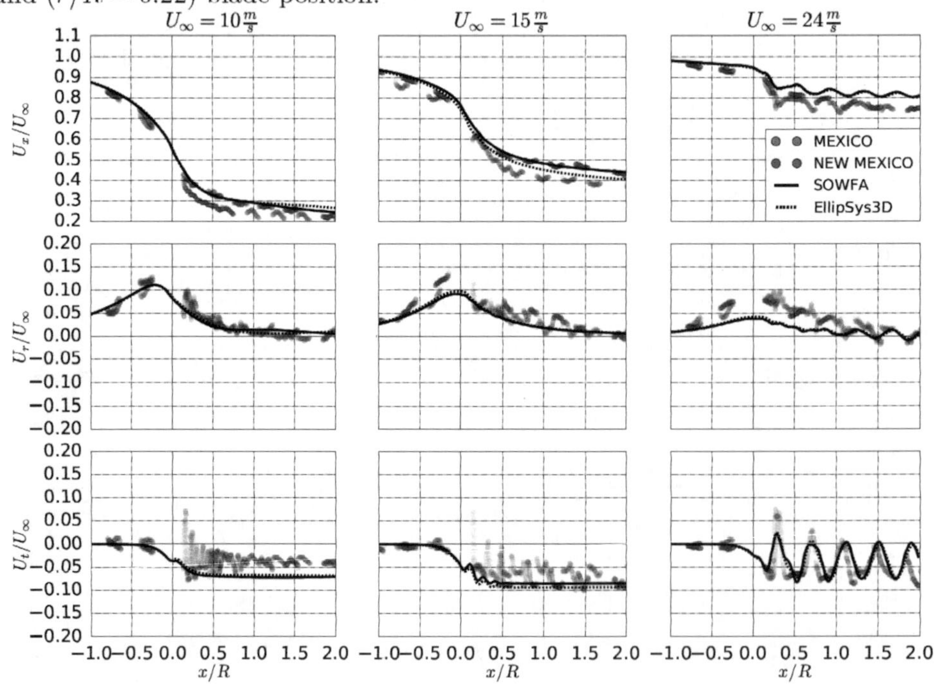

Figure 8. Axial profile of phase averaged ($\Psi = 0^o$) velocity components for different flow cases at outbound ($r/R = 0.67$) blade position.

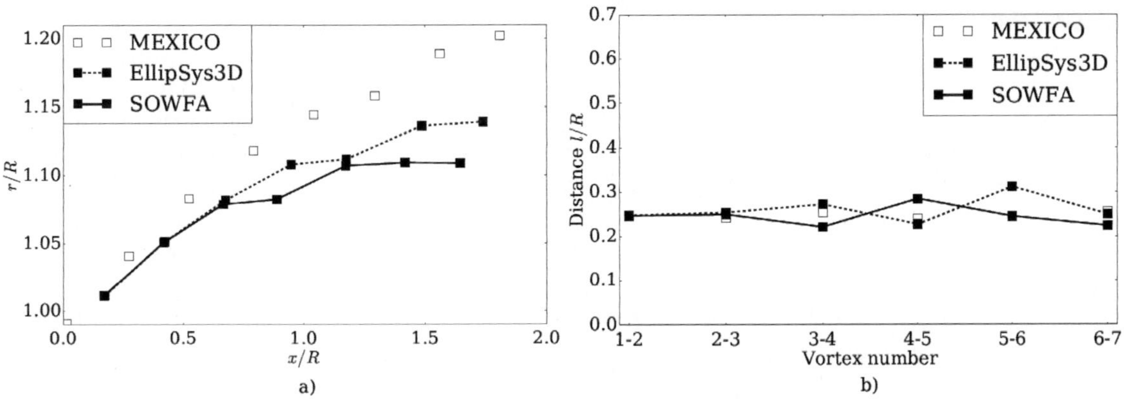

Figure 9. Normalized vortex locations (a) and distances (b) for $U_\infty = 15\,m/s$.

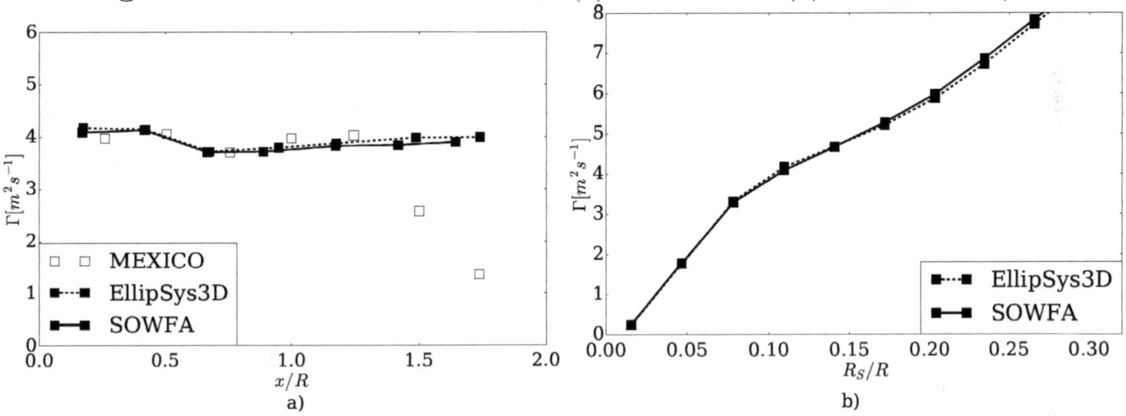

Figure 10. Circulation of vortices (a) and integration radius R_S/R (b) for $U_\infty = 15\,m/s$.

4. Conclusion

The two examined frameworks, EllipSys3D and SOWFA, compare very well in the near wake of the rotor used in the MEXICO and NEW MEXICO experiment. When comparing them against experimental data it can be seen that they predict well the experimental results from the NEW MEXICO experiment while there is an overprediction compared to MEXICO results due to the inlet velocity mismatch in this experiment. Even vortex properties matched among the simulations and predicted the same trend as in the MEXICO experiment, although the resolution in the rotor vicinity might be too coarse for well resolving the helicoidal vortex structure.

Both frameworks obtain very similar velocites and AOAs, and fail at the moment where 3D effects start to dominate. As this occurs only for a very high velocity at the inbound region, it is very encouraging for using only non-altered airfoil data in the presented case.

Future work could be to conduct simulations with a finer resolved region around the rotor and also to find a suitable method for including 3D effects in order to reproduce well the results for the high velocity case. Furthermore the influence of a turbulent inflow could be examined as well.

Acknowledgments

This work is partially supported the Canadian Research Chair on the Nordic Environment Aerodynamics of Wind Turbines, the Natural Sciences and Engineering Research Council

(NSERC) of Canada, the Innovation Fund Denmark (grant number 1305-00024B) and DTU Risø central computing facilities. Thanks for the great work done by Matthew Churchield and colleagues at National Wind Technology Center, Boulder, CO, by establishing the open source framework SOWFA. The data used have been supplied by the consortium which carried out the EU FP5 project Mexico: 'Model rotor EXperiments In COntrolled conditions'. The consortium received additional support to perform the New Mexico measurements from the EU projects ESWIRP and INNWIND.EU.

References

[1] Sørensen, J. N. and Shen, W. Z., 2002. Numerical Modeling of Wind Turbine Wakes. *Journal of Fluids Engineering*, 124(2), pp. 393.
[2] Schepers, J. G. and Snel, H., 2007, *Model Experiments in Controlled Conditions, Final report*, Report, ECN-E-07-042, Energy Research Center of the Netherlands, ECN.
[3] Schepers, J.G., Boorsma K. et al., 2014, *Final report of IEA Task 29, Mexnext (Phase 2)*, Report, ECN-E-14-060, Energy Research Center of the Netherlands, ECN.
[4] Troldborg, N., 2008, *Actuator Line Modeling of Wind Turbine Wakes*, PhD Thesis, Technical University of Denmark, Copenhagen.
[5] Churchfield M J, Lee S, Michalakes J and Moriarty P J 2012 A numerical study of the effects of atmospheric and wake turbulence on wind turbine dynamics *Journal of Turbulence* 13
[6] Rhie, C.M. and Chow, W.L., 1983. Numerical study of the turbulent flow past an airfoil with trailing edge separation, *AIAA journal*, 21(11), pp. 1525–1532.
[7] Sørensen, N. N., 1995, *General Purpose Flow Solver Applied to Flow over Hills*, PhD thesis, Technical University of Denmark, Copenhagen.
[8] Jasak, H., 1996, *Error analysis and estimation in the Finite Volume method with applications to fluid flows*, PhD Thesis, Imperial College, University of London.
[9] Shen, W. Z., Zhu, W. J., and Sørensen, J. N., 2012, Actuator line/NavierStokes computations for the MEXICO rotor: comparison with detailed measurements, *Wind Energy*, 15(5), pp. 811–825.
[10] Vermeer, L. J., Sørensen, J. N., and Crespo, A., 2003, Wind turbine wake aerodynamics, *Progress in Aerospace Sciences*, 39(67), pp. 467–510.
[11] Meneveau, C., Lund, T. and Cabot, W., 1996, A Lagrangian dynamic subgrid-scale model of turbulence, J. Fluid Mech, vol 319, pp. 353–385.
[12] Troldborg, N., Zahle, F., Rthor, P.-E. and Sørensen, N. N., 2015, Comparison of wind turbine wake properties in non-sheared inflow predicted by different CFD rotor models, *Wind Energy*, 18(7), pp. 1239–1250.
[13] Nilsson, K., Shen, W. Z., Sørensen, J. N., Breton, S.-P., and Ivanell, S., 2015, Validation of the actuator line method using near wake measurements of the MEXICO rotor, *Wind Energy*, 18(9), pp. 499–514.

IOP Conf. Series: Journal of Physics: Conf. Series **854** (2017) 012034 doi:10.1088/1742-6596/854/1/012034

An influence of the different incoming wake-like flows on the rotor vibrations

I V Naumov[1], I K Kabardin[1], R F Mikkelsen[2], V L Okulov[1,2], J N Sørensen[2]

[1]Kutateladze Institute of Thermophysics, SB RAS, Novosibirsk 630090, Russia
[2]Department of Wind Energy, Technical University of Denmark, 2800 Lyngby, Denmark

E-mail: naumov@itp.nsc.ru

Abstract. The aim of the current investigation is the rotor vibrations generated by the disturbances caused the different types of incoming wake-like flows. Those wakes arriving at the tested rotor were created by two ways: a passive wake generator (immobile disk) and an upstream rotating rotor as an active wake generator. The influence of both wakes on the tested rotor was studied in a water flume. A model of the tested three-bladed rotor designed using Glauert's optimum theory at an optimal tip speed ratio $\lambda = 5$ was placed in both "passive" and "active" wakes to recognize dissimilarities on the vibrations of the tested rotor. The distance from the wake generators to the tested rotor was varied from 4 to 8 rotor diameters. Also, the shift between the rotor axis and axis of the incoming wakes was changed to 0, 0.5 and 1 rotor diameters. The flow condition before rotor was measured with high temporal accuracy using LDA. The turbulent intensity of the incoming wake flows changed from 3 to 16% due to the types of the wake generators. Power and thrust characteristics and their pulsations of the tested rotor were measured by strain gauges. The dependences of power coefficients from tip speed ratios and positions of the wake generators were documented. The present study showed a strong influence of the initial flow from the two different wake generators on the rotor vibrations.

1. Introduction

Wind turbines are often sited near obstacles such as forests, ridges, hills and cliffs [1]. Some of them can be very favorable for producing the wind power but other ones must be strictly avoided since they can generate considerable stagnation of flow ("wind shade") in the rotor area (fig. 1). Next factor is that the obstacles appearing before the wind turbine rotor cause turbulence level increasing. The turbine efficiency depends on the nature of the turbulence. An effect of increasing the level of free-stream turbulence may lead to an earlier breaking vortex wake behind a wind turbine, which is very important for the arrangement of wind parks [2, 3].

With regard to the wind-farm developments, there is an increasing interest in studying the interaction and the power degeneration between two or more interacting wind turbines [4, 5]. The background is that the power degradation caused by the influence of the wakes has become a major subject in the design process of wind farms. Wind turbines are usually located in a wind farm to save space, to simplify the energy transfer and also to reduce the costs of an installation and maintenance of the wind turbines. The distance between the nearest wind turbines typically ranges from 3 to 10 rotor

IOP Conf. Series: Journal of Physics: Conf. Series **854** (2017) 012034 doi:10.1088/1742-6596/854/1/012034

diameters. The adverse effect appears when the inner wind turbines situated downstream of the forward ones and interact with strong rotor wakes. Numerous studies have reported a significant effect of external pulsations on power characteristics of wind turbines [6].

The present paper continues our previous experimental studies in the water flume of the flow structures around the different dual disk-rotor or rotor-rotor configurations to establish this new aspect of the rotor operations. The setup and experiment conditions are in detail described in [7-10] where only the flow properties and average torque and thrust of the tested rotor have been experienced. A model of the three-bladed rotor designed using Glauert's optimum theory at a tip speed ratio $\lambda = 5$ was placed in the different wakes behind the passive solid disk and the rotating upstream rotor as the active wake generator.

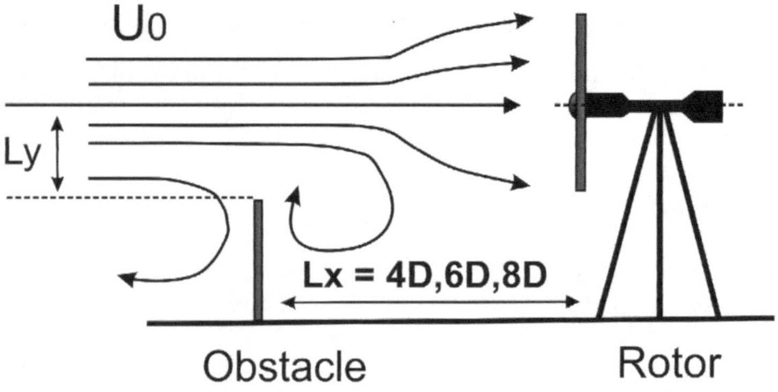

Figure 1. The schematic of the problem.

The purpose of the present investigation is to determine the rotor performance due to the interaction with the two types of the wakes. In our experiments, both active and passive wake generators (rotating rotor or immobile disk with the diameter close to the tested rotor) were used as the different oscillators in the upstream flow to recognize a dissimilarity in vibrations of the tested rotor. The wake influence has also been studied at the different distances between the tested rotor and wake generators or shifts of the rotor and wake axis. At first, by laser Doppler anemometry (LDA) method, we examined a structure and oscillations of the flow disturbances coming from both wake generators. The turbulent intensity of the incoming flow changed from 3 to 16 %. Then torque and thrust and their pulsation were measured by strain gauges. The dependences of power coefficients from tip speed ratio and the positions of the wake generators were studied. The present study also showed a strong influence of the types of the wake generators on the rotor vibrations. This conclusion is based on our comparison of the vibration of the tested rotor initiated by the Strouhal flow oscillations inherited to both disk or rotor wakes [11] and a blade frequency generated only in the rotor wakes. The detailed experimental description and main results are presented below.

2. Experimental Method and Results

The experiments were carried out in a water flume (fig. 2a, b). The length of the flume is 35 m, width is 3 m, and height is 1.0 m with 3 m transparent walls in the test section. A detailed characterization of the water flume flow can be found in [7-10]. The initial flow in the flume was subject to alow turbulence level within 2.5%, limiting the influence of external disturbances on the development of the inherent wake instability. The tested rotor was positioned downstream the wake generators. The three-bladed tested rotor has a diameter $D = 2R = 0.376$m, the blades consisting of SD7003 airfoil sections were specially designed for optimum operating conditions at the tip speed ratio $\lambda = 5$ [9, 10], where $\lambda = \Omega R/U_0$, and Ω is the angular speed of the rotor. The Reynolds numbers based on rotor diameter and the initial flow in the flume vary in the range $140.000 < Re < 240.000$.

The same rotor or disk with a diameter close to the rotor (0.9D) was used as the active and passive wake generators. The properties of the wake behind of the single rotor or disk are well known and they were studied in our previous work [7-10]. The setup location (fig. 2.a) at a height of 0.5 m from the flume bottom and at a distance of 1.5 m from the flume wall allows avoiding an influence of the boundary layer, whose thickness did not exceed 0.2 m from the bottom and the walls of the flume [9, 10]. The boundary layer influences on the entire flow pattern and changes the influence of the upstream obstacle on the investigated rotor; therefore, in some respect we here chose to research without the boundary layer influence. A ratio of the rotor area (0.111 m^2) and the area of the flume cross-section (3 m^2) is only 3.7%. For this reason, a blockage effect and other influences of the flume walls are actually very small and no corrections are made.

The wake generator and tested rotor are mounted on a platform displaced along the flume for different fixed distances between obstacle and rotor (fig. 2.a, b). The wake generator (disk or rotor 1) was axially placed at different positions Lx=4D, 6D and 8D infront of the wake affected rotor (fig. 2a). In other tests, the generator was placed at one fixed distance of 6 diameters upstream the rotor, withthe generator axis shifted from the rotor axis by Ly=0.5D and 1D (fig. 2b). The influence of the wake on the rotor performance and the thrust characteristics has been studied at the different positions (Lx and Ly) of the wake generator and tested rotor.

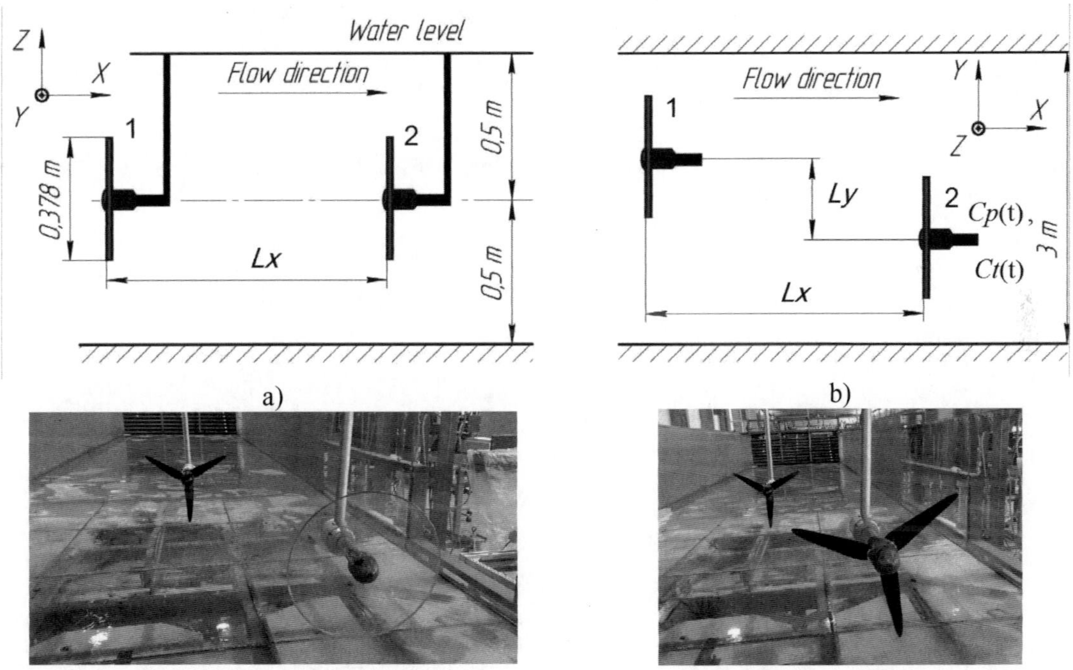

Figure 2. The schematic and photos of the experimental setup 1- obstacle disk or rotor, 2- three-bladed rotor.

At first, the velocity of the flow incoming to the tasting rotor was measured with high temporal accuracy using LDA. The initial free flow in the setup area of the flume has a uniform profile with velocity U_0 = 0.54 m/s and turbulence level 3 % [10]. The LDA measurements were carried out to find a deformation of the velocity profile by both wake generators. The local history of the axial velocity in each point was obtained using a Dantec 2-D Fiber flow LDA, based on a 1W Argon laser with a differential optical configuration and a frequency shift of 40 MHz. The diameter of the optical gauge is 112 mm and the focal length is 600 mm with a beam diameter of 1.35 mm. The wavelength of the laser beam is 514.5 nm (green light). The size of the probing optical field was 0.12×0.12×1.52 mm^3. These LDA measurements of both velocity and oscillations (RMS) are shown in fig.3 and fig.4.

The velocity profiles were measured at a distance of 0.5 D upstream of the tested rotor. The velocity is reduced up to 55 % with velocity oscillations growing up to 24% when the wake generator is an immobile disk and the tested rotor placed on the same axis (Ly = 0). The velocity was found to reduce up to 80 % when the rotating rotor was used as the active wake generator placed on the same axis. The level of RMS for the wake varied in the range from 3 to 16 % and becomes close to the turbulence intensity of the free flow when Lx=8D for both types of wake generators (fig. 3c). LDA velocity profiles of the rotor wake for different tip speed ratio at Lx = 6D (fig. 3d) show that the value of tip speed ratio almost does not influence the velocity attenuation in the wake.

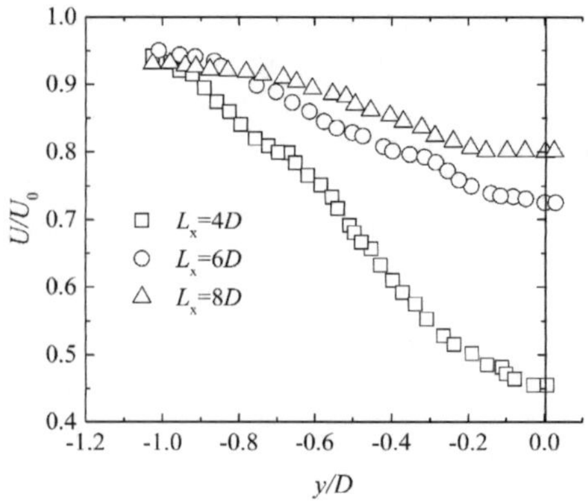

Figure 3a. LDA velocity profiles behind the passive wake generator – immobile disk.

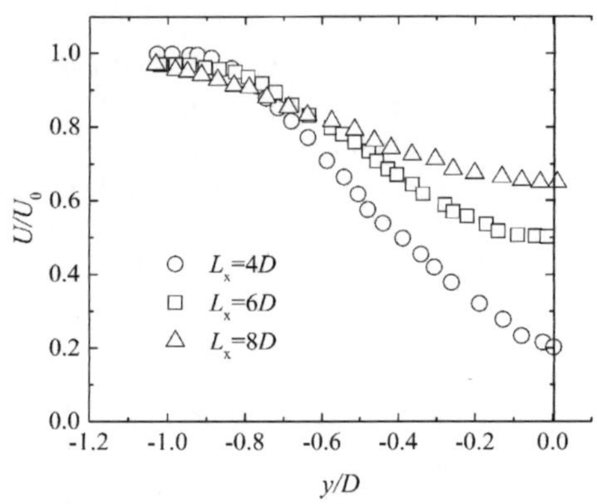

Figure 3b. LDA velocity profiles behind the active wake generator – rotating rotor.

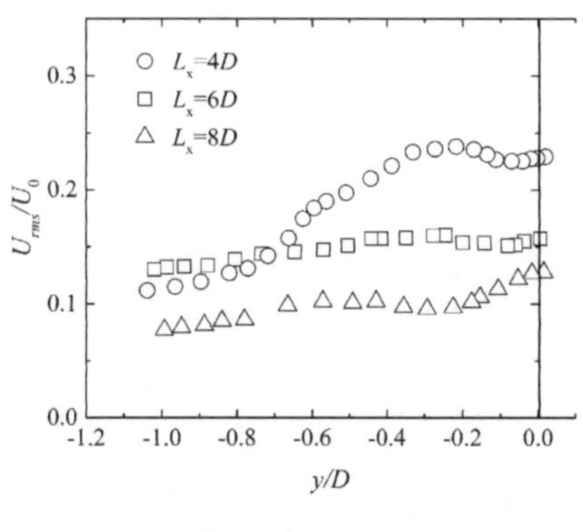

Figure 3c. RMS of LDA velocity behind the passive wake generator - immobile disk.

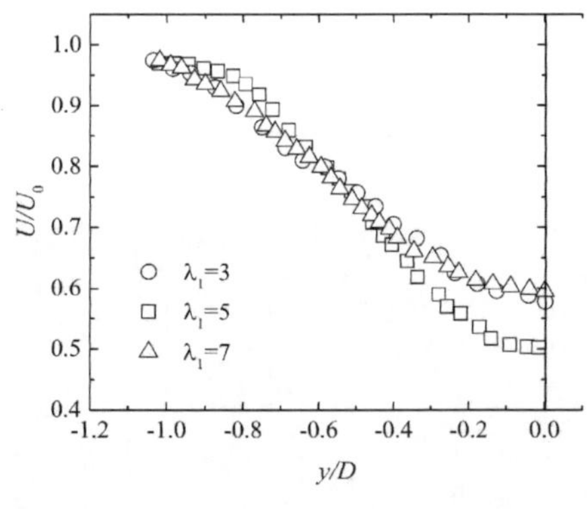

Figure 3d. LDA velocity profiles behind the rotating rotor for different tip speed ratio at Lx=6D.

Measured LDA velocity profiles with the axis of both passive and active wake generators shifted from the axis of the tested rotor at the different distances, are shown in fig. 4. From these dependencies, it can be seen that at Lx=4D both types of wake generators have a strong influence on the flow velocity profile and the active rotor extract more energy from the flow. It can be explained by the tip vortex system produced by the active rotor which does not inherit to the passive wake behind the disk. The velocity decrease in the wake is up to 40 % for the passive disk case and up to 20 % for the active rotor. The velocity attenuation in the rotor axis decreases only to 87 % for the disk and to 80 % for the rotor when Ly=0.5D. At Ly = 1D the velocity attenuation almost disappears in the rotor area for both passive and active types of wake generators (fig. 3a-c). For Lx=6D the velocity attenuation decreases up to two times.

IOP Conf. Series: Journal of Physics: Conf. Series **854** (2017) 012034 doi:10.1088/1742-6596/854/1/012034

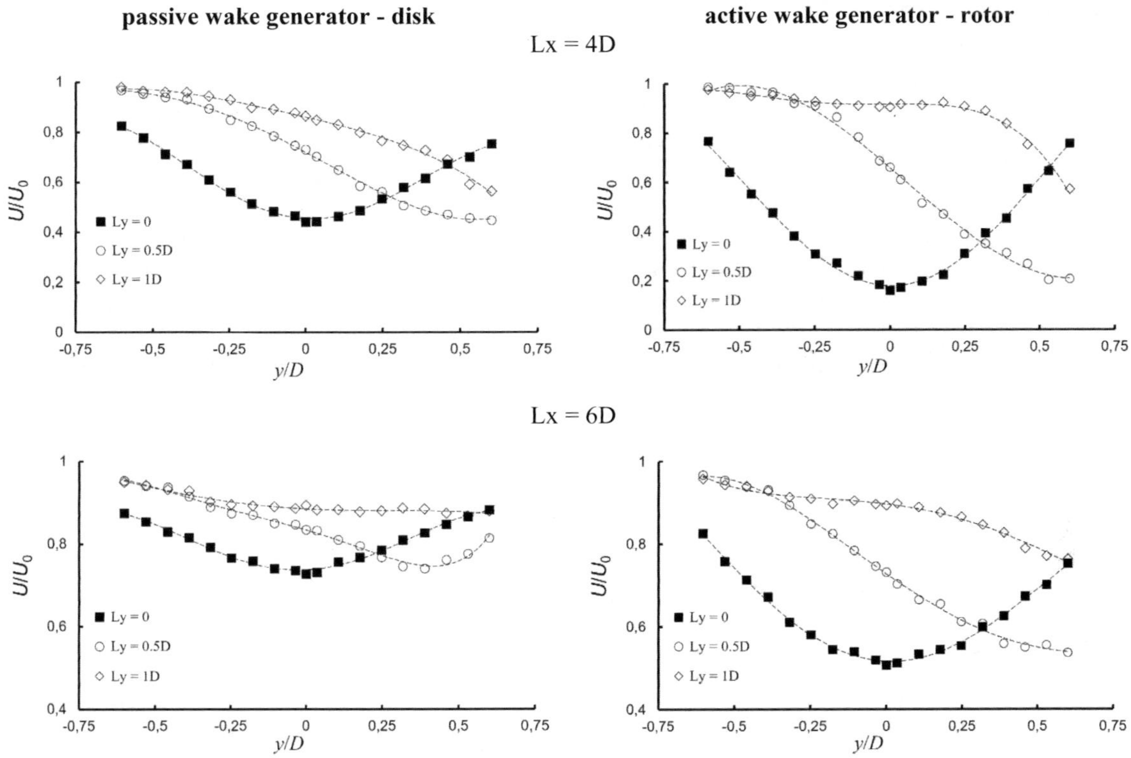

Figure 4. Wake velocity profiles at the different axis shifts behind passive (left) or active wake generators (right).

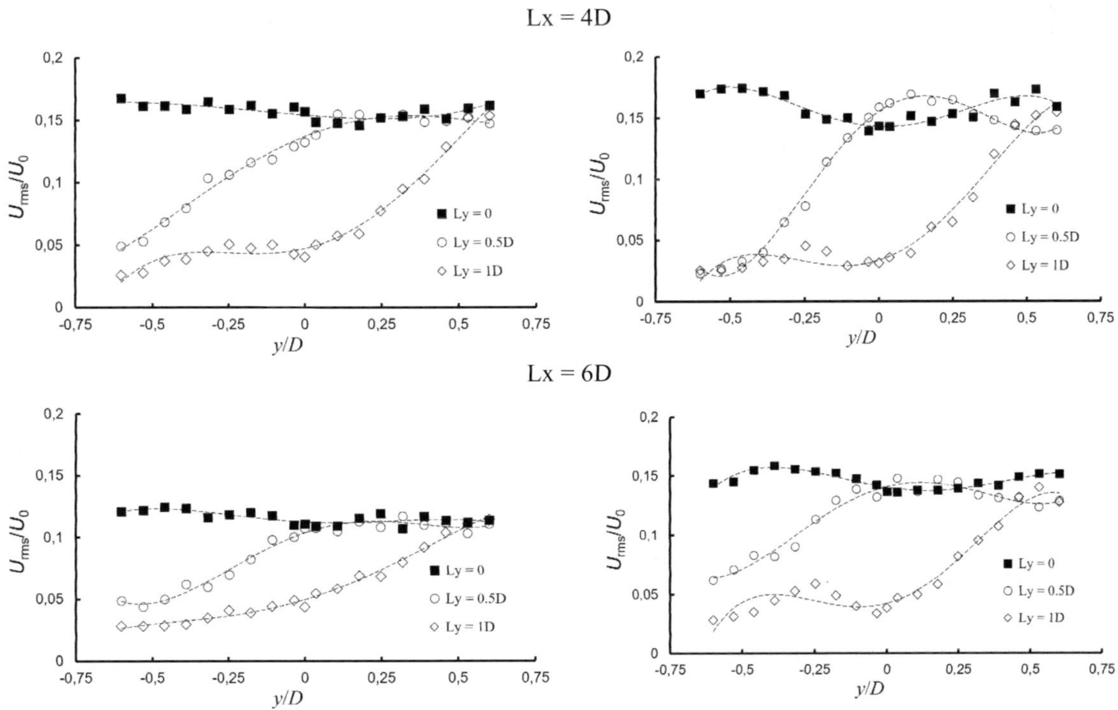

Figure 5. RMS in the wakes at the different axis shifts behind passive (left) or active wake generators (right).

3. Developments of the flow and rotor pulsations

The largest flow fluctuations (about 15 %) upstream of the tested rotor are observed at the minimum axial distance between the wake generators and tested rotor (Lx = 4D) when both rotors are coaxial (fig. 5). By increasing Lx to 6D and Ly=0, the pulsations decreases 3 % in the passive disk wake and only to 1% behind the active rotor wake. With increasing Ly, the wake pulsations also decreases to 5 % for Ly=0.5 and for Ly =1D they become less 4-5 % for both types of the wake generators and both values of Lx (fig. 5).

An influence of both wakes on the power and thrust of the tested rotor was examined too (fig. 6,7). Measurements of the rotor characteristics, torque (M) and thrust (T), were collected by strain sensors installed in the rotor mounting [9, 10]. The voltage of the sensors was amplified by a preamplifier Scout 55, produced by Hootinger Baldwin Messtechnik, and was digitized by the ADC produced by National Instruments Company. Both strain sensors were calibrated with an inaccuracy of less than one percent using reference weights. This system recorded the electrical signal of the strain sensors with a frequency of 120 Hz for 60 s. The obtained data served to calculate average values of the torque (M) and thrust (T) acting on the rotor axis. An evolution of the optimal operating regimes for the second tested rotor placed into both wakes generated by the passive immobile disk or the active rotating rotor were achieved (fig. 6).

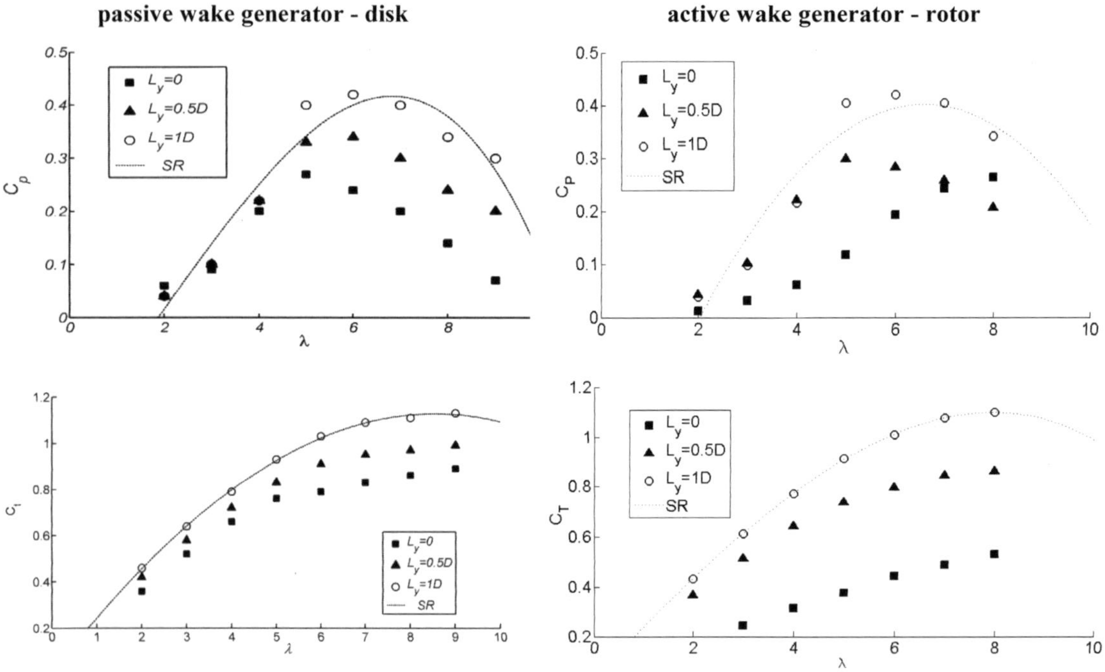

Figure 6. Power C_P and trust C_T coefficients of the tested rotor at Lx=6D and in different wakes.

For the single rotor, an angular velocity of the servo drive n_0 = 2.29 rps, coincides to the design TSR of the single rotor (λ_0 = 5) with the maximal power coefficient C_{P0}= 0.37 and the incoming free stream flow (U_0 = 0.54 m/s). The dependencies of Fig. 6 show that the wakes behind both the disk and rotor strongly influence on the rotor performance when the rotor axis coincides with the wake axis (Ly = 0) or contains in the wake (Ly = 0.5D). An impact of these wakes decreases also with an increasing of Lx (table 1). At Ly = 0.5D the rotor is situated in the half of the wake and C_P is rising. At Ly = 1D the rotor does not feel the influence of the wake and take values as the single rotor.

Table 1. Operating parameters of the dual setups.

Types of the setup	Parameters	Values						
	L_x - *distance*	4*D*			6*D*			8*D*
	Ly - distance	0	0.5*D*	1*D*	0	0.5*D*	1*D*	0
D-R (at λ_2= 5)	*max* C_P/ C_{P0}	0.48	0.8	1	0.69	0.85	1	0.83
	Ct/ Ct_0	0.65	0.87	1	0.83	0.9	1	0.95
	Ct_2, RMS,%	18.6	9.77	3.14	15.9	9.39	4.44	6.94
R-R (at λ_2= 5)	*max* C_{P2}/ C_{P0}	0.11	0.7	0.99	0.39	0.75	1	0.51
	Ct_2/ Ct_0	0.66	0.73	0.99	0.62	0.81	1	0.74
	Ct_2, RMS,%	18.54	10.08	3.08	14.29	9.54	3.57	7.95

The data's on figure 6 and table 1 shows that both the passive disk wake and the active rotor wake almost don't influence on the rotor performance when Ly = 1D. The power and trust coefficients in this case is almost equal to the single rotor value (solid curve).

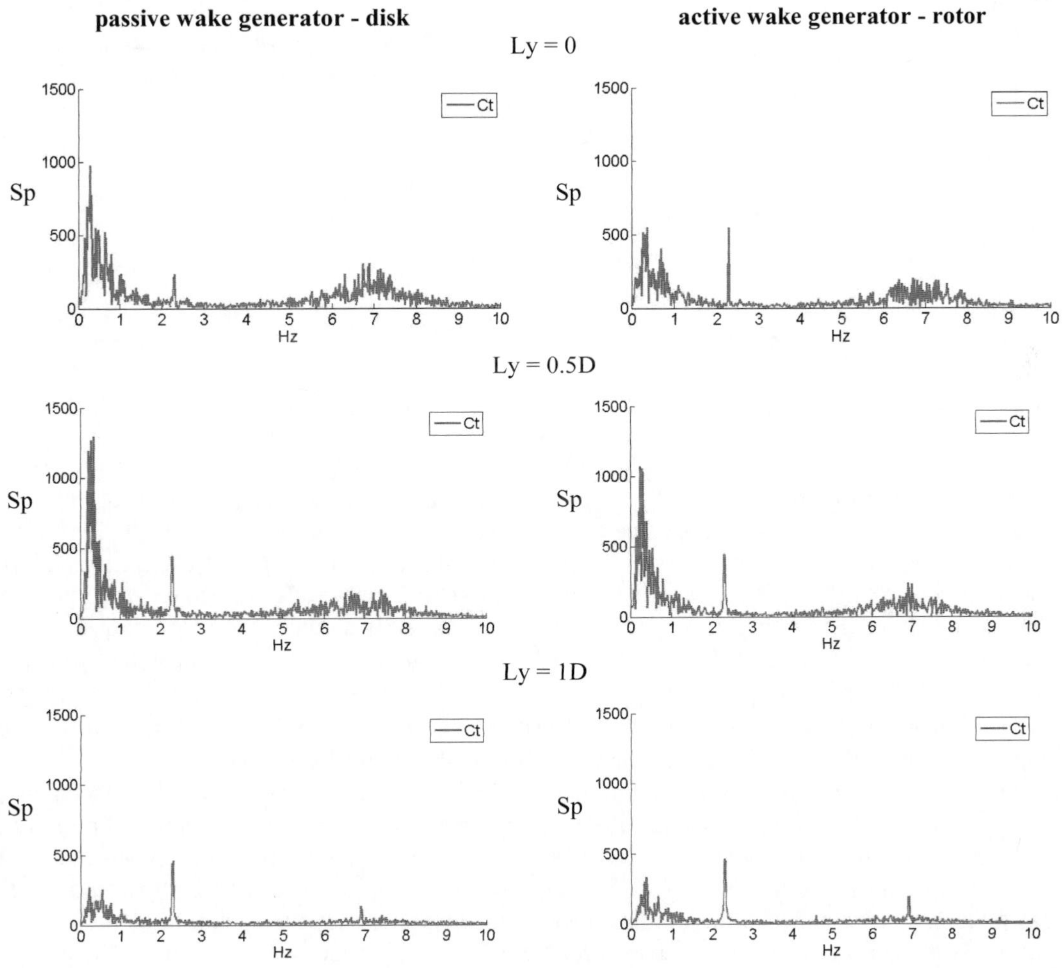

Figure 7. The spectra of C_T signal of rotor at λ_2=5 for of Lx=4 and different Ly, located in passive wake (left) or active wake (right).

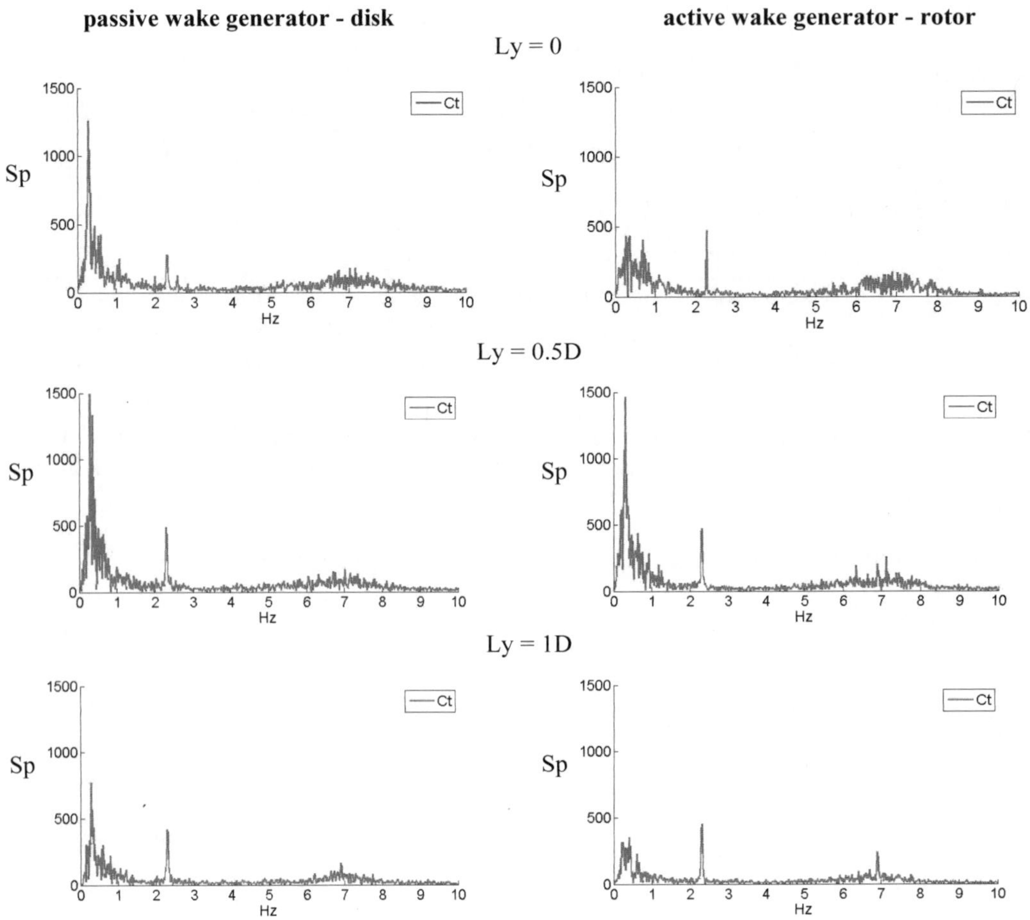

Figure 8. The spectra of C_T signal of the rotor at λ_2=5 Lx=6 and different Ly, located in the passive wake (left) or active wake (right).

The existence of strong oscillations (fig. 7, 8) was found on the same frequency of 0.27 Hz for all spectra of the C_T of the tested rotor, a value that correspond well to the Strouhal oscillation in both disk or rotor wakes [10]. Furthermore, computing spectra of the C_T of the tested rotor, a signal with frequencies of the rotation (2.29 Hz) and blade-rotation (6.87 Hz) itself was found too. For the Lx = 4D and Ly = 0 case, the spectra of C_T rotor frequency peak is greater than for the obstacle -disk case. This situation is caused by the upstream rotor wake influence, because the upstream rotational rotor speed was the same as for rotor one. However, for other cases of Lx and the none optimal regime of the first rotor operation, the spectra have not indicated any frequencies coming from the upstream rotor with the interactions of the active wake behind the first rotor.

The Strouhal frequency usually takes a dominant role in the spectra of the tested rotor placed inside of the passive wakes (Ly = 0 or 0.5D) and the value is about three times higher than the rotating or blade frequencies. In the active wake, the same behavior of the Strouhal frequency inherits only the case Ly = 0.5D. In the active wakes with the coaxial position of the tested rotor at Lx =4D and 6D, the Strouhal frequency does not take the dominant value in the spectra. An explanation of the different behavior of both spectra in the coaxial positions can be also connected with the different wake development in the case of the dual coaxial configuration from disks and rotors founded in [12].

At Ly =1D, when the tested rotor is not situated into the both wakes, the Strouhal frequency also become unimportant and both spectra becomes similar which coincides with the conclusion of an insignificant impact on the average values of the torque (M) and thrust (T) of the outside rotor position from the wakes (fig. 6 and table 1).

4. Conclusions

An interaction between the axial-flow turbine and wakes generated by the disk and rotor placed upstream of the tested rotor was experimentally studied to estimate an influence of both passive and active obstacles on the rotor vibrations. The LDA measurements of the incoming flow have shown that the velocity attenuation by the passive wake generator – immobile disk is less than for the active wake generator – spinning rotor. The rms measured by LDA before the tested rotor has shown that the flow pulsations behave in the same manner for both types of the obstacle. The power of the tested rotor placed in the disk wake is greater than one in the rotor wake. The spectra of the rotor vibrations have shown an existence of the Strouhal frequency, the rotor and blade frequencies of the rotation itself, but they have not indicated any frequencies inherited the upstream rotor.

The obtained results are of interest for further development of aerodynamics of the wind farm where the turbine interactions with the wakes behind other turbines.

5. Acknowledgments

The research was supported by the Russian Science Foundation (Project № 14-19-00487).

References

[1] Thomas A. Wind Power in Power systems. *John Wiley and sons, ltd.* 2005:1120.

[2] Port-Agel F,Wu YT, Chen CH. A numerical study of the effects of wind direction on turbine wakes and power losses in a large wind farm. *Energies*. 2013; 6(10): 5297–5313.

[3] Chamorro LP, Hill C, Neary VS, Gunawan B, Arndt REA, Sotiropoulos F. Effects of energetic coherent motions on the power and wake of an axial-flow turbine. *Physics of Fluids* 2015; 27(5): 055104.

[4] Troldborg N, Larsen GC, Madsen, HA, Hansen KS, Sørensen JN, Mikkelsen R. Numerical simulations of wake interaction between two wind turbines at various inflow conditions. *Wind Energy* 2011; 14: 859-876.

[5] Bartl J, Pierella F, Sætran L. Wake measurements behind an array of two model wind turbines. *Energy Procedia* 2012; 24:305-312.

[6] Breton SP, Nilsson K, Olivares-Espinosa H, Masson C, Dufresne L, Ivanell S. Study of the influence of imposed turbulence on the asymptotic wake deficit in a very long line of wind turbines. *Renewable Energy* 2014; 70: 153-163.

[7] Okulov VL, Naumov IV, Mikkelsen RF, Sørensen JN. Wake effect on a uniform flow behind wind-turbine model. *J. of Physics: Conference series*. 2015; 625: 012011.

[8] Naumov IV, Kabardin IK, Mikkelsen RF, Okulov VL, Sørensen JN. Rotor performance and wake conditions in non-uniform flow behind an obstacle. *Journal of Physics: Conference Series* 2016; 753: 032051

[9] Naumov IV, Mikkelsen RF, Okulov VL. Estimation of Wake Propagation behind the Rotors of Wind-Powered Generators. *Thermal Engineering* 2016; 63(3): 208–213.

[10] Okulov VL, Naumov IV, Mikkelsen RF, Kabardin IK, Sørensen JN. A regular Strouhal number for large-scale instability in the far wake of a rotor. *J. Fluid Mech.* 2014; 747: 369-380.

[11] Naumov IV, Litvinov IV, Mikkelsen RF, Okulov VL. Investigation of a wake decay behind a circular disk in a hydro channel at high Reynolds numbers. *Thermophysics and Aeromechanics* 2015; 22(6): 657-665.

[12] Okulov VL, Mikkelsen RF, Naumov IV, Litvinov IV, Gesheva E, Sørensen JN. Comparison of the far wake behind dual rotor and dual disk configurations. Journal of Physics: Conference Series 2016; 753: 032060.

Wake developments behind different configurations of passive disks and active rotors

V L Okulov [1,2], I V Litvinov[2], R F Mikkelsen[1], I V Naumov[2], J N Sørensen[1]

[1]Department of Wind Energy, Technical University of Denmark, Lyngby, Denmark
[2] Kutateladze Institute of Thermophysics SB RAS, Novosibirsk, Russia

E-mail: vaok@dtu.dk

Abstract. The present paper takes a broad view on our previous experimental studies of flows behind different single and dual configurations from passive disks or active rotors to establish new aspects of the wake development [1-4]. The aim of the present examination is to obtain a better understanding of the wake formations and interactions between wind turbines in wind farms. A correlation between independent investigations of the near [1] and far wakes behind single [2] and dual [3-4] systems will be established to the same operating regimes and flow conditions. New examinations of the old data need because two main differences in the wake behaviour for the disk-disk and the rotor-rotor systems were found: the wake intensity grows for the dual disks in comparison with the single one, but in contrast to this, wake intensity behind the dual rotor system is smaller than the one behind a single rotor. These differences may be explained by an influence of the rotor tip vortices which are absent in the disk-disk model. The present retesting of the near and far wake data should provide an evidence of this conclusion.

1. Introduction

In spite of the extensive investigations of turbine interaction in wind farms, the prediction of the available power and structural stability of wind turbines located in the wake of other turbines in a wind farm are still open questions [5-6].

In the present paper, we will compare the wakes behind solitary setups containing a solid disk (D) or a three-bladed HAWT rotor (R); dual coaxial models consisting of either two identical rotors (R-R) or two identical disks (D-D), and a mixing coaxial model including both single disk and rotor (D-R). In the comparison, we used the solid disk because both disk and rotor far wakes have inherited similar Strouhal oscillations [1]. All systems were studied at the same spatial positions in a uniform inflow of the water flume [1-4] and the dual systems were investigated for different interspatial distances along the setup axis, $L_x = 4$, 6 and 8D. The investigations were carried out in the water flume of length 35m, 3m width and an operative height of 0.9m. The 3m wide test section is fitted with transparent walls at a distance of 20 m from the channel inlet. The freestream flow speed in the flume was $U_0 = 0.54$ m/s. The axis of setup with the rotors or disks was positioned at a height of 0.5m from the channel bottom and 1.5 m away from the two walls of the flume. The experimental setup is basically the same as in [1-4] with the disk diameter $D_d= 0.3$ m and the three-bladed rotors have and a blade length of 0.159 m.

The purpose of the analysis is to reconstruct and compare the wake developments and decays behind all systems to recognize and describe factors impacting on the wake formation and evolution. The present retesting of both near and far wake data should provide an evidence of a significant role of the tip vortices in the rotor which are absent and could not effect on the wakes behind the disks.

2. Dissimilar behavior of far wakes behind different dual systems

New examinations of the old data need because two main differences in the far wake behavior for the D-D and R-R systems were found in [3]. The wake intensity grows for the dual disks in comparison with the single one, but in contrast to this, wake intensity behind the dual rotor system is smaller than the one behind a single rotor. Fig. 1 shows downstream variations of the maximum axial velocity deficit $\frac{\Delta U}{U_0}=\frac{U_0-U}{U_0}$ in the wake behind all possible configurations (U is velocity on the setup axis): the single D and R [2]; the D-D [3], D-R [4] and the R-R [3] setups for the different distances of the dual systems: L_x= 4, 6, 8D.

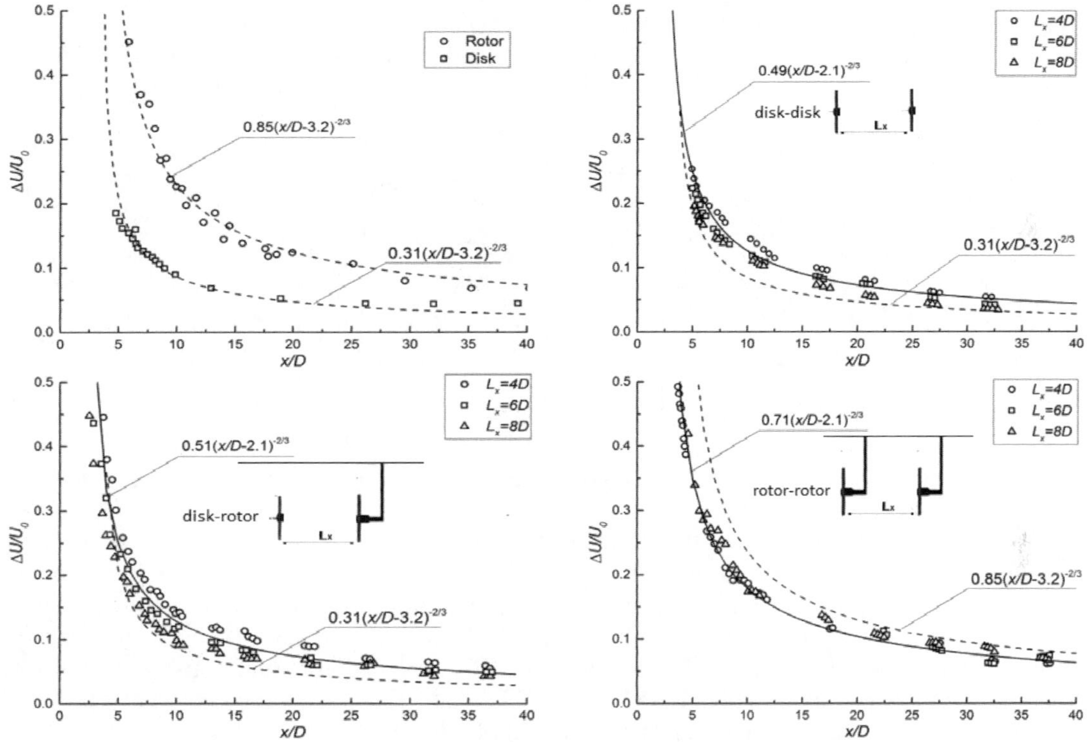

Figure 1. Far wakes of the different dual configurations. Dashed lines indicate both single cases.

The experimental data using regression techniques to fit all velocity profiles were approximated by a rational function $G(x)=\Delta U(x)/U_0=a(x-x_0)^{-2/3}$ (lines on Fig. 1), where the attenuation constants a and x_0 are collected in table 1.

Table 1. The attenuation constants for the single and dual configurations of the disk or rotor setups.

Type of the setup	a	x_0
Single disk	0.31	3.2
D-D	0.49	2.1
D-R (optimal R)	0.51	2.1
R-R (optimal R and R)	0.71	2.1
Single rotor (optimal R)	0.85	3.2

As seen from all graphs of Fig. 1, the velocity deficits of the far wakes from all single and dual configurations are well described by this rational function with the same exponent (-2/3) and the different values of the attenuation constants (Table 1). For clarity, the first graph shows the wake developments only for the single disk [8] or rotor [9]. The disk just as a passive flow damper and the rotor as an active energy conversion explain the difference in the strengths of both wakes.

For all dual systems in which the disk placed at a forward position, the data on the next two plots of Fig. 1 and the corresponding constants of Table 1 indicate some growth of the velocity deficits in the far wakes as compared with ones behind the single disk. Fig. 1 also shows that the total wake deficit increases in the both cases (D-D and D-R) when the distance between the first disk and second element decreases from $L_x = 8D$ to $4D$. A damping influence of the second disk or rotor on the initial wake behind the forward passive disk produces the extra velocity deficit for both cases (D-D and D-R) too.

The last plot of Fig. 1 shows the wake behavior behind two active wind turbine rotors (R-R) extracting power from the flow at the optimal operating conditions of both ones.

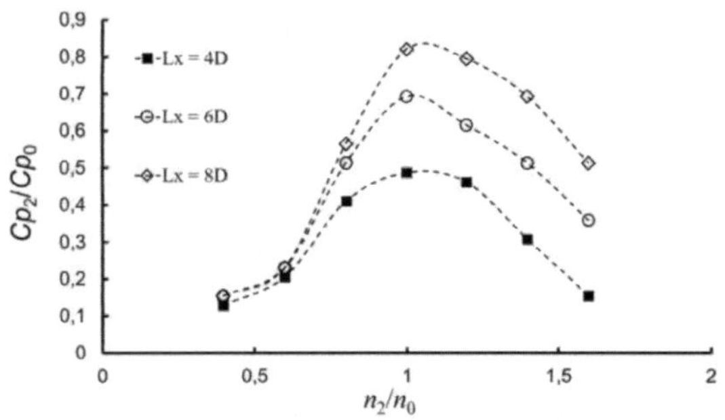

(a)

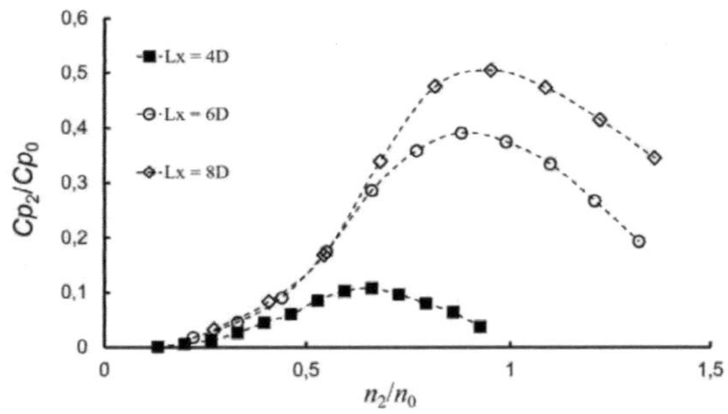

(b)

Figure 2. Variations of the power coefficients of the second rotors in the setups: (a) D-R and (b) R-R.

For the first rotor an angular velocity of the servo drive, $n_1 \equiv n_0 = 2.29$ rps, coincides to the design TSR of the single rotor ($\lambda_1 \equiv \lambda_0 = 5$) with the maximal power coefficient $C_{P0}= 0.37$ and the incoming freestream flow ($U_1 \equiv U_0 = 0.54$ m/s). An evolution of the optimal operating regimes for the second rotor in both D-R and R-R cases is not such a simple task since the velocity U_2 of the incident flow upstream from the second rotor is not known and not uniform behind either the forward disk or rotor. The optimal angular velocity of the servo drive (n_2) in the both cases were varied until optimum power conditions ($max\ C_{P2}$) were achieved (Fig.2 and Table 2) [4, 10].

Table 2. Operating parameters of the dual setups.

Types of the setup	Parameters	Values		
	L_x - *distance*	4*D*	6*D*	8*D*
D-D	U_2/U_0	0.41	0.58	0.61
	$(C_{T2}+ C_{T1})/C_{T0}$	1.69	1.86	1.97
D-R (optimal R)	U_2/U_0	0.45	0.73	0.80
	n_2/n_0	1	1	1
	$max\ C_{P2}/ C_{P0}$	0.48	0.69	0.83
	$(C_{T2}+ C_{T1})/C_{T0}$	1.59	1.62	1.64
R-R (optimal R and R)	U_2/U_0	0.33	0.55	0.68
	n_2/n_0	0.66	0.88	0.95
	$max\ C_{P2}/ C_{P0}$	0.11	0.39	0.51
	$(C_{T2}+ C_{T1})/C_{T0}$	1.32	1.65	1.74

Table 2 also shows estimations of the axial velocity component U_2 of the incident flow before the second disk or rotor and a ratio $(C_{T2}+ C_{T1})/C_{T0}$ of the total trust coefficients of all dual setups to the corresponding values of the single cases.

It is clearly seen from the R-R plot, just like any other case of Fig. 1, that the velocity deficit decreases monotonically and in a smooth manner for all the considered cases without large deviation from the fitting G - curve, shown by the solid line. Nevertheless, the wake deficit behind both active rotors (R-R) decreases, as compared to the single samples (R) in contrast to the former dual systems with the forward passive disk (D-D and D-R) when it grows. Next difference of the R-R case indicates by more concentrations of the measuring point along the fitting curve. Hence, we can conclude that the variations of the rotor positions do not have a significant effect on the wake deficit just like other cases with the forward passive disk (D-D and D-R).

3. Possible explanations: total thrust, turbulence or near wake impacts

The comparison between the dual passive disks, both passive disk and following active rotor and the dual active rotors shows a strong difference between the wake behaviors in the systems with the forward passive and active obstacles. The evaluation of both wake curves (D-D and D-R) for the passive forward disk indicates an increase in the wake deficit behind the second obstacle, as compared to the single samples (D). In contrast to this, the wake behind the active system of the dual rotors (R-R) decreases, as compared to the single samples (R).

Possible differences in the total thrust behaviors of the passive or active systems were tested and compared as our first idea for this explanation. The thrusts of the disk and rotor systems were measured in [10-11]. The ratio of the total thrust $(C_{T2}+ C_{T1})$ of these dual D-D, D-R and R-R systems to the single samples of D or R (C_{T0}) was calculated by varying the distance L_x (Table 2). For all cases, the flow velocity (U_2) in the wake before the second disk or rotor was also estimated and presented in the same table. From this investigation, we can conclude that the total thrust grows for all cases, as compared to the single samples, and it is not clear why the duplication of the passive or active systems results differently in the wake deficit.

Our next step for the explanation of the difference between D-D or D-R and R-R systems concerns the testing of possible differences in the turbulence characteristics. This was tested and Fig. 3 shows the comparison of the distribution of the RMS-values in the wake behind all the considered disk and rotor systems.

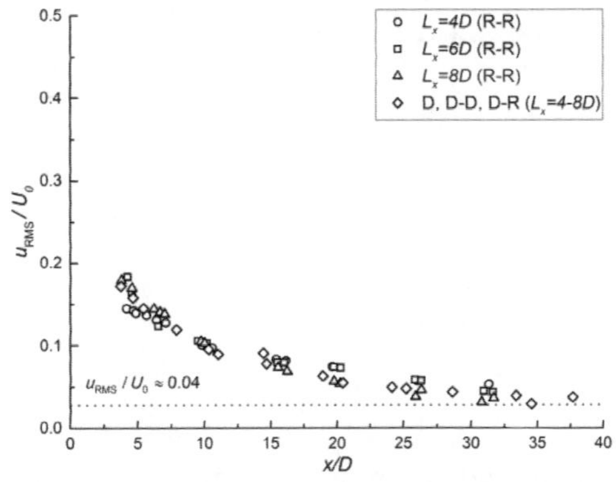

Figure 3. A comparison of RMS in the far wakes for both D-D, D-R, and R-R systems.

As seen in the figure the distributions are almost equal. Hence, it can be concluded that it is not the turbulence characteristics that cause the different wake behavior of the different configurations.

One can just guess that a strong influence of the helical tip vortices producing the active systems only to explain these differences. Indeed, the helical vortex system behind rotors plays an important role in a generation of the velocity deficit in the wake. For example, these vortices can reduce the velocity in the near wake with up to two times as much as in the rotor plane [12-14]. The helical tip vortices are not present when considering the flow about a disk [15]. Thus, the most likely candidate for explaining the different behavior is that the tip vortex in the near wake has an influence on the far wake characteristics in the rotor case.

The properties of the helical tip vortices behind a single rotor and the near wake of a single disk have been studied in many investigations (see e.g. [1, 15-17] and references hereby). In Fig 4 we reproduce profiles of the typical axial velocity distributions in the wakes behind both single rotor and disk to compare them with the ones generating behind the different dual systems on the same distance respectively. The examinations of the new data of the velocity profiles indicate a strong difference in the near wake behavior for the dual passive or active setups as it was found for the velocity deficit in the far wakes (Fig. 1).

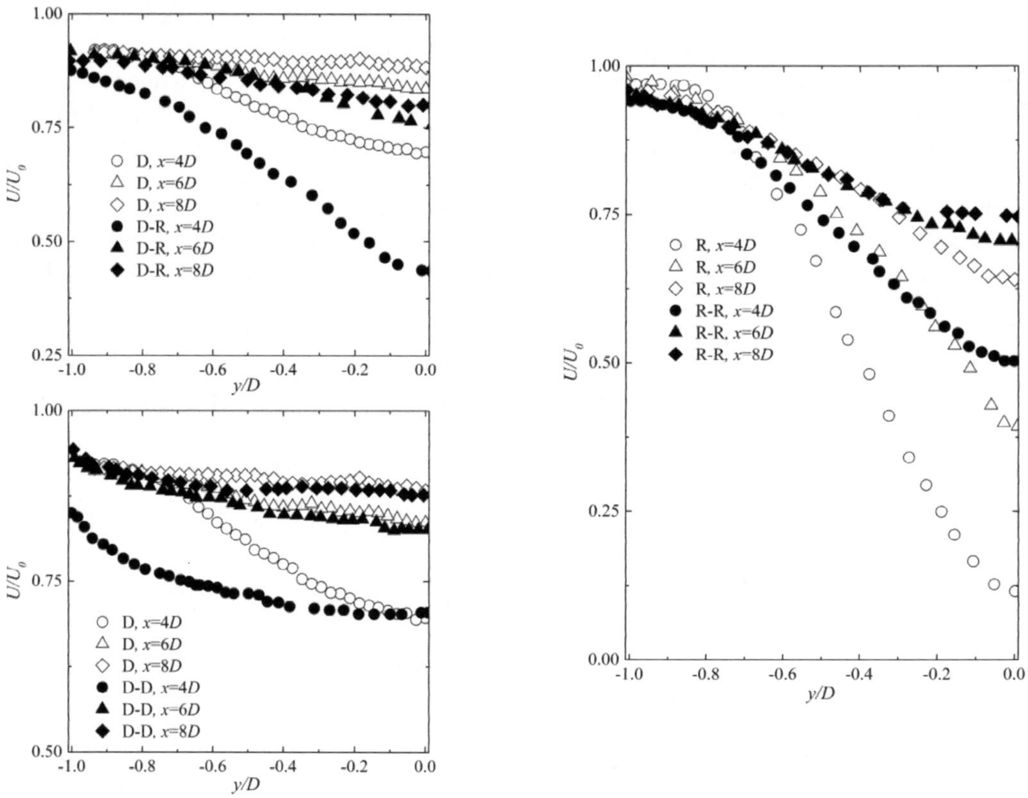

Figure 4. A comparison of the axial velocity profiles behind different single and dual testing systems with optimal TSR and L_x=8D.

Indeed the wake intensity grows for the dual systems when the disks put as the initial passive flow disturbance (for D-D) because the second disk produced an additional resistance in the flow. In contrast to this, the wake intensity behind the dual active rotor system (R-R) becomes smaller than it is behind the single rotor by a reducing of the tip vortex influence in the wake development when the second rotor put in a disturbance. It should also be mentioned that the disturbances coming from the first disk make a very similar "reduction" of the wake profiles behind the second rotor in the dual system D-R to compare with the single rotor (please compare the black symbols on the plots D-R and R-R of Fig. 4).

4. Conclusions

The far wake developments behind the passive and active dual systems (disk-disk; disk-rotor and rotor-rotor) have been studied for different interspatial positions and operating regimes. The measurements show that the velocity deficits in the far wakes for all the considered cases follow a rational dependence with the same rate, but with different wake attenuation coefficients. The resulting velocity deficit of the dual disks strongly depends on the distance between them and the wake intensity grows for the dual disks in comparison with the single one. In contrast to this, the decay of the velocity deficit behind dual rotors is nearly independent of the interspatial distance between them and the wake intensity behind the dual rotor system is smaller than the one behind a single rotor. The differences may be explained by the influence of the rotor tip vortices which are absent in the disk-disk model. Thus, our experiments indicate that a dual disk system cannot satisfactorily replace a system of dual rotors when analyzing wake properties behind wind turbines but the first disk in the disk-rotor system may be used as a method of the tip vortex control in the rotor wake.

5. Acknowledgments

The research was supported by the Russian Science Foundation (Project № 14-19-00487).

References

[1] Okulov VL, Naumov IV, Mikkelsen RF, Kabardin IK, Sørensen JN. A regular Strouhal number for large-scale instability in the far wake of a rotor. *J. Fluid Mech.* 2014; 747: 369-380.
[2] Okulov VL, Naumov IV, Mikkelsen RF, Sørensen JN. Wake effect on a uniform flow behind wind-turbine model. *Journal of Physics: Conference Series* 2015; 625: 012011.
[3] Okulov VL, Mikkelsen RF, Naumov IV, Litvinov IV, Gesheva E, Sørensen JN. Comparison of the far wake behind dual rotor and dual disk configurations. *Journal of Physics: Conference Series* 2016; 753: 032060.
[4] Naumov IV, Kabardin IK, Mikkelsen RF, Okulov VL, Sørensen JN. Rotor performance and wake conditions in non-uniform flow behind an obstacle. *Journal of Physics: Conference Series* 2016; 753: 032051
[5] Calaf M, Meneveau C and Meyers J Large eddy simulation study of fully developed wind-turbine array boundary layers. *Phys. Fluids* 2010; 22: 015110
[6] Porte-Agel F,Wu YT, Chen CH. A numerical study of the effects of wind direction on turbine wakes and power losses in a large wind farm. *Energies*. 2013; 6(10): 5297–5313.
[7] Nygaard N.G. Wakes in very large wind farms and the effect of neighbouring wind farms. *Journal of Physics: Conference Series* 2014; 524: 012162.
[8] Naumov IV, Litvinov IV, Mikkelsen RF, Okulov VL. Investigation of a wake decay behind a circular disk in a hydro channel at high Reynolds numbers. *Thermophysics and Aeromechanics* 2015; 22(6): 657-665.
[9] Naumov IV, Mikkelsen RF, Okulov VL. Estimation of Wake Propagation behind the Rotors of Wind-Powered Generators. *Thermal Engineering* 2016; 63(3): 208–213.
[10] Okulov VL, Mikkelsen RF, Sørensen JN. Naumov IV, Tsoy MA. Power Properties of Two Interacting Wind Turbine Rotors. *Journal of Energy Resources Technology* 2017; 139(5): 051210.
[11] Naumov I.V., Litvinov I.V., Mikkelsen R.F., Okulov V.L. Experimental investigation of wake evolution behind a couple of flat disks in the hydro channel. *Thermophysics and Aeromechanics* 2016; 23(5): 657–666.
[12] Okulov V.L. Limit cases for rotor theories with Betz optimization. *Journal of Physics: Conference Series* 2014; 524: 01212.
[13] Kuik GAM van, Sørensen JN, Okulov VL. The rotor theories by Professor Joukowsky: Momentum Theories. *Prog Aerospace Sci* 2015; 73: 1-18. DOI: 10.1016/j.paerosci.2014.10.
[14] Okulov VL, Sørensen JN, Wood DH. The rotor theories by Professor Joukowsky: Vortex Theories. *Prog Aerospace Sci* 2015; 73: 19-46. DOI: 10.1016/j.paerosci.2014.10.002.
[15] Naumov IV, Rahmanov VV, Okulov VL, Velte CM, Mayer KE, Mikkelsen RF. Flow diagnistics downstream of a tribladed rotor model. *Thermophysics and Aeromechanics* 2012; 19(2): 171-181.
[16] Naumov IV, Mikkelsen RF, Okulov VL, Sørensen JN. PIV and LDA measurements of the wake behind a wind turbine model. *Journal of Physics: Conference Series* 2014; 524 (1): 012168
[17] Bobinski T., Goujon-Durand S., Wesfreid J.E. Instabilities in the wake of a circular disk. *Physical Review E* 2014; 89 (5): 053021

Optimization Under Uncertainty for Wake Steering Strategies

Julian Quick[1], Jennifer Annoni[1], Ryan King[1], Katherine Dykes[1], Paul Fleming[1], Andrew Ning[2]

[1] National Renewable Energy Laboratory, 15013 Denver West Pkwy, Golden, CO 80401, USA
[2] Department of Mechanical University, Brigham Young University, Provo, UT 84602, USA

E-mail: julian.quick@nrel.gov

Abstract.
Wind turbines in a wind power plant experience significant power losses because of aerodynamic interactions between turbines. One control strategy to reduce these losses is known as "wake steering," in which upstream turbines are yawed to direct wakes away from downstream turbines. Previous wake steering research has assumed perfect information, however, there can be significant uncertainty in many aspects of the problem, including wind inflow and various turbine measurements. Uncertainty has significant implications for performance of wake steering strategies. Consequently, the authors formulate and solve an optimization under uncertainty (OUU) problem for finding optimal wake steering strategies in the presence of yaw angle uncertainty. The OUU wake steering strategy is demonstrated on a two-turbine test case and on the utility-scale, offshore Princess Amalia Wind Farm. When we accounted for yaw angle uncertainty in the Princess Amalia Wind Farm case, inflow-direction-specific OUU solutions produced between 0% and 1.4% more power than the deterministically optimized steering strategies, resulting in an overall annual average improvement of 0.2%. More importantly, the deterministic optimization is expected to perform worse and with more downside risk than the OUU result when realistic uncertainty is taken into account. Additionally, the OUU solution produces fewer extreme yaw situations than the deterministic solution.

1. Introduction

Wind power plant control can be used to efficiently increase wind energy production by maximizing power in wind plants that are already installed [1,2]. It can also be used to mitigate structural loads to maximize the lifetime of the turbines and better integrate wind energy into the energy market. One wind power plant control strategy, known as wake steering, has been explored extensively in the literature [3–5]. Wake steering involves intelligently offsetting wind turbine yaw alignments from the incoming wind so that wakes are deflected away from downstream turbines. This control strategy may allow wind plant operators to produce more power than by controlling each turbine to maximize its own energy capture by yawing directly towards the incoming flow.

Accurately measuring and controlling yaw misalignment can be an issue when designing wake steering schemes. Complex phenomena such as vorticity caused by the turbines' blades, the speed-up effect around the nacelle, and sensor errors introduce significant uncertainty in measuring each turbine's yaw angle relative to the incoming wind. Several studies have examined wind-power-plant-level control strategies and have reported significant gains in annual energy production [3, 4, 6]. These studies use deterministic formulations of the wind energy system, assuming perfect information on parameter inputs. The uncertainty associated with wind plant model parameters such as inflow direction, yaw misalignment, or average speed may cause a wake steering strategy in the field to perform significantly worse than anticipated. One potential solution to this problem is optimization under uncertainty (OUU). This technique has been used in a variety of studies to provide a robust solution to a problem under varying levels of uncertainty [7–9]. Although other studies have modeled flow within a wind power plant using uncertainty in wind direction [7, 10], the authors are unaware of any previous yaw steering optimization that has explicitly taken uncertainty into account.

In this paper, we propose a methodology to explicitly take uncertainty in yaw position into account during the process of optimizing turbines' yaw offsets for wake steering schemes. As a first step, we examined how large yaw misalignment uncertainty affects the optimal solution. We used an engineering wake model to simulate the flow within a wind power plant, described in Section 2.1. In Section 2.2, a stochastic optimization approach is formulated to determine the optimal turbine control settings under uncertain yaw misalignment. Two case studies are used to further understand the effects of yaw misalignment uncertainty, detailed in Section 2.3. The first case study uses two turbines, in which the front turbine directly wakes the downstream turbine. The second case study considers the Princess Amalia Wind Farm, wherein we assume a single wind speed for each direction. The Princess Amalia Wind Farm was used in a previous wake steering study [3]. The results of these case studies are discussed in detail in Section 3. Finally, Section 4 addresses some conclusions and future work.

2. Methodology, application, and approach

In this study, we applied the FLOw Redirection and Induction in Steady State (FLORIS) engineering wake model to a two-turbine test case and the Princess Amalia Wind Farm to broadly quantify potential benefits of explicitly taking yaw misalignment into account when designing wake steering schemes. For this study, we assumed a constant and extreme yaw position error $\omega \sim \mathcal{N}(0°, \sigma^2)$, which represents a Gaussian distribution of yaw misalignment errors centered at 0° offset and with a standard deviation σ equal to 20°. This random yaw position error was added to the design yaw offsets during the OUU uncertainty quantification routine, simulating 30,000 possible scenarios resulting from different misalignment errors in each optimization iteration. A convergence study was used to identify the number of samples required to sufficiently minimize the variance of the expected power production. The turbine yaw controls were bounded from -30° to 30° and the yaw misalignment uncertainty was unbounded.

2.1. Engineering Wake Model: FLORIS

The FLORIS model, developed in [4], is an extension of Jensen's model [11], which approximates wake deflection caused by yaw offsets [12]. Further modifications have been made to better model the wake velocity profile and effects of partial wake overlap. The steady-state power of each turbine i, denoted as P_i, is given by the FLORIS model as:

$$P_i = \frac{1}{2}\rho A_i C_P(a_i) U_i^3 \tag{1}$$

where ρ is the air density, A_i is the area of the rotor, C_P is the power coefficient, and U_i is the effective wind speed at turbine i. The effective velocity at the downstream turbine i is found by combining the effect of the wakes of the upstream turbine j, weighting the wake zones (near wake zone, far wake zone, and mixing wake zone) by their overlap with the rotor as:

$$U_i = U_\infty \left(1 - 2\sqrt{\sum_j \left[a_j \sum_{q=1}^{3} c_{j,q}(X_i) \min\left(\frac{A_{j,i,q}^{overlap}}{A_i}, 1\right)\right]^2}\right) \tag{2}$$

where U_∞ is the free-stream velocity, X_i is the x (streamwise direction) location of turbine i, $A_{j,i,q}^{overlap}$ is the overlap area of a wake zone q of a turbine i with the rotor of turbine j, and $c_{i,q}(x)$ is a coefficient that defines the recovery of a zone q to the free-stream conditions:

$$c_{i,q}(x) = \left(\frac{D_i}{D_i + 2k_e m_{U,q}(\gamma_i)[x - X_i]}\right)^2 \tag{3}$$

where $m_{U,q}$ is defined as:

$$m_{U,q}(\gamma_i) = \frac{M_{U,q}}{\cos(a_U + b_U \gamma_i)} \tag{4}$$

for $q = 1, 2, 3$ (corresponding to the three wake overlap zones), where a_U and b_U are tuned model parameters, D_i is the rotor diameter of turbine i, γ_i is the yaw offset of turbine i, and $M_{U,q}$ are tuned scaling factors that ensure that the velocity in the outer zones of the wake will recover to the free-stream conditions faster than in the inner zones. Additional details can be found in [4].

When a turbine is yawed, it exerts a force on the flow that causes the wake to deflect in a particular direction. Empirical formulas were presented in [12] and used in the FLORIS formulation [4]. In addition to wake deflection, there is a rotation-induced lateral offset that is caused by the interaction of the wake rotation with the shear layer [4]. The total lateral offset is represented as:

$$y_{w,i} = Y_i + \delta y_{w,yaw,i}(x) + \delta y_{w,rotation,i}(x) \tag{5}$$

where Y_i is the y (cross-stream direction) location of the turbine. $\delta y_{w,yaw,i}(x)$ and $\delta y_{w,rotation,i}(x)$ are defined in detail in [4].

A modification was made to the FLORIS model to allow for extreme yaw events in the uncertainty quantification routine. As $a_U + b_U\gamma_i$ approaches 90°, the denominator in Equation (4) approaches zero and $m_{U,q}$ approaches infinity. We introduce a simple modification: when $a_U + b_U\gamma_i$ is greater than 85°, we replace the $a_U + b_U\gamma_i$ term with 85°. Though we believe that this modification should be sufficient to capture general trends, future work is needed to model wakes caused by extreme yaws more rigorously.

2.2. Problem Formulation

Using the FLORIS model described earlier, the power production of a wind power plant can be computed for the whole year (i.e., the annual energy production). The annual energy production can be used in the cost function to optimize the yaw angles of the turbines in the wind plant with and without uncertainty. The remainder of this section formulates the yaw steering OUU problem using the average annual power production as the object of interest.

The average annual power production is a linear sum of each speed-and-direction-specific energy production value:

$$\Pi(\boldsymbol{Y}, \boldsymbol{\omega}) = \sum_{i=1}^{n} \sum_{j=1}^{m} P(\theta_i) P(u_{ij}) \pi(\boldsymbol{Y}_{ij} + \boldsymbol{\omega}_{ij}, \theta_i, u_{ij}) \tag{6}$$

where $P(\theta_i)$ is the discrete probability mass function of θ_i, $P(u_{ij})$ is the discrete probability mass function of u_{ij}, n is the number of directional bins, m is the number of wind speed bins,

$\boldsymbol{Y}$ is an m by n by T tensor, where T is the number of turbines and each $\boldsymbol{Y}_{ij}$ is a speed-and-direction-specific vector of T yaw offsets for controlling the entire wind power plant, and $\boldsymbol{\omega}_{ij}$ is a speed-and-direction-specific array of T random variables representing yaw misalignment errors: $\boldsymbol{Y}_{ij} = [\boldsymbol{Y}_{ij,1}, \boldsymbol{Y}_{ij,2}, \ldots, \boldsymbol{Y}_{ij,T}]$ and $\boldsymbol{\omega}_{ij} = [\boldsymbol{\omega}_{ij,1}, \boldsymbol{\omega}_{ij,2}, \ldots, \boldsymbol{\omega}_{ij,T}]$. Each $\boldsymbol{\omega}_{ijk}$ entry is a Gaussian distribution centered at zero with standard deviations of 20°. $\pi(\boldsymbol{x}, \theta, u)$ is the power production assuming unidirectional θ degree inflow with average wind speed u, and $\boldsymbol{x}$ is a speed-and-direction-specific steering scheme vector of length T, and $\Pi(\boldsymbol{Y}, \boldsymbol{\omega})$ is annual power production averaged across wind direction and speed frequencies.

In the OUU formulation [13], we optimize the expected value of Π:

$$\mathbb{E}[\Pi(\boldsymbol{Y}, \boldsymbol{\omega})] = \sum_{i=1}^{n} \sum_{j=1}^{m} P(\theta_i) P(u_{ij}) \int \ldots \int \pi(\boldsymbol{Y}_{ij} + \boldsymbol{\omega}_{ij}, \theta_i, u_{ij}) P(\boldsymbol{\omega}_{ij}) d\boldsymbol{\omega}_{ij} \quad (7)$$

where $\mathbb{E}$ is the expected value and $P(\boldsymbol{\omega}_{ij})$ is the probability density function of the multivariate normal vector of yaw misalignment, $\boldsymbol{\omega}_{ij}$.

Because of the intensive computational nature of the problem, we chose to only examine the average wind speed for 72 directional bins in the Princess Amalia Wind Farm wake steering strategy optimization. We plant to expand the problem to consider multiple speeds in each direction bin in future work. Thus, our problem is reduced to the following:

$$\mathbb{E}[\Pi(\boldsymbol{X}, \boldsymbol{\omega})] = \sum_{i=1}^{n} P(\theta_i) \int \ldots \int \pi(\boldsymbol{X}_i + \boldsymbol{\omega}_i, \theta_i) P(\boldsymbol{\omega}_i) d\boldsymbol{\omega}_i \quad (8)$$

where $\boldsymbol{X} = [\boldsymbol{X}_1, \boldsymbol{X}_2, \ldots, \boldsymbol{X}_n]$, where each $\boldsymbol{X}_i$ is an inflow-direction-specific steering scheme for each of the wind power plant's T turbines, so, for each i directional bin, $\boldsymbol{X}_i = [\boldsymbol{X}_{i,1}, \boldsymbol{X}_{i,2} \ldots \boldsymbol{X}_{i,T}]$.

In the OUU setup, we approximate the expected value using a sample average approximation with K samples:

$$\widehat{\mathbb{E}}[\Pi(\boldsymbol{X}, \boldsymbol{\omega})] = \sum_{i=1}^{n} P(\theta_i) \frac{1}{K} \sum_{K} \pi(\boldsymbol{X}_i + \boldsymbol{\omega}_i^k, \theta_i) \quad (9)$$

where $\widehat{\mathbb{E}}$ is an approximation of the true expected value and $\boldsymbol{\omega}_i^k$ is the kth evaluation of the random vector $\boldsymbol{\omega}_i$.

Thus, we formulate the wake steering OUU problem as follows:

$$\boldsymbol{X}^*_{OUU} = \arg\max_{\boldsymbol{X}} \widehat{\mathbb{E}}[\Pi(\boldsymbol{X}, \boldsymbol{\omega})] = \arg\max_{\boldsymbol{X}} \left(\sum_{i=1}^{n} P(\theta_i) \frac{1}{K} \sum_{K} \pi(\boldsymbol{X}_i + \boldsymbol{\omega}_i^k, \theta_i) \right) \quad (10)$$

Similarly, the deterministic wake steering optimization is formulated as:

$$\boldsymbol{X}^*_{det} = \arg\max_{\boldsymbol{X}} \Pi(\boldsymbol{X}, \mathbb{E}[\boldsymbol{\omega}]) = \arg\max_{\boldsymbol{X}} \left(\sum_{i=1}^{n} P(\theta_i) \pi(\boldsymbol{X}_i, \theta_i) \right) \quad (11)$$

Because a metric is needed to assess the quality of each solution, we introduce the value of the stochastic solution (VSS). Our definition is similar to the VSS metric introduced by Birge and Louveaux [13], but is expressed as a percent increase in production rather than an absolute value increase. We define the VSS as the ratio of the expected values of deterministic and OUU solutions, given yaw uncertainty:

$$VSS = \frac{\mathbb{E}[\Pi(\boldsymbol{X}^*_{\mathrm{OUU}}, \boldsymbol{\omega})]}{\mathbb{E}[\Pi(\boldsymbol{X}^*_{\mathrm{det}}, \boldsymbol{\omega})]} \quad (12)$$

where VSS is the value of the stochastic solution evaluated across the wind rose; $\boldsymbol{\omega}$ is an uncertain n by T matrix of yaw misalignment, normally distributed around zero with standard deviation equal to 20°; $\boldsymbol{X}^*_{OUU}$ is the optimal yaw strategy obtained via OUU, assuming unbiased, Gaussian yaw misalignment ω; and $\boldsymbol{X}^*_{det}$ is the optimal yaw strategy obtained via deterministic optimization, assuming perfect control of yaw alignment.

In the Princess Amalia Wind Farm case, it was useful to report direction-specific VSS values, defined as:

$$vss_i = \frac{\mathbb{E}[\pi_{\mathrm{i}}(\boldsymbol{X}^*_{\mathrm{OUU,i}} + \omega)]}{\mathbb{E}[\pi_{\mathrm{i}}(\boldsymbol{X}^*_{\mathrm{det,i}} + \omega)]} \tag{13}$$

where: vss_i is the value of the stochastic solution considered in the ith direction.

2.3. Case studies

We explored the benefits of OUU in a two-turbine test case as well as for the Princess Amalia Wind Farm, explicitly taking yaw misalignment uncertainty into account during the optimization process.

The two-turbine case has unidirectional flow with a spacing of 10 rotor diameters. The turbines are both 5 MW and the inflow is 10.6 m/s and orientated such that the front turbine directly wakes the back turbine. We investigated the two-turbine OUU setup for various levels of yaw misalignment uncertainty. We performed a parameter sweep across possible values of the front turbine with a nested Latin hypercube sampling routine—simulating yaw misalignment uncertainty in each turbine—to find the optimum steering strategy at various levels of uncertainty, recording the VSS and optimum front turbine angle associated with each level of yaw misalignment uncertainty.

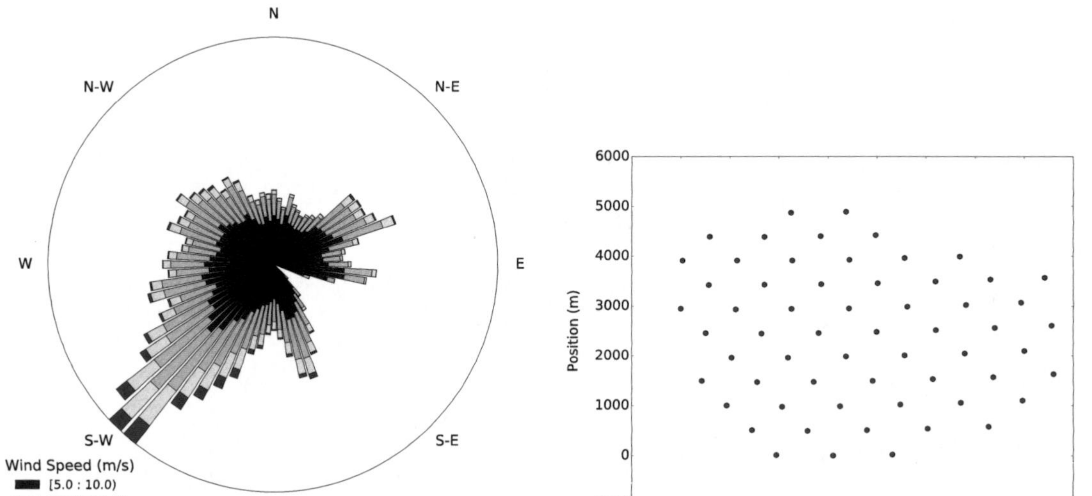

Figure 1. Princess Amalia Wind Farm characteristics. The wind rose is shown on the left. The turbine layout is shown on the right.

Next, we used the Princess Amalia Wind Farm specifications to compare wake steering schemes. We assumed a single wind rose with 72 directional bins (Figure 1), where each bin's wind characteristics are summarized by the average annual wind speed. The wind power plant has 60 five-MW turbines with rotor diameters of 126 m (Figure 1). We approached the Princess

Amalia Wind Farm wake stering OUU problem by nesting a semirandom sampling routine within an outer optimization. We chose to use the Latin hypercube sampling approach and the constrained optimization by using a linear approximation (COBYLA) optimization algorithm. The optimal yaw arrangement was found using each of the 72 recorded directions and the associated mean wind speeds in parallel optimizations with 10 multiple random optimization starts associated with each direction. The Latin hypercube samples the uncertain inputs semirandomly by dividing their bounded values into a grid to obtain sparsely selected random samples and avoid needlessly sampling the model [14]. The Latin hypercube sampling was highly parallelized using Sandia National Laboratories' DAKOTA analysis engine [14]. The COBYLA routine uses a sequential trust-region algorithm to create variable-sized linear approximations of the objective function and constraints to solve the optimization [14,15]. Thus, COBYLA creates a constructed optimum, as opposed to other metaheuristic algorithms that search for optimums by semirandomly perturbing the candidate solution. We ran COBYLA for 1,000 iterations with a nested Latin hypercube sampling routine with 30,000 samples. For each direction-specific yawing scheme, we maximized the expected power production using this nested sampling approach.

3. Results

First, we performed the wake steering OUU on the two-turbine case, varying the level of yaw uncertainty. Next, using the Princess Amalia Wind Farm, we assumed a single yaw misalignment uncertainty. We report the average power produced by the Princess Amalia Wind Farm using deterministic and OUU optimal strategies, assuming a single wind speed per direction. We also examine the flow geometry, distribution of yaw misalignment, and distributions of potential power generation resulting from possible yaw misalignment errors for a direction that yielded large VSS.

3.1. Two-turbine test case

We examined a two-turbine case to better understand how yaw misalignment uncertainty affects the optimal solution. We found that the OUU solution generally prefers less steering and that the optimal front turbine yaw offset decreased with increases in yaw misalignment uncertainty (Figure 2). As expected, the VSS increased with yaw misalignment uncertainty. When the standard deviation of the Gaussian distribution of errors in both turbines' yaw misalignment reached 20°(the uncertainty assumed for the Princess Amalia Wind Farm case), the OUU strategy performed 1.1% better than the deterministically formulated strategy when both strategies were implemented with noise in the yaw sensors. The optimal expected power decreased from 6.9 to 5.7 GW when the standard deviation of the turbines' yaw misalignment increased from 0° to 30°. As yaw misalignment uncertainty increased, the optimum strategy shifted to extract more power from the front turbine. Thus, the downside risk of offsetting the front turbine increased and there was less utility in wake steering.

3.2. Princess Amalia Wind Farm case

In this section, we present the results of the OUU problem using the Princess Amalia Wind Farm. We found the optimal yaw configuration for this case using the deterministic and OUU optimization formulations developed in Section 2.2. If there is perfect knowledge of the alignment of each wind turbine's yaw relative to the average direction of the incoming wind, there is no need for the OUU approach. Given uncertainty in yaw alignments, there appear to be various benefits from OUU depending on the incoming wind direction (Figure 3). For example, considering the northern inflow (0°) direction, the OUU solution produces 1.4% more power than the deterministic solution, under the yaw misalignment uncertainty introduced in Section 2.2.

We expect the northern direction OUU solution to yield 1.4% more power than the deterministic strategy [Equation (9)]. The probability distributions and deterministic values

IOP Conf. Series: Journal of Physics: Conf. Series **854** (2017) 012036 doi:10.1088/1742-6596/854/1/012036

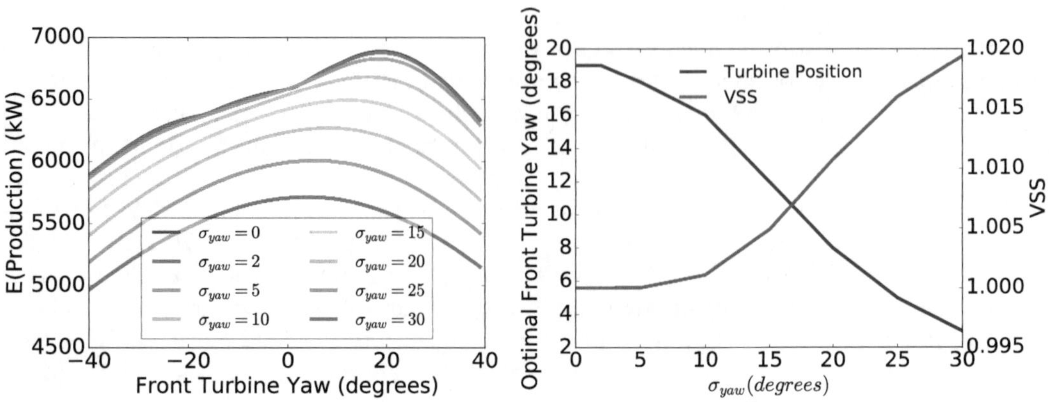

Figure 2. (Left) Power production resulting from different front turbine yaw angles. Contours of different yaw misalignment uncertainty values are shown. (Right) Optimal front turbine arrangement and VSS plotted against yaw misalignment uncertainty.

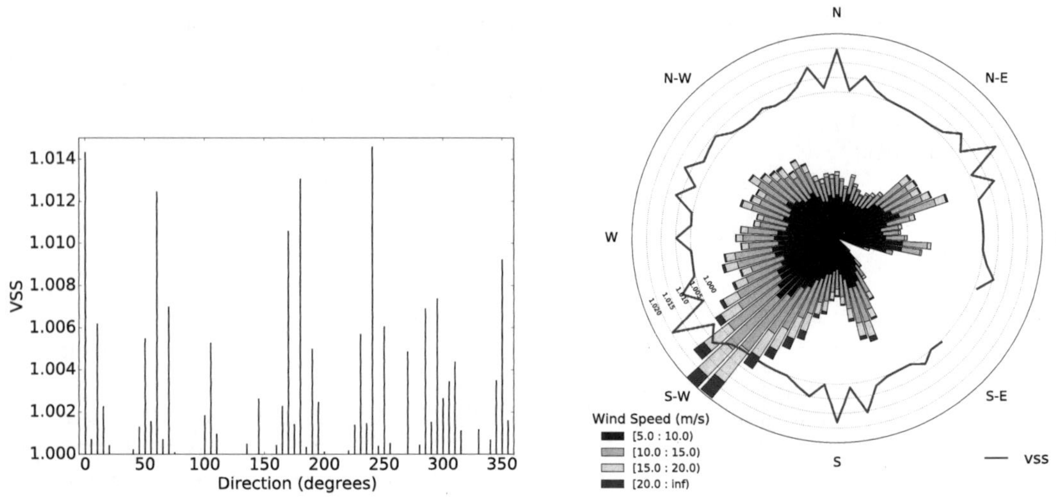

Figure 3. (Left) The vss for each direction considered. (Right) The vss plotted along the site wind rose. In some directions, the OUU solution outperformed the deterministic strategy's power production (under yaw misalignment uncertainty) by nearly 1.4%, whereas the OUUs associated with other directions yielded little or no improvements in expected power production.

of the OUU optimal strategy, deterministic optimal strategy, and baseline steering strategy are shown in Figure 4 (left) for the northern inflow case. The distribution of yaw offsets in the OUU and deterministic optimum strategies are shown in Figure 4 (right). Similarly to the two-turbine case, introducing uncertainty resulted in less utility to steering wakes. Consequently, the OUU strategy uses lesser offsets, which is attractive because lesser offsets reduce turbines' non-torque drivetrain loads and yaw moments. Figure 5 shows the flow fields for the northern inflow's baseline, deterministic, and OUU solutions, respectively. Figure 6 shows a close-up of the flow fields for the deterministic and OUU strategies to highlight the differences between the two solutions. The deterministic strategy steers wakes further away from downstream turbines, thus, the OUU solution prefers less yaw offsets.

IOP Conf. Series: Journal of Physics: Conf. Series **854** (2017) 012036 doi:10.1088/1742-6596/854/1/012036

We expect the robust design to produce 0.2% more average annual power than the deterministic design (Table 1). In each direction, the deterministic solution that performed best under certain yaw alignments was compared to the OUU solution that performed best under yaw alignment uncertainty. These results were generated using 10 random initial starting points per direction.

Table 1. Optimization results (GW).

	Deterministic Average Power Production	Stochastic Average Power Production
OUU Solution	110.4	100.6
Deterministic Solution	113.1	100.1
No Yaw Offsets	105.1	97.4

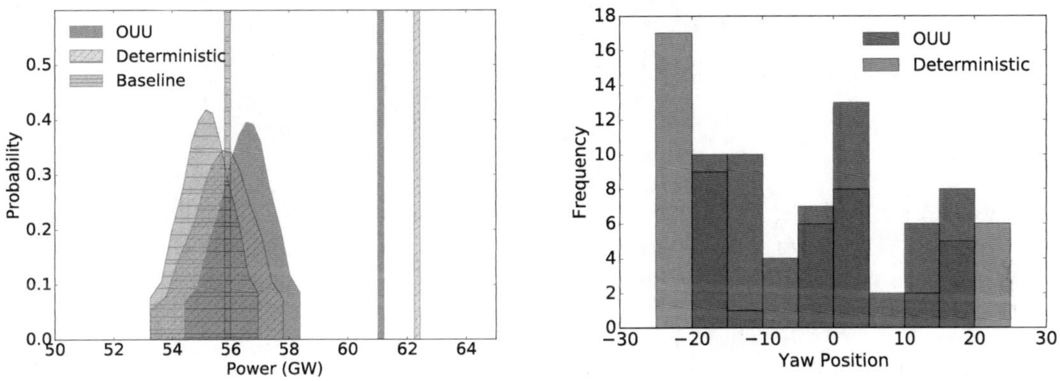

Figure 4. (Left) Empirical distributions of energy production under yaw misalignment uncertainty for OUU, deterministic, and baseline (no yaw offsets) steering strategies for the northern inflow case. The probability distributions show possible yaw misalignment. The straight bars show the forecasted production for each arrangement, assuming perfect knowledge of each turbine's yaw alignment. (Right) Frequency occurrence of deterministic and OUU solution yaw positions for the northern inflow. OUU solutions generally produced strategies with less extreme offsets.

4. Conclusions

Robust optimization offers an attractive solution to the yaw misalignment problem when designing wake steering strategies. Deterministically, there is a natural optimum steering strategy balancing the trade-offs of losing power from the steering turbine's offset and gaining power by steering the velocity deficit away from downstream turbines. When yaw misalignment uncertainty is considered, the optimal wake steering shifts towards lesser yaw offsets, extracting more power from the front turbines. Future work in this area could examine other sources of uncertainty. In addition, it may be useful to integrate wake steering strategies into the turbine placement design (this was done deterministically in [3]), explicitly taking measurement uncertainty into account during the design process.

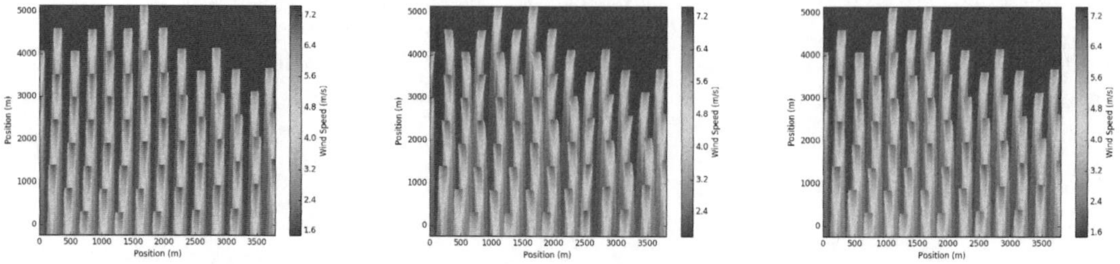

Figure 5. (Left) No steering with the northern inflow. (Middle) Deterministic optimal steering with the northern inflow. (Right) OUU optimal steering with the northern inflow.

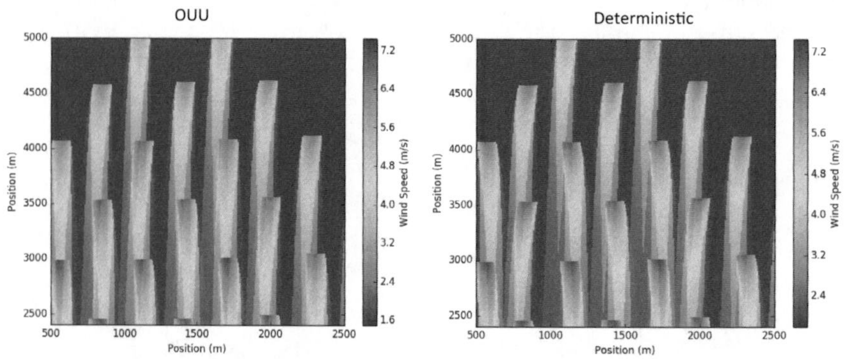

Figure 6. Close-up of the flow fields associated with the OUU and deterministic optimal yaw offset strategies. The deterministic strategy steers the wakes directly next to the downstream turbines, whereas the OUU solution requires less steering.

5. Acknowledgments

This work was supported by the U.S. Department of Energy under Contract No. DE-AC36-08GO28308 with the National Renewable Energy Laboratory. Funding for the work was provided by the DOE Office of Energy Efficiency and Renewable Energy, Wind Energy Technologies Office.

References

[1] Kathryn E Johnson and Naveen Thomas. Wind farm control: Addressing the aerodynamic interaction among wind turbines. In *American Control Conference, 2009. ACC'09.*, pages 2104–2109. IEEE, 2009.

[2] Jason R Marden, Shalom D Ruben, and Lucy Y Pao. A model-free approach to wind farm control using game theoretic methods. *IEEE Transactions on Control Systems Technology*, 21(4):1207–1214, 2013.

[3] Paul A Fleming, Andrew Ning, Pieter MO Gebraad, and Katherine Dykes. Wind plant system engineering through optimization of layout and yaw control. *Wind Energy*, 19(2):329–344, 2016.

[4] PMO Gebraad, FW Teeuwisse, JW Wingerden, PA Fleming, SD Ruben, JR Marden, and LY Pao. Wind plant power optimization through yaw control using a parametric model for wake effects–a CFD simulation study. *Wind Energy*, 19(1):95–114, 2016.

[5] Steffen Raach, David Schlipf, Friedemann Borisade, and Po Wen Cheng. Wake redirecting using feedback control to improve the power output of wind farms. In *American Control Conference (ACC), 2016*, pages 1387–1392. IEEE, 2016.

IOP Conf. Series: Journal of Physics: Conf. Series **854** (2017) 012036 doi:10.1088/1742-6596/854/1/012036

[6] Ervin Bossanyi and Tiago Jorge. Optimisation of wind plant sector management for energy and loads. In *Control Conference (ECC), 2016 European*, pages 922–927. IEEE, 2016.

[7] Javier Serrano Gonzalez, Manuel Burgos Payan, and Jess M. Riquelme-Santos. Optimization of wind farm turbine layout including decision making under risk. *IEEE Systems Journal*, 6(1):94–102.

[8] Nikolaos V Sahinidis. Optimization under uncertainty: state-of-the-art and opportunities. *Computers & Chemical Engineering*, 28(6):971–983, 2004.

[9] J. Quick, K. Dykes, P. Graf, and F. Zahle. Optimization under uncertainty of site-specific turbine configurations. *Journal of Physics: Conference Series*, 753(6):062012, 2016.

[10] M Gaumond, P-E Réthoré, Søren Ott, Alfredo Peña, Andreas Bechmann, and Kurt Schaldemose Hansen. Evaluation of the wind direction uncertainty and its impact on wake modeling at the horns rev offshore wind farm. *Wind Energy*, 17(8):1169–1178, 2014.

[11] I Katic, J Højstrup, and NO Jensen. A simple model for cluster efficiency. In *European Wind Energy Association Conference and Exhibition*, pages 407–410, 1986.

[12] Ángel Jiménez, Antonio Crespo, and Emilio Migoya. Application of a LES technique to characterize the wake deflection of a wind turbine in yaw. *Wind energy*, 13(6):559–572, 2010.

[13] John R. Birge and François Louveaux. *Introduction to Stochastic Programming*. Springer Series in Operations Research and Financial Engineering. Springer, July 1997.

[14] Brian M. Adams, Mohamed S. Ebeida, Michael S. Eldred, Gianluca Geraci, John D. Jakeman, Kathryn A. Maupin, Jason A. Monschke, Laura P. Swiler, J. Adam Stephens, Dena M. Vigil, Timothy M. Wildey. Dakota, a multilevel parallel object-oriented framework for design optimization, parameter estimation, uncertainty quantification, and sensitivity analysis: Version 6.5 theory manual, July 2014.

[15] M. J. D. Powell. A Direct Search Optimization Method that Models the Objective and Constraint Functions by Linear Interpolation. In Susana Gomez and Jean-Pierre Hennart, editors, *Advances in Optimization and Numerical Analysis*, number 275 in Mathematics and Its Applications, pages 51–67. Springer Netherlands.

Results of the GABLS3 diurnal-cycle benchmark for wind energy applications

J Sanz Rodrigo[1], D Allaerts[9], M Avila[4], J Barcons[4], D Cavar[5], RA Chávez Arroyo[1], M Churchfield[2], B Kosovic[3], JK Lundquist[2,6], J Meyers[7], D Muñoz Esparza[3], JMLM Palma[8], JM Tomaszewski[6], N Troldborg[5], MP van der Laan[5] and C Veiga Rodrigues[8]

[1]National Renewable Energy Centre, Sarriguren, Spain
[2]National Renewable Energy Laboratory, Boulder, CO, U.S.A.
[3]National Centre for Atmospheric Research, Boulder, CO, U.S.A.
[4]Barcelona Supercomputing Centre, Barcelona, Spain
[5]Technical University of Denmark, Roskilde, Denmark
[6]University of Colorado Boulder, Boulder, CO, U.S.A.
[7]University of Leuven, Leuven, Belgium
[8]University of Porto, Porto, Portugal

E-mail: jsrodrigo@cener.com

Abstract. We present results of the GABLS3 model intercomparison benchmark revisited for wind energy applications. The case consists of a diurnal cycle, measured at the 200-m tall Cabauw tower in the Netherlands, including a nocturnal low-level jet. The benchmark includes a sensitivity analysis of WRF simulations using two input meteorological databases and five planetary boundary-layer schemes. A reference set of mesoscale tendencies is used to drive microscale simulations using RANS k-ε and LES turbulence models. The validation is based on rotor-based quantities of interest. Cycle-integrated mean absolute errors are used to quantify model performance. The results of the benchmark are used to discuss input uncertainties from mesoscale modelling, different meso-micro coupling strategies (online vs offline) and consistency between RANS and LES codes when dealing with boundary-layer mean flow quantities. Overall, all the microscale simulations produce a consistent coupling with mesoscale forcings.

1. Introduction

The increasing growth of wind turbines, with rotor tip heights approaching 200 m, and wind farm clusters extending for tens of kilometers, is pushing the wind farm flow modeling community to consider effective ways of integrating forcing from large meteorological processes in the simulation of the flow at wind farm scale based on computational fluid dynamic (CFD) models. The dynamics of these forcings determine the interplay between the wind climatology, relevant for the assessment of the wind resource, and the wind conditions relevant for wind turbine siting.

A recent review of state-of-the-art methodologies for mesoscale-to-microscale modeling is provided in Sanz Rodrigo et al [1]. Outstanding challenges include: coupling of codes dealing with fundamental differences in terms of physical hypothesis and numerical methods, lack of suitable parameterization in the "terra incognita" [2] that links mesoscale and microscale turbulence processes, and lack of a systematic evaluation procedure that can identify the source of modeling errors. Additionally, for the wind energy community, design conditions are almost entirely based on idealized turbulence and boundary-layer models rather than recognizing and classifying wind conditions that account for mesoscale forcing in design standards.

With these challenges in mind, the boundary-layer meteorology community has conducted a series of model intercomparison GEWEX Atmospheric Boundary Layer Studies (GABLS) [3]. The third GABLS benchmark deals with a diurnal cycle observed at the Cabauw meteorological tower in the Netherlands, under relatively stationary synoptic conditions, that lead to the development of a strong nocturnal low-level jet (LLJ) [4][5]. Model intercomparison results for single-column models (SCM) are reported in [6].

This benchmark has been revisited by the wind energy community to help design meso-micro methodologies for resource assessment and design tools, in particular: incorporating time- and height-dependent mesoscale forcing in microscale models, turbulence modeling at varying atmospheric stability conditions, defining suitable surface boundary conditions for momentum and heat and characterization of the wind profile in (non-logarithmic) LLJ conditions [7]. Initial results, using a single-column model (SCM) as proxy for microscale 3D models, show how adding more realistic mesoscale forcing systematically leads to better performance considering cycle-aggregated rotor-based quantities of interest. Still, a large bias is observed in the hour-to-hour evolution of the vertical wind profile due to the inherent input uncertainty from the mesoscale model [8][9].

This paper summarizes the results of the Windbench/GABLS3 model intercomparison benchmark, discussing three topics: characterization of uncertainties in the input forcing, methodologies for meso-micro coupling, and consistency of turbulence models at different fidelity levels.

2. The Windbench/GABLS3 Benchmark

The GABLS3 original set-up is described in Bosveld et al. [5]. The case analyzes the period from 12:00 UTC 1 July to 12:00 UTC 2 July 2006, at the Cabauw Experimental Site for Atmospheric Research (CESAR), located in the Netherlands (51.971°N, 4.927°E). The elevation of the site is approximately -0.7 m, surrounded by relatively flat terrain characterized by grassland, fields and some scattered tree lines and villages. The roughness length for the wind direction sector of interest (60° - 120°) is 15 cm.

The CESAR measurements are carried out at a 200-m tower, free of obstacles up to a few hundred meters in all directions. The measurements include 10-min averaged vertical profiles of wind speed, wind direction, temperature and humidity at heights 10, 20, 40, 80, 140 and 200 m, as well as surface radiation and energy budgets. Turbulence fluxes are also monitored at four heights: 3, 60, 100 and 180 m. A RASS profiler measures wind speed, wind direction and virtual temperature above 200 m.

The original GABLS3 setup uses a simplified set of forcing terms that was obtained by piecewise linear approximations of the mesoscale tendencies in RACMO numerical weather prediction model to obtain a better agreement of the wind speed at 200m. In our model intercomparison we will use directly the input forcings derived from mesoscale, as described in the next section, without calibration against observations. This is in favor of testing a general methodology for offline coupling of mesoscale and microscale models.

The GABLS3 benchmark for wind energy is published as a Windbench repository in [7].

2.1. Reference WRF simulation: Input Data for Microscale Models

Input forcing for most of the microscale models (excepting VENTOS®/M) is derived from a mesoscale simulation using the Weather Research and Forecasting model WRF-ARW v3.8 [10]. Following previous work from Kleczek et al [11], the reference WRF simulation is based on a one-way nesting configuration of three concentric square domains centered at the Cabauw site, of the same size (Figure 1) based on a 181x181 points grid with 9, 3 and 1 km horizontal resolution. The vertical grid, approximately 13 km high, is based on 46 terrain-following (eta) levels with 24 levels in the first 1000 m, the first level at approximately 13 m, a uniform spacing of 25 m over the first 300 m and then stretched to a uniform resolution of 600 m in the upper part. The U.S. Geological Survey (USGS) land-use surface data, that comes by default with the WRF model, is used together with the unified Noah land-surface model to define the boundary conditions at the surface. Other physical parameterizations used are: the rapid radiative transfer model (RRTM), the Dudhia radiation scheme and the Yonsei University (YSU) first-order PBL scheme [12]. The simulation uses input data from

ERA-Interim [13] with a spin-up time of 24 hours. The input forcing of VENTOS®/M differs from the other microscale models and is discussed below.

Time and height-dependent mesoscale forcing is extracted directly from the momentum budget components (so-called "tendencies" in WRF) after filtering out fluctuations at the grid-resolution scale by spatially averaging data from a 9-km wide 3x3 grid around the site and using a 1-hour rolling window on the time series. More details about this meso-micro coupling method and a sensitivity analysis of rotor-averaged forcing on the filter set-up are discussed in [9].

2.2. Validation Data and Metrics

The performance of the models is based on quantities of interest relevant for wind energy applications. These quantities are evaluated across a reference rotor span of 160 m, between 40 and 200 m, characteristic of an 8-MW large wind turbine. Besides hub-height wind speed S_{hub} and direction WD_{hub}, it is relevant to consider the rotor-equivalent wind speed *REWS*, the turbulence intensity (not evaluated here), the wind speed shear α, and the wind direction shear or veer ψ.

The *REWS* is especially suitable to account for wind shear in wind turbine power performance tests [14]. The *REWS* is the wind speed corresponding to the kinetic energy flux through the swept rotor area, when accounting for the vertical shear:

$$REWS = \left[\frac{1}{A} \sum_i \left(A_i S_i^3 \cos \beta_i \right) \right]^{1/3} \tag{1}$$

where A is the rotor area and A_i are the horizontal segments that separate vertical measurement points of horizontal wind speed S_i across the rotor plane. The *REWS* is here weighted by the cosine of the angle β_i of the wind direction WD_i with respect to the hub-height wind direction to account for the effect of wind veer [15].

Wind shear is defined by fitting a power-law curve across the rotor wind speed points S_i:

$$S_i = S_{hub} \left(\frac{z_i}{z_{hub}} \right)^{\alpha} \tag{2}$$

Similarly, wind veer is defined as the slope ψ of the linear fit of the wind direction difference:

$$\beta_i = \psi \left(z_i - z_{hub} \right) \tag{3}$$

To evaluate simulations and measurements consistently, these quantities are obtained after resampling, by linear interpolation, velocity and wind direction vertical profiles at 10 points across the rotor area and then computing the *REWS* and the shear functional fits. While these fitting functions are commonly used in wind energy, their suitability in LLJ conditions is questionable. The regression coefficient from the fitting can be used to determine this suitability.

A rolling average with a window size of one hour is applied to simulation and observational data to remove the impact of high frequency fluctuations in the analysis.

Validation results are quantified in terms of the mean absolute error (MAE):

$$MAE = \frac{1}{N} \sum_{i=1}^{N} \left| \chi_{pred} - \chi_{obs} \right| \tag{4}$$

where χ is any of the above mentioned quantities of interest, predicted (*pred*) or observed (*obs*), and N is the number of samples evaluated in the time series.

3. Participating Models

Table1 shows a list of the models participating in the model intercomparison benchmark. The microscale models do not include humidity, which is not relevant in this case since the selection criteria for the cycle excluded wet conditions.

Table 1: Summary of model simulations. Monin-Obukhov similarity theory (MOST) surface boundary conditions use either heat flux (H), 2-m (T_2) or skin temperature (T_{SK}) from WRF

Name	Input	Turbulence	z-Levels	Surface B.C.
WRF-YSU (ref)	ERA Interim	YSU	46	Noah
WRF	ERA Interim, GFS	MYJ, MYNN, QNSE, TEMF, YSU	46	Noah
WRF-YSU_LES	ERA Interim	LES-TKE	101	Noah
WRF-VentosM_ke	ERA Interim	YSU/k-ε	70	MOST, H
CFDWind1D_ke	WRF (ref)	k-ε	301	MOST, T_2
Alya-CFDWind1D_ke	WRF (ref)	k-ε	500	MOST, T_2
Ellipsys1D_ke	WRF (ref)	k-ε	512	MOST, T_{SK}
Ellipsys3D_ke	WRF (ref)	k-ε	192	MOST, T_{SK}
Ellipsys3D_LES	WRF (ref)	*Smagorinsky*	128	MOST, T_{SK}
SP-Wind_LES	WRF (ref)	LES-TKE	500	MOST, T_2

3.1. WRF

We use the Advanced Research Weather forecasting model (WRF) version 3.8.1 [10] configured with three nested domains with grid sizes of 27, 9, and 3 km (Figure 1). All model domains have 61 x 61 horizontal grid points and are centered at the Cabauw tower. We use the same vertical grid as in the WRF-YSU (ref) simulation (see Section 2.1), with 46 eta levels defined to increase the vertical resolution in the lowest part of the planetary boundary layer (PBL). The boundary conditions at the surface are defined by the default USGS land-use surface data and the unified Noah land-surface model. The rapid radiative transfer model (RRTM) and the Dudhia radiation scheme are also used. All simulations are spun-up for 24 hours.

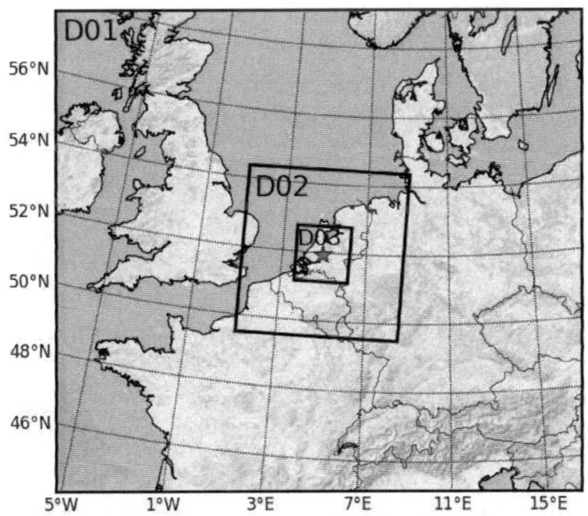

Figure 1: Model domain arrangement, with the red star indicating the center of each domain representing the Cabauw tower.

We perform a sensitivity analysis varying PBL schemes and global input data, similar to Kleczek et al. [11]. The first of the selected PBL schemes is Yonsei University (YSU) [12], a non-local, first-order closure scheme. Other PBL schemes used in this study are the Mellor–Yamada–Janjic (MYJ) [16], Mellor–Yamada–Nakanishi–Niino (MYNN2.5) [17], and Quasi-Normal Scale Elimination (QNSE) [18], which are classified as local, 1.5-order turbulent kinetic energy (*TKE*) closure schemes. Total Energy Mass Flux (TEMF) is the final PBL scheme tested and is classified as a hybrid local/non-local,1.5-order closure scheme [19]. We assess the sensitivity of WRF to input data by running each of the five PBL schemes with two different sources of input data: ERA-Interim [13] and GFS reanalysis. This ensemble of simulations is used in this study to quantify the spread of mesoscale solutions that could be input into the microscale simulations.

3.2. WRF-LES

Five additional telescopic nests are added to the reference mesoscale set-up of 3 nests to physically downscale, using one-way coupling, to a resolution of 12.5 m in the innermost domain using large-eddy simulation (LES). All domains use a 181x181 horizontal grid. The finest three nests are refined vertically to double the number of levels and obtain an isotropic grid of 12.5 m in the lowest 300 m. A prognostic equation for *TKE* is solved for the LES subgrid-scale model [23], using the WRF default constants [10]. The time step is reduced from 60 s in the outermost domain to 0.2 s in the innermost nest. Since the cycle starts from unstable atmospheric conditions, spin-up is deemed unnecessary. Hence, the first few hours of the cycle are transitioning to reach equilibrium with the mesoscale boundary conditions.

Mean flow profiles are obtained by spatial averaging over a 1-km wide horizontal box. Fluctuations about the mean flow within this box are used to compute the *TKE*.

3.3. VENTOS®/M

The VENTOS®/M computer code is an atmospheric flow solver employing a one-way dynamical coupling methodology, where WRF model simulation results are used as initial and boundary conditions. It solves the URANS equations assuming anelastic fluid, together with a transport equation for potential temperature and a k-ε turbulence model [20].

The WRF simulation follows the reference set-up described in Section 3.1 with few differences, namely the software version was WRF-ARW v3.6.1 and a fourth nesting level was added, increasing the downscaling to 1-km horizontal resolution. Physical parameterizations, grid dimensions, nesting grid ratios and the simulation spin-up time were kept equal to the reference simulation.

The VENTOS®/M domain encompasses 12x12 km^2. Centered on Cabauw, the horizontal resolution was kept constant at 160 m within a square area of 3x3 km^2, afterwards expanding towards the boundaries. At each vertical level the horizontal grid was composed of 47x47 control volumes. The vertical grid consisted of 10 km columns with 70 control volumes, expanding from 4 m above the surface to 730 m at the domain top, where a 4-km Rayleigh damping layer was set. The simulation time-step was 1.5 s and the boundary conditions were updated every 5 minutes. The simulation spin-up time was 12 hours, less than the WRF simulation. Similarly to WRF-LES, the results were smoothed through spatial averaging using 1-km wide horizontal boxes. The VENTOS®/M model surface boundary condition is based on Monin-Obukhov similarity theory (MOST) functions and a prescribed surface heat flux, as such quantity is expected to be less sensitive to height discrepancies between the mesoscale and microscale orographies. As heat-flux boundary conditions are not suitable to differentiate between intermittent and fully-turbulent regimes [21], the VENTOS®/M model always assumes the latter, which is not appropriate if stratification conditions are indeed very stable.

3.4. CFDWind

CFDWind1D is a python-based finite-difference code. The single-column model (SCM) is used as a prototype to design the CFDWind 3D model [9]. They are both based on unsteady Reynolds-Averaged Navier Stokes (URANS) equations using the k-ε model of Sogachev et al [23] with constants: $C_{\varepsilon 1}$ = 1.52, $C_{\varepsilon 2}$ = 1.833, σ_k = 2.95, σ_ε = 2.95 and C_μ = 0.03.

The SCM is solved on a 4-km long log-linear vertical grid with 301 levels using a time step of 1 s. Pressure gradient and advection forcings are lumped together as a time and height-dependent equivalent geostrophic wind that enters momentum equations as source terms. The advection temperature tendency is also added as a source term in the potential temperature equation. No-slip conditions are defined for momentum equations at the surface. Surface boundary conditions for potential temperature are defined based on MOST, inferring the surface temperature by prescribing the diurnal 2-m temperature from the mesoscale input data and using the dynamic surface-layer friction velocity and heat flux as described in [9].

3.5. Alya-CFDWind

Alya-CFDWind1D is a fortran-based finite-elements code equivalent to CFDWindSCM. Similarly, the model is used at early-stage design and as a precursor of the in-house 3D model Alya-CFDWind for numerical modelling of wind farms [24][25]. The same set of k-ε constants than CFDWind1D is used in this study.

The 1D mesh is 4-km tall using 500 vertical elements with a geometrically growing rate from 0.5 m up to 10 m. The time step is set to 10 s.

3.6. Ellipsys3D

The EllipSys3D code [26][27] is an in-house multiblock finite volume solver for the incompressible Navier-Stokes equations in general curvilinear coordinates. The code uses a collocated variable arrangement, where revised Rhie/Chow interpolation is used to avoid odd/even pressure coupling. The code is parallelized with Message Passing Interface (MPI) for executions on distributed/shared memory machines, using a non-overlapping domain decomposition technique.

The URANS case is based on k-ε model [22], with the same set of constants than the other URANS models of the benchmark. The problem is solved on a 6-km high vertical domain with 192 tanh stretched grid points using a time step of 1 s.

No-slip wall boundary condition at the bottom and a symmetry boundary condition at the top were applied. Periodic boundaries in both horizontal directions were used to mimic the 1D SCM basic setup. Pressure gradient and momentum advection forcings together with the temperature tendencies are included in the code in a way completely analogous to the procedure presented for the CFDWindSCM code [9].

In the LES case, the 4x4x4 km domain was discretized by 128x128x128 equidistantly spaced grid points, and Smagorinsky SGS model is used. Previously described URANS boundary conditions and time stepping are also applied in the LES case. EllipSys1D [28] is a recent "stripped down" version of EllipSys3D. The basic 3D URANS set-up was also applied in the 1D case. Only 512 tanh stretched grid points were used for discretizing the 6-km high vertical domain.

3.7. SP-Wind

SP-Wind is an in-house LES code developed at KU Leuven [29][30][31] for research on the simulation and optimization of turbulent flows. SP-Wind solves the Boussinesq form of the conservation equations for mass, momentum and potential temperature on a three-dimensional Cartesian grid. The horizontal directions are discretized with pseudo-spectral schemes, while the vertical direction uses a fourth-order energy-conservative finite difference scheme. Time integration is performed using a classic four-stage fourth-order Runge—Kutta scheme. The subgrid-scale stress and heat flux are computed with a prognostic TKE model [22], and the surface boundary conditions are imposed using classic Monin-Obukhov similarity theory [32] with standard stability correction functions [33].

The LES is solved on a horizontally periodic domain of 5x5x5 km with 400x400x500 grid points. The vertical grid has a uniform spacing of 5 m in the first 2 km, above which it is stretched to a maximum grid size of 40 m. A Rayleigh damping layer is added above 4 km. Initial profiles, mesoscale forcing and surface temperature (inferred from the 2-m temperature) are all extracted from mesoscale WRF simulations as in [9].

In view of the large computational cost, no spin-up time is used for the LES simulations. Analysis of the vertical profiles of turbulent kinetic energy shows that the positive surface heat flux at UTC 2006-07-01 12:00 leads to intense generation of turbulence, which rapidly fills up the entire boundary layer and reduces the transition time to less than one hour.

4. Results

4.1. Input uncertainties: Sensitivity Analysis at Mesoscale

We conduct a sensitivity analysis on the impacts of PBL scheme and global input data on WRF. Atmospheric profiles of potential temperature (θ), moisture (q) and winds (S, WD) over Cabauw at midnight LT on 2006-07-02 illustrate noticeable spread in the model output by the different configurations (Figure 2). Most of the configurations exhibit very similar temperature (Figure 2a) and wind (Figure 3c,d) profiles, while more spread is evident in the moisture profiles (Figure 2b). Further, the TEMF-GFS (orange open squares) appears to deviate most from the other configurations, most noticeably in the lowest layers of the potential temperature, moisture, and wind speed profiles. The YSU-ERA (green filled dots), the reference mesoscale model used as input for the microscale simulations in this study, shows good agreement to the low-level moisture observations from the Cabauw tower (black dots), but overestimates the winds and the slope of the temperature inversion to a similar degree as the other configurations. Aside from the large difference in the TEMF-GFS and TEMF-ERA output, input data appear to have little impact on the model solution in this particular case, suggesting the input datasets agree on the larger-scale forcing, unlike other cases which show more spread between input datasets.

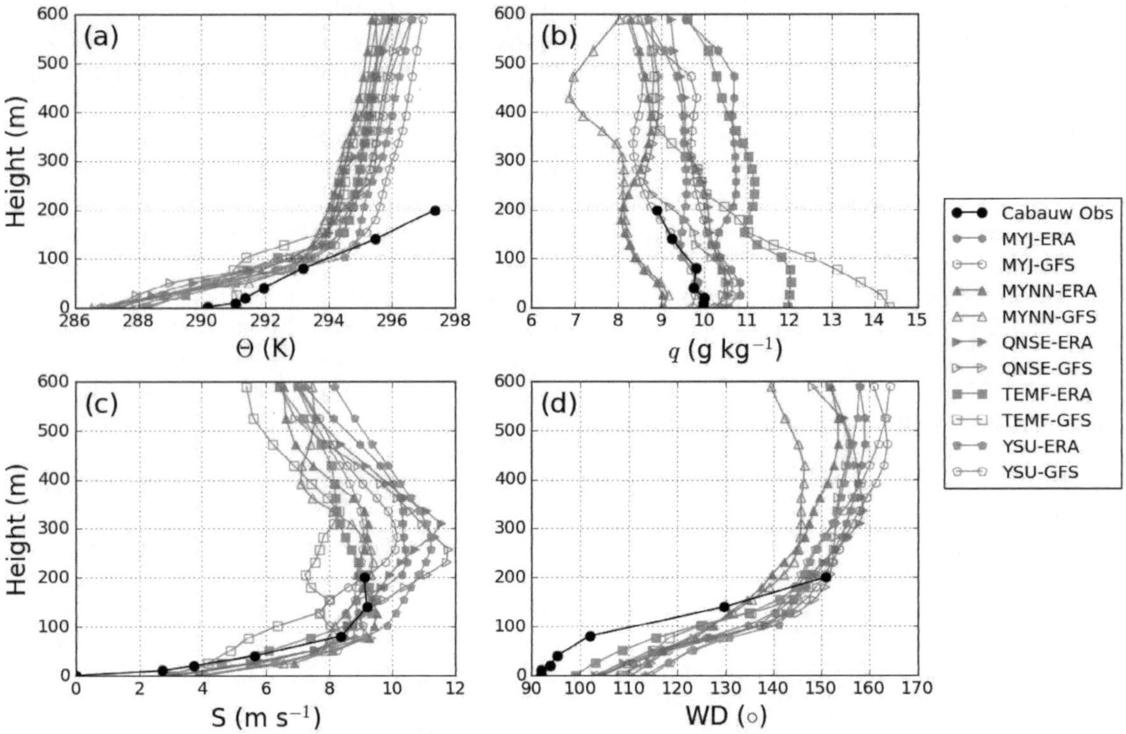

Figure 2: Atmospheric profiles of (a) potential temperature (θ), (b) water vapor mixing ratio (*q*), (c) wind speed (S), and (d) wind direction (WD) over Cabauw at midnight LT on 2006-07-02.

Time series of the same variables plotted over the 24-hour period between UTC 2006-07-01 1200 and 2006-07-02 1200 at 80 m show a similar spread between the PBL scheme and input data configurations (Figure 3). The TEMF-GFS again deviates most from the other simulations and the Cabauw observations, especially in the moisture (Figure 3b) and wind speed (Figure 3c) data. The other configurations form a generally narrow spread of output data. In particular, the time series of 80-m potential temperature (Figure 3a) exhibits a spread of ~2 K between the configurations, giving the whole ensemble a potential temperature bias consistently between -1 and -4 K. For the moisture and wind time series (Figure 3b-d), the ensemble of configurations frequently envelope the Cabauw observations during the 24-hour period, which may allude to minimal errors caused by choice of PBL scheme and input data on the microscale simulations. Similar conclusions may be drawn from time series of friction velocity ( ), PBL height (*PBL h*), latent heat flux (*LE*), and sensible heat flux (*H*) over the same 24-hour period at Cabauw (Figure 4). The greatest differences between the configurations is the TEMF-GFS friction velocity and the TEMF-ERA PBL height. The largest spread of output data is present in the PBL height, while the latent and sensible heat fluxes have the smallest spread. The ensemble of configurations generally overestimated latent and sensible heat fluxes during the day when compared to Cabauw observations, with fewer error during the night.

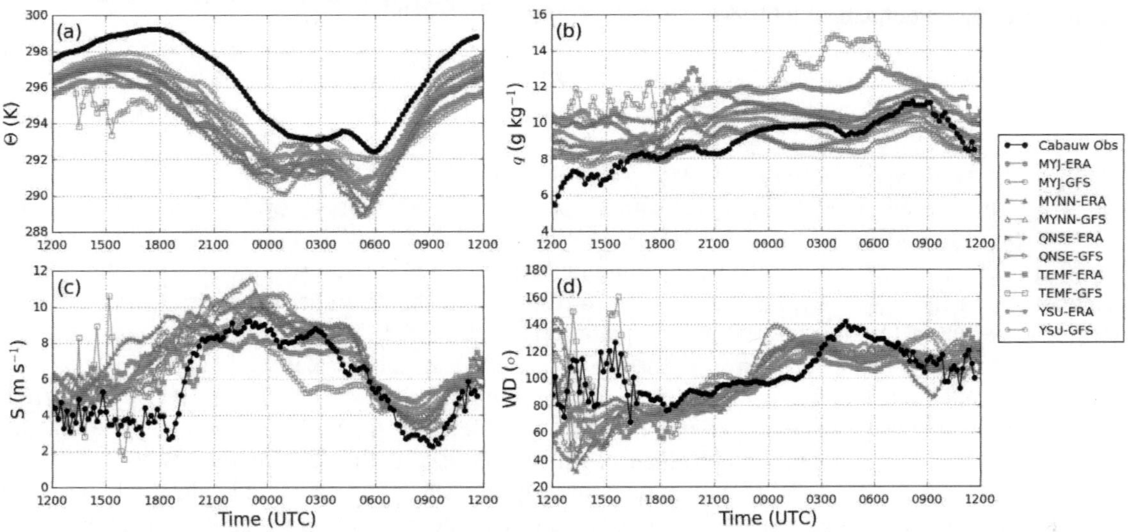

Figure 3: Time series of 80-m (a) potential temperature (Θ), (b) water vapor mixing ratio (*q*), (c) wind speed (*S*), and (d) wind direction (*WD*) over Cabauw during the 24-hour period between UTC 2006-07-01 1200 and 2006-07-02 1200.

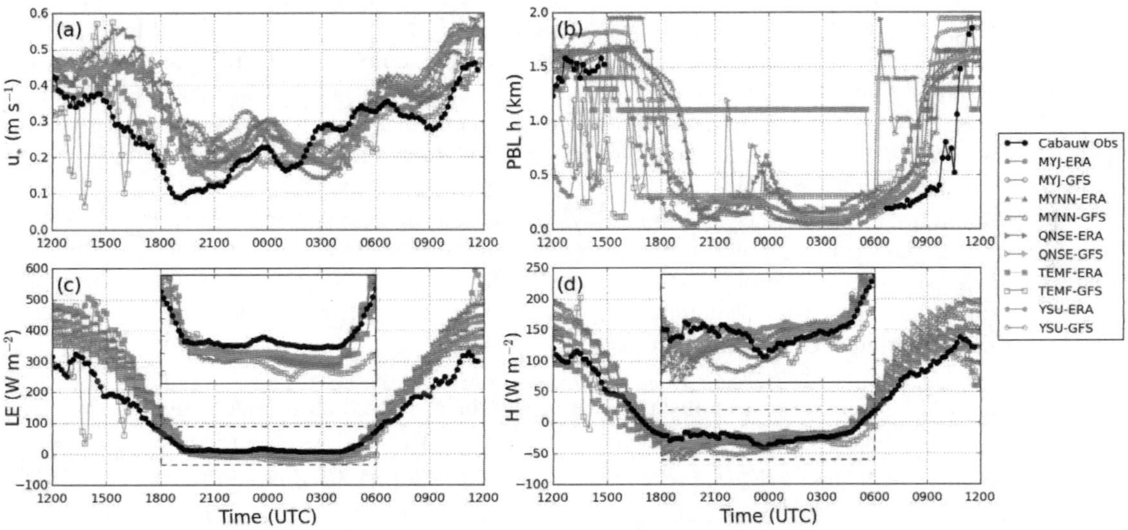

Figure 4: Time series of (a) friction velocity (u_*), (b) planetary boundary layer height (*PBL h*), (c) latent heat flux (*LE*), and (d) sensible heat flux (*H*) over Cabauw during the 24-hour period between UTC 2006-07-01 1200 and 2006-07-02 1200.

These results show that, excluding the TEMF, choice of input data has a near-negligible impact on the WRF output for this particular case, although input boundary condition datasets exhibit stronger influence in other cases [34]. Rather, differences in both the vertical profiles (Figure 2) and the time series (Figure 3 and Figure 4) between the tested configurations appear to be caused more so by choice of PBL scheme. Despite this dependence on PBL scheme, the ensemble of choices tends to maintain a consistent spread near the observations, with the only significant biases evident in the potential temperature and daytime latent and sensible heat fluxes. The greatest differences between PBL schemes occur in the PBL height calculations (Figure 4b). PBL heights were determined via direct output from each scheme, where each scheme uses a different approach that contributed to the scatter in the data. In the YSU and TEMF schemes, the PBL height is diagnosed from the bulk Richardson number [12] [19]. For the local PBL schemes (MYJ, MYNN, and QNSE), PBL height is diagnosed as the height where prognostic TKE decreases to a sufficiently small value (0.2 $m^2 s^{-2}$, 1.0 x 10^{-6} $m^2 s^{-2}$, and 0.01 $m^2 s^{-2}$, respectively) [16] [17] [18]. Overall, the nontrivial spread in mesoscale forcings should be considered when making a choice in PBL scheme for a simulation that will be used as forcing for a microscale simulation.

4.2. Consistency among LES and RANS Microscale Models

All of the microscale models produce similar patterns of the diurnal cycle, demonstrating the effectiveness of the offline coupling methodology. Figure 5 shows contour plots of the evolution of the vertical profile of mean flow quantities computed by the participating models compared to the observations on the top row. The second row corresponds to the reference mesoscale model that is used to derive input forcings and boundary conditions to drive the microscale simulations. Since no higher-resolution terrain or land cover information is added at microscale, we can assume that this mesoscale simulation is already a good reference for microscale models to verify a correct implementation of the meso-micro methodology. The differences in the mean flow arise from the different ways each model represents turbulence, which is noticed in the *TKE* contour plots.

The ensemble mean of the WRF simulations used in the sensitivity analysis is plotted in the third row. By ensemble averaging, we obtain a better match with the observations than by using any single WRF simulation – this result was also seen for the LLJ cases discussed in [34]. The ensemble here provides a better prediction of the LLJ with a distinct velocity maximum around midnight instead of a broader double-peak as in the reference WRF simulation.

The microscale models diverge in their estimates of rotor-based quantities of interest. Time series of rotor-based quantities of interest are shown in Figure 6. The spread of the models for *REWS* is around 2 m s^{-1} and around 15° in terms of WD_{hub}. The spread in terms of wind shear and wind veer is also large specially during nighttime stable conditions. Vertical profiles of wind speed and direction at midnight (Figure 7) show how the models capture the characteristics of the LLJ. The phase error in the input data dominates the bias in the simulations; this input error cannot be corrected at microscale by simply changing the turbulence model.

Table 2 summarizes the differences between the simulations and observations in terms of (4) with respect to observations (MAE_{obs}) and with respect to the reference WRF simulation (MAE_{ref}). These are differences integrated over the whole diurnal cycle; therefore mixing all kinds of surface stability and large-scale conditions into a single quantity. We shall focus on the MAE_{ref} to quantify the impact of choosing a different turbulence model at microscale.

Table 2: MAE with respect to observations and to the reference mesoscale simulation.

	***REWS* [m s^{-1}]**		**S_{hub} [m s^{-1}]**		**WD_{hub} [°]**		**α (shear)**		**ψ (veer)**	
	MAE_{obs}	**MAE_{ref}**	**MAE_{obs}**	**MAE_{ref}**	**MAE_{obs}**	**MAE_{ref}**	**MAE_{obs}**	**MAE_{ref}**	**MAE_{obs}**	**MAE_{ref}**
WRF-YSU (ref)	1.26	0.00	1.35	0.00	10.49	0.00	0.13	0.00	0.07	0.00
WRF ensemble	1.16	0.63	1.12	0.76	14.24	5.14	0.11	0.05	0.07	0.02
WRF-YSU_LES	1.51	0.45	1.60	0.54	10.67	4.09	0.15	0.05	0.06	0.03
WRF-VentosM_ke	1.56	0.69	1.59	0.72	10.74	6.34	0.12	0.05	0.06	0.03
CFDWind1D_ke	1.56	0.63	1.62	0.68	11.49	6.22	0.15	0.09	0.06	0.05
Alya-CFDWind1D_ke	1.48	0.73	1.42	0.70	11.30	2.90	0.14	0.07	0.07	0.02
Ellipsys1D_ke	1.37	0.55	1.50	0.61	11.51	5.22	0.16	0.10	0.06	0.04
Ellipsys3D_ke	1.36	0.66	1.52	0.74	10.61	5.17	0.16	0.09	0.06	0.04
Ellipsys3D_LES	1.38	1.13	1.37	1.27	11.90	11.99	0.18	0.09	0.09	0.06
SP-Wind_LES	1.47	0.63	1.38	0.69	8.79	3.25	0.13	0.04	0.07	0.02

IOP Conf. Series: Journal of Physics: Conf. Series **854** (2017) 012037 doi:10.1088/1742-6596/854/1/012037

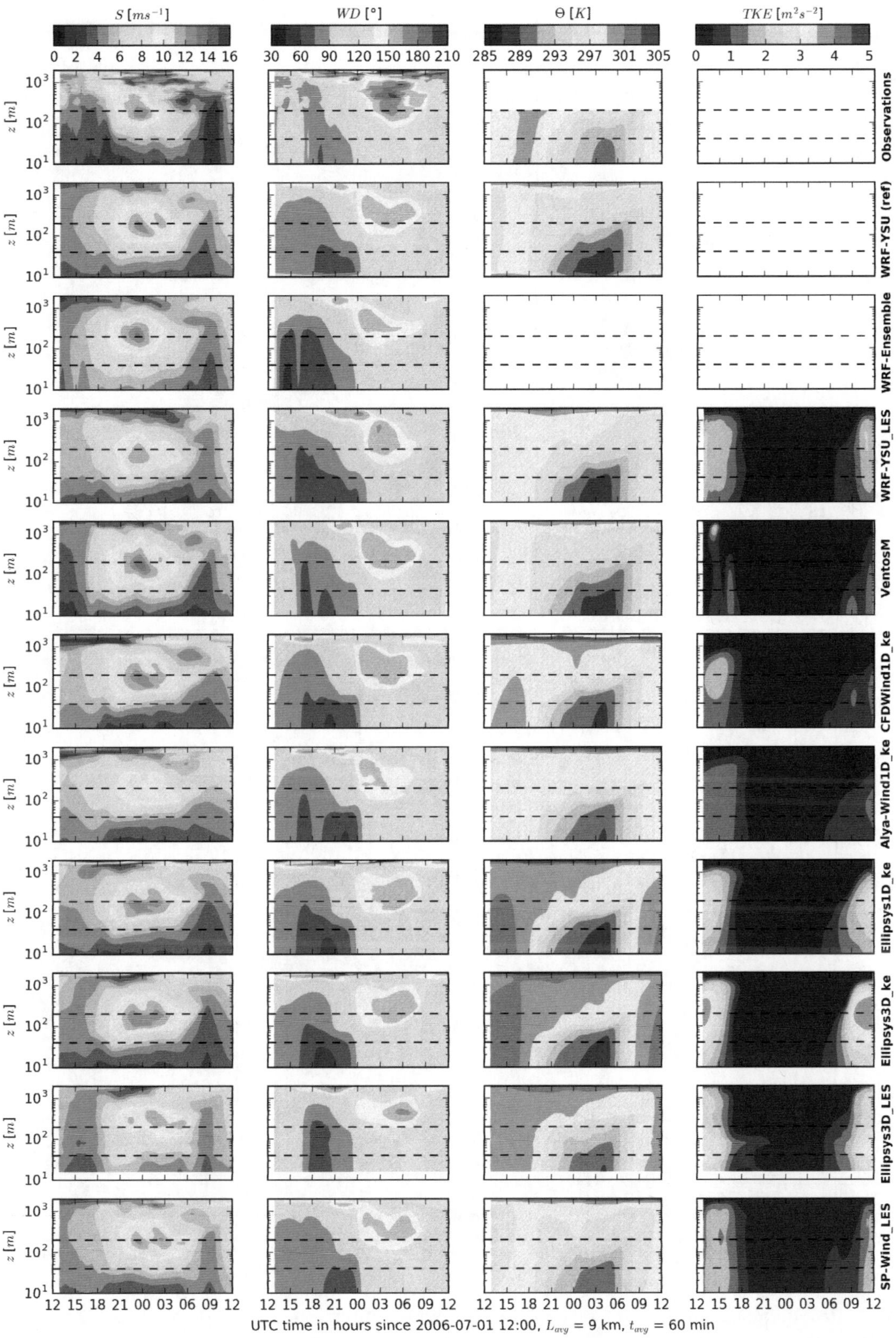

Figure 5: Time-height contours of horizontal wind speed *S*, direction *WD*, potential temperature Θ and turbulent kinetic energy *TKE*. Dotted lines denote a rotor diameter of 160 m at 120 m hub-height.

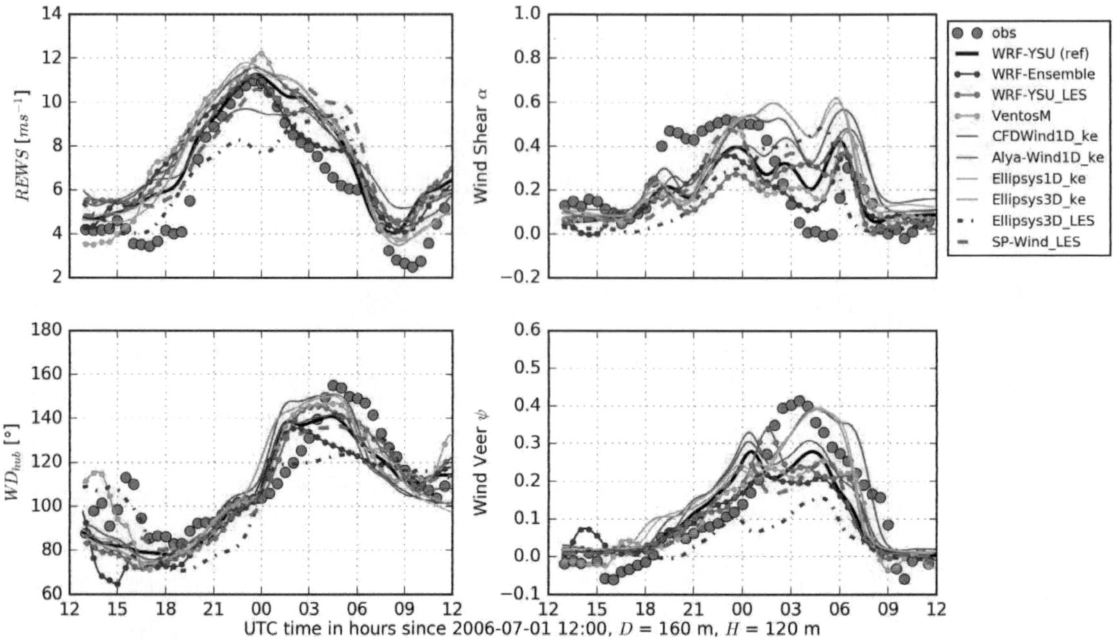

Figure 6: Time series of rotor-based quantities of interest used for validation.

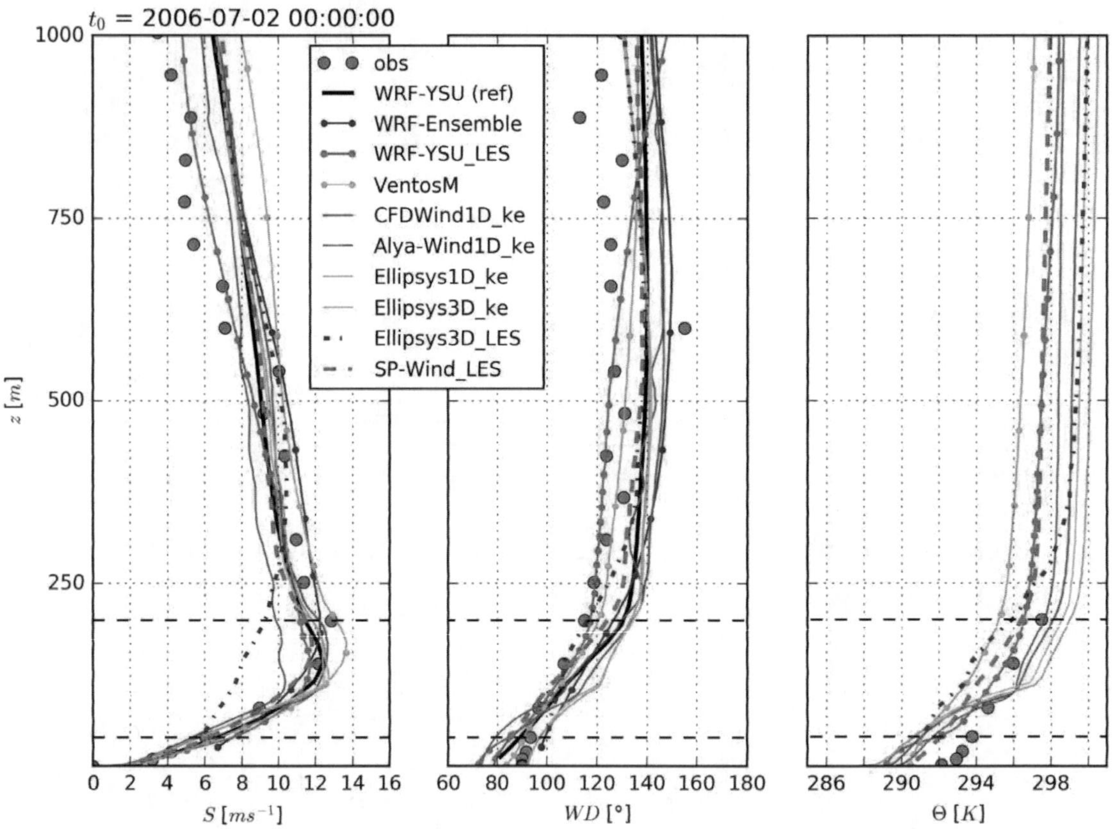

Figure 7: Vertical profiles of wind speed, wind direction and potential temperature at UTC 2006-07-02 00:00:00.

WRF-YSU_LES is the closest to the reference, which is to be expected since they are results from different nests of the same simulation. Still, differences are significant of around 0.5 m s^{-1} of wind speed and 4° of wind direction at hub-height. Microscale models increase the error by 0.2-0.7 m s^{-1} of wind speed and up to 7° of wind direction at hub-height with respect to the WRF-YSU_LES results. With respect to observations, all simulations show similar results with a *MAE* of 1.1-1.6 m s^{-1} of wind speed and up to 14° of wind direction at hub-height. Considering vertical wind speed shear and wind direction veer, SP-Wind and the WRF ensemble produce the closest results to the reference WRF simulation.

Different sets of k-ε constants have been tested (not shown) leading to changes that are within the spread shown in Figure 6, which is also comparable to that observed in the WRF sensitivity analysis when changing the planetary boundary-layer scheme.

Recently, capabilities of EllipSys3D have been extended to cover wall modeled LES of stratified flows and at the same time, a "striped" down 1D version of the code have been made operational [28]. Figures 5 and 6 and Tables 2 and 3 show a general good agreement between URANS based EllipSys1D and EllipSys3D computations. Some minor differences exist thought; a possible cause of them might be related to the fact that the vertical velocity (W) and the advection terms are implicitly assumed to be zero in the EllipSys1D. Further investigations are necessary to confirm this.

Initial LES computations based on coarse grid resolution and very basic Smagorinsky SGS model show that EllipSys3D LES is capable of reproducing the basic LLJ features observed at the Cabauw site, but Tables 2 and 3 MAE comparisons and Figures 2 and 3 indicate that significantly higher resolution and a more advanced approach to SGS turbulence modeling are needed in order to capture all main details relevant for its application in a wind energy context.

Regarding VENTOS®/M wind speed results, both Figures 5 and 7 show that the LLJ magnitude was reasonably well predicted. The diurnal cycle of the simulation wind speed shows over-predictions around 1.4 m s^{-1}, affecting the *REWS* and S_{hub} error values in Table 2 which are higher than the reference simulation. Good agreement was obtained for the wind direction, shear and veer regarding their integrated error. Analysis of Figure 6 indicates a generalized under-prediction of α for several nocturnal periods, analogous to a higher turbulent shape factor of the boundary-layer. These mismatches happen also with WRF and, despite the VENTOS®/M limitations of its heat-flux boundary condition, the microscale simulation predicts higher values of α and closer to the observations. The results further show a -2.5 K temperature bias that occurs in both WRF and VENTOS®/M results, as well as in the other microscale simulations, which originates from the ERA-Interim input data.

Table 3: MAE with respect to the reference mesoscale simulation for unstable ('*u*': z/L < -0.2), neutral ('*n*': -0.2 < z/L < 0.2) and stable ('*s*': z/L > 0.2) conditions.

	***REWS* [m s^{-1}]**			**S_{hub} [m s^{-1}]**			**WD_{hub} [°]**			**α (shear)**			**ψ (veer)**		
	U	***n***	***s***	***u***	***n***	***s***	***u***	***n***	***s***	***u***	***n***	***s***	***u***	***n***	***s***
WRF-YSU (ref)	0.00	0.00	0.00	0.00	0.00	0.00	0.00	0.00	0.00	0.00	0.00	0.00	0.00	0.00	0.00
WRF ensemble	0.45	0.51	0.82	0.51	0.60	1.02	5.79	8.98	3.97	0.03	0.09	0.06	0.01	0.02	0.03
WRF-YSU_LES	0.48	0.44	0.42	0.58	0.72	0.48	4.59	4.15	3.62	0.03	0.05	0.08	0.02	0.02	0.03
WRF-VentosM	0.64	0.75	0.74	0.68	0.80	0.73	9.74	7.02	3.10	0.04	0.08	0.05	0.04	0.02	0.03
CFDWind1D_ke	0.74	0.45	0.56	0.73	0.53	0.65	5.62	7.87	6.53	0.07	0.10	0.11	0.05	0.09	0.05
Alya-CFDWind1D_ke	0.79	0.39	0.72	0.78	0.51	0.66	4.66	3.16	1.24	0.09	0.03	0.06	0.02	0.02	0.03
Ellipsys1D_ke	0.58	0.40	0.55	0.57	0.40	0.69	4.30	7.47	5.75	0.06	0.12	0.13	0.03	0.10	0.05
Ellipsys3D_ke	0.67	0.47	0.68	0.67	0.60	0.83	4.13	8.03	5.70	0.04	0.14	0.13	0.02	0.10	0.04
Ellipsys3D_LES	0.56	1.27	1.64	0.59	1.06	1.93	11.69	8.36	12.81	0.07	0.10	0.11	0.02	0.05	0.11
SP-Wind_LES	0.48	1.36	0.66	0.50	1.38	0.76	3.00	1.72	3.71	0.02	0.02	0.06	0.01	0.02	0.03

Finally, Table 3 shows the MAE_{ref} is computed for different stability classes filtering with the observed stability parameter z/L, where L is the Obukhov length and $z = 10$ m: unstable ('*u*': z/L < -0.2), neutral ('*n*': -0.2 < z/L < 0.2) and stable ('*s*': z/L > 0.2). Not all the models behave similarly depending on stability. The WRF ensemble *REWS* is more sensitive in stable conditions. This is also the case for Ellipsys3D_LES probably due to the coarser resolution of the simulation compared to the other LES simulations that do not show this high sensitivity in stable conditions.

In general, it is difficult to extract more meaningful conclusions from Table 2 and Table 3 due to the limited statistical significance of the samples. The overall assessment would be richer if several diurnal cycles from uncorrelated synoptic conditions would have been tested. The ensemble WRF simulations for the same cycle already show significant improvement on mean flow quantities.

5. Conclusions

Results of the GABLS3 diurnal cycle benchmark with an emphasis on rotor-relevant values are presented. The main challenge for microscale models was to produce consistent flow fields with respect to the mesoscale model that was used to derive their input forcings. This consistency has been achieved by both LES and URANS models. The spread of the models is significant but of similar magnitude as that shown by WRF using different boundary-layer parameterizations. The input uncertainty coming from the mesoscale, even in relatively ideal conditions, is large and results in *MAE* of wind speed at hub-height of the order of 1.1-1.6 m s^{-1} over the whole cycle and hourly errors of up to 3 m s^{-1}. This is partly mitigated when using an ensemble average of several simulations which also lead to better results in terms of wind shear and wind veer.

By ensuring consistency of the microscale models at introducing input forcings we can proceed with further analysis on how RANS and LES models interpret the structure of turbulence in different stability conditions.

Acknowledgments

This article was produced with funding from the "MesoWake" Marie Curie International Outgoing Fellowship (FP7-PEOPLE-2013-IOF, European Commission's grant agreement number 624562). It also counts with contributions supported by the New European Wind Atlas (NEWA, FP7-ENERGY.2013.10.1.2, European Commission's grant agreement number 618122). WRF-LES simulations have been produced using high-performance computing resources from the PRACE-MesoWake project in the MareNostrum system, based in Barcelona, Spain. JMT's mesoscale ensemble simulations were conducted using the Extreme Science and Engineering Discovery Environment (XSEDE), which is supported by National Science Foundation grant number ACI-1053575. JKL's efforts were supported by the National Renewable Energy Laboratory under APUP UGA-0-41026-22. NREL is a national laboratory of the U. S. Department of Energy, Office of Energy Efficiency and Renewable Energy, operated by the Alliance for Sustainable Energy, LLC. The benchmark has been coordinated within the International Energy Agency IEA-Wind Task 31 "Wakebench".

References

[1] Sanz Rodrigo J, Chávez Arroyo R-A, Moriarty P, Churchfield M, Kosović B, Réthoré P-E, Hansen KS, Hahmann A, Mirocha JD and Rife D 2016 Mesoscale to microscale wind farm flow modelling and evaluation. *WIREs Energy Environ*, doi:10.1002/wene.214

[2] Wyngaard JC 2004 Toward Numerical Modeling in the "Terra Incognita". *J. Atmos. Sci.* **61**: 1816–1826, doi: 10.1175/1520-0469(2004)061<1816:TNMITT>2.0.CO;2

[3] Holtslag AAM, et al. (2013) Stable atmospheric boundary layers and diurnal cycles—challenges for weather and climate models. *Bull. Am. Meteorol. Soc.* **94**: 1691–1706, doi:10.1175/bams-d-11-00187.1

[4] Baas, P., Bosveld, F.C., Baltink, K. and Holtslag, A.A.M. 2009 A Climatology of Nocturnal Low-Level Jets at Cabauw, *J. Appl. Meteorol. Climatol.* **48**: 1627-1642, doi: 10.1175/2009JAMC1965.1, 2009

[5] Bosveld FC, Baas P, vanMeijgaard E, de Bruijn EIF, Steeneveld GJ and Holtslag AAM 2014 The third GABLS intercomparison case for evaluation studies of boundary-layer models, Part A: case selection and set-up. *Boundary-Layer Meteorol.* **152**: 133-156, doi:

IOP Conf. Series: Journal of Physics: Conf. Series **854** (2017) 012037 doi:10.1088/1742-6596/854/1/012037

10.1007/s10546-014-9917-3
[6] Bosveld FC et al. 2014 The third GABLS Intercomparison case for evaluation studies of Boundary-Layer Models Part B: Results and Process Understanding, *Boundary–Layer Meteorol.* **152**: 157-187, doi: 10.1007/s10546-014-9919-1
[7] Sanz Rodrigo J 2017 Windbench/GABLS3 benchmark. http://windbench.net/gabls-3, last accessed in March 2017
[8] Sanz Rodrigo, Churchfield M and Kosovic B 2016 A wind energy benchmark for ABL modeling of a diurnal cycle with a nocturnal low-level jet: GABLS3 revisited. *J. Phys. Conf. Ser.* **753**: 032024, doi:10.1088/1742-6596/753/3/032024
[9] Sanz Rodrigo J, Churchfield M and Kosović B 2017 A methodology for the design and testing of atmospheric boundary layer models for wind energy applications. *Wind Energ. Sci.* **2**: 1-20, doi:10.5194/wes-2-1-2017
[10] Skamarock WC, Klemp JB, Dudhia J, Gill DO, Barker DM, Duda MG, Huang X-Y, Wang W and Powers JG 2008 A description of the advanced research WRF version 3, Technical Note NCAR/TN-475+STR, NCAR, Boulder, CO, June 2008.
[11] Kleczek MA, Steeveneveld GL and Holtslag AAM 2014 Evaluation of the Weather Research and Forecasting Mesoscale Model for GABLS3: Impact on Boundary-Layer Schemes, Boundary Conditions and Spin-Up. *Boundary-Layer Meteorol.* **152**: 213-243, doi: 10.1007/s10546-014-9925-3
[12] Hong S-Y and Noh Y 2006 A New Vertical Diffusion Package with an Explicit Treatment of Entrainment Processes. *Mon. Wea. Rev.*, **134**: 2318–2341, doi: 10.1175/MWR3199.1
[13] Dee DP et al. 2011 The ERA-Interim reanalysis: Configuration and performance of the data assimilation system. *Quart. J. R. Meteorol. Soc.* **137**: 553-597, doi: 10.1002/qj.828
[14] Wagner R, Cañadillas B, Clifton A, Feeney S, Nygaard N, Martin CSt, Tüxen E and Wagenaar JW 2014 Rotor equivalent wind speed for power curve measurement – comparative exercise for IEA Wind Annex 32. *J. Phys. Conf. Ser.* **524**: 012108, doi: 10.1088/1742-6596/524/1/012108
[15] Choukulkar A, Pichugina Y, Clack CTM, Calhoun R, Banta R, Brewer A and Hardesty M 2015 A new formulation for rotor equivalent wind speed for wind resource assessment and wind power forecasting. *Wind Energy* **19**: 1439–1452, doi:10.1002/we.1929
[16] Janjić ZI 1994 The Step-Mountain Eta Coordinate Model: Further Developments of the Convection, Viscous Sublayer, and Turbulence Closure Schemes. *Mon. Wea. Rev.* **122**: 927–945, doi: 10.1175/1520-0493(1994)122<0927:TSMECM>2.0.CO;2
[17] Nakanishi M and H Niino 2006 An Improved Mellor–Yamada Level-3 Model: Its Numerical Stability and Application to a Regional Prediction of Advection Fog. *Boundary-Layer Meteorol* **119**: 397-407, doi:10.1007/s10546-005-9030-8.
[18] Sukoriansky S, Galperin B and Pero V 2005 Application of a New Spectral Theory of Stably Stratified Turbulence to the Atmospheric Boundary Layer over Sea Ice. *Boundary-Layer Meteorol.* **117**: 231-257. doi:10.1007/s10546-004-6848-4.
[19] Angevine WM 2005 An Integrated Turbulence Scheme for Boundary Layers with Shallow Cumulus Applied to Pollutant Transport. *J. Appl. Meteor.* **44**: 1436–1452, doi: 10.1175/JAM2284.1.
[20] Veiga Rodrigues C, Palma JMLM and Rodrigues ÁH 2016 Atmospheric Flow over a Mountainous Region by a One-Way Coupled Approach Based on Reynolds-Averaged Turbulence Modelling. *Boundary-Layer Meteorol.* **159**: 407–437, doi: 10.1007/s10546-015-0116-7
[21] van de Wiel BJH, Moene AF, Steeneveld GJ, Hartogensis OK and Holtslag AAM. 2007 Predicting the Collapse of Turbulence in Stably Stratified Boundary Layers. *Flow Turbulence Combust.*, **79**:251–274, doi: 10.1007/s10494-007-9094-2
[22] Sogachev A, Kelly M and Leclerc MY 2012 Consistent Two-Equation Closure Modelling for Atmospheric Research: Buoyancy and Vegetation Implementations. *Boundary-Layer Meteorol.* **145**: 307–327, doi:10.1007/s10546-012-9726-5
[23] Deardorff J W 1980 Stratocumulus-capped mixed layers derived from a three dimensional

model. *Boundary-Layer Meteorol.* **18**: 495-527, doi:10.1007/BF00119502

[24] Avila M, Folch A, Houzeaux G, Eguzkitza B, Prieto L and Cabezon D 2013 A Parallel CFD Model for Wind Farms. *Procedia Comput. Sci.* **18**: 2157 – 2166, doi:10.1016/j.procs.2013.05.386

[25] Vázquez M et al. 2015 Alya: Multiphysics engineering simulation toward exascale. *J. Comput. Scie.* **14**: 15–27, doi:10.1016/j.jocs.2015.12.007

[26] Michelsen JA 1994 Block structured Multigrid solution of 2D and 3D elliptic PDE's. Technical Report AFM 94-06, Technical University of Denmark, Department of Fluid Mechanics, May 1994.

[27] Sørensen NN 1995 General Purpose Flow Solver Applied to Flow over Hills. Risø-R-827-(EN), Risø National Laboratory, Roskilde, Denmark, June 1995

[28] van der Laan MP and Sørensen NN 2017 A 1D version of EllipSys. Technical Report, DTU Wind Energy E-0141, March 2017

[29] Meyers J, Meneveau C 2010 Large eddy simulations of large wind-turbine arrays in the atmospheric boundary layer. AIAA Paper No. 2010-827, doi:10.2514/6.2010-827

[30] Munters W, Meneveau C and Meyers J 2016 Turbulent inflow precursor method with time-varying direction for large-eddy simulations and applications to wind farms. *Boundary-Layer Meteorol.* **159**: 305-328, doi:10.1007/s10546-016-0127-z

[31] Allaerts D and Meyers J 2017 Boundary-layer development and gravity waves in conventionally neutral wind farms. *J. Fluid Mech.* **814**: 95-130, doi:10.1017/jfm.2017.11

[32] Moeng C-H 1984 A large-eddy-simulation model for the study of planetary boundary-layer turbulence. *J. Atmos. Sci.* **41**: 2052-2062, doi:10.1175/1520-0469(1984)041<2052:ALESMF>2.0.CO;2

[33] Dyer A J 1974 A review of flux-profile relationships. *Boundary-Layer Meteorol.* **7**: 363-372, doi:10.1007/BF00240838

[34] Vanderwende BJ, Lundquist JK, Rhodes ME, Takle ES and Irvin SL 2015 Observing and Simulating the Summertime Low-Level Jet in Central Iowa. *Mon. Weater Rev.* 143(6): 2319–2336, doi:10.1175/mwr-d-14-00325.1

IOP Conf. Series: Journal of Physics: Conf. Series **854** (2017) 012038 doi:10.1088/1742-6596/854/1/012038

Modelling of Wind Turbine Loads nearby a Wind Farm

B Roscher[1,2], A Werkmeister[1], G Jacobs[1], R Schelenz[1]

[1]Center for Wind Power Drives, RWTH Aachen, Campus Boulevard 61, 52074 Aachen, Germany

[2] Corresponding author: bjoern.roscher@cwd.rwth-aachen.de

Abstract. Each wind turbine experiences a variety of loads during its lifetime, especially inside a wind farm due to the wake effect between the turbines. This paper describes a possibility to observe a load spectrum while considering wake effects in a wind farm by through the turbulence intensity. The turbulence intensity is distributed along the wind rose of Alpha Ventus. For each turbulence intensity, a Weibull characteristic is calculated. The resulting wind fields are used to determine the loads through a multibody simulation of an imaginary wind turbine located at FINO-1, representing a closely placed wind turbine at the outer edge of a wind farm. These loads are analyzed and summed up. As expected, the change of the turbulence intensity due to the wake effect has an impact on the internal loading of a wind turbine inside a wind farm. Based on the assumed loading conditions, the maximum loads increased by a factor of almost 2.5.

1. Introduction

From 2010 till 2015, Germany installed more than 6900 wind turbines (WT) [1]. These WTs have to be placed somewhere. Unfortunately, almost 78% of the available onshore space cannot be used [2] either to logistical problems or social restrictions like the distance towards inhabited areas. One approach is to place the WT offshore, as there is more space available. But offshore space is regulated as well. Leading to the fact that space has to be used as beneficial as possible. This is mainly achieved by increasing the number of WT and therefore the annual energy production. The increased number of WTs means a closer spacing inside a defined area. Leading to an aerodynamic interaction between the WTs, also known as wake effects.

These wake effects result in higher turbulence intensities and greater dynamic inside the WT system. This could mean a faster damage accumulation and higher maintenance cycles, therefore an increase of downtime. This paper will inspect the influence of the wake effects on the damage accumulation on a single WT. As it is not possible to measure the direct influence on an existing turbine without an extensive measurement campaign, this paper will use the data of the research platform FINO-1 [3], where the wake effects can be observed by the wind farm Alpha Ventus. The damage accumulation will be determined through the use of a generic CWD reference turbine C3x126 [4], which is assumed to be located at the location of FINO-1. It needs to be stated that the FINO-1 is spaced rather close, which is occurs commonly at locations with a clear preferred wind direction. The turbine will be loaded in two different load situations and integrated over one year. To limit the displayed work, the considered WT components is the rotor hub as it observes all aerodynamical loads and is directly connected to the drive train.

IOP Conf. Series: Journal of Physics: Conf. Series **854** (2017) 012038 doi:10.1088/1742-6596/854/1/012038

2. Site Assessment of FINO-1 and Alpha Ventus

The measurement station FINO-1, as well as the wind park Alpha Ventus, are both located in the North Sea of Germany. The wind park consists out of 12 WTs orientated as indicated in figure 1. It can be noted the WT are spaced almost equally along the rectangular. The first 6 WT are REpower 5M and the last 6 are Areva Multibrid M5000, their configurations are given in table 1.

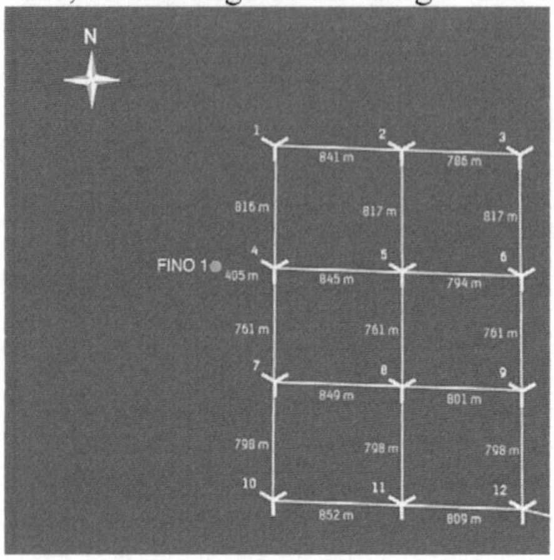

Figure 1. Spacing of WT inside Alpha Ventus, including location of FINO-1 [5]

Table 1. Wind turbine configurations inside Alpha Ventus, [5]

	Repower 5M	Areva Multibrid M5000
Wind turbine number	1-6	7-12
Rated power [MW]	5.0	5.0
Hub height over sea level [m]	92	90
Rotor diameter [m]	126	116
Rated wind speed [m/s]	14.0	12.5

The measurement station is included in figure 1 as well. It can be noted that FINO-1 is spaced close to WT 4, which leads to the biggest wake effect. However, FINO-1 will experience mostly a free stream inflow as the main wind direction is out of the southwest, observable in figure 2.

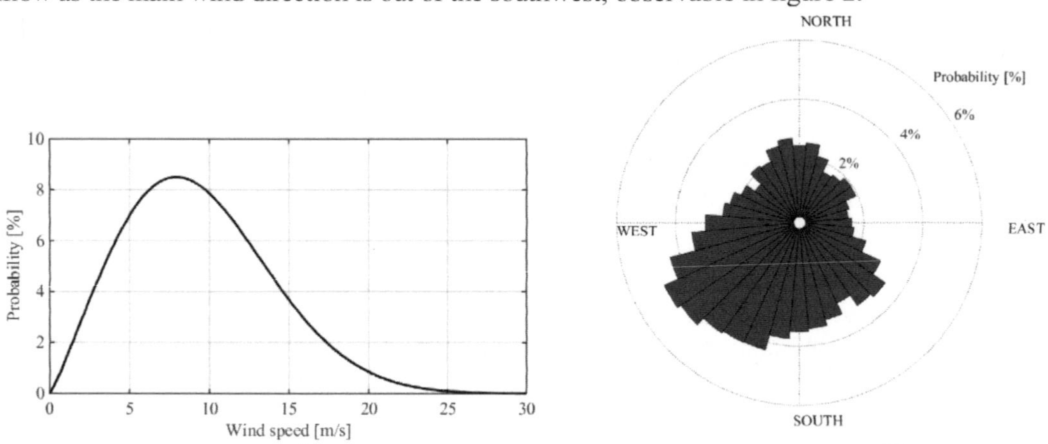

(a) **(b)**

Figure 2. (a) Weibull distribution (A = 10.61 and k = 2.15) and (b) wind direction distribution at FINO-1 in 80 m height [3]

As mentioned the nearby placed WT will shed a certain wake, which needs to be modeled. The wake effect will be approached by an increased turbulence intensity (TI), as done by Fradsen [9] and according to the IEC-61400 [6]. The TI can be calculated out of the FINO-1 database using equation (1), where σ represents the standard deviation of the measurement and U the mean free velocity.

$$TI = \frac{\sigma}{U} \quad (1)$$

In figure 3 the TI distribution is projected against the inflow direction. The black line is equal to the TI distribution after Alpha Ventus was erected. As a reference, the TI distribution before the erection of Alpha Ventus was included as well, as a gray dashed-dotted line. It needs to be noted that the displayed lines represent the mean value of the TI at the certain direction.

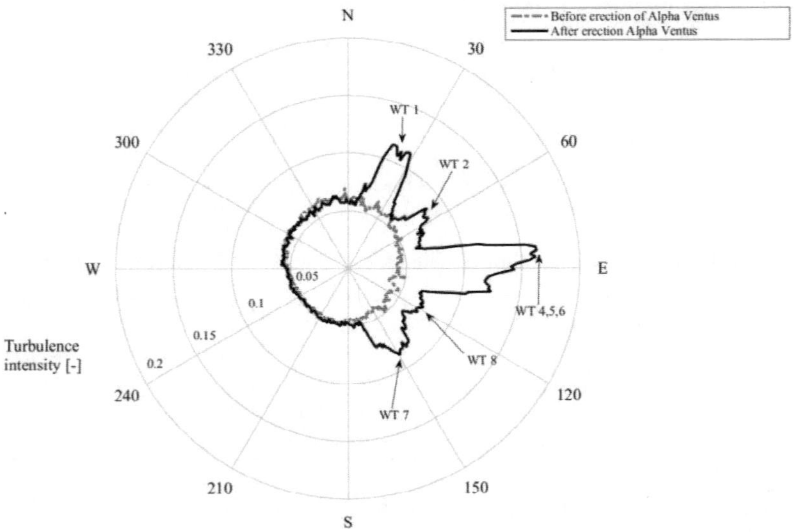

Figure 3. Turbulence intensity distribution

Out of figure 3, it can be observed that if the wind appears from the northeast, east and southeast the TI is increased significantly. This can be traced back to the wake effects of WT 1, 4 and 7, as these turbines are the closest. Additionally, the wake effects of WT 2 and 8 can be observed as well. However, their effect will be neglect, as the greater TI is not significantly with respect to the other WT. The highest TI in the distribution can be found while the wind occurs from the east, this is due two facts. Firstly, WT 4 is spaced closer than any other turbine. Secondly, the wake effects of WT 4, 5, and 6 are summing up at this inflow angle. A clear distinction between the two influences is not possible without any additional information. Additionally, the TI could also be calculated using equation (2) [6] and equation (3) [9], where D represents the distance normalized by the rotor diameter.

$$TI_{IEC} = \sqrt{\frac{0.9U_{hub}^2}{\left(1.5 + 0.3D \cdot \sqrt{U_{hub}}\right)^2} + I_{ref}^2} \quad (2)$$

$$TI_{Fradsen} = \sqrt{\frac{1}{\left(1.5 + \frac{0.8D}{C_T}\right)^2} + I_{ref}^2} \quad (3)$$

Figure 4, displays the TI distribution with respect to equations (2-3) as well as the one out of the FINO-1 data. Additionally, the design TI of the Repower 5M is included in the graph, showing that after a certain distance the TI is below the design point. It can be noted that the function of Fradsen [9], aligns with the measurements nicely. However, this alignment is based on an assumed thrust coefficient (0.644), which would introduce an additional uncertainty in this approach. Therefore, it was decided to use the TI distribution out of table 2 for the load calculations. This will neglect the effects of WT 2 and 8 but their effect is rather low with respect to the freestream inflow TI.

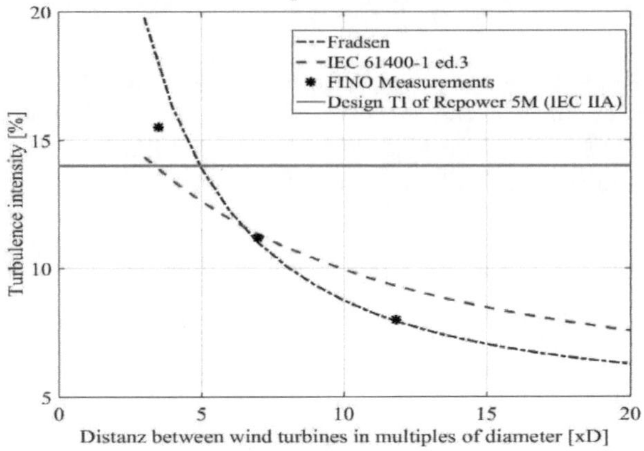

Figure 4: TI at various distances according to Fradsen [9] and IEC 61400-1 [6].

Table 2. Investigated inflow conditions, resulting out of the TI distribution

	TI = 5%	TI = 10%	TI = 15%
Represents:	Free stream	Wake effect of WT 1, 7	Wake effect of 4, 5, 6
Inflow angle span	Rest	[10°, 30°] [140°, 160°]	[75°, 115°]
Occurrence per year	79.9%	10.1%	10%

In order to evaluate the effect of the various TI, a Weibull distribution for each of them is determined by filtering the wind speeds with respect to their inflow angle, as done in figure 5. The solid black curve equals the Weibull distribution of the low TI. The dashed gray line corresponds to the Weibull distribution of TI equals 10%. The dashed-dotted line serves as the Weibull distribution of the high TI. Additionally, the overall Weibull distribution was included with a dotted line. It can be noted that through the increase of the TI, the mean wind speed is reduced, following from the fact that the previously WT that exploited the wake already extracted energy out of the flow.

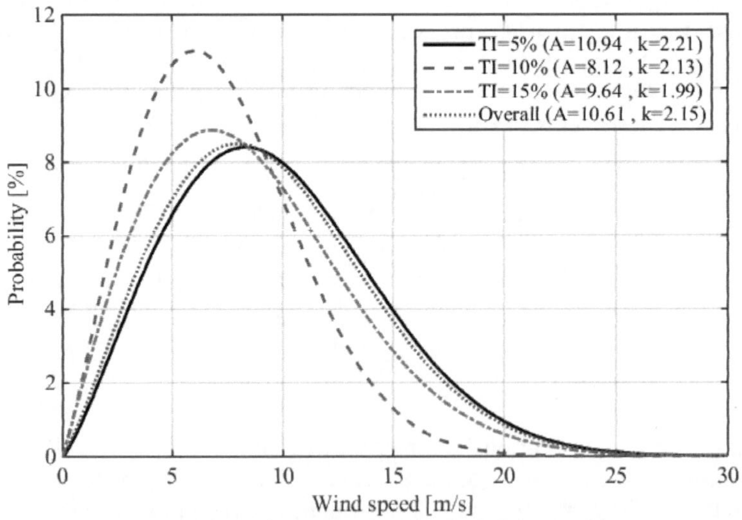

Figure 5. Weibull distribution for different inflow conditions

IOP Conf. Series: Journal of Physics: Conf. Series **854** (2017) 012038 doi:10.1088/1742-6596/854/1/012038

3. Load determination

In order to evaluate the wake effects without computational costly CFD calculations, it will be assumed that at the position of FINO-1 a wind turbine is located, that experiences the wind profiles according to the existing measurements. In this paper, the loads on this fictional WT will be determined by a Multibody Co-Simulation using SIMPACK with AERODYN [10] and MATLAB. The generic WT C3x126 is the reference WT of the Center for Wind Power Drives [4]. The turbine consists out of multiple flexible bodies such that deflection and displacement can be observed. The configuration of the WT can be found in table 3. The WT will be loaded with variations of turbulent wind velocities along the power curve, similar to the IEC 61400-1 DLC 1.1 [6] with a normal turbulence model. The WT produces power in a wind speed range between 3m/s and 25m/s. The loads will be cumulated over one year using the weighting of the probabilities, shown in figure 5, and then compared to each other.

Table 3. Wind turbine characteristic of C3x126 [4]

	Value	Unit
Rated power	3.0	MW
Tower height	112.0	m
Rotor diameter	126.0	m
Cut-in wind speed v_{in}	3.0	m/s
Cut-out wind speed v_{out}	25.0	m/s
Rated wind speed v_{rated}	11.0	m/s
Gearbox ratio	1: 92.28	
Main suspension	4-Point	

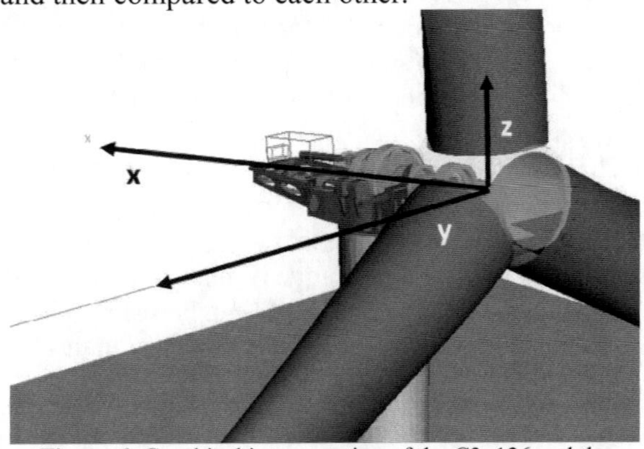

Figure 6. Graphical interpretation of the C3x126 and the coordinate system

The wind fields with the three different TI, 5%, 10% and 15% (see figure 5) are scaled toward the hub height using the logarithmic scaling law, used in the IEC 61400[6]. Sequentially, the corresponding 3D-wind fields are calculated with TurbSim [7] and step size of 0.5 m/s, such that a representative resolution was reached. This means for each TI distribution 45 wind fields and simulations covering ten minutes production time are calculated. All three TI distributions are generated with the same random seed values, such that the wind fields can be compared without the influences of random numbers. The results are analyzed at the hub coordinate system (see figure 6) in order to compare the loads at the drive train. This coordinate system is not rotating, meaning that the y-component equals an in-plane (rotor plane) component. The z-component represents gravitational effects and vertical upwards streams. For convenience, the moment around the y-axis will be called pitch moment and the moments around the z-axis yaw moment, inspired by the aeronautics terms.

4. Results

In order to evaluate the effect of the different inflow conditions, this paper will analyze the cumulated loads of two different conditions. At first, the influence of the TI has to be analyzed. For this two TIs are compared the 5% and the 15% with the Weibull distribution of FINO-1 (see figure 5, “overall”) in order to show the impact of the TI.

After each multibody-simulation, a rain flow counting is applied and multiplied with the Weibull frequency of the corresponding wind speed. Subsequently, the load cycle for each amplitude is calculated, given in subsection 4.1.

The second part (subsection 4.2) compares the design loads for a wind turbine which is sited at FINO before and after Alpha Ventus was build up. Before Alpha Ventus was build up a wind turbine sited at FINO would have seen a TI of 5% with the Weibull distribution (see figure 4, "5% TI”). The label “FINO after Alpha Ventus” is a combination of the three TIs (5%; 10% and 15%), their individual Weibull characteristics (see figure 5). The different TIs are cumulated and weighted with the occurrence

IOP Conf. Series: Journal of Physics: Conf. Series **854** (2017) 012038 doi:10.1088/1742-6596/854/1/012038

per year (see table 2). For the analysis, the hub coordinate system is used in order to quantify the effect on the wind turbine loads.

4.1. Impact of the turbulence intensity between 5% and 15%

It is expected that the impact of the turbulence intensity will result in a higher load at the yawing and pitching torque of the rotor, because of a higher imbalance of the inflow wind speed. The results are displayed in figure 7-12 and are as predicted. It can be noted that the presented results are partially discontinues and zigzagged, this is based on the bin-size used in rain flow counting. It could occur that at certain loadings no values are available. However, the main relevance in these plots is a trend with respect to the load cycles. The thrust at the rotor does not change significantly (see figure 7). The amplitudes of the side force for lower load cycles increases because of the higher turbulence effects in the rotor plane (see figure 8). According to this, also the torque in z-direction raises (see figure 12). The load in z-direction does not change much, which is logical as the biggest contributions origin from the rotor weight.

The main shaft torque has a different gradient for load cycles between 10^4- 10^6. The main shaft torque relates mainly to the control strategy. One would expect higher amplitudes for a higher turbulence intensity, but it seems to be that the control of the model is fast enough to compensate the higher TI. It might be that the results change if more random seeds for one wind speed are considered.

The biggest influence of the TI can be seen in figure 11 and figure 12 at the rotor pitching and yawing torque. The amplitude for load cycles lower than 10^5 is more than two times higher than before.

IOP Conf. Series: Journal of Physics: Conf. Series **854** (2017) 012038 doi:10.1088/1742-6596/854/1/012038

Comparison of a turbulence intensity of 5% and 15%

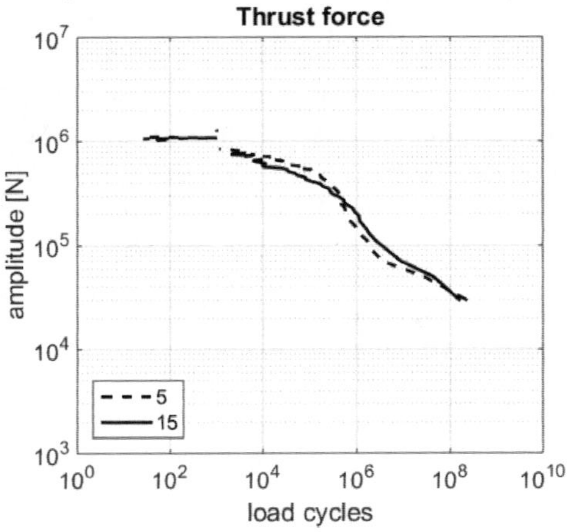

Figure 7. Thrust force of the rotor

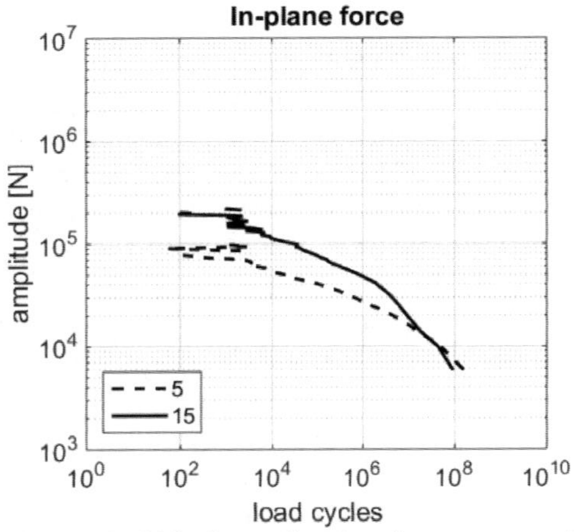

Figure 8. Side force (in-plane)

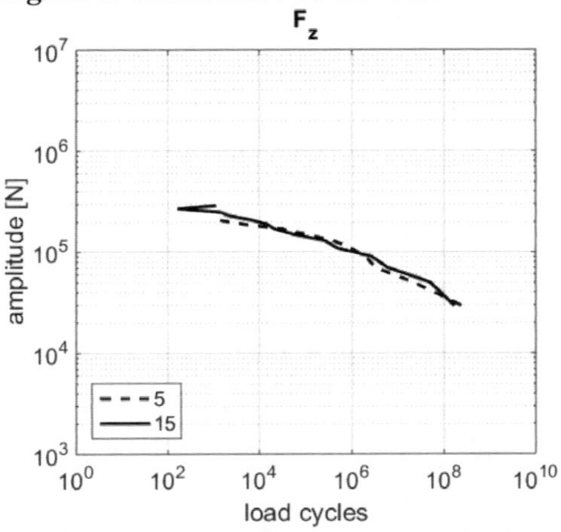

Figure 9. Force in z-direction (gravitation)

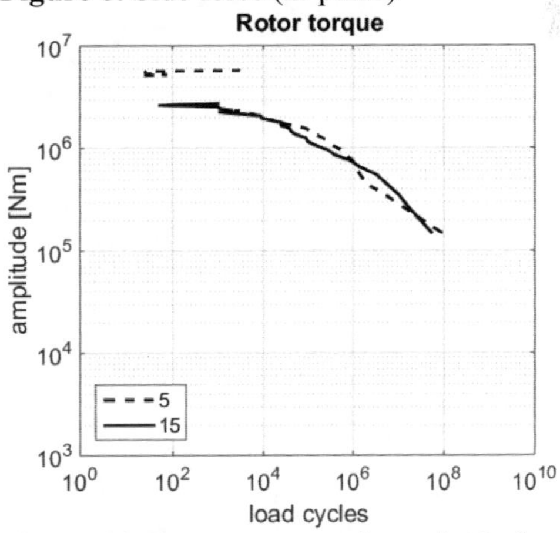

Figure 10. Rotor torque at the main shaft

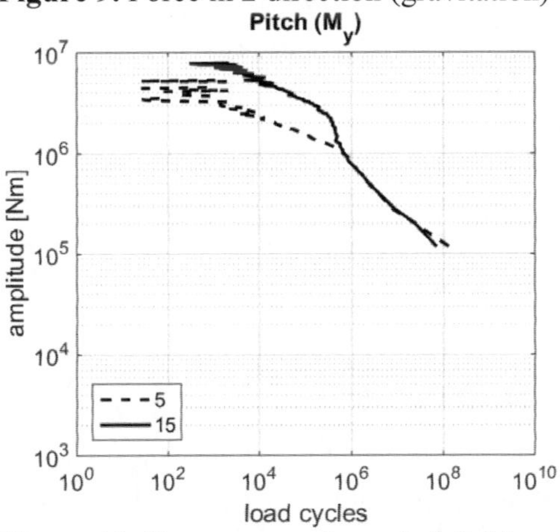

Figure 11. Torque around y-axis (pitch)

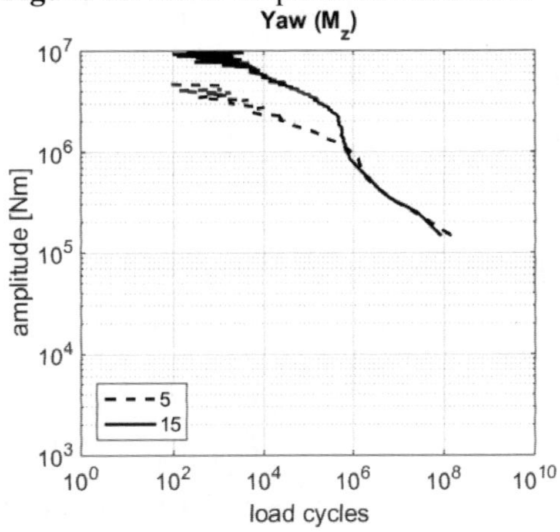

Figure 12. Torque around z-axis (yaw)

IOP Conf. Series: Journal of Physics: Conf. Series **854** (2017) 012038 doi:10.1088/1742-6596/854/1/012038

4.2. Before and After Alpha Ventus

As a next step, the influence of the different TIs for a wind turbine placed at the site of FINO-1 is quantified. The situation before Alpha Ventus is calculated with a 5% TI and the one at the position of FINO-1 after Alpha Ventus was built is called "FINO after Alpha Ventus" in the following figures. The simulations are done with all three TIs analyzed with a rain flow classification and their specific Weibull characteristic. After this, the load cycles are weighted by the occurrence per year of each TI. The results are shown in figure 13-15 and are as suspected. The influence of the TIs is visible for the yawing and pitching torque and the side force. In all three diagrams, the wind turbine experiences a higher load at higher TIs.

Comparison of a turbine in the free stream between one placed at FINO-1

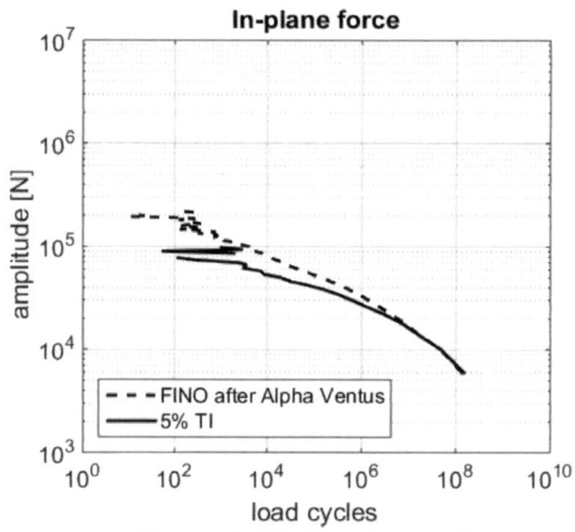

Figure 13. Side force (in-plane)

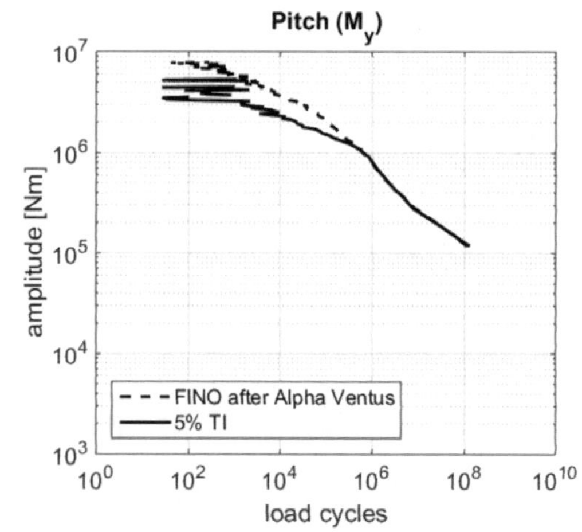

Figure 14. Torque around y-axis (pitch)

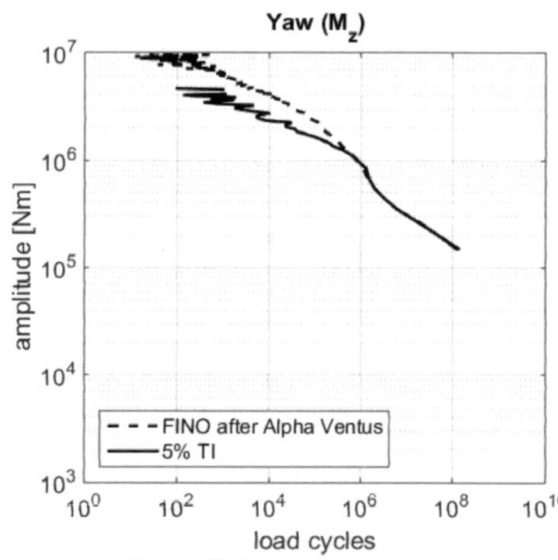

Figure 15. Torque around z-axis (yaw)

The differences are observable but only at low load cycles. Additionally, the presented results are reduced to a one level load cycle collective with equation (4) according to DIN 50100 [8].

$$DEL = \sum S_i^k \cdot \left(\frac{N_i}{N_{eq}}\right)^{\frac{1}{k}}, with\ k = 3 \tag{4}$$

Equation (4) is used to calculate a torque or force for an equivalent load cycle, therefore often is used 10^7 load cycles. This is determined by every combination of load cycles and amplitudes. Afterwards, all load cycles can be summed up to one amplitude with an equivalent load cycle. Comparing with respect to an equivalent load cycle was done, in order to observe the load variance due to the variation of the TI. The results are shown in table 4.

Table 4. Investigated inflow conditions, resulting out of the turbulence intensity distribution

	5% TI	FINO after Alpha Ventus	N_{eq}
Load cycle Pitching torque N_{My}	13 MNm	32 MNm	$1 \cdot 10^7$
Load cycle Yawing torque N_{Mz}	12 MNm	33 MNm	$1 \cdot 10^7$

The results show that the higher turbulence intensity results in approximately 2.5 times higher dynamic load at the bearings of the wind turbine and the supporting structure.

The results do not cover a complete system design and do not cover the statistical safety. For a design load analyses, one has to calculate at least six wind fields with six different random seeds to be sure that the results are statistically safe. But the results show that it is important to calculate the influence of the TI within a wind farm and that it is not enough to measure the free stream velocities before a wind farm is built. In the case of Alpha Ventus, the wind turbines are designed with a turbulence class of A or B according to the IEC 61400-1 and meaning that the mean TI is always above 10% (see Figure 4).

Before Alpha Ventus was built the TI was particularly low and it is possible that a turbine does not change the TI above 15% while the distance between two wind turbines is higher than 5 diameters. In order to validate the effect, more data of downtime statistics or data of a condition monitoring system of wind farms is needed.

5. Conclusion

Overall, this paper presented a method to determine wake effects out the FINO-1 measurements and integrating those into a load spectrum. Such a load spectrum was used to determine load cycles acting on a fictional wind turbine based on MBS simulation environment implemented in SIMPACK. The results show that a higher turbulence intensity has an impact on loads of a wind turbine and therefore on the cumulated load cycles, especially with respect to the in-plane side force as well as the moments around the y and z-axis. In further studies, it will be necessary to include actual measurements of an entire wind farm in order to validate the shown approach.

IOP Conf. Series: Journal of Physics: Conf. Series **854** (2017) 012038 doi:10.1088/1742-6596/854/1/012038

References

[1] German Wind Energy Association, Installed wind power capacity in Germany, *website*, (https://www.wind-energie.de/en/infocenter/statistiken/deutschland/installed-wind-power-capacity-germany) (accessed 02-13-2017)

[2] Pfaffel S et al 2011, IWES Annual Report, *Windenergie Report Deutschland 2011* 54

[3] Federal Maritime and Hydrographic Agency, FINO-1 Research Platform database, (http://www.fino1.de/en/) (accessed 02-01-2017)

[4] Werkmeister A, Schelenz R, Jacobs G, Calculation of the design loads with SIMPACK, Simpack User Meeting, *Conference, 09-2015*, Augsburg

[5] Alpha Ventus, Federal Ministry for Economic Affairs and Energy (https://www.alpha-ventus.de/english/) (accessed 02-01-2017)

[6] Internation Electrotechnical Commission standard DIN EN 61400-1, design requirements, Release 2015

[7] National Renewable Energy Laboratory (NREL), TurbSim A stochastic, full-field, turbulence simulator primarily for use with InflowWind/AeroDyn-based simulation tools, *website* (https://nwtc.nrel.gov/TurbSim) (accessed 02-17-2017)

[8] Internation Electrotechnical Commission standard DIN 50100:2015, Load controlled fatigue testing - Execution and evaluation of cyclic tests at constant load amplitudes on metallic specimens and components, Release 2015

[9] Fradsen S, Turbulence and Turbulence-generated structural loading in wind turbine cluster, Riso-R-1188, Roskilde, Release 2007

[10] János Zierath et al., Field test validation of Flex5, MSC.Adams, alaska/Wind and SIMPACK for load calculations on wind turbines, Wind Energy 2016, 19:1201-1222, DOI: 10.1002/we.1892

Dynamic Flow Model for Real-Time Application in Wind Farm Control

Andreas Rott[1]*, Sjoerd Boersma[2], Jan-Willem van Wingerden[2], Martin Kühn[1]

[1] Institute of Physics, ForWind, Carl von Ossietzky University of Oldenburg, Oldenburg, Germany

[2] Delft Center for Systems and Control, Delft University of Technology, Delft, the Netherlands

E-mail: `andreas.rott@uni-oldenburg.de`

Abstract. For short-term power predictions and estimations of the available power during curtailment of a wind farm, it is necessary to consider the flow dynamics and aerodynamic interactions of the turbines. In this paper, a control-oriented dynamic two-dimensional wind farm model is introduced that aims to incorporate real-time measurements such as flow velocities at turbine locations to estimate the ambient wind farm flow. The model is intended to derive flow predictions for real-time applications. Since fully resolved computational fluid dynamics are too CPU-intensive for such a task, the dynamic model presented in this paper relies on an approximation of the flow equations in a two-dimensional framework. A semi-Lagrangian advection scheme and a step-wise flow solver together offer fast calculation speed, which scales linearly with the number of grid points. In order to emulate effects of realistic three-dimensional wind farm flow, a relaxation of the two-dimensional continuity equation is presented. Furthermore, with little extra computational expense, additional dynamic state variables for various possible applications can be propagated along the wind flow. For instance, a dynamic confidence parameter can provide estimations of the accuracy of flow predictions, while a turbulence parameter adds the possibility to estimate wake induced loads on downstream turbines. In order to demonstrate the performance and validity of the new model it is compared with other models. At first a two turbine reference case is compared with a steady-state model and secondly with results obtained by the dynamic wind farm flow model WFSim. Finally a small wind farm is simulated in order to show the computational scaling of the model.

1. Introduction

Wind energy plays an important role in the increase of the renewable energies to meet the requirements of the Paris Agreement. Consequently, wind farms become larger and wake effects become even more prominent. In particular, offshore wind farms, which often are constructed with a regular mesh-like layout, are subject to aerodynamic interactions that cause significant power losses and wake induced loads [1, 2]. While onshore wind energy is a strong competitor in the energy market and has become at good sites, one of the cheapest power sources [3], the energy generation cost of offshore wind energy still has a lot of potential for improvement. One promising approach to achieve that, is optimized wind farm control with wake steering [4]. Therefore, the consideration of the aerodynamic interactions in the control of the individual turbines within a wind farm control scheme is an active field of research. One key point for those concepts is knowledge about the current ambient wind farm flow. In the past years, great effort was spend on the development and improvement of steady wake models and flow simulations to estimate and predict the power output of a wind farm, that are computationally fast enough to be used for optimized wind farm control. An extensive overview is provided in [5]. But only in the most recent years research also focuses on dynamic models that are fast enough for online predictions and for control design. These models are required to accurately represent the dynamic wind farm flow in large wind farms because of the high sensitivity of changes of the wind direction on the power output [6]. In [7] a propagation model for LiDAR measurements to correct the temporal distortion and to reconstruct the wind field between the measurements is introduced. In this paper this propagation model is extended to a complete 2D dynamic wind farm flow model that is able to simulate and predict the wind farm flow at hub-height with a relatively high reliability at relatively low computational costs. Additionally the semi-Lagrangian structure of the model allows to propagate any number of other parameters/properties with the flow, which can be utilized for various purposes. The model is designed so that it works for flow estimations from SCADA data or as a basis for wind farm control optimizations. Therefore the objectives of this papers are threefold: To introduce the structure of the dynamic model and describe its application (see section 2), demonstrate simulation results and the performance of the model (see section 3) and to discuss its capabilities (see section 4) and finally a brief conclusion is given in section 5.

2. Method

The wind farm flow model presented in this paper is a 2D medium fidelity flow and property propagation scheme. The algorithm solves the discretized Navier-Stokes equation for incompressible flow in a step-wise scheme, that includes a relaxation of the continuity equation to approximate 3D wind farm flow with relatively low computational expenses. Next to the longitudinal and lateral wind velocities (u and v, respectively) of the 2D domain, any number of additional parameters can be propagated with the flow by the presented method as explained in 2.2. These parameters are referred to as additional states in this paper, they can be used to simulate a variety of flow properties.

2.1. General flow equations

The Navier-Stokes equation and the continuity equation (for incompressible flow) are the governing flow equations illustrated in the vector equation (1) and the scalar equation (2)

$$\frac{\partial \vec{u}}{\partial t} = -(\vec{u} \cdot \nabla)\vec{u} - \frac{1}{\rho}\nabla p + \nu \nabla^2 \vec{u} + \vec{f} \tag{1}$$

$$\nabla \cdot \vec{u} = 0 \tag{2}$$

Here the vector $\vec{u} := (u, v)^T$ is the short form for the smooth function $\vec{u}(x, y, t) \in \mathcal{C}^\infty(\mathbb{R}^2 \times \mathbb{R})$ and it describes the 2D flow velocity, while the superscript T denotes the transpose of a matrix or vector. The variables $t, \rho, p, \nu \in \mathbb{R}$ are the time, density, pressure and the kinematic viscosity of the fluid, respectively. $\vec{f} := \left(\frac{f_x}{\rho}, \frac{f_y}{\rho}\right)$ (short for the vector function $\vec{f}(x, y, t) : \mathbb{R}^2 \times \mathbb{R} \rightarrow \mathbb{R}^2$) is the external force (over density) acting on the flow. The symbol $\nabla := \left(\frac{\partial}{\partial x}, \frac{\partial}{\partial y}\right)^T$ is a vector operator of the partial spatial derivatives with $x, y \in \mathbb{R}$ as the horizontal and vertical positions.

The widely used Helmholtz-Hodge Decomposition [8] is a useful mathematical tool that lets us define a projection, which allows to reduce (1),(2). Furthermore, this technique is used to build a relaxation of the continuity equation as will be described in section 2.8. The fundamental theorem of vector calculus states that every sufficiently smooth vector field is uniquely composed of a solenoidal vector field and a conservative vector field. In more mathematical terms: Let $\vec{\xi}(x, y) \in \mathbb{R}^2$ be a smooth vector field. Hence there exists a unique vector field $\vec{q}(x, y) \in \mathbb{R}^2$, with

$\nabla \cdot \vec{q} = 0$ (solenoidal, here ' $\cdot$ ' is the dot product) and a unique vector field $\vec{r}(x,y) \in \mathbb{R}^2$, with $\nabla \times \vec{r} = 0$ (rotational free, here ' $\times$ ' is the cross product), with:

$$\vec{\xi} = \vec{q} + \vec{r} \tag{3}$$

According to this decomposition the projection P can be defined, that projects a smooth vector field onto its divergence free component:

$$P(\vec{\xi}) := \vec{\xi} - \vec{r} = \vec{q} \tag{4}$$

The construction of the projection can be derived by taking the divergence of (3) and the linearity of the divergence operator:

$$\nabla \cdot \vec{\xi} = \underbrace{\nabla \cdot \vec{q}}_{=0} + \nabla \cdot \vec{r} = \nabla \cdot \vec{r} \tag{5}$$

by definition the vector field $\vec{r}$ is rotational free, which is equivalent to the existence of a scalar field $p(x,y) \in \mathbb{R}$, with $\vec{r} = \nabla p$. Therefore we can substitute $\vec{r}$ and get:

$$\nabla \cdot \vec{\xi} = \nabla^2 p \tag{6}$$

This is a Poisson-equation, which can be numerically solved to find p, from where $\vec{r}$ can be derived, which fully defines the projection according to (4). The use of the letter p here is on purpose, because this scalar field indeed resembles the pressure field of the Navier-Stokes equation.

Overall we have shown that the solution of the projection P fulfills the continuity equation. Therefore we can use this projection to combine (1) and (2) in one equation as it was done in [9]:

$$P\left(\frac{\partial \vec{u}}{\partial t}\right) = P\left(-(\vec{u} \cdot \nabla)\vec{u} - \frac{1}{\rho}\nabla p + \nu \nabla^2 \vec{u} + \vec{f}\right) \tag{7}$$

Due to the fact that $P(\vec{u}) = \vec{u}$, the linearity of the projection and $P(\nabla p) = 0$, since ∇p is rotational free, the final equation describing the flow field is:

$$\frac{\partial \vec{u}}{\partial t} = P\left(-(\vec{u} \cdot \nabla)\vec{u} + \nu \nabla^2 \vec{u} + \vec{f}\right) \tag{8}$$

2.2. General advection equation

The structure of the flow-solver allows to simulate additional states, which can represent flow properties, as long as they can be substituted by a scalar value. The propagation of an additional state is modeled by the following equation [9]:

$$\frac{\partial s}{\partial t} = -(\vec{u} \cdot \nabla)s + \kappa \nabla^2 s - \beta s + S \tag{9}$$

The function $s := s(x,y,t) : \mathbb{R}^2 \times \mathbb{R} \to \mathbb{R}$ describes the value of the state at every grid point, the variables $\kappa, \beta \in \mathbb{R}$ are parameters that model the diffusion and the dissipation, respectively. The function $S := S(x,y,t) : \mathbb{R}^2 \times \mathbb{R} \to \mathbb{R}$ is the source term of the respective state.

This formula is strongly related to the Navier-Stokes equation, aside from the dissipation term. In the following the algorithm is presented that is used to solve a discretized version of (8) and (9) resulting in the flow velocities and additional parameters.

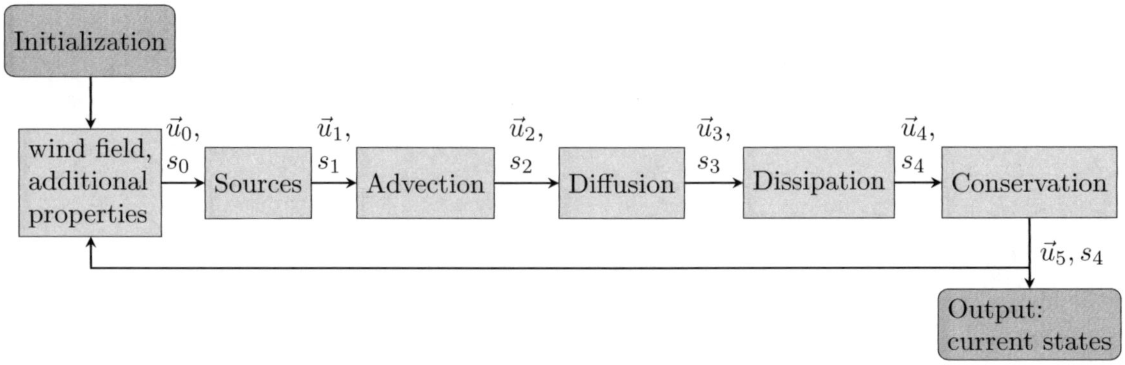

Figure 1: Structure of the flow-solver.

2.3. Structure of the solver

In order to solve the flow velocities at hub-height and additional flow properties (states), a fast, modular, step-wise approximation of the flow-equations has been implemented similar to the solver presented in [9]. The separate terms of the flow equations are solved successively as illustrated in Fig. 1.

In the initialization, the initial windfield $\vec{u}_0$ and the initial additional states s_0 are constructed. This can be done by assuming a uniform flow for the flow velocities and zero for the additional states. This initial wind field is passed to the iteration-phase of the algorithm, the individual steps of which are described in the following sections.

2.4. Source - Step

In the first step of the algorithm the sources are applied to the simulation domain. This is represented by the external forces $\vec{f}$ in terms of the flow velocities and by the source S for additional parameters. The step description of the algorithm yields:

$$\vec{u}_1 = \vec{u}_0 + \vec{f} \tag{10}$$

$$s_1 = s_0 + S \tag{11}$$

The index indicates the steps of the algorithm as illustrated in Fig. 1. The commonly used Actuator Disk model is used to calculate the thrust of a wind turbine. The rotor area is represented by a porous disk, that reduces the wind velocity. The magnitude of deceleration is calculated by the thrust T of a turbine:

$$T = \frac{1}{2}\rho A u_\infty^2 c_T = \frac{1}{2}\rho A u_\perp^2 \frac{1}{(1-a)^2} 4a(1-a) \tag{12}$$

$$= 2\rho A u_\perp^2 \frac{a}{1-a} \tag{13}$$

Where A is the swept rotor area, $u_\perp$ is the wind speed at the rotor that is perpendicular to the rotor area. According to momentum theory it holds $u_\perp = u_\infty(1-a)$, where u_∞ is the free-stream velocity perpendicular to the rotor area. $c_T := 4a(1-a)$ is the thrust coefficient, which is brought into relation with the axial induction factor $a \in \left[0, \frac{1}{2}\right)$ of the turbine. Finally the discretized volume force $\vec{f} = \Delta t \frac{T}{\Delta x \Delta y}\vec{n}$ is applied to the wind field. $\Delta x, \Delta y$ are the lateral and longitudinal grid size, Δt is the time-step size and $\vec{n}$ is the normal vector of the rotor area.

Modeling the source term of an additional state cannot be generalized. As an example, in order to model a turbine induced turbulence intensity a direct relation between the force, that is generated by the turbine, can be formulated. Here the relation is modeled as a proportional correlation:

$$S = c_{TI}\|\vec{f}\| \tag{14}$$

The coefficient $c_{TI} \in \mathbb{R}$ can be parameterized to scale the induced turbulence intensity properly. The source of confidence coefficient on the other hand can be modeled as a constant that overwrites the current confidence state at the locations close to the measurement point, whenever a new measurement is available, as described in section 2.9.

2.5. Advection step

The advection step involves the non-linear term of the Navier-Stokes equation and it is also the most computational intensive step of the algorithm. In this step, the following is solved:

$$\frac{\partial \vec{u}}{\partial t} = -(\vec{u}\cdot\nabla)\vec{u} \tag{15}$$

$$\frac{\partial s}{\partial t} = -(\vec{u}\cdot\nabla)s \tag{16}$$

The algorithm calculates the current values of the states u, v, s from the prior time-step according to the following scheme, which was derived from the method of characteristics:

Let $\vec{\lambda}(\tau) := (x(\tau), y(\tau))^T = (x_c, y_c)^T + \vec{u}_c\tau$ be a path in the domain, that passes the point $(x_c, y_c)^T$ at $\tau = 0$, where $\vec{u}_c = (u_c, v_c)^T$ is the velocity at $(x_c, y_c)^T$ and $\tau \in \mathbb{R}$. This is a characteristic of the wind field and it allows us to parameterize the spatial and temporal components with $t(\tau) = \tau$ such that $s(x(\tau), y(\tau), t(\tau)) = s(\lambda(\tau), t(\tau))$ is only depending on one variable τ. According to the chain rule it holds:

$$\begin{aligned}\frac{ds}{d\tau} &= \frac{\partial s}{\partial \vec{\lambda}}\frac{d\vec{\lambda}}{d\tau} + \frac{\partial s}{\partial t}\frac{dt}{d\tau} \\ &= \frac{\partial s}{\partial \vec{\lambda}}\vec{u}_c + \frac{\partial s}{\partial t}\end{aligned} \tag{17}$$

With (16) we can show, that this expression vanishes, since:

$$\begin{aligned}\frac{\partial s}{\partial \vec{\lambda}}\vec{u}_c + \frac{\partial s}{\partial t} &= \left(\frac{\partial s}{\partial x}, \frac{\partial s}{\partial y}\right)\cdot\vec{u}_c + \frac{\partial s}{\partial t} \\ &= (\vec{u}_c\cdot\nabla)s + \frac{\partial s}{\partial t} \\ &= 0\end{aligned} \tag{18}$$

Hence s is constant with respect to τ on the path λ. The temporarily discretized statement we derive from this is:

$$\begin{aligned} s\left(\vec{\lambda}(\tau), t(\tau)\right) &= s\left(\vec{\lambda}(\tau-\Delta\tau), \tau-\Delta\tau\right) \\ &= s\left(\vec{\lambda}(\tau) - \vec{u}_c\Delta\tau, \tau-\Delta\tau\right) \\ \Rightarrow s(x_c, y_c, t) &= s(x_c - u_c\Delta t, y_c - v_c\Delta t, t-\Delta t)\end{aligned} \tag{19}$$

where $\Delta\tau := \Delta t$ is chosen to resemble one time step. Since $\vec{\lambda}$ can be chosen arbitrarily, this is valid in the spatial discretization at every grid point. The proof for the velocity vector is analogous.

Tantamount to the above we can say that the current state at the location (x, y) can be derived from the state of the previous time-step at the location $(x, y) - (u, v)\Delta t$. The step description for the algorithm results in:

$$\vec{u}_2(x,y) = \vec{u}_1(x - u\Delta t, y - v\Delta t) \tag{20}$$

$$s_2(x,y) = s_1(x - u\Delta t, y - v\Delta t) \tag{21}$$

The algorithm approximates this by linear interpolation of the states to obtain the value at the position $(x_+, y_+) - (u_+, v_+)\Delta t$ for each grid point (x_+, y_+), where (u_+, v_+) is the respective velocity.

2.6. Diffusion step

The diffusion step simulates viscous stress and the blending of each state in one time step. In terms of flow velocity this term is defined by the viscosity term. According to [9] the straight forward explicit representation for the step description of the discrete algorithm is:

$$\vec{u}_3 = \vec{u}_2 + \Delta t \nu \nabla^2 \vec{u}_2 \tag{22}$$

$$s_3 = s_2 + \Delta t \kappa \nabla^2 s_2 \tag{23}$$

Here the parameter ν for the viscosity is a synthetic parameter, which needs to be adjusted to measurement or simulation data, in order to correctly represent 3D flow diffusion.

2.7. Dissipation step

The dissipation term reduces the value of a state in each iteration by a defined percentage $\beta \in [0, 1]$ per second. This simulates the disappearance of the respective state from the simulation domain. The step-wise term description for the dissipation is:

$$s_4 = s_3(1 - \beta)^{\Delta t} \tag{24}$$

The exponent Δt is necessary, since the dissipation parameter β is the dissipation rate per second. For the wind speeds u, v an adjusted dissipation term is implemented in the algorithm, although the Navier-Stokes equation does not include a dissipation term. The dissipation term for the wind speeds is set up not to reduce the wind speeds, but rather to reduce wake deficits to better fit to measurement data. The drawback of the 2D model is that the mixing with the free flow is underrepresented due to the missing wake recovery above and below the wake. To emulate the behaviour the following term is implemented in the model:

$$\vec{u}_4 = (1 - \gamma)^{\Delta t} \vec{u}_3 + (1 - (1 - \gamma)^{\Delta t}) \vec{u}_\infty \tag{25}$$

The parameter $\gamma \in [0, 1]$ defines the magnitude of the wake dissipation, the vector $\vec{u}_\infty \in \mathbb{R}^2$ represents the free-flow velocity. This term lets the entire flow field converge to the free-stream wind velocity, while the change is proportional to the difference to the free-stream flow.

2.8. Conservation step

The conservation step is the last step of the flow solver an it is only used for the flow velocities u and v. It represents the continuity equation (2) for incompressible fluids, which ensures that in every time-step the amount of mass entering a finite volume equals the amount of mass leaving it. As already explained before, it is possible to combine the continuity equation for incompressible fluids with the Navier-Stokes equation into one equation by using the Helmholtz-Hodge decomposition. The presented method utilizes a relaxation of the continuity equation with the help of the described projection as explained in the following.

Regarding the conservation of mass a 2D model does not accurately represent a 3D domain, since the application of the continuity equation ensures that all the mass is conserved within the 2D horizontal plane. A turbine in such a model is comparable to a wind turbine model in a wind tunnel that occupies the entire height of the wind tunnel, since there is no possibility for the wind to flow above or below the turbine. Therefore, the jet effect on the sides of the turbine and its wake is higher than expected. Additionally the effect of lateral forces (e.g. caused by a yawed turbine) is damped by the 2D mass conservation, because a shift in the flow direction has to transfer directly to the boundaries of the domain.

To mitigate these "2D effects" a relaxation parameter is introduced into the model. With this the implementation of the relaxed continuity Equation is as follows:

$$\vec{u}_5 = P_\delta(\vec{u}_4) = \vec{u}_4 - \delta\vec{r} \tag{26}$$

Where P_δ is the adjusted version of the projection P explained above. $\delta \in [0, 1]$ is the relaxation parameter. $\vec{r} := \nabla p$, where p is the solution of the Poisson-Equation $\nabla \cdot \vec{u}_4 = \nabla^2 p$.

In the case of $\delta = 1$ the full continuity equation is applied in each time-step. For smaller $\delta \in [0, 1)$ the wind field is not completely projected onto its solenoidal component, but "pushed" into this direction i.e not all the pressure is resolved to fulfill the continuity equation.

2.9. Online measurements

The embedding of online measurements into the dynamic flow model is an extension of the flow solver and will be introduced here briefly. It is a crucial part for the application of the model in order to predict the wind farm flow derived from measurements. The challenge is to mix online information about the current flow, gathered by measurements at distinct point wise locations, into the model, so that a coherent flow field can be established. To achieve this, the confidence coefficient $c(x, y, t) : \mathbb{R}^2 \times \mathbb{R} \to \mathbb{R}$ is introduced and is treated by the solver as an additional state. It is initialized as $c(x, y, 0) \equiv 0$. When new measurements are available at the locations $(x_{m,i}, y_{m,i})$ at the time instance t_m a continuous flow field $\vec{m}(x, y, t_m)$ can be derived from these measurements by the shepard-interpolation [10]. Now the prevailing flow field $\vec{u}$ is mixed with $\vec{m}$ according to:

$$\vec{u}_{\text{new}} = c \cdot \vec{u} + (1 - c) \cdot \vec{m} \quad \forall x, y \text{ and } t = t_m \tag{27}$$

This results in the new flow field $\vec{u}_{\text{new}}$. The confidence coefficient gets updated to c_{new}, when a measurement is embedded according to :

$$c_{\text{new}}(x, y, t_m) = \begin{cases} 1 & \text{if there exists an } i \text{ with } (x - x_{m,i})^2 + (y - y_{m,i})^2 \leq d^2 \\ c(x, y, t_m) & \text{otherwise} \end{cases} \tag{28}$$

Where d is a configurable distance, that influences the confidence reach of a measurement.

Since c is a state, its values are propagated downstream with the flow by the flow solver, which means, that the influence of new measurements is predominantly upstream.

3. Results

To demonstrate the performance of the model, comparisons with two other models are presented in this section. In the first part, a comparison is made with a steady-state model, which is based on the well-known Jensen Park model [11]. The second part validates the model presented here with the dynamic model WFSim [12]. For these comparisons an area of 1400 m × 2400 m is used on which two turbines of the type NREL 5MW [13] are placed at a distance of approximate 6 rotor diameters. The grid spacing is 50 m × 50 m and the time step is 1 s. The wind speed is set to $8\,\text{ms}^{-1}$, coming from the south with uniform boundary conditions, so that the turbines build a full wake situation. The parameters are set to minimize the RMS error of the wind fields in a domain around the wakes, as illustrated in by the red box in Fig. 2 (right image).

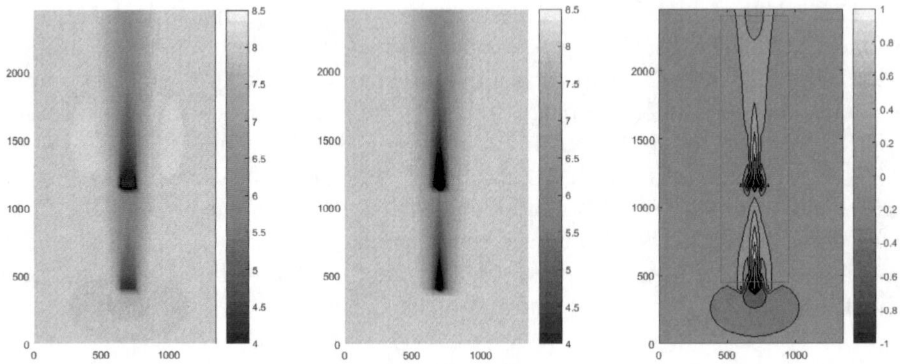

Figure 2: Test Case 1a: The plots show the longitudinal flow velocities in [ms^{-1}] over the simulation domain computed by the presented dynamic model (left) and the steady-state model (center). The contour plot (right) shows the difference

Table 1: Dynamic model parameters

Parameters	
ν	$11.5\,m^2s^{-1}$
γ	0.0029
δ	0.03

Table 2: Difference in the wake area

Results	Test case 1a	Test case 2a
RMS	$0.244\,ms^{-1}$	$0.230\,ms^{-1}$
avg. abs. err	$0.149\,ms^{-1}$	$0.144\,ms^{-1}$

3.1. Steady-State Model comparison

For the comparison with a steady-state model the dynamic model has to run for a short time under stationary inflow conditions so that the flow converges to an equilibrium state. To make sure that this equilibrium is reached and that there is no influence of the propagation the simulation time for the convergence was set to 625 s, which equals the time the free-stream flow needs to pass the whole domain twice. The steady-state model, which was used for comparison, calculates the wake deficit based on the Jensen-Park model [14], where wake interactions are modeled according to [15]. Furthermore a Gaussian overlay was used according to [16].

The parameters of the dynamic model are adjusted accordingly in order to minimize the differences in the wake area.

As a performance indicator the average absolute error and the RMS error are calculated for a domain that includes the wakes as depicted by the red box in Fig 2. The first case (Test Case 1a) we look at is the full wake situation, where both turbines run at their local optimum, which means an axial induction factor of $a_{1,2} = \frac{1}{3}$. The determined parameters and the results are shown in Table 1 and Table 2.

In the second case (Test Case 2a) the results of the two models are compared, where the axial induction of the turbines is reduced to $a_{1,2} = \frac{1}{6}$.

3.2. Dynamic model - WFSim

Next the dynamic model will be compared to WFSim [12]. For the comparison we look at the equilibrium state of both models. But this time the relaxation factor δ for the continuity equation was set to 1, since WFSim solves the 2D continuity equation completely. Again in the first test case (Test Case 1b) the models are compared for a standard situation, where the axial induction factors are $a_{1,2} = \frac{1}{3}$.

In the second test case (Test Case 2b) the axial inductions of the turbines are reduced again to $a_{1,2} = \frac{1}{6}$.

Table 3: Parameters for comparison with WFSim

Parameters	
ν	$5.2\,\mathrm{m^2s^{-1}}$
ε	0.0068
δ	1

Table 4: Difference between dynamic model and WFSim

Results	Test Case 1b	Test Case 2b
RMS	$0.345\,\mathrm{ms^{-1}}$	$0.187\,\mathrm{ms^{-1}}$
avg. abs. error	$0.204\,\mathrm{ms^{-1}}$	$0.124\,\mathrm{ms^{-1}}$

Table 5: Performance Test: The CPU-Time is computed for a flow-simulation with two states per grid point (u, v), the values in the brackets represent a flow-simulation with one additional state i.e. three states per grid point (u, v, s).

Resolution	grid points	avg. CPU Time ± std	avg. CPU Time per iter.
50 m × 50 m	3136	4.48 s ± 0.14 s (5.08 s ± 0.16 s)	0.007 s (0.008 s)
25 m × 25 m	12319	13.41 s ± 0.09 s (14.38 s ± 0.13 s)	0.022 s (0.024 s)
20 m × 20 m	19118	19.55 s ± 0.21 s (21.55 s ± 0.19 s)	0.033 s (0.036 s)
10 m × 10 m	76156	75.21 s ± 0.40 s (84.25 s ± 1.09 s)	0.125 s (0.140 s)

3.3. Performance test

In order to examine the scalability of the algorithm, 600 s of a virtual wind farm consisting of 9 turbines are simulated on a standard PC[1]. The size of the domain is fixed to 2400 m × 3150 m, while the resolution is increased according to Table 5. The experiment is performed for a standard flow-simulation with two states per grid point (u, v) and for three states at each grid point (u, v, s). Each resolution is simulated 10 times and the average calculation time plus standard deviation is shown in Table 5. The results of the CPU time show a clear linear dependence to the number of grid points with a coefficient of determination of $R^2 = 0.9988$. The simulation of an additional state increases the CPU-Time by approximately 12 %. In the most complex test case 228468 states have been simulated, while the average calculation time of one time-step took 0.14 s.

4. Discussion

Although the comparisons of the presented model with the steady-state model and WFSim show a relatively good agreement for the equilibrium state, the propagation of the model still needs to be verified. In order to do this, validations with high-fidelity simulations will be carried out in the future.

The configuration variables $(\nu, \varepsilon, \delta)$, allow the model to be parameterized for varying atmospheric conditions. However it needs to be validated how accurately the two-dimensional flow simulation can describe three dimensional flow. As long as the vertical mixing of the flow is sufficiently strong the differences over the height of the rotor area are expected to be neglectable. But especially in stable atmospheric conditions the vertical mixing is very limited, which can lead to significant differences of wind speed and wind direction (wind shear and wind veer) over the height of the rotor area. It will be investigated if separate parallel instances of the flow-solver, that represent different heights over the rotor area, can be used to improve the power estimation. Also for this, validations with high-fidelity simulations will be done in the future.

Overall the dynamic model shows relatively good scalability and the possibility of simulating additional flow properties can be utilized for a range of applications. E.g. turbine induced turbulence intensity (cf. equation (14)) can be used to estimate wake induced loads at downstream turbines or the confidence coefficient (cf. equation (28)) is necessary to integrate

[1] Intel(R) Core(TM) i5-3340M CPU @2.70 GHz

online measurements into the simulation as explained in section 2.9. But the parameterization and validation of these additional states is a future task.

5. Conclusion

In this work a dynamic flow model for real-time application is introduced. The model has relatively low computational complexity and scales linearly with domain size (i.e. number of grid points). The advantage of the model is, that next to the flow velocities, additional states can be added to the simulation for a variety of applications, without increasing the computational complexity significantly. In future work validations with high-fidelity simulations will be made to further verify the model and its dynamic properties.

Acknowledgements

This work was partially funded by the German Federal Ministry for Economic Affairs and Energy (BMWi) in the scope of the OWP Control project (FKZ 0324131A). The research is supported by the German Academic Exchange Service (DAAD) with funds from the Federal Ministry of Education and Research (BMBF).

References

[1] M. Méchali, R. Barthelmie, S. Frandsen, L. Jensen, and P.-E. Réthoré. Wake effects at Horns Rev and their influence on energy production. In *Proceedings of the European Wind Energy Conference and Exhibition, Athens, Greece*, 2006.

[2] P. Frohboese and C. Schmuck. Thrust coefficients used for estimation of wake effects for fatigue load calculation. In *European Wind Energy Conference*, pages 1–10, 2010.

[3] C. Kost, J. N. Mayer, J. Thomsen, N. Hartmann, C. Senkpiel, S. Philipps, S. Nold, S. Lude, N. Saad, and T. Schlegl. Levelized cost of electricity renewable energy technologies. *Fraunhofer Institute for Solar Energy Systems ISE*, 2013.

[4] D. Medici and J. Dahlberg. Potential improvement of wind turbine array efficiency by active wake control (awc). In *European Wind Energy Conference, Madrid.*

[5] S. Boersma, B. M. Doekemeijer, P. M. O. Gebraad, P. A. Fleming, J. Annoni, A. K. Scholbrock, J. A. Frederik, and J. W. van Wingerden. A tutorial on control-oriented modelling and control of wind farms. *ACC*, 2017.

[6] R. J. Barthelmie, S.C. Pryor, S. T. Frandsen, K. S. Hansen, J.G. Schepers, K. Rados, W. Schlez, A. Neubert, L.E. Jensen, and S. Neckelmann. Quantifying the impact of wind turbine wakes on power output at offshore wind farms. *Journal of Atmospheric and Oceanic Technology*, 27(8):1302–1317, 2010.

[7] H. Beck, A. Rott, D. Trabucchi, and M. Kühn. Volumetric wind field measurements of wind turbine wake with long-range lidar. In *EWEA2015, 17-20 November 2015, Paris, France*, 2015.

[8] H. Bhatia, G. Norgard, V. Pascucci, and P.-T. Bremer. The Helmholtz-Hodge decompositiona survey. *IEEE Transactions on visualization and computer graphics*, 19(8):1386–1404, 2013.

[9] J. Stam. Stable fluids. In *Proceedings of the 26th annual conference on Computer graphics and interactive techniques - SIGGRAPH 1999.* Association for Computing Machinery (ACM), 1999.

[10] D. Shepard. A two-dimensional interpolation function for irregularly-spaced data. In *Proceedings of the 1968 23rd ACM national conference*, pages 517–524. ACM, 1968.

[11] A. Rott and M. Kühn. Combining Model-Based and Data-Driven Optimization of Wind Farm Operation in a Learning Database. In *EAWE 10th PhD Seminar on Wind Energy in Europe, Orléans, France*, 2014.

[12] S. Boersma, P.M.O. Gebraad, M. Vali, B.M. Doekemeijer, and J.W. van Wingerden. A control-oriented dynamic wind farm flow model: WFSim. *Journal of Physics: Conference Series*, 753:032005, 2016.

[13] J. Jonkman, S. Butterfield, W. Musial, and G. Scott. Definition of a 5-mw reference wind turbine for offshore system development. *National Renewable Energy Laboratory, Golden, CO, Technical Report No. NREL/TP-500-38060*, 2009.

[14] N. O. Jensen. A note on wind generator interaction. *Risø-M-2411 Risø National Laboratory Roskilde*, pages 1–16, 1983.

[15] I Katic, J Højstrup, and NO Jensen. A simple model for cluster efficiency. *European wind energy association conference and exhibition*, (October):407–410, 1986.

[16] B. F. Lund, P. N. Oma, and T. Jakobsen. Wind farm production maximization through joint optimization of turbine efficiency and yaw. simulated case studies. *Poster EWEA*, 2014.

IOP Conf. Series: Journal of Physics: Conf. Series **854** (2017) 012040 doi:10.1088/1742-6596/854/1/012040

Large Eddy Simulation of an SD7003 Airfoil: Effects of Reynolds number and Subgrid-scale modeling

Hamid Sarlak

Fluid Mechanics Section, Department of Wind Energy, Technical University of Denmark, 2800 Kongens Lyngby, Denmark.

E-mail: hsar@dtu.dk

Abstract. This paper presents results of a series of numerical simulations in order to study aerodynamic characteristics of the low Reynolds number Selig-Donovan airfoil, SD7003. Large Eddy Simulation (LES) technique is used for all computations at chord-based Reynolds numbers 10,000, 24,000 and 60,000 and simulations have been performed to primarily investigate the role of sub-grid scale (SGS) modeling on the dynamics of flow generated over the airfoil, which has not been dealt with in great detail in the past. It is seen that simulations are increasingly getting influenced by SGS modeling with increasing the Reynolds number, and the effect is visible even at a relatively low chord-Reynolds number of 60,000. Among the tested models, the dynamic Smagorinsky gives the poorest predictions of the flow, with overprediction of lift and a larger separation on airfoils suction side. Among various models, the implicit LES offers closest pressure distribution predictions compared with literature.

1. Introduction

Over the past few years, studies have proved the capability of LES for airfoil simulations (cf. Mellen et al. [3]). Recently, a study was performed with the aim of investigating potential of LES in predicting the airfoil characteristics at high Reynolds numbers [1]. Uranga et al. [12] performed LES of the flow over a Selig-Donovan SD7003 airfoil for a range of Reynolds numbers between 10,000 and 60,000 at $\alpha = 4^o$, resulting in generation of laminar and fully turbulent flows over the airfoil, respectively. Recently, Sarlak et al. [8, 9] investigated aerodynamics of an S826 airfoil at low and moderate Reynolds numbers using LES and RANS methods.

While there has been a number of numerical studies for the flow past airfoils using LES techniques, an investigation of various subgrid scale models has not been studied extensively in the literature. The current paper investigates characteristics of the SD7003 airfoil at low Reynolds numbers and examines a number of SGS models in predicting the separation location, pressure distribution as well as airfoil polars.

2. Numerical Modeling

2.1. Governing equations

The flow is governed by Navier-Stokes equations which reads as

$$\frac{\partial \mathbf{v}}{\partial t} + \mathbf{v} \cdot \nabla \mathbf{v} = -\frac{\nabla p}{\rho} + \nabla \cdot [(\nu + \nu_{sgs}) \nabla \mathbf{v}], \quad (1)$$

where

$$\begin{aligned} \nu_{sgs} &= 0 && \text{No model: implicit LES} \\ \nu_{sgs} &= c_s \Delta^2 |\bar{S}| && \text{Mix-}S\text{: Mix-S} \\ \nu_{sgs} &= c_{ms} \Delta^{1.5} q_c^{0.25} |\bar{S}|^{0.5} && \text{Mix-}\omega\text{: Mix-O Dynamic} \\ \nu_{sgs} &= c_{mo} \Delta^{1.5} q_c^{0.25} |\bar{\Omega}|^{0.5} && \text{Smagorinsky: DynSmag,} \end{aligned}$$

and ρ and ν are the fluid density and molecular viscosity, respectively. Also, $\mathbf{v}$ represents the filtered velocity vector, for simplicity called "valocity" from now on, and $p = p + (1/2)\rho u_j u_j$, is the modified pressure (it is pressure combined with the isotopic part of the subgrid-scale stress tensor which acts as a pressure). ν_{sgs} is the eddy viscosity to be specified by the SGS model. Four different SGS models are used to evaluate ν_{sgs}, as described in equation 1. Here, *NO model* refers to the case in which there is no explicit representation for turbulent viscosity and the only effects of kinetic energy dissipation are those arising from numerical dissipation. This case is included here as a measure of the relative impact of the SGS models. *Smagorinsky* refers to the standard Smagorinsky model, and *Mix-S* and *Mix-ω* represent the two variants of the mixed-scale model [5]. $q_c = (\bar{\tilde{u}}_i - \tilde{u}_i)^2$ is the sub-filter scale kinetic energy obtained by an explicit filtering (shown by bar) of larger size than the grid size, δ being the grid size, $\tilde{S}_{ij}(\mathbf{x}, t)$ and $\Omega = \nabla \times \tilde{u}(\mathbf{x}, t)$ are the resolved strain rate and vorticity, respectively. $c_s = 0.01$, $c_{mo} = 0.01$ and $c_{ms} = 0.06$ are (fixed) model constants used in the present study. The mixed-scale model is chosen in the existing code (see below) because of its low computational cost and its performance. Formally, it depends on the small scales through the term q_c (as a result of scale similarity) and on the resolved large scales through the resolved velocity gradient tensor. As a result, the model is able to predict a laminar flow close to the solid wall without a damping function.

2.2. Numerical solver

For the numerical simulations of the governing incompressible Navier-Stokes equations, the block structured general purpose flow solver, EllipSys3D [4, 11], is used in which the equations are discretized using a finite volume method for the primitive variables. Diffusive and convective terms are discretized using 2nd order central differencing schemes (CDS) and a blend of CDS and 3rd order QUICK scheme. Time is discretized using a second order backward Euler scheme and the solution is marched in time using inner time stepping. Pressure checker-boarding is prevented by using Rhie-Chow interpolation on a collocated grid arrangement and the pressure correction equation is solved using PISO algorithm.

3. Results

LES is used for all simulations and that four different subgrid scale (SGS) models are used for all simulations: a no-model (or implicit LES), dynamic Smagorinsky, as well as two variants of a mixed-scale SGS model referred to as Mix-O and Mix-S in [7] (for brevity, details of the models are not discussed in this paper). Following the meshing study [6] a numerical grid consisting of $1024 \times 256 \times 128$ cells in the streamwise direction, normal to the airfoil surface and along the span is used for the simulations. The span to chord ratio is set to 0.4. A non-dimensional time step of $t^* = dt.c/U_0 = 0.001$ is used in 40000 time steps and the statistics are time-averaged for the last 20000 time steps.

To validate current simulations, experimental results of [10] are used for the polars while the LES work of [2] is used to compare the pressure dstribution over airfoil. Figure 1 shows a comparison of lift polars where the simulations using the four various SGS models are being compared at angles of attack $\alpha = 4^o$ and $\alpha = 8^o$. As can be seen, at $\alpha = 4^o$ the agreement between LES and experiments is good for LES cases except the dynamic Smagorinsky which over-predicts the lift by a higher margin. At $\alpha = 8^o$, results of implicit LES and Mix-O model are nearly identical but dynamic Smagorinsky and Mix-S models overpredict and underpredict the lift by around 10%, respectively. To further investigate the effect of SGS models, comparisons of flow structures using various models is presented in the following section.

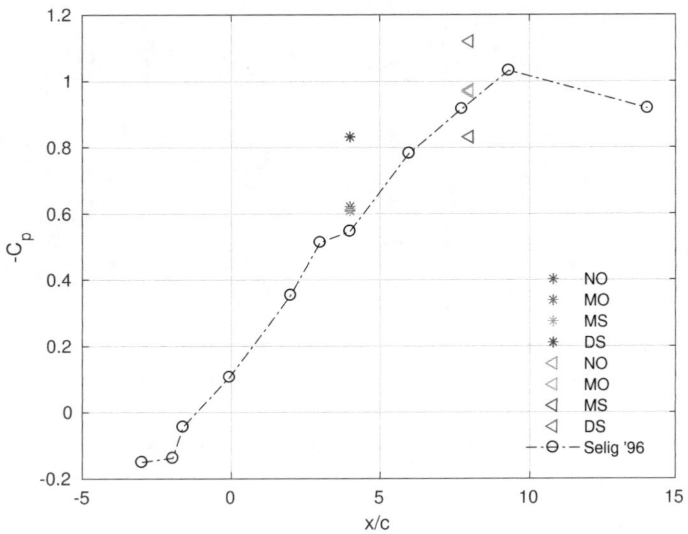

Figure 1. Lift polars for the SD7003 airfoil at Reynolds number 60,000. Comparison against experiments by Selig [10] are performed at $\alpha = 4^o$ and $\alpha = 8^o$.

3.1. Effect of angle of attack on velocity distribution over airfoil

To get an overview of the flow structures over the airfoil at various configurations, snapshots of flow past the airfoil at various angles of attack are plotted in figure 2. As can be seen, at $\alpha = 4^o$, the separation –shown by blue color, occurs at the middle of the chord and continues towards the trailing edge. By increasing angle of attack to $\alpha = 8^o$, laminar separation and flow re-attachment happens over the airfoil and at the two higher angles of attack, the flow undergoes leading edge separation which becomes stronger at $\alpha = 16^o$.

3.2. Effect of Reynolds number on the flow structures

Figure 3 presents the effect of Reynolds number on the flow attachment over the airfoil and subsequent aerodynamic loads. Two flow regimes at chord-based Reynolds numbers of Re=24,000 and Re=60,000 are investigated and flow past airfoil at angles of attack $\alpha = 4^o$, 8° and 12° are illustrated respectively. As can be seen, for all cases, the boundary layer gets thinner as a result of increasing the Reynolds number. At $\alpha = 4^o$ the Re=24,000 flow gets separated at nearly x/c=0.1 and exhibits recirculation further downstream whereas for the higher Reynolds number flow, a visible separation point is hardly seen. This is due to the dominance of higher momentum and its entrainment to the boundary layer to oppose adverse pressure gradients. At both $\alpha = 8^o$ and $\alpha = 10^o$, increasing the Reynolds number results in moving the separation point further upstream, with such phenomenon being more significant at $\alpha = 10^o$ where a separation bubble spanning the whole airfoil turns into one spanning about 15% of the chord.

3.3. Effect of SGS modeling on turbulence intensities and pressure and velocity distributions over airfoil

In order to quantify the effect of SGS modeling, one can first compare magnitude of the normalized viscosity, that is, eddy viscosity relative to the molecular viscosity. Figure 4 shows

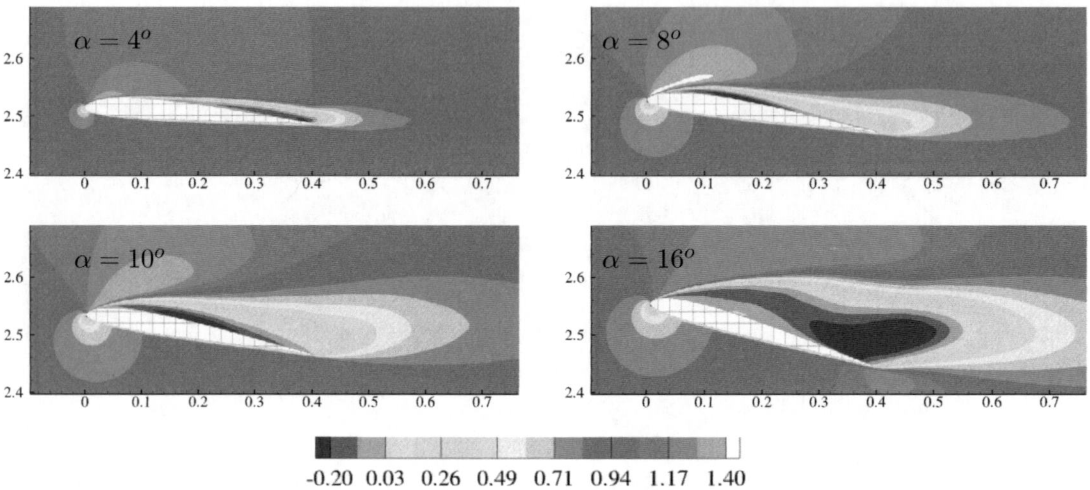

Figure 2. Mean streamwise velocity contours on the SD7003 airfoil for Reynolds number 24,000 at $\alpha = 4^o$, $\alpha = 8^o$, $\alpha = 10^o$, and $\alpha = 16^o$, respectively. x-axis: x/c, y-axis: y/c.

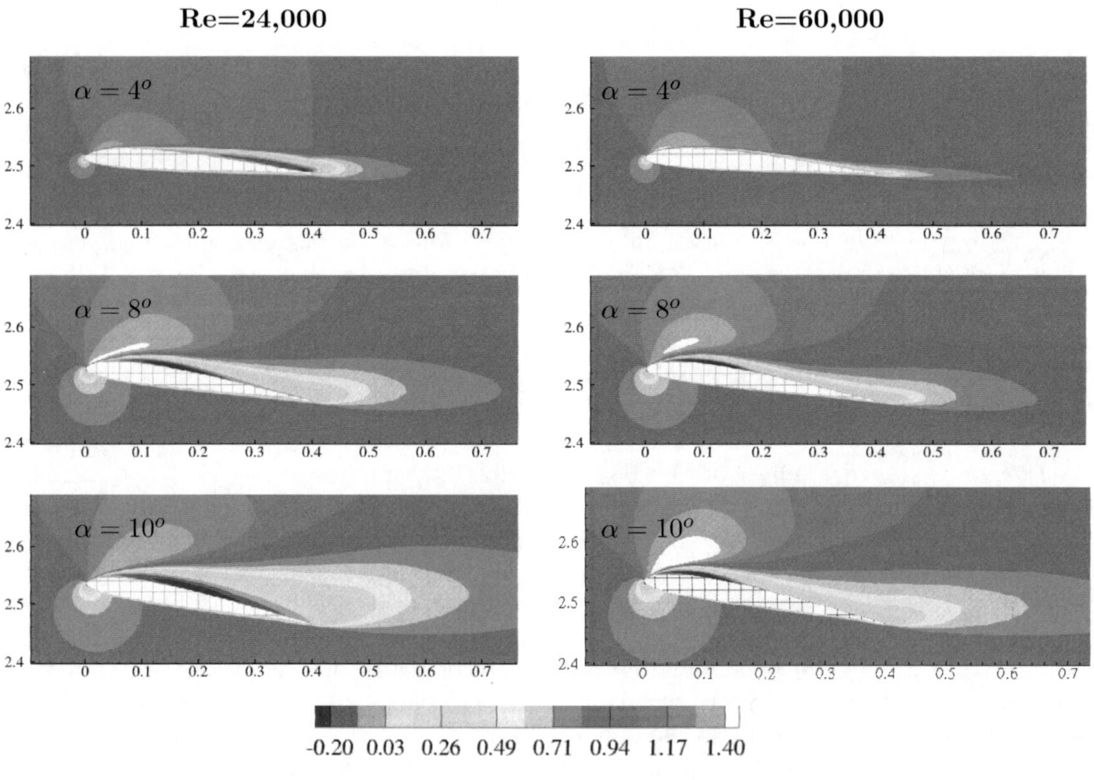

Figure 3. Mean streamwise velocity contours on the SD7003 airfoil for Reynolds numbers 24,000 (left) and 60,000 (right) at $\alpha = 4^o$ (top), $\alpha = 8^o$ (middle), and $\alpha = 10^o$ (bottom). Results are obtained using "MO" SGS model. x-axis: x/c, y-axis: y/c.

such quantities for the three Reynolds numbers (10,000, 24,000 and 60,000) at $\alpha = 4^o$. As can be seen, the normalized eddy viscosities are of the order of O(1), meaning that the contribution

IOP Conf. Series: Journal of Physics: Conf. Series **854** (2017) 012040 doi:10.1088/1742-6596/854/1/012040

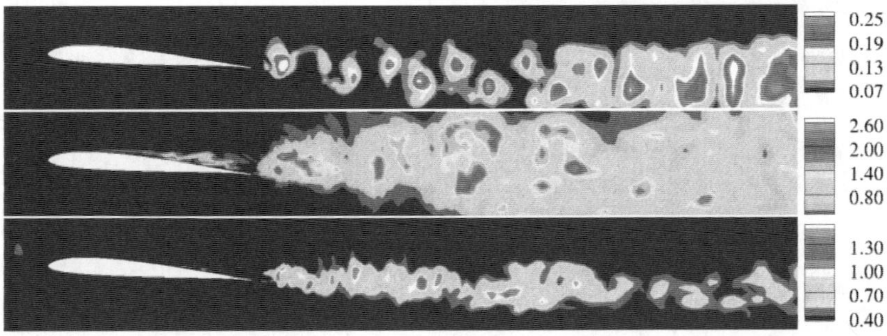

Figure 4. Effect of Reynolds number on turbulent viscosities. Normalized SGS viscosity contours on the SD7003 airfoil for Reynolds numbers 10,000 (top) and 60,000 (bottom) at $\alpha = 4^o$.

of the LES-generated viscosity is of the same order as the molecular viscosity. Looking at the color bar, note that the normalized values for Re=60,000 are approximately 6 times higher than Re=10,000. The trend at Re=24,000, however, suggests generally higher values of eddy viscosity than both lower and higher Reynolds numbers. While one explanation could be that at Re=24,000 flow starts exhibiting more violent fluctuations due to a transient nature, this trend needs to be investigated further.

To compare the different SGS models in predicting the flow past the airfoil, two sets of results at Reynolds numbers 10,000 and 60,000 are presented. Figure 5 illustrates the streamwise turbulence intensity $\sqrt{< u'u' >}/U_0$ contours obtained using the four SGS models for the case of Re=10,000 and figure 6 presents the same results for Re=60,000. Both cases demonstrate an angle of attack of $\alpha = 4^o$. As can be seen, even the effects of SGS model for the case of Re=10,000 are hardly noticeable. On the other hand, the SGS model for Re=60,000 can have a large impact on the quality of the results obtained. This is an interesting observation since the normalized eddy viscosity is of the order of unity for all of these cases and therefore one would imagine a much less impact of SGS models involved.

Comparison of various SGS models in streamwise velocity contours is also informative as it contains information about separation and recirculating flows on the suction side of the airfoil. Figures 7 and 8 illustrate the comparisons of velocity contours. Again, as can be imagined, at Re=10,000, results are insensitive to the SGS model whereas results of the Re=60,000 flow exhibit more significant differences. In particular, it is observed that among the four investigated models, the dynamic Smagorinsky presents a stronger separation point and recirculating flow towards the leading edge.

In order to better quantify the differences and compare against available references, pressure distribution over the airfoil at Re=60,000 is plotted in figure 9 for various SGS models at $\alpha = 4^o$. As can be noticed, there is a sizeable difference between model predictions. Interestingly, among various models, the implicit LES version predicts the flow separation and peak pressures almost dentical to the LES computations of Galbraith and Visbal [2]. The overprediction of lift using the dynamic Smagorinsky model that was observed in figure 1 can now be explained by the separation behaviour (particularly delayed reattachment) and the corresponding pressure distribution which results in higher area below the Cp curve. The wiggles on the curves are representing spanwise variation of pressure and are intentionally plotted to compare 3D effects visually. Clearly, all models except the Dynamic Smagorinsky predict quite a 2D pressure distribution over the airfoil on the suction side. It is also worthy to note how different is the behaviour of various models in capturing the laminar separation bubble - size and location.

IOP Conf. Series: Journal of Physics: Conf. Series **854** (2017) 012040 doi:10.1088/1742-6596/854/1/012040

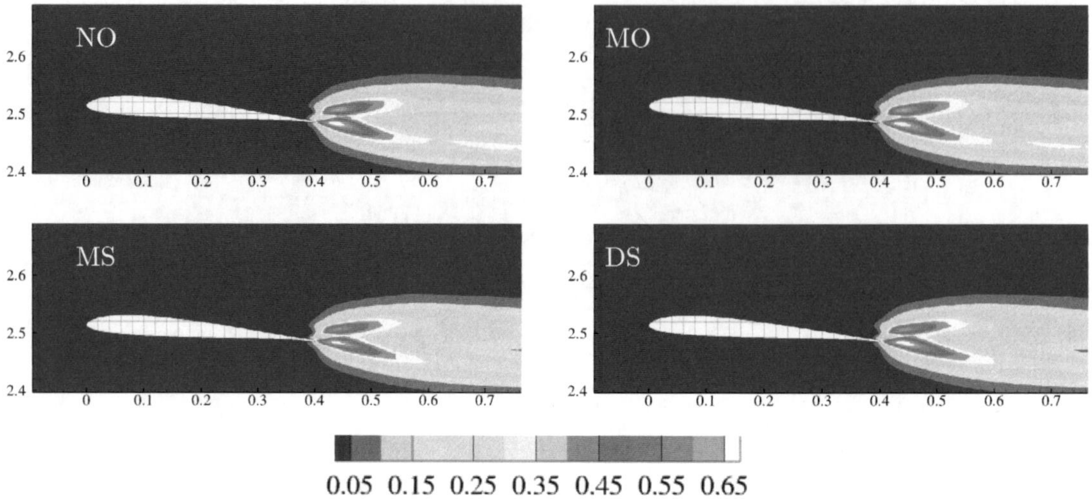

Figure 5. Effect of SGS modeling on the predicted streamwise turbulence intensity contours on the SD7003 airfoil for Reynolds number 10,000 obtained using various SGS models at $\alpha = 4^o$. NO: implicit (no model) LES, MO: Mix-Ω, MS: Mix-S, DS: dynamic Smagorinsky, as in [7]. x-axis: x/c, y-axis: y/c.

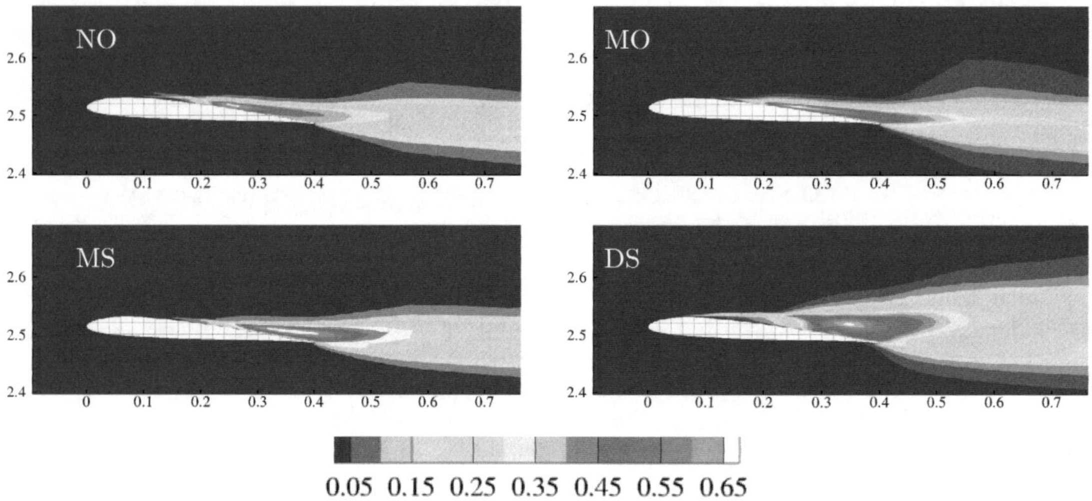

Figure 6. Effect of SGS modeling on the predicted streamwise turbulence intensity contours on the SD7003 airfoil for Reynolds number 60,000 and angle of attack $\alpha = 4^o$ using various SGS models. NO: implicit (no model) LES, MO: Mix-Ω, MS: Mix-S, DS: dynamic Smagorinsky, as in [7]. x-axis: x/c, y-axis: y/c.

4. Conclusions

Large Eddy Simulation of an SD7003 airfoil at low Reynolds numbers was performed at various angles of attack at chord-Reynolds numbers of Re=10,000, Re=24,000 and Re=60,000 in order to assess the applicability of four SGS models and investigate their role in predicting turbulence at different flow regimes.

The effect of Reynolds number on the flow behavior was investigated and it was seen that

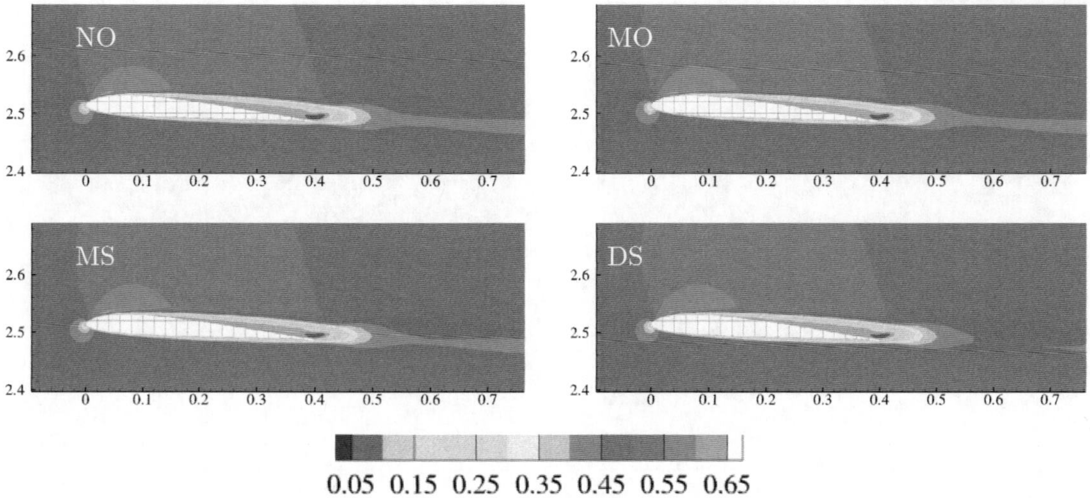

Figure 7. Effect of SGS modeling on the predicted mean streamwise velocity contours on the SD7003 airfoil for Reynolds number 10,000 obtained using various SGS models. NO: implicit (no model) LES, MO: Mix-Ω, MS: Mix-S, DS: dynamic Smagorinsky, as in [7]. x-axis: x/c, y-axis: y/c.

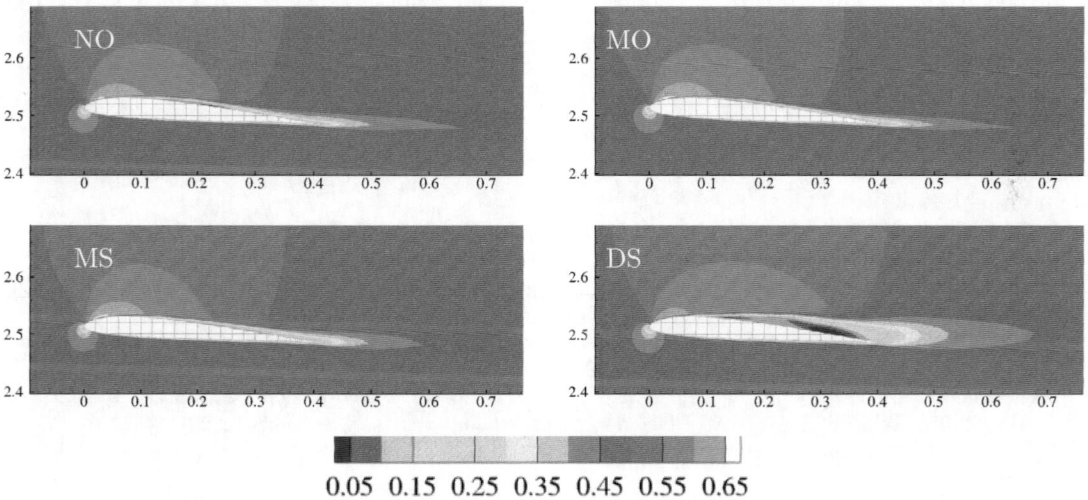

Figure 8. Effect of SGS modeling on the predicted mean streamwise velocity contours on the SD7003 airfoil for Reynolds number 60,000 obtained using various SGS models. NO: implicit (no model) LES, MO: Mix-Ω, MS: Mix-S, DS: dynamic Smagorinsky, as in [7]. x-axis: x/c, y-axis: y/c.

increasing Reynolds number results in shrinking the separation bubble that is formed on airfoil's suction side and moving the separation point towards the leading edge.

Through the study of four different SGS models at different Reynolds numbers, it was seen that at Re=10,000 the SGS models predict almost identical flow structures, whereas with increasing the Reynolds number to 60,000, the results vary among the different models. This is in contraty to the less-sensitive results of actuator line simulations presented in [7], mainly due to

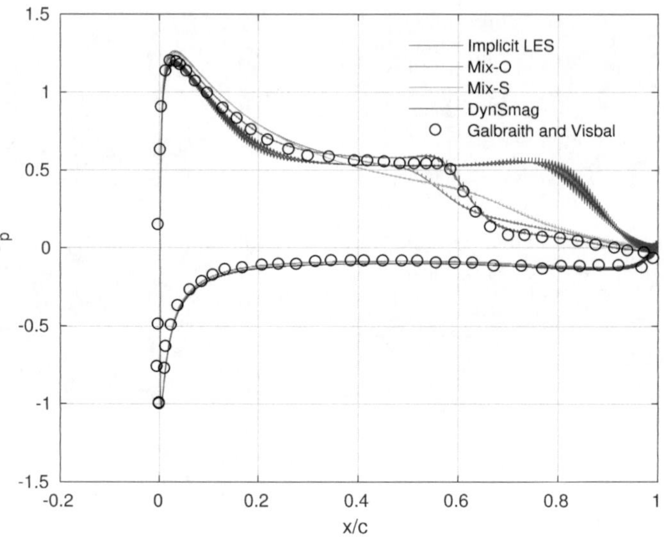

Figure 9. Pressure coefficient distribution over the SD7003 airfoil at Reynolds number 60,000 using $\alpha = 4^o$ and comparison against Galbraith and Visbal [2].

the presence of solid surfaces. In particular, the dynamic Smagorinsky model seemed to predict an earlier and stronger separation followed by a stronger wake and higher turbulence levels which eventually led to overprediction of lift. A more detailed comparison of the instantaneous results for spectral analysis of the signals would shed more light on the origin of the discrepancies. This will be performed in a future study.

References

[1] L. Davidson. *LESFOIL: Large Eddy Simulation of Flow Around a High Lift Airfoil: Results of the Project LESFOIL Supported by the European Union 1998-2001*, volume 83. Springer, 2003.

[2] Marshall Galbraith and Miguel Visbal. Implicit large eddy simulation of low-reynolds-number transitional flow past the sd7003 airfoil. In *40th Fluid Dynamics Conference and Exhibit*, page 4737.

[3] CP. Mellen, Fröhlich J., and W. Rodi. Lessons from lesfoil project on large-eddy simulation of flow around an airfoil. *AIAA journal*, 41(4):573–581, 2003.

[4] JA. Michelsen. *Basis3D-a platform for development of multiblock PDE solvers.* PhD thesis, Technical Note AFM92-05, Technical University of Denmark, Department of Fluid Mechanics, 1992.

[5] P. Sagaut. *Simulations numériques d'écoulements décollés avec des modèles de sous-maille.* PhD thesis, 1995.

[6] H. Sarlak. *Large eddy simulation of turbulent flows in wind energy.* PhD thesis, Technical University of Denmark, 2014.

[7] H Sarlak, C Meneveau, and JN Sørense. Role of subgrid-scale modelling in large eddy simulation of wind turbine wake interactions. *Renewable Energy*, 77:386–399, 2015.

[8] H. Sarlak, R. Mikkelsen, S. Sarmast, and JN. Sørensen. Aerodynamic behaviour of nrel s826 at re=100,000. *Journal of Physics, Conf. Series*, 524(1):012027, 2014.

[9] H Sarlak, T Nishino, Jens Nørkær Sørensen, Theodore Simos, and Charalambos Tsitouras. Urans simulations of separated flow with stall cells over an nrel s826 airfoil. In *AIP Conference Proceedings*, volume 1738, page 030039. AIP Publishing, 2016.

[10] MS. Selig, CA. Lyon, P. Giguere, C. Ninham, and JJ. Guglielmo. *Summary of low-speed airfoil data*, volume 2. SoarTech Publications, 1995.

[11] NN. Sørensen. *General purpose flow solver applied to flow over hills.* PhD thesis, Technical University of Denmark, Risø National Laboratory for Sustainable Energy, 1995.

[12] A. Uranga, PO. Persson, M. Drela, and J. Peraire. Implicit large eddy simulation of transitional flows over airfoils and wings. *Proceedings of the 19th AIAA Computational Fluid Dynamics, no. AIAA*, 4131, 2009.

IOP Conf. Series: Journal of Physics: Conf. Series **854** (2017) 012041 doi:10.1088/1742-6596/854/1/012041

Verification and Calibration of a Reduced Order Wind Farm Model by Wind Tunnel Experiments

J Schreiber[1], E M Nanos[1], F Campagnolo[1], C L Bottasso[1,2]

[1] Wind Energy Institute, Technische Universität München, Boltzmannstraße 15, D-85748 Garching bei München, Germany

[2] Dipartimento di Scienze e Tecnologie Aerospaziali, Politecnico di Milano, Via La Masa 34, I-20156 Milano, Italy

E-mail: {johannes.schreiber, em.nanos, filippo.campagnolo, carlo.bottasso}@tum.de

Abstract. In this paper an adaptation of the FLORIS approach is considered that models the wind flow and power production within a wind farm. In preparation to the use of this model for wind farm control, this paper considers the problem of its calibration and validation with the use of experimental observations. The model parameters are first identified based on measurements performed on an isolated scaled wind turbine operated in a boundary layer wind tunnel in various wind-misalignment conditions. Next, the wind farm model is verified with results of experimental tests conducted on three interacting scaled wind turbines. Although some differences in the estimated absolute power are observed, the model appears to be capable of identifying with good accuracy the wind turbine misalignment angles that, by deflecting the wake, lead to maximum power for the investigated layouts.

1. Introduction

In a wind farm environment wind turbine wakes, which are characterized by a lower wind speed and higher turbulence intensity than the free steam, can adversely affect other turbines. This in turn may lead to higher fatigue loads and a significantly reduced power output on affected turbines.

To increase total wind farm power and/or reduce fatigue loads, several techniques have been proposed [1, 2]. At present, one of the most promising approaches seems to be a technique where the wake is deflected by operating the wind turbine in yaw-misalignment condition with respect to the incoming wind [3]. In fact, as the wind turbine is yawed out of the wind, its wake is laterally deflected, which may reduce its interaction with downstream machines. Wind farm control strategies based on wake deflection have been proposed to increase the total energy capture and/or decrease fatigue loading [3, 4]. In this context, reduced order wind farm models as the FLORIS (FLOw Redirection and Induction in Steady-state) approach [4] may be used to enable model-based wind farm control.

This paper describes first a FLORIS-like wind farm model. Next, its parameters are calibrated based on wake measurements of a scaled wind turbine in a wind tunnel environment. Finally, the tuned model is used to estimate the wind farm power output for several different wind farm layouts comprising three scaled interacting wind turbines. For each layout, a variety of different yaw-misalignment combinations are tested and the turbine power is compared to the model-predicted one. Results and the causes for the observed mismatches are discussed.

2. Wind farm model

In this section a reduced order wind farm model is presented, following the work of Gebraad et al. [4]. In the present study, the model has been re-implemented with some modifications. First, the model describing expansion, reduction and deflection of a single wind turbine wake is presented. Next, the models that describe wind turbine power extraction and multiple wake interactions are presented. Finally, the process of calculating wind farm flow and power is summarized.

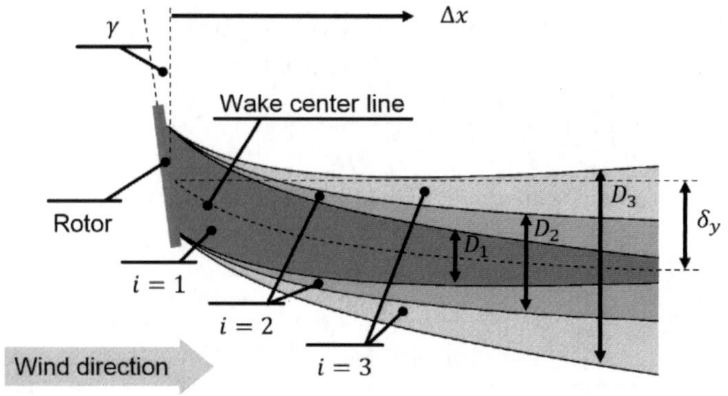

Figure 1. Wake model with three zones.

It is assumed that every wind turbine wake consists of three wake zones, as depicted in figure 1. In the generic wake zone i, the Jensen model [5] is used to describe wake expansion in terms of coefficient $k_{\mathrm{e},i}$. Experimental results suggest that the wake diameter of a turbine operated in a misaligned condition with respect to the wind direction is reduced (see the results section later on in this work). Taking this effect into account, the wake zone outer diameter is defined as

$$D_i(\Delta x, \gamma) = \max\left(0, (D + 2\Delta x k_{\mathrm{e},i}) \cos(\gamma)^{k_{\mathrm{e},\gamma}}\right), \tag{1}$$

where Δx is the distance downstream of the wind turbine, D the rotor diameter, γ the wind turbine yaw-misalignment angle and $k_{\mathrm{e},\gamma}$ a parameter that describes the reduced wake expansion due to wind turbine yaw-misalignment. As the wake zones can have a negative expansion coefficient, the wake diameter has to be limited to positive values.

The wake velocity in each wake zone is described by the Jensen wake model as

$$U_i(\Delta x) = U_\infty\Big(1 - r_i(\Delta x)\Big), \tag{2}$$

where U_∞ is the ambient free stream velocity and r_i the reduction factor defined as

$$r_i(\Delta x) = 2a\left(\frac{D}{D + 2\Delta x k_{\mathrm{r},i}}\right)^2, \tag{3}$$

where a is the wind turbine induction and $k_{\mathrm{r},i}$ the wake reduction parameter of wake zone i.

The wake center line deflection due to yaw-misalignment is taken into account as described in [4], leading to

$$\delta_y(\Delta x, \gamma) = \frac{C_\mathrm{T}(\gamma)\left(15(2k_\mathrm{d}\frac{\Delta x}{D} + 1)^4 + C_\mathrm{T}(\gamma)^2\right)}{30\frac{k_\mathrm{d}}{D}(2k_\mathrm{d}\frac{\Delta x}{D} + 1)^5} - \frac{C_\mathrm{T}(\gamma)D\left(15 + C_T(\gamma)^2\right)}{30k_\mathrm{d}}, \tag{4}$$

where k_{d} is the single parameter describing the recovery of the wake flow direction and C_{T} the turbine thrust coefficient, which in turn is defined as a function of the rotor induction a and the yaw-misalignment as

$$C_T(\gamma) = \frac{1}{2}\cos(\gamma)^2 \sin(\gamma)\Big(4a(1-a)\Big). \tag{5}$$

For calculating the power extracted by a wind turbine, its rotor disk is split into m discrete elements e. The turbine power is obtained by summing up the power extracted in each element

$$P = \sum_{e=1}^{m} \frac{1}{2}\rho A_e C_{\mathrm{P}}(\gamma) V_e^3, \tag{6}$$

where ρ is the air density, A_e the element area, V_e the wind velocity at the discrete element and $C_{\mathrm{P}}(\gamma)$ the power coefficient expressed as a function of yaw-misalignment as

$$C_{\mathrm{P}}(\gamma) = C_{\mathrm{P},\gamma=0}\cos(\gamma)^{k_{\mathrm{P}}}. \tag{7}$$

Furthermore, $C_{\mathrm{P},\gamma=0}$ is the power coefficient of the turbine operating aligned with the wind, while k_{p} is the parameter taking into account power reduction due to yaw-misalignment. Speed V_e is calculated based on the wake deficits of all upstream turbines

$$V_e = U_\infty\left(1 - \left(\sum_{w=1}^{n} r_w^2\right)^{\frac{1}{2}}\right), \tag{8}$$

where n is the number of wake zones overlapping with the turbine rotor, while r_w the reduction factor of a wake zone impinging the element. In case $n = 0$, no wake is impinging on the element and therefore $V_e = V_\infty$.

The implemented algorithm is organized as follows. First, the power and wake characteristics of the first upwind turbine is calculated. In a second step, the next wind turbine is considered and the wake position, reduction, and expansion of all upwind turbines are interpolated at the given downwind position. Based on this, the turbine power can readily be computed by equation (6) and (8). Finally, the turbine wake is also computed, using equations (1,2) and (4), and the second step is repeated until the last turbine is reached.

3. Results

3.1. Experimental setup

The experiments described in this section were conducted with a scaled wind farm (see figure 2) composed of three identical scaled wind turbine models, longitudinally spaced 4 diameters (D) apart, whose rotor diameter is equal to 1.1 m (in the following named G1s for Generic 1 m diameter rotor), which were already used in other research projects [3, 6, 7]. The models were operated in the boundary-layer test section of the wind tunnel of the Politecnico di Milano, which has a cross-sectional area of 13.84 m by 3.84 m and a length of 36 m. Atmospheric boundary-layer conditions were simulated by the use of spires placed at the chamber inlet. The vertical profile of the longitudinal wind speed was measured prior to testing, resulting in the following best-fitted exponential law

$$U(z) = U_H\left(\frac{z}{z_H}\right)^{0.088}, \tag{9}$$

where $U_H \approx 5.7$ m/s and $z_H = 0.825$ m are the free-stream wind speed at hub height and the elevation of the rotor axis from the ground, respectively. The turbulence intensity (TI) at hub

Figure 2. Wind farm layout in the wind tunnel.

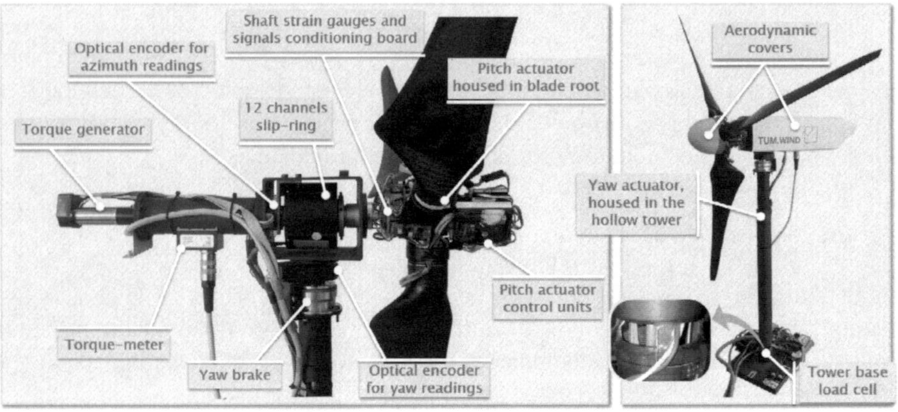

Figure 3. `G1` rotor-nacelle assembly (left) and overall model layout (right).

height was circa 5%. The undisturbed wind speed was measured by means of a pitot tube, also shown in figure 2, placed at hub height and 3D in front of the upstream model.

The dimensions of the model (see figure 3) are a compromise among the need for miniaturization, wind tunnel blockage, Reynolds effects and the need to realize multiple wind turbine interference conditions typical of wind farm operations. The scaled wind turbine is characterized by realistic aerodynamic performance, both at the airfoil and rotor levels, and generates a wake with shape, deficit and recovery that match closely the ones of a full scale machine. Moreover, the model features active individual pitch, torque and yaw control that, together with a comprehensive onboard sensorization of the machine (including measures of shaft and tower loads), enables the testing of modern control strategies.

Each model is controlled by a `M1 Bachmann` module that hard-real-time executes, similarly to what is done on real wind turbines, control laws similar to the ones described in [8] and references therein. In the present study, only operation below rated wind speed was considered. Therefore, the turbines were torque controlled according to a precomputed quadratic relation between rotor speed and torque.

3.2. Model parameter identification

For the identification of the model parameters, the wake velocity of an isolated G1 wind turbine was measured with hot wire probes and compared to model-predicted velocities. In the wind conditions described above (TI of circa 5%), wake measurements were available at hub height at several downwind distances (3D, 4D, 6D, 9D). In this first set of experiments, no wake measurements in yawed condition had been conducted ($\gamma = 0$). Therefore, a second set of experiments had to be used for identifying the parameters that influence the wake in case of yaw-misalignment ($k_{\mathrm{e},\gamma}$ and k_d). In this second set of experiments, TI was much lower (TI of circa 1%), but the wind turbine was operated with yaw-misalignments between $-20° < \gamma < +20°$. The wake velocity was measured at a distance of 4D downwind of the wind turbine, again at hub height.

For the parameter identification, the hot wire velocity measurements V_HW were utilized to solve the minimization problem

$$\min_p \int \left(V_\mathrm{HW}(x) - V_\mathrm{M}(x,p)\right)^2 \,\mathrm{d}x, \tag{10}$$

where x is the lateral position of the measurement, V_M the model-predicted wake velocity for a set of model parameters p. The problem is solved by the Nelder-Mead simplex direct search algorithm implemented in the MATLAB function `fminsearch`. The wind turbine induction was obtained from a G1 blade element momentum simulation and set to $a = 0.35$ for operation below rated wind speed.

For the first set of experiments, the model parameters to be identified are defined as

$$p_1 = \{k_{\mathrm{e},1}, k_{\mathrm{e},2}, k_{\mathrm{e},3}, k_{\mathrm{r},1}, k_{\mathrm{r},2}, k_{\mathrm{r},3}\}. \tag{11}$$

Figure 4 shows the hot wire measurements $V_\mathrm{HW}(x)$ (red dashed line) and the model-predicted wake velocity $V_\mathrm{M}(x,p)$ (blue solid line) for the identified set of parameters. Only measurements at 4D were utilized for the identification (red solid line). The good quality matching of the profiles at 3D, 6D and 9D, since they were not used for calibrating the model, demonstrate its good generality.

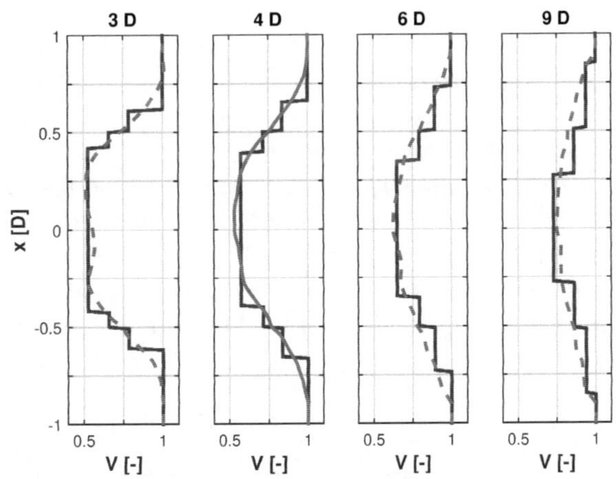

Figure 4. First measurement set (TI circa 5%), modeled (blue solid line) and measured (red solid and dashed lines) wake deficit for different distances behind the wind turbine.

To identify the wake parameters that play a role in turbine yaw-misalignment, the second set of experiments was used and the parameters to be identified by equation (10) were set to be

$$p_2 = \{k_{\mathrm{e},\gamma}, k_\mathrm{d}\}. \tag{12}$$

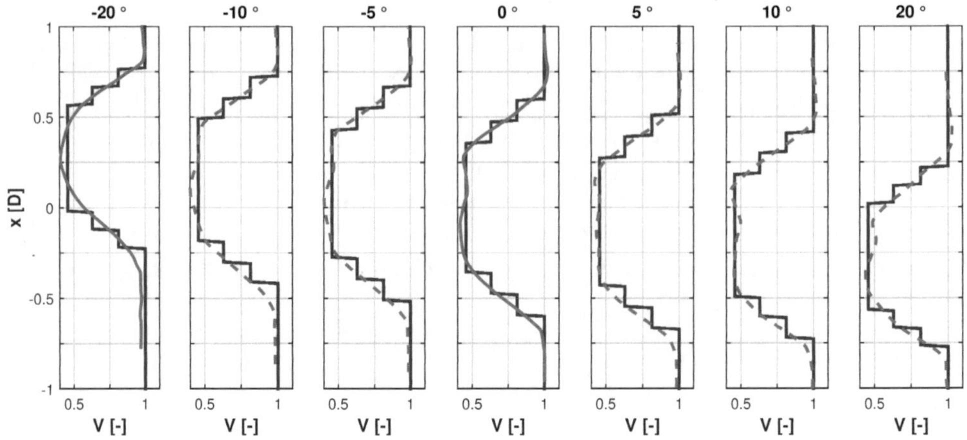

Figure 5. Second measurement set (TI circa 1%), modeled (blue solid line) and measured (red solid and dashed lines) wake deficit for different yaw-misalignments.

The parameters p_1 were re-identified for the low TI case to take the slower wake recovery of this different flow condition into account. Figure 5 shows the hot wire measurements $V_{\mathrm{HW}}(x)$ and the model-predicted wake velocity $V_{\mathrm{M}}(x,p)$ for the identified set of parameters. Again, the generality of the wake model is demonstrated by only taking the measurements of $\gamma = -20°$ and $\gamma = 0°$ into account during the identification. Note the wake diameter reduction in the cases characterized by larger yaw-misalignment. The modeled maximum wake diameter for $\gamma = 0°$ is 1.2D, whereas for $\gamma = \pm 20°$ the wake diameter is only 1D, giving a good fit with the wake measurements.

It is assumed that the parameters $k_{\mathrm{e},\gamma}$ and k_{d} are independent of TI. Therefore, they can be used to describe the wake also for the higher TI cases used in the wind farm experiments.

Parameter k_{p} was identified based on a subset of the wind farm experiments in which the first turbine yaw-misalignment was $-36° < \gamma < 0°$. In this subset, the first wind turbine experimental power coefficient $C_{\mathrm{P,Exp}}(\gamma)$ was calculated based on turbine power measurements and pitot tube measurements 3D in front of the hub, as shown in figure 6 (red circles). The squared error between the measured and modeled power coefficients, given by equation (7), was minimized with respect to the free parameters k_{p} and $C_{\mathrm{P},\gamma=0}$, yielding the modeled power coefficient shown in figure 6 (blue solid line). Coefficient $C_{\mathrm{P},\gamma=0}$ was included in the free parameters to account for a low precision in the pitot tube measurements.

The full set of identified model parameters is reported in Table 1. For the sake of completeness, Table 1 also reports the identified parameters for low TI, which correspond to Figure 5.

Table 1. Identified model parameters.

	$k_{\mathrm{e},1}$	$k_{\mathrm{e},2}$	$k_{\mathrm{e},3}$	$k_{\mathrm{r},1}$	$k_{\mathrm{r},2}$	$k_{\mathrm{r},3}$	$k_{\mathrm{e},\gamma}$	k_{d}	k_{p}
high TI	-0.0251	0.0011	0.0386	0.0320	0.0669	0.2130	2.8808	0.1219	1.7870
low TI	-0.0363	-0.0062	0.0236	0.0140	0.0433	0.1864	2.8808	0.1219	1.7870

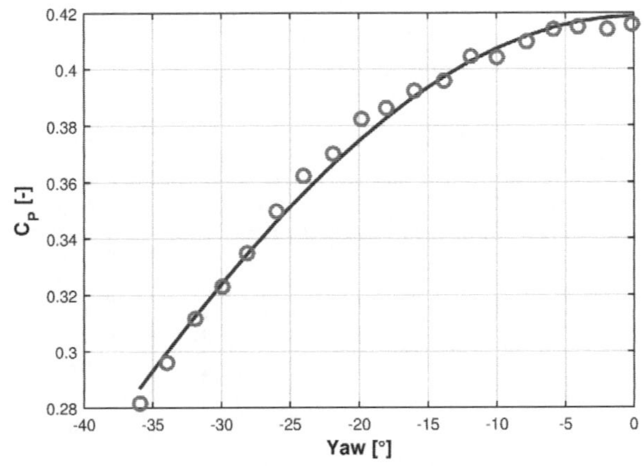

Figure 6. Measured (red circles) and modeled (blue solid line) power coefficients plotted as functions of wind turbine yaw-misalignment.

3.3. Wind farm experiments

In the wind farm experiments, three G1s were operated in the wind tunnel in three different layouts (noted A, B and C) as shown in figure 7. In each layout, the most upwind wind turbine is labeled WT1, the middle turbine is noted WT2, while the most downwind wind turbine is termed WT3. The turbines were torque controlled below rated wind speed, and the reduced order model assumed a constant operation of all turbines with $C_{\mathrm{P},\gamma=0} = 0.419$ and $a = 0.35$.

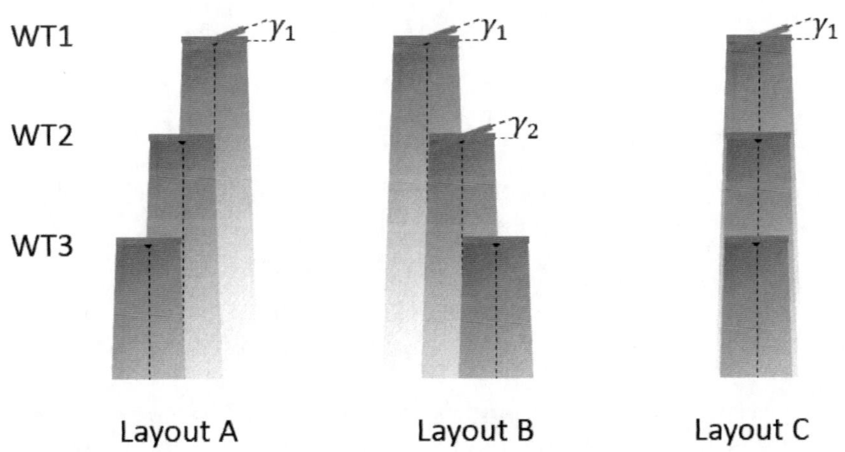

Figure 7. Wind farm layouts A, B, C with lateral displacements of ±0.5D and 0D. The longitudinal displacement is approximately 4D. Note that the sketch is not to scale.

First, for every layout several experiments were conducted with different yaw-misalignments of WT1 (γ_1), in order to deflect the wake away from the downstream turbines. For these experiments, figure 8 shows the power coefficient of all three wind turbines and the total wind farm in layouts A, B and C, as indicated by the column title. The red circles indicate the measured and the blue solid line the modeled power coefficient, respectively.

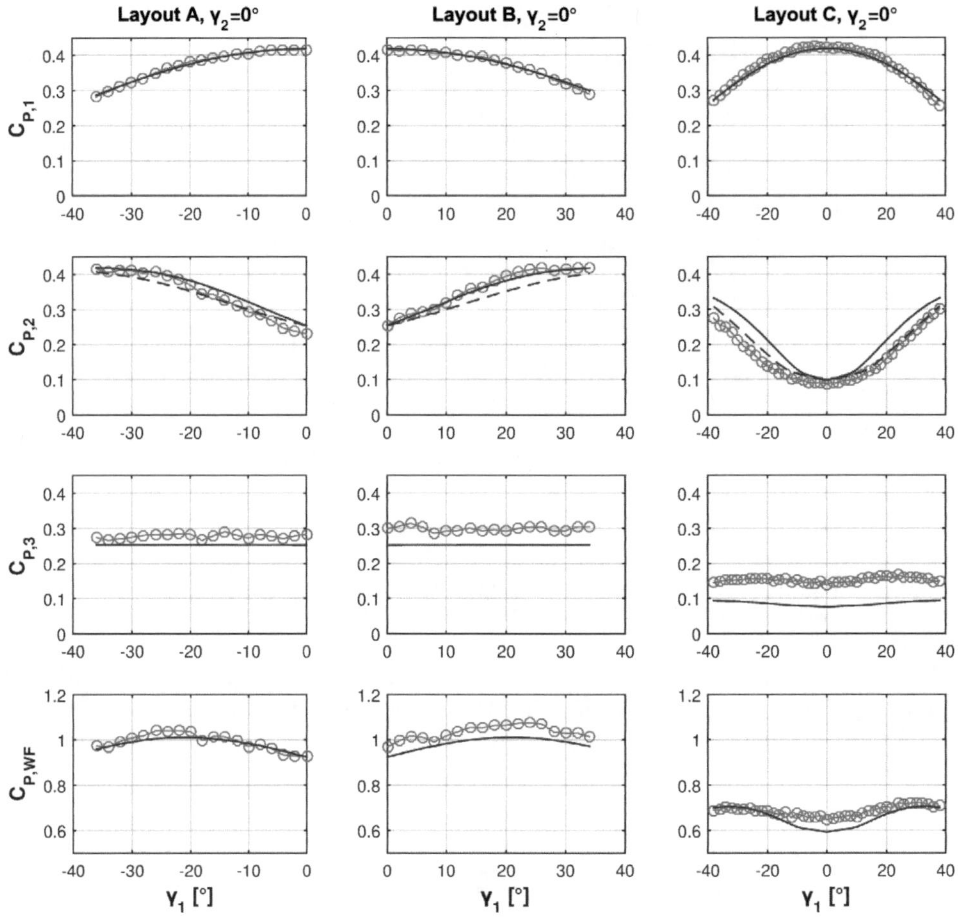

Figure 8. Measured (red circles) and modeled (blue solid line) power coefficient for WT1, WT2, WT3 and the total wind farm, for layouts A, B and C.

The first row of subplots shows the power coefficient of WT1 ($C_{\mathrm{P},1}$) in each layout. As expected from the identification of parameter k_p, the model-predicted and measured results correlate very well.

The second row of subplots shows the power coefficient of WT2 ($C_{\mathrm{P},2}$). For $\gamma_1 = 0°$, the model predicts well the downwind turbine power for all layouts. However, a small asymmetric behavior in the measurements can be observed between the symmetric layouts A and B — perhaps due to asymmetric wake behavior or due to a small horizontal variation in the wind tunnel inflow speed. In the experiments of layout B, maximum power is reached already at $\gamma_1 \approx 25°$ and further upwind turbine yawing does not influence the second wind turbine power anymore. In the full wake case of layout C, the model significantly over-predicts power for $|\gamma_1| > 10°$. A first hypothesis that could explain this behavior is that the model predicts a slightly inaccurate wake location in the yawed cases (which is possible, given that the corresponding parameter was identified at a much lower TI). Indeed, simulations with a larger wake deflection parameter ($k_\mathrm{d} = 0.28$, see blue dashed line) show improved results for layout C, but the new parameter

IOP Conf. Series: Journal of Physics: Conf. Series **854** (2017) 012041 doi:10.1088/1742-6596/854/1/012041

affects also the results in layouts A and B, where now the modeled power coefficient exhibits an increased error. Results in layout C might improve by adjusting the model in such a way that in yaw-misalignment conditions the wake speed decreases or the wake diameter reduces further than currently predicted, but this would worsen the results in layouts A and B. The previously mentioned slight lateral variation in the inflow speed, which cannot be captured by the use of a single pitot tube, could also be a partial cause of the mismatch seen here. It is also possible that the wake deflection position is affected by the downwind turbine position. Understanding the reasons causing these discrepancies will be part of further studies, involving additional wind farm flow measurements.

The third row of subplots shows the power coefficient of WT3 ($C_{\mathrm{P},3}$). In layout A and B, no significant γ_1-dependency can be observed in the experiments as well as in the model. Nevertheless, the power is constantly under-predicted. The cause might be a faster wake recovery of the wake of WT2 that, operating within the wake of WT1, experiences a higher turbulence intensity, leading to a faster wake mixing. Conversely, a strongly deflected wake (i.e. $|\gamma_1| > 25°$) should in that case also lead to lower power at WT3 — an effect that however is not observed in the experiments. Again, further studies and measurements are necessary to better explain this contradiction. For layout C, the power of WT3 is clearly affected by γ_1. The model correctly predicts an increase of power for increased γ_1. However, above $|\gamma_1| > 20°$ the experimental results show a decrease of power, which might be caused by a slower wake recovery of the WT2 wake compared to the full wake case ($\gamma_1 = 0$).

The total power coefficient of all three wind turbines ($C_{\mathrm{P,WF}} = C_{\mathrm{P},1} + C_{\mathrm{P},2} + C_{\mathrm{P},3}$) is shown in the last row of figure 8. Taking into account the discrepancies noted above, the overall correlation is rather good. In layouts A and B the predicted power achieves a maximum for $|\gamma_1| = 20°$, which correlates well with the experimental data. In the full wake case, the predicted power is maximum at $|\gamma_1| = 34°$, which again correlates well with the experimental data.

For layout B, experiments were also conducted in which WT2 is operating in yaw-misalignment (γ_2). Figure 9 shows experiments in which WT2 is yawed by $\gamma_2 = 8°$ (first column), $\gamma_2 = 18°$ (second column) and $\gamma_2 = 28°$ (third column). The power of WT1 is again well predicted. For WT2 the modeled and measured power decreases for increased γ_2, even though the model under-predicts this effect slightly at higher γ_2. The power of WT3 increases as expected with increased γ_2. Surprisingly, for $\gamma_2 = 28°$ the WT3 power exceeds the maximum power coefficient in four experiments — possibly due to a flow acceleration just outside of the wakes of WT1 and WT2. The total power of all three turbines follows the trend of the experiments, but the previously observed power over-prediction becomes smaller for increased γ_2, mainly due to the under estimation at WT2 and the smaller error at WT3.

The full set of experiments, not shown here for brevity, includes all combinations of γ_1 and γ_2 in steps of 2° around the point of maximum power. By using this data, the experimental optimum yaw configuration could be readily identified and was found to be at $\gamma_1 = 20°$ and $\gamma_2 = 16°$. The model-predicted point of maximum wind farm power was on the other hand found to be located at $\gamma_1 = 20°$ and $\gamma_2 = 22°$.

4. Conclusions and outlook

In this paper, the parameters of a reduced order wind farm model were identified with the help of wake measurements along a hub-height horizontal line for an isolated wind turbine. The modeled wakes are in good agreement with the measurements, even though only a small subset of the measurements were taken into account for the model identification procedure.

For three different wind farm layouts, including a variety of yaw angles of the two upstream wind turbines, the model-predicted wind turbine power coefficient was compared with experimental measurements. The comparison shows a good correlation in the overall trends, but the absolute values are not always well predicted especially for the last downstream wind turbine.

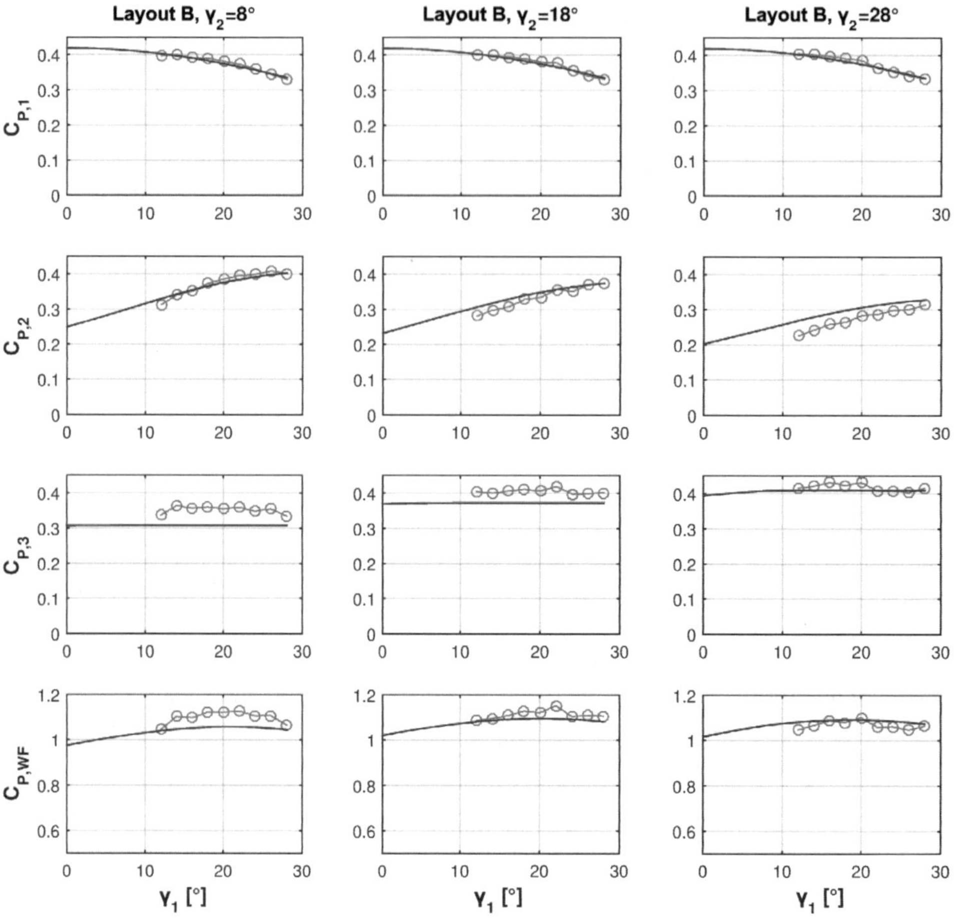

Figure 9. Measured (red circles) and modeled (blue solid line) power coefficient for WT1, WT2, WT3 and the total wind farm, for layout B for different γ_2 values.

The causes are believed to be a combination of various effects, including a faster recovery of wakes shed by waked turbines, flow acceleration outside the wakes, lateral flow speed variations in the wind tunnel, Reynolds number effects in the experiments, the assumption of a uniform power coefficient on each rotor, the assumption of axisymmetric wakes and certainly additional not modeled and fully understood effects like the influence of the downwind turbines on the upstream wake development.

Regardless of these open questions, it is important to note that the employed rather simplistic wind farm model, after a tuning by wake measurements, predicts well the operating point of maximum wind farm power. This is a promising result in view of the use of the model for wind farm control. It is also worth noting that the power gradient is very small around the optimum yaw angles for all studied layouts, implying that for wind farm control purposes a rough estimation of the optimum yaw configuration should be enough to harvest most of its potential.

Future work will try to clarify the open points and improve the model, taking into account the various deficiencies stated above. In addition, rotor load-based wind estimation and wake detection techniques [9] will be coupled with the model for the development of robust model-based closed-loop wind farm control.

Acknowledgments
The authors wish to thank Dr. Vlaho Petrović for his help in the preparation and realization of the experiments, and Mr. Amr Balbaa for the assistance with the parameter identification.

This research was partially supported by the European Union Horizon 2020 research and innovation program under the Marie Skłodowska-Curie grant agreement No. 642108, and it was also partially supported by the German Federal Ministry for Economic Affairs and Energy (BMWi) within the CompactWind project (FKZ: 0325492D).

References
[1] Wang J, Bottasso C L and Campagnolo F 2016 Wake redirection: comparison of analytical, numerical and experimental models *Journal of Physics: Conference Series* **753** 32064
[2] Fleming P A, Gebraad P M O, Lee S, van Wingerden J-W, Johnson K, Churchfield M, Michalakes J, Spalart P and Moriarty P 2014 Evaluating techniques for redirecting turbine wakes using SOWFA *Renewable Energy* **70** 211-8
[3] Campagnolo F, Petrović V, Schreiber J, Nanos E M, Croce A and Bottasso C L 2016 Wind tunnel testing of a closed-loop wake deflection controller for wind farm power maximization *Journal of Physics: Conference Series* **753** 32006
[4] Gebraad P M, Teeuwisse F W, Van Wingerden J W, Fleming P A, Ruben S D, Marden J R and Pao L Y 2016 *Wind Energy* **19** 95–114 ISSN 10954244
[5] Jensen N O 1983 A note on wind generator interaction (Roskilde, Denmark: Risø National Laboratory)
[6] Campagnolo F, Petrović V, Bottasso, C L and Croce A 2016 Wind tunnel testing of wake control strategies *Proceedings of the American Control Conference* 513–518
[7] Campagnolo F, Petrović V, Nanos E M, Tan C W, Bottasso C L, Paek I, Kim H and Kim K 2016 Wind tunnel testing of power maximization control strategies applied to a multi-turbine floating wind power platform *Proceedings of the International Offshore and Polar Engineering Conference* 309–316
[8] Bossanyi, E 2000 The design of closed loop controllers for wind turbines *Wind Energy*, **3** 149–163.
[9] Schreiber J, Cacciola S, Campagnolo F, Petrović V, Mourembles D and Bottasso C L 2016 Wind shear estimation and wake detection by rotor loads - First wind tunnel verification *Journal of Physics: Conference Series* **753** 32027

IOP Conf. Series: Journal of Physics: Conf. Series **854** (2017) 012042 doi:10.1088/1742-6596/854/1/012042

Wind-farm simulation over moderately complex terrain

Antonio Segalini[1] and Francesco Castellani[2]
[1]STandUP for Wind, Linné FLOW Centre, KTH Mechanics, Stockholm, Sweden
[2]Department of Engineering, University of Perugia, Perugia, Italy
E-mail: segalini@mech.kth.se

Abstract. A comparison between three independent software to estimate the power production and the flow field in a wind farm is conducted, validating them against SCADA (Supervisory, Control And Data Acquisition) data. The three software were ORFEUS, WindSim and WAsP: ORFEUS and WAsP are linearised solvers, while WindSim is fully nonlinear. A wake model (namely a prescribed velocity deficit associated to the turbines) is used by WAsP, while ORFEUS and WindSim use the actuator-disc method to account for the turbines presence. The comparison indicates that ORFEUS and WAsP perform slightly better than WindSim in the assessment of the polar efficiency. The wakes simulated with ORFEUS appear more persistent than the ones of WindSim, which uses a two-equation closure model for the turbulence effects.

1. Introduction

In the recent years the number of wind turbines installed onshore increased drastically due to the demand of clean renewable energy. Ideally, it is preferable to install turbines over flat terrains with no roughness, but such ideal sites are rarely available for energy harvesting [1]. However, there are several sites where the terrain is not flat and the roughness affects the wind but the wind resource is sufficient to make the wind-farm project profitable. Such complex terrains have the advantage that they are not very populated and therefore wind farms have a limited social impact. Furthermore, being onshore installations, their cost is approximately half of their equivalent offshore version.

The industrial methods used today to evaluate wind farms were developed for simpler cases such as offshore farms (or onshore over plains) where the inflow profile is approximately homogeneous, with low shear and low turbulence intensity, together with wake models [2, 3] where the wind turbine wakes are superimposed in post-processing to the velocity field estimated over the terrain without the turbines: by means of an assumed wake behaviour (often associated to empirical parameters) and to a law describing how multiple turbine wakes interact, it is possible to estimate rapidly the expected power production. However, a recent study [4] identified that wake models are subjected to significant uncertainty already in offshore cases, while even larger uncertainty is expected onshore since the wakes evolution will be influenced by the terrain condition through vertical displacement and enhanced diffusion. Regarding complex terrain, an interesting test case has recently provided useful indications: in [5], the free flow over the terrain was addressed, while in [6, 7, 8] the combination of wakes and terrain effects

IOP Conf. Series: Journal of Physics: Conf. Series **854** (2017) 012042 doi:10.1088/1742-6596/854/1/012042

was investigated and the significant amount of wake distortion due to the terrain quantified. Consequently, wake-models physics should be improved to properly estimate the wind resource for wind farms placed in complex terrains, although only few studies have been devoted to this issue despite its importance.

An industrial alternative to wake models is provided by computational fluid dynamics (CFD) where the flow-motion equations are solved in a discretised domain at every grid point. Usually, a Reynolds-averaged approach is used, where turbulence affects the mean wind field through the Reynolds stresses, which are modelled in several ways. A two-equation approach ($k - \epsilon$, $k - \omega$ with all their variants) is generally considered to be the one that will provide the most accurate results, although there are several evidences where simpler models have been equally successful, if not even better than the more complex two-equations models [9, 10, 11, 12].

Pushed by the growth of the wind-energy sector, a number of international collaborations have begun to determine the accuracy of the currently-used tools and their approximations by comparing predictions with measurements performed offshore or onshore, although the latter cases are quite rare. The present work is aimed at comparing three numerical models (ORFEUS, a recently-developed linearised code for wind-farm assessment in complex terrain, WindSim, a CFD solver based on finite volumes, and WAsP, an analytical solver with the Park wake model integrated for wind turbines) with experimental data measured in a small wind farm located over a moderately-complex terrain.

2. Numerical codes

2.1. ORFEUS

The new ORFEUS code, developed in 2016 at KTH, is based on a linearised approximation of the Navier-Stokes equations around the undisturbed incoming boundary-layer profile [12]. The Boussinesq approximation is used to model the Reynolds stresses, where the eddy viscosity is prescribed as a linear function of the height from the ground, similarly to the approach of Ott *et al.* [10]: the model does not require the boundary-layer approximation and can indeed account for turbulent diffusion in all directions. Wind turbines and forestry are introduced in the simulation by means of body forces related to the thrust coefficient and to the leaf-area density, respectively. The terrain topography is introduced by means of a coordinate-mapping approach, similarly to Jackson & Hunt [9]. The linearised equations are solved through a spectral approach in a Cartesian grid by means of the Fourier transform in the horizontal plane and Chebyshev polynomials in the vertical non-homogeneous direction, providing a fast and accurate numerical solution of the linearised partial differential equations. Additional details on the equations and the numerical implementation can be found in [13]. In contrast with what claimed in [12, 13], here no thrust enhancement correction has been adopted: the latter was originally proposed in [12] to cope for the discrepancy between linearised and non-linear actuator disc by forcing a higher thrust force to get the same disc velocity. Since the power estimated by ORFEUS was much lower than what measured by the four turbines, it was decided to perform the simulations by using the actual thrust coefficient, so that the applied force is globally the same as in the real farm. This appeared as an easy fix to the problem as the wakes would become weaker and decay faster, although other possibilities might explain the negative bias of ORFEUS, such as the neglect of the wake-meandering phenomenon, discussed in [10].

2.2. WindSim

The numerical CFD simulation of the wakes was implemented using the WindSim model, namely a numerical code based on PHOENICS which solves the Reynolds Averaged Navier-Stokes

(RANS) equations [14]. The General Collocated Velocity method was preferred for the solution of the momentum equations while the RNG-$k-\epsilon$ model was used for turbulence closure. This setup was preferred to have a good convergence but the time for calculation was longer (more than 3 hours for each direction). The turbine wakes are simulated with a simple actuator-disc model, where the axial force applied on the disc surface depends on the wind speed, according to the thrust-coefficient curve of the turbine. Since PHOENICS works on a Cartesian orthogonal grid, the rotor surface could be interpolated only using squared elements.

Wake models could be also used in WinSim, although here the actuator-disc approach was preferred, simulating wind turbines together with the terrain in order to improve the accuracy of the simulations: although the implementation of the actuator disc in WindSim is constrained by the grid orientation (as only body forces orthogonal to the grid can be imposed), the flow field is expected to be well captured in the far wake. This problem does not exist in ORFEUS since the grid is automatically rotated according to the incoming wind direction, and the turbine forces are imposed in a generic volume before the Fourier transformation.

2.3. WAsP

The WAsP (Wind Atlas analysis and application Program [15]) model was additionally used for wakes calculation with the standard Park model [2]. The wake calculations were performed with the standard wake decay constant $k = 0.075$ that is considered appropriate for onshore cases.

3. Wind-farm description

The wind farm studied in the present work is located in the south part of Italy and it is characterised by mild elevation changes, as depicted in figure 1 where the terrain elevation (normalised by the rotor diameter) and the turbines position are reported. Table 1 lists the position of the turbines and their name: the turbine T1 is here used as the origin of the adopted reference frame. The ground is covered by olive trees with an equivalent roughness height of $z_0 = 0.3$ m.

The four turbines have a rated power of 2 MW and a rotor diameter of $D = 100$ m. The thrust coefficient in the investigated range of wind speeds is $C_T \approx 0.81$. SCADA (Supervisory, Control And Data Acquisition) data is available for the wind turbines for more than one year, where the wind speed and direction are measured at the nacelle of every turbine. A SCADA system is collecting and storing measurements on a 10 minutes basis. Thanks to the availability of such operational data, a comparison and validation of the numerical results is possible. Nevertheless, filtering of the database is necessary in order to have a fair comparison with the models. Data filtering was done through the following steps:

(i) data were at first selected when all the four turbines were running for more than 90% of each 10 minute interval.

(ii) a second filter was applied on the wind speed in order to have a nacelle wind speed for each machine below 9 m/s. The SCADA data were therefore filtered within the region in which the thrust coefficient is 0.81 ± 0.08 (below the rated wind speed), ensuring that a consistent statistics of farm efficiency can be observed in similar conditions;

(iii) finally, the database was prepared for the directional analysis by filtering on twelve sectors with an overall width of 15 degrees, as a good compromise between statistical convergence and direction sensitivity. The direction signal of the vane mounted on the nacelle of the most-upstream wind turbine was used as the reference flow direction.

Since no information was available on the atmospheric stratification condition, the incoming boundary layer was assumed to be neutral.

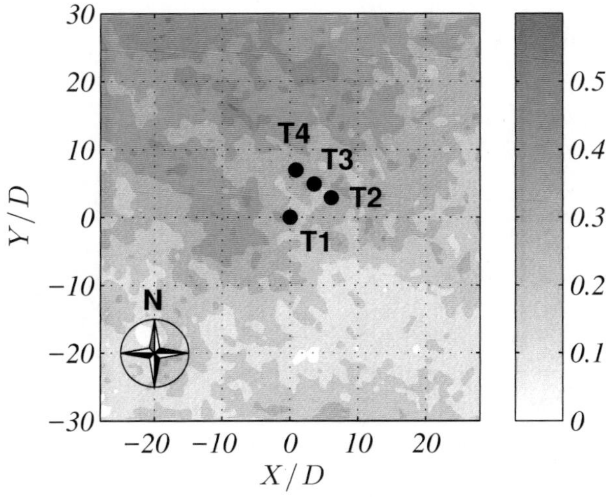

Figure 1. Topography of the investigated terrain and position of the four turbines present. The colorbar indicates the terrain elevation normalised by the turbines diameter.

Table 1. Position of the available turbines.

Name	X/D	Y/D	Z_{hub}/D
T1	0	0	0.8
T2	6.05	2.90	0.8
T3	3.52	4.94	0.8
T4	0.88	6.98	0.8

In ORFEUS the terrain is discretised by 1024×1024 grid points distributed over an area of $160D \times 120D$. 80 logarithmically spaced grid points were used in the vertical direction up to 50D. For the sake of simplicity, a homogeneous roughness of $z_0/D = 0.003$ was assumed, neglecting the slightly roughness variation of the real terrain. The simulation of one wind sector took less than 5 minutes. WindSim had a grid composed by 1.9 million cells (197 × 214 × 45 in the streamwise, spanwise and vertical direction, respectively) with a minimum horizontal grid spacing of 5.5 m.

For each of the three software, the output of the simulation for each direction sector was used to estimate the polar efficiency as well as the normalised power for each turbine.

4. Results

The efficiency of the farm is one of the most important parameters to be quantified for a given layout. For an onshore wind farm it can be defined as [16]

$$\eta(\theta) = \frac{1}{N_t P_{\max}} \sum_{i=1}^{N_t} P_i \,, \tag{1}$$

where P_i is the power of the ith turbine, and $P_{\max}$ is the maximum power of the $N_t = 4$ turbines. θ indicates the wind direction. The advantage of this definition is that the efficiency remains a number below 1 for any terrain and wind-farm arrangement. It can be expected that the

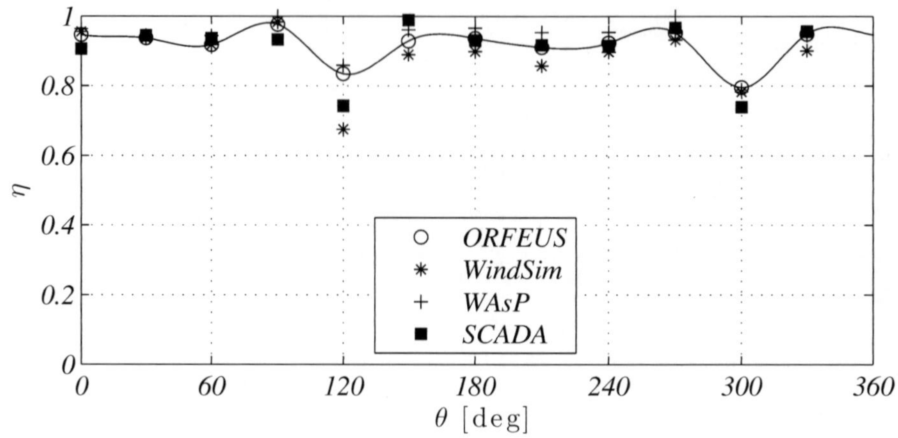

Figure 2. Polar efficiency of the investigated farm estimated from ORFEUS, WindSim and WAsP against SCADA data.

polar efficiency should weakly depend on the wind-speed magnitude, facilitating the averaging process of the SCADA data and reducing the number of simulations needed: this might not be the case if one or more turbines operate above rated speed, as the thrust coefficient might reduce significantly, but these events were removed by the adopted filtering procedure.

Figure 2 shows the polar efficiency obtained from the three available codes against the filtered SCADA data. The experimental data indicate that the polar efficiency is nearly constant to $\eta \approx 0.85$, although two deeps exist at $\theta \approx 120°$ and $300°$, corresponding to the wind directions where the turbines T2, T3 and T4 are aligned and, consequently, affected by significant wake effects. All codes identify correctly these two deeps, although with large errors for $\theta = 120°$ while the specular case with $\theta = 300°$ does not show such large discrepancies. WindSim seems to be more pessimistic than ORFEUS and WAsP in the polar-efficiency estimation. The overestimation of WAsP is a consequence of the wake-model approach, since the upstream field is not affected by the turbines presence, and no blockage effect is accounted for. The opposite happens for WindSim and ORFEUS that are elliptic codes and can account for upstream effects.

In order to quantify the systematic error, table 2 reports the relative error in the estimation of the polar efficiency of the three investigated software against the available SCADA data. Interestingly, it is generally observable that both ORFEUS and WAsP perform better than WindSim, although the discrepancy might also be due to uncertainty in the processing of SCADA data. In any case, an average error of 3.5%, 5.0% and 4.7% affects ORFEUS, WindSim and WAsP, respectively.

In order to understand the origin of the various deeps, and to see how the different wakes are simulated by ORFEUS and WindSim, figure 3 shows the comparison for several wind directions between the two software: a reasonably good agreement between the two software is observed both qualitatively and quantitatively. The wakes simulated by ORFEUS appear stronger and more persistent than the ones of WindSim, probably due to the $k-\epsilon$ closure scheme of the latter software. The sectors $\theta = 120°$ and $\theta = 300°$ are characterised by the overlap of the wakes of T2, T3 and T4, although the wake deficit does not increase in WindSim as it does for ORFEUS.

The normalised power produced by each turbine is reported in figure 4 for the same 6 wind directions analysed in figure 3: ORFEUS and WAsP provide in general a good estimation, although slightly positively biased. WindSim appears more conservative with the largest error for $\theta = 120°$.

IOP Conf. Series: Journal of Physics: Conf. Series **854** (2017) 012042 doi:10.1088/1742-6596/854/1/012042

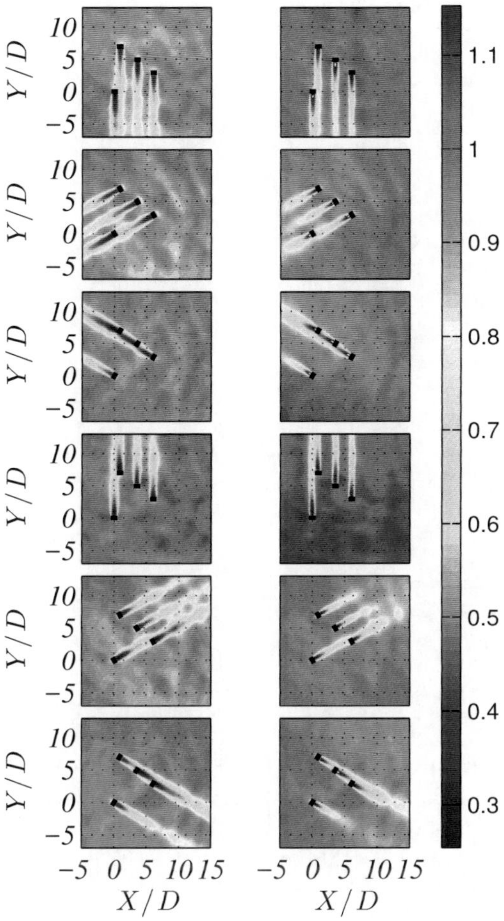

Figure 3. Wind speed (normalised by the reference speed of $U_{\text{ref}} = 5.9$ m/s) on a plane parallel to the ground located at hub height for the wind directions 0°, 60°, 120°, 180°, 240° and 300° (increasing from top to bottom of the figure). The positions of the turbines and their radial extension are indicated by the black thick lines. (*Left*) Field obtained with ORFEUS. (*Right*) Field obtained with WindSim.

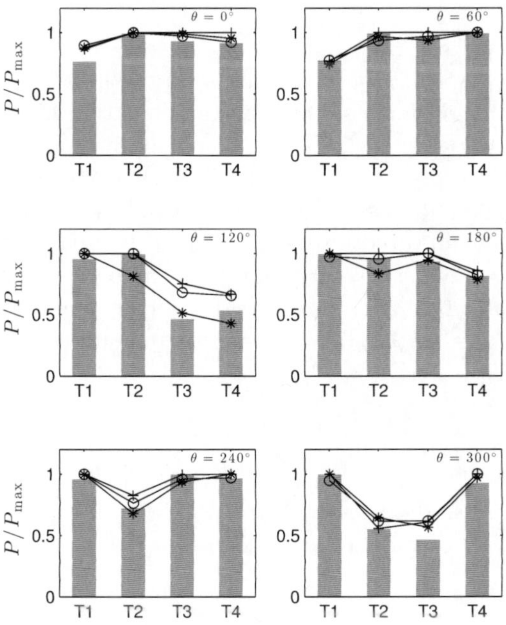

Figure 4. Normalised powers for different wind directions. (*bars*) SCADA data, (*circles*) ORFEUS, (*asterisks*) WindSim, (*plusses*) WAsP.

Table 2. Relative error (in percent) in the polar efficiency between the three software and the SCADA data.

θ [deg]	ORFEUS	WindSim	WAsP
0	4.4	5.3	6.6
30	-0.8	-0.7	0.3
60	-1.9	-2.6	0.6
90	4.6	5.0	7.2
120	12.3	-7.3	15.3
150	-6.1	-10.5	-2.9
180	0.7	-4.2	3.6
210	-0.7	-6.1	3.9
240	0.9	-1.3	4.4
270	-1.5	-3.2	3.4
300	7.6	7.6	7.1
330	-0.8	-6.0	-1.1

5. Conclusions

In the present work a comparison between three independent software to assess the energy yield of a wind farm placed over a moderately complex terrain has been performed. Two linearised software (ORFEUS and WAsP) were compared against a non-linear tool (WindSim): being linearised, the first two allow for a faster and numerically-efficient solution (ORFEUS is based on spectral elements, providing a high-accuracy numerical solution of the involved equations, while WAsP is based on an analytical solution) compared to WindSim, that fully accounts for the nonlinearity of the flow-motion equations with a solution performed with the finite-volume method. The turbulence closure schemes of the three codes is substantially different as WAsP is based on the mixing-length approach [15], ORFEUS on a prescribed linear eddy viscosity and WindSim on a $k - \epsilon$ model. WAsP uses the Jensen wake model to account for the turbines, while WindSim and ORFEUS simulate the turbines by means of the actuator-disc approach.

According to the present results, ORFEUS and WAsP perform slightly better than WindSim in the assessment of the polar efficiency, although positively biased. Despite the simplifications, the velocity field estimated by the ORFEUS code is close to the one simulated with WindSim, with wind-turbine wakes that are slightly more persistent (and less diffusive) than what WindSim provides. Interestingly, while all software for almost all wind directions were affected by an error of $\pm 5\%$, they all showed significant deviations for $\theta = 120°$, corresponding to a waked sector, although the specular waked sector $\theta = 300°$ was not affected by the same error. The reason for this might be due to the processing of the SCADA data or to the different upstream conditions, although more data is maybe required to sort this out properly.

Acknowledgments
The authors acknowledge the Italian company Lucky Wind spa for providing the production data and all the technical information used in the validation of the methods.

Bibliography

[1] Alfredsson P H and Segalini A 2017 *Phil. Trans. R. Soc. A* **375** 20160096
[2] Katić I, Høstrup J and Jensen N O 1986 A simple model for cluster efficiency *Proceedings of the European Wind Energy*
[3] Bastankhah M and Porté-Agel F 2014 *Ren. Energy* **70** 116–123
[4] Walker K, Adams N, Gribben B, Gellatly B, Nygaard N G, Henderson A, Jimémez M M, Schmidt S R, Rodriguez J R, Paredes D, Harrington G, Connell N, Peronne O, Cordoba M, Housley P, Cussons R, Håkansson M, Knauer A and Maguire E 2015 *Wind Energy* **19** 979–996
[5] Castellani F, Astolfi D, Burlando M and Terzi L 2015 *Journal of Wind Energy and Industrial Aerodynamics* **147** 320–329
[6] Politis E, Prospathopoulos J, Cabezon D, Hansen K, Chaviaropoulos P and Barthelmie R 2012 *Wind Energy* **15** 161–182
[7] Castellani F, Astolfi D, Piccioni E and Terzi L 2015 Numerical and experimental methods for wake flow analysis in complex terrain *Journal of Physics: Conference Series* vol 625 (IOP Publishing) p 012042
[8] Castellani F, Astolfi D, Mana M, Piccioni E, Becchetti M and Terzi L 2017 *Wind Energy* **10.1002/we.2094**
[9] Jackson P S and Hunt J C R 1975 *Quart. J. R. Met. Soc.* **101** 929–955
[10] Ott S, Berg J and Nielsen M 2011 Linearised CFD Models for Wakes Tech. rep. Risø National Laboratory
[11] Segalini A, Nakamura T and Fukagata K 2016 *Boundary-Layer Meteorol.* **161** 439–460
[12] Ebenhoch R, Muro B, Dahlberg J Å, Hägglund P B and Segalini A 2016 *Wind Energy* **DOI: 10.1002/we.2067**
[13] Segalini A 2017 *Phil. Trans. R. Soc. A* **375** 20160099
[14] Pep M, Gravdahl A and Romero M 2003 Wind flow over complex terrain: application of linear and CFD models *European wind energy conference and exhibition, 16-19 June, Madrid, Spain*
[15] Troen I and Lundtang Petersen E 1989 European wind atlas Tech. rep. Roskilde: Risø National Laboratory
[16] Castellani F, Astolfi D, Terzi L, Hansen K and Rodrigo J 2014 Analysing wind farm efficiency on complex terrains *Journal of Physics: Conference Series* vol 524 (IOP Publishing) p 012142

IOP Conf. Series: Journal of Physics: Conf. Series **854** (2017) 012043 doi:10.1088/1742-6596/854/1/012043

Influence of Tip Speed Ratio on Wake Flow Characteristics Utilizing Fully Resolved CFD Methodology

M. Salman Siddiqui[1], Adil Rasheed[2], Trond Kvamsdal[1,2] and Mandar Tabib[2]

[1] Department of Mathematical Sciences, Norwegian University of Science and Technology, NO-7491 Trondheim, Norway

[2] SINTEF ICT, Department of Applied Mathematics, Postboks 4760 Sluppen, NO-7465 Trondheim, Norway

E-mail: muhammad.siddiqui@ntnu.no, adil.rasheed@sintef.no, trond.kvamsdal@ntnu.no, mandar.tabib@sintef.no

Abstract. Dominant flow structures in the wake region behind the turbine employed in the Blind Test campaign [1], [2] is investigated numerically. The effect on the wake configuration at variable operating conditions are studied. The importance of the introduction of turbine tower inside the numerical framework is highlighted. High-fidelity simulations are performed with Multiple Reference Frame (MRF) numerical methodology. A thorough comparison among the cases is presented, and the wake evolution is analyzed at variable stations downstream of the turbine. Streamlines of flow field traveled towards ground adjacent to turbine tower and strongly dependent on the operating tip speed ratio. Wake is composed of tower shadow superimposed by rotor wake. Shadow of the tower varies from x/R=2 until x/R=4 and breaks down into small vortices with the interaction of rotor wake. This study also shows that the wake distribution consists of two zones; inner zone composed of disturbances generated by blade root, nacelle and the tower, and an outer zone consisting of tip vortices.

1. Introduction and objective

The complex wind turbine wake dynamics, especially in the near wake, is hindering the denser layout of wind turbines inside a farm [3], [4]. It is, therefore, vital to resolve the wake to study the complex physics inside wind farms, depending on the rotor size and the blade aerodynamic characteristics. To perform such task, there are various types of methods available depending on the level of complexity to evaluate the aerodynamic behavior of wind turbines [5]. One of the most widely used analytical technique is the use of the Blade Element Momentum (BEM) theory. Experimentation, on the other hand is rigorous and complex, however, it is considered important to check the validity of numerical simulations and BEM tools [6], [7]. Due to an ever increasing size of wind turbine, it is sometimes not possible to validate the results at full scale under standard conditions [8]. Recent advancement of computers nowadays, the Computational Fluid Dynamics (CFD) is becoming a vital alternative[9]. It also offers certain advantages concerning qualitative analysis, visualizations of pressure distribution on turbine blade surfaces and evaluation of the dynamic behavior under variable inflow conditions can be explicitly formulated [5]. It is not unique in the sense of computational cost. However, the results from the latest numerical methods give a clear depiction of flow field under real time [10], [11]. However, still, CFD methods rely on correct experimental data for the initial validation.

Therefore, to study the wake dynamics we utilized the experimental results conducted by the fellow researchers at Norwegian University of Sciences & Technology (NTNU) [12]. These experimental campaigns, known as *Blind Test*, have become popular for benchmarking and validating the performance evaluation of small wind turbines [1], [2], [13], [14], [15]. Previous studies also incorporated the effect of tower and tunnel blockage using actuator line modeling [16], [17]. To this end, we start by comparing MRF computational methodology for modeling of the flow field in wind turbines [18], [19]. Exact experimental conditions are formulated, and the ability of the present method to predict the turbine's performance under various operating conditions is evaluated. Moreover, near wake dynamics is explicitly represented with the qualitative plots of vorticity and velocity distribution in different positions in the downstream direction.

2. Governing Equations

The flow around wind turbines, even the very large ones, is still essentially incompressible with Mach numbers, hardly exceeding 0.25 based on blade tip speed. Hence, application of the incompressible form of Navier Stokes equation using a Multiple Reference Frame (MRF) is reasonably justified. Adopting the MRF approach, the computational domain is divided into two concentric zones: rotational and stationary with the former containing the rotating turbine. The governing equations are slightly modified and added with the source terms to be compatible with the reference frame in which they are solved [20]. In the stationary zone the following set

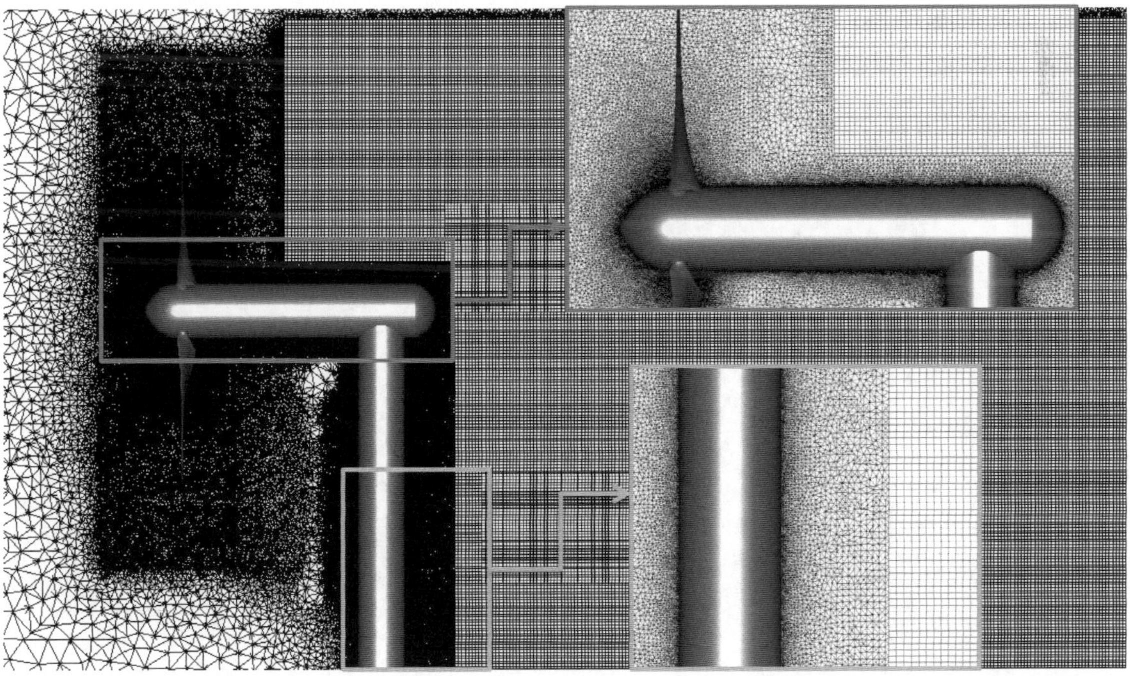

Figure 1. NTNU Blind Test T1: Computational mesh consisting of hexahedral cells in the wake block and tetrahedral cells in the rest of domain. Prism extrusion is employed to allow finer mesh grading near the turbine surface. Mesh consist of 40×10^6 elements.

of equations for mass and momentum are solved:

$$\nabla \cdot \mathbf{u_a} = 0 \tag{1}$$

$$\nabla \cdot (\mathbf{u_a} \otimes \mathbf{u_a}) = -\nabla p + \nabla \cdot (\nu + \nu_t)\nabla(\mathbf{u_a} + (\nabla \mathbf{u_a})^{\mathbf{T}}) \tag{2}$$

where $\mathbf{u_a}$ is the absolute velocity as seen from a stationary reference frame. In the rotating rotational zone the same equations are rewritten in terms of the relative speed $\mathbf{u_r}$ (relative to the rotating frame of reference) given by:

$$\nabla \cdot \mathbf{u_r} = 0 \tag{3}$$

$$\nabla \cdot (\mathbf{u_r} \otimes \mathbf{u_r}) + \mathbf{2\Omega} \times \mathbf{u_R} + \mathbf{\Omega} \times (\mathbf{\Omega} \times \mathbf{r}) = -\nabla p + \nabla \cdot (\nu + \nu_t)\nabla(\mathbf{u_r} + (\nabla \mathbf{u_r})^{\mathbf{T}}) \tag{4}$$

$$\mathbf{u_a} = \mathbf{u_r} + \mathbf{\Omega} \times \mathbf{r} \tag{5}$$

Where $\mathbf{\Omega}$ is the rotational speed of the reference frame with a stationary observer (here it will also be equal to the rotational speed of the turbine), p is pressure, ν is the kinematic viscosity of the air and ν_t is the turbulent kinetic viscosity. Since MRF computes a steady state solution, the temporal derivatives are ignored in the equations. However, the certain loss in accuracy is compensated with the substantial decrease in the solution time. The computational efficiency is considered ideal for the solution of wind turbine simulations at various operating conditions. For turbulence modeling, the $k-\omega SST$ model [21] with standard wall functions are used.

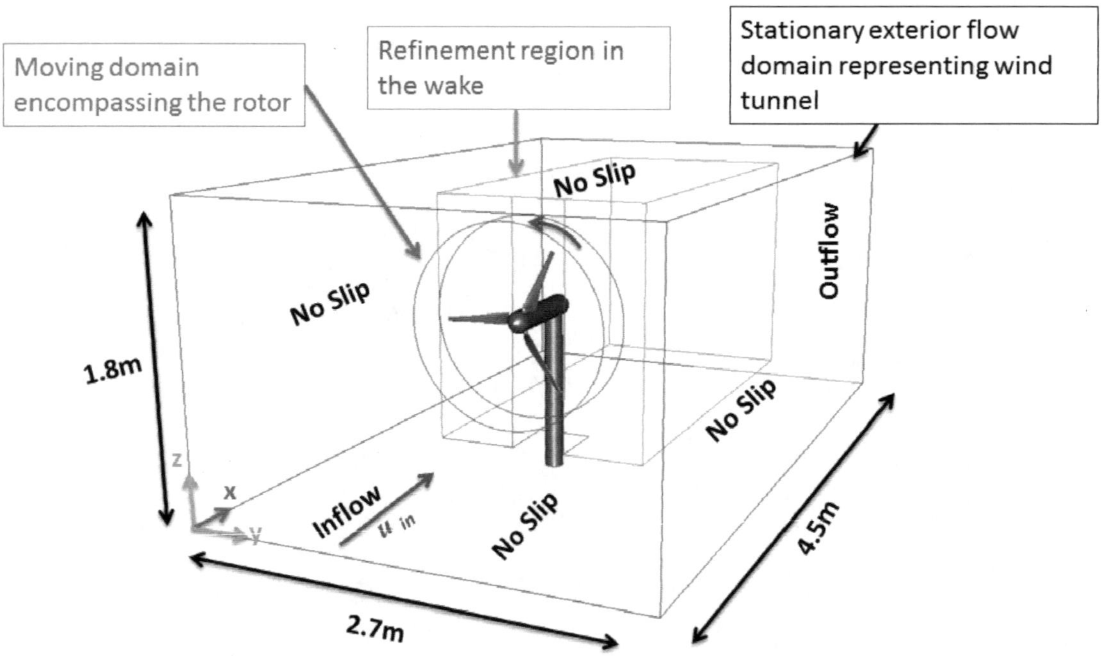

Figure 2. NTNU Blind Test T1: Computational setup and boundary conditions of full problem domain. The extent of the domain is $2.7 \times 4.5 \times 1.8$. Two zone approach is used with stationary and moving regions.

IOP Conf. Series: Journal of Physics: Conf. Series **854** (2017) 012043 doi:10.1088/1742-6596/854/1/012043

3. Approach and methods

3.1. CAD model

The geometry of the blind test turbine blade consists of 14% thick airfoil made from the standard NREL S8265MW profile with a constant twist along the spanwise direction [22]. Three such blades of 0.89m length each with very high stiffness values constitute a rotor. The high stiffness of the blades allows for neglecting the deformation in the analysis. Exact representation of experimental turbine geometry is followed; however, sharp corners in the actual geometry at the junction between blade root and nacelle are smoothened to avoid the need of excessive mesh refinements, thereby, lowering the computational cost considerably. It is assumed that such modification will not influence the analysis of wake behavior undertaken in this work. However, any modification of the sharp blade tip is avoided because of its sensitivity towards the blades aerodynamic behavior. The CAD model of the turbine is presented in Figure 1.

3.2. Domain and Mesh

A hybrid finite element mesh is generated with a smooth transition from hexahedral cells close to the blade surface to the tetrahedral elements away from it. While the former is used to model boundary layer accurately and to have better control over the y^+ (used according to the standard wall function $y^+ \approx 30$), the latter is employed because of the associated convenience towards local refinements (Figure 1). To capture the wake evolution with better accuracy, a wake block is generated with highly refined structured elements from x/R=1.5 until x/R=7. This way an accurate representation of near wake can be evaluated with an overall mesh size of around 40 $\times 10^6$ cells. The generated grid resolution is approximately two times higher than the most refined test case submitted in Blind Test campaign (CMR Prototech 26 $\times 10^6$ [1]). To construct the computational setup, we have adopted a two zones approach: stationary zone (4.5m×2.7m×1.8m) representing the wind tunnel and rotating region encompassing the turbine rotor. The relative tolerance between the zone is kept low (≈ 0.0001) to allow for accurate flux transfer. The computational domain is presented in Figure 2

3.3. Boundary Conditions

Simulations are conducted for uniform inlet velocity profile, applied on the inlet face and for different Tip Speed Ratio ($TSR = R\omega/U_{inf}$). The TSR is changed by adjusting the rotational speed of the turbine. At the outlet face, a standard outlet boundary condition is employed. The coupling between the rotating and stationary zone is distinguished with an interface. The walls of the wind tunnel along with the turbine surface is applied with the no-slip boundary condition, where the wall functions are employed to work with relatively coarse mesh [23]. Figure 2 represents the boundary conditions used on different faces of the computational domain, which are chosen based on the boundary conditions imposed in the experimental setup [2], [14], [22].

3.4. Computational setup and Solver

The computational setup is based on the experimental investigation described in [2][22]. Fluid parameters are selected to comply with standard operating conditions with the reference fluid density $\rho = 1.225 kg/m^3$, dynamic viscosity $\mu = 1.82 \times 10^{-5} kg/m.s$, and incoming wind velocity $U_\infty = 10 m/s$ and under varying turbine rotational speed $\Omega(rad/s) = 67, 134, 233$, which corresponds to the TSR of 3, 6, 10 respectively. The parameters of turbulent kinetic energy (k) and rate of dissipation (ω) are varied until the desired representation of turbulent intensity at the turbine location is achieved [7]. The solver is created in OpenFOAM-2.3.0 (OF) [20]. To ensure continuity, OF uses an elliptic equation for the modified pressure which involves combining the continuity equation with the divergence of the momentum equation. This elliptic equation along with the momentum equation and turbulence equation are solved in a segregated manner using the Semi-Implicit Method for Pressure-Linked Equations (SIMPLE) algorithm. The OF uses a finite volume discretization technique, and all the equations are integrated over control volumes (CV) using Green-Gauss divergence theorem [4]. The Gauss divergence theorem converts the volume integral of the divergence of a variable into a surface integral oft the (normal component for vector valued) variable over faces comprising the CV. Thus, the divergence term defining the convection terms can simply be computed using the face values of variables in the CV. The face values of variables are obtained from their neighboring cell centered values by using convective scheme. In this work, all the equations (except k and turbulence equations) use second order linear discretization scheme, while the turbulent equations use upwind convection schemes. Similarly, the diffusion term involving Laplacian operator (the divergence of the gradient) is simplified to compute the gradient of the variable at the faces. The gradient term can be split into the contribution from the orthogonal and the non-orthogonal parts, and both these contributions are accounted for. The computations are performed on 48 cores on a 3.2 GHz Intel(R) Xeon (R) CPU machine and achieved convergence in approximately 1.5 weeks for each tip speed.

4. Result and Discussion results

4.1. Validation study

Validation against the experiment is performed to check the credibility of the simulation methodology. For the comparison of wake deficit profiles, data along the horizontal and vertical lines behind the turbine center at a location of $z = -0.95m$ to $z = +0.95m$ are extracted. Wake deficit profiles (U = 1 - U_{wake}/U_{ref}) shown in Figure 3 corresponds to a TSR=6. Hence we may infer that we have achieved a closer agreement between the given experimental data and numerical computed results. The better agreement obtained in the current investigation can be attributed to the inclusion of the nacelle and the tower in the simulation. Most of the previous studies did not include the nacelle and the tower due to computational complexity. On the other hand the MRF methodology used herein ignores the physics associated with the unsteadiness in the flow, and that can partially explain the differences between the given experimental data and the numerical simulated results. The use of the Sliding Mesh Interface (SMI) instead of MRF would enabled a better representation of the unsteadiness in the flow. The horizontal asymmetry across rotor center is successfully captured by the present simulations, the inherent reason of which is still investigated but currently unknown at present.

Having developed a sound basis through a quantitative comparison, the opportunity is

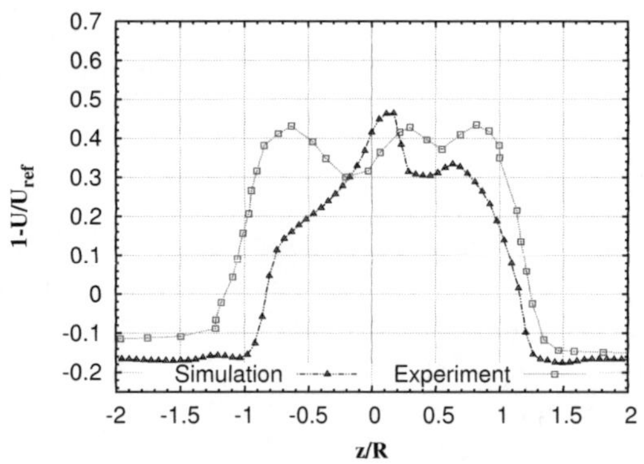

(a) Horizontal

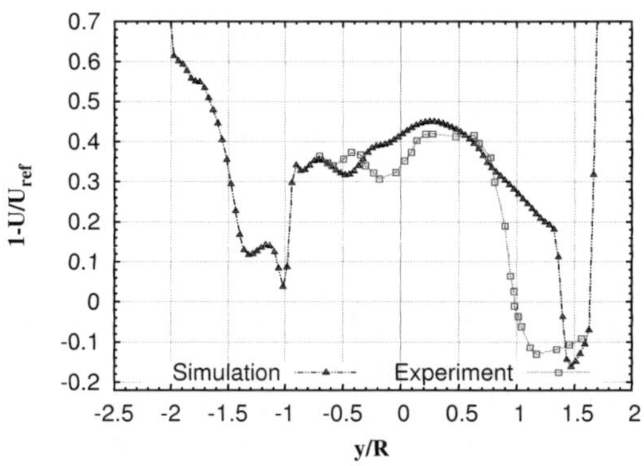

(b) Vertical

Figure 3. NTNU Blind Test T1: Validation against experimental results of Krogstad *et al.* [1] at design operating conditions of TSR=6 and at the downstream station x/R=6

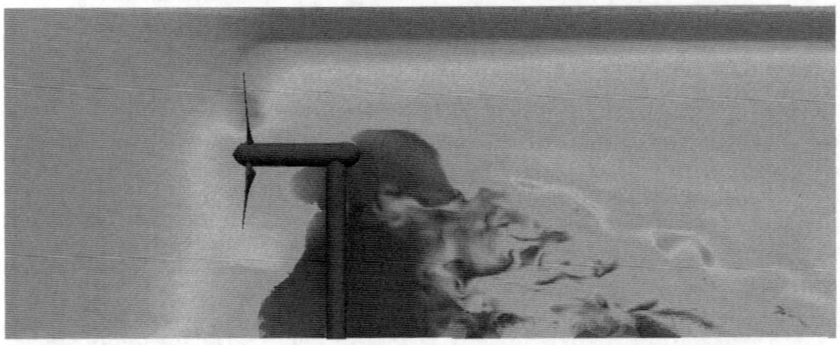

(a) TSR=3

(b) TSR=6

(c) TSR=10

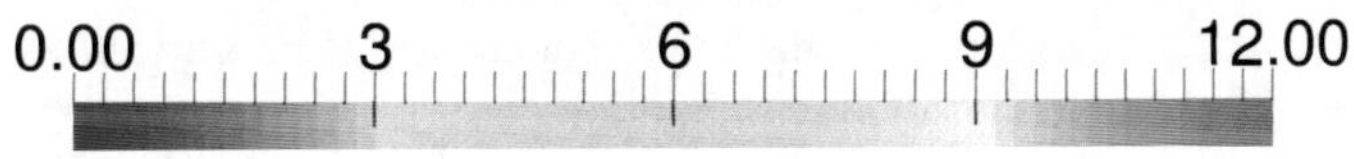

Figure 4. NTNU Blind Test T1: Velocity magnitude (m/s) (inertial reference frame) at three tip speed across the turbine. Tower shadow is clearly visible along with flow acceleration at near the wind tunnel walls. Rapid flow acceleration is also observed near the hub at higher tip speeds. *units:(m/s)

exploited to evaluate the qualitative behavior of wakes at different operating conditions in the next section.

4.2. Flow dynamics versus Tip Speed Ratio (TSR)

Figure 4 presents the contour of wind speed at a 2D vertical plane passing through the center of the nacelle and aligned in the streamwise direction for the three TSRs. There are two mechanisms by which complicated flow structures are generated: One due to the movement of air around the cylindrical turbine tower including the nacelle and second due to the churning action of the rotor. These flow structures though independently generated soon interacting with each other resulting in smaller structures. This interaction causes turbulent mixing resulting in faster diffusion of the flow structures ultimately leading to full wake recovery. The restoration pattern of velocity deficit in the presence of turbine tower, therefore, depends on the operating tip speed. Wake recovery length at $TSR = 3$ due to the presence of tower is observed to be $x/R = 4$, whereas the length increases to $x/R = 5$ at the $TSR = 10$. For the designed operating conditions of $TSR = 6$, the length lies somewhere in between the corresponding values at $TSR = 3$ and $TSR = 10$. The presence of the tower results in highly unpredictable wake behavior on the lee-side of the turbine also reported in [22].

Streamlines of velocity are found to deviate towards the ground after the interaction with the turbine structure as shown in Figure 5. This phenomenon was also reported by the authors who performed the wind tunnel experiments [2]. A similar trend is observed at each tip speed, and a complex helical streamline pattern is identified. The streamlines are found to be significantly more complex at lower TSR where the major portion of the blade is inside the stall regime [1]. On the other hand, less complicated wake structure is observed at the design TSR=6, and the streamlines are concentrated behind the turbine tower. In contrast, at TSR=10, an identical wake structure similar to TSR=3 is observed, however, the streamlines have lost their concentration and expanded indicating an increase in the overall wake width.

Constant velocity deficit is recognized in the rotor wake for the tip speed of 6. However, under off-design conditions, a significantly larger variation of flow distribution near the rotor blades is discovered. One of the interesting observation is the generation of high velocities near the turbine root at TSR of 10. It depicts that the turbine has started to work like a propeller increasing the flow velocity in the wake. Moreover, an acceleration in the flow is also observed in the streamwise direction between the rotor edge and the wind tunnel walls. The magnitude of this velocity is found to increase with the increasing tip speed. It is attributed to the blockage effect present inside the wind tunnel. This effect becomes significant at higher tip speeds where the rotor produces resistance equivalent to a solid wall to the incoming flow. Due to which a tremendous increase in the velocity between the rotor and wind tunnel walls is observed.

4.3. Formation of wake structures vs. TSR

High turbulence level pockets are generated on the lee-side of the turbine due to the presence of turbine structure as shown in Figure 6. These pockets are more significant in near wake region, where large disturbances are observed inside the wake core behind the nacelle and tower in comparison to the rotor. Wake generated from the stationary and rotating components of the turbine is distinguishable up to $x/R = 2$, where rotor shadow is separately identifiable. However, the wake starts to lose its character in the downstream direction and breaks down into

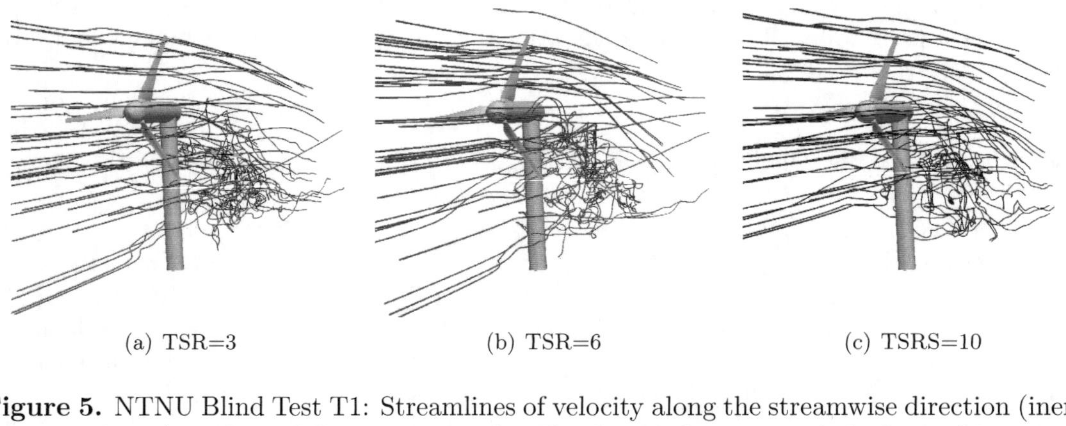

(a) TSR=3 (b) TSR=6 (c) TSRS=10

Figure 5. NTNU Blind Test T1: Streamlines of velocity along the streamwise direction (inertial reference frame) at three different tip speeds. The flow deflection towards the turbine tower is clearly visible.

IOP Conf. Series: Journal of Physics: Conf. Series **854** (2017) 012043 doi:10.1088/1742-6596/854/1/012043

smaller vortices at $x/R = 4$. Spanwise diffusion becomes active at this stage which makes the flow more homogeneous. While moving further downstream at $x/R = 6$ the difference rather becomes insignificant, and the two wakes started to diffuse together to form a combine wake.

The contours of velocity deficit recognized the asymmetry in the wake with stronger velocity reported on the left-hand side as compared to the right hand side (see Figure 6). Hence the contours also identify asymmetry in the horizontal wake profile across rotor, which is mentioned in the section 4.1. The tower wake rotates in the opposite direction to the rotor and is the prime reason for this behavior. Whereas for the off-design conditions similar lateral distribution of velocity is observed. However, the wakes overall width increases and the wake asymmetry is found to get reduced at higher TSR as compared to lower values. Moreover, the wake recovery length is examined to be considerably faster at higher tip speed as compared to smaller ones, the reason of this behavior is explained in the upcoming section.

4.4. Tower effect versus Tip Speed Ratio (TSR)

The turbine tower introduces significant variations in the flow field which enhance turbulence mixing. The distribution of fluctuations is such that intermittent spiral shape structures are

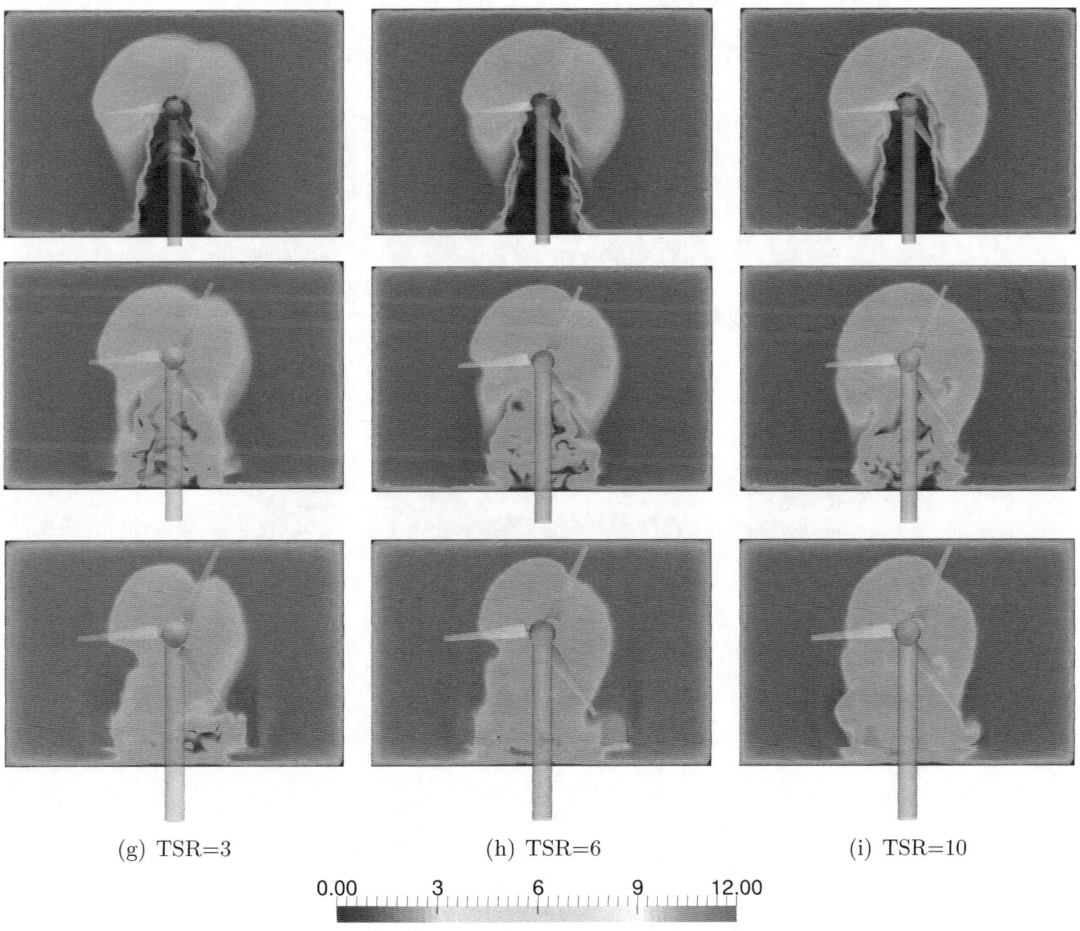

Figure 6. NTNU Blind Test T1: Wake structure characterization (m/s) on x/R=2(top–row), x/R=4(intermediate–row) and x/R=6(bottom–row) at various TSR. Rotor and tower wake is clearly identified with wake expansion at higher tip speeds.

IOP Conf. Series: Journal of Physics: Conf. Series **854** (2017) 012043 doi:10.1088/1742-6596/854/1/012043

found to emerge across the cylindrical tower. It later develops into a vortex pairing in the downstream direction, where the magnitude of vorticity is found to be more behind the central region as compared to the outer region consisting of blades tip. The complex inner flow configuration is primarily composed of vortices generated by turbine root, nacelle, and the tower. The flow structures travel downwards and begin to diffuse together with the interaction and breaks into smaller vortices at x/R=3. The tip vortices, on the other hand, remains undisturbed until x/R=6, thereafter they begin to interact with the central vortex street. Overall wake profile obtained at this station have more thickness and become slightly shifted leftward from the turbine axis. It happens mainly because of the opposite directions of the tower and the rotor

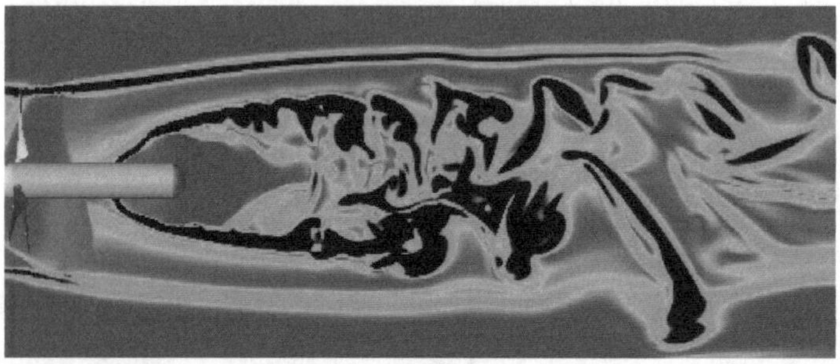

(a) TSR=3

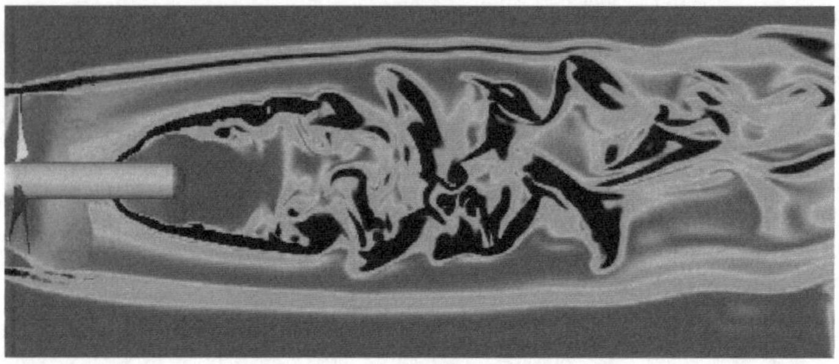

(b) TSR=6

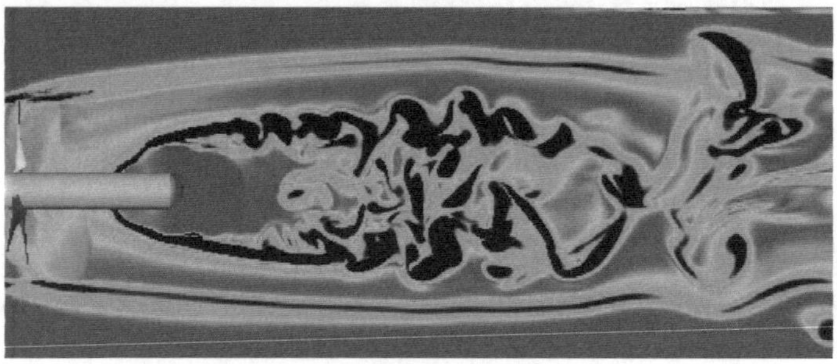

(c) TSR=10

Figure 7. NTNU Blind Test T1: Vorticity (1/s) distribution from turbine tower and blade tips (inertial reference frame) at three different tip speeds. Significant vortices emanating from the turbine tip and central tower can be observed. The characteristics of wake vorticity vary with the operating tip speed.

wake.

Gradients in the flow field are lower in the central wake at x/R=6 as compared to x/R=2. This uneven distribution causes molecular diffusion and turbulent mixing which enhance wake recovery. From the Figure 7, a significant drop in the wakes strength is observed at x/R=6 for all tip speed. This region is classified as the transition zone from near wake to the complicated, far wake. Tower shadow is also recognized from the plots and is determined by a region of zero vorticity distribution near the tower. Whereas different configuration of vortices emanating from turbine structure are observed corresponding to a particular tip speed. Significantly large vortex structures are observed to originate across the cylindrical tower at lower tip speed (TSR=3). The formation evolves and break down into smaller structures while advecting downwards in the wake. However, they require more downstream distance before the fluctuations get dissipated. Whereas the vortex street at larger speed (TSR=10) results into smaller vortex structures, which disappear quick and require less distance downstream. The design condition (TSR=6) are corresponding to an average vortex pairing behavior in comparison to the off-design operating conditions.

5. CONCLUSION

Numerical investigation of wake distribution behind NTNU blind test turbine 1 was conducted under various operating conditions. Dominant components inside the near wake were identified. Reynolds Average Navier Stokes (RANS) simulations were performed with Multiple Reference Frame (MRF) methodology and k-ω SST turbulence model. The full turbine was modeled, and the boundary layer was fully resolved throughout the turbine geometry. Close agreement of mean velocity profiles in the horizontal and vertical direction was observed in the experiments. The results presented in the paper highlights the following significant observations.

- Fast recovery of velocity was observed due to the presence of turbine tower inside the computational framework. It caused an increased turbulent kinetic energy level in the wake adjacent to the turbine.
- Tower shadow comprised of region with significant higher velocity deficit. It got broken with the interaction of rotor wake and started to move towards the ground.
- Tower wake became insignificant and lost its character at x/R=6 and overall diffused wake was observed. This distance was also known as the end of near wake region.
- Vortices generated from the tower comprised of large structures at lower tip speed, where significantly smaller vortices were recognized at higher values.
- Flow streamlines followed a downward trend towards the ground at every tip speed. More concentrated streamline configurations were distinguished at lower tip speed as compared to higher values.
- Region of high velocity was recognized near blade root at higher tip speed values, which showed that turbine was enhancing flow velocity by working as a propeller.

Acknowledgments

The authors acknowledge the financial support from the Norwegian Research Council and the industrial partners of NOWITECH: Norwegian Research Centre for Offshore Wind Technology (Grant No.:193823/S60) (http://www.nowitech.no) and FSI-WT (Grant No.:216465/E20)(http://www.fsi-wt.no). Furthermore, the authors greatly acknowledge the Norwegian metacenter for computational science (NOTUR-reference number: NN9322K/1589)(www.notur.no) for giving us access to the Vilje high performance computer at the Norwegian University of Science and Technology (NTNU).

References

[1] Krogstad P A and Eriksen P E 2013 *Renewable Energy* **50** 325 – 333
[2] Krogstad P A and Saetran L 2012 *Wind Energy* **15** 443–457
[3] Siddiqui M S, Durrani N and Akhtar I 2013 *ASME Power Conference* **2** V002T09A020
[4] Cavar D, Rthor P E, Bechmann A, Sørensen N, Martinez B, Zahle F, Berg J and Kelly M 2016 *Wind Energy Science Discussions* **1** 5570
[5] Siddiqui M S, Durrani N and Akhtar I 2015 *Renewable Energy* **74** 661–670
[6] Heinz J C, Sørensen N N and Zahle F 2016 *Wind Energy* **19** 22052221 ISSN 1095-4244
[7] Siddiqui M S, Rasheed A, Kvamsdal T and Tabib M 2015 *Energy Procedia* **80** 312–320
[8] Siddiqui M S, Durrani N and Akhtar I 2013 *ASME Power Conference* **2** V002T09A020(11)
[9] Sørensen N N and Zahle F 2016 *Journal of Physics: Conference Series (Online)* **753** ISSN 1742-6596
[10] Nordanger K, Holdahl R, Kvarving A M, Rasheed A and Kvamsdal T 2015 *Computer Methods in Applied Mechanics and Engineering* **284** 664–688
[11] Nordanger K, Holdahl R, Kvamsdal T, Kvarving A M and Rasheed A 2015 *Computer Methods in Applied Mechanics and Engineering* **290** 183 – 208
[12] Eriksen P E and ge Krogstad P 2012 *Energy Procedia* **24** 378 – 384
[13] Oggiano L 2014 *Energy Procedia* **58** 111 – 116
[14] Pierella F, Krogstad P A and Saetran L 2014 *Renewable Energy* **70** 62–77
[15] Sreenivas K, Mittal A, Hereth L, Taylor L K and Hilbert C B 2016 *Journal of Wind Engineering and Industrial Aerodynamics* **157** 145 – 157
[16] Sarlak H, Nishino T, Martnez-Tossas L, Meneveau C and Sørensen J 2016 *Renewable Energy* **93** 340352
[17] Sarmast S, HSChivaee, Ivanell S and Mikkelsen R 2014 *In Journal of Physics: Conference Series* **524, No.1** 012137
[18] Siddiqui M S, Rasheed A, Tabib M and Kvamsdal T January 5-11, 2017 *35th Wind Energy Symposium AIAA SciTech Forum*
[19] Siddiqui M S, Rasheed A, Tabib M and Kvamsdal T 2016 *Journal of Physics: Conference Series* **753** 032059
[20] Jasak H 5-8 January, 2008 *47th AIAA Aerospace Sciences Meeting* **52** (AIAA 2009–341)
[21] Menter F 2009 *International Journal of Computational Fluid Dynamics* **23** 305–316
[22] Krogstad P A, Saetran L and Adaramola M S 2015 *Journal of Fluids and Structures* **52** 65–80
[23] Wilcox D 1994 *AIAA Journal. 32* 247–255

Wake Conference 2017 IOP Publishing
IOP Conf. Series: Journal of Physics: Conf. Series **854** (2017) 012044 doi:10.1088/1742-6596/854/1/012044

Near wake region of an industrial scale wind turbine: comparing LES-ALM with LES-SMI simulations using data mining (POD)

Mandar Tabib[1], M. Salman Siddiqui[1,2], Eivind Fonn[1], Adil Rasheed[1], Trond Kvamsdal[1,2]

[1] SINTEF Digital,Mathematics and Cybernetics, Postboks 4760 Sluppen, NO-7465 Trondheim, Norway

[2] Department of Mathematical Sciences, Norwegian University of Science and Technology, NO-7491 Trondheim, Norway

E-mail: mandar.tabib@sintef.no,adil.rasheed@sintef.no

Abstract. Accurate prediction of power generation capability needs proper assessment of blade loading and wake behavior. In this regard, the Sliding Mesh Interface (SMI) approach and the Actuator Line Model (ALM) are two diverse computational fluid dynamics (CFD) based approaches of simulating the turbine behavior, each having its own merits and demerits. The SMI technique simulates the unsteady flow by explicitly modeling the blades and their rotation using a dynamic mesh, while in Actuator Line Model, the blades are not modeled explicitly but each blade is resolved as a rotating line (made of N actuator segments), over which the forces are computed. The current work focuses on simulating an industrial scale reference turbine and in differentiating the near wake dynamics predicted by these two approaches using Large Eddy Simulation (LES) and Proper Orthogonal Decomposition (POD) technique (a data mining tool). Initially, the ALM is compared with FAST model for the prediction of variation of power coefficient with the Tip Speed Ratio (TSR). The ALM is able to capture the varying trend and it predicts a similar optimum tip speed ratio as the FAST model. At this optimum TSR condition, the ALM is compared with the SMI method for a study limited to the near wake region. Comparisons between SMI and ALM shows that : (a) The SMI is predicting more complex 3D nature of the flow, and (b) the POD shows that ALM captures the shear regions of wake but it does not capture the vast compendium of length and time scales of eddies as SMI does. However, despite these limitations, the ALM has been able to capture the qualitative trend in wake deficit and the power coefficient variation with tip speed.

1. Introduction and objective

Wake dynamics have been shown to influence the power production capabilities of downstream turbines depending upon the inter-turbine distance in a wind farm layout [1–5]. Actuator Line Model (ALM) with Large Eddy Simulation (LES) turbulence model has been used popularly to understand the wake dynamics in wind farms as it is computationally tractable to do so for multi-turbine set-ups. However, it is well-known that ALM does not explicitly resolve turbine blades and hence might not be expected to be accurate enough. On the other hand, the Sliding Mesh Interface (SMI) approach resolves the blade and is expected to be more accurate but it is computationally intractable to perform for a set-up with multiple turbines. In this regard, the current work compares the SMI and ALM methods for one industrial scale turbine. The NREL 5MW reference turbine [6] is chosen for this work, as it is a realistic and standardized industrial-scale off-shore turbine model. NREL 5 MW turbine [6] [7] consists of three $63m$ long blades defined in terms of eight cross sectional profiles (DU21, DU25, DU30, DU35, DU40 and NACA64) and twist angles at different locations away from the hub (as described in [6]). These multiple-sections with diverse angles of attack provide an ideal opportunity to test and benchmark models and methodologies that can later be applied to solve a bigger range of industrial challenges. Hence, it is popularly used by leading groups (several of USA DOEs Wind and Hydro-power Technologies Programs, EU's UpWind research program, and the International Energy Agency (IEA)'s Wind Annex XXIII Subtask Offshore Code Comparison Collaboration) to test methodologies. Such industrial scale wind-turbine blades comprising of multiple sections exhibit complex flow-patterns along the blade length [8] and generate wakes influencing wind-farm operations. In this regard, a comparison of wake dynamics and power production simulation results between sliding mesh approach (SMI) and Actuator line model will be useful to understand the comparative predictive performance of model. Thus, the objectives of this work are :

1.1. Objectives

(i) Compare the performance of ALM-LES turbulence model with FAST model in predicting power coefficient variation with varying tip speed ratio.

(ii) Compare the performance of ALM-LES turbulence model with SMI-LES approach in capturing wake dynamics at the optimum tip speed ratio.

2. Approach and Methods

The study is limited to the near wake region. The approach involves comparison of ALM and SMI approaches. Initially, the ALM is compared with FAST model for the prediction of variation of power coefficient with Tip Speed Ratio (TSR). This is done for the NREL 5 MW reference turbine. Then at the optimum TSR, ALM and SMI are compared for predicting wake structure and flow patterns. The wake dynamics simulated by SMI and ALM for the optimim $TSR = 7.5$ are compared by analyzing mean wake deficit profiles and by using Proper Orthogonal Decomposition (POD). Due to lack of experimental data of wake deficit velocity profiles for NREL 5 MW reference turbine, it is not possible to do validation of the models, so we resort to a verification exercise (i.e. comparison of different models). The governing equations and application details of SMI, ALM and POD is explained next section.

2.1. Sliding Mesh Interface (SMI) approach

The conservation equations are solved on a moving mesh (a sliding mesh) to account for the impeller motion and to capture unsteady interactions between rotating part and stationary bafes. This approach involves dividing the computational domain into a rotating and a stationary zone using a sliding interface. The approach assumes that the rotating computational mesh moves (or slides) relative to the stationary frame. The terms of the governing equations are integrated on a control volume, and the effect of the moving control volume is accounted by including the mesh motion flux in the computation of face mass flux in the convective terms (in other words the convective terms in the governing equation are modied). Thus, the face mass fluxes in the convective terms have to be computed relative to the mesh motion flux (ie. the volume swept by moving cell face during its movement with the cell face velocity). This mesh motion flux is computed based on the space (or volume) conservation law, which ensures that the moving face velocity is calculated from the face-centered positions, such that the surface vectors as well as calculation volumes inside rotating part remain constant. The mesh position in the rotating domain is updated after every time step (as it changes with impeller rotation) and so are the cell face positions at the sliding interface. At the sliding interface, a conservative interpolation is used for both mass and momentum using a set of fictitious control volumes. The governing equations are solved only in an inertial reference frame. Following equations are solved throughout the domain for SMI approach using LES turbulence model.

$$\nabla \cdot \mathbf{u_a} = 0 \tag{1}$$

$$\frac{\partial \mathbf{u_a}}{\partial t} + \nabla \cdot (\mathbf{u_a} \otimes \mathbf{u_a}) = -\nabla p + \nabla \cdot (\nu + \nu_t)\nabla(\mathbf{u_a} + (\nabla \mathbf{u_a})^{\mathbf{T}}) \tag{2}$$

2.2. Actuator Line Model (ALM)

The turbine is modeled using actuator line model (ALM) approach, which was first developed by Sørensen and Shen [9]. The actuator line model (Equations 3, 4 and 5) uses the velocity field input from the CFD model and outputs body force, which are used as the sink term in the momentum equation. The ALM approach resolves each blade of the turbine as a rotating line (made of N actuator segments), over which the forces are computed. The forces at each segment comprise of lift force and drag forces (Equations 3, 4), which are computed from the local relative velocity (V_{rel}), local twist angle, blade chord (c), local actuator width (w) and local angle of attack (α) at a given actuator segment. The local angle of attack is computed from the tangential and normal component of relative velocity at the segment. The lift coefficient (C_l) and drag coefficient (C_d) at each segment (in Equations 3, 4 and 5) are a function of local angle of attack, and this dependency is provided as an input (blade airfoil data) to the ALM model. The force at an actuator segment i (f_i) is a point force and it is translated on to the fluid domain as a volumetric body force (F_i) using Gaussian projection (Equation 5). The regularization parameter (ε) in Equation 5 represents the width of the Gaussian and determines the concentration of the force. Larger the parameter, more smoothed out the force is on the flow field. The negative sign in Equation 5 accounts for the fact that the force exerted by turbine on the flow field is equal and opposite to the force experienced by it due to the flow. At the location (x, y, z) of the fluid domain, the overall body force is summation of force over all N actuator segments of the turbine, where (x_j, y_j, z_j) is the location of the jth segment and r_j is the distance between segment j and the fluid domain location.

$$L = \frac{1}{2} C_l(\alpha) \rho V_{rel}^2 cw \tag{3}$$

$$D = \frac{1}{2} C_d(\alpha) \rho V_{rel}^2 cw \tag{4}$$

$$F_i^T(x, y, z) = -\sum_{j=1}^{N} f_i^T(x_j, y_j, z_j, t) \frac{1}{\varepsilon^3 \pi^{3/2}} exp\left[-\frac{r_j^2}{\varepsilon}\right] \tag{5}$$

2.2.1. Numerical implementation - Computational domain, mesh and parameters The turbulence is modeled using a one-equation sub-grid scale (SGS) turbulent kinetic energy LES model for both SMI and ALM. Figure 1 shows the computational domain, mesh and suitability of mesh. The adequacy of mesh for LES Simulations can be judged through the contour of the ratio of sub-grid-scale-kinetic-energy k_{sgs} to total-kinetic-energy k_{total} (as seen in figure 1(c)-1(d)). In LES simulations, the mesh-size determines spatial filtering and establishes cut-off between resolved and unresolved (modeled sub-grid scale) parts of flow. Finer the mesh, more part of flow is resolved and more accurate are the simulation. As per criteria of Pope (2000) [10], for a well-resolved LES, less than 20% of the total kinetic energy should be modeled sub-grid-scale part (i.e. k_{sgs}/k_{total} ratio should be less than 0.2). Figure 1(c)-1(d) shows this criteria being satisfied for ALM and for most regions in SMI case. The domain size for SMI and ALM simulations (figure 1(a) and 1(b)) are kept similar: $200m \times 120m(streamwise) \times 200m$ for SMI and $200m \times 190m(streamwisedirection) \times 200m$ for ALM. SMI uses a mesh of 8 million cell elements comprising of mostly tetrahedral cells with few layers of prismatic cells to capture the blade boundary layer region, while the ALM uses a grid of 1.5 Million cells with about 50 grid points across the rotor diameter (thus minimum grid length is about $2.5m$). This grid should be sufficient to resolve wake tip vortex [11–13]. For SMI, the blade surface has been treated as wall with no-slip boundary and employs a wall function based on Spalding's law [14] that gives a continuous kinematic viscosity profile to the wall over wide range of y^+. This is required because the average y^+ value near blade wall is 667 while the minimum y^+ value is 11 and in terms of grid size - the minimum wall normal grid size is $0.008m$ near blade. Simulations are conducted with a uniform inlet wind velocity of $9m/s$ applied on the inlet face at different TSRs. The TSR is changed by adjusting the rotational speed of the turbine while keeping the inlet wind velocity constant. At the outlet face a standard outlet boundary condition is used (fixed pressure value and zero normal gradient for rest). A free slip boundary condition is applied on the rest of the surfaces. For the actuator line method, about 40 actuator segments (i.e parameter N in equation 5) were used in the simulation. The regularization parameter (ε in equation 5)in ALM is chosen to be about two times the cube root of grid volume size in that region. This is selected so as to ensure that the force is not overly concentrated to cause numerical oscillations / solver instability, and neither does the force becomes too smoothed so as to cause no resistance to the wind flowing through the turbine. For both the cases, the LES runs are initialized from the results from Reynolds Averaged Navier Stokes (RANS) simulations. For SMI, the initial RANS results are run for a total time corresponding to about four revolutions. For ALM, the RANS results are run for $100s$ of simulation time (which corresponds to the passage of 17 revolutions of the turbine and approximately 5 times flow passage through domain at wind free stream velocity). For ALM, the LES results are run for $300s$ of simulation time

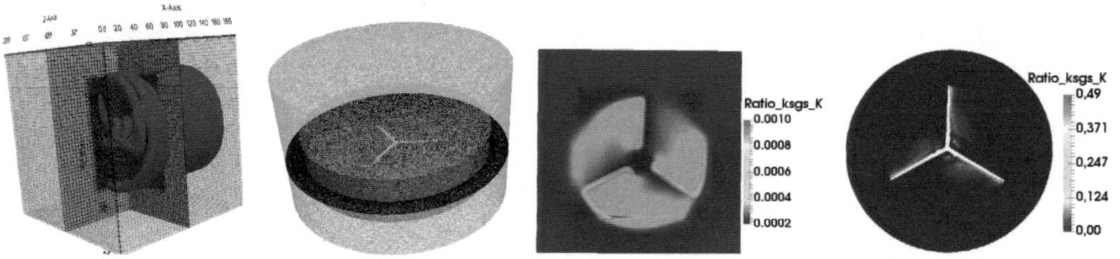

(a) ALM (b) SMI (c) ALM-Mesh Adequacy (d) SMI-Mesh Adequacy

Figure 1. Domain and Mesh information for SMI and ALM

(which corresponds to passage of 51 revolutions of turbine and approximately 14 times flow passage through domain at wind free stream velocity). The ALM simulations run at a time step of around $0.007s$ (corresponding to courant number of 0.1 which is constrained by the tip speed). The averaging was done for last $100s$. The ALM simulations are run with 16 cores on a 3.2 GHz Intel(R) Xeon (R) CPU machine and took one week of computational time. For SMI, the LES results are run for a total time of $32s$ corresponding to nearly six revolutions (and the flow developed after 4 revolutions). All the SMI LES computations are performed on 256 cores of a 2.6GHz Intel(R) Xeon(R) CPU machine on $Vilje$, the high performance computational facility at Norwegian University of Science and Technology. The simulation time-step reached a lower value of 6e-05 s in-order to maintain the Courant number of 1. The simulation took around 3 weeks of computational time on the *Vilje*. Regarding solver, OpenFOAM was used for both SMI and ALM simulations. The equations are solved in a segregated manner using the Semi-Implicit Method for Pressure-Linked Equations (SIMPLE) algorithm. In this work, all the equations (except k_{sgs}) use second order linear convection discretization scheme. Similarly, the gradient term computation at cell faces accounts for both the orthogonal and the non-orthogonal parts.

2.3. Proper Orthogonal Decomposition (POD)

For the computation of the POD modes two dimensional snapshots of any variable (velocity components here) is required. The N snapshots are represented by $\mathbf{U} = [\mathbf{u^1}, \mathbf{u^2} \cdots \mathbf{u^N}]$ which is used to compute the covariance matrix given by $\mathbf{C} = \mathbf{U^T U}$. After this an eigenvalue problem $\mathbf{C}\mathbf{A}^i = \lambda_i \mathbf{A}^i$ is solved to obtain the eigenvalues λ^i and eigen vectors $\mathbf{A}^i$ which are sorted in a decreasing order as $\lambda_1 > \lambda_2 > \cdots > \lambda_N$. POD modes are then computed as

$$\phi^i = \frac{\sum_{n=1}^{N} A_n^i \mathbf{u}^n}{||\sum_{n=1}^{N} A_n^i \mathbf{u}^n||}, i = 1, \cdots, N \tag{6}$$

With POD modes arranged as $\Psi = [\phi^1 \phi^2 \cdots \phi^N]$. POD coefficients a_i can be found from the snapshot n as $\mathbf{a}^n = \Psi^T \mathbf{u}^n$. From this a snapshot can be reconstructed as $\mathbf{u}^n = \Psi \mathbf{a}^n$. Relative energy given by any i^{th} mode is given by $\lambda_i / \sum_{j=1}^{N} \lambda_j$. POD has been successfully applied to understand wake dynamics [15, 16].

3. Result and Discussion results

3.1. Verification study: Power coefficient (C_p) Vs tip speed ratio(TSR)

For the NREL 5 MW turbine, Jonkman et.al. ([6]) studied the evolution of the power coefficient as a function of the TSR and blade-pitch surface by running FAST [17] with AeroDyn simulations. From these simulations, they found that the peak power coefficient of 0.482 occurred at a TSR of 7.55 and a rotor-collective blade-pitch angle of 0.0°. Similar study was carried out by Rannam Chaaban (data available at https://wind.nrel.gov/forum/wind/viewtopic.php?t=582) and he predicted the optimum C_p to occur at $TSR = 8$. Furthermore. Rannam's study has been used for the verification of our ALM and SMI model in Figure 2. The ALM method follows a similar trend and gives optimum power coefficient at a the TSR of 7.5 but it over-predicts the optimum power coefficient value (around 0.56). Figure 2 shows a comparison between the variation in power coefficient predicted by ALM and FAST methods. It is possible to obtain similar results as FAST by tuning regularization parameter ϵ. However, this is not done in the current work as it's a verification exercise and not a validation exercise, and all models can differ from the exact experimental observation (which are currently not available for NREL 5 MW reference turbine). The SMI method predicts a $C_p = 0.54$ at $TSR = 7.5$. The next verification is done for wind deficit.

3.2. Wake dynamics with tip speed ratios

The ALM has also been used to compare the impact of TSR on mean wind deficit (figure 4). Figure 4 compares normalized mean wind deficit $(\Delta U/U_{inflow}) = (U_{inflow} - U)/U_{inflow}$ variations at four locations downstream of the turbine ($0.3R, 0.45R, 0.9R$ and $1.3R$, where R stands for the radius of turbine = $63m$). In all the four figures, the wind deficit at $TSR = 7.5$ is higher in the region $1.1 > z/R > 0.1$ as compared to $TSR = 9$ and $TSR = 6$. The lowest wind deficit values $(\Delta U/U_{inflow})$ in the region $(1.1 > z/R > 0.1)$ at all the four downstream locations is for $TSR = 6$. This could be explained on the basis of the energy extraction by the turbine. At

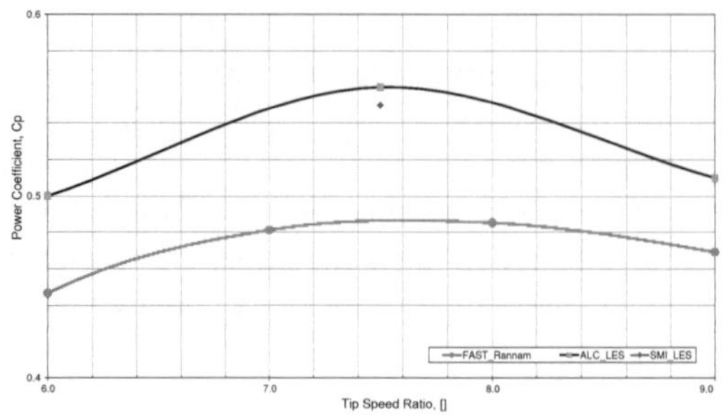

Figure 2. Verification - power coefficient versus tip speed ratio for ALM, SMI and FAST.

an optimum TSR of 7.5, more energy is extracted by the blades in that region leading to higher wind deficit and also higher C_p, while as we increase or decrease the tip speed ratio (towards 9 or 6), the energy extracted is lesser resulting in lower wind deficit and lower C_p. The wake deficit by SMI for TSR 7.5 is compared with ALM in (figure 4), and it differs in two ways, firstly SMI predicts a higher wind deficit in the near hub region ($0.1 > z/R$) at locations $0.3R$, $0.9R$ and $1.3R$ downstream of turbine. This is because the hub is explicitly modeled in case of SMI, whereas in ALM, the hub region experiences higher wind velocity and a lower negative wind deficit ($0.1 > z/R$), secondly in regions near tip vortex ($0.9R > z/R > 1.1$), ALM shows a sharp velocity gradient corresponding to the presence of shear region accompanying tip vortex and it shows higher wake velocity deficit in the core wake region ($0.2R > z/R > 0.9$). Either, it could be because of higher turbulence that the wakes are decaying faster in SMI or because of mesh size. Despite these differences in wake velocity deficit, the power coefficient in SMI is slightly lower than ALM because higher energy gets extracted at the tip edge region (which produces most of the torque) which compensates for the lower energy extracted at the core wake region in SMI, as compared to the ALM case.

Figure 3 compares the flow structures and flow pattern captured by SMI and ALM Approach. The wake structures obtained in the near wake region by SMI and ALM are shown in figure 3(a) and in figure 3(d) respectively. The vorticity contours show that both the methods (and the mesh used) are able to capture the helical wake structures downstream of blade at an altitude along the periphery of blade tip. A higher vorticity region is captured by SMI at regions downstream at altitude between the tip periphery and root, which could be because of a chaotic flow induced in this region by explicit rotating blade of SMI as shown in streamlines (figure 3(e)). Figure 3(b) and figure 3(e) compares the flow path (streamlines) predicted by SMI and ALM, with streamlines colored by magnitude of velocity. SMI captures a higher 3D chaotic flow behavior in regions behind the turbine as compared to the ALM (which does not show this much chaotic 3D flow). Further, the highest velocity in ALM streamline is about 11 m/s while SMI streamline captures a highest velocity (65 m/s) almost equivalent to the blade tip speed ($\omega r = 1.071 * 63 = 67.4 m/s$), which could be due to non-slip boundary on the rotating blade geometry in SMI. For sake of comparison, the color-bar range for streamlines is set at the highest velocity magnitude captured by the SMI. Further, figure 3(c) and figure 3(f) shows the flow pattern at the DU40 segment (located near the hub) in terms of mean velocity vectors captured by the SMI and ALM respectively. In figure 3(f), the velocity vectors colored by velocity magnitude is imposed on a pressure contour. At TSR 7.5, SMI shows the wind approaching the blade at an angle of attack and the explicit rotating bluff body airfoil segment has major influence on the velocity vector direction and magnitude as it crosses the blade, while with ALM, the velocity vectors seemed to only deviate slightly as it crossed the similar blade region (note that ALM do not have explicit blade and this DU40 region was identified using

IOP Conf. Series: Journal of Physics: Conf. Series **854** (2017) 012044 doi:10.1088/1742-6596/854/1/012044

vorticity iso-surface). The flow vector differences can be attributed to the distributed nature of force projection in ALM methodology along with the non-explicit resolution of blade. However, despite these limitations, ALM has been able to predict the trends in wake deficit and power coefficients. Thus, the performance of ALM method is compared with the SMI model. The impact of these two diverse methodologies on wake dynamics is further captured by POD as described in section 3.3 below.

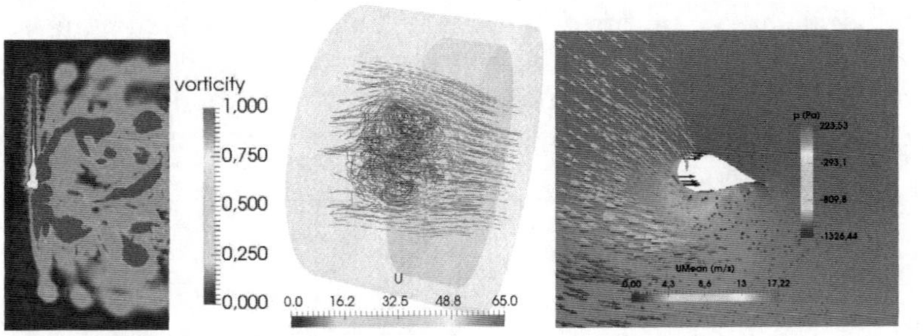

(a) SMI Wake (b) SMI Streamline (c) SMI Velocity vector at DU40

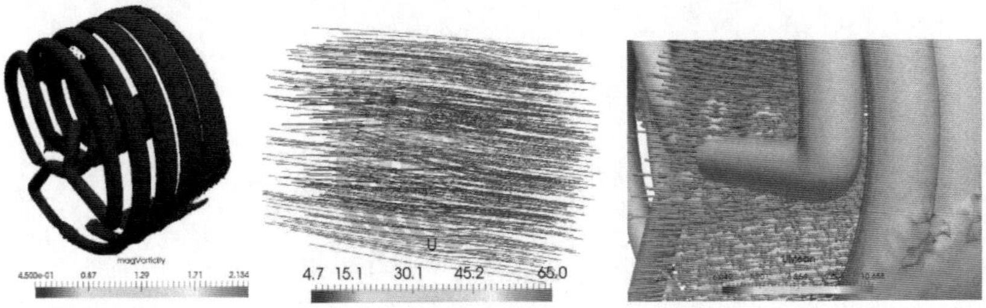

(d) ALM Wake (e) ALM Streamline (f) ALM Velocity vector at DU40

Figure 3. Impact of tip speed ratio (TSR) on flow pattern

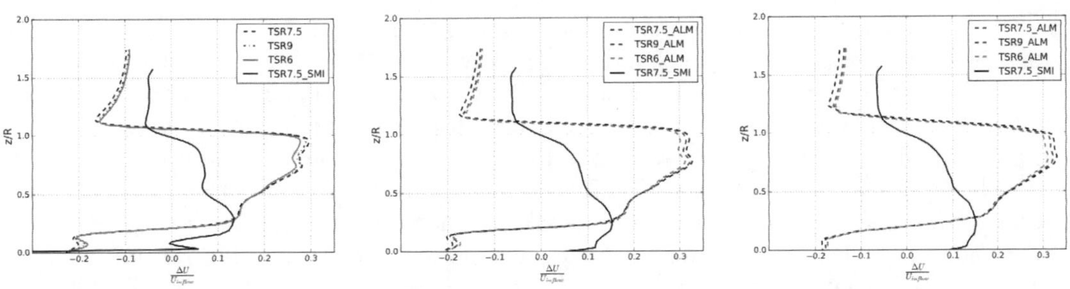

(a) 0.3R distance behind turbine (b) 0.9R distance behind turbine (c) 1.3R distance behind turbine

Figure 4. Verifying impact of tip speed ratio on normalized mean velocity deficit.

3.3. POD comparison

To conduct POD the simulation data from ALM and SMI was first interpolated onto a rectilinear grid. For ALM, the data sampled at 7 Hz, is interpolated on a uniform rectilinear grid measuring

380×400 elements, with a grid-size of $0.5m \times 0.5m$. The POD is applied on 720 planes over a total time of $100s$. For SMI, the data sampled at 166 Hz, is interpolated on a uniform rectilinear grid measuring 240×400 elements, with a grid-size of $0.5m \times 0.5m$. The POD is applied on 1000 planes over a total time of $6s$ (corresponding to nearly one rotation). It is ensured that most of the flow realizations are captured in the samples used for conducting the POD. Figure 6 and figure 7 shows the contours of decomposed velocity modes. Their values do not physically represent the velocity. These values represent distribution of energy content in the mode, and gives idea about overall energy captured by the mode and the flow structure associated with it. High values (red color regions) signifies higher energy content and blue color regions signifies lower energy content. The color bars are not relevant to the study here and are not shown, only the total energy captured by a mode and associated flow structure is relevant. Figure 6

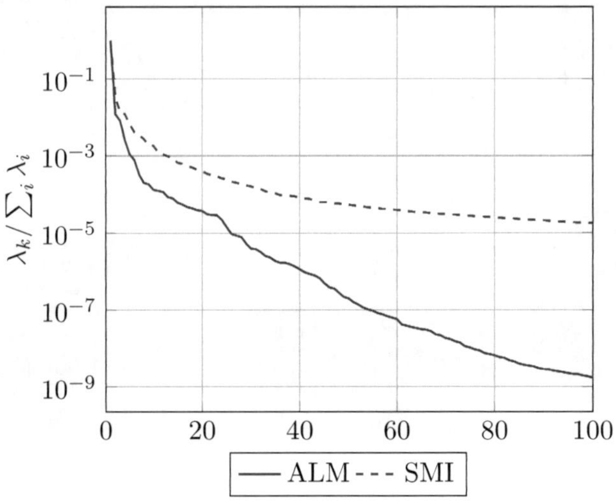

Figure 5. Energy spectra of SMI and ALM simulations

gives the flow structures represented by the first four modes for the ALM simulations conducted for at $TSR = 7.5$. It can be seen from the figure that the first mode represents the most energetic (97.5% of the total energy) and probable realization of the flow which looks like the mean wind field. As we move to higher modes we discover that the associated energy decrease rapidly. Together, the first three modes account for 99.5% of the total energy. The spatial scales associated with flow structures also decrease. It is also clear from the figures that the first four modes capture distinct kinds of flow structures hinting at a clear scale separation at least in space. On the other hand, in the case of SMI, the first mode corresponds to only 90% of the total energy. It hints at a spreading of the larger flow structures to higher modes. The explicit churning motion of the blades generate large eddies (Figure 7) which break down into smaller eddies and no clear scale separation is observed when compared to the ALM simulation as a result of which the energy drop across modes is more gradual. However, together the first 30 modes capture 99.5% of the total energy. The need for slightly more number of modes in the case of SMI suggests that it is able to capture flow with more unsteadiness and wider range of spatio-temporal scales either owing to explicit blade resolution or perhaps the faster decay of wakes through the energy cascading process (and hence there are more length and time scales in the given domain). It is worth highlighting here that although the effects of hub vortex is quantitatively captured using both the approach, the same is not true for the tip-vortices. A reasonable explanation can be that the current POD is conducted on 2D snapshots where the tip

IOP Conf. Series: Journal of Physics: Conf. Series **854** (2017) 012044 doi:10.1088/1742-6596/854/1/012044

vortices are only represented as dots contributing very little energy compared to the other flow structures. In order to capture the helical tip vortices, POD using three dimensional field should be conducted. This was not possible during the course of this work but remains an investigation for the future.

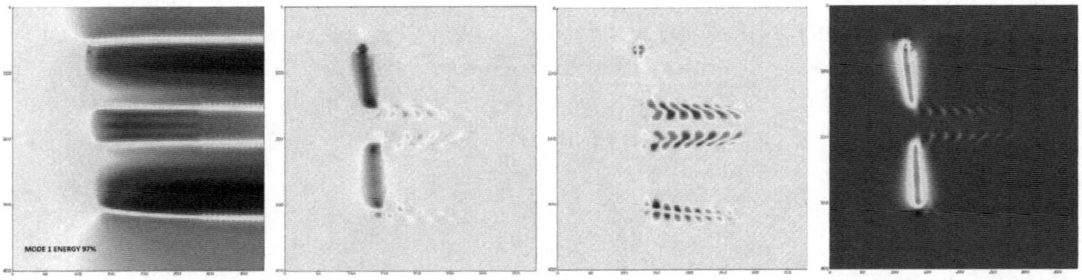

(a) Mode1 Energy 97% (b) Mode2 Energy 1.21% (c) Mode3 Energy 0.8% (d) Mode4 Energy 0.2%

Figure 6. First four modes extracted from the ALM simulations using POD. Contours of Decomposed velocity modes.Red color region represents high values and blue color region represents low values.

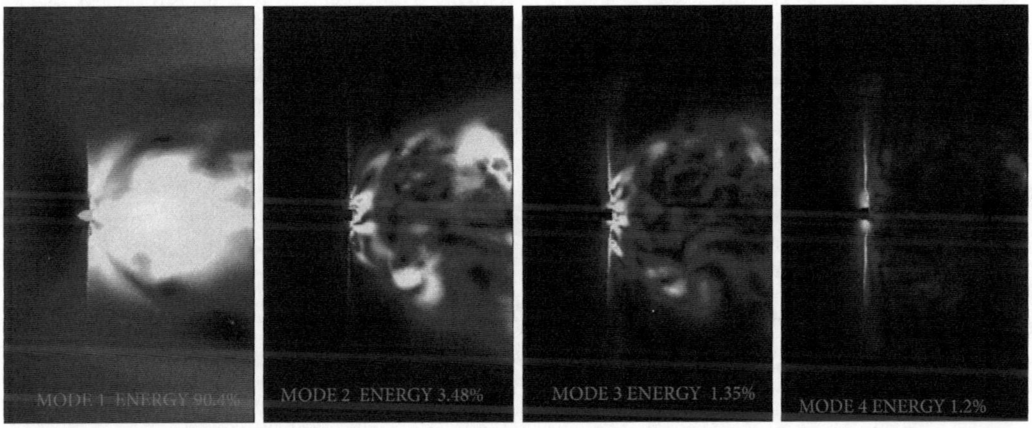

Figure 7. First four modes extracted from the SMI simulations using POD.Contours of Decomposed velocity modes.Red color region represents high values and blue color region represents low values.

4. Conclusion and future work

The paper aims to use LES and POD technique to differentiate wake dynamics captured by two diverse methods (ALM and SMI techniques) for an industrial scale wind turbine. The paper has been able to show that:

- ALM method is compared with the FAST model in its ability to compute power coefficient with varying tip speed ratios for an industrial scale turbine, and ALM is able to capture the trend in a similar way as FAST.
- The POD analysis also reveals that SMI has been able to capture a compendium of eddies with variable time scales. As a result, its first mode is not as dominant (90% of energy) as

the first mode from ALM (which captures 97% of energy). The ALM first mode shows the shear along the tip periphery and hub periphery dominantly and most energy lies in this mode itself, while the rest of structures (scales) have less energy and are in the remaining modes.

- A comparison in prediction of flow-pattern and wake structure by ALM and SMI shows that the SMI predicts a more complex 3D nature of flow. Despite these differences, the ALM has been able to follow the qualitative trend in wake deficit and power coefficient variation with tip speed, owing to the line source and the distributed nature of force projection. The current work is focussed on near wake region study. In future, it will be interesting to study the wakes further downstream.

Acknowledgments
The authors acknowledge the financial support from the Norwegian Research Council and the industrial partners of NOWITECH: Norwegian Research Centre for Offshore Wind Technology (Grant No.:193823/S60) (http://www.nowitech.no) and FSI-WT (Grant No.:216465/E20)(http://www.fsi-wt.no).

References

[1] Churchfield M, Lee S, Moriarty P, Martnez L, Leonardi S, Vijayakumar G and Brasseur J 2012 *50th AIAA Aerospace Sciences Meeting Including the New Horizons Forum and Aerospace Exposition* 124 (2) 393–399
[2] Porté-Agel F, Wu Y, Lu H and Conzemius R 2011 *Journal of Wind Engineering and Industrial Aerodynamics* 99(4): 154–168
[3] Tabib M V, Rasheed A, and Kvamsdal T 2015 *Journal of Physics: Conference Series* 625: 012032
[4] Tabib M V, Rasheed A and Fuchs F 2016 *Journal of Physics: Conference Series* 753 (3), 032063
[5] Tabib M V, Rasheed A, and Kvamsdal T 2015 *Energy Procedia, 80* 302311
[6] Jonkman J, Butterfield S, Musial W and Scott G 2009 *National Renewable Energy Laboratory* **Tech. Rep. NREL/TP-500e38060**
[7] Kooijman H, Lindenburg C, Winkelaar D and Hooft E 2003 *DOWEC Dutch Offshore Wind Energy Converter 1997-2003 Public Reports , DOWEC 10046-009, ECN-CX–01-135, Petten, the Netherlands: Energy Research Center of the Netherlands*
[8] Tabib M V and Rasheed A 2017 *EERA Deepwind Conference, Trondheim*
[9] Sørensen J and Shen W 2002 *Journal of Fluids Engineering* 124 (2) 393–399
[10] Pope S 2000 *Cambridge University Press*
[11] Troldborg N, Sorensen J and Mikkelsen R 2010 *Wind Energy* **13** 86–99
[12] Martinez L, Leonardi S, Churchfield M and Moriarty P 2012 A comparison of actuator disk and actuator line wind turbine models and best practices for their use
[13] Jha P, Churchfield M, Moriarty P and Schmitz S 2014 *Journal of Solar Energy Engineering, Transactions of the ASME* **136**
[14] Spalding D 1961 *Transactions ofth ASIDE, Series E: Journal of Applied Mechanics* 28,455–458
[15] Bastine D, Witha B, Wchter M and Peinke J 2014 *Journal of Physics: Conference Series* **524**
[16] Braud C, Heitz D, Braud P, Arroyo G and Delville J 2004 *Experiments in Fluids* **37** 95–104
[17] Jonkman J, Marshal L and Buhl J 2005 *National Renewable Energy Laboratory, Golden, CO,* **Tech. Rep. NREL/EL-500-38230**

Wake meandering and its relationship with the incoming wind characteristics: a statistical approach applied to long-term on-field observations

E. Torres Garcia[1*], S. Aubrun[1], M. Boquet[2], P. Royer[2], O. Coupiac[3] and N. Girard[3]

1 Univ. Orleans, PRISME EA4229, 45072 Orleans, France

2 Leosphere, Jean Rostand 14-16, 91400 Orsay, France

3 Maia Eolis, Tour de Lille, Bvd de Turin 59777 Lille, France

*eulalio.torres-garcia@univ-orleans.fr

Abstract. In several papers, the importance of the atmospheric flow in the wake development of wind turbines (WT) has been pointed out, making it clear that it is necessary to have long-term on-field observations for an appropriate description of the wake development, its structure and dynamics. This work presents a statistical approach to wake meandering, y_ω, and the relationship that this behavior has with the incoming wind conditions and neighboring wakes. The work was developed in the framework of the French project SMARTEOLE. The study is based on a 7-month measurement campaign in which a pulsed scanning LiDAR system was used. The ground based LiDAR, measures the flow field in a segment such that the wake of two wind turbines can be captured quasi-horizontally. The analysis filters the incoming wind conditions according to the thermal stability, wind direction and wind velocity at hub height; therefore, the wakes that are developed in periods with similar wind conditions are expected to be analogous, hence meandering can be tracked and statistically analyzed. A well-defined wake evolution was found and the uncertainty analysis made on the wake meandering uncovered some interesting characteristics, including the number of samples required to reach a statistical uncertainty on the mean wake position between 2×10^{-2} D and 8×10^{-2} D for a confidence interval of 95%.

1. Introduction

In wind farms (WFs), some of the undesired effects of wind turbine (WT) wakes are power losses and a decrease in the lifetime of structures embedded within the wakes [1]. Many simulation tools, numerical [2–5] and experimental [6,7], focus on the wake's development in order to better understand the wake physics and reduce such undesired effects; the final objective is to design efficient wind farm layouts. The problem arises when the physics exceeds the simulation tool limitations, i.e. for numerical simulations the number of operations is beyond current calculation capabilities and for experimental simulation, the changing environmental variables make it impossible to simulate them all. As a result, from a wind farm design to its operation, there is a power output difference of up to 20% (mostly negative) [8]. Detailed observations of the atmospheric flow and its interaction with operational WTs (upwind and downwind) are required to calibrate numerical models and corroborate wind tunnel experiments.

When doing field measurements, the expected outcome is the understanding of the complex atmospheric flow field behavior, which in turn generates a rule that can be extended to any real case (see [9–11] for some studies dealing with wakes in on-field measurements). The problem to generate such outcome is the atmospheric stability, which depends on variables such as mean wind velocity ($\overline{U}$), mean wind direction (*WD*), Monin-Obukhov Length (L) and turbulence intensity (I_u). Moreover, the wake topology is related to many other aspects such: wind farm layout, landscape and neighboring wakes [11], that make it a hard target to describe. Detailed observations over long periods will help to understand the synergy between atmospheric flow and wakes, to characterize the impact of the wakes on other wind turbines and also, as Rhodes et al. [12] suggested, the impact on the local environment.

One of the promising measuring techniques to characterize WT wakes is Light Detection and Ranging (LiDAR) sensors. LiDAR remotely senses the Doppler shift of laser light backscattered from particles carried by the wind in order to measure a line-of-sight (LOS) radial wind velocity. The system considers measurements of this LOS velocity in multiple radial directions in order to estimate wind characteristics [13]. LiDARs have been used to estimate wind characteristics of boundary layer dynamics [14,15] and also for wake dynamics [12,16].

In the present study a LiDAR measurement campaign, lasting 7 months, is analyzed. The objective was to analyze the unsteady wake behavior of two wind turbines. The measurements were performed by a ground based scanning LiDAR WINDCUBE 200S. The LiDAR was programmed to acquire 30 LOS to form a plane at a specific elevation angle; three elevation angles were programmed. Details of the LiDAR and experimental set up are presented in section 2. In order to analyze the wake physics, each plane was treated individually, assuming conditions of homogenous flow along the plane and also assuming that the turbulent structures move as frozen entities transported by the mean wind [17]. The methods used to categorize measurements and the methods to analyze the categories are presented in section 3. The unsteady wake behavior was assessed statistically, the outcome of the statistical analysis and the discussion are presented in section 4. Finally, conclusions are drawn in section 5.

2. Experimental set up

The measurement campaign took place between 15/11/2015 and 30/05/2016. More than 4300 hours of measurements were collected and analyzed.

2.1. The wind farm

The WF belongs to Maïa Eolis (Engie Green) and it is located in the north of France, on the western limit of the municipality of Ablaincourt-Pressoir, see Figure 1. The WF is composed of seven SENVION MM82 wind turbines, whose diameter (D) and hub height are 82 m and 80 m, respectively. The turbines are sited from north to south and spaced approximately 3.5 D apart. The WTs are named SMV1 to SMV7, SMV1 being the northernmost; for a detailed description see [18]. The measurement campaign focused on SMV5 and SMV6; these turbines are aligned to a 207° reference line (orientated in meteorological coordinates).

2.2. The instrumentation

A pulsed scanning LiDAR WINDCUBE 200S, provided by LEOSPHERE, was located at ground level on the outskirts of Ablaincourt-Pressoir (see B, in Figure 1 right); this location is 1320 m east from the line of WTs. The system has a velocity accuracy of 0.1 $m \cdot s^{-1}$ [19] and was programmed to acquire 115 measurement points along a line of sight (LOS). 30 LOSs form a plane position indicator (PPI). Each PPI was measured between azimuth 243° and 273° with respect to the geographical north and in meteorological coordinates (see C, in Figure 1 right). The acquisition resolution along each LOS is 25 m and the displacement between LOS is 1° at $2° \cdot s^{-1}$ acquisition rate, therefore a full PPI is obtained every 15 s. The LiDAR was programmed to acquire PPIs at three elevation angles: α_1=2.5°, α_2=3.8° and α_3=5.2°, with respect to the ground level. The selection of azimuths and elevations for acquiring measurements was based on the geographical location of the turbines, their geometrical parameters and the historical wind rose; hence, SMV5 and SMV6 wakes were captured within the specified PPIs. In addition, because of the location of the LiDAR and the selected elevation angles, the PPIs crossed SMV5 at heights of 58 m for α_1, 79 m for α_2 and 120 m for α_3.

An 80 m high lattice met mast is located 1.6 km northeast of SMV6, see A in Figure 1. An 80 m anemometer and a 40 m wind vane were operational throughout the measurement campaign.

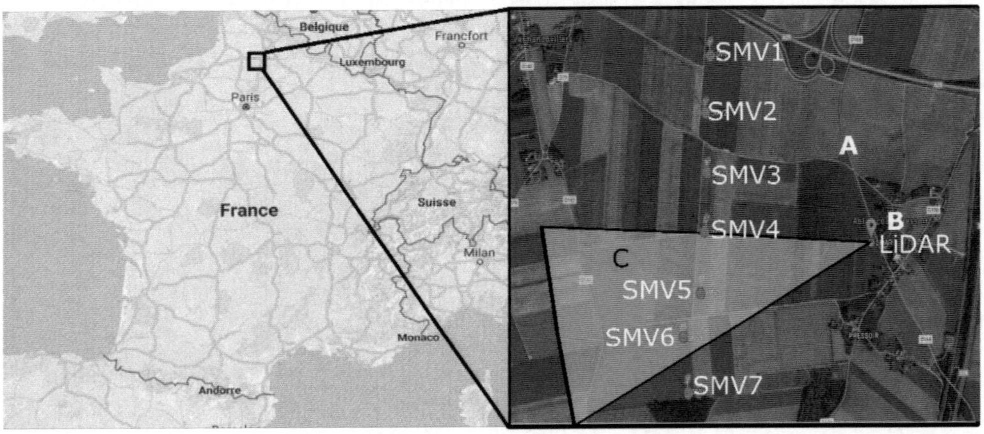

Figure 1. Map of France indicating the wind farm site (left). On the right, the wind farm composed of seven WTs marked with red dots. A meteorological mast (A) is located in the surrounding area. The location of the LiDAR is marked with a B. The shadowed area C is the LiDAR scanned area.

2.3. External resources

The Modern-Era Retrospective analysis for Research and Applications version 2 (MERRA-2) dataset is a global reanalysis to assimilate space-based observations of aerosols and represent their interactions with other physical processes in the climate system [20]. The MERRA-2 spatial resolution is 1/2° (latitude) and 2/3° (longitude) and the temporal resolution is 1 hour. MERRA-2 has proven to be a valuable dataset for the determination of the overall stability conditions of a site, as long as the surrounding grid points are representative for the site [21]. The data were extracted from the closest MERRA-2 grid point located 2.5 E- 49.5 N, 44 km South West from the wind farm. The MERRA-2 data provide information about wind direction, wind velocity, pressure and temperature. The Monin-Obukhov Length (L) can be calculated from MERRA-2 [21].

3. Methods

In order to ensure the right statistical description the atmospheric wind conditions were filtered. The three filter criteria applied were: 1) Neutral atmosphere corresponding to $|L| \geq 1000$. In this case, the WT wake recovers more slowly than in unstable/convective conditions [22]; 2) three wind directions (WD) associated with three different wake patterns; 3) two wind speeds (U) at different WT operational set points.

The stability periods were obtained from the MERRA-2 dataset. As MERRA data are averaged values over a 1/2° × 2/3° tile, the representativeness of the 2.5 E- 49.5 N grid point was assessed by comparing to a bilinear interpolation between the four grid points closest to the WF and the position of the SMV5 WT. No change in the stability periods was found.

Three WDs that produce three different wake patterns were chosen: 1) two wakes inline; 2) two distinct wakes separated 1 D in the crosswise direction; 3) two distinct wakes separated 2 D in the crosswise direction. The WDs considered in each case are limited by the turbulence intensity (I_u). An $I_u = \sigma_U / \overline{U} \approx 0.2$ was determined using information from the meteorological mast. Knowing that the transversal fluctuations are defined as $I_v = 0.75\ I_u$ [23], then the angle of variation is close to $\tan^{-1}(I_v)$, leading to WD bins of ± 8.53°. The historical wind rose (most common WD from the South-West) and the WTs geographical positions were used to determine the center of the *WD* bins. Each post-processed PPI provides the environmental *WD*, where this filter was applied.

Based on the power coefficient of the wind turbine SENVION MM82 (Figure 2) two significant wind speeds were selected, 7 m·s^{-1} and 11 m·s^{-1}, represented by circles on the C_P curve. The standard deviation (σ_U) for the two selected velocities was calculated from the turbulence intensity I_u (σ_U is also presented with error bars in Figure 2). The limits for classifying U data are therefore 7 ± 1.4 m·s^{-1} and 11 ± 2.2 m·s^{-1}. This filter was applied to each single PPI, by extracting the undisturbed wind velocity at hub height.

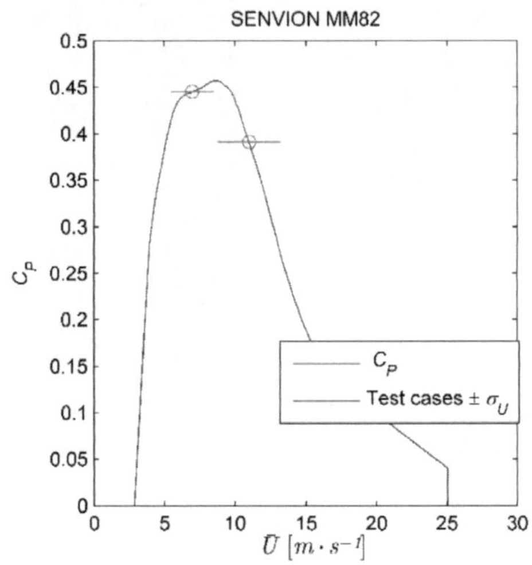

Figure 2. Power coefficient (C_P) of the wind turbine SENVION MM82 as a function of the incoming wind velocity ($\bar{U}$).

3.1. LiDAR data processing

A tool, provided by LEOSPHERE, converts the radial wind velocity measurements into estimated horizontal wind velocities by applying a correction from the cosine of the averaged wind direction. The averaged LiDAR *WD* is retrieved from locations outside of the wake influence.

No measurement was obtained at certain locations due to the lack of environmental aerosols; data at these locations were interpolated. The measurements taken at WT positions were also interpolated. An 80% data availability criterion was established for each PPI scan. If a certain PPI scan had more than 80% of available data, the interpolation was applied, otherwise the PPI was not taken into account. In order to interpolate the measurements, the Laplacian operator was applied over each PPI. Those parts of the PPI that were not in contact with unknown values were dropped and the unknown values were computed by the least squares approach. The measured points were not modified.

In order to validate the *WD* deduced from scanning LiDAR measurements, the alignement of WT wakes with this *WD* was checked. To do so, the inertial coordinates were translated to wind related coordinates, thus each PPI was rotated clockwise according to the *WD* around the ground normal component, to end up with a *WD* orientated system $[X_{WD}, Y_{WD}, Z_{WD}]$. The LiDAR location is defined by $[X_{WD}, Y_{WD}, Z_{WD}]=(0,0,0)$. If the *WD* is correctly determined, the rotated PPI should present wakes aligned with the X_{WD} axis. The velocity field was interpolated on a regular grid. It was verified that the instantaneous velocity profile at 4 D downstream had the minimum velocity aligned with the WT hub ± 0.6 D (transversal variation caused by I_v). If the PPI fulfilled this condition, the *WD* is correct and the PPI was taken into account for the statistics.

3.2. Wake tracking

The wake tracking procedure was applied to each PPI (distributed along time, t) in the *WD* system, following the procedure proposed in [24]. The wake tracking trajectory is based on the calculation of the wake center coordinate, y_ω, which is a weighted average of the locations Y_{WD} distributed in the crosswise direction $Y_{WD} \pm 0.75$ D from the wind turbine position. The weighting is the exponential of the instantaneous local velocity deficit Du_i, for the N profile points:

$$y_\omega(X_{WD},t) = \sum_{i=1}^{N} \exp\left(Du_i(X_{WD},Y_{WD},t) \times Y_{WD^i}\right) \Big/ \sum_{i=1}^{N} \exp\left(Du_i(X_{WD},Y_{WD},t)\right), \quad (1)$$

where Du_i, is defined as the difference between the local velocity and the maximum velocity at a constant downstream distance (X_{WD}); the Du_i is located at position Y_{WD^i}. This method was designed to work in the far wake area, where the averaged velocity deficit profile is expected to be characterized by only one local minimum.

3.3. Statistical analysis

Since the positions of the wake center in each time step are independent (PPI distributed over 7 months) and the process is considered statistically stationary due to the imposed constraints, all the accessible microstates (captured by each PPI) are equi-probable over a long period of time, and can then be analyzed as an ensemble. The collection of PPIs gives a collection of wake center positions (y_ω) and the statistical analysis was therefore carried out on these wake center positions. The mean and the standard deviation from the wake center location were calculated, then, the relative statistical uncertainty convergence (ϵ) was calculated as a function of the cumulative standard deviation of the wake center position ($C\sigma_{y\omega}$). The statistical uncertainty for the mean wake position is a function of the wake center dispersion and the number of samples, $\epsilon = 1.96 \cdot \sigma_{y\omega} / \sqrt{N}$ where $\sigma_{y\omega}$ is the standard deviation of the wake center position along the number of independent samples N [25]. The factor 1.96 stands for a confidence level of 95%.

4. Results and discussion

Table 1 presents the number of samples treated per case and per PPI elevation. The two digits at the beginning of the case name refer to the *U* and the next three to the *WD*. The minimum number of samples was found in the case 07244, where wakes are parallel, while the maximum number of samples was found in cases 07207 and 11207 (wakes overlapped). When determining the mean wake position, more samples means greater accuracy. There are two facts involved in the determination of the amount of data: one is the recurrence of the wind direction and the other is the availability of quality measurements.

Table 1. Number of PPI-samples (N) available for each case and for each elevation angle. The two digits at the beginning of the case name refer to the *U* and the next three to the *WD*.

		Cases					
		07207	11207	07225	11225	07244	11244
elev.	α_1	284	248	155	255	121	149
	α_2	324	258	167	248	107	152
	α_3	242	184	131	258	105	153

Figure 3 presents the ensemble-averaged velocity fields for α_2, $U = 7 \pm 1.4$ m·s^{-1} and 3 *WD*s (one per panel). In the three cases, the wake development aligns with the *WD*, confirming the ability to determine the *WD* accurately from LiDAR measurements. In the left panel, case 07207, SMV6 and SMV5 are aligned one in front of the other. The SMV6 wake is well identified, presenting a velocity deficit directly downstream of the WT hub. This velocity deficit decreases as the distance from the WT increases. The SMV6 wake is directly in line with turbine SMV5; hence, the downstream velocity of SMV5 has a further decrease that is also noticeable on the plot. On the left side of SMV6 and SMV5 wakes, the signature of the SMV7 wake can be observed. This WT is outside of the measured PPI. The expansion of the SMV7 wake meets with the wake of SMV5 at around [X_{WD} , Y_{WD}]= (900,750). In the panel in the middle, case 07225, the crosswise separation between hubs SMV7 to SMV6 and SMV6 to SMV5 is roughly 3 D and 1 D, respectively. Although SMV7 is outside of the measured PPI, some of its effects are still visible. At around $[X_{WD},Y_{WD}]$=(1200,500) the SMV7 wake meets the wake of SMV6. The SMV6 wake expansion is very clear. The right limit of the SMV6 wake reaches the position of SMV5, then the SMV5 wake experiences a further velocity decrease; the process of expansion is also noticeable. At the position of SMV4 and upstream, a velocity reduction is measured. One should bear in mind that the measurements at all WT positions were interpolated. In the right panel, case 07244, the crosswise spacing between SMV7 and SMV6 is 3 D and between SMV6 and SMV5 it is 2 D. The velocity decreases just in front of SMV6 and downstream, a further reduction is observed. The velocity deficit remains for a greater distance in comparison with the previous two cases. The SMV6 wake and SMV5 wake have a similar behavior. Upstream of SMV4, the velocity decreases in agreement with SMV5 and SMV6. The specific value of the velocity deficit is beyond of the scope of the present paper.

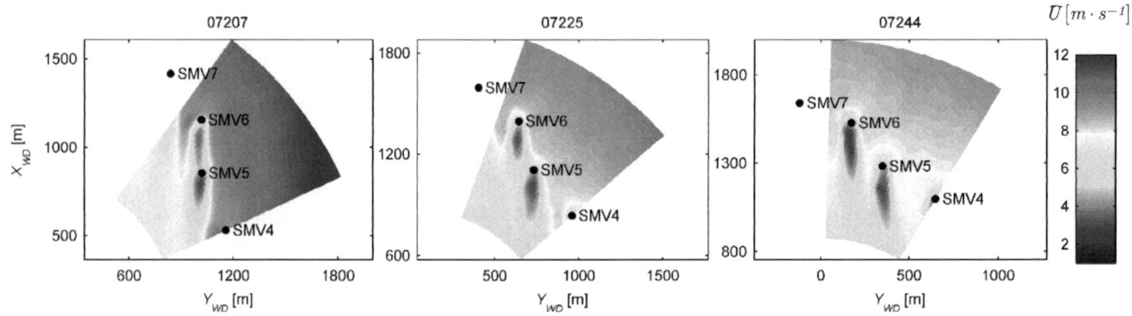

Figure 3. Mean PPI at $\alpha_2 = 3.8°$, for $U = 7 \pm 1.4$ m· s^{-1} and three *WD*s ± 8.53°.

In the subsequent plots, the coordinates $[X_{WD}, Y_{WD}, Z_{WD}]$ were nondimensionalised by D and the origin was set to the position of the SMV6 hub.

Figure 4 presents the mean position of the wake center (y_ω) and its standard deviation. The y_ω blue color stands for α_1, the green for α_2 and the red for α_3. SMV6 and SMV5 positions are marked with a black circle and a black square, respectively. The gray area corresponds to the domain where the algorithm for the y_ω computation was applied. Only cases with $U = 7 \pm 1.4$ m·s^{-1} were analyzed, since between the two incoming wind velocities selected (U = 7 m·s^{-1} and U = 11 m·s^{-1}) no remarkable differences were observed.

The first plot of Figure 4 shows the case 07207. The wake center starts at the SMV6 position and is displaced slightly in the -y_{WD}/D direction. This displacement can be attributed to the SMV7 wake, but also to the inability of the method to separate single wake effects. At the SMV5 position, the center starts at hub position for α_1 and α_2. For α_3, the wake center remains aligned with the SMV6 wake. The standard deviation increases with the distance downstream with a noticeable increase after SMV5, due to the sum of the effects of the two wakes. A similar effect was reported by Muller et al. [24].

In the central plot of Figure 4, the SMV6 wake shows the effects of two neighboring wakes interacting in the same domain of calculation. On the -y_{WD}/D side, the SMV7 wake is developed and on the +y_{WD}/D side the SMV5 wake, both at different distances streamwise. Since SMV7 is located at y_{WD}/D ≈ -1, and the wake tracking process weights all the velocities, the center of SMV6 is attracted in this direction. From x_{WD}/D = 3.5, SMV6 y_ω is displaced towards the direction of SMV5 and the standard deviation increases; this deviation diminishes at approximately 3 D downstream of SMV5 and, farther downstream, returns to its initial position (aligned with the WT hub). The SMV5 wake shows a slight displacement towards the direction of SMV6, which is smaller in comparison with SMV6. The standard deviation in SMV5 is similar to case 07207 and higher than SMV6.

The lower plot in Figure 4 presents the case 07244. The y_ω is centered to its respective WT hub. The standard deviation in the SMV6 wake increases with the distance. Following the same curve, 1 D downstream of SMV5 the standard deviation diminishes with the distance. The SMV5 wake shows a higher standard deviation, because the wind is already turbulent (SMV6 wake upstream).

IOP Conf. Series: Journal of Physics: Conf. Series **854** (2017) 012045 doi:10.1088/1742-6596/854/1/012045

Figure 5 shows the y_ω standard deviation ($\sigma_{y\omega}$) with respect to the location downstream from the source (SMV6 and SMV5 WTs). Each row represents a *WD* range and each column a *U* range. As in the previous plot, blue stands for $\alpha_1 = 2.5°$, green for $\alpha_2 = 3.8°$ and red for $\alpha_3 = 5.2°$; the solid lines represent $\sigma_{y\omega}$ calculated from the collection of wake center positions for SMV6, and the dotted lines SMV5. In most of the cases, from the WT source and up to 1 D downstream, the $\sigma_{y\omega}$ diminishes and then increases with the distance. The interpretation of the results in the near wake should be done carefully, since all the values measured at the WT location were interpolated and because of the limitations of the wake tracking method.

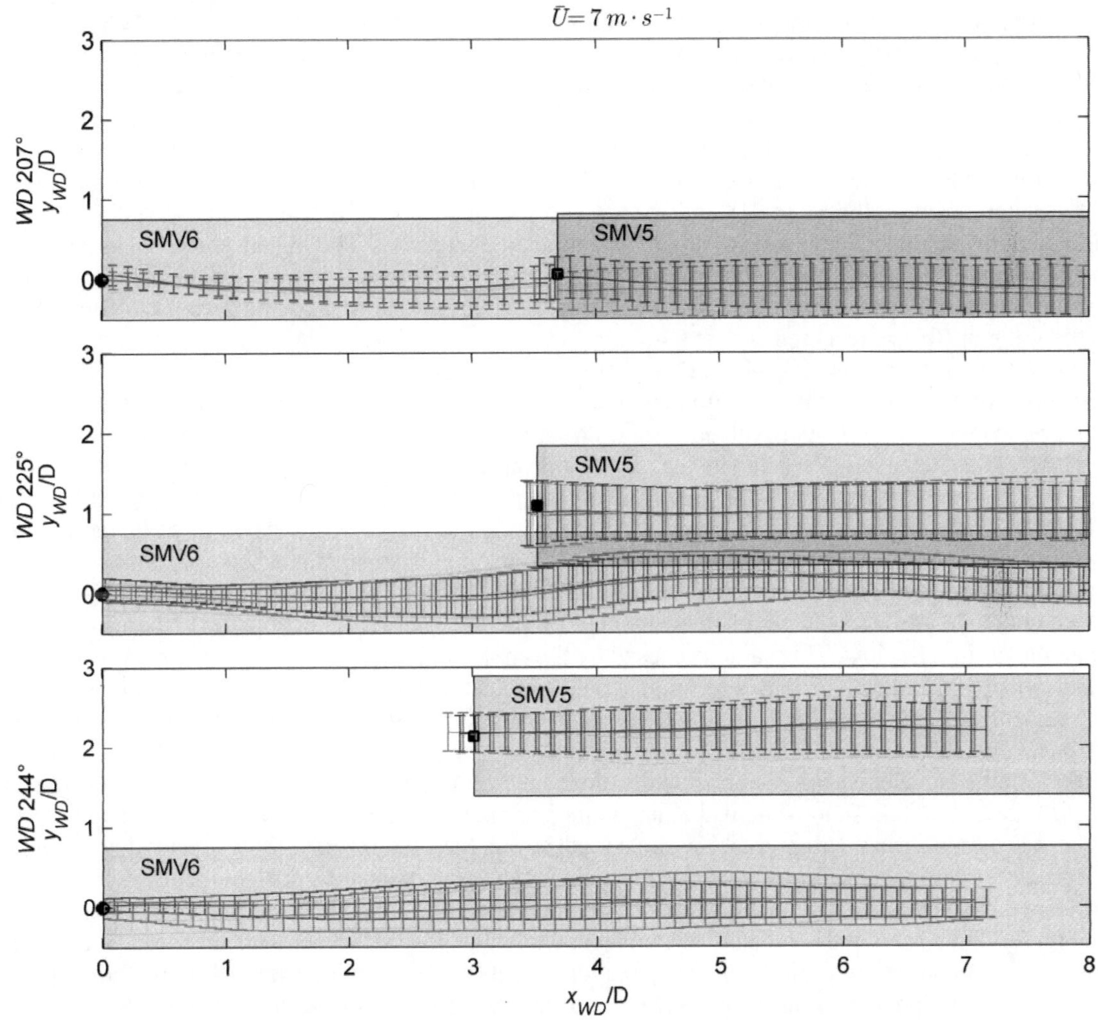

Figure 4. Ensemble average of the wake center locations (y_ω) and its standard deviation ($\pm\ \sigma_{y\omega}$) at different downstream distances. Blue stands for α_1, green for α_2 and red for α_3. SMV6 and SMV5 positions are marked with a black circle and a black square, respectively. The gray area corresponds to y_{WD} /D= WT ± 0.75, the area of wake tracking calculation.

In Figure 5a (case 07207), SMV6's $\sigma_{y\omega}$ increases from 1 D until 3 D downstream and then remains quasi-constant with a value of 0.19. Results from on-field measurements [26] report a value of $\sigma_{y\omega}$ = 0.17 at x/D = 3 for a I_u similar to the one used in the present paper. The SMV5 $\sigma_{y\omega}$ starts with a jump of around 0.05 with respect to the ending of SMV6 and then has a positive slope; this jump is associated to the perturbation due to the WT rotation, and as the flow goes downstream, the accumulation of the wakes of SMV6 and SMV5 is clear since the curve continues to show a positive slope. By increasing the velocity (Figure 5b, case11207), in SMV6 the same initial positive slope is observed, but around 2 D downstream the $\sigma_{y\omega}$ appears to be constant (0.21). In SMV5 the same behavior as in SMV6 is observed but with an increase of $\sigma_{y\omega}$ = 0.15. Note that the trend changes at around 2 D downstream of each WT. This position could mark the end of the near wake and the beginning of the far wake.

In Figure 5c (case 07225), the $\sigma_{y\omega}$ for SMV6 increases from ≈ 0.13 at 1 D downstream up to 0.28 for α_1 or 0.38 for α_3 at approximately 4 D downstream. This difference can be attributed to the limited number of samples; α_3 was calculated with only 131 samples (Table 1). Downstream of SMV6, at around 2 D, there is a change in the tendency; the same position (2 D) and value (0.21) are verified in all the cases. The wake tends to follow the main wind stream, but reaching the position of SMV5, the rotor induces a local modification of the flow and then the wake center is attracted in this direction. In the SMV6 wake, this effect increases the $\sigma_{y\omega}$ at around the position of SMV5. This feature disappears between $x_{WD/D}$= 7 and $x_{WD/D}$=8, where the SMV6 wake reaches the nominal value of 0.2. There is more dispersion in the upper part of the wake that in the lower part. Unfortunately, the number of samples is smaller in the higher elevation. Figure 5d (case 11225) was calculated with almost the same amount of samples per elevation (between 248 and 258). The effect of suction of the wake is observed, increasing the $\sigma_{y\omega}$ value up to 4 D downstream of SMV6 and then decreasing. The higher plane, α_3, experiences more fluctuations. From SMV5, the three elevations give similar results but, with double the number of samples, the congruency between planes is better. The values appear to take into account the turbulence of the wake upstream, reaching almost 0.4.

Figure 5e presents case 07244. A clear difference between PPI elevations is seen, with α_3 the one with a higher $\sigma_{y\omega}$, and also the one with the fewest samples (105). The wake that started upstream shows an increase in the $\sigma_{y\omega}$ up to approximately 1.5 D downstream of the wake that started farther downstream, then there is a decrease in the $\sigma_{y\omega}$. The measurements close to the ground have the same tendency as for case 07225, reaching the nominal value of 0.2 at the end. Case 11244 is presented in Figure 5 f, where SMV6 experiences fluctuations due to disturbance by SMV5. Downstream the value of 0.2 is reached. SMV5 presents a similar positive slope to case 11207 and 3 D downstream it becomes almost constant but with different values between planes. These differences will be explained from Figures 6 to 8.

To conclude, the $\sigma_{y\omega}$ starts at a value between 0.1 and 0.2, providing that upstream there is no other wake disturbing the flow. This value is caused by the environmental I_u. Then, $\sigma_{y\omega}$ diminishes at around 1 D downstream and then increases again. It should be remembered, however, that, as mentioned before, the tracking method might not be suitable for the near wake region. If there are no neighboring wakes, the $\sigma_{y\omega}$ will reach a value of 0.2. If there are neighboring wakes, the second wake will show the sum of effects. In the parallel wakes case, they will influence each other (at least for the transversal distances tested here): the dispersion of the wake that starts first increases up to 1 D downstream of the second wake. The second wake will present the same tendency as the first one but with a greater dispersion. In all the cases at around 2 D downstream, a change in the tendency was observed; this position could mark the end of the near wake and the beginning of the far wake.

Figures 6 to 8 present the cumulative statistical uncertainty (ϵ) convergence of the mean wake center position as a function of the distance downstream from SMV6. In the plots, the vertical axis corresponds to the streamwise position of the wake, while the horizontal axis corresponds to the cumulative statistical uncertainty for a number of samples. In order to facilitate comparison between cases the same length of horizontal axis was used. The three PPI elevations are presented in columns; the first three columns are for U = 7 m·s^{-1} and the last 3 for U = 11 m·s^{-1}.

When the *WD* is 207° (Figure 6, case 07207), at the closest position to the SMV6 hub (x_{WD}/D =0) the three elevations need only a few samples to converge, as is expected since the source of the wake is fixed and there is no flow modification upstream. Even at the higher elevation, ϵ appears to be constant from a very few samples. At this position the PPI measures above the wake. Moving streamwise, there is more meander and then, to reach convergence, more samples are needed for the three elevations. From SMV6, 60 samples are more than enough to reach a quasi-constant value. The flow leaving SMV6 reaches SMV5, and then in the SMV5 wake, 200 samples are needed to reach the same value of ϵ, due to the turbulence created upstream. In case 11207 in the three elevations, from x_{WD}/D =0, more samples are needed to converge in comparison with case 07207. In α_1 and α_2, a similar convergence with the distance downstream is observed; these two elevations at approximately 150 samples reached the same value of convergence as at 60 samples with U of 7 m·s^{-1}. The higher elevation appears to be more fluctuating but constant along the wake. There is no change in the convergence with the distance downstream. The behavior in SMV5's wake is the sum of SMV6 and SMV5 wakes: more fluctuations are measured and, to reach the same statistical uncertainty, more samples are needed. With the distance downstream, the meandering is increased. Comparing cases 07207 and 11207, it can be seen that the higher the velocity, the larger the number of samples required to reach convergence. It follows, therefore, that the higher elevation also needs more samples to converge (effect of the velocity profile of the atmospheric boundary layer).

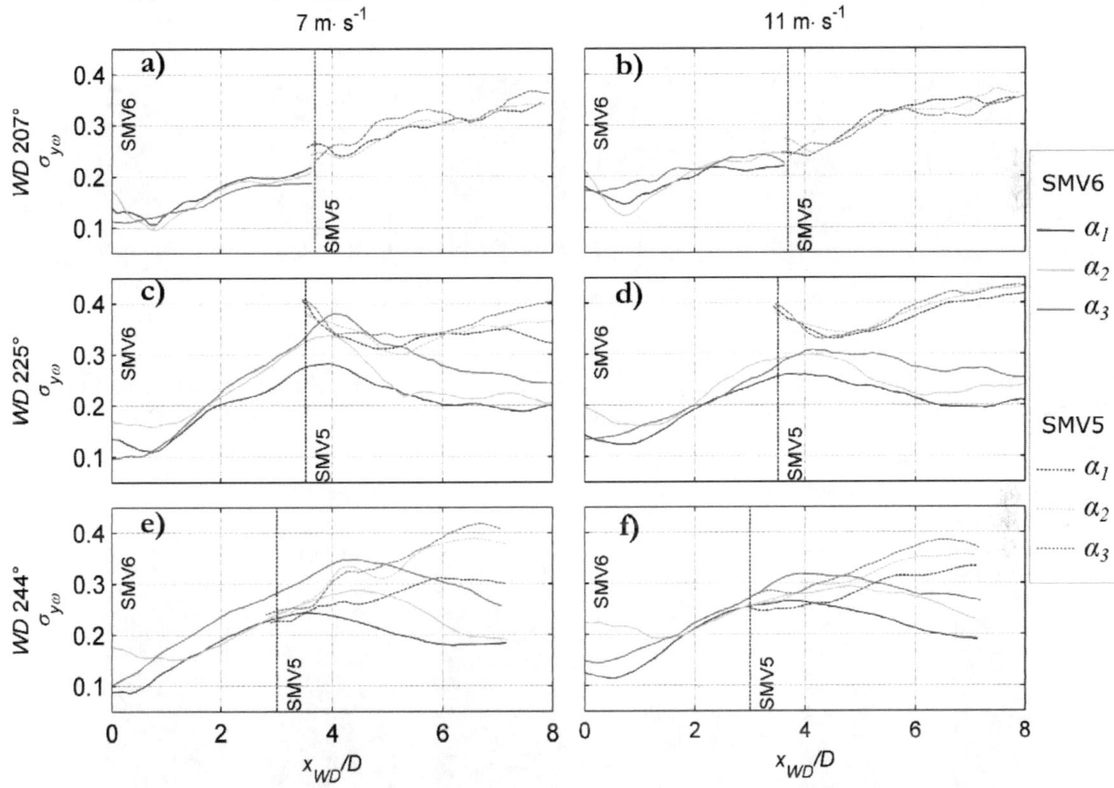

Figure 5. Standard deviation from wake center position ($\sigma_{y\omega}$) as a function of the distance. Blue stands for $\alpha_1 = 2.5°$, green for $\alpha_2 = 3.8°$ and red for $\alpha_3 = 5.2°$. The position x_{WD}/D=0 is the position of SMV6 while the position of SMV5 is referenced with a vertical black dotted line; solid colored lines present the $\sigma_{y\omega}$ for SMV6's wake while the dotted lines correspond to SMV5's wake.

In the cases 07225 and 11225 (Figure 7), SMV6 can be divided into two: the first part, $0 < x_{WD}/D < 3.5$ where the wake behaves in a similar fashion to SMV6 in Figure 6, and the part from $x_{WD}/D = 3.5$ onwards where an influence of SMV5 on SMV6 is seen. In α_1 and α_2 of case 07225, the increase in meandering with the distance is clear, with a break at the position SMV5 and the tendency downstream follows as at the beginning. In α_3 the convergence is slower, mainly in the region of SMV5, where a "hump" appears, and from $x_{WD}/D > 3.5$ the ϵ appears to decay constantly all along the wake with the number of samples. α_1 of case 11225 is similar to the α_3 case 07225, but requires more samples to converge. The α_2 of case 11225 presents the same tendency as α_3 of case 07225 but the "hump" is upstream of SMV5. From 200 samples the ϵ looks constant. In the α_3 of case 11225, more environmental fluctuations are measured and the ϵ value decays slowly. In both cases, the wake on SMV5 is more affected by the turbulence generated by SMV6. Notice that convergence is more difficult to achieve than in the previous case, where SMV5 was directly in line with the wake of SMV6.

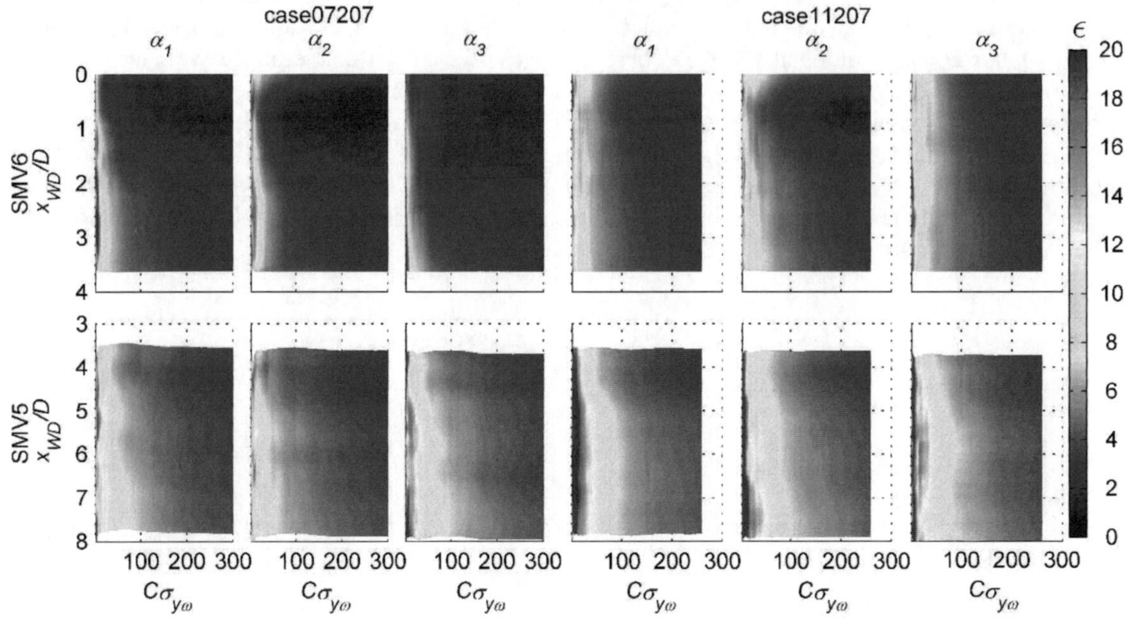

Figure 6. Statistical uncertainty convergence ($\epsilon \times 10^{-2}$ D) as a function of the streamwise position and the cumulative standard deviation of the wake center position ($C\sigma_{y\omega}$). Cases with *WD* = 207°.

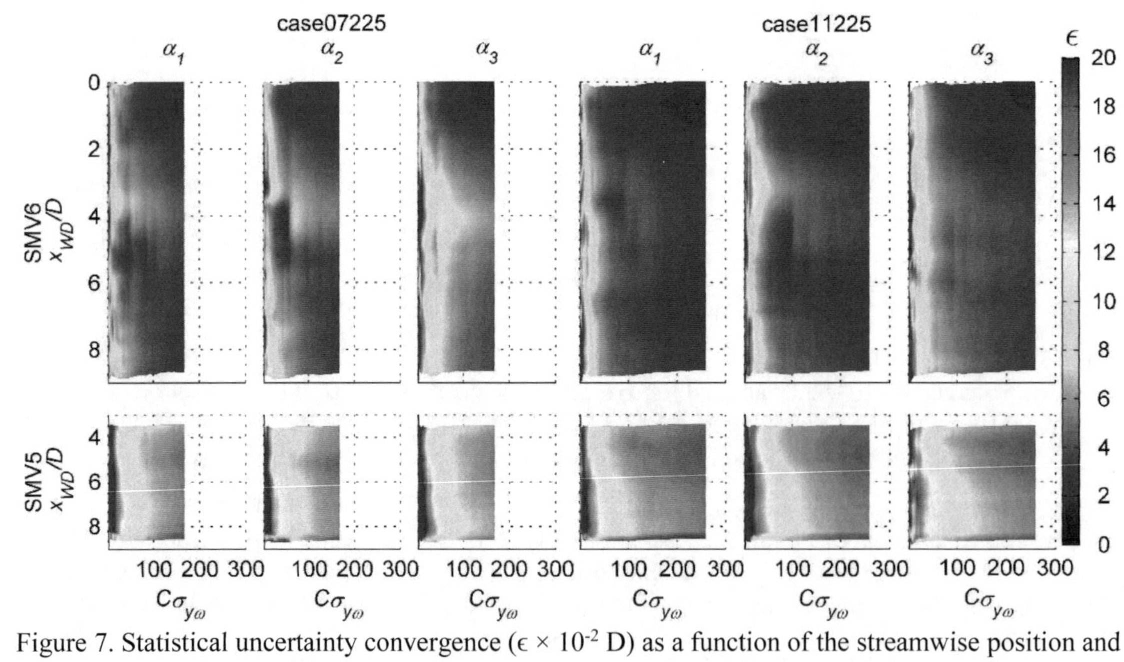

Figure 7. Statistical uncertainty convergence ($\epsilon \times 10^{-2}$ D) as a function of the streamwise position and the cumulative standard deviation of the wake center position ($C\sigma_{y\omega}$). Cases with *WD* 225°.

Figure 8 shows the statistical uncertainty convergence for cases 07244 and 1244. The spacing between hubs' centers is 2 D, even though the influence of SMV5 on SMV6 can be observed. Hence in

SMV6 at x_{WD}/D = 4, more samples are needed to converge. The zone of influence starts approximately 1.5 D upstream of SMV5, making it harder to reach convergence. As the wake propagates downstream (at around 2.5 D downstream of SMV5), the fluctuations are smaller and the error converges faster. The fluctuations appear to be bigger in the higher elevation. Besides the crosswise spacing between WTs, SMV5 receives a turbulent wind and the wind speed does not make any difference. In other words, the statistical uncertainty convergence of SMV5 is similar for the two cases 07244 and 11244.

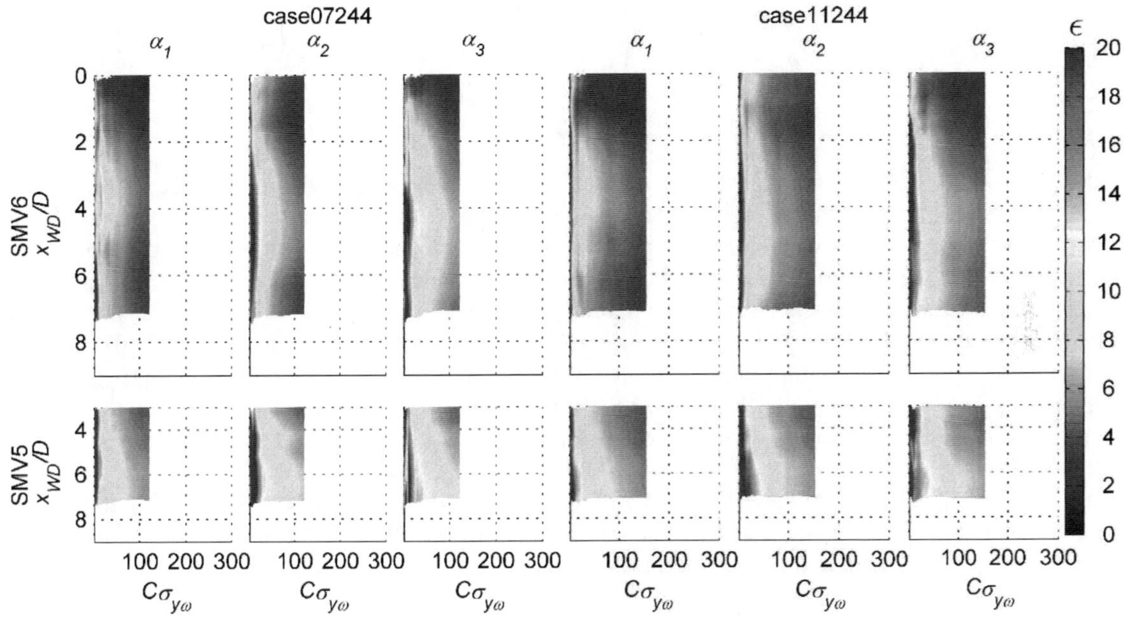

Figure 8. Statistical uncertainty convergence ($\epsilon \times 10^{-2}$ D) as a function of the streamwise position and the cumulative standard deviation of the wake center position ($C\sigma_{y\omega}$). Cases with *WD* 244°.

Table 2 summarizes the statistical uncertainty reached at various positions downstream of SMV6 for each case and for each elevation. As shown in the graphs, the higher values are concentrated where there is interaction between the wake SMV6 and that of SMV5 (shaded in gray). The lower values are reached when the incoming flow is undisturbed (in SMV6 at x_{WD}/D =1 at any elevation). The lateral spacing between the wind turbines does not reduce the fluctuations, and for the cases treated here, a spacing of 2 D produces more fluctuations than if the WT were directly in line with the wake of a WT upstream.

Table 2. Statistical uncertainty ($\epsilon \times 10^{-2}$ D) reached at different distances downstream for the six cases tested. In grey the maximum values at positions of interference of SMV5 on SMV6.

		SMV6 Cases						SMV5 Cases					
	x_{WD}/D	07 207	11 207	07 225	11 225	07 244	11 244	07 207	11 207	07 225	11 225	07 244	11 244
α_1	1	1.4	1.9	1.7	1.5	1.7	1.8						
	2	2	2.5	2.7	2.1	3	2.8						
	3.5	2.3	2.7	3.7	3	4.2	4.1						
	4.5			4.3	3.2	4.2	4.1	2.6	3.1	5	4.1	4.7	4.1
	6			3.3	2.7	3.4	3.6	3.3	4.1	5.1	4.5	5.4	4.9
	7			3.1	2.4	3.2	3.1	3.5	4.3	5.3	5	5.5	5.1
α_2	1	1.3	1.8	2.4	2	2.8	3.4						
	2	1.9	2.6	3.1	2.1	3.1	3.2						

	3.5	2.2	3	4.6	3.4	4.5	4.3						
	4.5			4.8	3.7	5.1	4.5	2.6	3	5.3	4.3	6	4.7
	6			3.4	2.9	4.3	4.4	3.4	4	5	4.7	6.6	5.4
	7			3.3	2.7	3.5	3.8	3.5	4.1	5.4	5.1	6.9	5.6
	1	1.4	2.3	2.2	1.8	2.5	2.6						
	2	1.8	2.6	3.8	2.3	3.7	3.3						
α_3	3.5	2	2.9	5.5	3.1	5.1	4.6						
	4.5			5.1	3.7	6.1	5	2.9	3.2	5.1	4	5.5	4.7
	6			4.1	3.4	5.6	4.5	3.6	3.9	5.3	4.7	6.6	5.8
	7			3.9	3.3	5	4.3	3.5	3.9	5.9	5.1	7.5	6

5. Conclusion

In the present study a 7-month measurement campaign, in which a LiDAR ground based system was used, has been analyzed. The target was to analyze the unsteady wake behavior of two wind turbines. The LiDAR measurements were categorized according to atmospheric conditions. To determine the categories, three filter criteria were applied: 1) Neutral atmosphere corresponding to $|L| \geq 1000$. In this case, the WT wake recovers more slowly than in unstable/convective conditions [22]; 2) three wind directions (*WD*) associated with three different wake patterns; 3) two wind speeds (U) at different WT operational set points. The atmospheric stability was extracted from an external resource (MERRA-2 dataset). The *WD* was determined from LiDAR measurements and corroborated by the wake analysis process. In this process, the PPIs were rotated according to the *WD* and the minimum velocity 4 D downstream of the WT hub was verified, hence confirming or not the *WD*. Only PPIs with the right *WD* were used for the statistical process. To each PPI within the same category, a wake tracking procedure was applied and the wake center was determined. In each category the mean and the standard deviation from the wake center were calculated, and the statistical uncertainty was determined.

Analyses showed that the mean wake center position remains streamwise aligned from 1 D downstream onwards. If parallel wakes develop, the centers of both wakes are displaced, moving closer to each other and the fluctuation increases in both wakes. In the cases analyzed here, the maximum crosswise spacing between wind turbines was 2 D. This close spacing affects the neighboring wakes. Since the WTs are also separated streamwise, the effects seen at the wake center depend on its position. The wake center of the first WT deviates towards the second WT and this effect disappears with the distance downstream as the wake aligns with its WT hub origin. The wake that starts second does not deviate from its center but experiences more fluctuations (effect of the wake upstream).

Wake meandering, determined by the standard deviation from the wake's center, increases from 0.1 D at 1 D downstream of the WT location to 0.19 at 3 D downstream. A similar value was found in wind tunnel experiments [26], but this value is conditioned by the nonexistence of neighboring wakes. The statistical uncertainty convergence analysis showed that 120 samples are enough to reach a statistical uncertainty of 8×10^{-2} D for conditions of wakes' interaction, while for non-disturbed incoming wind conditions, the same number of samples provide a statistical uncertainty of 2×10^{-2} D.

Acknowledgements

This research was funded by the French National Research Agency, under the project SMARTEOLE, reference ANR-14-CE05-0034.

References

[1] L.J. Vermeer, J.N. Sørensen, A. Crespo, Wind turbine wake aerodynamics, Prog. Aerosp. Sci. 39 (2003) 467–510. doi:10.1016/S0376-0421(03)00078-2.
[2] M. Bastankhah, F. Porté-Agel, A new analytical model for wind-turbine wakes, Renew. Energy. 70 (2014) 116–123. doi:10.1016/j.renene.2014.01.002.
[3] T.J. Larsen, H.A. Madsen, G.C. Larsen, K.S. Hansen, Validation of the dynamic wake meander model for loads and power production in the Egmond aan Zee wind farm, Wind Energy. 16

(2013) 605–624. doi:10.1002/we.1563.
[4] Y.-T. Wu, F. Porté-Agel, Atmospheric Turbulence Effects on Wind-Turbine Wakes: An LES Study, Energies. 5 (2012) 5340–5362. doi:10.3390/en5125340.
[5] A. Jimenez, A. Crespo, E. Migoya, J. Garcia, Advances in large-eddy simulation of a wind turbine wake, J. Phys. Conf. Ser. 75 (2007) 12041. doi:10.1088/1742-6596/75/1/012041.
[6] G. España, S. Aubrun, S. Loyer, P. Devinant, Wind tunnel study of the wake meandering downstream of a modelled wind turbine as an effect of large scale turbulent eddies, J. Wind Eng. Ind. Aerodyn. 101 (2012) 24–33. doi:10.1016/j.jweia.2011.10.011.
[7] L.P. Chamorro, F. Porté-Agel, A Wind-Tunnel Investigation of Wind-Turbine Wakes: Boundary-Layer Turbulence Effects, Boundary-Layer Meteorol. 132 (2009) 129–149. doi:10.1007/s10546-009-9380-8.
[8] R.J. Barthelmie, S.T. Frandsen, M.N. Nielsen, S.C. Pryor, P.-E. Rethore, H.E. Jørgensen, Modelling and measurements of power losses and turbulence intensity in wind turbine wakes at Middelgrunden offshore wind farm, Wind Energy. 10 (2007) 517–528. doi:10.1002/we.238.
[9] J.K. Lundquist, E.S. Takle, M. Boquet, B. Kosovic, M.E. Rhodes, D. Rajewski, R. Doorenboos, S. Irvin, M.L. Aitken, K. Friedrich, P.T. Quelet, J. Rana, C. St. Martin, B. Vanderwende, W. Rochelle, Lidar observations of interacting wind turbine wakes in an onshore wind farm, Ewea 2014. (2014) 33–36.
[10] V.-M. Kumer, J. Reuder, B. Svardal, C. Sætre, P. Eecen, Characterisation of Single Wind Turbine Wakes with Static and Scanning WINTWEX-W LiDAR Data, Energy Procedia. 80 (2015) 245–254. doi:10.1016/j.egypro.2015.11.428.
[11] G.C. Larsen, E. Machefaux, A. Chougule, Wake meandering under non-neutral atmospheric stability conditions - theory and facts, J. Phys. Conf. Ser. 625 (2015) 12036. http://stacks.iop.org/1742-6596/625/i=1/a=012036.
[12] M.E. Rhodes, J.K. Lundquist, The Effect of Wind-Turbine Wakes on Summertime US Midwest Atmospheric Wind Profiles as Observed with Ground-Based Doppler Lidar, Boundary-Layer Meteorol. 149 (2013) 85–103. doi:10.1007/s10546-013-9834-x.
[13] A. Peña, C.B. Hasager, M. Badger, R.J. Barthelmie, F. Bingöl, J.-P. Cariou, S. Emeis, S.T. Frandsen, M. Harris, I. Karagali, S.E. Larsen, J. Mann, T. Mikkelsen, M. Pitter, S. Pryor, A. Sathe, D. Schlipf, C. Slinger, R. Wagner, Remote Sensing for Wind Energy DTU Wind Energy E-Report Remote Sensing for Wind Energy, 2015.
[14] C.J. Grund, R.M. Banta, J.L. George, J.N. Howell, M.J. Post, R.A. Richer, A.M. Weickmann, High-resolution doppler lidar for boundary layer and cloud research, J. Atmos. Ocean. Technol. 18 (2001) 376–393. doi:10.1175/1520-0426(2001)018<0376:HRDLFB>2.0.CO;2.
[15] S.C. Tucker, C.J. Senff, A.M. Weickmann, W.A. Brewer, R.M. Banta, S.P. Sandberg, D.C. Law, R.M. Hardesty, Doppler Lidar Estimation of Mixing Height Using Turbulence, Shear, and Aerosol Profiles, J. Atmos. Ocean. Technol. 26 (2009) 673–688. doi:10.1175/2008JTECHA1157.1.
[16] J.-J. Trujillo, F. Bingöl, G.C. Larsen, J. Mann, M. Kühn, Light detection and ranging measurements of wake dynamics. Part II: two-dimensional scanning, Wind Energy. 14 (2011) 61–75. doi:10.1002/we.402.
[17] D. Schlipf, D. Trabucchi, O. Bischoff, M. Hofsäß, J. Mann, T. Mikkelsen, A. Rettenmeier, J.-J. Trujillo, M. Kühn, Testing of Frozen Turbulence Hypothesis for Wind Turbine Applications with a Scanning LIDAR System, … Symp. …. 12 (2010) 5410.
[18] S. Aubrun, E. Torres-Garcia, M. Boquet, O. Coupiac, N. Girard, Wind turbine wake tracking and its correlations with wind turbine monitoring sensors. Preliminary results, J. Phys. Conf. Ser. 753 (2016) 32003. doi:10.1088/1742-6596/753/3/032003.
[19] R. Wagner, Comparison test of WLS200S-22 (Final), DTU Wind Energy, Denmark, 2014.
[20] A. Molod, L. Takacs, M. Suarez, J. Bacmeister, Development of the GEOS-5 atmospheric general circulation model: evolution from MERRA to MERRA2, Geosci. Model Dev. 8 (2015) 1339–1356. doi:10.5194/gmd-8-1339-2015.

[21] D. Li, C. Schmitt, C. Meissner, A. Vignaroli, Stability classification for CFD simulations in complex terrain, EWEA Resorce Assess. (2013).
[22] G.V. Iungo, F. Porté-Agel, Volumetric Lidar Scanning of Wind Turbine Wakes under Convective and Neutral Atmospheric Stability Regimes, J. Atmos. Ocean. Technol. 31 (2014) 2035–2048. doi:10.1175/JTECH-D-13-00252.1.
[23] VDI guideline 3783/12, Physical modelling of flow and dispersion processes in the atmospheric boundary layer - Application of wind tunnels, (2000).
[24] Y.-A. Muller, S. Aubrun, C. Masson, Determination of real-time predictors of the wind turbine wake meandering, Exp. Fluids. 56 (2015) 53. doi:10.1007/s00348-015-1923-9.
[25] L.H. Benedict, R.D. Gould, Towards better uncertainty estimates for turbulence statistics, Exp. Fluids. 22 (1996) 129–136. doi:10.1007/s003480050030.
[26] S. Aubrun, T.F. Tchouaké, G. España, G. Larsen, J. Mann, F. Bingöl, Comparison between Wind Tunnel and Field Experiments on Wind Turbine Wake Meandering, in: M. Oberlack, J. Peinke, A. Talamelli, L. Castillo, M. Hölling (Eds.), Prog. Turbul. Wind Energy IV Proc. iTi Conf. Turbul. 2010, Springer Berlin Heidelberg, Berlin, Heidelberg, 2012: pp. 223–226. doi:10.1007/978-3-642-28968-2_47.

IOP Conf. Series: Journal of Physics: Conf. Series **854** (2017) 012046 doi:10.1088/1742-6596/854/1/012046

Challenges in simulating coastal effects on an offshore wind farm

M. P. van der Laan[1], A. Peña[1], P. Volker[1], K. S. Hansen[2], N. N. Sørensen[1], S. Ott[1], C. B. Hasager[1]

[1]Technical University of Denmark, DTU Wind Energy, Risø Campus, DK-4000 Roskilde, Denmark

[2]Technical University of Denmark, DTU Wind Energy, Lyngby Campus, DK-2800 Kgs. Lyngby, Denmark

E-mail: plaa@dtu.dk

Abstract. The effect of a coastline on an offshore wind farm is investigated with a Reynolds-averaged Navier-Stokes (RANS) model. The trends of the RANS model compare relatively well with results from a mesoscale model and measurements of wind turbine power. In addition, challenges of modeling a large domain in RANS are discussed.

1. Introduction

Wind turbine wakes can cause energy losses in wind farms [1] and their effect needs to be modeled in order to design energy efficient wind farm layouts [2]. Wake losses in offshore wind farms are often modeled by assuming offshore conditions for all wind directions [3, 4, 5, 6, 7]; however, many offshore wind farms are build in the vicinity of a coastline. Dörenkämper et al.[8] used Large Eddy Simulations (LES) of an offshore wind farm to show that a varying distance between the coast and a wind farm can have a large influence on the wind turbine power. In addition, coastal effects are clearly visible in measurements of wind turbine power (Supervisory Control and Data Acquisition [SCADA]) of the Danish Anholt offshore wind farm [9]. For westerly and south westerly winds, the distance from the coastline (Danish Peninsula Djursland) to the Anholt wind farm varies between 15 and 50 km. This causes a horizontal wind speed gradient that results in a variation of wind turbine power along the north-south oriented rows, which is clearly visible in the measurements.

The effect of wind turbine wakes can be modeled with different levels of model fidelity [10], ranging from simple engineering wake models up to high fidelity Computational Fluid Dynamics (CFD), e.g. LES. In previous work, we have shown that Reynolds-averaged Navier-Stokes (RANS) simulations of wind turbine wake interaction in a neutral atmospheric surface layer can predict power deficits that compare well with LES [11] and SCADA measurements [7], if a proper turbulence model is used [12]. In this study, we want to investigate if the effect of the Djursland coastline on the Anholt offshore wind farm can be modeled by a RANS setup. The main model challenge is how to model the inflow of the Anholt wind farm including the horizontal wind speed gradient. In the LES of Dörenkämper et al.[8] this model challenge was handled by selecting an inflow profile from a library of precursor simulations representing different distances to the coast. In the current article, we can afford to include the coastline in a large flow domain and model it as a roughness change since the grid requirements of RANS are not as demanding as in LES of atmospheric flows. The resulting horizontal wind speed gradient from RANS is compared with mesoscale simulations of Djursland using the Weather, Research and Forecasting (WRF) Model without the Anholt wind farm [13]. In addition, the horizontal wind speed gradient derived from the SCADA measurements using the official power curve is also compared with RANS. Finally, the simulated power deficits from RANS (including the coastal effects) are validated with the SCADA measurements.

The Anholt offshore wind farm test case is presented in Section 2. The methodology of the RANS simulations are described in Section 3. The comparison of the RANS simulations with the mesoscale simulations and SCADA measurements is discussed in Section 4.

2. Test case

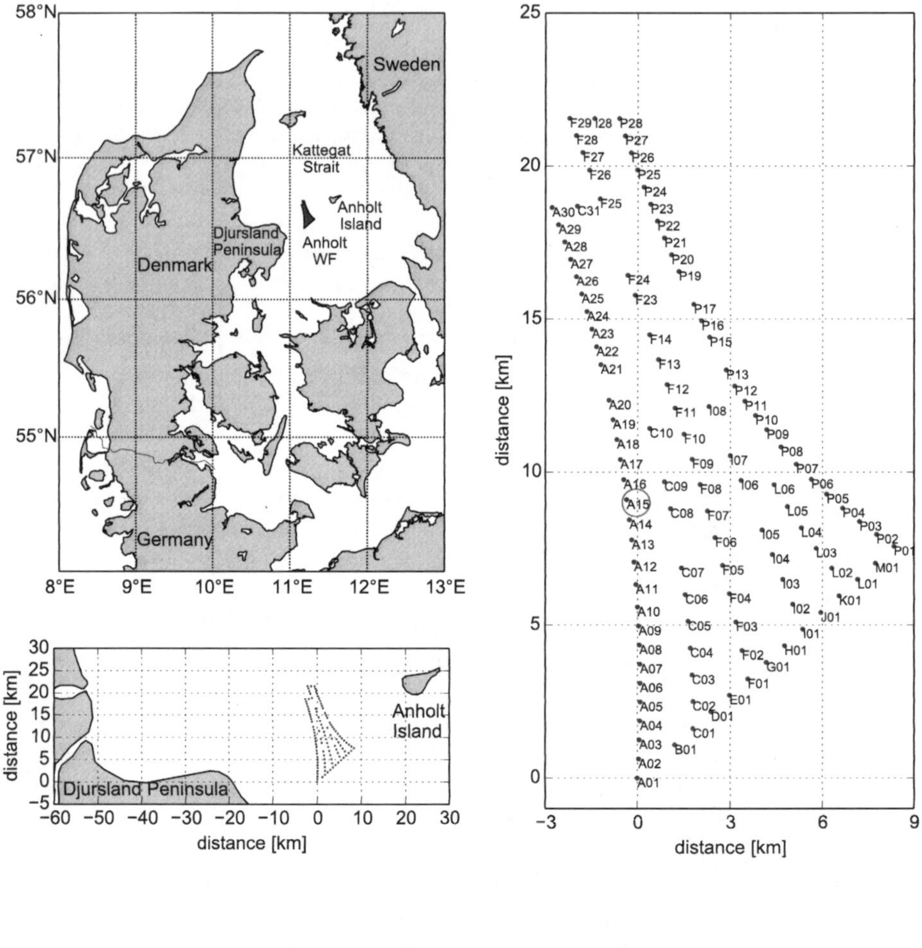

Figure 1. Location and layout of the Anholt offshore wind farm. Top left plot: Denmark, bottom left and right plots are zoomed views showing relative distances. Blue dots represent wind turbines. Red circle marks wind turbine A15 that is used as a reference wind turbine in Figures 6, 7 and 8.

The location and layout of the Anholt offshore wind farm is shown in Figure 1. The Anholt wind farm is located in the Kattegat Strait, 15 km east of the Djursland Peninsula, and it fits in a rectangular box of 12 km by 22 km. Anholt consists of 111 Siemens 3.6 MW wind turbines with a rotor diameter and hub height of 120 and 81.6 m, respectively. The smallest distance between the wind turbines is 4.9 rotor diameters. For westerly wind directions, the distance from the Anholt wind farm to the coast varies between 15 and 50 km. In Figure 2, a wind rose at the middle of the Anholt wind farm is extracted from WRF simulations of Denmark [13], for the year 2014. The Anholt wind farm is affected by the wake of the Djursland Peninsula between 200-270°, which is about 30% of the wind rose. We are interested in a wind direction sector where the distance of the coast to the Anholt wind farm strongly varies for the north-south rows, i.e. 240-270°.

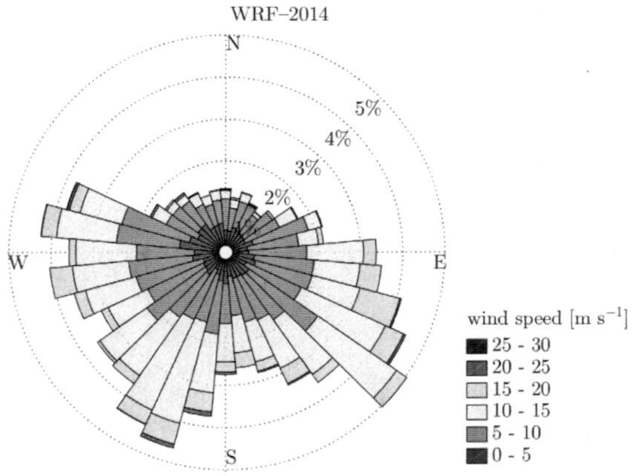

Figure 2. Wind rose of Anholt wind farm for the year 2014, simulated by WRF.

DONG Energy has provided raw SCADA measurements of all wind turbines, measured for 2.5 years between 2013-01-01 and 2015-07-01. The SCADA measurements are filtered for a wind speed bin of 9 ± 1 m/s, where the reference wind speed is determined from the power of a group of 4 wind turbines using the official power curve. For the wind direction sectors 230-270°, 270-280° and 280-310°, wind turbines A22-A25, A17-A20, A23-A26 are used to calculate the reference wind speed, respectively. The reference wind direction is determined from wind turbines A01, A16, A30, F29, P28 and P01 using yaw sensors that are calibrated with the peak deficit of neighboring wind turbines. One should keep in mind that it is difficult to obtain a global reference wind speed and wind direction for westerly wind directions because the inflow conditions vary in the north-south oriented rows. In addition, the SCADA measurements have been filtered for neutral and near-neutral atmospheric stability [14] using the WRF results.

3. Methodology

A brief description of the RANS setup is presented in this Section, while a full description can be found in previous work [15]. The RANS equations are solved in EllipSys3D, an incompressible flow solver developed at DTU Wind Energy [16, 17]. The turbulence is modeled by an alternative k-ε model that is suited to model wind turbines wakes (k-ε-f_P model [12]) and produces a finite boundary layer height by limiting the global turbulent length scale [18]. In addition, Coriolis forces are added to the momentum equations, which means that wind veer is also present in the flow domain. Only neutral atmospheric stability is considered. The effect of the coastline is modeled by roughness only and terrain heights are not modeled.

The numerical domain is plotted in Figure 3, where every 16th grid line is shown. The domain has a size of $300 \times 300 \times 0.9$ km^3 in x, y and z (height) direction, respectively. A green area is

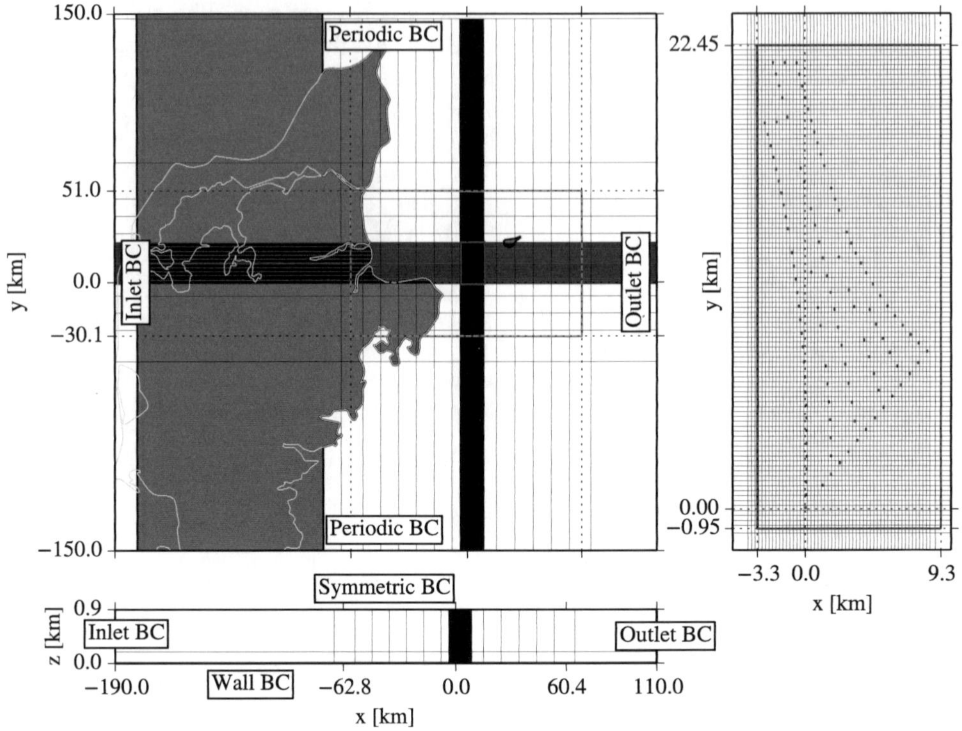

Figure 3. Grid, boundary conditions and roughness of wind farm simulations. Left top plot: top view, Left bottom plot: side view, right plot is a zoomed view of the left top plot. Every 16th grid line is plotted. Green area in left top plot indicates the modeled land with a roughness of 0.05 m, while cyan line indicates the real coastline. Red and blue boxes are regions with uniform horizontal spacing, blue squares in the right plot are actuator disks.

shown in the top left plot of Figure 3, which indicates where a land roughness is applied. Both the east coast of Denmark and the Anholt island are included. The land areas are modeled by a constant land roughness of 0.05 m, representing farmland with open appearances. The sea roughness is set to 10^{-4} m. The real coastline is plotted in the top left Figure 3 as a cyan line. The modeled land roughness deviates from the real roughness for two reasons. In order to use periodic conditions at the boundaries at $y = \pm 150$ km, the coastline is changed to a vertical line at $x = -78$ km for 120 km $<$ y $<$ -90 km. In addition, the land roughness area is 100 km extended upstream (in x direction), up to $x = -178$ km. A sea roughness is used in the first 12 km of domain such that the roughness at inlet boundary condition (BC) ($x = -190$ km) is in balance with the roughness at the outlet BC ($x = 110$ km), where fully developed conditions are assumed. If land roughness would be used at the inlet, the inlet Atmospheric Boundary Layer (ABL) profile based on land roughness would not develop to an ABL profile based on sea roughness (calculated by a precursor simulation based on sea roughness), after the roughness change, for $x \to \infty$, because of mass conservation. To illustrate this issue, a RANS simulation of a 100 km long domain has been carried out, where the first 20 km is land roughness and the remaining 80 km is based on sea roughness. The inlet profile is based on a precursor simulation with land roughness. In Figure 4, the ABL profiles at three downstream distances are compared with the precursor profiles based on land and sea roughnesses. Figure 4 shows that the ABL profile from the inlet does not develop to the precursor profile based on the sea roughness because the total mass flow of the land profile does not match the total mass flow of sea profile.

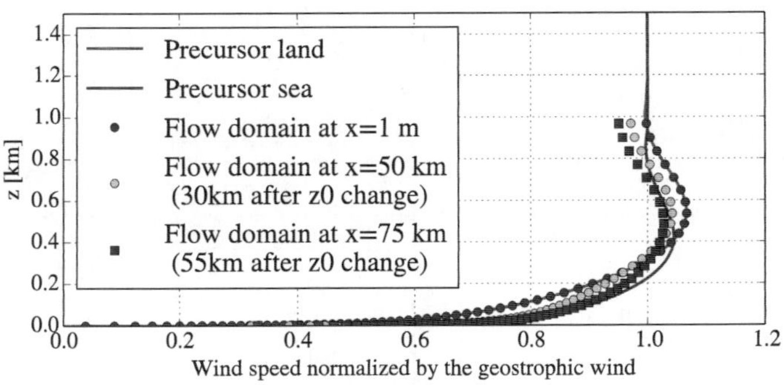

Figure 4. Downstream development of an ABL profile after a roughness change from land to sea using an inlet profile based on land that is not in balance with the sea roughness used at the outlet.

The inlet profiles that are applied to the inlet BC are calculated with a precursor simulation using a geostrophic wind of 10.5 m/s, the sea roughness (10^{-4} m), a Coriolis parameter of $f_c = 1.22 \times 10^{-4}$ 1/s (latitude of 56.6°) and a maximum turbulent length scale of $\ell_{t,\max} = 0.00027\frac{G}{f_c} = 23.3$ m, following Blackadar [19]. The results of the precursor simulation are shown in Figure 5, where the bottom figures are zoomed views of the top figures. Note that we do not use the additional max limiter on ε that we have used in previous work [15] to avoid an overshoot in the turbulent length scale because its effect is negligible. The other BCs in the numerical domain of Figure 3 are a wall boundary at $z = 0$ suited to satisfy a logarithmic profile [20] and a symmetric BC at the $z = 0.9$ km is applied. Ideally, one would prefer to use a taller domain height; however, the combination of the large domain with roughness changes and an ABL profile with low eddy viscosity above the ABL height ($z \approx 1$ km) can result in numerical instabilities. When a low domain height of 0.9 km is used, these numerical instabilities disappear.

In Figure 3, red and blue rectangles indicate regions where a uniform horizontal spacing of 600 m and 15 m is applied, respectively. Around the blue rectangle, the grid spacing grows outwards from 15 m to 600 m with a grow ratio of approximately 1.2. A spacing of 15 m, or $D/8$, is based on grid study of a single wind turbine wake [12]. The 600 m ($5D$) spacing is based on a grid study presented in Section 4.1. The total amount of cells is 120 million. Without modeling the coastline the total amount of cells would be around 100 million, which means that the coastline adds 20% of cells to the grid. The first cell height is set to 0.05 m.

The wind turbines are modeled as actuator disks [21] with variable forcing, where a fixed force distribution is scaled by the disk averaged velocity and a thrust coefficient that is function of the disk averaged velocity [11].

4. Results and Discussion

4.1. Wind farm inflow conditions and grid sensitivity study

Figure 6 shows the wind speed at row A calculated by RANS and WRF (both without the wind farm present in the simulations). The WRF data is binned in 10 degree bins at the location of wind turbine A15 and it is extracted from the same period as the SCADA measurements. The WRF data is filtered for neutral conditions, where only data points with a Monin Obukhov length larger than 500 m are kept. The Monin Obukhov length is calculated at the first cell

IOP Conf. Series: Journal of Physics: Conf. Series **854** (2017) 012046 doi:10.1088/1742-6596/854/1/012046

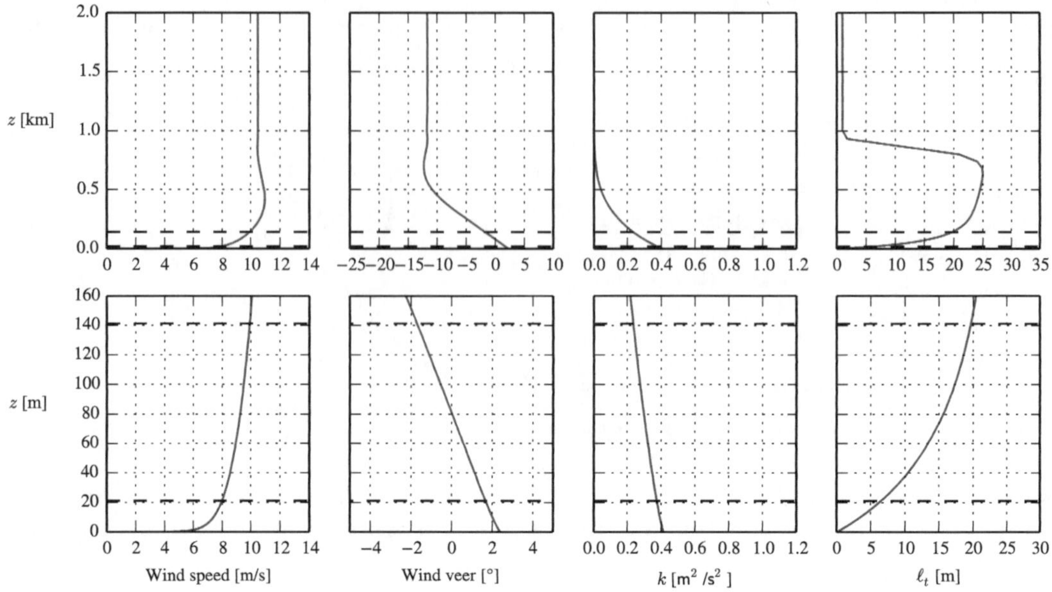

Figure 5. Inlet profiles calculated by the precursor and rotated for zero wind veer at hub height. Bottom plots are a zoomed view of the top plots. Rotor area is shown as black dashed lines.

located at 14 m. Note that the WRF data is not filtered for wind speed because the number of data points becomes too small if wind speed filtering is applied. The RANS results are made a follows. First a parametric run of 41 different freestream wind directions between 245° and 285° with an interval of 1° is carried out, without the wind farm present in the flow domain. Secondly, the wind direction at the location of wind turbine A15 is probed and it is used to bin all RANS simulations in 10° intervals from which a linear bin average is calculated. The re-binning is necessary because the freestream wind direction at the inlet changes downstream due to the interaction of two roughness changes with the Coriolis forces (sea → land → sea, while going from the inlet towards wind turbine A15). For example, the RANS simulation for a freestream wind direction of 285° calculates a wind direction of 281° at the location of wind turbine A15.

Three RANS results are shown in Figure 6 that are made by three different horizontal spacings in the red rectangle from Figure 3, namely 300 m, 600 m and 1200 m. Note that the finer spaced area in the blue rectangle from Figure 3 is not present in this grid sensitivity study. Figure 6 shows that there is only a small difference between a spacing of 1200 m and 600 m (mainly for the wind direction sectors 260° ± 5° and 270° ± 5°), while there is no noticeable difference between a spacing of 600 m and 300 m. Therefore, a grid spacing of 600 m is sufficient to model the effect of the coastline.

The measured trends in the horizontal wind speed gradient are reasonably well captured by the RANS and WRF models, although there are differences. The observed mean wind speed gradient in the SCADA measurements is larger than those calculated by RANS and WRF for 250° ± 5° and 260° ± 5°. For a wind direction of 275° and larger, the wake of Djursland does not affect the wind farm, which is observed as a nearly flat curve for the wind direction bin 280° ± 5° in Figure 6, for the RANS results and the SCADA measurements (where a small

IOP Conf. Series: Journal of Physics: Conf. Series **854** (2017) 012046 doi:10.1088/1742-6596/854/1/012046

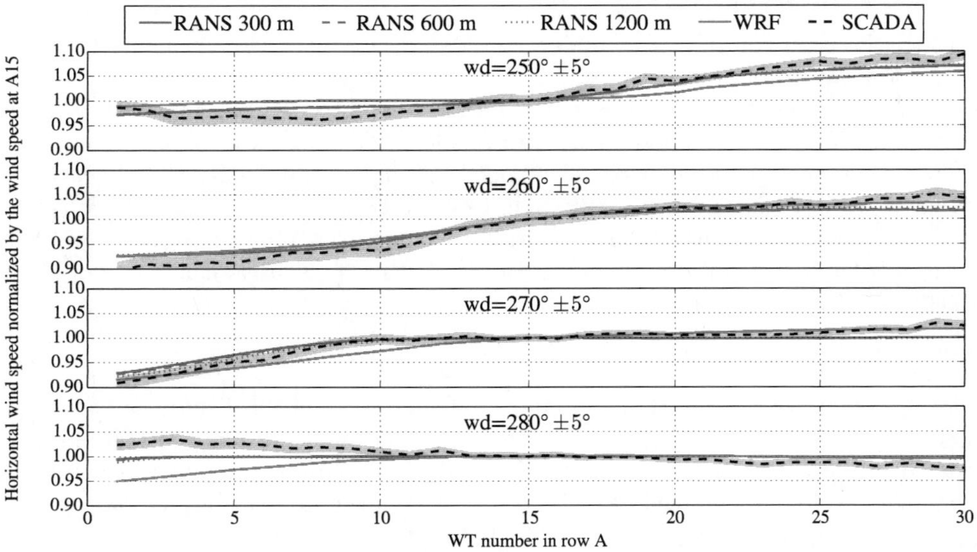

Figure 6. Wind speed at row A for westerly wind directions. RANS and WRF simulations are carried out without wind farm. The three RANS results are made with a horizontal spacing of 300 m, 600 m and 1200 m. The gray region around the SCADA measurements represents the uncertainty of the mean.

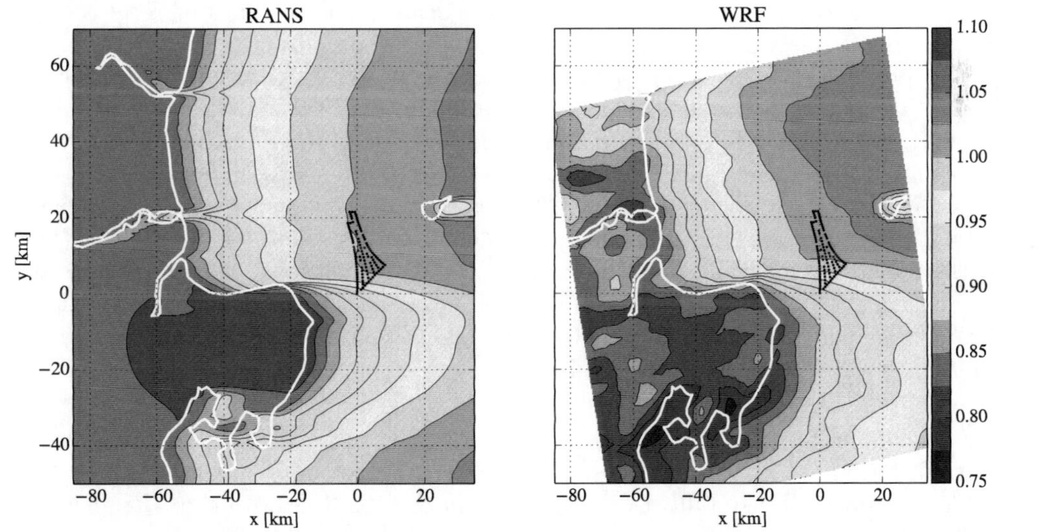

Figure 7. Contours of the horizontal wind speed at hub height for a wind direction of $270° \pm 5°$, calculated by RANS and WRF. Coastlines and wind turbines are shown as a white line and black filled circles.

gradient is observed over the entire row A in the opposite direction). However, the WRF results still show a gradient for the same wind direction bin for wind turbines A1-A9. This indicates that the wake of Djursland is more wide spread in WRF compared to the RANS results and

SCADA measurements. The horizontal wind speed gradient of the other three wind direction bins from WRF show similar trends compared to RANS, although the gradients in WRF are shifted towards the south for $250° \pm 5°$ and $270° \pm 5°$.

Figure 7 depicts wind speed contours at hub height calculated by RANS and WRF, for the wind direction bin $270° \pm 5°$. The wind speed contours from WRF show a complex flow over land because a non uniform roughness is used. The speed up effects from the fjords is clearly visible in the RANS results, while is absent in the results from WRF, possibly because the WRF model is run with a horizontal resolution of 2 km. Overall, the comparison between RANS and WRF is difficult because RANS does not include any mesoscale effects (e.g. a varying geostrophic wind) and the WRF results are not filtered for wind speed.

4.2. Wake effects
Figure 8 shows a comparison of the wind farm simulation from RANS with the SCADA measurements, for 4 wind directions bins of 10°. The left two columns are contour plots of horizontal wind speed calculated by RANS. The right two columns are plots of wind turbine powers normalized by the power of wind turbine A15, where the most right plots are taken from the SCADA measurements. RANS simulation including the wind farm are carried out for every 5° between 230° and 290°, where the wind direction is measured at the location of wind turbine A15. The RANS results (both wind speed contours and powers) are Gaussian averaged with a standard deviation of 5° to represent the wind direction uncertainty in SCADA measurements [22]. Both the RANS results and SCADA measurements show a gradient in wind turbine power along the north-south rows that changes with wind direction. For the wind directions $250° \pm 5°$ and $260° \pm 5°$, the gradient in wind turbine power along the wind turbine rows A26-A30 and F25-F29 is stronger in the SCADA measurements compared to the RANS simulations, which is also visible in Figure 6. As a result, the downstream wind turbines of rows A26-A30 and F25-F29 also produce less power in the RANS simulations compared to the measurements. The wind turbine power from RANS compares well with the SCADA measurements for the other two wind direction bins $270° \pm 5°$ and $280° \pm 5°$, where the influence of Djursland Peninsula is small. While we have shown that RANS setup can model the effect of a coastline on a offshore wind farm, the comparison of the RANS results with the SCADA measurements could be improved by changing the input parameters of the RANS simulations, such as the roughness length of the coastline and the maximum turbulent length scale. A higher roughness of the coastline would increase the horizontal wind speed gradient in the RANS simulations. Setting a lower maximum turbulent length scale (ignoring the relation of Blackadar [19]) would result in a lower ABL height, a reduced eddy viscosity, and a higher wind shear and wind veer, that could affect the magnitude of the horizontal wind speed gradient.

4.3. Model challenges in RANS
In the current RANS setup, it is not possible to use a numerical domain with a height taller than 0.9 km due to numerical instabilities, as described in Section 3. Above the ABL height, a layer of low eddy viscosity exists, which can create numerical waves if the domain is tall enough to include this layer (> 0.9 km) and large enough in the horizontal dimensions. This problem needs to be solved because wind turbines wakes can be affected by the symmetric BC at the top of the domain if the domain height is shallow compared to the rotor diameter. One could possibly damp the fluctuations in the W velocity component with a source term in the W-momentum equation that is only active in the region above the ABL height.

5. Conclusions
A horizontal wind speed gradient along the north-south rows is observed in SCADA measurements of the Anholt wind farm for western wind directions. The wind speed gradient is

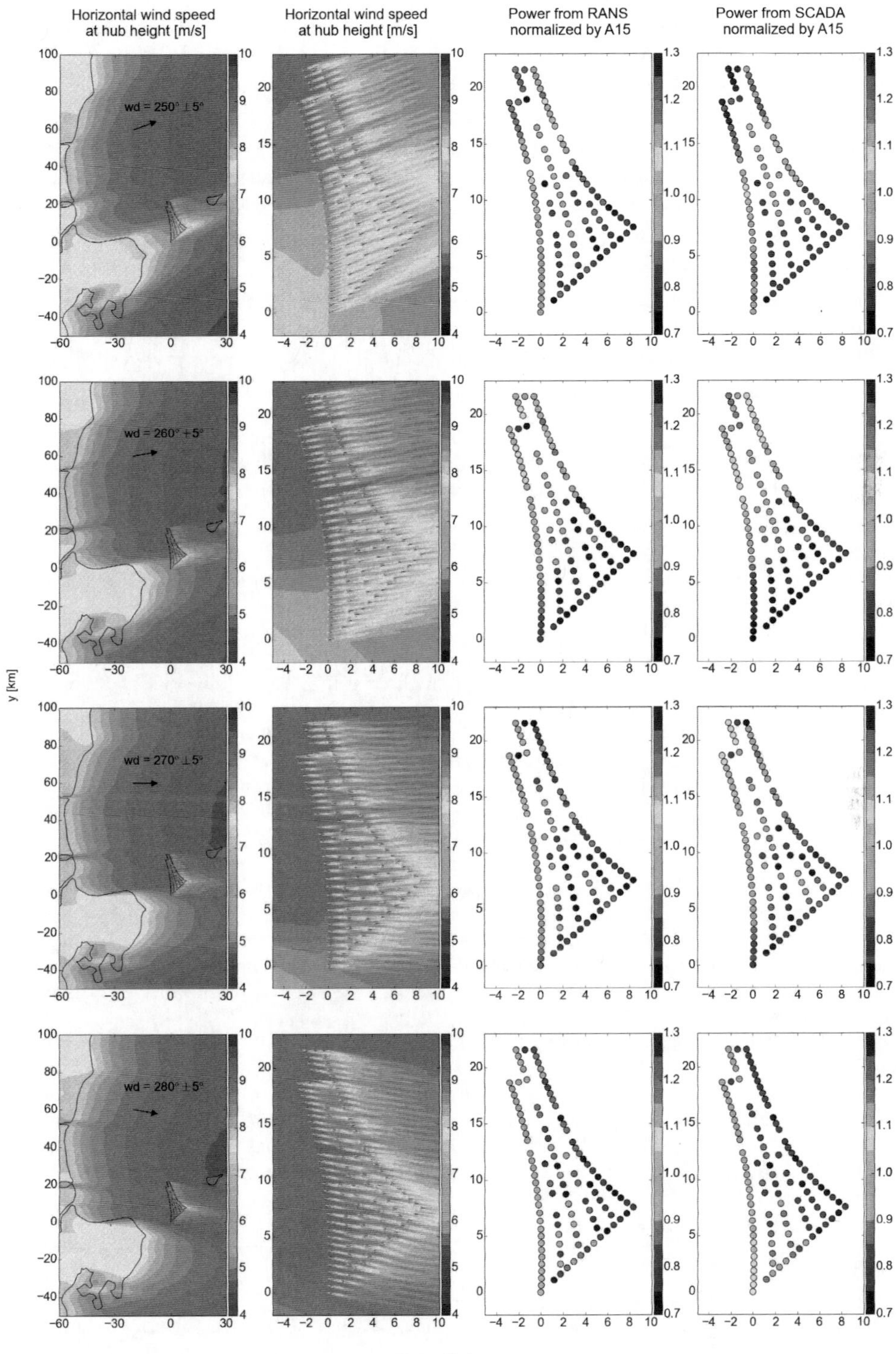

Figure 8. Contours of horizontal wind speed at hub height and normalized wind turbine power, for a wind direction of $250 - 280° \pm 5°$, calculated by RANS and observed in SCADA measurements. Coastlines and wind turbines are shown as a black line and black filled circles, respectively.

caused by a varying distance of the coastline to the wind turbines. The effect of the coastline on the Anholt offshore wind farm is investigated by a RANS model that can simulate wind turbines wakes in a neutral ABL including Coriolis forces. The trends of the horizontal wind speed gradient at the wind farm is reasonably well captured by the RANS model compared to a WRF model and the SCADA measurements. The RANS modeled effect of the coastline on the wind turbines wakes and corresponding power deficits shows similar trends as the SCADA measurements, although a higher gradient in wind turbine power along the north-south rows is observed for the wind direction sectors $250° \pm 5°$ and $260° \pm 5°$. The comparison could be improved by changing the input parameters of the RANS simulations. Overall, we shown that the RANS setup can model the effect of the coastline on the Anholt offshore wind farm.

The results from the RANS model are obtained using a large domain of 300×300 km^2, with a relative low domain height of 0.9 km. It is preferred to use taller domains; however, numerical instabilities are observed for taller domains due to the low eddy viscosity region above the ABL height. This problem needs to be solved in order to improve the RANS setup.

Acknowledgments
We would like to thank DONG Energy and partners for providing the SCADA measurements.

References

[1] Barthelmie R J, Frandsen S T, Nielsen N M, Pryor S C, Réthoré P E and Jørgensen H E 2007 *Wind Energy* **10** 217
[2] King R N, Hamlington P E, P G and Dykes K 2016 *American Institute of Aeronautics and Astronautics, 34th Wind Energy Symposium* 2016-2199
[3] Churchfield M J, Lee S, Moriarty P J, Martinez L A, Leonardi S, Vijayakumar G and Brasseur J G 2012 *AIAA conference 2012, Nashville, USA*
[4] Storey R C, Norris S E, Stol K A and Cater J E 2013 *Wind Energy* **16** 845
[5] Porté-Agel F, Wu Y T and Chen C H 2013 *Energies* **6** 5297
[6] Nilsson K, Ivanell S, Hansen K S, Mikkelsen R, Sørensen J N, Breton S P and Henningson D 2015 *Wind Energy* **18** 499
[7] van der Laan M P, Sørensen N N, Réthoré P E, Mann J, Kelly M C, Troldborg N, Hansen K S and Murcia J P 2015 *Wind Energy* **18** 2065
[8] Dörenkämper M, Witha B, Steinfeld G, Heinemann D and Khn M 2015 *Journal of Wind Engineering and Industrial Aerodynamics* **144** 146–153
[9] Damgaard S 2015 Open issues in wake model validation `http://www.windpower.org/download/2639/08_open_issues_in_wake_model_validationpdf` Vindkraftnet meeting on wakes
[10] Göçmen T, van der Laan M P, Réthoré P E, Peña Diaz A, Larsen G C and Ott S 2016 *Renewable and Sustainable Energy Reviews* **60** 752
[11] van der Laan M P, Sørensen N N, Réthoré P E, Mann J, Kelly M C and Troldborg N 2015 *Wind Energy* **18** 2223
[12] van der Laan M P, Sørensen N N, Réthoré P E, Mann J, Kelly M C, Troldborg N, Schepers J G and Machefaux E 2015 *Wind Energy* **18** 889
[13] Peña A and Hahmann A N 2017 30-year mesoscale model simulations for the noise from wind turbines and risk of cardiovascular disease project Tech. Rep. DTU Wind Energy-E-Report-0055(EN) DTU Wind Energy
[14] Hansen K S, Barthelmie R J, Jensen L E and Sommer A 2012 *Wind Energy* **15** 183
[15] van der Laan M P, Hansen K S, Sørensen N N and Réthoré P E 2015 *Journal of Physics: Conference Series* **524** 1
[16] Sørensen N N 1994 *General purpose flow solver applied to flow over hills* Ph.D. thesis DTU
[17] Michelsen J A 1992 Basis3d - a platform for development of multiblock PDE solvers. Tech. rep. DTU
[18] Apsley D D and Castro I P 1997 *Boundary-Layer Meteorology* **83** 75
[19] Blackadar A K 1962 *Journal of Geophysical Research* **67** 3095
[20] Sørensen N N, Bechmann A, Johansen J, Myllerup L, Botha P, Vinther S and Nielsen B S 2007 *Journal of Physics: Conference series* **75** 1
[21] Mikkelsen R 2003 *Actuator Disc Methods Applied to Wind Turbines* Ph.D. thesis DTU
[22] Gaumond M, Réthoré P E, Ott S, Peña A, Bechmann A and Hansen K S 2014 *Wind Energy* **17** 1169

IOP Conf. Series: Journal of Physics: Conf. Series **854** (2017) 012047 doi:10.1088/1742-6596/854/1/012047

Large-Eddy Simulation of Waked Turbines in a Scaled Wind Farm Facility

J Wang[1], D McLean[1], F Campagnolo[1], T Yu[1] and C L Bottasso[1,2]

[1] Wind Energy Institute, Technische Universität München, Boltzmannstraße 15, D-85748 Garching bei München, Germany

[2] Dipartimento di Scienze e Tecnologie Aerospaziali, Politecnico di Milano, Via La Masa 34, I-20156 Milano, Italy

E-mail: jesse.wang@tum.de

Abstract.
The aim of this paper is to present the numerical simulation of waked scaled wind turbines operating in a boundary layer wind tunnel. The simulation uses a LES-lifting-line numerical model. An immersed boundary method in conjunction with an adequate wall model is used to represent the effects of both the wind turbine nacelle and tower, which are shown to have a considerable effect on the wake behavior. Multi-airfoil data calibrated at different Reynolds numbers are used to account for the lift and drag characteristics at the low and varying Reynolds conditions encountered in the experiments. The present study focuses on low turbulence inflow conditions and inflow non-uniformity due to wind tunnel characteristics, while higher turbulence conditions are considered in a separate study. The numerical model is validated by using experimental data obtained during test campaigns conducted with the scaled wind farm facility. The simulation and experimental results are compared in terms of power capture, rotor thrust, downstream velocity profiles and turbulence intensity.

1. Introduction

Wind farms are collections of wind turbines, often operating in close proximity of one another. During operation, several major phenomena take place within a wind farm: there is an interaction between the atmospheric boundary layer and each individual wind turbine (as well as a more global interaction between the boundary layer and the whole wind farm), and an interaction among upstream and downstream wind turbines through their wakes. The wake produced by the upstream wind turbines, coupled with the atmospheric flow, has a significant influence on the performance of downstream operating machines. In fact, the wake-affected inflow results in lower power output and increased loading for the downstream wind turbines when compared to an isolated condition. A number of control strategies are currently being investigated to optimize the operation of wind farms, including power derating and wake deflection [10].

Our group has developed scaled wind turbine models, operated in a large boundary layer wind tunnel [4]. The machines are actively controlled, and optionally aeroelastically scaled, and are designed to generate realistic wakes notwithstanding the low Reynolds numbers at which they operate. This unique facility, which has been previously used within other research projects [5–7], enables experiments to study aeroservoelasticity, turbine wakes, machine-to-machine interactions, and wind farm control for power maximization and load mitigation. The facility is highly instrumented, allowing for the collection of a wide range of high quality data, both regarding flow conditions and the mechanical response of the machines.

A parallel on-going research effort of our group aims at developing a digital copy of this experimental facility, including the wind turbines and the wind tunnel itself. State-of-the-art numerical methods are used to model the wind tunnel generated flow, the wind turbines (including blades, nacelle, tower), and their shed wakes interacting with the other machines and the wind tunnel. Initial efforts towards the validation of this simulation tool for the single wind turbine case were described in previous work [21]. This paper describes recent improvements in the capabilities of the numerical model for the single wind turbine case, and it also considers multiple waked setups in a number of operating conditions, including the effects of changes in axial induction (power derating).

2. Numerical model

2.1. CFD formulation

The current simulation model is developed within SOWFA [8], a simulation tool based on a standard incompressible solver in the OpenFOAM repository. The current version of the code uses an actuator line method embedded in a large-eddy simulation (LES) environment, coupled with the aeroservoelastic simulator FAST [13]. An immersed boundary (IB) formulation [11] is used to model the wind turbine nacelle and tower. A blended connective scheme algorithm, coupled with the Gamma [12] and center differencing schemes, is also implemented in the solver to minimize numerical diffusion (Gamma is used at near wake, while center differencing scheme is used at far wake). The employed numerical methods are described in detail in Ref. [22].

2.2. Immersed boundary method

An IB method is employed to model the turbine geometry, as this approach does not require a complex body-conformal mesh, which usually implies a tedious process and significant effort in the present structured-grid context [15]. Furthermore, the immersed boundary method can usually achieve an adequate local resolution with minimum overall mesh density, and eliminate the risk of obtaining deteriorated local cells.

The adopted IB approach uses a direct imposition of the boundary conditions (discrete forcing approach). To introduce the characteristics of this method and explain its advantages over the alternative indirect imposition method, the principles of both are briefly reviewed next. The present formulation is derived based on Mittal et al. [15] and Bandringa's work [1]. The incompressible governing equations are given as

$$\nabla \cdot \vec{u} = 0, \tag{1a}$$

$$\frac{\partial \vec{u}}{\partial t} + \vec{u} \cdot \nabla \vec{u} + \frac{1}{\rho} \nabla p - \frac{\mu}{\rho} \nabla^2 \vec{u} = 0, \tag{1b}$$

where $\vec{u}$ is the fluid velocity, p the pressure, ρ the density and μ the dynamic viscosity. The governing equations are first discretized without considering the immersed boundary and are written as

$$[L]\{\underline{U}^*\} = 0, \tag{2}$$

where $\underline{U}^* = (\vec{u}, p)$ specifically indicates the solution of the velocity field without considering the effects of the immersed boundary, while L is the operator representing the discretized Navier-Stokes equations. Next, the immersed boundary condition (Γ_b) is imposed on the flow-field $\underline{U} = \underline{U_\Gamma}$, resulting into Eq. (2) being reformulated as

$$[L + (I - L)\delta(|\vec{x_k} - \vec{x_{ib}}|)]\{\underline{U}\} = \{\underline{U_\Gamma}\}d(|\vec{x_k} - \vec{x_{ib}}|). \tag{3}$$

Here $[I]$ is the identity matrix and δ is the Dirac delta function, which allows for the formula to switch between Eq. (2) and $\underline{U} = U_\Gamma$. However, the cell center of a specified volume mesh location (vector), noted $\overrightarrow{x_k}$, in general does not coincide with the immersed boundary point, $\overrightarrow{x_{ib}}$. The two numerical methods — direct and indirect — differ in the way they address this problem. The indirect method tries to replace δ with a smear function d, which writes

$$[L]\{\underline{U}\} = \{\underline{U_\Gamma} - \underline{U}^*\}d(|\overrightarrow{x_k} - \overrightarrow{x_{ib}}|) + [L]\{\underline{U}^*\}d(|\overrightarrow{x_k} - \overrightarrow{x_{ib}}|). \quad (4)$$

The forcing term resulting from this smooth distribution implies that sharp edges of the geometry cannot be exactly represented. Therefore, the forcing term will smear in space, which is an undesirable effect when one wants to resolve the surface boundary layer of the geometry at high Reynolds numbers.

On the other hand, the direct imposition approach preserves the sharpness of the geometry. No smeared force will be generated in the neighboring cells, but instead an interpolation technique is used to treat the IB cell center using its neighboring cells. The current solver uses a two-dimensional polynomial function with second order accuracy [11]. The Dirichlet immersed boundary condition imposed at location k is formulated as

$$\underline{U_k} = \underline{U_{ib}} + C_0(x_k - x_{ib}) + C_1(y_k - y_{ib}) + C_2(x_k - x_{ib})(y_k - y_{ib}) + C_3(x_k - x_{ib})^2 + C_4(y_k - y_{ib})^2 \quad (5)$$

where all the coefficients C_0, C_1, C_2, C_3 and C_4 are computed using weighted least squares on the extended stencil [11]. x_k and x_{ib} represent x coordinate (scalar) of the specified points. With this interpolation technique, the direct imposition method is able to locate the boundary surface, identify each individual solid and fluid cell, and assign proper boundary conditions to the fluid-solid interface. This way, the sharpness of the geometry surface is preserved and no extra stability constraints are introduced. With the sharp representation of the IB, one can directly impose boundary conditions and a wall model on the surface boundary, which typically results in an improved quality of the solution for high Reynolds number viscous flows.

3. Computational setup

The computational setup is similar to the one described in Ref. [22], and it is shown in figure 1. Zone 1 represents the background mesh, with isotropic grid spacing of $\Delta x = \Delta y = \Delta z = 0.08$ m. Zones 2 and 3 represent progressively denser grids, the latter having a cell size of 0.01 m. The wind turbines, whose diameter D is 1.1 m, are aligned with each other in the streamwise direction (i.e. there is no lateral offset), and longitudinally spaced by 4.1 rotor diameter. Two layouts are considered, comprising either two or three aligned wind turbines. During the experiments, each wind turbine operated under its own pitch and torque control system, and therefore it adjusted automatically to the local wind conditions. On the other hand, the simulated wind turbine rotors spun at the constant angular velocity measured in the experiments. It should be noticed that this has also the effect of limiting the impact of errors in the inflow conditions for downstream wind turbines, which clearly deviate from the real ones on account of inaccuracies in the modeling of the wakes. The closed-loop operation of the simulated wind turbines would amplify such errors, and it will be the subject of specific investigations once the validation and tuning of the present simulation environment has been completed.

A static velocity inflow map, measured experimentally using scanning LiDARs [20], is imposed at the inlet of the computational domain to represent the non-uniform inflow generated by the wind tunnel. This inflow map offers a more realistic input compared to a uniform inlet velocity, as described more in detail in Ref. [22]. In fact, it was observed that matching the inflow non-uniformity ultimately increases the precision of the numerical simulations. In particular, the simulated wake profiles more closely match the experimental data. This is a nontrivial

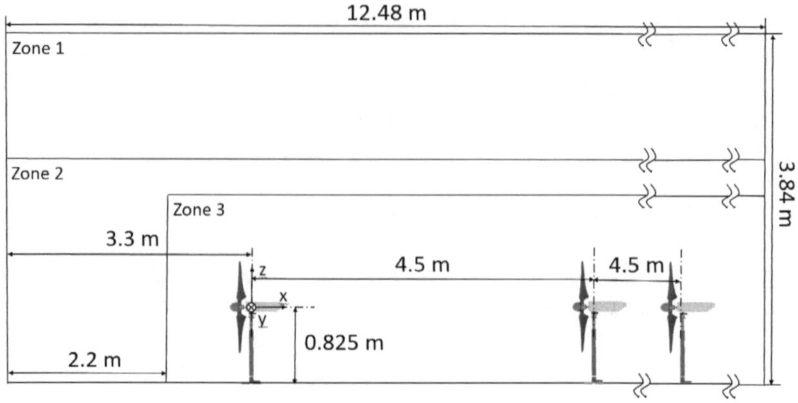

Figure 1: Experimental layout and computational domain.

accomplishment in a relatively large size wind tunnel, where a discrete number of fans and the internal structural arrangement of the channel induce some non-uniformities in the flow field.

Four different axial induction factors (increasing levels of power derating) are considered in the simulations, matching corresponding experimental conditions. All simulations use the same value of the actuator line Gaussian width parameter, ϵ. A blended connective interpolation scheme is employed in the domain, with a deferred correction Gamma bounded scheme in the wind turbine region (β_m=0.45) and an unbounded central differencing scheme in the wake region.

4. Results and analysis

4.1. Single wind turbine case

4.1.1. Integral rotor quantities Table 1 reports the power (P) and thrust (T) for a single wind turbine setup at four different power ratings, namely 1.0, 0.975, 0.95, 0.925. The machine was operating in region II (partial load condition) at a wind speed of 5.97 m/sec measured at hub height, 3D in front of the rotor disk, while the air density was approximately 1.19 kg/m^3. The wind speed was employed to normalize the velocities measured within the wake. Due to the inflow non-uniformity, the rotor-disk average wind speed was however slightly lower, i.e. approximately 5.83 m/sec [22]. Derating is accomplished by holding the rotor angular velocity constant above the wind speed that makes the power production exceeding its derated level at optimal pitch setting, and then using the pitch controller to maintain the power production as the wind speed increases. This means that reducing the demanded power requires either pitching the blades to feather while also slightly decreasing the rotor angular velocity. The model includes the presence of the nacelle and tower, although this has only a very marginal effect on the rotor performance.

Results show that power and thrust are in general in reasonable agreement. However, for increasing derating the error in the power increases, while the error in the thrust decreases. This is probably due to the accuracy of the airfoil polars. In the present simulations, the airfoil lift and drag are computed at each instant of time based on the local angle of attack by interpolating tabulated values at Reynolds 75000 and 90000. Future efforts will try to improve the quality of the polars by identification techniques [3], as these have clearly a crucial importance in a LES-lifting-like formulation.

4.1.2. Wake properties The quantitative verification of the wake characteristics is essential for the validation of simulation tools, as the matching of rotor quantities does clearly not guarantee an accurate modeling of the wake. For the non-derated case, figure 2 shows the time-averaged

Table 1: Rotor power and thrust for a single wind turbine configuration at 1.0, 0.975, 0.95 and 0.925 derating.

	Type	$1.0P$	$0.975P$	$0.95P$	$0.925P$
Power	Exp. $[W]$	45.79	44.36	43.20	42.11
	Sim. $[W]$	45.09	42.88	41.03	39.35
	ΔP %	-1.50	-3.34	-5.02	-6.52
Thrust	Exp. $[N]$	15.18	14.24	13.62	13.10
	Sim. $[N]$	15.79	14.44	13.55	12.82
	ΔT %	4.02	1.48	-0.19	-1.72

velocity profile and turbulence intensity (TI) at hub height at 3D, 4D and 7D. The blue x symbols indicate results obtained by including nacelle and tower in the model (noted CASE 1 in this paper), while the red crosses correspond to the case without nacelle and tower (noted CASE 2). Circles are used to report the experimental values, obtained by triple hot-wire probes [20].

First of all, it is important to note that the wake deficit measured immediately downstream of the rotor agrees with 1D momentum theory expectations, since the BEM-based axial induction factor of the rotor for the non-derated case is approximately 0.31. Moreover, CASE 1 performs better than CASE 2 in terms of velocity profile for all three downstream positions. This highlights the importance of including nacelle and tower in the simulation models (as already noted by other authors [18]), especially for the present scaled wind turbine for which these two components are relatively bigger than in a multi-MW wind turbine. Indeed, the sum of the frontal area of the nacelle and of the portion of tower located within the rotor swept area is $0.037A$, with $A = \pi\frac{D^2}{4}$, while it is $0.023A$ for the 5MW NREL wind turbine [14]. Although this parameter is larger for the G1, we expect that the effects of nacelle and tower on wake evolution might not be negligible even for typical multi-MW wind turbines.

The mean-over-the-rotor velocity difference $\overline{\mathbf{\Delta U}}$ between simulation and experiment is -0.4%, 2.9% and 2.9% for CASE 1 at 3D, 4D and 7D, respectively, while for CASE 2 it is 6.9%, 17.5% and 6.4% at the same respective positions. Moreover, the wake profiles of CASE 1 are closely correlated with the experimental curves for all three downstream positions. Matching is also good at 3D, a position in the near wake. Reference [22] also shows results at 0.56D, a position far closer to the rotor, where results are reasonable but not as accurate as here. While it is still very challenging to achieve high accuracy very close to the rotor (LES-lifting-line doesn't allow fully capturing the near wake behavior), the present results show that relatively good results can be obtained further downstream, where mixing generates a more elementary, smooth and quasi-Gaussian shape of the wake speed deficit.

An accurate modeling of turbulence intensity at the near wake by LES is still difficult to achieve, with some good quality results shown by Ref. [17] and [16]. For the present work, figure 2 in the bottom row reports the time-averaged TI profiles. The mean-over-the-rotor-disk difference ΔTI between simulation and experiment is -18.8%, -17.3% and -25.9% for CASE 1 at 3D, 4D and 7D, respectively, while ΔTI for CASE 2 is -52.1%, -41.2% and -41.3% at the same respective positions. Clearly, the accuracy of the mean TI for CASE 1 is significantly superior to CASE 2, despite still having a relatively large margin of error with respect to the experimental results. Here again the inclusion of nacelle and tower in the model proves to be of great importance.

It could appear surprising that a good match between experimental and numerical wake deficits is observed, while the prediction of the turbulence intensity is unsatisfactory. However, the following considerations could potentially explain this behaviour. First, unlike Pesudo-

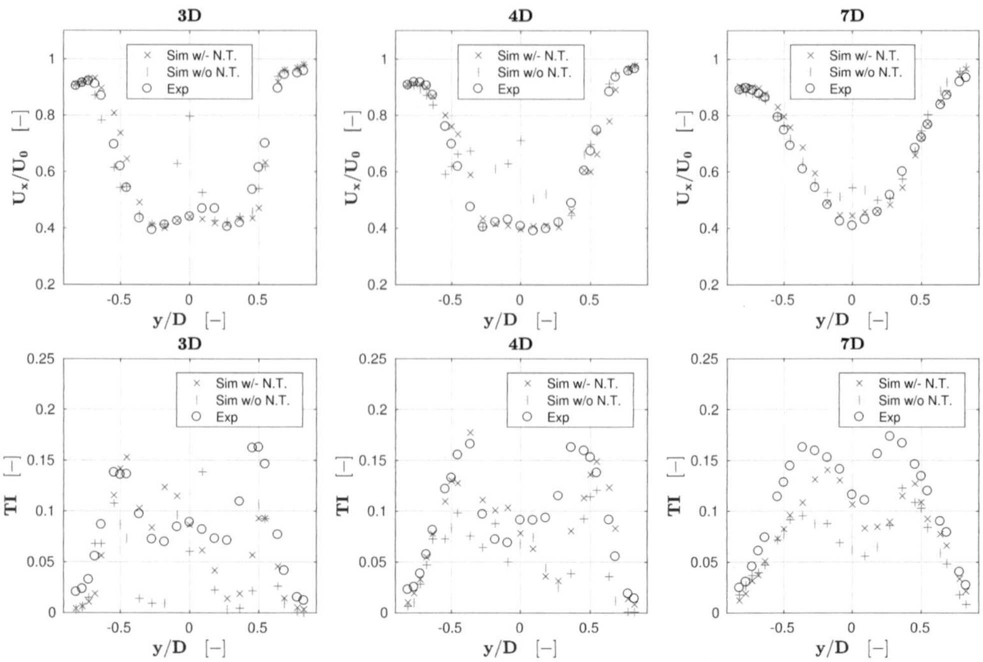

Figure 2: Normalized time-averaged streamwise velocity and TI profiles at hub height for the single wind turbine configuration at 3D, 4D and 7D. Blue x symbols: model includes nacelle and tower (CASE 1); red crosses: model without nacelle and tower (CASE 2); black circles: experimental results.

spectral and Energy-conserving discretization method, the adopted finite-volume method cannot guarantee zero numerical dissipation [2], which can only be limited by employing a fine spatial and time grids, as well as by using dedicated differencing schemes. As described in section 2.1, the low-diffusion Gamma scheme is used in the present LES framework. The results shown in this section were obtained by manually tuning the parameter β_m, a weighing factor for the Gamma Convection bounded scheme that controls the amount of numerical dissipation [12]. By decreasing β_m, one may observe the appearance of numerical oscillations in proximity of the immersed boundaries.These may increase the tendency of the simulation to crash and induce artificial fluctuations that contaminate the velocity profile whose accuracy, in turn, drops significantly. Increasing β_m, on the other hand, increases the upwind-like behavior of the scheme, which results in more turbulent kinetic energy being dissipated, while leaving the velocity profile largely unaffected. Indeed, the numerical dissipation is mainly affecting second order quantities, like TI, while first order quantities, such as velocity, are less influenced by the intrinsic numerical dissipation.

Second, in this paper a spatially non-uniform but temporally steady inflow map was used as inlet condition for the simulations. Such inflow differs from experimental inlet flow, which is characterized by a 2% of turbulence intensity. Such a difference can also be observed in figure 2, where the simulated TI outside the wake is almost null. This implies that the experimental wake recovery should be slightly faster than the simulated one, which in turn means that the experimental wake deficit should be slightly lower than the simulated one. On the other hand, the intrinsic numerical diffusion of the adopted numerical method makes the wake to recover slightly faster than what is observed in reality. This means that, on one side, the lower simulated TI implies a stronger wake deficit, while the intrinsic numerical diffusion implies a lower wake deficit. These two opposite effects thus compensate each other, leading to the good observed matches between the experimental and numerical wake deficits.

4.2. Two aligned wind turbines

4.2.1. Integral rotor quantities Next, we consider the case of two wind turbines, where the upstream one (termed wind turbine No.1) is located 4.1D in front of the second (termed wind turbine No.2) (cf. figure 1). As shown earlier in section 4.1, the wake profile and velocity magnitude for the single wind turbine case appears to be of a good quality. Therefore, it is expected that the inflow conditions for the donwstream machine in the two wind turbine case will also be accurate.

Table 2 reports the power P and thrust T for wind turbine No.2. In the non-derated condition for CASE 1, there is a reasonable agreement for P with respect to the experimental value, with only an about 4% difference. Such accuracy for power is non-trivial to obtain, as its absolute value is much lower than that of wind turbine No.1 and there is also a strong sensitivity of this result to even small variations in the wake shed by the upstream machine. On the other hand, CASE 2 shows a much larger discrepancy in power, which is about 20% larger than the experimental value. In fact, neglecting the nacelle and tower from the model will cause a substantial change in the downstream flow mixing, resulting in the strong differences in the velocity profile and TI shown in figure 2.

Table 2: Comparison of rotor power and thrust for wind turbine No.2 at 1.0, 0.975, 0.95 and 0.925 derating values of wind turbine No.1. Values in brackets represents results for the model without nacelle and tower (CASE 2).

	Type	$1.0P$	$0.975P$	$0.95P$	$0.925P$
Power	Exp. $[W]$	4.41	5.53	6.40	7.31
	Sim. $[W]$	4.23 (5.51)	6.14 (6.83)	7.76 (7.54)	7.69 (8.11)
	ΔP %	-4.08 (22.66)	11.11 (23.65)	18.45 (15.10)	4.17 (9.86)
Thrust	Exp. $[N]$	4.14	4.58	5.00	5.29
	Sim. $[N]$	3.65 (4.16)	4.30 (4.37)	4.78 (4.58)	4.49 (4.73)
	ΔT %	-11.84 (0.57)	-6.09 (-4.46)	-4.43 (-8.40)	-15.23 (-10.56)

The table also illustrates that the accuracy of the results decreases with increasing derating of wind turbine No.1. The reason of this behavior is still under investigation. On the other hand, the results show that the model is able to appreciate the expected power increase in wind turbine No.2 that follows from derating wind turbine No.1.

4.2.2. Wake properties Figure 3 shows the wake properties in terms of time-averaged velocity profile and TI downstream of the two aligned wind turbines. Similarly to the wake analysis for the single wind turbine configuration, results for CASE 1 are well correlated with the experimental values. The mean velocity difference between simulation and experiment across the rotor disk, $\overline{\mathbf{\Delta U}}$, is 1.5%, 2.7% and 2.6% for CASE 1 respectively at 8D, 10D and 11D downstream of wind turbine No.1, while $\overline{\mathbf{\Delta U}}$ for CASE 2 is -8.1%, -10.1% and -8.4% at the same positions. Unlike the single wind turbine setup, the mean wind speeds for CASE 1 are consistently greater than for CASE 2, with the latter showing a larger wake deficit than the former at all three downstream positions. This might be due to a lack of flow mixing when nacelle and tower are not present in the model, and to the larger power capture of wind turbine No.2. The good comparison of wake properties at 11D (slight higher wake recovery), a distance quite far from the upstream wind turbine, demonstrates again the goodness of the present LES formulation based on the Gamma approach. In fact, excessive numerical dissipation has always been an issue for the numerical modeling of wakes for finite volume method.

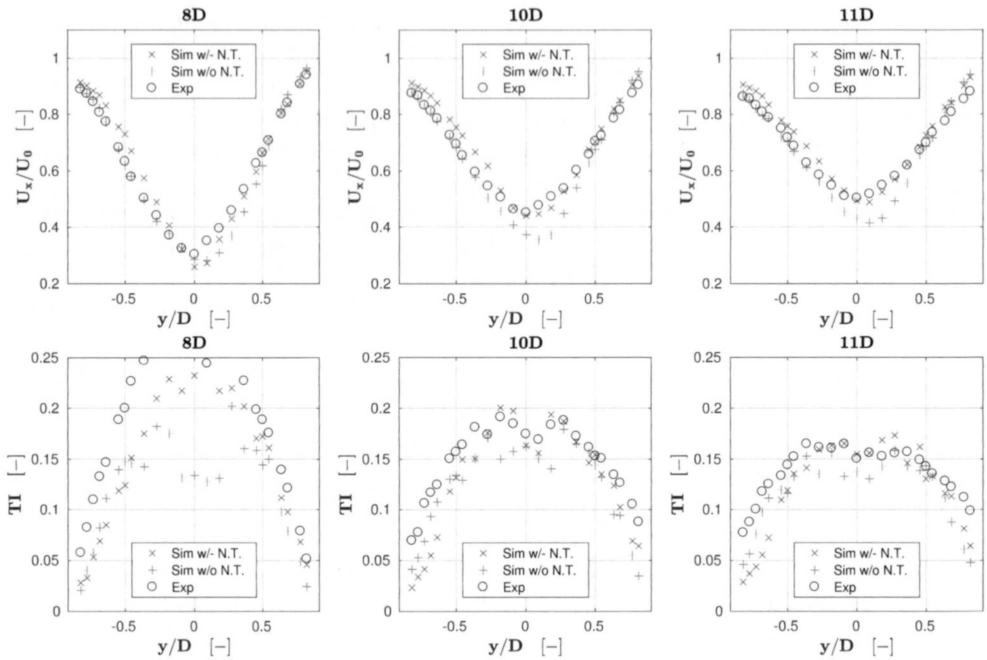

Figure 3: Normalized time-averaged streamwise velocity and TI profiles at hub height for wind turbine No.2 at 8D, 10D and 11D downstream of wind turbine No.1. Blue x symbols: model includes nacelle and tower (CASE 1); red crosses: model without nacelle and tower (CASE 2); black circles: experimental results.

The behavior of TI for the double wind turbine configuration is well predicted in terms of its shape and mean value across the rotor. Indeed, the effects produced by the intrinsic numerical diffusion become less dominant as the upcoming flow becomes turbulent, as shown by Troldborg et al. [19]. Since the upstream wind turbine adds substantial turbulence to the flow impinging on the second wind turbine, the results for the double wind turbine setup appear to be better than in the single wind turbine case by at least 10% on average.

Moreover, by comparing TI for CASE 1 and CASE 2 at 8D, a position that is located 3.9D downstream of wind turbine No.2, it appears that neglecting nacelle and tower from the model dramatically underestimates TI by over 40%. This provides an explanation for the consistent over-estimation of the velocity deficit seen for CASE 2, as discussed earlier in this section.

4.3. Three aligned wind turbines

4.3.1. Integral rotor quantities The three wind turbine configuration is considered only in the non-derated condition. Table 3 reports power capture and thrust for all three operating wind turbines. The power capture for the two front wind turbines shows a slight difference with respect to the values in Table 1 and 2. This is due to a slight different inflow speed in the wind tunnel during the two sets of experiments. In fact, since each wind turbine was operated by its own closed-loop controller, a variation of inflow speed will result in the wind turbines operating at a different set point. The power capture of wind turbine No.3 for CASE 1 shows a 17% difference with respect to the experimental value. This can be attributed to a lack of background

IOP Conf. Series: Journal of Physics: Conf. Series **854** (2017) 012047 doi:10.1088/1742-6596/854/1/012047

turbulence intensity, differences in the airfoil polars, or other effects not yet understood. Here again, CASE 2 shows a much larger error, indicating the necessity of modeling nacelle and tower. Similarly to the two aligned wind turbines cases, we observed that the accuracy of the results decreases as the derating of the upstream wind turbine increases.

Table 3: Comparison of rotor power and thrust for three aligned wind turbines.Values in brackets represents results for the model without nacelle and tower (CASE 2).

	Type	WT No.1	WT No.2	WT No.3
Power	Exp. $[W]$	45.32	4.55	6.86
	Sim. $[W]$	44.94 (45.44)	4.14 (5.23)	5.88 (3.52)
	ΔP %	-0.91 (0.19)	-3.04 (22.33)	-17.92 (-50.88)
Thrust	Exp. $[N]$	14.99	3.48	4.81
	Sim. $[N]$	15.75 (15.79)	3.62 (4.08)	4.11 (3.45)
	ΔT %	5.07 (5.34)	3.98 (17.25)	-14.52 (-28.28)

4.3.2. Wake properties Figure 4 shows normalized instantaneous $\mathbf{U_x}$ and TI fields at a streamwise vertical plane for CASE 1 and CASE 2. The plots reveals the dramatic difference in the flow evolution which results from including or not nacelle and tower in the model. In fact, nacelle and tower generate a substantial amount of turbulence in the field, which in turn induces an enhanced downstream flow mixing. Plots of Reynolds stress, vorticity and turbulent kinetic energy all demonstrate a similar dependence on the presence of nacelle and tower. At the same time, these model features also cause considerable changes in the air flow velocity in the wakes, creating large discrepancies in power output and loading.

Furthermore, the TI plots also indicate that turbulence starts to accumulate after approximately 2D downstream of wind turbine No.2 and 3, whereas it is relatively low immediately downstream of the rotors. As expected, TI is higher downstream of the blade tip and root regions than for the rest of the blade span. The rotor appears to be initially damping, just immediately downstream from it, the incoming TI, which then bursts up again further downstream.

5. Conclusion and outlook

In this paper, the results from a LES-lifting-line simulation formulation were discussed with reference to the experimental measurements obtained on scaled wind turbine models operating in a wind tunnel. Three different setups were considered: a single isolated machine, as well as double and triple aligned and fully waked configurations.

From the comparison of power capture, thrust, wake profile and other flow quantities, it is concluded that the results of the simulation model are in general in a good agreement with the measured data. The quality of the results, however, appears to be decreasing for increasing derating of the front machine, an issue still needs to be properly investigated.

The paper also considered the effects of the inclusion of nacelle and tower in the model. In all cases, results always consistently indicate the crucial importance of modeling the effects of nacelle and tower, as their absence has a strong effect on the mixing of the flow. The nacelle and tower of the scaled models are in general larger than the ones of a real wind turbine, because of the need to house all necessary mechanical and electronic components, including pitch, torque and yaw actuators and various sensors. Because of this, the effects of neglecting nacelle and tower

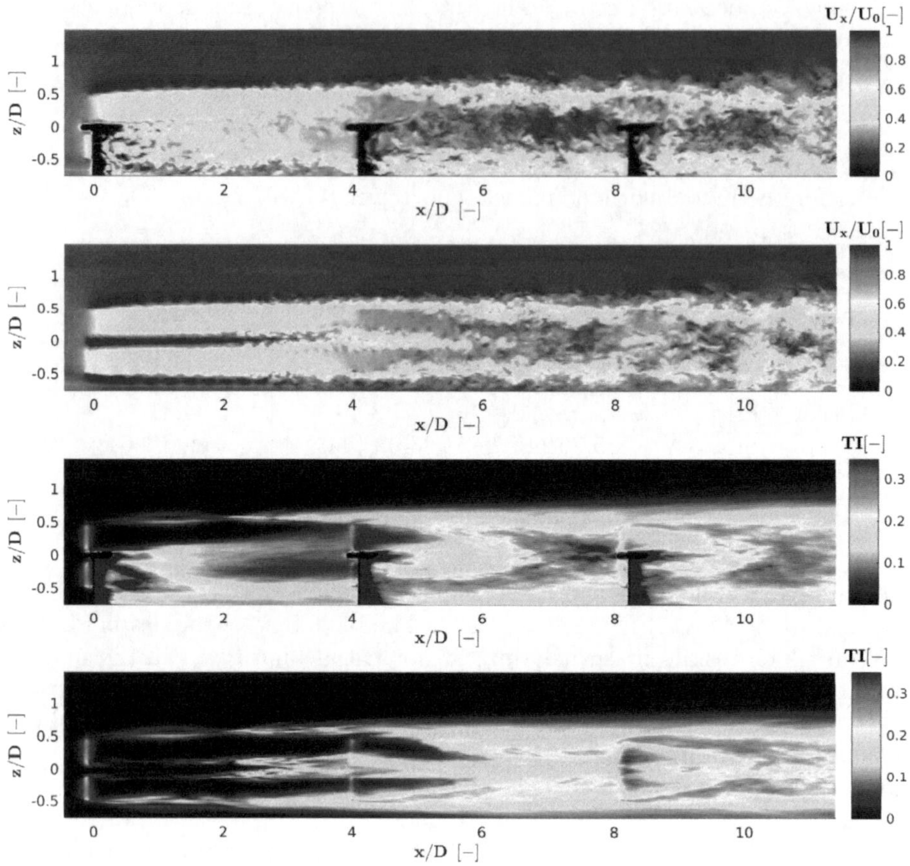

Figure 4: Normalized instantaneous streamwise velocity and TI fields in a vertical plane for the triple wind turbine configuration. From top to bottom: velocity field for CASE 1, velocity field for CASE 2, TI for CASE 1 and TI for CASE 2.

from the model might be more pronounced in this than in a full scale case. Similar conclusions on the importance of a complete model have also been reported by other authors [18].

A parallel study is considering a similar validation activity for cases of higher ambient turbulence. Turbulence generating devices were used in the wind tunnel to generate a turbulent sheared profile, in which the wind turbines were operated in different configurations and at different operating points, as done in the present work. A large-eddy simulation of the turbulence generation will be used to produce a turbulent wind field (same way as in the experiment), which will then be used as input for the subsequent LES-lifting-line simulations of the interacting wind turbines. The study of these conditions, in addition to the low turbulence case considered here, will be used to further the understanding of the present simulation framework, with the final goal of obtaining a digital model of the scaled wind farm facility with predictive capabilities.

References

[1] Bandringa H 2010 *Immersed boundary methods* (Netherlands: University of Groningen)

[2] Barakos G N 2016 An Overview of the CFD Analyses in the MARE-WINT Project *Springer International Publishing*, 275-6

[3] Bottasso C L, Cacciola S, and Iriarte X 2014 Calibration of wind turbine lifting line models from rotor loads *Journal of Wind Engineering and Industrial Aerodynamics*, **124** 29-45

[4] Bottasso C L, Campagnolo F and Petrović V 2014 Wind tunnel testing of scaled wind turbine models: Beyond aerodynamics *Journal of wind engineering and industrial aerodynamics* **127** 11-28

[5] Campagnolo F, Croce A, Manos E, Petrović V, Schreiber J and Bottasso C L 2016 Wind tunnel testing of a. closed-loop wake deflection controller for wind farm power maximization *Journal of Physics: Conference Series* **753**

[6] Campagnolo F, Petrović V, Bottasso C L and Croce A 2016 Wind tunnel testing of wake control strategies *Proceedings of the American Control Conference* 513-18

[7] Campagnolo F, Petrović V, Nanos E, Tan C, Bottasso C L, Paek I, Kim H, and Kim K 2016 Wind tunnel testing of power maximization control strategies applied to a multi-turbine floating wind power platform *Proceedings of the International Offshore and Polar Engineering Conference* 309-16

[8] Churchfield M and Lee S 2012 *Nwtc design codes-sowfa* (USA: http://wind. nrel. gov/designcodes/simulator S./SOWFA)

[9] Van Dooren M, Khn M, Petrović V, Bottasso C, Campagnolo F, Sjholm M, Angelou N, Mikkelsen T, Croce A and Zasso A 2016 Demonstration of synchronised scanning lidar measurements of 2d velocity fields in a boundary-layer wind tunnel *Journal of Physics: Conference Series* **753**

[10] Fleming P, Gebraad P, Lee S, van Wingerden J W, Johnson K, Churchfield M, Michalakes J, Spalart P and Moriarty P 2014 Evaluating techniques for redirecting turbine wakes using sowfa *Renewable Energy* **70** 211-18

[11] Jasak H and Rigler D 2014 Finite volume immersed boundary method for turbulent flow simulations *9th OpenFOAM Workshop*

[12] Jasak H, Weller H and Gosman A 1999 High resolution nvd differencing scheme for arbitrarily unstructured meshes *International journal for numerical methods in fluids* **31(2)** 431-49

[13] Jonkman J and Buhl M L 2004 *Fast user guide* (USA: Technical report, EL-500-29798, National Renewable Energy Laboratory, Golden, CO)

[14] Jonkman J, Butterfield S, Musial W and Scott G 2009 Definition of a 5-MW reference wind turbine for offshore system development *US: NREL Technical Report No. NREL/TP-500-38060*

[15] Mittal R and Iaccarino G 2005 Immersed boundary methods *Annu. Rev. Fluid Mech.* **37** 239-61

[16] Porté-Agel F, Lu H and Wu Y T 2014 Interaction between large wind farms and the atmospheric boundary layer *Procedia Iutam* **10** 307-18

[17] Porté-Agel F, Wu Y T, Lu H and Conzemius R J 2011 Large-eddy simulation of atmospheric boundary layer flow through wind turbines and wind farms *Journal of Wind Engineering and Industrial Aerodynamics* **99(4)** 154-68

[18] Santoni C, Carrasquillo K, Arenas-Navarro I, Leonardi S 2017 Effect of tower and nacelle on the flow past a wind turbine *Wind Energy, preprint (under review)*

[19] Troldborg N, Zahle F, Rthor P E and Sørensen J 2015 Comparison of wind turbine wake properties in non-sheared inow predicted by different computational uid dynamics rotor models *Wind Energy* **18** 1239-50

[20] Van Dooren M, Khn M, Petrović V, Bottasso C, Campagnolo F, Sjholm M, Angelou N, Mikkelsen T, Croce A and Zasso A 2016 Demonstration of synchronised scanning lidar measurements of 2d velocity fields in a boundary-layer wind tunnel *Journal of Physics: Conference Series* **753**

[21] Wang J, Bottasso C L and Campagnolo F 2016 Les modeling of a scaled wind farm facility in a boundary layer wind tunnel *4th Symposium on OpenFOAM in Wind Energy*

[22] Wang J, Foley S, Nanos E M, Yu T, Campagnolo F and Bottasso C L 2017 Numerical and experimental study for wake redirection techniques in a boundary layer wind tunnel *Journal of Physics: Conference Series, preprint (under review)*

Numerical and Experimental Study of Wake Redirection Techniques in a Boundary Layer Wind Tunnel

J Wang[1], S Foley[1], E M Nanos[1], T Yu[1], F Campagnolo[1], C L Bottasso[1,2], A Zanotti[2] and A Croce[2]

[1] Wind Energy Institute, Technische Universität München, Boltzmannstraße 15, D-85748 Garching bei München, Germany

[2] Dipartimento di Scienze e Tecnologie Aerospaziali, Politecnico di Milano, Via La Masa 34, I-20156 Milano, Italy

E-mail: jesse.wang@tum.de

Abstract.
The aim of the present paper is to validate a wind farm LES framework in the context of two distinct wake redirection techniques: yaw misalignment and individual cyclic pitch control. A test campaign was conducted using scaled wind turbine models in a boundary layer wind tunnel, where both particle image velocimetry and hot-wire thermo anemometers were used to obtain high quality measurements of the downstream flow. A LiDAR system was also employed to determine the non-uniformity of the inflow velocity field. A high-fidelity large-eddy simulation lifting-line model was used to simulate the aerodynamic behavior of the system, including the geometry of the wind turbine nacelle and tower. A tuning-free Lagrangian scale-dependent dynamic approach was adopted to improve the sub-grid scale modeling. Comparisons with experimental measurements are used to systematically validate the simulations. The LES results are in good agreement with the PIV and hot-wire data in terms of time-averaged wake profiles, turbulence intensity and Reynolds shear stresses. Discrepancies are also highlighted, to guide future improvements.

1. Introduction

Wake redirection is currently one of the most promising wind farm control methodologies, which may lead to the improved power capture and reduced structural loading. The current research on wake redirection is very active, covering high-fidelity numerical simulations [14], scaled experiments in the wind tunnel [10], direct measurements in the field [1], reduced order models [2] and control methods [9]. In a previous paper [27], we used a three-pronged approach based on the combination of an analytical formulation, a LES-lifting-line CFD method [26] and wind tunnel scaled models to investigate two wake redirection methods, yaw misalignment (noted YM in this paper) and cyclic pitch control (noted CyPC). The analysis showed that, as already noticed by other authors, YM is more effective than CyPC in displacing the wake laterally. CyPC, however, was shown to have some effect on the speed of recovery of the wake, although the practical use of this concept is put in serious doubt by the large loads that are generated on the wind turbine.

In the present paper, we use additional wind tunnel measurements to further characterize the behavior of the CFD approach in the context of these two wake redirection techniques. Our long term goal is to develop the ability to accurately simulate our wind tunnel experiments by a validated numerical environment, and to upscale the validation to full scale. The present paper considers both integral rotor quantities such as power and thrust, as well as local flow measurements obtained by particle image velocimetry (PIV) and hot-wire probes. Scanning LiDARs are used to create a map of the wind tunnel inlet, to account for some spatial variability of the generated flow conditions, which is useful for a more accurate verification of the numerical results. The analysis highlights the general good qualitative agreement between experiments and simulations. The results also show that some local improvements are still necessary, highlighting some weakness of the numerical approach that motivate further refinements.

2. Simulation model

2.1. CFD formulation

The simulation model is implemented in the SOWFA code [12], a CFD simulation tool based on an incompressible solver within the OpenFOAM repository, employing the Boussinesq approximation to include buoyancy-driven flow. Large scale transient simulations are conducted using the PIMPLE time marching algorithm. The grid-collocated formulation computes quantities at cell centers. The interpolation of Rhie and Chow [24] is employed to remove numerical oscillations caused by pressure-velocity decoupling induced by grid collocation. The time marching scheme uses a second order backward implicit discretization. The momentum and potential temperature equations use a diagonal incomplete preconditioned bi-conjugate gradient linear solver, while pressure is solved using a geometric agglomerated algebraic multi-grid solver. Parallel computing is based on the Message Passing Interface (MPI) [13]. In addition, SOWFA uses the lifting-line method embedded in a large-eddy simulation (LES) environment, coupled with the aeroservoelastic simulator FAST [13].

Apart from the numerical methods used for wind farm simulations, other techniques are employed to perform more robust and accurate large-eddy simulations of the scaled wind farm facility. The immersed boundary formulation is used for the modeling of geometries, like nacelle and tower, within the control volume. The convection vector-field scheme uses a deferred correction Gamma-bounded interpolation method, which belongs to the family of categorized normalized variable diagram (NVD) schemes [18]. This Gamma scheme contains a pre-specified constant, β_m, which allows one to limit the numerical diffusion and to minimize numerical dispersion. In general, a larger value of β_m implies a lower dispersion and a higher diffusion, vice versa. Therefore, β_m=0.35 is employed within the near wake to stabilize the simulations, since actuator line body forces and immersed boundary possibly increment the numerical stability issues, while β_m=0.05 is used for the far wake to minimize numerical diffusion. This algorithm represents an important improvement with respect to normal unbounded and non-flux-limited central differencing schemes. In fact, it is capable of minimizing the numerical dispersion otherwise often observed at large yaw misalignment angles, where the wake is oblique to the grid. Moreover, this approach performs better than other low diffusive schemes such as the one of Vanleer in terms of limiting the numerical diffusion in the wake region, where small-scale eddies are dissipated due to such flux limiters [16]. A tuning-free Lagrangian scale-dependent dynamic approach is used for the Smagorinsky sub-grid scale modeling [23]. This approach improves the modeling of inhomogeneous flow fields such as in the case of wind turbine wakes, where first and second-order flow quantities vary significantly within the domain.

2.2. Actuator line method

The LES formulation is augmented with an actuator line method. The complete geometry of the blades is not resolved in the present LES formulation, due to its excessive computational costs.

Instead, blades are represented by a given number of points on an actuator line. Aerodynamic forces are calculated based on the local angle of attack by means of look-up tables storing the local airfoil lift and drag coefficients. The resulting forces acting at each blade section are then smoothly mapped to a body force field by a Gaussian projection characterized by a width parameter ε. The determination of the ε parameters is dependent on several factors. According to Ref. [13], ε should account for cell size, chord width and other geometrical parameters, but cannot be lower than a minimum value to avoid spurious oscillations of the velocity field. A comprehensive study of the effect of the projection width on the simulation quality is reported in Ref. [20].

2.3. Immersed boundary method

An immersed boundary (IB) formulation [17] is used to model the wind turbine nacelle and tower. The IB method is employed to avoid the use of surface conforming meshes to resolve the shape of the body [22]. The present IB approach, based on a discrete forcing method, uses a direct imposition of the boundary conditions. This way, the sharpness of the body shape is preserved and no extra stability constraints are introduced. In fact, with this approach boundary conditions and wall models can be directly imposed on the IB surfaces, yielding a significant improved solution quality for higher Reynolds number viscous flows [3], such as the present case of wind turbines operating in an atmospheric boundary layer. Details of the formulation are described in Ref. [28].

3. Experimental methods

Experiments were conducted in a 36 m $\times$ 16.7 m $\times$ 3.84 m boundary layer wind tunnel at Politecnico di Milano (POLIMI) [7]. Tests were performed with a scaled wind turbine model, shown in figure 1, together with the reference frame adopted to express the flow velocity components, which has a rotor diameter (D) of 1.1 m and a rated rotor speed of 850 RPM, and which has already been used within other research projects [9–11].

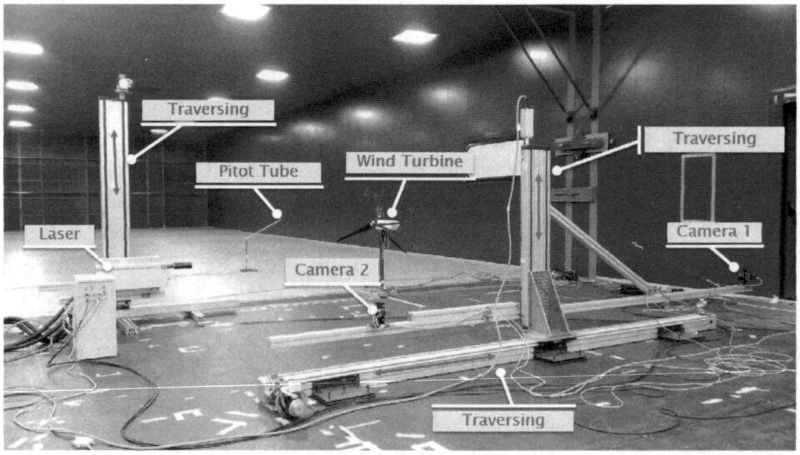

Figure 1: Experimental setup within the boundary layer wind tunnel at Politecnico di Milano.

The model is characterized by a realistic aerodynamic performance, both at the airfoil and rotor levels, and its wake is characterized by shape, deficit and recovery that are in good accordance with those of full-scale machines. Moreover, the model features active individual pitch, torque and yaw control that, together with a comprehensive onboard sensorization of the machine (including measures of shaft and tower loads), enables the testing of modern control strategies similar to the ones described in [5] and references therein.

The wind tunnel can be operated with different configuration of spires and pyramids at the inlet of the test section, to realize varying vertical shear and turbulence intensity conditions. Hot-wire probes, LiDAR (Light Detection And Ranging) and PIV (Particle Image Velocimetry) are used to perform high quality measurements of aerodynamic quantities, which will then be used for the validation of the LES-lifting-line model.

3.1. Inflow measurements

Previous experimental testing in the POLIMI facility has revealed that the inflow generated by the wind tunnel is non-uniformly distributed, with variations in flow velocity of up to 6%. The inhomogeneity of the flow is non-negligible, and it will influence the operation of the wind turbine models and of the associated flow-field. To address this problem, instead of imposing a uniform inflow for the inlet boundary condition, a non-uniform inflow was used to more accurately represent the incoming flow generated in the wind tunnel. Properly capturing the inflow velocity requires the full cross-sectional wind map to be recorded upstream of the wind turbines, rather than sampling a single point (or several scattered points in space) at hub height. The measured experimental wind map can then be implemented as the inlet boundary condition to the CFD simulation.

In order to obtain a reliable high resolution wind map, two synchronized scanning LiDARs were employed. LiDARs are instruments equipped with a steering laser beam that fires rapid pulses and, by measuring the frequency shift of the backscattered light from the aerosols present in the air, are able to record the line-of-sight component of the wind speed. In this work, LiDAR data was obtained from the collaborative efforts of ForWind-Oldenburg, TUM, Technical University of Denmark and POLIMI [25]. Figure 2 shows the time-averaged velocity scanned inflow map. The figure clearly shows the lack of uniformity of the inflow, probably due to the presence of a discrete number of fans and of the structural arrangement of the tunnel just upstream of the test section. The black circle indicates the location of the rotor, which appears to be placed in correspondence with a local peak of the flow velocity.

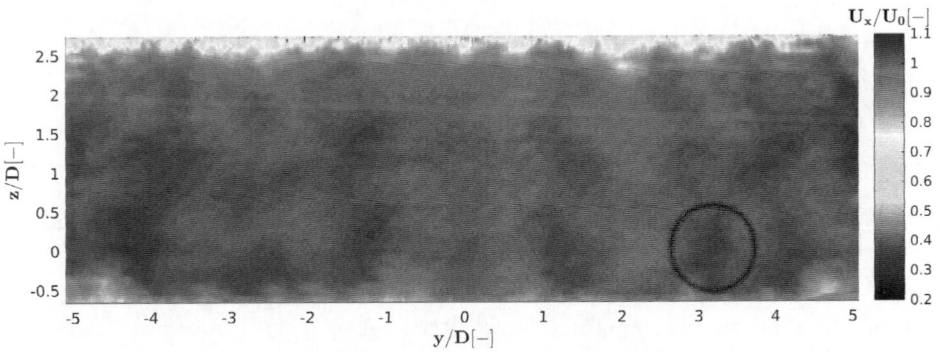

Figure 2: Inflow velocity profile (facing upstream) as measured by two scanning LiDARs at a 3D distance upstream of the wind turbine model.

3.2. Wake measurement

The stereo PIV instrumentation, shown in figure 1, was used to measure the three velocity components at cross-flow planes located at two different positions at 0.56D and 6D downstream from the rotor. The measurement planes cover quite a broad fraction of the wind turbine wake. In order to achieve a higher spatial resolution of the velocity field, the measurement area was divided into several smaller windows with slight overlapping areas between them. A rapid scanning of the entire measurement area was achieved by the use of an automated traversing system, moving both the laser and the cameras. In particular, the two cameras, equipped with a 1952 × 1112 pixel array, were connected to a metallic arm and mounted on a 4 m × 2 m double axis traversing system. The Nd:Yag double pulsed laser, with a 200 mJ output energy, was also mounted on a traversing system to move the laser sheet simultaneously with the measurement window. The measuring windows were divided into 32 × 32 pixel interpolation areas, which resulted in an approximately 15 mm spatial resolution. For each measuring window, 200 pairs of images were acquired (per camera) without any phase lock (time-averaged flow field). Additional details concerning the PIV instrumentation are given in [29]

4. Computational setup

The computational domain, as shown in figure 3, has a height of 3.84 m height, equal to the height of the wind tunnel test section. The inlet is located 3.3 m upstream of the wind turbine, in the same plane where the LiDAR inflow map is measured. The streamwise length is 10D. The lateral dimension is reduced to 4 m, to limit the domain size and hence the computational cost. Multiple simulation tests have shown that the reduction of the lateral dimension has no noticeable effect on the results. After determining the size of the computational domain, grid resolution requirements are investigated through a mesh independence study. Three levels of refinement are used: zone 1 defines a cube-shaped background mesh ($\Delta x = \Delta y = \Delta z$) with a cell size of 0.08 m, zone 2 is the first level of refinement with a cell size of 0.04 m, followed by zone 3 that uses a size of 0.01 m.

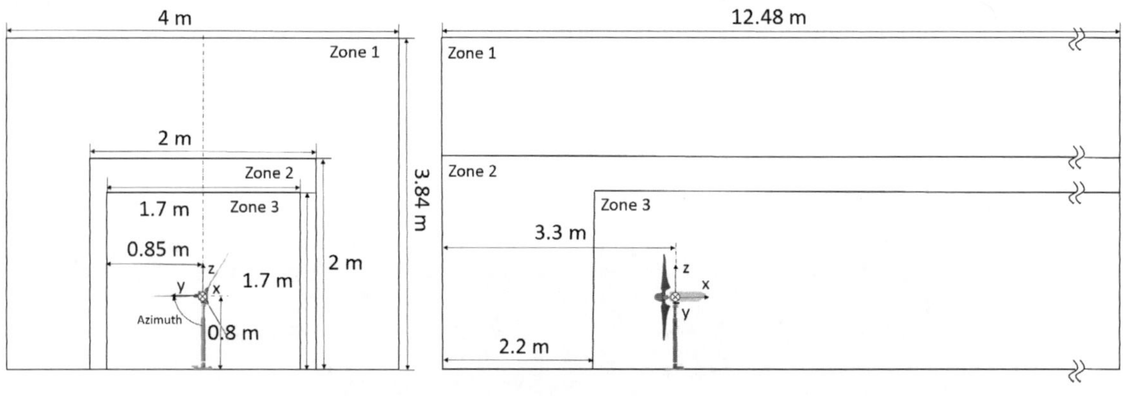

Figure 3: Views of the computational domain. Left: cross-section. Right: lateral view.

The inflow velocity measured by LiDARs is used at the domain inlet, to capture the lack of uniformity of the wind tunnel generated flow. Apart from this, the inflow of the numerical simulation is steady and non-turbulent. This creates a difference between the numerical and experimental flow conditions, because the latter has a turbulence intensity equal to about 2%. This should be kept in mind when comparing numerical and experimental results, because the different ambient turbulence conditions may have some effects.

Slip boundary condition is applied at the left and right walls, while a wall model equivalent to a no-slip condition is used for the floor and ceiling of the tunnel. A wall model is also used for the immersed boundary surface that models the nacelle and tower. Neumann type boundary condition for the pressure at all walls is also imposed. A modified version of the mesh partitioning algorithm was developed, which tries to allocate the same number of cells to all processors without cutting the mesh at the immersed boundary surfaces.

A fixed time-step of 0.0002 s is used for baseline and two CyPC setups, while 0.0003 s is used for YM setup. The time-step equates to a maximum Courant number C_{max} of 0.17 and a displacement of 0.99 cells per time-step at the tip of the blades for the baseline case. In all cases, the machines are operating in the partial load region at an averaged wind speed of 5.83 m/sec across the rotor, while the rotating speed varies from 780 to 840 RPM due to different redirection control methods. The Initial blade pitch were kept to be constant for all 4 setups. The ratio $\frac{\varepsilon}{\Delta x}$ between the Gaussian width and the cell size has been kept equal to 2.2 for the baseline and two CyPC simulations, while it has increased to 3.4 for 20 deg yaw misalignment case. Most simulations were performed on an Intel Haswell-EP computing cluster. With these computational resources, 5.2 s of physical time (≈26,000 time steps) require about 13,440 cpu hours. The solution converges to steady state after about 4 s of physical time.

5. Results and analysis

5.1. Power and thrust

Four operating conditions are considered: a baseline condition (constant pitch setting, no misalignment), cyclic pitching with two different phase angles, and a yaw misalignment of 20 deg. For CyPC, the blade is pitched according to $\theta_i = \theta_0 + \theta_c \cos(\psi_i + \gamma)$, where θ_0 is the collective pitch constant, θ_c the 1P pitch amplitude, ψ_i the blade azimuth angle (counter-clockwise looking upstream, and null when the blade is pointing vertically up along the tower), and γ is the phase angle (clockwise looking upstream). Two values for γ are considered: 52 and 270 deg. The wind tunnel was operated without spires at the test chamber inlet, resulting in low-turbulence (≈2%) flow conditions.

Table 1 shows the wind turbine power and thrust at the four operating conditions. A good agreement between the simulations and experiments is observed for the baseline case, with differences of 3% and 6% for power and thrust, respectively. The match is less accurate for the CyPC cases. For both cases, power is underestimated by the numerical simulations, while thrust is underestimated in one case and overestimated in the other.

Table 1: Comparison of rotor power and thrust between simulation and experiment for the baseline case, cyclic pitch amplitude of 5.3 deg with phase shifts of 52 and 270 deg, and yaw misalignment φ=20 deg.

	Type	Baseline	CyPC γ=52 [deg]	CyPC γ=270 [deg]	YM φ=20 [deg]
Power	Exp. $[W]$	44.55	35.19	35.04	40.98
	Sim. $[W]$	43.29	32.24	31.63	40.63
	ΔP %	-2.82	-8.39	-9.74	-0.85
Thrust	Exp. $[N]$	15.93	14.03	15.42	13.69
	Sim. $[N]$	16.85	14.78	14.53	13.59
	ΔT %	5.76	5.34	-5.79	-0.73

Several possible effects may help explain the larger errors in the CyPC cases. First, unsteady airfoil aerodynamic effects (including dynamic stall) are completely neglected, as the present lifting-line formulation only uses look-up-tables of steady lift and drag coefficients. This could be improved by using the suitable unsteady aerodynamics model for the airfoils. Some of these models, such as the Beddoes-Leishman [21], are available within the current LES framework. However, this model requires the definition of several airfoil-dependent governing parameters that were unknown to us at the time when simulations were performed, and would need to be specifically calibrated. Second, the CyPC experiments were conducted at a lower RPM than the baseline (780 RPM instead of 840 RPM, corresponding to a reduction of about 7%, due to the shift of operating point under cyclic pitch control), which induces a reduction in the operating Reynolds of the airfoils. The airfoil lift and drag coefficients, obtained during previous research activities [10], are tabulated for two values of the Reynolds number (namely, Re=90000 and Re=75000), and they are interpolated during the simulation based on the local instantaneous conditions [21]. However, the lower Reynolds coefficients may be less accurate than the higher ones, which might induce a larger discrepancy in the results in this case. This might be improved by tuning the aerodynamic coefficients, as suggested in Ref. [6].

The same table shows also good results for the misalignment case. However, various numerical experiments showed a significant sensitivity of the results to the Gaussian width parameter ε. In fact, the results of the table were obtained with a parameter ε 54% higher than the one used for the baseline case, as addressed in section 4. This problem deserves further careful investigation. Here again, a possible reason might be the lack of modeling of unsteady aerodynamic effects at the level of the airfoils, effects that increase for increasing misalignment. Clearly, finding a specific value of the Gaussian width parameter to correct for this deficiency of the method is not a sensible option, and it is certainly not suggested as a general solution.

5.2. Wake properties

5.2.1. Cyclic pitch case Normalized time-averaged velocity plots are shown in figure 4 and 5. The figures report results just behind the rotor (x/D=0.56) and in the far wake (x/D=6), for the baseline and the two considered CyPC cases.

The analysis of the U_x component at 0.56D downstream of the rotor is particularly interesting, as this is a position where few results have been previously reported. The images show that the use of CyPC has a strong effect on the wake structure, leading to marked unsymmetrical shapes. Measurements are missing from two areas left and right of the rotor disk, where, due to the close proximity of the measuring plane with the wind turbine, part of the nacelle (which is of a white color) was on the background, leading to bad correlation between the images. A comparison between experimental and numerical results shows that there is in general a good qualitative agreement, and that the main distortion effects caused by CyPC are reasonably captured. The difference $\mathbf{\Delta U_i}$ between experiment and simulation velocities was evaluated at several locations within the rotor swept area (black dot circles in figure 4). Based on these quantities, the average velocity difference $\overline{\mathbf{\Delta U}}$ and its root mean square $\mathbf{\Delta U_{RMS}}$ were calculated. $\overline{\mathbf{\Delta U}}$ is equal to -1.36%, 2.69% and 5.05% for the baseline, CyPC γ=52 and 270 deg cases, respectively, while $\mathbf{\Delta U_{RMS}}$ is 0.43 m/sec, 0.79 m/sec and 0.72 m/sec for the same three cases. Positive value of $\overline{\mathbf{\Delta U}}$ means that the average simulated velocity is higher.

Figure 5 shows the wake plots in the far wake region at 6D. $\overline{\mathbf{\Delta U}}$ in this case is 4.9%, 9.4% and 4.9% for the baseline, CyPC γ=52 and 270 deg cases, respectively. $\mathbf{\Delta U_{RMS}}$ is 0.33 m/sec, 0.46 m/sec and 0.44 m/sec for the same three cases. $\mathbf{\Delta U_{RMS}}$ are slightly improved at 6D compared to the near wake for all three cases, while $\overline{\mathbf{\Delta U}}$ at 6D show in general higher wake recovery. Lower $\mathbf{\Delta U_{RMS}}$ implies a higher correlation between LES and PIV wake profile. In fact, compared with the near wake region, the flow in the far wake is further mixed and less structured, exhibiting a velocity profile that is typically well approximated by a single Gaussian. Lastly, it should be remarked that CyPC leads to a faster recovery of the wake than in the baseline case, as already noticed in Ref. [27]. In principle, this could be of interest for wind farm control, although the very large loads exerted on the rotor by the use of CyPC severely limits the possible applicability of this control concept.

Next, the hub-height velocity profile is considered. Figure 6 reports the time-averaged streamwise velocity profiles at hub height for the three considered operating conditions. In the near wake (0.56D), matching for the baseline case is reasonable, except the region close to the hub. Here, the effects of the nacelle and tower are not captured accurately. A visualization

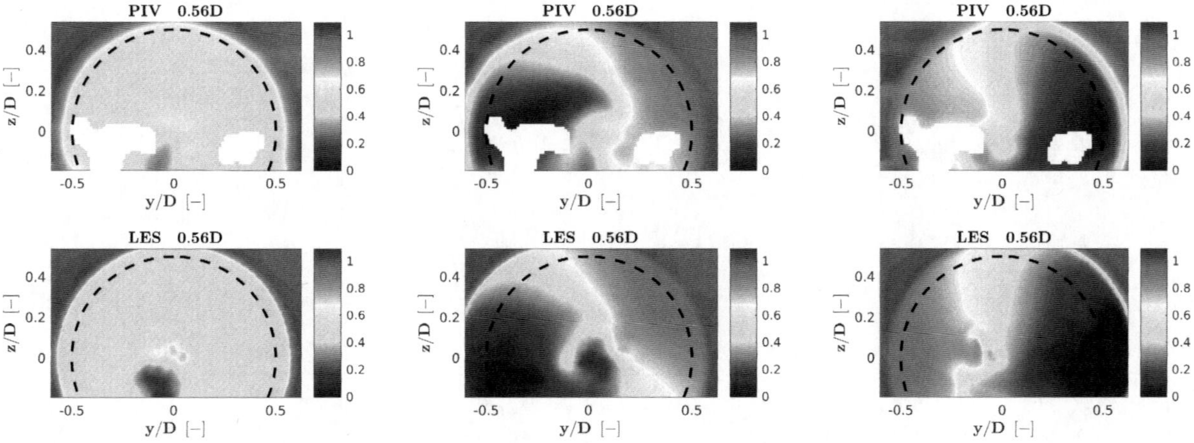

Figure 4: Normalized time-averaged streamwise velocity on the yz-plane at 0.56D downstream of the wind turbine (facing upstream). From left to right, baseline condition and cyclic pitch amplitude of 5.3 deg with phase shift γ of 52 and 270 deg. Top: PIV experimental measurements; bottom: simulation results.

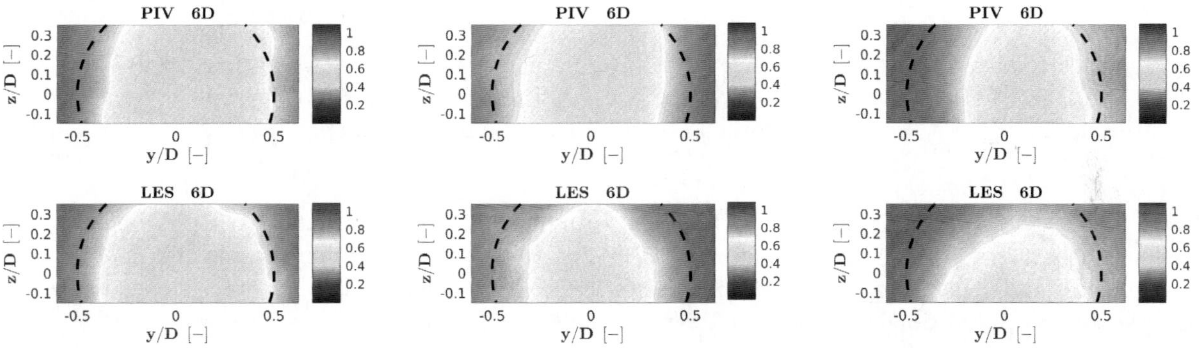

Figure 5: Normalized time-averaged streamwise velocity on the yz-plane at 6D downstream of the wind turbine (facing upstream). From left to right, baseline condition and cyclic pitch amplitude of 5.3 deg with phase shift γ of 52 and 270 deg. Top: PIV experimental measurements; bottom: simulation results.

of the numerical results seems to indicate a strong tower shedding. Accuracy might be affected by a lack of background turbulence in the simulations. In fact, as previously explained, the wind tunnel flow is characterized by a turbulence intensity of 2%, while the numerical simulations are conducted in uniform non-turbulent inlet conditions. For the CyPC case, matching is much poorer, with significant differences both close to the hub and especially at the tip of the low speed region. As previously noted, differences might be due to limits of the lifting-line formulation, and/or because of a lack of grid resolution.

The situation is improved for the 6D case, as expected. Here good results are obtained for the baseline case, while small discrepancies are visible for the CyPC conditions. Note that the wind speed velocities at the left and right ends of the rotor are not identical, which, as previously explained, is a phenomenon due to the non-uniform inflow conditions of the wind tunnel.

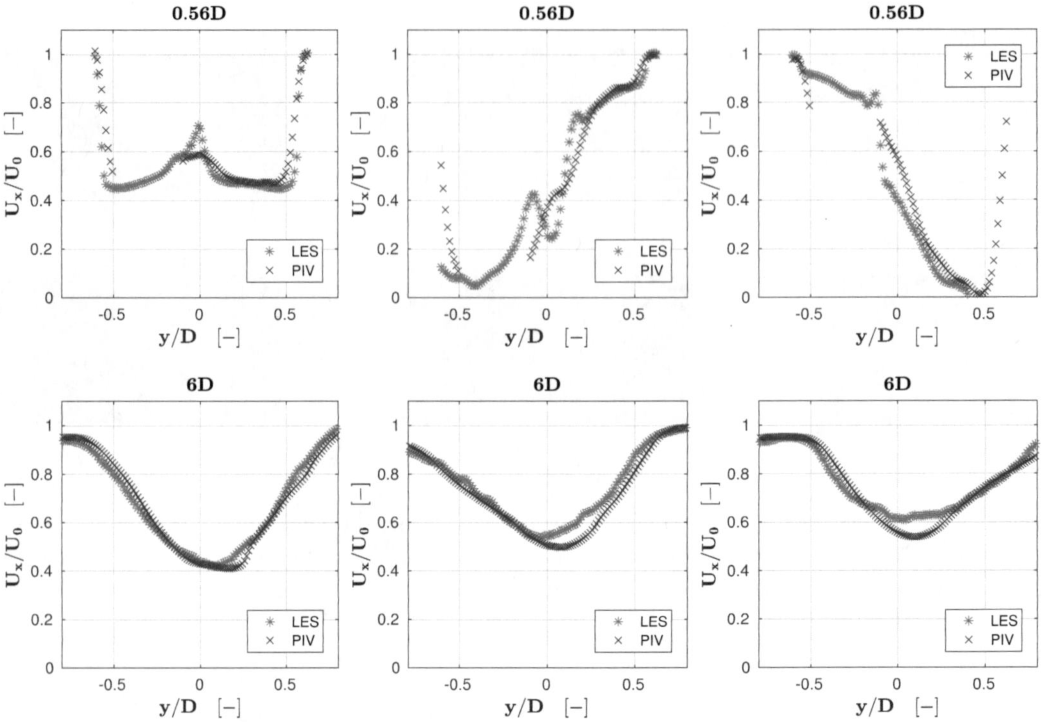

Figure 6: Normalized time-averaged streamwise velocity along a hub-height horizontal line at 0.56D (top) and 6D (bottom) downstream of the wind turbine. From left to right, baseline condition and cyclic pitch amplitude of 5.3 deg with phase shift γ of 52 and 270 deg.

5.2.2. Active yaw control Next, we consider the wake velocity profiles for the 20 deg YM case. As previously mentioned, the ε parameter was adjusted to better approximate the experimental results, while the other parameters remained identical to the baseline case. Figure 7 shows the numerically computed time-averaged streamwise velocity at three downstream locations; no PIV measurements are available in this case. The wake structure exhibits a characteristic 'kidney' shape, in accordance with the observations of other authors [4, 15].

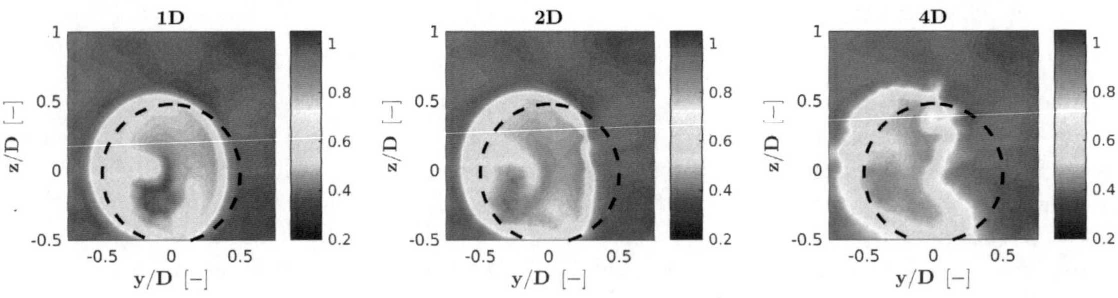

Figure 7: Simulated normalized time-averaged streamwise velocity (facing upstream) at 1D, 4D, and 7D downstream of a wind turbine with 20 deg yaw misalignment.

Figure 8 reports the first order (velocity profile) and second order (turbulence intensity TI and Reynolds shear stress $\overline{u'v'}$) flow quantities at 4D for the baseline case (top row of plots) and YM case (bottom row of plots). Experimental measurements were obtained with triple hot-wire probes [25], while simulations are reported with and without the presence of nacelle and tower. The baseline wake profile with nacelle and tower included correlates well with the measurement data. It appears that the presence of the nacelle and tower in the model is indispensable for an accurate solution ($\overline{\mathbf{\Delta U}} = 2\%$ in this case), as already noticed by other authors [8]. Although the situation is not as good for the YM case, matching with the experimental measurements is still quite reasonable ($\overline{\mathbf{\Delta U}} = 4\%$). In addition, the prediction of the displacement of the wake center, computed as the location of the minimum of the Gauss fit to the wake, appears to be highly accurate and essentially identical to -0.26D for both the simulation and experiment.

Similar effects can be observed for the two second order quantities, TI and $\overline{u'v'}$. Regarding the former quantity, the plots show that ambient TI, as seen at the end of the scan line, is nearly 0% for the simulation case and 2% for the experiments, as previously noted. Although there is more scatter for TI than for $\overline{u'v'}$, the plots show once again the importance of including tower and nacelle in the model to improve accuracy.

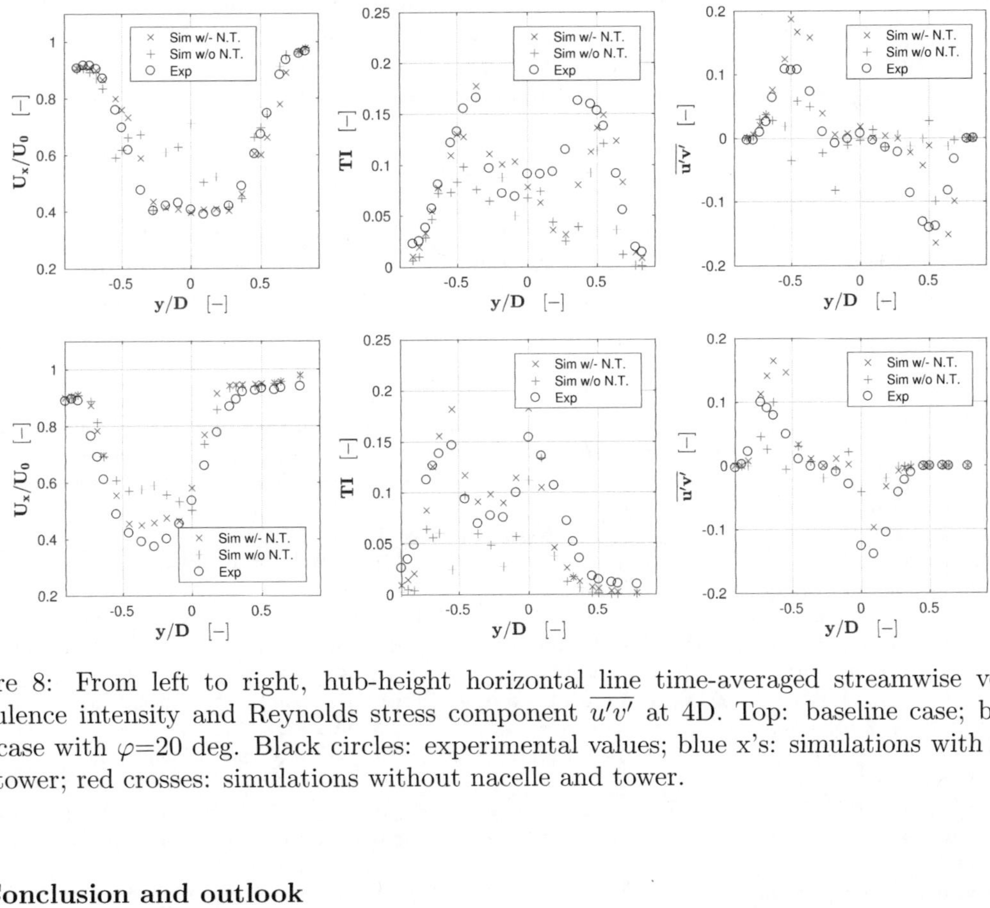

Figure 8: From left to right, hub-height horizontal line time-averaged streamwise velocity, turbulence intensity and Reynolds stress component $\overline{u'v'}$ at 4D. Top: baseline case; bottom: YM case with φ=20 deg. Black circles: experimental values; blue x's: simulations with nacelle and tower; red crosses: simulations without nacelle and tower.

6. Conclusion and outlook

This paper has presented the comparison of the results of a LES-lifting-line solver with wind tunnel measurements obtained with scaled wind turbine models. The analysis has considered a baseline flow-aligned case, the use of cyclic pitching to affect the wake, and the misalignment of the wind turbine with the incoming wind. Comparisons included integral rotor quantities as power and thrust, and local flow quantities measured by PIV and hot-wire probes. Measurements were performed in the near wake region, and after vortex breakdown in the far wake. Two scanning LiDARs were used to generate a map of the velocity at the inlet of the test section, to account for the non-uniformity of the wind tunnel generated flow.

Based on the results shown in the paper, it appears that the CFD formulation is able to capture both the integral and local quantities with reasonable accuracy. In general, the baseline case exhibits the best results, while both CyPC and YM are qualitatively acceptable but not always very precise.

Overall, results appear to be very encouraging, although there are several effects that may potentially further improve the level of accuracy of the current numerical model. The LES framework can be improved on three fronts: using more accurate airfoil polars, which may be achieved with the existing polar-identification procedures, and introducing unsteady effects in the airfoil model. Results presented herein showed in a very clear manner the importance of including the nacelle and tower in the CFD model. However, the same results also show that the accuracy of such components is still not sufficient, and improvements may be obtained by the use of denser grids and possibly an improved formulation. Finally, the present simulation did not match the ambient turbulence of the experiment, which might be the source of additional discrepancies. For higher turbulence cases (which are closer to full scale operating conditions), eddies in the wind tunnel flow are generated by the use of spires at the inlet. For such conditions we have already developed a CFD model that can be used as a precursor of the lifting-line simulations. In the future, we will work on expanding that method to also account for the non-uniformity of the mean velocity, a problem that was considered here using a LiDAR-measured inflow map.

Acknowledgments

This project was partly funded by the EU Horizon 2020 research and innovation programme under the Marie Skłodowska-Curie grant agreement No. 642108. The authors wish to thank G. Campanardi and D. Grassi from the Politecnico di Milano for their contribution to the PIV measurements.

References

[1] Aitken M L, Banta R M, Pichugina Y L and Lundquist J K 2014 Quantifying wind turbine wake characteristics from scanning remote sensor data *Journal of Atmospheric and Oceanic Technology* **31** 765-87

[2] Annoni J, Gebraad P, and Seiler P 2016 Wind farm flow modeling using an input-output reduced-order model *American Control Conference* 506-12

[3] Bandringa H 2010 *Immersed boundary methods* (Netherlands: University of Groningen)

[4] Bastankhah M and Porté-Agel F 2016 Experimental and theoretical study of wind turbine wakes in yawed conditions *Journal of Fluid Mechanics* **806** 506-41

[5] Bossanyi E 2000 The design of closed loop controllers for wind turbines *Wind Energy,* **3** 149-63

[6] Bottasso C L, Cacciola S, and Iriarte X 2014 Calibration of wind turbine lifting line models from rotor loads *Journal of Wind Engineering and Industrial Aerodynamics,* **124** 29-45

[7] Bottasso C L, Campagnolo F and Petrović V 2014 Wind tunnel testing of scaled wind turbine models: Beyond aerodynamics *Journal of wind engineering and industrial aerodynamics* **127** 11-28

[8] Santoni C, Carrasquillo K, Arenas-Navarro I, Leonardi S 2017 Effect of tower and nacelle on the flow past a wind turbine *Wind Energy, preprint (under review)*

[9] Campagnolo F, Croce A, Manos E, Petrović V, Schreiber J and Bottasso C L 2016 Wind tunnel testing of a. closed-loop wake deflection controller for wind farm power maximization *Journal of Physics: Conference Series* **753**

[10] Campagnolo F, Petrović V, Bottasso C L and Croce A 2016 Wind tunnel testing of wake control strategies *Proceedings of the American Control Conference* 513-18

[11] Campagnolo F, Petrović V, Nanos E, Tan C, Bottasso C L, Paek I, Kim H, and Kim K 2016 Wind tunnel testing of power maximization control strategies applied to a multi-turbine floating wind power platform *Proceedings of the International Offshore and Polar Engineering Conference* 309-16

[12] Churchfield M and Lee S 2012 *NWTC design codes-sowfa* (USA: http://wind. nrel. gov/designcodes/simulator S./SOWFA)

[13] Churchfield M, Lee S, Moriarty P, Martinez L A, Leonardi S, Vijayakumar G, and Brasseur J 2012 large-eddy simulation of wind-plant aerodynamics *AIAA paper* **537** 2012
[14] Fleming P, Gebraad P, Lee S, van Wingerden J W, Johnson K, Churchfield M, Michalakes J, Spalart P and Moriarty P 2014 Evaluating techniques for redirecting turbine wakes using sowfa *Renewable Energy* **70** 211-18
[15] Howland M F, Bossuyt J, Martinez-Tossas L A, Meyers J and Meneveau C 2016 Wake structure of wind turbines in yaw under uniform inflow conditions *arXiv* (*preprint* arXiv:1603.06632)
[16] Jasak H, 1996 *Error analysis and estimation for finite volume method with applications to fluid flow* (UK: Imperial College London)
[17] Jasak H and Rigler D 2014 Finite volume immersed boundary method for turbulent flow simulations *9th OpenFOAM Workshop*
[18] Jasak H, Weller H and Gosman A 1999 High resolution nvd differencing scheme for arbitrarily unstructured meshes *International journal for numerical methods in fluids* **31(2)** 431-49
[19] Jonkman J and Buhl M L 2004 *Fast user guide* (USA: Technical report, EL-500-29798, National Renewable Energy Laboratory, Golden, CO)
[20] Martinez L A, Leonardi S, Churchfield M, and Moriarty P 2012 comparison of actuator disk and actuator line wind turbine models and best practices for their use. *AIAA Paper* **1** 2012-0900
[21] Moriarty P L and Craig H 2005 *Aerodyn user's guide* (USA: National Renewable Energy Laboratory)
[22] Mittal R and Iaccarino G 2005 Immersed boundary methods *Annu. Rev. Fluid Mech.* **37** 239-61
[23] Porté-Agel F, Wu Y T, Lu H and Conzemius R J 2011 Large-eddy simulation of atmospheric boundary layer flow through wind turbines and wind farms *Journal of Wind Engineering and Industrial Aerodynamics* **99(4)** 154-68
[24] Rhie C and Chow W 1983 Numerical study of the turbulent flow past an airfoil with trailing edge separation *AIAA journal* **21(11)** 1525-32
[25] Van Dooren M, Khn M, Petrović V, Bottasso C, Campagnolo F, Sjholm M, Angelou N, Mikkelsen T, Croce A and Zasso A 2016 Demonstration of synchronised scanning lidar measurements of 2d velocity fields in a boundary-layer wind tunnel *Journal of Physics: Conference Series* **753**
[26] Wang J, Bottasso C L and Campagnolo F 2016a Les modeling of a scaled wind farm facility in a boundary layer wind tunnel *4th Symposium on OpenFOAM in Wind Energy*
[27] Wang J, Bottasso C L and Campagnolo F 2016b Wake redirection: comparison of analytical, numerical and experimental models *Journal of Physics: Conference Series* **753**
[28] Wang J, Mclean D, Campagnolo F, Yu T and Bottasso C L 2017 Large-eddy simulation of waked turbines in a scaled wind farm facility *Journal of Physics: Conference Series, preprint (under review)*
[29] Zanotti A, Ermacora M, Campanardi G and Gibertini G 2014 Stereo particle image velocimetry measurements of perpendicular bladevortex interaction over an oscillating airfoil *Experiments in Fluids* **55** 1-13

Comparison of the Actuator Line Model with Fully Resolved Simulations in Complex Environmental Conditions

Pascal Weihing, Christoph Schulz, Thorsten Lutz, Ewald Krämer
University of Stuttgart, Institute of Aerodynamics and Gas Dynamics, Pfaffenwaldring 21, 70569 Stuttgart

E-mail: weihing@iag.uni-stuttgart.de

Abstract. In the present paper the actuator line method is compared with fully resolved wind turbine simulations in offshore and complex terrain applications. In such flow fields, which are characterized by non-homogeneous and unsteady velocity distributions in the rotor plane, unsteady aerodynamic effects are likely and it is unclear how these characterize the wake development and load behavior of the wind turbine. The wake properties and loads are therefore compared for the case of a 5 MW wind turbine operating in a typical maritime atmosphere and a 2.4 MW onshore turbine located at a complex terrain site downstream of an escarpment. It was found that the actuator line predicts the wake structure, wake deflection and wake deficit in good agreement with the fully resolved simulation. However, an overestimation of velocity fluctuations was observed.

1. Introduction

The massive extension of wind power use calls for clustering of wind turbines in wind farms and for establishing sites of relatively lower wind speed in complex terrain. Especially for the latter, as summarized by Alfredsson and Sagalini [1], the aerodynamic modeling is multidisciplinary and challenging, involving effects of orography, roughness or stratification on the turbulent atmospheric boundary layer or the question on how to model the turbine itself. The present paper shall contribute to the assessment of turbine modeling by evaluating the actuator line method (ACL) in offshore and complex terrain conditions. The ACL has been widely used in the wind turbine community for the prediction of turbine performance, wake and the loads [2, 3]. Particularly, with focus on wind farms it seems to be reasonable that details from the rotor geometry as for example viscous wake effects of the flow around the airfoil are probably irrelevant for the development of the wake state multiple rotor diameters downstream of the turbine. This gives the rationale to shift the time and length scales to the order of the airfoil's chord by representing those with an appropriate source term distribution. This follows both a significant reduction of computational effort, as well as a massive simplification in the preparation of a fully resolved (FR) simulation. In close to design conditions, the blade element momentum theory which serves as the foundation of the ACL provides reliable results in the industrial framework. However, in complex environmental conditions like in the turbulent atmosphere or in the case of misalignments of the wind vector with the rotor axis, the spatial and temporal distribution of the velocity components in the rotor plane is far from uniform and unsteady aerodynamic effects are likely [4, 5, 6]. These effects which can be summarized as a phase shift between the change in angle of attack and the responding loads are naturally taken into account when conducting a FR turbine simulation. It is not clarified how these flow fields affect the wake properties. Besides of inter and intra code comparisons conducted in MexNext projects [7] where the wake is compared in controlled uniform conditions, the work of Troldborg et al. [8] contrasts actuator techniques with FR simulations in non-sheared inflow conditions and obtain an overall good agreement for turbulent inflow. The current study, as the latter, shall use both models within the same code and shed more light into the scope of application of the ACL, particularly, for non-uniform sheared turbulent flows. Therefore, the wake development, as well as the load prediction are compared with FR simulations for an offshore case with atmospheric turbulence and a case in complex terrain.

In the next section the numerical modeling and code is described, followed by the presentation of the test cases, the meshing and the numerical settings in Sec. 3. The results are discussed thereon in Sec. 4 and conclusions are drawn in Sec. 5.

2. Numerical modeling

The simulations in the present work have been performed using the code FLOWer which has been developed by the *German Aerospace Center* (DLR) and has been extended by the *Institute of Aerodynamics and Gas Dynamics* (IAG). The code solves the compressible Navier-Stokes equations on block-structured grids using the finite volume method. Convergence is accelerated by a three-level multi-grid method. For unsteady simulations, time discretization is conducted using the dual-time stepping algorithm by Jameson [9], whereas the convective fluxes are discretized by default with the central Jameson-Schmidt-Turkel (JST) scheme [10]. In order to improve conservation of vortical structures, the latter has been extended in Schäferlein [11] by the fifth-order *weighted essentially non-oscillatory* (WENO) scheme in conjunction with a HLLC or l^2-*Roe* [12, 13] Riemann solver, respectively. For the FR rotor simulations the relative movements of the body conforming grids are applied utilizing the overset grid technique and taking into account the corresponding whirl fluxes.

The basic implementation of the actuator line method is aligned with literature [14, 15], where the source term distribution representing the two dimensional lift and drag forces of each blade element are distributed over the surrounding grid cells using a three dimensional Gaussian smearing with a radially constant factor ϵ. Tait-Bryan angles transform the inertially sampled velocities into the airfoil coordinate system by taking into account the turbine kinematics of yaw, tilt, cone and the current azimuthal position of the blade. It is to be noted, that for the approximation of the angle of attack (AoA), the sampling points of the velocities do not coincide with the locations of the actuator points. Theoretically, the up- and down-wash in the center of a potential vortex is discontinuous. In reality, at least high gradients can be expected. In order to circumvent the problem that the sampled velocity might be deteriorated by the induced velocity of the bound vortex, the sampling is applied at some distance upstream of the airfoil and is then corrected iteratively by the induced velocity of the bound vortex applying Biot-Savart's law. The procedure is as follows: The initial value of the AoA is determined from the velocity triangle in the sample point, from which the lift distribution Γ is calculated. With Biot-Savart's

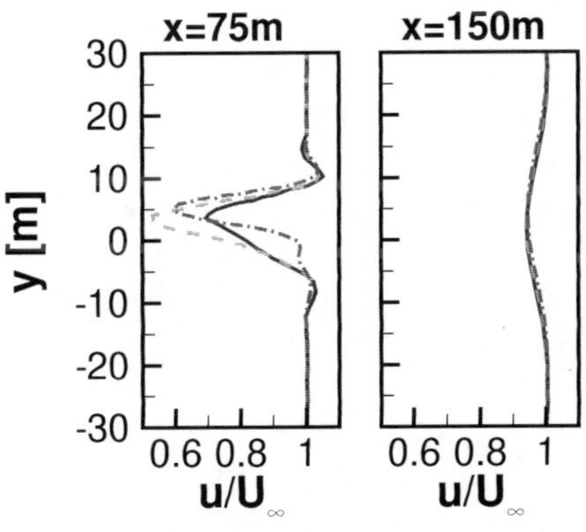

Figure 1. Wake of an isolated nacelle at two different downstream positions. Solid: FR, dashed: ACL with $c_d = 0.8$, $s_r = 0.2, k_f = 0.005$, dashed-dot: ACL with $c_d = 0.8$, $s_r = 0.16, k_f = 0.004$

law the induced velocity from the actuator point on the sampling point is determined and the effective velocity is obtained from which the inflow angle is recalculated. After five to ten iterations usually a converged AoA value is found. The nacelle can be modeled as an extraction of momentum within the volume of the nacelle. In FLOWer, this has been implemented based on the nacelle geometry and a user specified drag coefficient for that shape. From which the integrally estimated drag force of the body volume force increments are calculated and imposed to the flow field in those cells found within the nacelle. It was refrained from using the Gaussian smearing again, as the compressible framework of FLOWer does not show pressure velocity decoupling. For bluff nacelle shapes, where vortex shedding of the wake is likely, an orthogonally to the local wind vector acting side force is superposed which follows a simple sinusoidal signal $F_{sf} = k_f F_{D,nc} \sin 2\pi \frac{Sr U_\infty}{L} t$. where $F_{D,nc}$ is the drag force of the nacelle. The frequency is calculated from the Strouhal number the length of the nacelle L and the free stream velocity. The amplitude k_f and can calibrated using literature values [16]. Figure 1 shows the wake behind an isolated nacelle simulation in uniform inflow. The geometry is a cuboid of $20 \times 6.5 \times 6m^3$ that is later employed in the offshore turbine simulation. The actuator type simulation is compared with a FR representation of the nacelle and shows overall good agreement for the near and farther wake. Particularly, the deflection can be mimicked. The tower is represented as a single actuator line with cylindrical profile. Here again the mentioned side force is applied.

All simulations have been carried out using the Spalart-Allmaras turbulence model and a zonal detached eddy simulation approach, where the ambient turbulence is resolved in LES mode and the actual turbine meshes of the FR simulations are treated in URANS mode [13, 17, 18].

3. Test cases and numerical setup

3.1. The wind turbines and their operating points

In the present work two cases are investigated that cover important environmental conditions of wind farms, namely the maritime atmospheric boundary referred to as *caseOffshore*, as well as high turbulence and shear found in complex terrain (*caseTerrain*). For *caseOffshore* a 5 MW wind turbine of 126 m diameter and a hub height of 95 m is considered. The tilt angle is 6° and the cone angle −4°. The turbine is operating in partial load at a constant rotational speed of 10.3 rpm which corresponds to a tip-speed ratio (TSR) of 8.5. The pitch angle is −1°.

For *caseTerrain* a turbine of 2.4 MW with a rotor diameter of 109 m, and tilt and cone angles of 5° and −2°, respectively is considered. The turbine is sited on a plateau downstream of an escarpment that is located close to Stötten in southern Germany (cf. Fig. 3). Details on the site can be found in [19]. The TSR is 6.64 and the pitch angle 4.41°.

It should be noted, that both turbines are commercial, hence the scaling of the load results has to be suppressed.

3.2. Inflow conditions

The atmospheric conditions for *caseOffshore* are fed from a precursor simulation conducted at the University of Oldenburg using the meteorologic LES solver PALM [20]. This run was driven by WRF and LiDAR data that were obtained upstream of the turbine. The atmosphere was almost neutrally stratified with an averaged wind speed at hub height of ≈ 8 m/s and a turbulence intensity of 4.5 %. A time series of 58 s was extracted and ramped according to Meister et al.[21] to obtain temporal periodicity. The inflow data set was prescribed more than 2.5 D upstream of the turbine using a Dirichlet Boundary condition.

For *caseTerrain* a steady, homogeneous boundary layer profile is superposed with synthetic turbulence generated from the Mann model [22]. The underlying boundary layer follows power law with an exponent $\alpha = 0.14$. The model parameters for the Mann box were $L = 40$ m, $\Gamma = 3.9$, $\alpha\varepsilon^{2/3} = 0.035$. The standard deviation of the stream-wise velocity component was $\sigma_u = 0.9$ m/s. The turbulence is introduced with body forces similar to Troldborg [23] ca. 1 km upstream of the turbine. At the virtual turbine position this results in an inflow velocity of around 11 m/s at a turbulence intensity of about 10%. Details on the inflow generation and the propagation through the orography can be found in Schulz [19].

3.3. Meshing

For the FR simulations the relevant turbine components as tower, nacelle, spinner and the blades are meshed with body conforming grids. The blade meshes were created automatically using the IAG script *Automesh* in a C-H topology with locally adapted y_p^+ spacing of ≈ 1 in the boundary layer. The blade of *caseOffshore* is discretized by $256 \times 100 \times 192$ cells in circumferential, wall normal and span-wise direction. For *caseTerrain* $220 \times 100 \times 140$ cells were used. All other component grids were created manually in *Pointwise* with resolved boundary layers. Each component was then assembled within FLOWer using the overset grid technique as shown in Fig. 2. In both cases the area of the wake and the inflow is discretized by $1\,\mathrm{m}^3$ volumes. Away from that, the mesh is coarsened stepwise using hanging nodes. In total *caseOffshore* consists of ca. 80 M cells, whereas in *caseTerrain* ca. 180 M cells were used.

For the ACL simulations the blades are represented in both cases by 100 points. The airfoil polars were generated in *XFOIL* in the the AoA range $-180° < \alpha < 180°$ using Viterna's extrapolation. For sections near the root interpolation is conducted between the airfoils. For *caseOffshore* the tower and the nacelle are modeled as actuators as described in sec. 2. In *caseTerrain* the tower and nacelle meshes of the FR simulation were used. In both cases no tip corrections have been applied.

Figure 2. *caseOffshore*: Assembly of the component meshes with the overset grid technique.

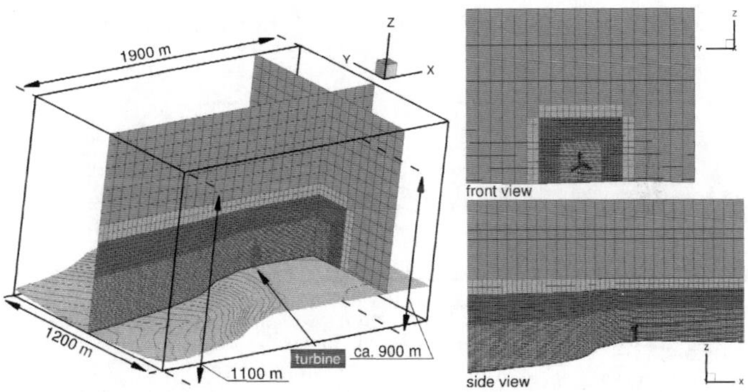

Figure 3. *caseTerrain*: Integration of the Turbine in the background domain [24].

3.4. Numerical Parameters

The low dissipative resolution of turbulent structures in the atmosphere and the wake is important in order to capture unsteady loads as well as the turbulent fluctuations induced by the turbine. Therefore, the fifth order WENO scheme is applied in the background mesh, while the more robust second order JST scheme is applied in the component structures. For the ACL simulations of *caseOffshore* the low low-Mach number Riemann solver l^2-*Roe* was employed [12, 13]. All other simulations use the HLLC Riemann solver. The time step in *caseOffshore* and *caseTerrain* was set equivalent to 1.5° and 2° azimuth increment, respectively, for both the FR and ACL simulations. Each physical time step was iterated towards a pseudo steady state using 40 sub-iterations. The simulation were run until periodicity of the flow field and the thrust was obtained. Data was then extracted for 10 revolutions in *caseOffshore* and 12 in *caseTerrain*.

4. Results

To accurately predict wakes by means of actuator type methods the generation of appropriate airfoil polars is crucial. To be able to interpret the wake results presented below, the sectional thrust loading of the rotor shall be discussed in Fig. 4 before. In each case the force is normalized by the span-wise mean value of the FR simulation. For *caseTerrain* the ACL distribution coincides very well with the results from the FR, except for a slight overestimation between $0.2 \leq {}^{r}/_{R} \leq 0.5$. Compared to *caseTerrain*, the thrust distribution in *caseOffshore* is linear over a wide range with a highly loaded tip, which is significantly overestimated by the ACL. There, it could be observed that the airfoils operate close to the beginning of flow separation revealing thick boundary layers and *XFOIL* predicts higher lift coefficients compared to CFD. Therefore, it can be expected, that in *caseOffshore* the ACL predicts a higher induction compared to the FR simulation.

To assess whether unsteady aerodynamic effects play a role, time series of the angle of attack are plotted v.s. the lift coefficient at different radial positions of *caseOffshore* in Fig. 5. In the inner part of the rotor, a strong non-linearity of the load response is present, which reduces further outboard, but is still pronounced in the mid portion of the rotor. As expected, the ACL reproduces the 2D airfoil polar. The AoA range is similar as for the FR. For the outer region the mean value of the AoA in the ACL is even slightly smaller compared to the FR. Nevertheless, the predicted lift coefficient is higher which stresses the fact that inaccurate airfoil polars are responsible for the higher loading of the ACL rotor.

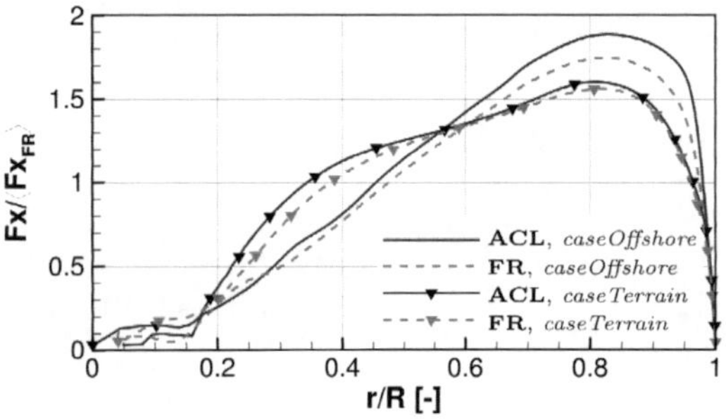

Figure 4. Time averaged sectional thrust. Each normalized by the spanwise mean value of the corresponding FR simulation.

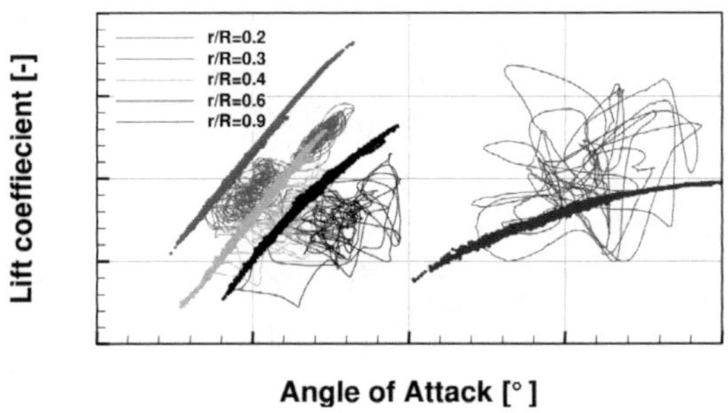

Figure 5. Characterization of unsteady aerodynamics for *caseOffshore*. Solid: FR; Symbols: ACL

Turning to the characterization of the wake, first the instability mechanisms shall be discussed for the different cases. For *caseOffshore* the parameters that define te wake state are of meteorologic as well as turbine specific kind. The shear of the wind profile as well as the tilt angle of the turbine result in an inhomogeneous azimuthal loading of the rotor. The tilt results in higher thrust for the 270° position compared to the 90° position as shown for the considered turbine by Meister et al. [21]. Further, the blade digs deeper into the wake when passing the upper rotor half. Particularly, at high TSR this results that the blade is affected by higher induction there. This effect is superposed by the shear of the wind profile which increases the loading with increasing height. The ambient turbulence adds to these effects that result in an azimuthally inhomogeneous strength and distance of the trailing vortices and favors the relative movement of those. As the tip vortices are already close to each other due to the high

TSR mutual inductance becomes very likely as shown in Fig. 6. Vortex pairing starts about one rotor radius downstream of the turbine in the ACL and the FR simulations. However, the tip vortices of ACL are more elliptic compared to the FR, which is related to the smearing of the force terms. From $2 \leq x/R \leq 4$ the vortices roll up to larger sub-structures that are finally soaked into the wake center at around $x/R = 3.25$. Due to the presence of the ground, this process occurs with preference in horizontal direction constricting the wake. In reaction to that, fresh outer fluid of higher kinetic energy fills the deficit and longitudinal vortices emerge. As the velocity of the incoming fluid is inhomogeneous a large scale motion of the entire wake is triggered. Once, the shear layer is broken up the mixing with the surrounding air is enhanced. Therefore, with increasing distance the coherent structures loose their directionality and become more and more isotropic. The instability process is captured by the ACL and the FR in similar way. The wake structure in the hub region is closely mimicked by the hub model. Further, it can be observed, that the ACL simulation resolves smaller structures, which is a consequence of the lower dissipation of the employed l^2-*Roe* Riemann solver compared to the HLLC.

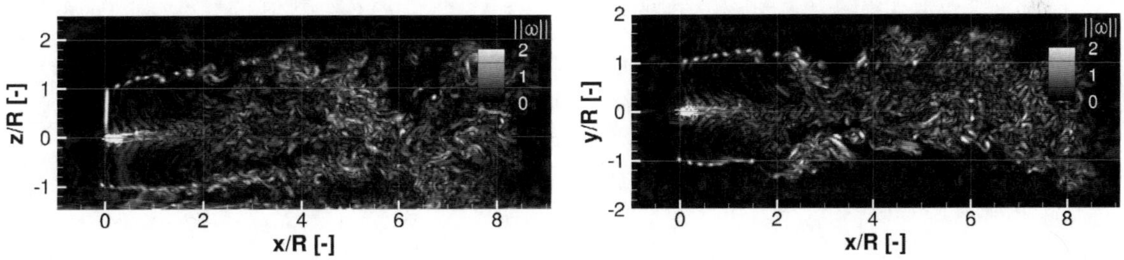

(a) *caseOffshore*: ACL vertical slice $y/R = 0$ (b) *caseOffshore*: ACL horizontal slice $z/R = 0$

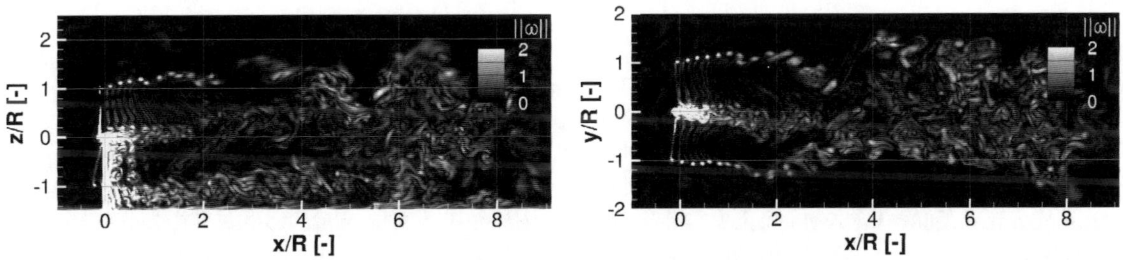

(c) *caseOffshore*: FR vertical slice $y/R = 0$ (d) *caseOffshore*: FR horizontal slice $z/R = 0$

Figure 6. Wake instability mechanisms visualized by vorticity magnitude for *caseOffshore*.

The vorticity contours of the flow around the turbine in *caseTerrain* are shown in Fig. 7. Compared to *caseOffshore* the considerably higher turbulence intensity can immediately be noticed showing vorticity levels of the same order as the tip and root vortices generated by the turbine. Further it can be observed that the incoming flow is rather homogeneous. Both facts result in a quicker disordering of the coherent structures in the wake. Vortex pairing can be observed at about one rotor diameter downstream of the turbine showing brake-up of the wake at around $x/R \geq 4$, especially in vertical direction. Hereby the upward deflection is larger as the inclination angle of the incoming flow. Overall, the role up process of the tip vortices that leads to global instability is less pronounced as for *caseOffshore*. When comparing the ACL with the FR, the wake structure is similar. However, stronger smearing of the tip vortices is present compared to the FR. It should be noted that as opposed to *caseOffshore* both simulations use the same spatial discretization scheme.

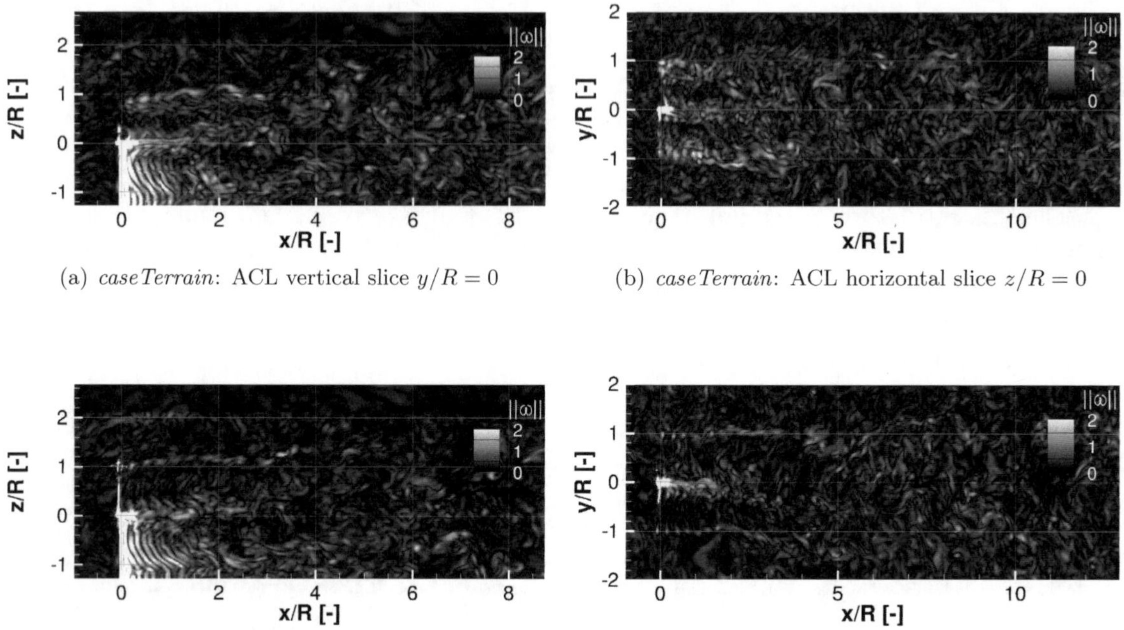

(a) *caseTerrain*: ACL vertical slice $y/R = 0$

(b) *caseTerrain*: ACL horizontal slice $z/R = 0$

(c) *caseTerrain*: FR vertical slice $y/R = 0$

(d) *caseTerrain*: FR horizonal slice $z/R = 0$

Figure 7. Wake instability mechanisms visualized by vorticity magnitude *caseTerrain*.

In order to evaluate the development of the wake more quantitatively the profiles of the mean velocity and their standard deviation σ are shown for *caseOffshore* in Fig. 8. The near wake is characterized by a linearly increasing velocity deficit in radial direction which is closely aligned with the rotor loading observed in Fig. 4. The high axial induction which reduces the free stream velocity by up to 50% in the outer part indicates that the turbine is operating close to the turbulent wake state [25]. This supports the findings made before regarding the entrainment of fluid from outside into the wake. The recovery process of the wake coincides well with the instability discussed before. When looking at the local maximum of the velocity in the wake center, this is phased out towards $x/R = 4$ showing a fairly uniform distribution and can be explained by the horizontal contraction of the wake. At $x/R = 6$, this shape is also present in the vertical profile, whereas in horizontal direction the wake is further rounded. Overall, the wake is deflected in positive z-direction and negative y-direction. The reasons for that is a slight negative mean v-component and positive w-component, as well as a rotor loading which is maximum in the fourth azimuthal quadrant. The ACL compares quite well with the FR capturing the shape and the levels of the wake deficit. The slightly higher deficit is connected with the mismatch of the loads in the outer part discussed in Fig. 4.

When looking at the velocity fluctuations it can be observed, that the ACL brings in higher perturbations into the flow field. In the near wake the FR simulation shows only peaks in the surrounding of the tip- and root vortices, whereas the ACL indicates enhanced fluctuations also in the middle part of the rotor. Similar observations were also made in Troldborg et al. [8] for turbulent non-sheared inflow. A possible explanation, is that due to the force smearing the bound vortex which trails helically into the flow field is larger than the viscous wake of the airfoil of the

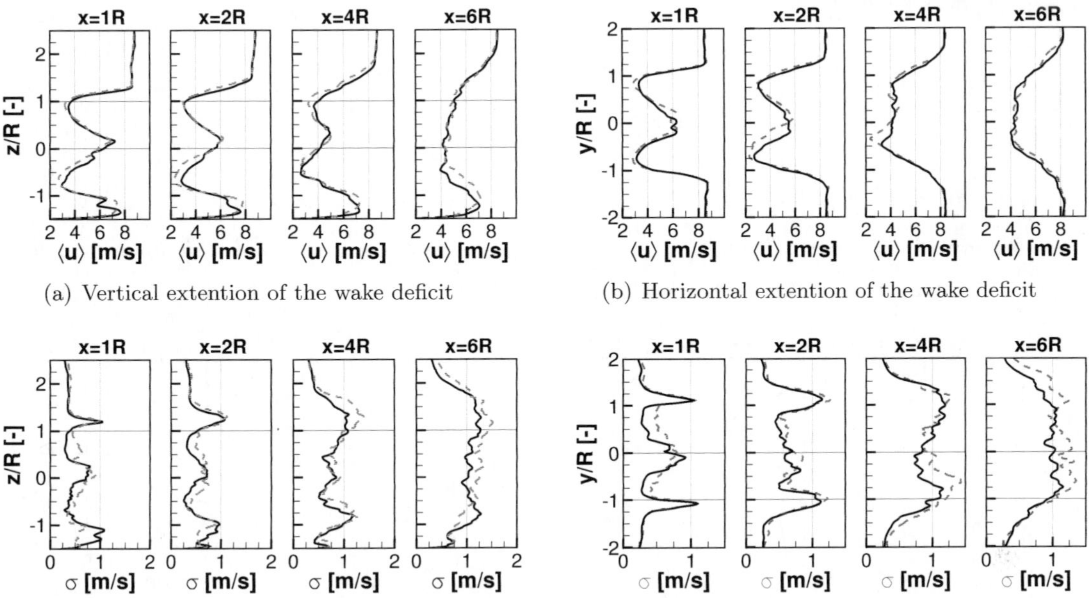

(a) Vertical extention of the wake deficit

(b) Horizontal extention of the wake deficit

(c) Vertical variation of the velocity fluctuations

(d) Horizontal variation of the velocity fluctuations

Figure 8. Profiles of mean axial velocity and standard deviation for *caseOffshore*. Solid: FR; dashed: ACL

FR simulation. Both lead to velocity variations that travel over the fixed observer. However, the thin viscous wake of the FR dissipates quicker as it requires a significantly finer mesh to be preserved. When looking further downstream, both turbine models predict a spatially blurring of the velocity fluctuations in the tip region which is a consequence of the breaking shear layer and mixing with the surrounding turbulent boundary layer. The entrainment of fluid into the wake center raises the level of turbulence there and diminishes the turbine foot-print.

For *caseTerrain*, the corresponding profiles are shown in Fig. 9. At first glance it is apparent, that the wake is more rounded compared to *caseOffshore* which is consistent to the higher loading in the mid portion of the rotor shown in Fig. 4. The escarpment in the terrain yields higher velocities near the ground and a negative shear with increasing height [19, 24]. This results in a stronger tower effect compared to *caseOffshore*. Particularly, in the very near wake the velocity fluctuations do not decay towards the ground. Still further downstream the fluctuations are more pronounced in the lower part of the wake which leads to a faster recovery of the wake there. Compared to *caseOffshore* the higher turbulence intensity is responsible for the fact that the footprint of the turbine is getting blurry after around two rotor diameters behind the turbine resulting in a Gaussian shaped wake deficit. Regarding the averaged deflection of the wake, the vertical profiles indicate an upward deflection in the near wake due to the inclination of the inflow, followed by a successive alignment to the orography further downstream. In horizontal direction the wake is slightly deflected to the negative y-axis which follows from a terrain induced small yaw angle in combination to an azimuthally asymmetric rotor loading due to the turbine tilt and cone angle. A more detailed discussion on the wake development in *caseTerrain* can be found in [24]. Comparing the wake deficit between the modeling techniques, good agreement of both can be found with a slight tendency of the ACL to extract more momentum from the flow field as the FR, although both had very similar force distributions. The comparison of the velocity fluctuations shows as for *caseOffshore* mostly higher standard deviations for the ACL compared to the FR simulation.

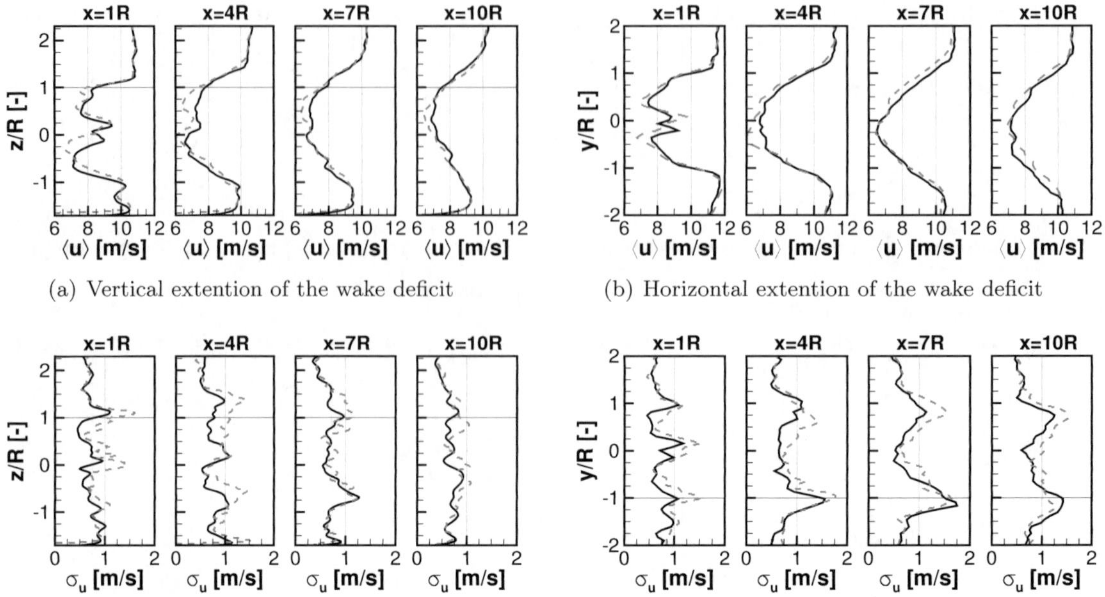

(a) Vertical extention of the wake deficit

(b) Horizontal extention of the wake deficit

(c) Vertical variation of the velocity fluctuations

(d) Horizontal variation of the velocity fluctuations

Figure 9. Profiles of mean axial velocity and standard deviation for *caseTerrain*. Solid: FR; dashed: ACL [24]

For the determination of fatigue loads on a fictive turbine located in the wake, it is important to break down the velocity fluctuations into the different frequency shares. Hence, spatially averaged velocity spectra of the axial component are displayed in Fig. 10 for the different models and cases. In *caseOffshore* the fluctuations show similarly to Meister [26] distinct peaks at the $3P$ frequency and its higher harmonics. As for the ACL only a single vortex system is injected into the flow field the energy is concentrated in the rotor frequency where the peaks are somewhat sharper and the amplitudes slightly higher. In the FR simulation velocity fluctuations are more diverse resulting for example from the viscous wake of the airfoil or flow separation and therefore resulting in a more broadband spectrum. In *caseTerrain* the situation is different. As already seen in Fig. 7 the wake structure is less dominated by the turbine features than in *caseOffshore* (Fig. 6) but more by the higher amount of atmospheric turbulence that directly enters the near wake. It seems that here the vortices induced by the ACL get disintegrated more easily as in the FR case, yielding the spreading of the $3P$ peaks. This seems to hold also further downstream as shown in the higher amplitudes of small scale turbulence at $x/R = 7$. In *caseOffshore* the high frequency regime at $x/R = 6$ is further amplified for the ACL, since a lower dissipation scheme was employed.

5. Conclusions

Simulations on the wake properties have been conducted for typical wind turbines sited in on- and offshore wind farms with different turbine modeling techniques, namely the actuator line method and the fully resolved representation of the rotor. The objective was to assess the modeling accuracy of the actuator line method in complex environmental conditions where different levels of turbulence, shear and orography effects apply. Therefore, two cases have been investigated, an offshore case where a 5 MW wind turbine was subject to the maritime boundary layer and a 2.4 MW turbine sited in complex terrain. For the offshore case precursor LES data from the PALM code has been used revealing positive shear and a mean velocity of 8 m/s with a

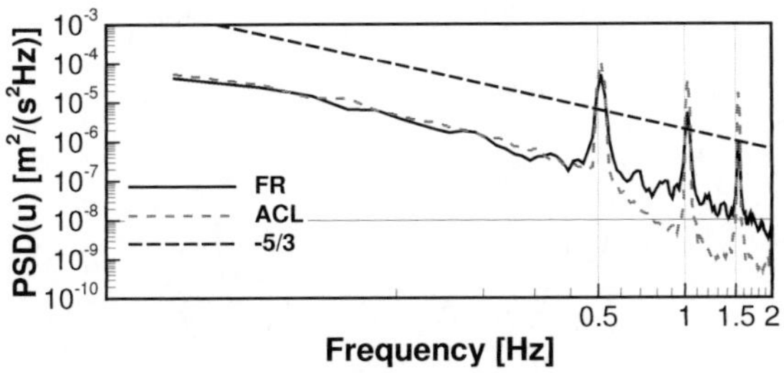

(a) *caseOffshore*: $x/R = 0.1$

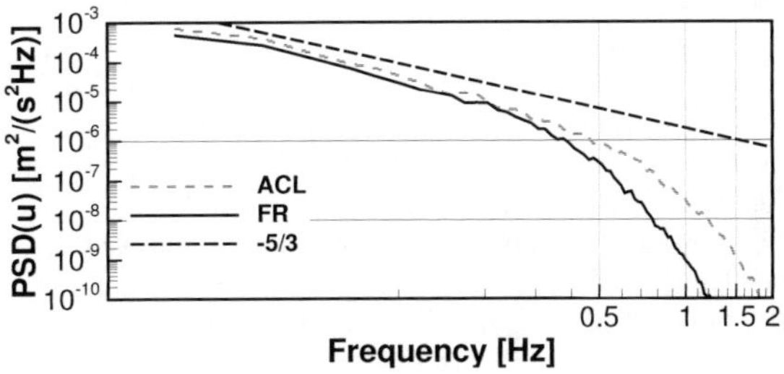

(b) *caseOffshore*: $x/R = 6$

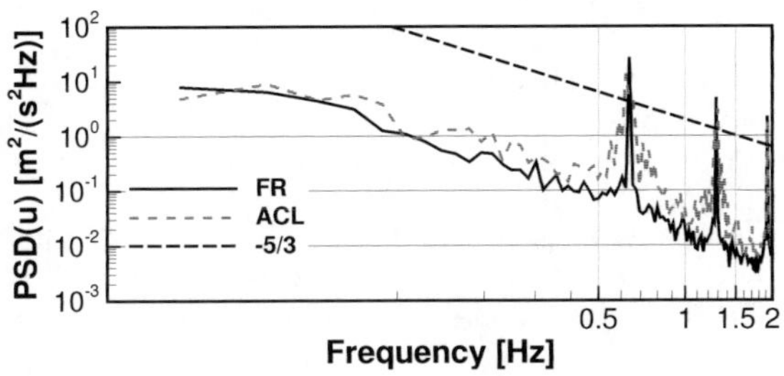

(c) *caseTerrain*: $x/R = 0.1$ [24]

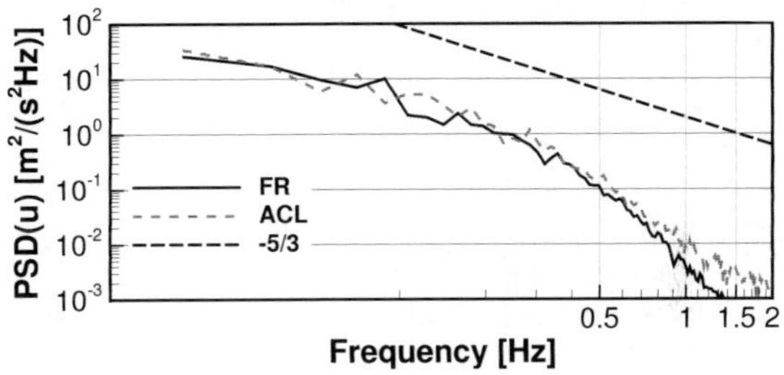

(d) *caseTerrain*: $x/R = 7$ [24]

Figure 10. Spatially averaged velocity spectra of the axial velocity component at different downstream positions.

moderate Ti of 4.5% at hub height. For the complex terrain case synthetic turbulence has been generated with the *Mann* model. The orography yields a negatively sheared wind profile at the turbine position with wind velocities of 11 m/s and turbulence intensities of about 10%.

It could be shown that in the cases considered unsteady aerodynamic effects are present for the rotor loads. Regarding the wake instability, distinct wake meandering was observed for the high induction offshore rotor. In complex terrain no such large scale motion of the wake was observed. The actuator line simulations showed good agreement for the mean wake properties but over predicted velocity fluctuations in the mid portion of the wake. It can be concluded that for wind farm optimization and power forecasts the actuator line seems accurate enough, even in complex terrain. However, the specification of accurate airfoil polars is essential. If the interest lies on the detailed load behavior of a turbine within a wind farm the actuator line model could certainly provide the disturbed inflow for the turbine of interest which is then simulated in the fully resolved fashion.

5.1. Acknowledgments

The authors acknowledge the *German Federal Ministry for Economic Affairs and Energy* for funding this study and the *High Perfomance Computing Center Stuttgart* for the contribution of the computational resources.

References

[1] Alfredsson P and Segalini A 2017 Introduction wind farms in complex terrains: an introduction
[2] Nilsson K, Shen W Z, Sørensen J N, Breton S P and Ivanell S 2015 *Wind Energy* **18** 499–514
[3] Troldborg N 2008 *Actuator Line Modeling of Wind Turbine Wakes* Ph.D. thesis Technical University of Denmark
[4] Huyer S A, Simms D and Robinson M C 1996 *AIAA journal* **34** 1410–1419

[5] Leishman J G 2002 *ASME 2002 Wind Energy Symposium* (American Society of Mechanical Engineers) pp 141–167
[6] Schepers J G 2012 *Aerospace Engineering. Delft University of Technology*
[7] Schepers J and Snel H 2007 *ECN report*
[8] Troldborg N, Zahle F, Réthoré P E and Sørensen N N 2015 *Wind Energy* **18** 1239–1250
[9] Jameson A 1991 *AIAA paper* **1596** 1991
[10] Jameson A, Schmidt W, Turkel E *et al.* 1981 *AIAA paper* **1259** 1981
[11] Kowarsch U, Oehrle C, Keßler M and Krämer E 2013
[12] Oßwald K, Siegmund A, Birken P, Hannemann V and Meister A 2015 *International Journal for Numerical Methods in Fluids*
[13] Weihing P, Letzgus J, Bangga G, Lutz T and Krämer E 2017 *Notes on Numerical Fluid Mechanics and Multidisciplinary Design (to be published)* (Springer)
[14] Mikkelsen R 2003 *Actuator Disc Methods Applied to Wind Turbines* Ph.D. thesis Technical University of Denmark
[15] Tossas L A M and Leonardi S 2013 Wind turbine modeling for computational fluid dynamics Tech. Rep. NREL/SR-5000-55054 National Renewable Energy Lab., Golden, CO (US)
[16] Limley J 2016 Implementierung eines modells zur berücksichtigung von gondeleffekten bei der actuator line methode im strömungslöser flower
[17] Weihing P, Meister K, Schulz C, Lutz T *et al.* 2014 *Journal of Physics: Conference Series* vol 524 (IOP Publishing) p 012143
[18] Schulz C, Klein L, Weihing P and Lutz T 2016 *Journal of Physics: Conference Series* vol 753 (IOP Publishing) p 032016
[19] Schulz C, Hofsäß M, Anger J, Rautenberg A, Lutz T, Cheng P W and Bange J 2016 *Journal of Physics: Conference Series* vol 753 (IOP Publishing) p 082017
[20] Raasch S and Schroter M 2001 *Meteorologische Zeitschrift* **10** 363–372
[21] Meister K, Lutz T and Krämer E 2014 *Journal of Physics: Conference Series* vol 555 (IOP Publishing) p 012071
[22] Mann J 1994 *Journal of fluid mechanics* **273** 141–168
[23] Troldborg N, Sørensen J N, Mikkelsen R and Sørensen N N 2014 *Wind Energy* **17** 657–669
[24] Schulz C 2017 *Numerische Untersuchungen des Verhaltens von Windenergieanlagen im komplexen Gelände unter turbulenter atmosphärischer Zuströmung* Ph.D. thesis Institute of Aerodynamics and Gas Dynamics, University of Stuttgart
[25] Hansen M O 2015 *Aerodynamics of wind turbines* (Routledge)
[26] Meister K 2015 *Numerische Untersuchung zum aerodynamischen und aeroelastischen Verhalten einer Windenergieanlage bei turbulenter atmosphärischer Zuströmung* (Shaker Verlag)

A wind turbine wake in changing atmospheric conditions: LES and lidar measurements

L Vollmer[1], J C-Y Lee[2], G Steinfeld[1], J K Lundquist[2]
[1]ForWind, Carl-von-Ossietzky Universität Oldenburg, Oldenburg, DE
[2]Department of Atmospheric and Oceanic Sciences, University of Colorado, Boulder, CO, USA
E-mail: lukas.vollmer@forwind.de

Abstract. This work aims to reproduce the measured atmospheric conditions during one day of the CWEX-11 campaign, with a transient LES. The selected period includes several interesting atmospheric conditions for wind power generation such as a nocturnal low-level jet, a highly turbulent convective daytime boundary layer, as well as a distinct evening transition between daytime and nocturnal boundary layers. To include synoptic conditions, large-scale forcing profiles for the LES were derived from a mesoscale simulation with the WRF model. A comparison with lidar measurements shows that the trend of the wind conditions and the diurnal cycle is well replicated by the model chain. Selected periods of the day are simulated with the NREL 5MW turbine model, followed by a qualitative comparison of measured and simulated wakes. We find a strong dependency of the meandering and the shape of the wake on wind profile and turbulence, while a categorization by Obukhov length is less representative for the different conditions. As the veer in the wind profile increases, the deviation of the wind direction at hub height from the direction of the largest wake impact also increases.

1. Introduction

The use of ground-based or nacelle-based lidars has rapidly increased knowledge about wind turbine wakes in recent years [1, 2, 3]. Lidars are mostly cheaper and are more mobile than meteorological masts, can be easier deployed in complex terrain and the remote technique allows measurements of the wake at a range of heights and distances. Still a single lidar cannot completely probe the three-dimensional structure of the wake as it evolves in time, leading to uncertainties in our understanding of wake evolution.

One method to complement lidar measurements of wind turbine wakes are numerical experiments [4, 5, 6]. Numerical experiments can deliver a temporally-resolved three-dimensional wind field to study the flow regions that were not observed. The possible benefit of the complementary simulations depends on the ability to reproduce, in the numerical environment, the aerodynamic behaviour of the turbine as well as the state of the atmospheric boundary layer.

Mesoscale models, with typical horizontal grid sizes in the order of 1-10 kilometers, can simulate the mesoscale and synoptic development of the atmospheric boundary layer. These models enable studies of the changing wind conditions at a wind farm site but are not highly resolved enough to model the aerodynamics of turbines and the interaction of turbines by means of their wakes. An approach for further downscaling towards smaller scales has been presented in Vollmer et al. [7, 8] using a Large-Eddy-Simulation model (LES), driven by the data of

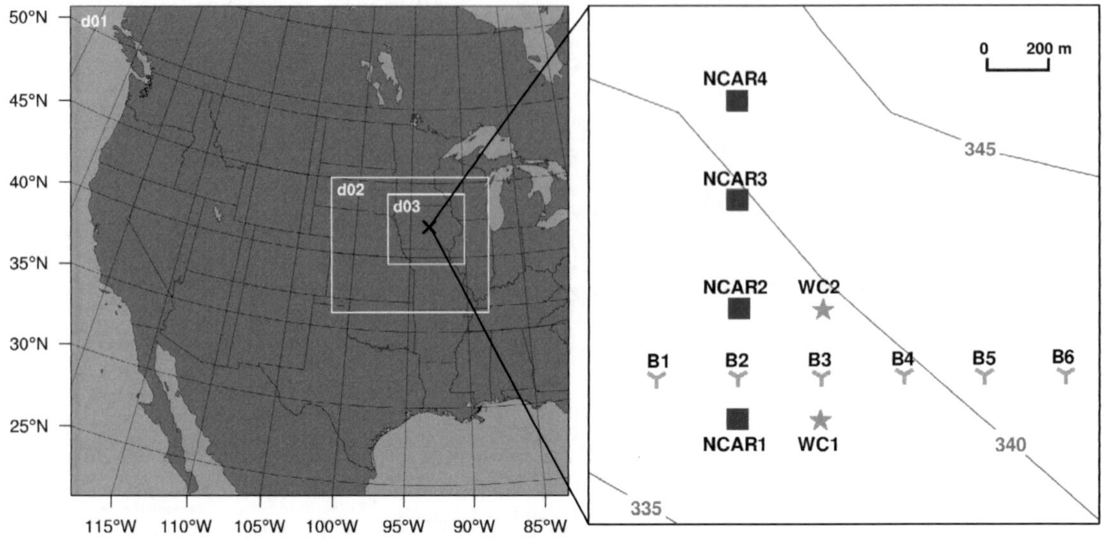

Figure 1. WRF model domain with the wind farm marked as cross in the innermost domain and topographic map showing the setup of the CWEX-11 campaign. Wind turbines (B1-B6), lidars (WC1-2) and surface flux stations (NCAR1-4).

a mesoscale model. The study showed that it is possible to replicate observed offshore wind conditions with a transient LES and that the simulated wakes show similar behaviour to the measured wakes.

In this paper we use the approach to simulate a case that is more interesting regarding the influence of atmospheric stability on wakes, as it represents an onshore day characterized by a strong diurnal cycle and a well-pronounced evening transition. The case study selected is one day (9 July) of the CWEX-11 measurement campaign in 2011 in Iowa, USA. From the selected day, lidar measurements from two vertical profiling lidars, placed upstream and downstream of a 1.5 MW turbine, and measurements from an upwind flux station are used. A part of the chosen day has been studied by means of quasi-stationary LES in Mirocha et al. [4]. Here we use the transient approach to simulate almost the whole day. During selected time periods, that represent different atmospheric stability conditions, wind turbine wake simulations with the NREL 5MW model turbine [9] are performed. Due to the different dimensions of the turbines in measurement and simulation, we performed a qualitative comparison in this study, and focused on the potential surplus of information that can be extracted from the accompanying simulations.

2. Methods

2.1. Observations

The Crop/Wind Energy EXperiment 2011 (CWEX-11) took observations along a row of wind turbines of a 200-turbine wind farm in Iowa, USA, from June to August 2011 [10, 2]. The turbines are 1.5 MW turbines from General Electric with a rotor diameter (D) of 77 m and a hub height of 80 m. The wind farm is situated in a flat region characterized by extensive agriculture (corn).

Two WindCube v1 lidars (WC1-2) measured the vertical wind profile at altitudes from 40 m to 220 m AGL. The lidars were placed 2.1 D south and 3.2 D north of one turbine of the southernmost row of the wind farm to capture free flow as well as wake flow for the prevailing southerly wind. Surface flux stations were also sampled upwind and downwind in similar fashion but for the next turbine to the west (Fig. 1). The original measured wind components from the lidars at 0.25 Hz were averaged every 2 minutes. In this paper we do not further analyse the uncertainty of the wind profile measurements that come from the velocity azimuth display (VAD) technique. For the effect of the individual beams and the volume averaging on the flow reconstruction by the lidar, especially inside wind turbine wakes, the reader is referred to Lundquist et al. [6].

Five-minute averages of the 20-Hz measurements from the sonic anemometers and gas analysers were made available from the National Center of Atmospheric Research (NCAR). The instruments were operating at 4.5 m above ground level (AGL). The Obukhov length was calculated as

$$L = -\frac{u_*^3\, T_v}{k\, g\, \overline{w'T_v'}} \tag{1}$$

using the friction velocity u_*, derived from the vertical fluxes of momentum in the streamwise and spanwise directions, the vertical kinematic heat flux $\overline{w'T_v'}$ and the virtual temperature T_v measured by the sonic anemometer. The factors k and g are the von-Kármán constant and the gravitational acceleration, respectively. The bulk Richardson number (Ri) was calculated as an alternative measure for atmospheric stability following Stull [11].

$$Ri = \frac{g\, \Delta\Theta_v\, \Delta z}{\Theta_v\, [(\Delta u)^2 + (\Delta v)^2]} \tag{2}$$

with $\Delta\Theta_v$, Δu and Δv the difference between the virtual potential temperature Θ_v, the streamwise velocity u and the cross-stream velocity v at two different measurement heights separated by Δz. As temperature measurements were not available above 10 m, Ri could only be calculated from the model results. The bulk Richardson number was calculated between the surface and the hub of the simulated turbine (Ri_0^{90}) and between the lower and upper rotor tip (Ri_{30}^{150})

2.2. Models

The model chain for the simulation of the case consisted of two models. The Weather Research and Forecasting (WRF) model [12] was used to simulate the changing wind conditions at the site during the day, required as large-scale forcing input [8] for the PArallelized Large-Eddy Simulation Model (PALM) [13].

Model version 3.6.1 of WRF was used for a 30 h simulation starting at 00 UTC (five hours ahead of local standard time LST) on 9 July [14]. Initial and boundary conditions for the one-way-nested domains were derived from the Global Forecast System (GFS) reanalysis, using the standard procedures as defined in Skamarock et al. [12]. The smallest domain covered the entire state of Iowa with 990 m horizontal resolution and a dimension of 571 x 511 grid points. Vertical levels were stretched towards the top with 70 levels in total and a vertical spacing of about 22 m in the lower 300 m AGL. The MYNN Level 2.5 scheme [15] was used for the TKE closure in the planetary boundary layer. A more detailed description and analysis of the WRF simulation is presented in Lee and Lundquist [14].

To create the large-scale forcing profiles for the LES, the hourly model results from WRF were horizontally averaged on a domain of 80 km x 80 km to average out pressure perturbations on smaller scales (See Vollmer et al. [8] for a discussion on the necessary size of the averaging domain). The domain was centered around the turbine, where the lidar measurements were made. The mean geostrophic wind vector at this location was calculated from the large-scale pressure gradients over the domain. The vertical profiles of the horizontal advection

IOP Conf. Series: Journal of Physics: Conf. Series **854** (2017) 012050 doi:10.1088/1742-6596/854/1/012050

of temperature, humidity and horizontal momentum as well as the geostrophic wind speed components were used for the large-scale forcing of PALM. To prevent the LES from drifting from the WRF results, a nudging scheme was used with a relaxation time constant of $\tau = 4$ h. Nudging was applied to the domain averaged vertical profiles of potential temperature, humidity and the horizontal wind components. More details about the preparation and application of the large-scale forcing can be found in Vollmer et al. [8].

Revision 1928 of PALM [13] was used with the same numerical schemes as in Vollmer et al. [16]. Simulations were run on a 640 x 640 x 160 grid with 5 m resolution. Above 600 m, vertical stretching of the grid was applied. The highest grid point is centered at 1715 m. Monin-Obukhov similarity theory is used at the bottom boundary with a surface roughness length of momentum of $z_0 = 0.14$ m, which represents the average roughness length inside of the WRF averaging domain. Roughness lengths of scalars were a factor of ten times smaller.

Sufficient information to implement the GE turbine was not available. However, to allow for a qualitative comparison of the wake behaviour, an ADM-R [17] representation of the NREL 5MW [9] research turbine model was used for wake simulations. The turbine has a rated power of 5 MW, a rotor diameter of 126 m and its hub is located at 90 m.

Adaptation of the rotor speed of the NREL 5MW turbine to the changing wind is ensured by a generator torque controller below rated wind speed (11.4 ms^{-1}) and a collective pitch controller above rated wind speed. The turbine orientation is controlled by a yaw controller based on a 30-second running average of the wind direction at the turbine hub. The yaw actuator is activated for a misalignment of more than 5 degrees. As the turbine influences the turbulence and fluxes inside the cyclic LES domain, the maximum length of individual turbine simulations was 35 minutes, of which the first 5 minutes were discarded as the wake was still developing. From the total 30 h of simulation, the first eight hours were discarded as they are used for the spin-up of both the WRF model and the LES.

3. Results

In the following we will discuss the performance of the model chain to simulate the state and development of the atmospheric boundary layer, followed by an analysis of the wake simulations. Figure 2 compares the results of the simulation of the ambient flow with the large scale forcing input data, the lidar measurements and the flux measurements. The wind speed and wind direction follow the trend of the input data, but deviations from the measurements can reach more than 2 ms^{-1} and 20° for wind speed and wind direction, respectively. The measured surface kinematic heat flux (Fig. 2g) is well replicated by the WRF simulation with a negative heat flux during nighttime and a positive one during daytime between 08:00 LST and 18:00 LST. The LES kinematic heat flux almost equals that of WRF. The diurnal cycle of the 2-min turbulence intensity (I) is well reproduced in magnitude, and the moment of restratification, quantified by the sign change of L, during the evening transition matches very well. The largest deviation of the LES from the WRF input is found in the measures for the vertical wind profile (Fig. 2e,f). The LES does a better job to simulate the veer of the wind profile of the nocturnal boundary layer than the mesoscale model, which appears to be limited to a maximum veer of $\Delta WD = 10°$ between the two selected heights. The shear of the wind profile on the other hand is well replicated by WRF with the exception of a period at the end of the night where the measured wind shear increases in the measurements. A low-level jet occurs in the simulation between 03:00 LST and 07:00 LST with a core height between 200 m and 300 m (not shown), thus clearly above the maximum rotor height. The core height is consistent with the lidar measurements, as no local maximum of the wind speed can be found in the measured profiles below 220 m.

Based on the results of the simulation of the ambient flow, three periods of 1.5 hours, marked by the shaded areas in Fig. 2, were selected in which a simulation with a turbine was conducted.

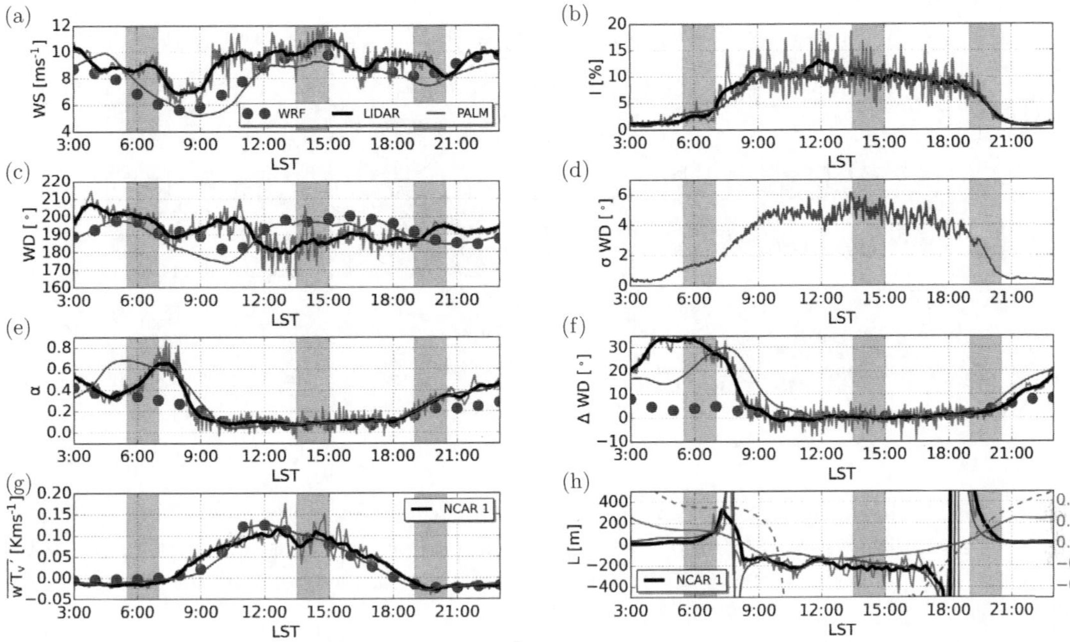

Figure 2. Comparison between one-hour running means (black bold), short-term averages (grey) of measurements, the large-scale forcing input from WRF (dots) and the domain averaged LES results (blue thin). (a) Two-minute hub-height wind speed, (b) hub height turbulence intensity, (c) hub-height wind direction and (d) hub-height standard deviation of the wind direction. (e) Two-minute vertical power law coefficient and (f) change of wind direction, both between 40 m and 140 m. (g) Five-minute surface kinematic heat flux, (h) Obukhov length and bulk Richardson number between surface and hub height (solid) and between lower (30 m) and upper (150 m) rotor tip (dashed). Shaded areas mark the selected periods for the turbine simulation. From left to right: SBL, CBL and NNBL.

The periods (Tab. 1) were selected to represent different meteorological conditions. During the first period, the turbine operates in a strongly sheared and veered wind profile with low turbulence that is created by the strong nocturnal stable stratification (SBL). The second period during the day is characterized by a convective boundary layer (CBL) with nearly no vertical shear but vigorous short scale fluctuations of wind speed and direction (Fig. 2b,d). The evening transition from the daytime to the nighttime layer with a near neutral stratification (NNBL) is selected as the third period. While the measures of the wind profile differ considerably between SBL and NNBL, the mean Obukhov length is almost equal and the negative heat flux is even lower during the NNBL period (Fig. 2g,h, Tab. 1). The long build-up of the strong shear of the nocturnal boundary layer can thus not be captured with the instantaneous measure of Obukhov length and heat flux at the surface. Using the bulk Richardson number as an alternative measure for the classification of the stability conditions during the periods allows to differentiate between the SBL and NNBL period only when using measurements at the heights of the rotor tips.

The wakes in the three situations exhibit very different characteristics, that appear in both, the measurements and the simulations. Fig. 3 shows the two-minute averaged profiles from the lidar measurements and from the simulations at the upwind and downwind lidar position. Even though the wind turbines in reality and in the LES model are different, and the ambient flow fields are not exactly equal, either, the defining wake characteristics are identical. During the

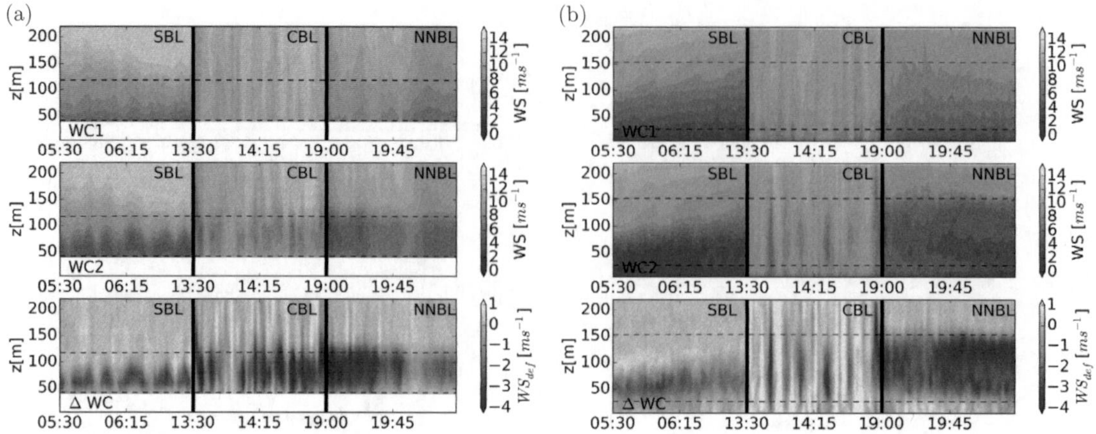

Figure 3. Temporal development of the horizontal velocity profile at the position of WC1 (top row) and WC2 (middle row) and difference between the two profiles (bottom row), (a) from the lidar measurements and (b) from the LES. The dashed lines mark the vertical boundaries of the respective rotor.

SBL period the wake is weaker and at a lower height than during the NNBL period, while the wake during the CBL period is just intermittently discernible. However, the vertical profiles of course only visualise a fraction of the full wake. Especially in the strong fluctuating wind conditions of the convective boundary layer it is impossible to localise the position inside the wake where it is sliced by the scan.

To gain further insight into the wake characteristics and behaviour, vertical cross-sections of the wake derived from the simulations were used. For this purpose the wind field was averaged along a half-circle downstream of the wind turbine with the radius of 3.22 D, corresponding to the distance of the downstream lidar from the turbine (Fig. 3b). This procedure was chosen to retrieve sections of the wake at the same distance for every wind direction. Example cross-sections for the three time intervals are shown in Fig. 4. The strong veer of the nocturnal boundary layer leads to a strong cross-stream stretching of the wake. Thus, the vertical profile at the lidar position only captures the lower left corner of the wake. The example wakes during the CBL and NNBL both have a more circular shape.

Table 1. Average quantities from the PALM simulations during the three periods for which wake simulations were run. Mean and turbulent quantities are at hub height. Vertical temperature flux and Obukhov length are calculated at the vertical center of the lowest grid cell at z = 2.5 m.

	Start [LST]	End [LST]	$\overline{WS}$ [ms^{-1}]	$\overline{WD}$ [°]	$\overline{I}$ [%]	$\overline{\sigma WD}$ [°]	$\overline{\Delta WD}$ [°]	$\overline{\alpha}$ [-]	$\overline{w'T'}$ [K ms^{-1}]	$\overline{L}$ [m]	$\overline{Ri}_0^{90}$ [-]	$\overline{Ri}_{30}^{150}$ [-]
SBL	05:30	07:00	7.5	194	3.4	1.5	22.8	0.65	-0.006	87	0.13	0.35
CBL	13:30	15:00	9.0	197	10.7	5.0	0.5	0.08	0.095	-172	-0.12	-0.75
NNBL	19:00	20:30	7.6	186	4.9	2.0	3.3	0.26	-0.025	93	0.11	0.16

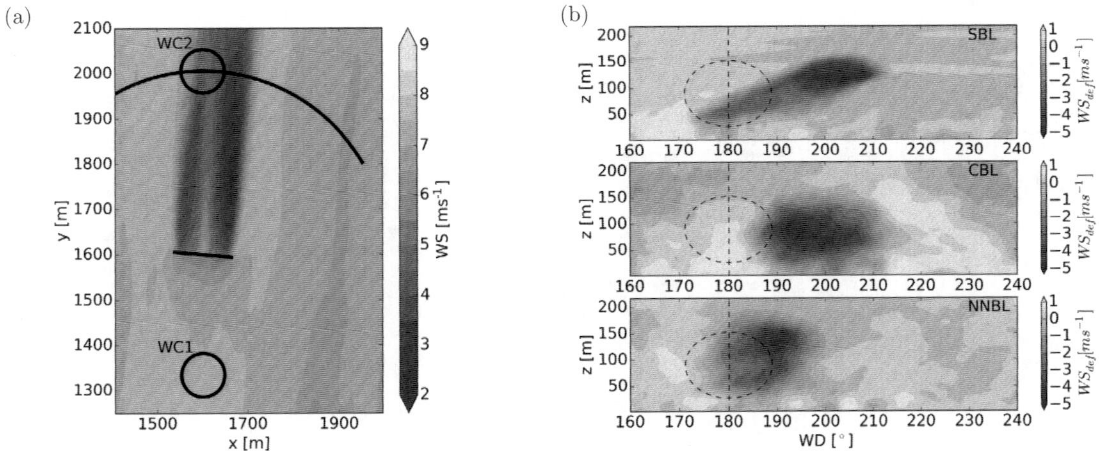

Figure 4. (a) LES wind speed average at hub height at 19:45 LST during the SBL period with the positions of the lidars marked as circles. Marked as black line is the line of constant distance to the turbine along which the velocities are analysed in Fig. 4 (b) Examples of the 2-min averaged wind speed along a constant distance (3.22 D) downstream of the turbine during the three simulation intervals. The vertical line denotes the position of WC2. The circle denotes the surface used for the calculation of the REWS at WD = 180°.

A direct comparison of the wakes based on the wind speed is difficult as the wakes can not be easily fitted to any geometric structure, e.g. a Gaussian-like shape, due to their asymmetry. To simplify the wake representation, the wind speeds downstream of the turbines were converted to rotor equivalent wind speeds (REWS) [18] as a function of the cross-stream coordinate x as in Vollmer et al. [16]. For this purpose the cube of the wind speed is averaged on virtual rotor surfaces of the diameter D of the NREL 5MW turbine and converted back to a wind speed.

$$\mathrm{REWS}(x) = \left(\frac{1}{A}\int_{x1}^{x2}\int_{z1}^{z2} u^3(x',z')\cos(\beta(x',z'))dz'dx'\right)^{1/3}, \; (x'-x)^2+(z'-z_h)^2 \leq (D/2)^2 \quad (3)$$

The wind veer β over the rotor area A is defined as the difference between the wind direction at hub height and at the specific coordinate [19]. Here we use the assumption that the virtual rotor is aligned with the wind direction at hub height. By calculating the REWS the wakes are reduced to two-dimensional structures and can be compared easier.

The REWS for all individual 2-minute averages is shown in Fig. 5a. Figure 5b shows a time series of the deficit position defined as the minimum of the REWS and the 2-minute average of the LES wind direction at the rotor center of the turbine. The analysis shows that the REWS shows a strong asymmetry in the SBL, with a sharp gradient on the right side of the wake and a flat slope on the left side. As the simulated wind veer increases during the SBL period (Fig. 2f), the difference between the wind direction of the REWS minimum and the hub height wind direction also increases. This difference reaches a value of 10 degrees by the end of the stable simulation. Though the effect is regarded to depend on both shear interacting with the wake rotation and the veer of the wind profile [16], it is mainly dependent on wind veer in the analysed case as shown in Fig. 5c.

The wake during the NNBL period appears to be more symmetric. While the wake shapes of NNBL and CBL appear similar in Fig. 4a, the main difference between the two periods is the strong fluctuation of the wind direction during the CBL period which leads to a permanent

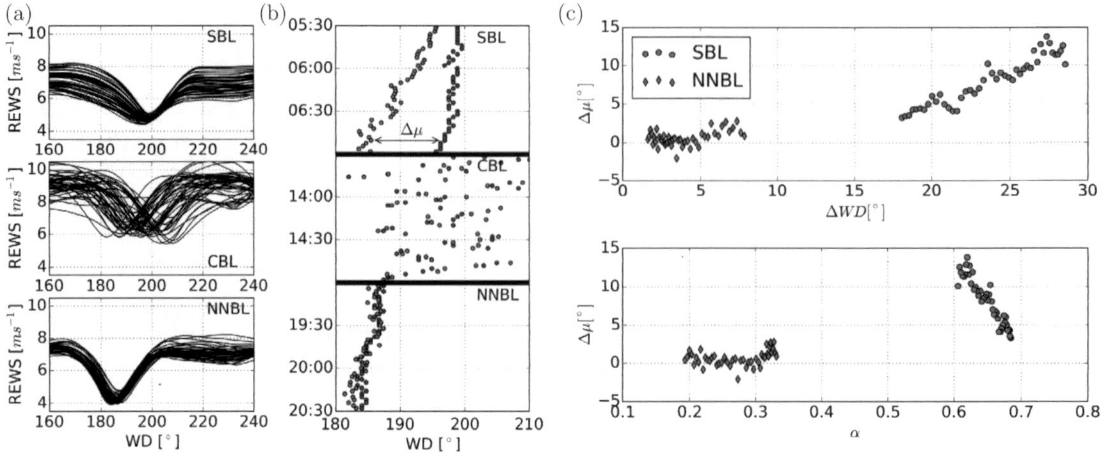

Figure 5. (a) REWS as function of wind direction of each 2-min timeframe. (b) Position of the minimum of the REWS (blue) and LES wind direction at the rotor center (red). (c) Difference between wake position and hub height wind direction $\Delta\mu$ plotted over the wind veer and the wind shear coefficient.

cross-stream meandering of the wake. During the NNBL period the REWS minimum generally coincides with the measured wind direction at hub height.

4. Discussion

Mesoscale to microscale flow modeling is regarded as one of the main research topics for the next generation of wind condition assessment and wind farm design tools [20]. The methodology to use a background state derived from mesoscale simulations for cyclic LES, presents an alternative to the so-called nesting of LES in mesoscale simulations. The advantage of the method is that only a limited amount and only low resolved mesoscale data is necessary for the setup of the LES boundary conditions. Thus, the simulations can be run from a database of temporally resolved mesoscale model data like planned e.g. in the New European Wind Atlas (NEWA) project. The disadvantage of the method is that the spatially resolved advection of turbulent structures, for example from orography or from upstream wind farms can not be included in the simulation. Here, nesting presents an advantage, even though the best practices of turbulence instigation and mesh refinement from the meso- to the microscale regime are still under discussion [21].

The results of this paper indicate that the model chain of WRF and PALM is generally able to replicate the state of the atmospheric boundary layer during the selected day. As also concluded in Vollmer et al. [8], the more important model for a good replication of wind speed and wind direction is the mesoscale model. However, in the given case with a strong diurnal cycle, we find the LES model capable of improving the vertical wind speed and wind direction profile of a strongly-stratified nocturnal boundary layer. The reason may be primarily related to the lower vertical resolution of the mesoscale model and the consequential parametrization of the boundary layer fluxes. To improve the match between measurements and simulations it might be worth incorporating the measurements into the large-scale forcing and nudging data as in Rodrigo et al. [19].

The wake simulations performed in the transient flow conditions present a method to enhance the understanding of the measured wake flow. This procedure can be regarded as an alternative or a supplement to long-term measurements that are typically classified in classes of wind speed, wind direction and stability [4, 22, 23]. While we were not able to replicate the wake in this study due to the lack of sufficient details about the wind turbine, we were still able to show the fundamental differences in wake behaviour and shape in the three periods of different atmospheric stability.

The differences can be related to the vertical profile of wind speed and direction and to the fluctuations of these parameters. The wind veer reaches values of up to 35° over the rotor surface during stable stratification, which leads to a wide stretching of the wake. During the daytime convective period the fluctuation of wake position and intensity is dominating the wind field. The bulk Richardson number, calculated from the vertical wind shear and the temperature difference over the rotor surface allows a categorization of the three periods of different wake behaviour. Atmospheric stability measured in the constant flux layer close to the surface, i.e. characterized by the Obukhov length, on the other hand, appears to be insufficient to differentiate between the periods as it is not able to capture the time history of the evolving flow. Thus, for a classification of wake measurements or performance measurements of wind turbines in stability classes the Richardson number is more suitable.

Abkar et al. [24] have shown in a simulation of an idealized diurnal cycle in LES how the wind farm power varies with the wind conditions influenced by atmospheric stability. For more sophisticated wind farm control like curtailment or wake steering, a precise knowledge about the wake deficit position is also crucial. Our results show that wind veer needs to be considered as it can indeed influence the wake deficit position.

Acknowledgments
The authors gratefully acknowledge the efforts of those who carried out the CWEX experiment, including the CU Boulder team of Dr. Matt Aitken and Mr. Michael Rhodes, the Iowa State Team of Dr. Gene Takle, Dan Rajewski, Russ Doorenbos, Kris Spoth, Jimmy Cayer, and the NCAR team including Dr. Steve Oncley and Dr. Tom Horst. We also extend appreciation to the wind farm operators and the landowners who permitted the deployment of the instruments. The DAAD is thanked for granting Lukas Vollmer a two-month stay at the Department of Atmospheric and Oceanic Sciences of the University of Colorado, Boulder, CO, USA. The work presented in this study is funded by the ministry of science and culture of Lower Saxony within the project ventus efficiens (ZN3024, MWK Hannover). Computer resources have been partly provided by the North German Supercomputing Alliance (HLRN).

References

[1] Trujillo J J, Bingöl F, Larsen G C, Mann J and Kühn M 2011 Light detection and ranging measurements of wake dynamics. Part II: two-dimensional scanning *Wind Energy* **14** 61–75

[2] Rhodes M E and Lundquist J K 2013 The Effect of Wind-Turbine Wakes on Summertime US Midwest Atmospheric Wind Profiles as Observed with Ground-Based Doppler Lidar *Boundary-Layer Meteorol.* **149** 85–103

[3] Machefaux E, Larsen G C, Troldborg N, Gaunaa M and Rettenmeier A 2015 Empirical modeling of single-wake advection and expansion using full-scale pulsed lidar-based measurements *Wind Energy* **18** 2085–2103

[4] Mirocha J D, Rajewski D A, Marjanovic N, Lundquist J K, Kosovic B, Draxl C and Churchfield M J 2015 Investigating wind turbine impacts on near-wake flow using profiling lidar data and large-eddy simulations with an actuator disk model *J. Renew. Sust. Energ.* **7** 043143

[5] Machefaux E, Larsen G C, Koblitz T, Troldborg N, Kelly M C, Chogoule A, Hansen K S and Sanz Rodrigo J 2015 An experimental and numerical study of the atmospheric stability impact on wind turbine wakes *Wind Energy* Published online

[6] Lundquist J K, Churchfield M J, Lee S and Clifton A 2015 Quantifying error of lidar and sodar doppler beam swinging measurements of wind turbine wakes using computational fluid dynamics *Atmos. Meas. Tech.* **8** 907–920

[7] Vollmer L, van Dooren M, Trabucchi D, Schneemann J, Steinfeld G, Witha B, Trujillo J and Kühn M 2015 First comparison of LES of an offshore wind turbine wake with dual-Doppler lidar measurements in a German offshore wind farm *J. Phys. Conf. Ser.* vol 625 (IOP Publishing) p 012001

[8] Vollmer L, Steinfeld G and Kühn M 2017 Transient LES of an offshore wind turbine *Wind Energy Science Discussions* **2017** 1–19

[9] Jonkman J M, Butterfield S, Musial W and Scott G 2009 Definition of a 5-MW reference wind turbine for offshore system development Technical Report NREL/TP-500-38060 National Renewable Energy Laboratory 1617 Cole Boulevard, Golden, Colorado 80401-3393

[10] Rajewski D A, Takle E S, Lundquist J K, Oncley S, Prueger J H, Horst T W, Rhodes M E, Pfeiffer R, Hatfield J L, Spoth K K and Doorenbos R K 2013 Crop wind energy experiment (CWEX): Observations of surface-layer, boundary layer, and mesoscale interactions with a wind farm *Bull. Amer. Meteor. Soc.* **94** 655–672

[11] Stull R B 1988 *An Introduction to Boundary Layer Meteorology* 1st ed (Kluwer Academic Publishers)

[12] Skamarock W, Klemp J, Dudhia J, Gill D, Barker D, Duda M, Huang X, Wang W and Powers J 2008 A description of the advanced research WRF version 3 Tech. Rep. NCAR/TN475+STR NCAR - National Center for Atmospheric Research Boulder, Colorado, USA

[13] Maronga B, Gryschka M, Heinze R, Hoffmann F, Kanani-Sühring F, Keck M, Ketelsen K, Letzel M O, Sühring M and Raasch S 2015 The parallelized large-eddy simulation model (palm) version 4.0 for atmospheric and oceanic flows: model formulation, recent developments, and future perspectives *Geosci. Model Dev.* **8** 2515–2551

[14] Lee J Y and Lundquist J 2017 Observing and simulating wind turbine wakes during the evening transition *in review in Boundary-Layer Meteorol.*

[15] Nakanishi M and Niino H 2006 An Improved MellorYamada Level-3 Model: Its Numerical Stability and Application to a Regional Prediction of Advection Fog *Boundary-Layer Meteorol.* **119** 397–407

[16] Vollmer L, Steinfeld G, Heinemann D and Kühn M 2016 Estimating the wake deflection downstream of a wind turbine in different atmospheric stabilities: an LES study *Wind Energy Science* **1** 129–141

[17] Witha B, Steinfeld G and Heinemann D 2014 *Wind Energy - Impact of Turbulence* Research Topics in Wind Energy 2 ed Hölling M, Peinke J and Ivanell S (Springer Berlin Heidelberg) pp 175–181 iSBN: 978-3-642-54696-9

[18] Wagner R, Canadillas B, Clifton A, Feeney S, Nygaard N, Poodt M, Martin C S, Tüxen E and Wagenaar J W 2014 Rotor equivalent wind speed for power curve measurement comparative exercise for iea wind annex 32 *J. Phys. Conf. Ser.* **524** 012108

[19] Rodrigo J S, Churchfield M and Kosovic B 2016 A wind energy benchmark for ABL modelling of a diurnal cycle with a nocturnal low-level jet: GABLS3 revisited *J. Phys. Conf. Ser.* **753** 032024

[20] Sanz Rodrigo J, Chvez Arroyo R A, Moriarty P, Churchfield M, Kosovi B, Rthor P E, Hansen K S, Hahmann A, Mirocha J D and Rife D 2017 Mesoscale to microscale wind farm flow modeling and evaluation *Wiley Interdisciplinary Reviews: Energy and Environment* **6** e214–n/a ISSN 2041-840X e214

[21] Muñoz-Esparza D, Kosovic B, Mirocha J and van Beeck J 2014 Bridging the Transition from Mesoscale to Microscale Turbulence in Numerical Weather Prediction Models *Boundary-Layer Meteorol.* **153** 409–440

[22] Hansen K S, Barthelmie R J, Jensen L E and Sommer A 2012 The impact of turbulence intensity and atmospheric stability on power deficits due to wind turbine wakes at Horns Rev wind farm *Wind Energy* **15** 183–196

[23] Dörenkämper M, Tambke J, Steinfeld G, Heinemann D and Kühn M 2014 Atmospheric Impacts on Power Curves of Multi-Megawatt Offshore Wind Turbines *Journal of Physics: Conference Series* **555** 012029

[24] Abkar M, Sharifi A and Port-Agel F 2016 Wake flow in a wind farm during a diurnal cycle *Journal of Turbulence* **17** 420–441

Theoretically optimal turbine resistance in very large wind farms

Alejandro Zapata, Takafumi Nishino*, Pierre-Luc Delafin

Cranfield University, Cranfield, Bedfordshire MK43 0AL, United Kingdom

*E-mail: t.nishino@cranfield.ac.uk

Abstract. In this paper we further investigate and validate the novel theoretical model of very large wind farms proposed recently by Nishino (*J. Phys.: Conf. Ser.* 753, 032054, 2016). One of the key features of the Nishino model is that a theoretically optimal turbine resistance (as well as optimal 'turbine-scale' and 'farm-scale' wind speed reduction rates) can be predicted analytically as a function of the farm density and the natural bottom friction observed before constructing the farm. To validate this theoretical model, a new set of 3D Reynolds-averaged Navier-Stokes (RANS) simulations are performed of a fully developed wind farm boundary layer over an aligned and staggered array of actuator discs with various disc resistance, inter-disc spacing and bottom roughness values. The results show that the theoretical model, which employs only one empirical model parameter, can be easily calibrated to predict very well the performance of various staggered arrays of actuator discs. This suggests the usefulness of the theoretical model not only for providing an upper limit to the performance of ideal large arrays but also for predicting the performance of realistic large arrays. The results also highlight the important fact that the optimal turbine resistance can be significantly smaller in a dense wind farm than in a sparse wind farm.

1. Introduction

The size of offshore wind farms has been dramatically increasing in recent years, causing a renewed interest in the modelling of very large wind farms. For example, in 2006, Frandsen et al. [1] reviewed a theoretical model for the interaction between a hypothetical very large wind farm and the planetary boundary layer (PBL), which was developed mainly by Frandsen and Emeis in the 1990's [2, 3]. Later on, Emeis [4, 5] enhanced the model by considering the effect of atmospheric stability. Calaf et al. [6] also improved the model using results of large-eddy simulations (LES) of a fully developed wind turbine array boundary layer (WTABL). More recently, Meneveau [7] further extended the model by taking into account the development of an internal boundary layer starting from the entrance of a wind farm. This family of theoretical models, which employ the concept of 'effective wind farm roughness' and are often referred to as the 'top-down' models, predict the reduction of wind speed by a large number of turbines with a given (i.e. empirically known or assumed) thrust coefficient.

Last year, a new type of theoretical model was proposed by Nishino [8] for the performance of a very large wind farm. Similarly to the traditional 'top-down' models, this new model also considers a fully developed boundary layer over a very large array of wind turbines. A noteworthy difference, however, is that the turbine thrust coefficient is not arbitrarily given but modelled using the classical (Lanchester-Betz type) actuator disc theory, which allows us to predict a practical upper limit to the efficiency of a very large wind farm together with 'optimal' wind speed reduction rates. The validity of this new theoretical model was only partially confirmed in [8]. Therefore in this study, we further validate this theoretical model by means of a comparison with a series of Reynolds-averaged Navier-Stokes (RANS) simulations of a fully developed boundary layer over an aligned and staggered array of actuator discs, for a range of inter-disc spacing, disc resistance and bottom roughness values.

2. Theoretical model

2.1. Nishino model

The theoretical model of very large wind farms proposed by Nishino [8] is based on a simple 'two-scale coupled' momentum conservation argument. The 'farm-scale' wind speed reduction factor, β, is calculated from the momentum balance between the driving force of the atmospheric boundary layer (ABL) and the resistance due to the turbines' thrust and bottom friction, whereas the 'turbine-scale' wind speed reduction factor, α, is calculated from the classical actuator disc theory. Eventually, these two wind speed reduction factors α and β satisfy the following momentum conservation equation (see [8] for further details of the derivation):

$$1-\beta^{\gamma} = \lambda \cdot \frac{1}{C_{f0}} \cdot \beta^2 \cdot 4\alpha(1-\alpha) \qquad (1)$$

where $\lambda = A/S$ is the 'farm density' (or the ratio of the turbine swept area, A, to the land area, S) and $C_{f0} = \tau_{\mathrm{w0}}/\frac{1}{2}\rho U_{F0}^2$ is the natural friction coefficient defined using the natural wind speed averaged across the so-called 'farm layer', U_{F0}. Note that $\beta = U_F/U_{F0}$ (≤ 1) is the velocity ratio that indicates how much the farm-layer wind speed changes from its natural state, whereas $\alpha = U_T/U_F$ (≤ 1) is the ratio of the average wind speed through the turbine swept area to that through the farm layer.

Although this model is essentially one-dimensional (1D; as with the classical actuator disc theory) and there are several possibilities in how to define the height of the farm layer, H_F, in real three-dimensional (3D) problems, Nishino [8] suggested that H_F can be defined such that

$$U_{F0} \equiv \frac{\int_0^{H_F} U_0 \mathrm{d}z}{H_F} = U_{T0} \equiv \frac{\int U_0 \mathrm{d}A}{A} \qquad (2)$$

where $U_0(z)$ is the natural wind speed observed before constructing the wind farm. As will be shown later, this definition of H_F tends to result in a very good agreement between the 1D theoretical model and 3D RANS simulations of actuator discs.

The only empirical parameter employed in the Nishino model is γ in (1). This parameter describes how the bottom friction (shear stress) changes in response to the reduction of the average wind speed through the farm layer. Specifically, γ is defined such that

$$\frac{\langle \tau_{\mathrm{w}} \rangle}{\tau_{\mathrm{w0}}} = \beta^{\gamma} = \left(\frac{U_F}{U_{F0}}\right)^{\gamma} \qquad (3)$$

where $\langle \tau_{\mathrm{w}} \rangle$ is the bottom shear stress averaged across the farm. Nishino [8] assumed $\gamma = 2$ to predict an upper limit to the performance of 'ideal' very large wind farms. For 'real' very large wind farms, however, the value of γ is expected to be less than 2 since the installation of turbines tends to increase the friction coefficient defined as $C_f^* = \langle \tau_{\mathrm{w}} \rangle/\frac{1}{2}\rho U_F^2$. It will be shown later that the γ value in our 3D RANS simulations of actuator discs (for a range of conditions) is around 1.5 to 1.6.

Based on the combination of α and β that satisfy the momentum conservation equation (1) for a given set of farm parameters λ and C_{f0} and a given empirical parameter γ, we can calculate the thrust and power coefficients of an ideal turbine (actuator disc) in a very large wind farm as

$$C_T \equiv \frac{\text{Thrust}}{\frac{1}{2}\rho U_{F0}^2 A} = \beta^2 \cdot 4\alpha(1-\alpha) = \beta^2 \cdot C_T^* \qquad (4)$$

and

$$C_P \equiv \frac{\text{Power}}{\frac{1}{2}\rho U_{F0}^3 A} = \beta^3 \cdot 4\alpha^2(1-\alpha) = \beta^3 \cdot C_P^* \qquad (5)$$

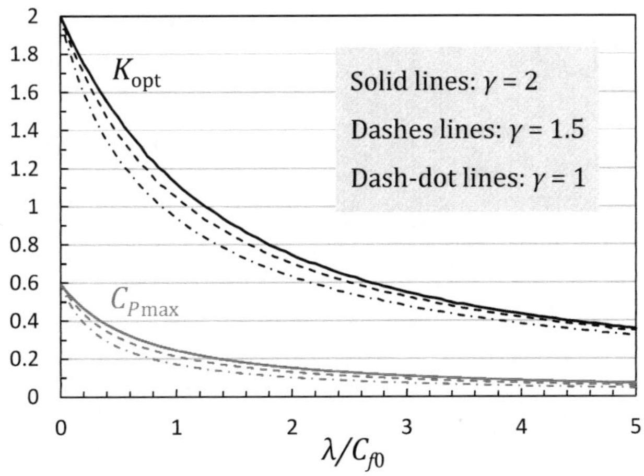

Figure 1. Theoretically optimal turbine resistance (K_{opt}) required for ideal turbines (actuator discs) to achieve the maximum power coefficient ($C_{P\max}$) in a very large wind farm.

where C_T^* and C_P^* are the 'local' or 'turbine-scale' thrust and power coefficients that are defined using U_F instead of U_{F0}. As demonstrated in [8], this theoretical model allows us to estimate very easily the 'optimal' set of wind speed reduction factors (α_{opt} and β_{opt}) to maximise the power output of a very large wind farm (for a given set of λ, C_{f0} and γ).

2.2. *Additional notes on the Nishino model*

A particularly interesting point here (which was not explicitly mentioned in [8]) is that, since the flow around each turbine has been modelled using the classical actuator disc theory, the 'turbine-scale' wind speed reduction rate α can be further linked to the so-called 'turbine resistance' K [9] (which is equivalent to another 'local' thrust coefficient C_T' [6] that is defined using U_T instead of U_F or U_{F0}). Specifically, the relationship between K and α can be described as

$$K = C_T' = \frac{C_T^*}{\alpha^2} = \frac{4(1-\alpha)}{\alpha} \qquad (6)$$

This means that the Nishino model also allows us to estimate an 'optimal' turbine resistance (K_{opt}) to maximise the power of a very large wind farm (for a given set of λ, C_{f0} and γ) as shown in Figure 1. As can be seen from the figure, the optimal turbine resistance decreases exponentially as the value of λ/C_{f0} increases (i.e. as the farm density increases relative to the natural friction coefficient).

3. RANS simulations

3.1. *Geometry and flow conditions*

Similarly to [6] and [8], for all simulations in this paper, a boundary-layer flow driven by a constant streamwise pressure gradient is considered over an 'infinitely large' wind farm. This 'infinitely large' farm is modelled as a doubly periodic array of actuator discs. Only one disc is placed in the middle of a relatively small (but periodic) rectangular computational domain since we employ (steady) RANS simulations instead of LES.

For all simulation cases, the disk diameter, D, is 100m. The disc is located at the centre of the domain with a vertical gap between the disc edge and the ground (bottom boundary) of 0.5D. The height of the computational domain, L_y, is fixed at 10D for all cases, whereas a square horizontal section of either (L_x, L_z) = (6D, 6D), (7D, 7D) or (14D, 14D) is employed, i.e. always keeping the streamwise (L_x) and spanwise (L_z) dimensions of the domain equal.

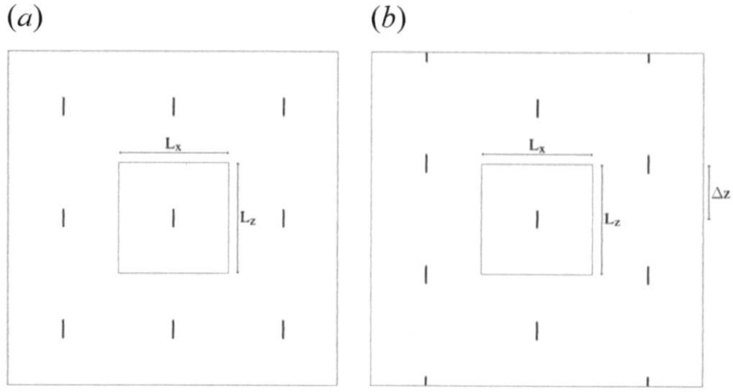

Figure 2. Schematic of array configurations: (*a*) aligned and (*b*) staggered ($\Delta z = 0.5L_z$).

Table 1. Summary of flow conditions obtained from 'empty box' simulations

Roughness Height, K_s [m]	Pressure Gradient [Pa/m]
0	-6.43×10^{-5}
1	-25.98×10^{-5}
5	-37.88×10^{-5}

The top boundary of the domain is treated as a symmetry boundary, whereas the side boundaries are treated as periodic boundaries. Two different types of array configurations are examined, namely 'aligned' and 'staggered' configurations (Figure 2). For the aligned cases, the discs are fully aligned (spacing between neighbouring discs is 1 domain-length in both streamwise and spanwise directions, i.e. 6*D*, 7*D* or 14*D*). For the staggered cases, the discs are arrayed in a fully staggered arrangement (spanwise displacement between two consecutive rows, Δz, is 0.5 domain-length, i.e. 3*D*, 3.5*D* or 7*D*). The bottom boundary is treated as a wall with three different nominal roughness heights of $K_s = 0$m, 1m and 5m (further details will be described later in Section 3.2 and in the Appendix).

To determine the streamwise pressure gradient across the computational domain (to be given as a boundary condition for the farm simulations), a preliminary 'empty box' simulation (i.e. setting the turbine resistance $K = 0$) was performed for each of the three different roughness cases with a fixed mass flow rate yielding a cross-sectional average velocity of $U_{av0} = 10$m/s. The results of these 'empty box' simulations are summarised in Table 1. For all simulations, the density and viscosity of air are assumed to be $\rho = 1.2$kg/m^3 and $\mu = 1.8 \times 10^{-5}$kg/m-s, respectively.

3.2. Computational methods

All computations, to solve numerically the 3D incompressible RANS equations, are performed using the commercial package 'ANSYS FLUENT 16.2' with its User Defined Functions (UDF) module for modifications. The Reynolds stress terms are modelled using the Standard *k*-ε model of Launder and Spalding [10] and the numerical method used is nominally second-order accurate in space for pressure, momentum, *k* and ε equations. The SIMPLE algorithm is used for pressure-velocity coupling.

The actuator disc is modelled as a stationary permeable disc (or porous disc) similarly to [11, 12] but with a local correction (or suppression) to the turbulent viscosity around the disc edge [8]. The loss of momentum in the streamwise (*x*) direction (per unit disc area) is calculated as

$$M_x = K \cdot \frac{1}{2}\rho U_d^2 \tag{7}$$

where U_d is the local (rather than disc averaged) streamwise velocity through the disc. Since the disc-averaged thrust and power can be calculated as $\int M_x \mathrm{d}A$ and $\int M_x U_d \mathrm{d}A$, respectively, the thrust and power coefficients of the disc are calculated as

$$C_T \equiv \frac{\text{Thrust}}{\frac{1}{2}\rho U_{F0}^2 A} = K \cdot \frac{\int U_d^2 \mathrm{d}A}{U_{F0}^2 A} \tag{8}$$

and

$$C_P \equiv \frac{\text{Power}}{\frac{1}{2}\rho U_{F0}^3 A} = K \cdot \frac{\int U_d^3 \mathrm{d}A}{U_{F0}^3 A} \tag{9}$$

For the disc resistance parameter K, two different values are used: $K = 0.5$ and 2, representing a low-resistance and high-resistance turbine, respectively. Note that $K = 2$ is the theoretically optimal value for an isolated ideal turbine, regardless of whether the flow is sheared or not [9].

The effect of roughness of the bottom boundary (or ground) is modelled using a modified wall function for 'fully rough' walls available in ANSYS FLUENT 16.2. Specifically, the flow condition at the bottom boundary (for the cases with K_s = 1m and 5m) is determined based on the following wall function equations [13]:

$$\frac{U_P u^*}{\tau_w/\rho} = \frac{1}{\kappa}\ln\left(E\frac{\rho u^* y_P}{\mu}\right) - \Delta B \qquad (10)$$

$$u^* = C_\mu^{0.25} k^{0.5}, \quad \Delta B = \frac{1}{\kappa}\ln(1 + C_S K_S^+) \qquad (11a,b)$$

$$K_S^+ = \frac{\rho K_S u^*}{u_\tau}, \quad u_\tau = \sqrt{\frac{\tau_w}{\rho}} \qquad (12a,b)$$

where U_P is the velocity at the centroid of the wall-adjacent cell (i.e. at $y = y_p$), $\kappa = 0.4187$ and E = 9.793 are the von Karman constant and an empirical constant used in ANSYS FLUENT, respectively, $C_\mu = 0.09$ is the eddy viscosity constant used in the Standard k-ε model, and $C_S = 0.5$ is the roughness constant used in this study. It should be noted that the nominal roughness height K_s used in this study is for 'sand-grain' type roughness (often used in engineering flow problems) but this is related to the aerodynamic roughness parameter y_0 (or 'z_0' often used in ABL flow problems) as $K_s = (E/C_s)y_0$ [14]. Therefore, the three different roughness heights considered in this study, K_s = 0m, 1m and 5m, correspond to y_0 = 0m, 0.051m and 0.255m, respectively. Traditionally, this wall roughness model in ANSYS FLUENT had a limitation that the distance from the wall to the centroid of the wall-adjacent cell needed to be greater than the roughness height K_s, which often caused difficulties in applying this model to ABL flow problems [14]. However, this limitation has been overcome in recent versions of ANSYS FLUENT by employing a remedial approach called 'virtually shifting the wall' [13], which, when the non-dimensional roughness height K_S^+ is in the 'fully rough' regime ($K_S^+ \geq 90$), allows to virtually shift the wall by 50% of the roughness height, i.e.

$$y^+_{p(\text{corrected})} = y^+_{p(\text{original})} + \frac{K_S^+}{2} \qquad (13)$$

This virtual wall shifting is justified on the assumption that roughness elements create a 'blockage' of approximately 50% of the roughness height [13]. This approach makes it possible to use a fine mesh near a rough wall with a large K_s value; therefore a sufficiently fine mesh designed for the case with K_s = 0m (to be described in detail in Section 3.3) can be used for K_s = 1m and 5m as well without modifications. A validation study is presented in the Appendix to confirm that the wall shear stress τ_w calculated using the above model is insensitive to the size of wall-adjacent cells.

All simulations are steady-state simulations and they are run for a very large number of iterations (approximately one million) until residuals and key variables, such as the bottom shear stress, are fully converged. This large number of iterations (to obtain a fully converged solution) is required due to the use of doubly periodic boundary conditions. Figure 3 shows examples of the convergence of spatially averaged bottom shear stress, $\langle\tau_w\rangle$, and the streamwise velocity averaged over the disc area, U_T.

3.3. *Computational grids*

Multi-block structured grids with hexahedral cells are used for all simulations. A 2D (cross-sectional) mesh is created first at one of the vertical (y-z) planes in the domain. This mesh has a minimum grid spacing of $0.01D$ inside and around the disc, and $0.001D$ above the bottom boundary (Figure 4a). This 2D mesh is then extruded in the streamwise (x) direction to create the final 3D mesh. For different domain sizes, the same 2D mesh around the disc is used as a base but with increasing the horizontal dimension before performing the streamwise extrusion. For the streamwise direction, a uniform grid spacing of $0.01D$ is used between $x = -0.1D$ and $0.5D$, with the disc located at $x = 0$, and allowing a

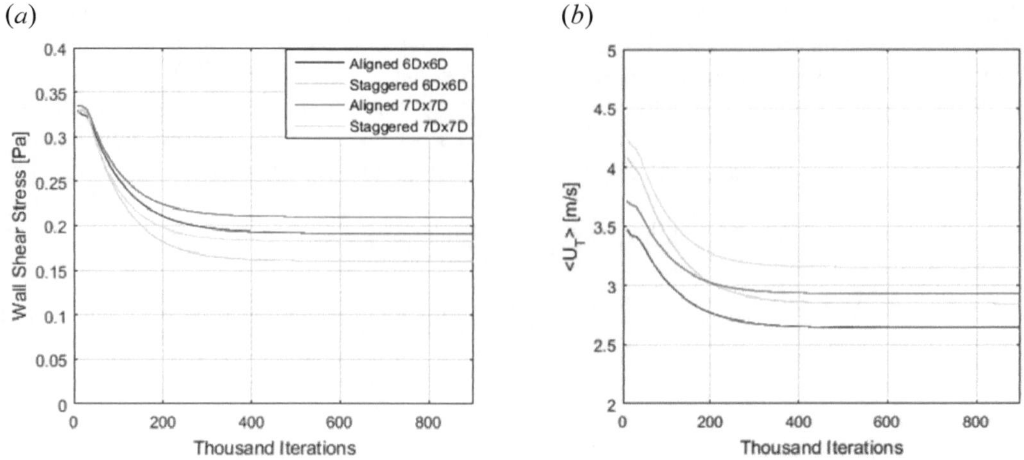

Figure 3. Examples of the convergence of solutions (for 4 cases with K_s = 5m and K = 2): (*a*) horizontally averaged bottom shear stress $\langle \tau_w \rangle$; and (*b*) disc-averaged streamwise velocity U_T.

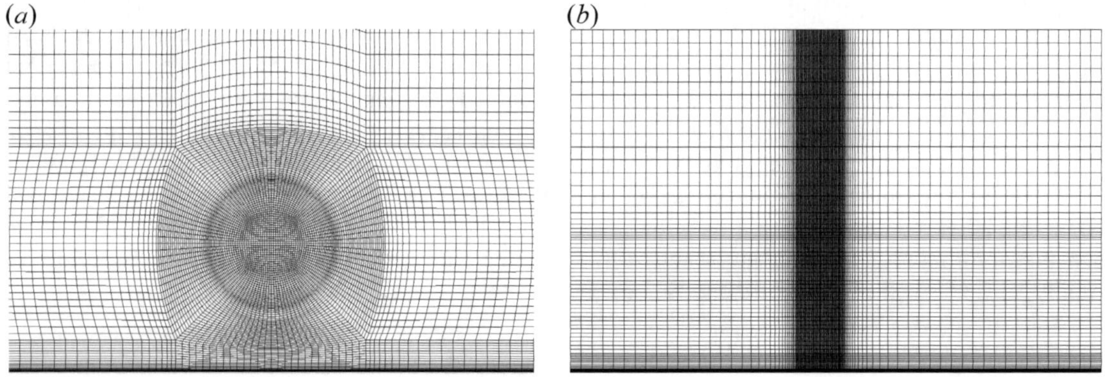

Figure 4. Computational mesh: (*a*) cross-sectional view around the disc (*y*-*z* plane); and (*b*) lateral view (*x*-*y* plane).

larger spacing outside this region, with a maximum spacing of approximately $0.15D$ (Figure 4b). This methodology ensures that all simulations are performed with the same high-quality mesh around the disc and near the bottom boundary. The total number of cells for the 3D domains varies between 1.64×10^6 for the $6D \times 6D$ case and 3.13×10^6 for the $14D \times 14D$ case.

3.4. Results

Figure 5 shows the 'natural' boundary layer profiles (streamwise velocity and turbulent kinetic energy or TKE) obtained with no disc resistance (K = 0) for the three different ground roughness cases. Note that, although not presented here, these 'natural' profiles are horizontally homogeneous across the entire domain due to the periodic boundary conditions. As explained in Section 3.1, these 'empty box' simulations are used to determine the 'natural' pressure gradient (to be used in the farm simulations with non-zero K values) as summarised in Table 1. Also, as described in Section 2.1, these 'natural' boundary layer profiles can be used to calculate parameters for comparison with the Nishino model, such as the farm-layer height (H_F), the natural farm-layer wind speed (U_{F0}) and the natural bottom shear stress (τ_{w0}). The farm-layer height, defined by Equation (2), was found to be $H_F = 2.6D$ for all three different roughness cases in this study.

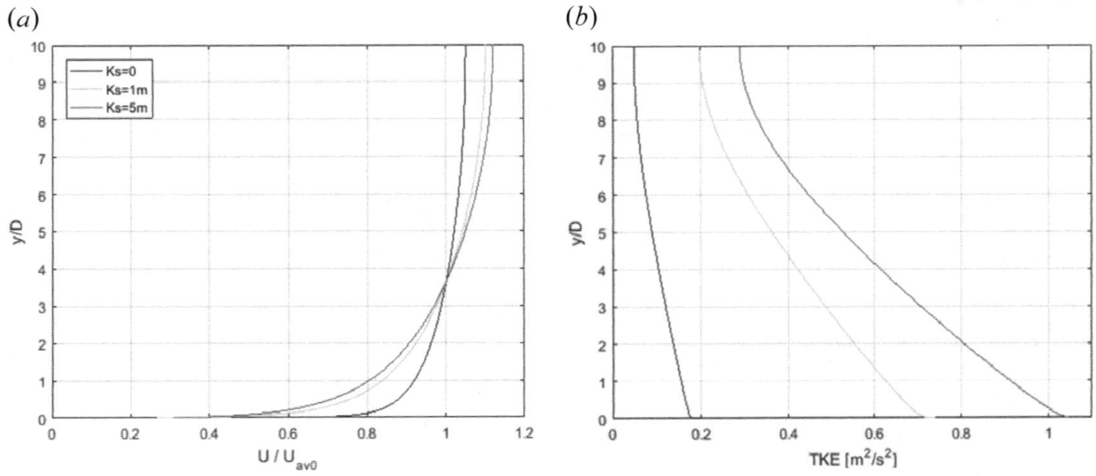

Figure 5. Boundary layer profiles for the 'natural' ($K = 0$) cases with three different bottom roughness heights: (*a*) streamwise velocity; and (*b*) turbulent kinetic energy.

Table 2. Summary of computational results for $K_s = 0$ m ($U_{F0} = 9.18$ m/s, $\tau_{w0} = 0.0643$ Pa)

Configuration	λ/C_{f0}	K	α	β	C_T	C_P	τ_w/τ_{w0}	γ
Aligned, $6D \times 6D$	17.5	2	0.503	0.307	0.0478	0.0074	0.165	1.53
Aligned, $7D \times 7D$	12.9	2	0.520	0.341	0.0631	0.0112	0.191	1.54
Staggered, $6D \times 6D$	17.5	2	0.641	0.247	0.0503	0.0080	0.120	1.52
Staggered, $7D \times 7D$	12.9	2	0.637	0.286	0.0666	0.0122	0.144	1.55

Table 3. Summary of computational results for $K_s = 1$ m ($U_{F0} = 8.34$ m/s, $\tau_{w0} = 0.2597$ Pa)

Configuration	λ/C_{f0}	K	α	β	C_T	C_P	τ_w/τ_{w0}	γ
Aligned, $6D \times 6D$	3.58	2	0.517	0.551	0.163	0.0466	0.417	1.47
Aligned, $7D \times 7D$	2.63	2	0.537	0.595	0.205	0.0655	0.463	1.48
Staggered, $6D \times 6D$	3.58	2	0.644	0.472	0.185	0.0564	0.337	1.45
Staggered, $7D \times 7D$	2.63	2	0.646	0.527	0.233	0.0794	0.389	1.48
Staggered, $14D \times 14D$	0.66	2	0.642	0.778	0.499	0.250	0.673	1.58
Staggered, $6D \times 6D$	3.58	0.5	0.866	0.626	0.147	0.0799	0.473	1.60
Staggered, $7D \times 7D$	2.63	0.5	0.869	0.681	0.175	0.104	0.539	1.61
Staggered, $14D \times 14D$	0.66	0.5	0.872	0.878	0.293	0.225	0.808	1.64

Table 4. Summary of computational results for $K_s = 5$ m ($U_{F0} = 8.01$ m/s, $\tau_{w0} = 0.3787$ Pa)

Configuration	λ/C_{f0}	K	α	β	C_T	C_P	τ_w/τ_{w0}	γ
Aligned, $6D \times 6D$	2.26	2	0.525	0.628	0.218	0.0722	0.507	1.46
Aligned, $7D \times 7D$	1.66	2	0.546	0.670	0.268	0.0982	0.555	1.47
Staggered, $6D \times 6D$	2.26	2	0.649	0.549	0.254	0.0904	0.425	1.43
Staggered, $7D \times 7D$	1.66	2	0.649	0.607	0.310	0.123	0.484	1.45
Staggered, $14D \times 14D$	0.42	2	0.649	0.834	0.587	0.318	0.756	1.54
Staggered, $6D \times 6D$	2.26	0.5	0.867	0.708	0.189	0.116	0.572	1.62
Staggered, $7D \times 7D$	1.66	0.5	0.872	0.757	0.218	0.144	0.638	1.61
Staggered, $14D \times 14D$	0.42	0.5	0.876	0.916	0.322	0.259	0.866	1.64

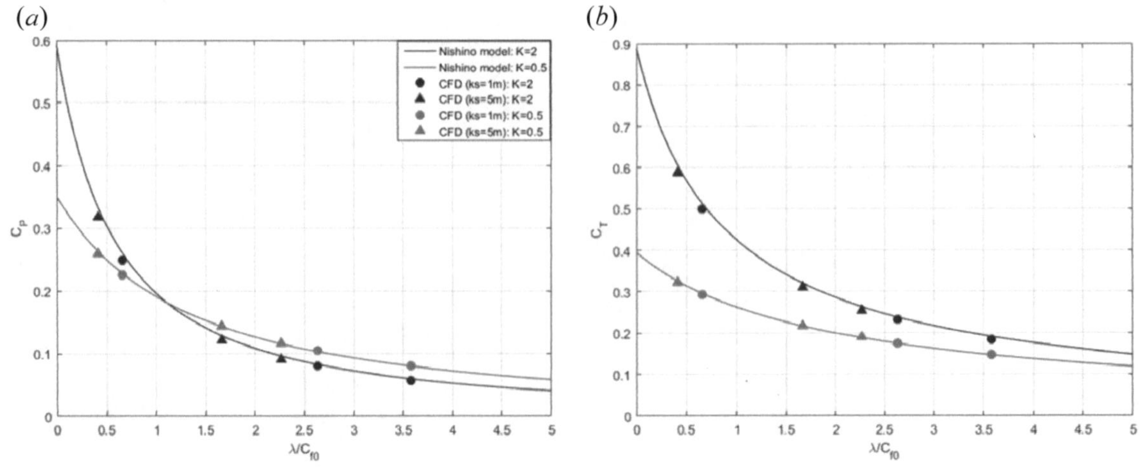

Figure 6. Comparisons between the 3D RANS simulations of staggered arrays and the Nishino model predictions (with $\gamma = 1.5$): (*a*) power coefficient; and (*b*) thrust coefficient.

Once these 'empty box' simulations have been conducted, it is possible to run the farm simulations (i.e. pressure-driven boundary layer simulations with non-zero K values) and calculate the parameters described in the Nishino model, such as α, β, C_T, C_P and γ, to evaluate the farm performance. The results are summarised in Tables 2, 3 and 4 for all scenarios investigated in this study, i.e. for three different roughness heights, three different domain sizes, two different array configurations (aligned and staggered) and two different disc resistance values. Note that the (normalised) farm density λ/C_{f0} depends only on the roughness height and domain size and is therefore already known before running the farm simulations. Of particular interest is that the values of γ obtained from the farm simulations performed in this study are all around 1.5 to 1.6. This agrees with the prediction made by Nishino [8] that the value of γ should be less than 2 for 'real' arrays (such as the 'aligned' and 'staggered' arrays considered here) and also suggests that this parameter γ (which is the only empirical parameter in the model) can be calibrated relatively easily for a wide range of practical array configurations.

Figure 6 shows a comparison between the farm simulations of 'staggered' arrays and the Nishino model [8] for the power coefficient C_P and thrust coefficient C_T. Note that $\gamma = 1.5$ has been employed in the Nishino model for this comparison. As can be seen from the figure, the agreement is excellent for both high and low disc resistance cases ($K = 2$ and 0.5). This shows the usefulness of the Nishino model not only for providing an upper limit to the performance of 'ideal' arrays (as described in [8]) but also for predicting the performance of 'staggered' arrays with wide ranges of K and λ/C_{f0} values. Most importantly, both simulations and theoretical predictions show a clear tendency that the optimal turbine resistance is smaller in a 'dense' wind farm than in a 'sparse' wind farm, i.e. low-resistance turbines (with $K = 0.5$ in this example) perform better than high-resistance turbines (with $K = 2$ in this example) when the normalised farm density λ/C_{f0} is large.

Although not presented in Figure 6, the simulations of 'aligned' arrays tend to give approximately 10% to 20% lower C_P values compared to the 'staggered' array simulations (see Tables 2, 3 and 4) and the Nishino model. To predict the performance of such 'less efficient' arrays accurately, the negative effect of turbine wake interaction would need to be taken into account in the theoretical model.

4. Concluding remarks

The results of 3D RANS simulations of 'infinitely large' wind farms presented above show that the simple 1D theoretical wind farm model proposed by Nishino [8] can be a good first approximation to the solution of this apparently complex 3D flow problem. In particular, it has been shown that the power coefficient of a number of ideal turbines (or actuator discs) arranged in a 'staggered' manner can be predicted very well as a function of the turbine resistance K and the normalised farm density λ/C_{f0} by adjusting the only empirical parameter γ in the theoretical model. It has also been shown that the value of γ should be approximately 1.5 for all scenarios considered in this study. It should be noted that, although the 3D RANS simulations performed in this study rely on the Standard k-ε turbulence model (with a local modification around the disc edge) and therefore the results are not as realistic as LES (for example) in terms of the details of turbulent mixing behind the disc, the simulations have been performed with care regarding the mesh quality and convergence to obtain sufficiently accurate numerical solutions of the governing equations. The excellent agreement between the 3D simulation results (for 'staggered' arrays) and the 1D theoretical model shown earlier in Figure 6 suggests the usefulness of the 1D theoretical model. As an example of how the theoretical model can be useful, it has also been shown that the optimal turbine resistance can be predicted easily as a function of the normalised farm density (and this prediction seems almost as accurate as what can be made using 3D RANS simulations). This analysis could be extended further in future studies to explore, for example, an optimal turbine rotor design as a function of the normalised farm density.

Although all simulation results presented in this paper suggest that the theoretical farm model is promising, there are still many important issues to be addressed in future studies. For example, the size and 'hub-height' of the rotors (relative to the thickness of the ABL) have been fixed for all scenarios in this study; these geometrical parameters, however, may affect the value of γ in the theoretical model. Another physical factor that may affect the value of γ is the stability of the ABL. Hence a more comprehensive set of numerical simulations would be necessary to calibrate the empirical parameter γ in the theoretical model for a wider range of problems. Apart from the calibration of γ, it would also be important to further examine the validity of using the classical actuator disc theory as the local-scale flow model in this two-scale coupled wind farm model. The present study has shown that this approach works well for 'fully staggered' arrays with a range of inter-disc spacing (from 6D to 14D); however this would not be the case when the inter-disc spacing is much smaller and there are strong (usually negative) interactions of individual turbine wakes. For such dense wind farms, some sort of corrections to the local-scale flow model would be required to account for the effect of turbine wake interactions. In addition, another type of correction to account for the influence of farm size would be necessary for the theoretical model to be applied to real (not 'infinitely' large) wind farms. Finally, but not least, the present study employs the assumption that the flow over the farm is driven by a constant streamwise pressure gradient; this is a commonly used assumption but not the case in the real world (as discussed in e.g. [6]). Most of these issues are currently under investigation.

Acknowledgments
Part of this study was conducted as part of the MULTI-WIND (integrated multi-scale modelling for optimal wind energy extraction) project, funded by EPSRC through the Institutional Sponsorship Fund allocated to Cranfield University (EP/P511134/1).

Appendix
As mentioned in Section 3.2, in this study we have used the 'sand-grain' type wall roughness model available in ANSYS FLUENT 16.2 to simulate the effect of bottom (ground) roughness. In order to use this approach, the near-wall flow must be in the 'fully rough' regime. This condition is satisfied if $K_s^+ \geq 90$. Blocken et al. [14] noted that in most cases, an ABL flow is in this 'fully rough' regime as the roughness is usually large enough to neglect the viscous sublayer. This is the case in the present study as well; it has been confirmed that $K_s^+ \geq 90$ in all simulations with K_s = 1m and 5m.

To confirm that the wall shear stress calculated using this wall function approach is insensitive to the size of wall-adjacent cells, we performed (separately from the farm simulations presented in the main part of this paper) a set of 2D RANS simulations of boundary layer flows. A short and tall 2D domain of (L_x, L_y) = (100m, 1000m) was employed with streamwise (x) periodic boundary conditions. Four different near-wall grid resolutions (0.001m, 0.01m, 0.1m and 1m) and seven different roughness heights (k_s= 0.005m, 0.01m, 0.05m, 0.1m, 0.5m, 1m and 5m) were considered, leading to 28 different simulation cases. Note that some of these cases are designed such that the size of wall-adjacent cells is larger than the roughness height, whereas the other cases are not. Figure 7 summarises the results of bottom shear stress obtained in these 2D boundary layer simulations. As can be seen from the figure, the sensitivity of the bottom shear stress to the wall-adjacent cell size is low. This supports the validity of the roughness modelling approach employed in the farm simulations in this study.

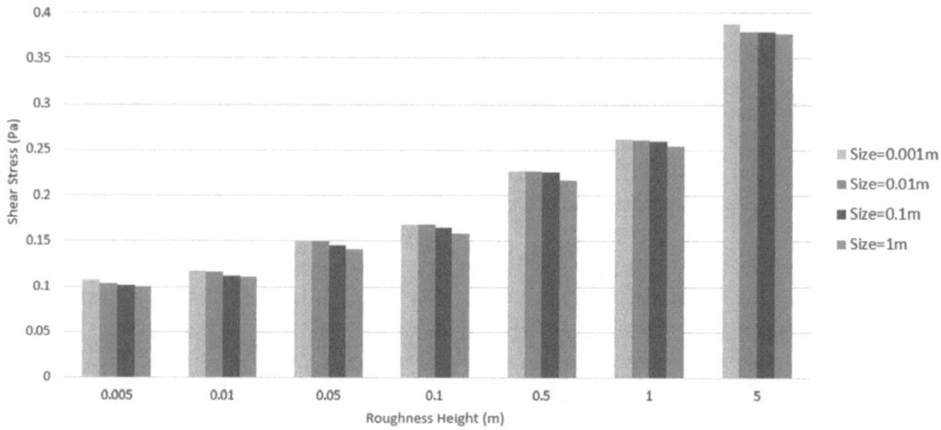

Figure 7. Sensitivity of the bottom shear stress to the size of wall-adjacent cells in 2D periodic boundary layer simulations (for seven different roughness heights).

References

[1] Frandsen S, Barthelmie R, Pryor S, Rathmann O, Larsen S, Højstrup J and Thøgersen M 2006 Analytical modelling of wind speed deficit in large offshore wind farms *Wind Energ.* **9** 39-53

[2] Frandsen S 1992 On the wind speed reduction in the center of large clusters of wind turbines *J Wind Eng. Ind. Aerodyn.* **39** 251-265

[3] Emeis S and Frandsen S 1993 Reduction of horizontal wind speed in a boundary layer with obstacles *Bound.-Layer Meteor.* **64** 297-305

[4] Emeis S 2010 A simple analytical wind park model considering atmospheric stability *Wind Energ.* **13** 459-469

[5] Emeis S 2012 *Wind Energy Meteorology: Atmospheric Physics for Wind Power Generation* (Berlin, Heidelberg: Springer) chapter 6

[6] Calaf M, Meneveau C and Meyers J 2010 Large eddy simulation study of fully developed wind-turbine array boundary layers *Phys. Fluids* **22** 015110

[7] Meneveau C 2012 The top-down model of wind farm boundary layers and its applications *J Turbul.* **13** N7

[8] Nishino T 2016 Two-scale momentum theory for very large wind farms *J. Phys.: Conf. Ser.* **753** (TORQUE 2016, Munich, Germany) 032054

[9] Draper S, Nishino T, Adcock T A A and Taylor P H 2016 Performance of an ideal turbine in an inviscid shear flow *J. Fluid Mech.* **796** 86-112

[10] Launder B E and Spalding D B 1974 The numerical computation of turbulent flows *Comput. Methods Appl. Mech. Eng.* **3** 269-289

[11] Nishino T and Willden R H J 2012 Effects of 3-D channel blockage and turbulent wake mixing on the limit of power extraction by tidal turbines *Int. J. Heat Fluid Flow* **37** 123-135

[12] Nishino T and Willden R H J 2013 Two-scale dynamics of flow past a partial cross-stream array of tidal turbines *J. Fluid Mech.* **730** 220-244

[13] ANSYS Inc. 2015 ANSYS Fluent User's Guide, Release 16.2

[14] Blocken B, Stathopoulos T and Carmeliet J 2007 CFD simulation of the atmospheric boundary layer: wall function problems *Atmospheric Environment* **41(2)** 238–252

IOP Conf. Series: Journal of Physics: Conf. Series **854** (2017) 012052 doi:10.1088/1742-6596/854/1/012052

Characterizing the coherent structures in large eddy simulations of aligned windfarms

Mengqi Zhang and Richard J. A. M. Stevens

Department of Physics, Mesa+ Institute and J.M. Burgers Centre for Fluid Dynamics, University of Twente, 7500AE Enschede, The Netherlands

E-mail: mengqi.zhang@utwente.nl, r.j.a.m.stevens@utwente.nl

Abstract. The present work studies the large coherent structures in large eddy simulations of windfarms using proper orthogonal decomposition (POD) method. In order to evaluate the effect of wind turbines on coherent structures, we consider three cases. One is a reference flow of a neutral atmospheric boundary layer and the other two are periodic and developing aligned windfarms. The number of wind turbines is large, 16×12 for periodic windfarm, and 12×12 for developing windfarm. The simulations are run for a long time in order to generate a sufficient database for POD analysis. In all cases, elongated streamwise counter rotating roll structures, covering 1 or 2 turbines in spanwise direction, are identified as the dominant POD mode. Another pattern, varying in streamwise direction, also appears in all the three cases.

1. Introduction

Research on renewable and clean energy receives more and more attentions nowadays due to the deficiency of resources available and the global warming effect. Wind energy is considered as one of the most promising future energy forms [1, 2]. However, the overall efficiency of energy harvesting function in windfarms located around the world remains low because of a lack of an explanation of the detailed physical mechanism at play. In order to better understand the flow dynamics of wind turbine wakes in windfarms, the large coherent structures in windfarms are studied in this work. We simulate the flow using large eddy simulation (LES) and extract the large coherent structures by means of proper orthogonal decomposition (POD) [3] and characterize their features. By doing so, the physical mechanism in windfarms can be further unveiled, paving the way for a better understanding of the energy harvesting process in windfarms.

Windfarm experiments provide authentic and trustworthy data for real cases. A benefit of simulations compared to experiments is that these provide full access to the entire flow field in order to study the main flow structures that are formed. Numerical simulations have been applied to studying the fluid dynamics in windfarm. Windfarm simulations based on Reynolds-Averaged Navier-Stokes (RANS) models are attractive because of the low computational cost [4]. LES on the other hand provide a compromise between computational cost and a more detailed account of the flow dynamics. In LES, the small scale motions are modelled by a subgrid-scale model and the large scale motions are solved directly according to the filtered Navier-Stokes equation. Turbines are modelled using an actuator disk approach or an actuator line approach. Recently, LES has become increasingly popular in the community [2, 4]. In the present paper, an in-house LES code [5, 6] is used to simulate aligned periodic and developing windfarms to investigate the effect of wind turbines on the flow structures formed inside the windfarm.

In order to disclose and characterize the most important flow structures in windfarms, we apply proper orthogonal decomposition (POD) to the turbulent flow fields obtained from LES. Fully developed turbulent flow possesses tremendously many degrees of freedom and a complete account of flow dynamics in all the spatiotemporal scales will be a daunting task. Therefore, a wiser strategy is to describe only the dynamically important degrees of freedom and discard the inessential ones. This is the idea behind mode decomposition methods, such as POD. This technique has been used widely and successfully in many other flow systems, such as wall-bounded flow, free shear flow, and convection, to identify the coherent flow structures, see Refs. [7,8].

In wind energy research, there are already several works on the POD analysis of flow fields in windfarms or flows behind wind turbines [9–15]. VerHulst & Meneveau [9] presented a comprehensive 3D POD analysis of aligned and staggered windfarms in addition to a neutral atmospheric boundary layer (ABL). They found streamwise counter rotating rolls as the dominant coherent structures creating strong ejection and sweep regions in the flow. It is believed that this mode contributes significantly to the kinetic energy re-distribution above and below the turbines and therefore are strongly related to the energy harvesting process. Other modes, such as streamwise varying structures and modes varying in vertical direction, are also identified. Hamilton et al. [13, 14], when analyzing experimental data of a wind turbine array, applied a double POD (DPOD) analysis to the measurements undertaken at separate streamwise stations. The idea of DPOD is to apply an additional POD analysis to the original POD modes with the same index. One of the merits of DPOD is that the effect of streamwise development, otherwise difficult to be taken into account in the POD analysis of separate measurements, can now be included in DPOD coefficients, but reduced to minimum in DPOD modes. Newman et al. [10] studied the entrainment process in a 3×5 array of wind turbines. The Reynolds shear stress is re-constructed at the turbine tip height via POD modes. A characteristic wavelength is identified for each POD mode. They found that lower POD modes (with larger kinetic energy) are associated with larger wavelength. Andersen et al. [15] performed LES of extended windfarms and analyzed POD in vertical planes. They showed that dominant length scales responsible for the entrainment process grow further inside the windfarm, but are restricted by the streamwise turbine spacing. Above the windfarm they observe larger structures.

Our methodology is closer to [9] with a longer simulation time and a larger computational domain. The results presented here complement the literature with new significances and are believed to be interesting to the researchers in windfarm, particularly those using POD as a probing tool. The present paper is organized as follows. In section 2 we introduce the used POD formalism and describe the LES setup. Section 3 presents the averaged velocity fields and the corresponding POD analysis. Finally in section 4, we conclude the paper and discuss possible future works.

2. Problem formulation

2.1. Proper orthogonal decomposition (POD)

POD is a powerful post-processing tool for revealing coherent structures in a flow system [3,9,16]. It can be interpreted as a singular value decomposition of the flow fields in a series of snapshots, obtained either numerically or experimentally. We present below its mathematical formulation. Other interpretations of POD and its mathematical formulations also exist. One can refer to [9, 17] in similar contexts of numerically simulated ABL and windfarms or [7, 18] for a general introduction to this technique. As noted by VerHulst & Meneveau [9], even though periodic boundary conditions are assumed in periodic windfarm, its POD modes need not be Fourier modes since the presence of turbines renders the flow inhomogeneous. This issue has also been discussed in general by George [19].

From the numerical results of LES, we first evoke Reynolds' decomposition of a certain flow variable $g_{total}(\boldsymbol{x}, t) = G(\boldsymbol{x}) + g(\boldsymbol{x}, t)$ to subtract the mean from total fields. $g(\boldsymbol{x}, t)$ can represent either velocity fields or flux fields. In the following, we take the streamwise, spanwise and vertical velocities (u, v, w, respectively) as an example. Using the flow data, we formulate a matrix **A** as

$$\mathbf{A} = \begin{bmatrix} \cdot & \cdot & \cdot & \cdot & \cdots & \cdot & \cdot & \cdot \\ \cdot & \cdot & \cdot & \cdot & \cdots & \cdot & \cdot & \cdot \\ \cdot & \cdot & \cdot & \cdot & \cdots & \cdot & \cdot & \cdot \end{bmatrix}_{3N_{xyz} \times N_t} \quad (1)$$

where $N_{xyz} = N_x \times N_z \times N_z$ and N_x, N_y, N_z are the grid resolutions in Cartesian coordinate and N_t is the number of snapshots recorded. The multiplier 3 accounts for three velocity components considered. By this arrangement, each row of **A** represents a snapshot of the flow fields and each column of **A** stands for the time history of a certain flow variable at a grid point in the flow. We apply singular value decomposition to matrix **A**, yielding

$$\mathbf{A} = \mathbf{U\Sigma V}^{\dagger}. \quad (2)$$

The superscript † represents transpose conjugate of a matrix. The unitary matrices **U** (of size $N_t \times N_t$) and **V** (of size $3N_{xyz} \times 3N_{xyz}$) are respectively the left and right singular vectors of **A**. $\mathbf{\Sigma}$ is a rectangular diagonal matrix whose diagonal entries are the singular values. Their physical interpretations will be explained below. In order to obtain these matrices, one can resort to the eigenvalue analysis by forming the correlation matrix $\mathbf{C} = \mathbf{A}^{\dagger}\mathbf{A}$, whose dimension is $3N_{xyz} \times 3N_{xyz}$ in this case. The symmetric matrix **C** satisfies the following relation

$$\mathbf{C} = \mathbf{A}^{\dagger}\mathbf{A} = \mathbf{V\Sigma}^2_{3N_{xyz}\times 3N_{xyz}}\mathbf{V}^{\dagger}. \quad (3)$$

We can easily see that each column of **V** spans the whole flow fields (u, v, w) and it is called POD mode of the flow. The eigenvalues in $\mathbf{\Sigma}^2_{3N_{xyz}\times 3N_{xyz}}$ (equal to the singular values squared λ^2) represent the kinetic energy contained in each POD mode of velocity fields. When N_{xyz} is relatively small, the above procedure is feasible; however, for large-scale numerical simulations as we do in this paper, N_{xyz} is of order of tens of millions, rendering the computation of equation (3) almost impossible due to the memory constraints. In practice, the correlation matrix **C** is usually formulated alternatively as $\mathbf{C} = \mathbf{A}\mathbf{A}^{\dagger}$ and one poses the corresponding eigenvalue problem as

$$\mathbf{C} = \mathbf{A}\mathbf{A}^{\dagger} = \mathbf{U\Sigma}^2_{N_t\times N_t}\mathbf{U}^{\dagger}. \quad (4)$$

After solving this eigenvalue problem, we can obtain **U** and $\mathbf{\Sigma}_{N_t\times N_t}$. In addition, by recalling equation (2), we obtain POD modes in **V** according to $\mathbf{\Sigma}^{-1}\mathbf{U}^{\dagger}\mathbf{A} = \mathbf{V}^{\dagger}$. From this equation, it is clear that **U** can be viewed as temporal coefficients for the snapshots in **A**, constituting the POD modes in **V**. Finally, as mentioned above, **U** and **V** are unitary matrices; hence, the columns in **U** (or **V**) form an orthonormal set, a property which we use as an *a posteriori* check for all the results shown below.

2.2. Setup in large eddy simulation

In this work we simulated a neutral ABL, an aligned periodic windfarm, and an aligned developing windfarm, see table 1. Turbines are modelled using an actuator disk model [20]. We use a constant thrust coefficient $C_T = 0.75$ for all turbines. It is assumes that turbines are operating in regime II [21], in which the thrust coefficient is constant as function of the wind speed. The constant thrust coefficient is a result of the blade rotation frequency that is adjusted appropriately to the incoming wind. According to Porté-Agel et al. [22] the turbines operate most of the time in this regime.

The resolution for all simulations is $512 \times 256 \times 128$ in respectively streamwise (x), spanwise (y) and vertical (z) directions in a fixed computational domain of $4\pi H \times 2\pi H \times H$, where H is

Types of flow	Turbine layout	s_x	s_y	s_{xy}	No. of snapshots
ABL	–	–	–	–	5300
Aligned periodic	16×12	$7.85D$	$5.24D$	$6.41D$	5467
Aligned developing	12×12	$7.85D$	$5.24D$	$6.41D$	6248

Table 1. Cases considered for ABL and windfarms in the present paper. Horizontal area $4\pi H \times 2\pi H$ is fixed for all the simulations. Grid resolution is also the same $512 \times 256 \times 128$.

the reference length in the vertical direction. This resolution is considered sufficient to simulate the large scale flow structures accurately in LES. The turbine diameter is $D = 0.1H$. As a result, each turbine has 5 grid points along its diameter. This grid resolution for turbines is comparable to the simulations performed in [9], and similar resolution guideline is put forward in Refs. [22, 23].

We simulated a periodic windfarms with 16×12 turbines in streamwise and spanwise directions, respectively. The used turbines spacing in streamwise direction is $s_x = 7.85D$ and in spanwise direction $s_y = 5.24D$. This gives a turbine density $s_{xy} = \sqrt{s_x s_y} = 6.41D$. For the developing windfarm, we use the concurrent precursor inflow method, introduced in [23], to simulate simultaneously the flow dynamics in both inflow field and windfarm. In this case, we consider a windfarm of 12×12 turbines. The turbine spacings are the same as the periodic windfarm, see table 1. In this study we only consider aligned wind farms.

The numerical methods employed in LES are explained in [5,6]. Time step in the computations is adjusted according to the CFL condition. We checked that the used sampling frequency is higher than the frequency scale of most energetic events in the auto-spectral density of streamwise velocity u. This ensures that all the essential dynamics are captured. The databases generated in the simulations are relatively large. For example, the size of database for the aligned periodic windfarm is about 2 terabytes.

3. Results

In this section, we present the POD analysis results. We first show the averaged fields which are subtracted from the total field. Then the results of POD analysis are given including the eigenvalue distributions and the shapes of POD modes for ABL and periodic and developing windfarms. The convergence has been checked by applying POD to 75% of the snapshots used in the main analysis and we found no significant differences.

3.1. Averaged fields

In figure 1, the averaged fields of streamwise velocity U, spanwise velocity V and vertical velocity W are shown for periodic windfarm with 16×12 turbines and in figure 2 for developing windfarm with 12×12 turbines. The flow velocity is normalized by $U_a = \sqrt{U_{\text{hub}}^2 + V_{\text{hub}}^2 + W_{\text{hub}}^2}$, where $U_{\text{hub}}, V_{\text{hub}}, W_{\text{hub}}$ are the averaged velocities in the horizontal plane at hub height ($z/H = 0.1$). Figure 1(a) shows the mean U field in the horizontal xy plane at hub height and reveals the structure of wind turbine wakes. Panel (b) displays V field, where one can observe regions of positive and negative velocities around the turbine positions. Physically, the wind turbines, modelled as actuator disks, act as blockage in the flow field. Due to the continuity of velocity field, fluid will flow around the turbines. Therefore, in the xy plane, some fluid has a positive V velocity and some fluid has a negative velocity. This results in the flow pattern shown in figure 1(b). The same figure also displays blue (small negative V) and green (small positive V) regions behind the turbines, showing up as patches in larger regions of small positive and negative V velocities, respectively. When a blue wake region lies in the larger region of negative V, it disappears due to the superposition of two small negative V values in the visualization. The same applies to the green positive wake regions. We note that the velocity in these regions

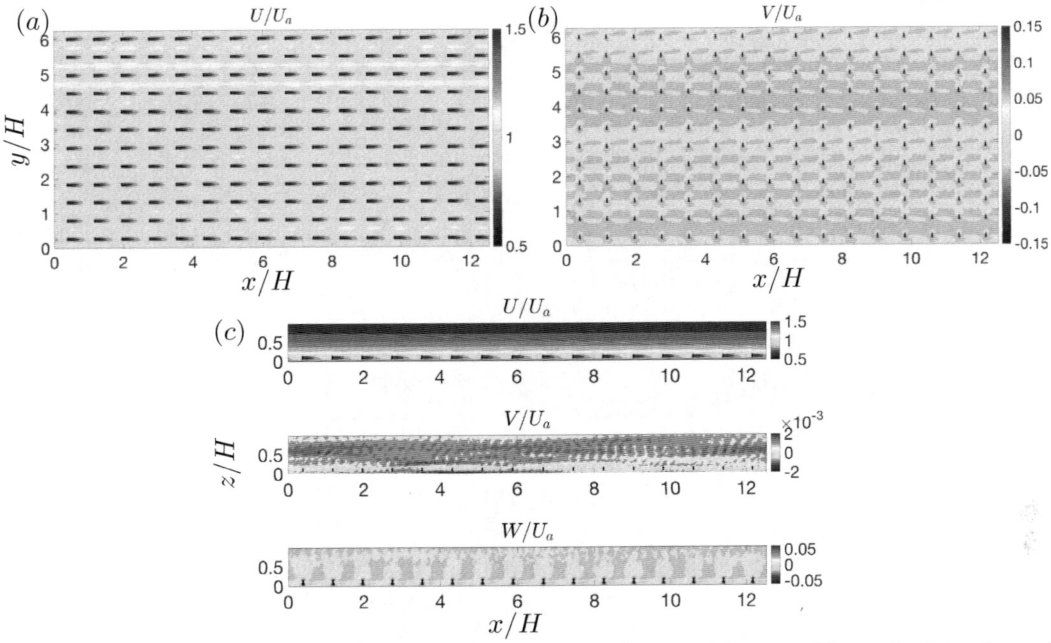

Figure 1. Averaged velocity fields for aligned periodic windfarms. The velocities have been normalized by U_a (see the text for its definition). (a) normalized U field in xy plane at the hub height of $z/H = 0.1$. (b) normalized V field in xy plane at the hub height of $z/H = 0.1$. (c) normalized U, V, W fields averaged over all the xz planes intersecting the turbines. The colorbars have been adjusted to facilitate comparison and show the mean features.

is small, i.e. about 1% of the reference velocity U_a.

By the same argument of flow continuity, in xz plane of W velocity, some fluid flows upwards and some downwards around the turbines. This is illustrated in the lower panel of figure 1(c), which is averaged over all the xz planes intersecting the turbines. The other two panels in figure 1(c) plot the velocity components U and V. Because of the imposed periodic condition in spanwise direction, the averaged V field is supposed to be vanishing. The V field in panel (c) reveals a small (around 0.1% of U_a) residual velocity, which is probably caused by numerical uncertainty, a relatively large sampling intervals, and the limited simulation length.

Figure 2 shows averaged velocity fields for a developing windfarm. The number of wind turbines in this case has been reduced to 12×12 due to a fringe region in the rear part of windfarm domain, which is used to create the inflow condition [23]. This part of the computational domain has also been discarded in POD analysis and is thus not shown in the figures below. One of the differences between the periodic and developing windfarms is the non-periodicity in streamwise direction in the developing case. In panels (a,b), the U and V velocities in xy plane manifest similar flow patterns in turbine scale as those in periodic windfarm, but in the scale of ABL, one can clearly see the streamwise development of U velocity. This non-periodicity in the streamwise direction is also clear in xz plane, see the figures in panel (c), which is again related to the lack of periodicity condition in the streamwise direction in developing windfarm.

3.2. POD analysis

In this section, we show large coherent structures educed by POD analysis from LES results. Compared to [9], whose computation domain is around $\pi H \times \pi H \times H$ with 4×6 turbines, our domain for the aligned case is 8 times larger ($4\pi H \times 2\pi H \times H$) with more turbines ($16 \times 12$). Therefore, by comparing the results in these two works, one can note the effects of computational

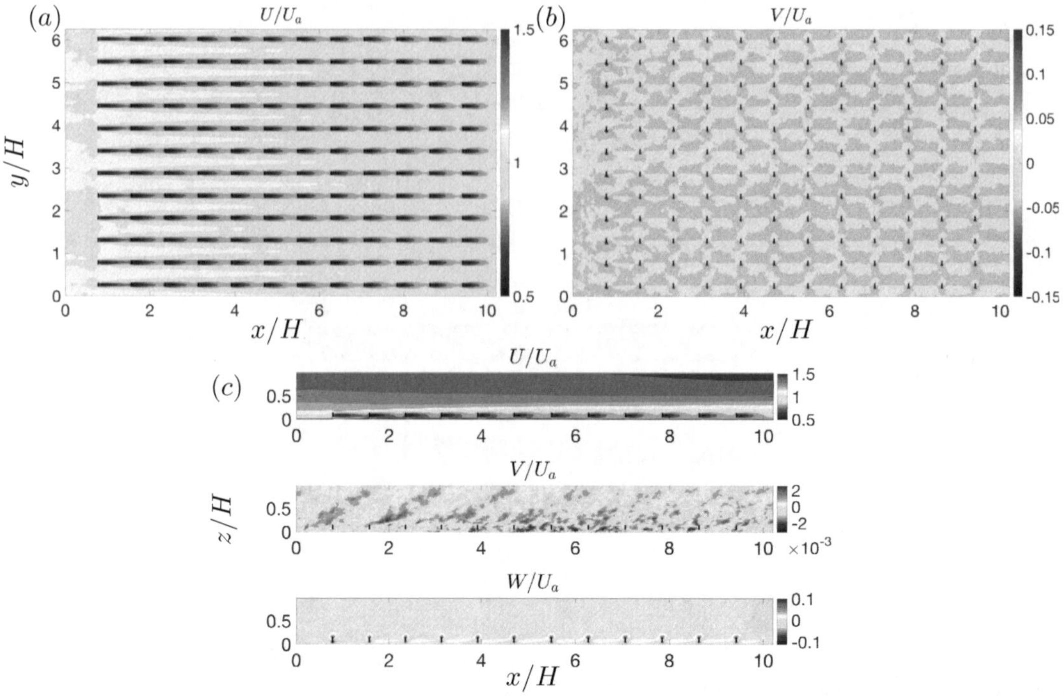

Figure 2. Averaged velocity fields for aligned developing windfarms. The velocities have been normalized by U_a (see the text for its definition). (a) normalized U field in xy plane at the hub height of $z/H = 0.1$. (b) normalized V field in xy plane at the hub height of $z/H = 0.1$. (c) normalized U, V, W fields averaged over all the xz planes intersecting the turbines. The colorbars have been adjusted to facilitate comparison.

domain and turbine layout on the POD results of windfarms and hopefully find new features of POD structures. Besides, we do not shift the velocity fields in spanwise direction to expand the database as [9] did, even though such a practice is reasonable based on the periodicity in the problem. Instead, we simulate the flow system long enough to create sufficiently many snapshots in terms of convergence and sampling. As mentioned above, we will present POD results of ABL, periodic and developing aligned windfarms below.

Firstly, the eigenvalues in POD analysis, which represent the kinetic energy of POD modes, are shown in figure 3. They are normalized by the largest eigenvalue in (a) and cumulated with a normalization of the maximum total sum in (b). From panel (a), we can see that the three cases all show a trend of decaying magnitudes of eigenvalues as the index of POD mode increases. The smallest eigenvalues are 2 orders smaller than the largest one, indicating that these higher POD modes bear less dynamic significance, at least from the point of view of kinetic energy composition. We thus concentrate on the first POD modes which have more kinetic energy. Panel (b) shows the cumulative eigenvalues normalized by the maximum sum. The trend is similar to the results in VerHulst & Meneveau [9].

Figure 4 (a) displays POD modes for the streamwise velocity in a horizontal xy plane at $z/H = 0.1$ for the ABL case. One can easily see that the dominant pattern of POD modes is the elongated streamwise rolls that vary slightly in spanwise direction. The vertical extent of these streamwise rolls can be of the order of the whole computational height as shown in panel (b). Aside from this structure, we see that POD modes 6 and 7 show another pattern which varies in the streamwise direction. In panel (b), the first POD modes of u, v, w velocities are plotted in the yz plane at $x/H = 1.963$. For u, the above discussed roll structures are clearly seen in this figure. For the v and w components, one can observe that there exists a counterclockwise

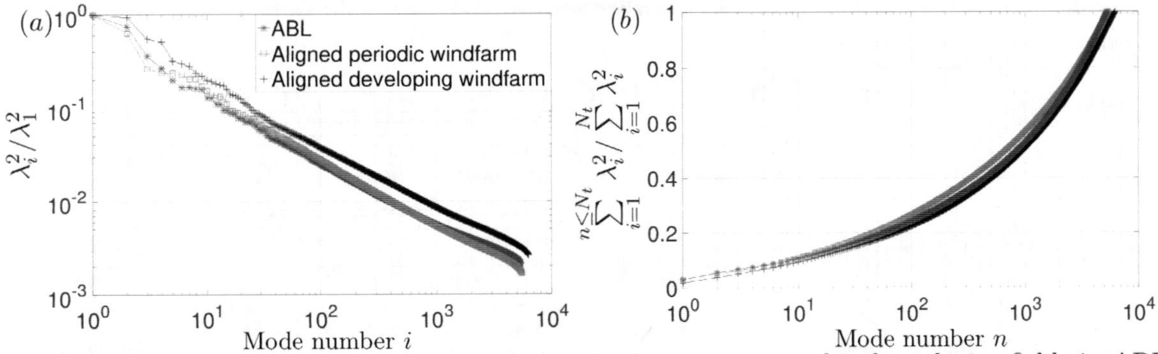

Figure 3. The distributions of eigenvalues solved in equation 4 for the velocity fields in ABL and aligned periodic and developing windfarms, respectively. (a) normalized eigenvalues by the largest one (λ_1^2), (b) normalized cumulative eigenvalues by the maximum total sum.

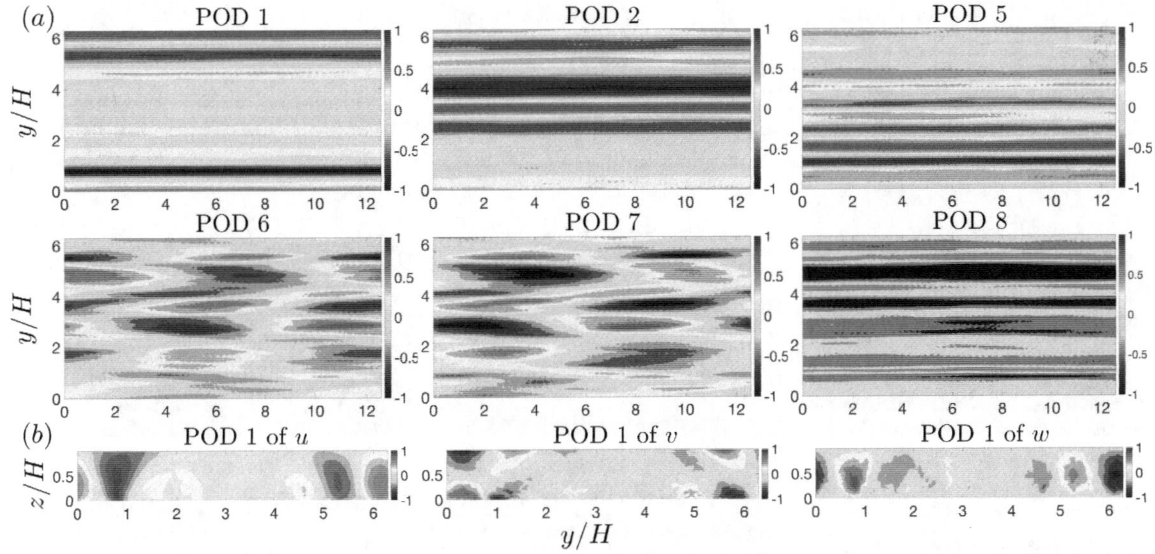

Figure 4. (a) POD modes of u component for ABL at $z/H = 0.1$. (b) The first POD modes for u, v, w in yz plane at $x/H = 1.963$. Each POD mode is normalized by its maximum absolute value.

roll at around $0 < y/H < 1.5$ and a clockwise roll at around $4.5 < y/H < 6$ and these two counter rotating rolls are at the same time modulated in the streamwise direction. These results altogether depict the rolls pattern seen in POD mode 1 of panel (a).

Figure 5 shows the POD results for an aligned periodic windfarm. Unlike ABL, now the roll structures in panel (a) are more clearly delimited. We can see that each roll structure covers one or two turbine spacing in spanwise direction, spanning roughly $2\pi H \times 1/12 = 0.523H$ to $2\pi H \times 2/12 = 1.047H$. Another feature of aligned periodic windfarm that differs from ABL is that the streamwise varying pattern appears at an earlier POD mode. This indicates that turbines can redistribute the kinetic energy across POD modes and lead to rearrangement of POD modes. This streamwise varying pattern has also been discussed in [9] for periodic windfarms. However, the structures shown here, covering $4-6H$ in the streamwise direction (like the ones in POD mode 5), are much larger than theirs in size. Their streamwise varying structures seem to be confined between the turbines in the streamwise direction, while ours span across several turbines. Here we also mention that Newman et al. [10] and Andersen et al. [15] showed with POD analysis

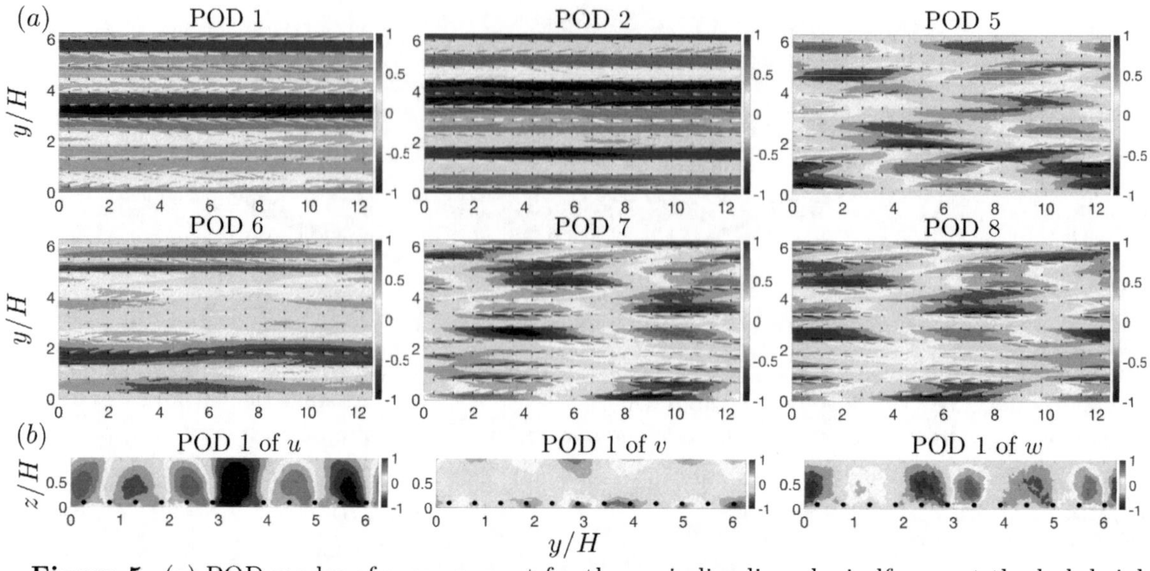

Figure 5. (a) POD modes of u component for the periodic aligned windfarms at the hub height $z/H = 0.1$. (b) The first POD modes for u, v, w in yz plane at $x/H = 1.963$ where the cross sections intersect the turbines. Each POD mode is normalized by its maximum absolute value.

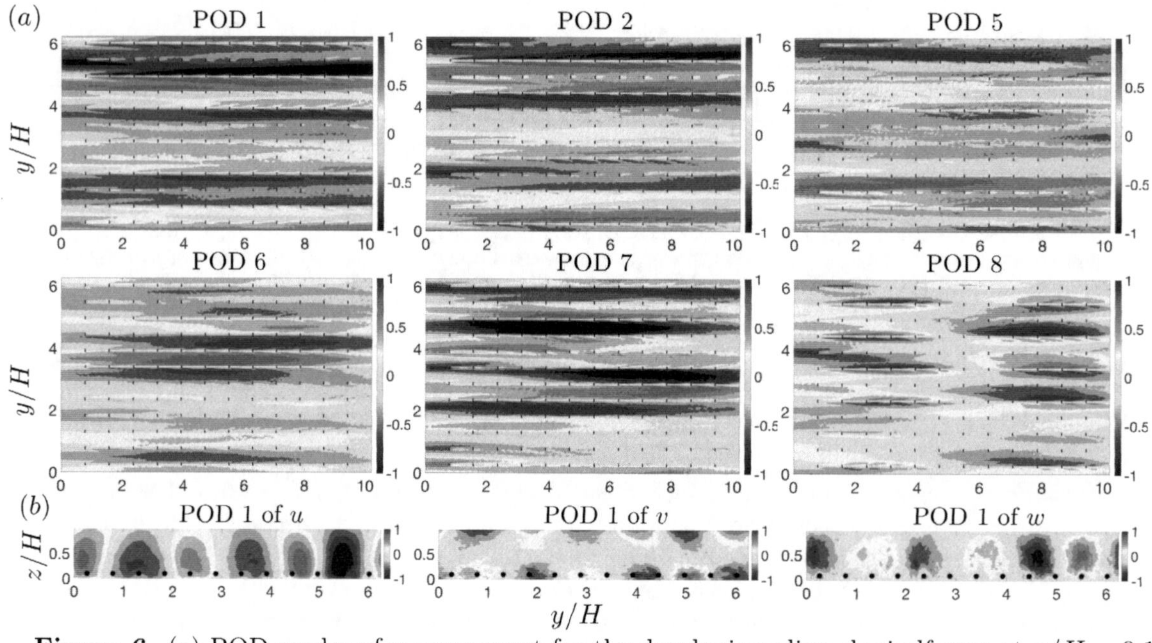

Figure 6. (a) POD modes of u component for the developing aligned windfarms at $z/H = 0.1$. (b) The first POD modes for u, v, w in yz plane at $x/H = 0.7854$ where the cross sections intersect the turbines. Each POD mode is normalized by its maximum absolute value.

in a vertical 2D plane that inside the windfarm the important characteristic length scales are restricted by the turbine spacing, but that this is not the case above the turbine tip height. When we look at the flow structures at the turbine tip height we also find shorter characteristic length scales, in agreement with the aforementioned studies.

Finally, figure 6 shows the POD analysis results for developing aligned windfarm. From this figure, due to the development of flow in x, one can note the relatively stronger non-periodicity of POD structures in the streamwise direction contrary to the results shown above. Nevertheless, the dominating pattern in this case still assumes similar roll structures as in the ABL and the periodic windfarm. Also, the spanwise size of the rolls appears similar to the ones observed in the periodic windfarm, covering 1 or 2 turbines in spanwise direction. The streamwise rolls start to diminish around POD mode 5 and for POD mode 8, its pattern resembles the streamwise varying pattern appearing in the other two flow cases. The shapes of POD modes 6 and 7 seem to lie in between the elongated streamwise structures and the streamwise varying structures.

4. Conclusion and discussions

In this work, we have performed LES of an ABL and aligned periodic and developing windfarms, resulting in databases of a very long simulation time and sufficient number of snapshots. We then applied POD analysis to the LES results and depicted the features of the dominant flow patterns in turbulent flow of windfarms.

We first showed the averaged fields for both periodic and developing windfarms. The turbines, modelled as actuator disks facing the atmospheric flow, drag the velocity fields and force the fluid particles to flow around them. This results in a U velocity field that clearly reveals the wind turbine wakes, a V velocity that has small positive and negative regions in xy plane, and a W velocity with positive and negative velocities around the turbines in xz plane. The developing windfarm clearly demonstrates streamwise development compared to the periodic windfarm, whose averaged fields are periodic in the streamwise direction.

By comparing POD results of windfarm to ABL, one can easily see the influence of wind turbines on the large coherent structures. For example, the turbines confine the coherent structures in between them, such as the streamwise constant rolls structures and the streamwise varying structures. The roll structures are thus more clearly shaped in the windfarms and consistently span one or two turbine spacing (that is $0.523H$ to $1.047H$) in spanwise direction. In [9], whose computation domain is smaller than ours, the spanwise length of the rolls in POD mode 1 is around $\pi H/2 = 1.571H$, indicating that the size of roll structures is dependent on computational domain and turbine layout. Also, one recognizes that the streamwise varying structures seem not to be exclusive for windfarms and also appear in ABL.

By comparing periodic windfarm to developing windfarm, we conclude that the structures do not change significantly between the two except for the streamwise non-periodicity expected for the developing windfarms. Also, there seem to be more elongated roll structures in the first several POD modes of developing windfarms. In the periodic windfarm the shorter roll structures appear at a lower POD mode. This indicates that these shorter roll structures are more important for the periodic windfarm.

The results presented in this paper concern only aligned windfarms. It is interesting to study in addition the effect of staggering of the wind turbines on the large coherent structures in both periodic and developing windfarms. Besides, other quantities, such as energy flux, and their coherent structures deserve to be further investigated. Finally, it is planned to explore the coherent structures in the spectral space after the database of physical flow is Fourier Transformed in order to describe their features at a certain frequency, which might be interesting for the frequency-wise flow control in windfarm. These analyses are currently undertaken and will be reported properly in the future.

Acknowledgement

This work is part of the Shell-NWO/FOM-initiative Computational sciences for energy research of Shell and Chemical Sciences, Earth and Live Sciences, Physical Sciences, FOM and STW and an STW VIDI grant (No. 14868). This work was carried out on the national e-infrastructure of SURFsara, a subsidiary of SURF cooperation, the collaborative ICT organization for Dutch education and research.

References

[1] J. N. Sørensen. Aerodynamic aspects of wind energy conversion. *Annu. Rev. Fluid Mech.*, 43(1):427–448, 2011.

[2] R. J. A. M. Stevens and C. Meneveau. Flow structure and turbulence in wind farms. *Annu. Rev. Fluid Mech.*, 49(1):311–339, 2017.

[3] P. Holmes, J.L. Lumley, and G. Berkooz. *Turbulence, Coherent Structures, Dynamical Systems and Symmetry*. Cambridge Monographs on Mechanics. Cambridge University Press, 1998.

[4] B. Sanderse, S.P. van der Pijl, and B. Koren. Review of computational fluid dynamics for wind turbine wake aerodynamics. *Wind Energy*, 14(7):799–819, 2011.

[5] M. Calaf, C. Meneveau, and J. Meyers. Large eddy simulation study of fully developed wind-turbine array boundary layers. *Physics of Fluids*, 22(1):015110, 2010.

[6] M. Calaf, M. B. Parlange, and C. Meneveau. Large eddy simulation study of scalar transport in fully developed wind-turbine array boundary layers. *Physics of Fluids*, 23(12):126603, 2011.

[7] G. Berkooz, P. Holmes, and J. L. Lumley. The proper orthogonal decomposition in the analysis of turbulent flows. *Annu. Rev. Fluid Mech.*, 25(1):539–575, 1993.

[8] C. W. Rowley. Model reduction for fluids, using balanced proper orthogonal decomposition. *International Journal of Bifurcation and Chaos*, 15(03):997–1013, 2005.

[9] C. VerHulst and C. Meneveau. Large eddy simulation study of the kinetic energy entrainment by energetic turbulent flow structures in large wind farms. *Physics of Fluids*, 26(2):025113, 2014.

[10] A. J. Newman, D. A. Drew, and L. Castillo. Pseudo spectral analysis of the energy entrainment in a scaled down wind farm. *Renewable Energy*, 70:129 – 141, 2014.

[11] S. J. Andersen, L. E. M. Lignarolo, D. Ragni, C. J. Simão Ferreira, J. N. Sørensen, R. F. Mikkelsen, and G. J. W. van Bussel. Comparison between PIV measurements and computations of the near-wake of an actuator disc. *J. Phys. Conf. Ser.*, 524(1):012173, 2014.

[12] J. N. Sørensen, R. F. Mikkelsen, D. S. Henningson, S. Ivanell, S. Sarmast, and S. J. Andersen. Simulation of wind turbine wakes using the actuator line technique. *Phil. Trans. R. Soc. A*, 373(2035), 01 2015.

[13] N. Hamilton, M. Tutkun, and R. B. Cal. Wind turbine boundary layer arrays for cartesian and staggered configurations: Part II, low-dimensional representations via the proper orthogonal decomposition. *Wind Energy*, 18(2):297–315, 2015.

[14] N. Hamilton, M. Tutkun, and R. B. Cal. Low-order representations of the canonical wind turbine array boundary layer via double proper orthogonal decomposition. *Physics of Fluids*, 28(2):025103, 2016.

[15] S. J. Andersen, J. N. Sørensen, and R. F. Mikkelsen. Turbulence and entrainment length scales in large wind farms. *Phil. Trans. R. Soc. A*, 375(2091), 2017.

[16] P. J. Holmes, J. L. Lumley, G. Berkooz, J. C. Mattingly, and R. W. Wittenberg. Low-dimensional models of coherent structures in turbulence. *Physics Reports*, 287(4):337 – 384, 1997.

[17] S. Shah and E. Bou-Zeid. Very-large-scale motions in the atmospheric boundary layer educed by snapshot proper orthogonal decomposition. *Boundary-Layer Meteorology*, 153(3):355–387, 2014.

[18] L. Sirovich. Turbulence and the dynamics of coherent structures, part 1: Coherent structures. *Q. Appl. Maths*, 1987.

[19] W. K. George. *Some thoughts on similarity, the POD, and finite boundaries*, pages 117–128. Birkhäuser Basel, Basel, 1999.

[20] J. N. Sørensen and A. Myken. Unsteady actuator disc model for horizontal axis wind turbines. *Journal of Wind Engineering and Industrial Aerodynamics*, 39(1):139 – 149, 1992.

[21] K. E. Johnson, L. J. Fingersh, M. J. Balas, and L. Y. Pao. Methods for increasing region 2 power capture on a variable-speed wind turbine. *J. of Solar Energy Engineering*, 126(4):1092–1100, 11 2004.

[22] F. Porté-Agel, Y. Wu, and C. Chen. A numerical study of the effects of wind direction on turbine wakes and power losses in a large wind farm. *Energies*, 6(10):5297–5313, 2013.

[23] R. J. A. M. Stevens, J. Graham, and C. Meneveau. A concurrent precursor inflow method for large eddy simulations and applications to finite length wind farms. *Renewable Energy*, 68:46 – 50, 2014.

International Conference and Exhibition in Advanced Materials and Microscopy (ICEAMM 2017)

IOP Conference Series: Materials Science and Engineering
Volume 343

Kuta, Indonesia
25 - 26 October 2017

Institute of Physics